*FUSARIUM* SPECIES

# *FUSARIUM* SPECIES:
## THEIR BIOLOGY AND TOXICOLOGY

**Abraham Z. Joffe**
Department of Botany, Laboratory of Mycology and Mycotoxicology
The Hebrew University of Jerusalem
Jerusalem, Israel

A WILEY-INTERSCIENCE PUBLICATION
**JOHN WILEY & SONS**
**New York · Chichester · Brisbane · Toronto · Singapore**

*Library of Congress Cataloging in Publication Data:*

Joffe, Abraham Z.
    Fusarium species.

    "A Wiley-Interscience publication."
    Bibliography: p.
    Includes index.
    1. Fusarium—Toxicology.   2. Fusarium—Classification.
3. Mycotoxins.   I. Title.
RA1242.F86J64   1986          615.9′52924          85-26494
ISBN 0-471-82732-0

Printed in the United States of America
10 9 8 7 6 5 4 3 2 1

*To my wife Rada and in memory of my Mother and Father*

# PREFACE

This monograph is a summary of nearly 50 years' research and study of the taxonomy of all the *Fusarium* species and their toxicity.

I was Director of the Laboratory of Mycology of the Institute of Epidemiology and Microbiology of Orenburg, USSR from 1943 to 1950. I established the etiology of alimentary toxic aleukia (ATA) and the role of overwintered cereals infected by cryophilic *Fusarium* species in specific climatic and environmental conditions that produced toxins causing thousands of deaths.

This research was later complemented by studies on the toxicity to plants and animals of the authentic *Fusarium* species associated with the fatal outbreaks of ATA, which I continued as Head of the Laboratory of Mycology and Mycotoxicology in the Department of Botany at the Hebrew University of Jerusalem and on sabbatical as Distinguished Researcher at the Division of Microbiology, Food and Drug Administration, Washington, D.C., in 1978–1979.

The idea of writing this book was conceived during the Conference of Mycotoxins in Human and Animal Health at the University of Maryland in 1976, and it crystallized later, thanks to many colleagues who contributed their support, encouragement, and stimulation during the preparation of the book.

This book is intended to provide a comprehensive survey of fusariotoxins and fusariotoxicoses. It covers toxic *Fusarium* species from all aspects and includes original material of great biological importance. Considerable work has been done on specific aspects of toxic *Fusarium* species, as may be seen from the table of contents and the 1800 references, which include monographs, reviews, and papers by research workers throughout the world, including the USSR. No definitive work exists, however, covering the entire field. This work covers sources of

*Fusarium* species; the conditions in which they grow in the field and laboratory; the toxic properties of cryophilic fungi; climatic, environmental, and ecological factors in toxin production in grain and soil; the symptoms produced by these toxins in animals and plants; and, most important, their danger to humans.

From the biological and chemical points of view, interest in fusariotoxins has concentrated upon their ability to produce trichothecene metabolites, which cause a variety of disorders in humans and animals. The role of *Fusarium* species and their toxins as causative agents in human disease has acquired considerable significance after many indications that four potent mycotoxins of the trichothecene group (T-2 toxin, diacetoxyscirpenol, deoxynivalenol, and nivalenol) may have been used in chemical warfare in Southeast Asia, as described in Chapter 11.

The book includes original studies on the taxonomy and distribution of toxic *Fusarium* species from various sources in Israel. Many hundreds of *Fusarium* cultures were isolated from various types of soil (light, medium, and heavy and from uncultivated desert-type soils from southern Israel) and also from cereals, from fields, and after storage. These cultures were incubated at different temperatures in order to examine their phytotoxic effects and their toxicity to animals. Several original investigations of fusariotoxicoses in humans and animals are also described.

Information is also presented on endemic Urov (or Kashin-Beck) disease in humans, cases of onychomycosis, keratomycosis, cutaneous ulcers and skin lesions, and possible involvement of *Fusarium* toxins in tumors, all similarly correlated with various *Fusarium* isolates.

A chapter on modern taxonomy has been included to facilitate the correct identification of trichothecene and other mycotoxin-producing *Fusarium* species and varieties.

This book should be of particular interest and use to mycologists, plant pathologists, soil and food scientists, agronomists, biologists, toxicologists, biochemists, chemists, oncologists, physicians, veterinarians and pharmacologists, and public health authorities.

It is my hope that this book may make a contribution to a wider understanding of the many problems connected with fusariotoxins and fusariotoxicoses.

I am deeply indebted to Dr. Sergei Gavrilovich Mironov (1893–1946), with whom I began my work on alimentary toxic aleukia at Orenburg. I also wish to acknowledge my indebtedness to Dr. Regina Schoental (of the Royal Veterinary College, University of London), with whom I have collaborated in research on the carcinogenic effects of toxic *F. poae* and *F. sporotrichioides*, for reading the manuscript and for her constructive advice.

I wish to thank Drs. Nicholas Petrus Jacobus Kriek, Henry Ungar,

Theodore Nobel, Irving Lutsky, and Natan Mor for their contributions and description of some histological preparations.

I am grateful to Dr. L. Stoloff of the FDA (Washington, D.C.) for allowing me to use their reference-card index and the many colleagues who have offered suggestions and valuable criticism during the writing of this treatise.

I would like to express special thanks to the Administration of the Faculty of Science of the Hebrew University for financial support in the preparation of this manuscript, to Eva Glikstein-Gaber for her technical assistance, and to Israel Baldinger for his line drawings.

I also thank Dr. Gruenberg-Fertig, from the Department of Botany at The Hebrew University, for her skill and general cooperation in preparing the subject index and for her critical reading of the manuscript.

ABRAHAM Z. JOFFE

Jerusalem, Israel
June 1986

# CONTENTS

# *FUSARIUM* SPECIES

# INTRODUCTION

In the 1940s and chiefly during World War II a very severe disease was widespread in the USSR. At that time I had the good fortune to work as principal and head of the Laboratory of Mycology at the Institute of Epidemiology and Microbiology of Orenburg, Russia.

The occurrence of the disease could be related to the near-famine conditions prevailing in rural regions of the Soviet Union. Under these conditions the population was forced to collect in spring even the over-wintered cereal grains (that had been left in the field); naturally, these were mainly contaminated by species of *Fusarium*. It was thus assumed that there was a correlation between the diet of these rural peasants and the development of the disease known as alimentary toxic aleukia (ATA).

The fusaria isolated from overwintered grains and soil have been shown to produce toxins under unfavorable climatic and environmental conditions, especially at low fluctuating temperatures.

The seasonal occurrence of ATA, its endemism, and the composition of the affected population suggested the importance of climatic and ecological factors in producing toxins that were found in field grains naturally infected by *Fusarium* species. This study enabled us to understand more clearly the role of environmental conditions in the mechanism of *Fusarium* toxin production in overwintered cereal grains and the etiology of the fatal ATA disease in humans.

This research was later complemented by studies on the toxicity to plants and animals of the authentic *Fusarium* species previously associated with the severe outbreaks of ATA, which I have continued since then, also as Head of the Laboratory of Mycology and Mycotoxicology in the Department of Botany of the Hebrew University of Jerusalem. The general emphasis of this monograph is on collection, isolation, and

**1**

identification of the *Fusarium* species from various subtropical and temperate regions and their cultivation and maintenance under special laboratory conditions. We also paid special attention to the fusariotoxins occurring naturally in the field. From the biological, chemical, and toxicological points of view, the interest in fusariotoxins has been centered mainly on the ability to produce trichothecene metabolites, chiefly T-2 toxin, by many *Fusarium* strains isolated from cereal grains (mainly from overwintered cereals) and foods and feeds which have adverse toxic effects on animals and caused a variety of health disorders in people in the USSR. The majority of investigations on toxic trichothecenes have been carried out on *Fusarium* strains under laboratory conditions. It is known that *Fusarium* isolates under laboratory conditions produced toxins different from those in the field crops, due to the dissimilarity of the environmental ecological conditions. Therefore studies of toxic principles in naturally contaminated cereal grains are most important and useful as they can throw some light on the causal relationship between cereal toxicosis and those *Fusarium* strains which produce mycotoxins that are associated with the diseases in domestic animals and humans.

Species and varieties which produce the most toxic trichothecene metabolites, mainly T-2 toxin, belong to *F. sporotrichioides, F. sporotrichioides* var. *tricinctum,* and *F. poae* (of the Sporotrichiella section), isolated only from temperate regions and not from Southeast Asia or Israel.

These naturally occurring trichothecene compounds and mainly T-2 toxin caused a variety of mycotoxicoses such as bean-hulls poisoning in Japan, moldy corn toxicosis in the United States, and ATA in the USSR. They caused several symptoms in animals, including skin inflammation, diarrhea, hemorrhages, vomiting, feed refusal, depression of bone marrow and hematopoietic system, and nervous disorders.

Toxic strains of *F. sporotrichioides* and *F. poae* occurring in the field and isolated from overwintered cereal grains and soil produced great quantities of T-2 toxin depending on the *Fusarium* strain's origin and culturing conditions. These *Fusarium* strains are described in this book in relation to their distribution, origin, incidence, phytotoxic action, mycotoxin production, their toxicity to laboratory and domestic animals, and mainly their implication in several human mycotoxicoses.

The importance of geographic distribution and origin of toxigenic *Fusarium* strains is clearly shown in the comparison of trichothecene T-2 toxin production by authentic ATA strains (isolated in the USSR) and strains used by researchers in the United States. Rabbit skin and brine shrimp test and physicochemical analysis have shown that the strains isolated in the USSR produced greater amounts of T-2 toxin than those from the United States. *F. sporotrichioides* and *F. poae* have been reported by the author as some of the most prominent and important fungi associated with ATA disease in the USSR. Therefore more attention has been given to the toxicity of these *Fusarium* strains and their mycotoxins.

TABLE I.1  *Fusarium* Strains of the Sporotrichiella Section in the Collection of the Laboratory of Mycology and Mycotoxicology, The Hebrew University of Jerusalem, Israel

| *Fusarium* strains | Overwintered Cereals | Locality | No. of Strains Isolated |
|---|---|---|---|
| F. sporotrichioides | Rye | USSR | 858 |
| F. sporotrichioides | Millet | USSR | 339 |
| F. sporotrichioides | Barley | USSR | 243 |
| F. sporotrichioides | Wheat | USSR | 85 |
| F. sporotrichioides var. tricinctum | Millet | USSR | 79 |
| F. sporotrichioides var. chlamydosporum | Millet | USSR | 31 |
| F. poae | Millet | USSR | 289 |
| F. poae | Wheat | USSR | 287 |
| F. poae | Barley | USSR | 127 |
| F. sporotrichioides var. chlamydosporum | Wheat | Canada | 54 |
| Total | | | 2392 |

In our laboratory of Mycology and Mycotoxicology we have the largest and most important collection of toxic and lethal single-spore *Fusarium* strains belonging to the Sporotrichiella section. They were isolated by the author from overwintered cereal grains and soil (in the Orenburg district of the USSR) which were involved in the fatal ATA disease in humans.

The isolated single-spore *Fusarium* strains of the Sporotrichiella section are presented in Table I-1.

Various single-spore cultures of *F. sporotrichioides, F. sporotrichioides* var. *tricinctum,* and *F. poae* differ from one another in the ability to produce trichothecene compounds, mainly T-2 toxin, even under the same laboratory conditions.

Recently we studied 308 random crude extracts from our *Fusarium* collection for their dermatitic toxicity on rabbits and brine shrimp larvae, and also by physicochemical analyses (*see* Table 2.1).

The crude extracts of T-2 toxin isolated from *Fusarium* species and varieties of the Sporotrichiella section have not only a localized dermatitic reaction characterized by irritation, inflammation and desquamation, and histopathological lesions (Fig. I.1 and Fig. I.2) but also have a general effect manifested by a loss of appetite, sleepiness, complete food refusal, loss of body weight, hemorrhages in different organs (mainly in the gastrointestinal tract and tissues), respiratory disorders, and other syndromes.

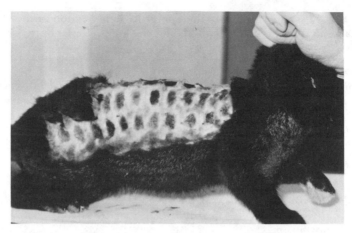

**Figure I.1**  General view of a rabbit treated with toxic extracts of *F. sporotrichioides* strains.

In all cases the rabbits died 18–48 hr after application of the toxic crude extracts.

The obtained results revealed that the strains of *F. sporotrichioides*, *F. sporotrichioides* var. *tricinctum*, and *F. poae* isolated in the 1940s, have retained their high levels of T-2 toxin production, indicating a remarkable persistence of toxigenic capability under special environmental conditions. We have reproduced in cats all characteristic clinical and histopathological symptoms of ATA disease with T-2 toxin isolated from these strains.

The literature review for all toxic *Fusarium* strains, belonging to the Sporotrichiella and other sections of the genus *Fusarium*, was completed at the end of 1984.

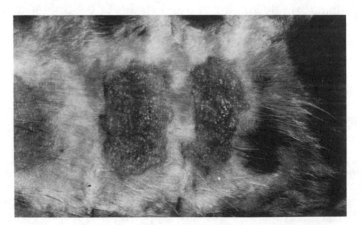

**Figure I.2**  Necrotic reactions on rabbit skin 24 hours after applications of extracts of *F. sporotrichioides* strains grown at 12°C for 21 days.

# CHAPTER 1

## BRIEF HISTORICAL BACKGROUND OF FUSARIOTOXICOSES

Fusariotoxicoses are diseases associated with the consumption of agricultural products infected by some *Fusarium* fungi causing intoxication in humans. Palchevski (1891) first described outbreaks of the fusariotoxicoses, called "toxic bread" intoxication, in people from the Far East. At the same time Woronin (1891), who studied the mycoflora of scabby cereals in the Ussuri district of Eastern Siberia where the disease was widespread, reported on the "inebriant bread" intoxication.

According to Sorokin (1890), Palchevski (1891), and Woronin (1891), the etiological agent of the "toxic bread" or "inebriant bread" intoxication or Taumelgetreide (staggering grain) in the Ussuri district was *F. roseum* and *Gibberella saubinetii*. These researchers also isolated other fungi, such as *Helminthosporium* sp. and *Cladosporium herbarum*.

Rozov (1889) noted that the "intoxicated bread" disease in the Ussuri district attacked not only rye but also wheat, oats, and other cereals and grasses. After eating bread prepared mainly from contaminated rye, rural peasants were severely affected. Even domestic animals such as dogs, poultry, pigs, and horses were also affected and refused to eat the moldy cereals. Some of the animals died after consumption of the intoxicated cereals.

Chemical investigations of the affected cereal grain from the Ussuri district were carried out by Gabrilovich (1906), Uglov (1913), and Pomasski (1915). Gabrilovich claimed that the toxin which caused the "inebriant bread" disease was one of the nitrous glucosides. Uglov stated that wheat grain from the Ussuri district had a higher acidity level than normal high quality grain. He also found that such grain contained

**5**

less protein than unaffected normal grain. Pomasski proved that *Fusarium* fungi destroyed the protein and changed the quantity of starch in the grain while also forming nitrous glucosides, amino acids, and ammonia.

In the far east of the USSR, Yachevski (1904), Naumov (1916), and later on Dounin (1926a), Agronomov *et al.* (1934), and Abramov (1939) studied the mycoflora of wheat, rye, oats, and barley in relation to the "inebriant bread" intoxication in man and mainly isolated *F. graminearum*. They suggested that the *F. roseum* isolated by Sorokin (1890), Palchevski (1891), and Woronin (1891) were in fact *F. graminearum* and *G. saubinetii* (*G. zeae*).

Dounin (1926a,b), Agronomov *et al.* (1934), and Abramov (1939) stated that the fusariotoxicosis of wheat and other cereal crops appeared in certain definite years marked by low temperatures, considerable precipitation, and high relative humidity (typical of a temperate climate). They also proved that the type of soil also influenced the formation of toxic *Fusarium* fungi in cereal grains causing diseases in humans and farm animals.

Contamination of grain by *F. graminearum* and the subsequent ingestion of bread made from the grain can produce "drunken bread syndrome," which is indigenous to the USSR (Gajdusek, 1953b). Inhabitants of the Ussuri district, after consumption of such bread (mainly prepared from infected rye), fell ill and showed the following disorders: headache, weakness, dizziness, exhaustion, nausea, vomiting, and shivering. Louria *et al.* (1968) also reported symptoms which included abdominal pain and marked ataxia.

According to Bilai and Pidoplichko (1970), fusariotoxicosis mostly appeared yearly in the Ussuri district, occasionally in Northern and Central Russia, and also in some years in Sweden, Germany, France, and Italy. Similar fusariotoxicoses have occurred in Finland, Central Europe, North America, China, and Japan (Uraguchi, 1971).

In 1932–1941 in the same areas of the USSR there was a disease of unknown etiology mainly among the rural population. The symptoms included headache, nausea, stomatitis, dermal necrosis, vomiting, hemorrhages, nose bleeding, and severe necrotic angina. The disease was named "septic angina" by Chilikin (1944, 1945); Mironov (1944, 1945a,b); Mironov and Fok (1944); Mironov and Joffe (1947a,b); Nesterov (1945); and Yefremov (1945); and later named ATA by Chilikin (1947) and Yefremov (1948).

One of the well-documented fusariotoxicoses in humans and animals since the 1940s is ATA (Joffe, 1960a, 1963a, 1965, 1971, 1974c, 1978b, 1983b, and in Chapter 6 of this volume).

The same syndromes were obtained on cats and monkeys with toxic overwintered grains (wheat, millet) contaminated by some *Fusarium* species (Alisova, 1947b; Joffe, 1960a; Rubinstein and Lyass, 1948; Sarkisov 1948, 1954; Sarkisov and Kvashnina, 1948).

In the 1960s a toxic compound named trichothecene was isolated from *Trichothecium roseum* (and some other compounds—as trichodermin from *Trichoderma viride,* verrucarin from *Myrothecium roridum,* and crotocin from *Cephalosporium crotocinigenum* and from various *Fusarium* species).

The mycotoxins isolated from the above-mentioned fungi are named (from the first isolated trichothecin compound) the "trichothecene" mycotoxins.

The etiology of ATA is connected with a trichothecene compound, mainly T-2 toxin, which was isolated from authentic outbreak samples of cereal grains infected by *F. poae* and *F. sporotrichioides* (Joffe, 1960a, 1971, 1974c, 1978b, 1983b; Joffe and Yagen, 1977, 1978; Lutsky and Mor, 1981a,b; Lutsky *et al.,* 1978; Sato *et al.,* 1975; Yagen *et al.,* 1977).

Another probable fusariotoxicosis appears to be the chronic Urov or Kashin-Beck disease, characterized by marked skeletal deformation in children, which occurred in endemic regions of Ussuri and elsewhere (*see* Chapter 6).

In the United States, fusariotoxicoses, during the first 30 years of the twentieth century, were associated with outbreaks in pigs and other domestic animals after feeding with scabby barley. This scabby barley, when exported to Germany, also caused poisoning in pigs (Miessner and Schoop, 1929).

Mundkur and Cochran (1930) isolated *G. saubinetii* (*G. zeae*), mainly from scabby barley which had caused intoxication in hogs. Dickson *et al.* (1930); Mains *et al.* (1930); Popp (1930); Popp and Contzen (1930); Roche *et al.* (1930); and Roche and Bohstedt (1930) reported that scabby barley, infected with *G. saubinetii,* caused toxicosis in mice, guinea pigs, chickens, dogs, and horses.

Christensen and Kernkamp (1936) and Hoyman (1941) described outbreaks among pigs fed with scabby barley infected by *F. roseum* and *G. saubinetii.* Similar outbreaks were reported in Ireland by McErlean (1952) when *Fusarium*-infected barley was fed to sows.

Later on a crystalline anabolic and uterotropic compound which caused vulvovaginitis in animals was isolated from corn infected by *G. zeae* by Stob *et al.* (1962). Christensen *et al.* (1965) noted that *F. graminearum* (*G. zeae*) caused weight increase of the uterus in female rats.

In Japan a red mold or wheat scab named Akakabibyo, affecting wheat, barley, oats, and rye, caused serious disorders in humans and domestic animals. At the beginning of the twentieth century Miyake (1909), Hara (1910), and Ito (1912a,b) isolated *G. saubinetii* from rice, scabby wheat, and oats, respectively, and stated that these infected cereals were the cause of disease in animals. Tochina (1933a,b) reported that scabby wheat caused loss of body weight, hemorrhages in various organs, and death in domestic animals.

Urakura (1933) reported cases of vomiting and diarrhea in humans, after eating products made from scabby wheat. He also described the outbreaks of disease linked with scabby wheat in horses and other farm animals. According to Urakura, the incidence of red mold-contaminated cereals in Japan was associated with heavy rains and cold weather.

Ikeda *et al.* (1964) and Kurata *et al.* (1964) also studied the toxicity of scabby wheat and barley contaminated by *F. roseum* and *F. graminearum*, known as "red mold" in Japan.

During the 1950s and 1960s research on fusariotoxins in Japan was carried out by Hirayama and Yomamoto (1948, 1950); Nakamura *et al.* (1951a,b); Takeda *et al.* (1953); Kurata *et al.* (1964); and Kurata (1978). They all reported outbreaks of food poisoning in humans in postwar Tokyo, accompanied by fever, vomiting, diarrhea, and headache, thought to be related to wheat flour contaminated by *F. graminearum*.

In 1955 rice contaminated by *G. zeae* was responsible for an outbreak of vomiting in Hoya and Tokyo, as well as in Ibaragi, Tochigi, Kanagawa, and Kochi Prefectures (Tsunoda, 1970).

Cho (1964) described an outbreak of barley scab in Korea in 1963 caused by *F. graminearum*. People affected showed signs of nausea, vomiting, abdominal pain, and diarrhea.

Yoshizawa (1983b) described the red mold diseases and natural occurrence of trichothecene mycotoxins in Japan, including historical background, human and animal toxicoses, and characterization of trichothecenes (nivalenol, deoxynivalenol, and its derivatives produced by *F. roseum*). The historical background of trichothecene producing *Fusarium* species and some other fungi and their role in mycotoxicosis was described by Ueno (1983a).

Problems associated with fusariotoxins and fusariotoxicoses and other mycotoxins have been reviewed by Austwick (1975); Bamburg (1969, 1972, 1976); Bamburg and Strong (1971); Bamburg *et al.* (1968a,b); Chu (1977); Ciegler (1978); Hesseltine (1976a,b); Hidy *et al.* (1977); Joffe (1971, 1974c, 1978a,b, 1983b); Marasas *et al.* (1984); Marth and Calanog (1976); Mirocha and Christensen (1982); Mirocha *et al.* (1977a,b, 1980); Pomeranz (1964); Saito and Ohtsube (1974); Saito *et al.* (1971); Smalley and Strong (1974); Ueno (1977a, 1980a,b); Uraguchi (1971); Uraguchi and Yamazaki (1978); Wilson and Hayes (1973); and Wogan (1975).

# CHAPTER 2

## TOXIC *FUSARIUM* SPECIES AND VARIETIES IN NATURE AND IN LABORATORY CONDITIONS

*Fusarium* fungi produce metabolites belonging to a number of chemical groups (e.g., trichothecenes, zearalenone, moniliformin, and butenolide) under favorable environmental conditions. The occurrence of these metabolites produced by toxic *Fusarium* species, in nature and under laboratory conditions, is described in this chapter, which deals only with those sections of the genus *Fusarium* that include toxic species.

The role of *Fusarium* strains and their toxins as causative agents in human disorders has acquired considerable significance after evidence that some potent mycotoxins of the trichothecene group (T-2 toxin, deoxynivalenol, and nivalenol) have been used in chemical warfare in Southeast Asia. The authentic *Fusarium* isolates associated with fatal ATA disease in humans have important scientific and practical value, and it will be useful to better understand their effective role in toxin production.

It is therefore of great significance to study the toxic *Fusarium* species mainly of the Sporotrichiella section, especially with regard to their temperature relationships, growth associations, and their capability for toxin production, as well as other toxic fusaria belonging to corresponding sections of the genus *Fusarium* investigated by different researchers.

Our studies (Joffe, 1960a,b, 1962a, 1965, 1974b) proved that the highest tolerance to low temperatures is displayed by toxic *Fusarium* species and varieties isolated from cereal grains overwintered under the cover of snow (Joffe, 1963a, 1971, 1983c).

Owing to the great variety of taxonomy systems used by the various authors quoted, it has been decided to give the name assigned by the original author to any given *Fusarium* species first, followed in parentheses by the name according to the classification of the author, as outlined in Chapter 10. Where no changes in classification are required, only the original name will be quoted. In some cases no reclassification is possible and the original names will be given separately at the end of the corresponding section, under their original classification. This applies in particular to *F. roseum* and *F. tricinctum*, according to Snyder and Hansen (1945).

## NATURALLY OCCURRING OUTBREAKS

### Sporotrichiella Section

*F. poae; F. sporotrichioides; F. sporotrichioides* var. *tricinctum;* and *F. sporotrichioides* var. *chlamydosporum.* The Sporotrichiella section, according to the taxonomy of the author (Joffe, 1974a, 1977), contains the following most toxic species of fusaria: *F. poae, F. sporotrichioides, F. sporotrichioides* var. *tricinctum,* and *F. sporotrichioides* var. *chlamydosporum.* These fungi produce toxic trichothecenes which widely occur in nature, mainly in temperate regions, and which have been associated with several outbreaks of intoxication in humans and animals.

One such case of intoxication is the well documented ATA in humans. This very severe and in most cases lethal disease is characterized by skin inflammation, diarrhea, extreme leukopenia, multiple hemorrhages, sepsis, and exhaustion of bone marrow (Joffe, 1971, 1974c, 1978b, 1983b).

The fusaria belonging to this Sporotrichiella section and implicated in this toxicosis have been described by the following researchers: Akhmeteli (1977); Akhmeteli *et al.* (1972a,b, 1973); Bilai (1947, 1948, 1952a,b,c, 1953, 1960, 1965, 1970a,b, 1977, 1978a,b); Bilai and Pidoplichko (1960, 1970); Ellison and Kotsonis (1973); Ermarkov *et al.* (1978); Kotik *et al.* (1979); Kotsonis *et al.* (1975a); Pidoplichko (1953); and Pidoplichko and Bilai (1946).

They were also described by Joffe (1960a,b, 1962a, 1963a, 1965, 1971, 1974c, 1978b, 1983b,c); Joffe and Palti (1975); Korpinen and Ylimaki (1972); Korpinen and Uoti (1974); Kvashnina (1948, 1968, 1976, 1978, 1979); Marasas *et al.* (1984); Mirocha and Pathre (1973); Rubinstein (1948, 1950a, 1951a,b, 1960a,b); Rubinstein and Lyass (1948); Rubinstein *et al.* (1961); Sarkisov 1944, 1945, 1948, 1950, 1954, 1960, 1961); Sarkisov and Kvashnina (1948); Sarkisov *et al.* (1948a,b,c); Ueno *et al.* (1972b, 1973d); and Szathmary *et al.* (1976).

Joffe and Yagen (1977, 1978); Lutsky and Mor (1981a,b); Lutsky *et al.* (1978); and Yagen *et al.* (1977, 1980) offered evidence that T-2 toxin isolated from *F. poae* and *F. sporotrichioides* was the compound causing ATA disease.

Outbreaks of "moldy corn toxicosis" and other fusariotoxicoses associated with *F. tricinctum* (*F. sporotrichioides* var. *tricinctum*), *F. poae*, and *F. sporotrichioides* were characterized by food refusal, hyperemia, hemorrhagic lesions, gastrointestinal and neurological disorders, and abortions and death among cattle and livestock. These fusaria produced T-2 toxin and related trichothecenes and have been found to occur naturally in animal feeds in the United States, Canada, the USSR, Japan, Korea, South Africa, Hungary, Scotland, Finland, and other countries (Bamburg *et al.*, 1968; Bamburg *et al.*, 1969; Burmeister, 1971; Ciegler, 1978; Dahlgren and Williams, 1972; Korpinen and Ylimaki, 1972; Marasas *et al.*, 1979b; Mirocha *et al.*, 1976a,b; Petrie *et al.*, 1977; Scott, 1978; Scott *et al.*, 1980; Shreeve *et al.*, 1975; Smalley, 1973; Spesivtseva, 1967; Szathmary *et al.*, 1976; Tookey *et al.*, 1972; Ueno, 1977a,b, 1980a,b,c, 1983b,c; Ueno *et al.*, 1972a,b; Wyatt *et al.*, 1972a; Yoshizawa, 1978, 1983b; Yoshizawa *et al.*, 1980a,b,c; 1981, 1982a,b; 1984).

Outbreaks of cardiomyopathies, which killed 40 people in Canada, the United States, and Europe (Belgium), among heavy beer drinkers (Morin and Daniel, 1967) were accompanied by gastrointestinal disorders (Bonnenfant, *et al.*, 1967). According to Schoental (1980a,b) these outbreaks were probably caused by *Fusarium* of the Sporotrichiella section contaminating grains used in beer production.

Hintikka (1983) listed the most frequent species of *Fusarium* (*F. avenaceum, F. culmorum, F. poae,* and *F. tricinctum*), which occurred in Finland, on oats, barley, wheat, and rye, and their toxins (trichothecene and zearalenone), and also their cytotoxic effects. The author noted that *Fusarium*-contaminated barley is not suitable for beer production and that T-2 toxin inhibited the germination, for malting, of barley grain by 40–80%.

Vesonder (1983) reported the natural occurrence of vomitoxin, nivalenol, diacetoxyscirpenol, and T-2 toxin in corn and feedstuffs in North America which were associated with *F. tricinctum* and *F. graminearum* (*Gibberella zeae*).

Scott (1983) described some incidences associated with trichothecenes in Canada related to *F. graminearum* (*G. zeae*), *F. poae,* and *F. sporotrichioides*.

## Martiella Section

*F. solani.* Bean-hulls, later seen to be contaminated by *F. solani*, have been implicated in widespread poisoning in horses in Hokkaido (Asami, 1932; Konishi and Ichijo, 1970a,b; Morimoto, 1936; Ohira, 1938; Oya and Morimoto, 1942; and Ueda *et al.*, 1967).

Sixty-nine adult cattle (out of a herd of 275) died in Georgia in 1969 after ingesting moldy sweet potatoes (*Ipomaea batatas*). The clinical signs and post mortem findings were characteristic of atypical interstitial pneumonia (AIP). *F. solani* (*F. javanicum*), isolated from the implicated sweet potatoes, reproduced the disease when cultured on sweet potatoes and fed to cattle (Peckham *et al.*, 1972).

Many cases of mycotic ulcers of the cornea have been found to be caused by *F. solani* (Jones *et al.*, 1969a,b,c, 1972; Joffe, unpublished data).

## Elegans Section

*F. oxysporum.*   This has been associated with urinary tract infections, cutaneous infections, corneal ulcers, mycotic keratitis, and also venous ulcers and other diseases (*see* Chapter 7).

## Liseola Section

*F. moniliforme.*   Van der Walt and Steyn (1943, 1945) reported a field outbreak associated with a Leukoencephalomalacia syndrome in South Africa in which horses displayed brain lesions accompanied by hemorrhage, edema, and liver damage. They related this toxicosis to sugar bean hay (*Phaseolus vulgaris*) contaminated with fungi, especially *F. moniliforme.*

Nelson and Osborne (1956) reported that swine died after eating moldy maize contaminated by *F. moniliforme*, which was the cause of death.

*F. moniliforme* was isolated from feed suspected of causing infertility in dairy cattle (Mirocha *et al.*, 1968c). *F. moniliforme* was one of the *Fusarium* strains isolated from sweet potatoes. One of the *F. moniliforme* isolates caused 20% or more weight reduction when fed to chicks (Doupnik *et al.*, 1971b).

Zearalenone, produced by *F. moniliforme* and *F. graminearum*, has been responsible for the hyperestrogenism seen in Yugoslavian pigs since 1963 (Ozegovic and Vukovic, 1972).

Maize or barley infected with *F. moniliforme* (*G. fujikuroi*) together with *F. graminearum* (*G. zeae*) was implicated in an outbreak of hyperestrogenism in pigs in Moldavia in the winter of 1972–1973 (Kurasova *et al.*, 1973).

*F. moniliforme* was isolated from maize associated with leukoencephalomalacia (LEM) in Egyptian donkeys (Badiali *et al.*, 1968; Wilson, 1971; Wilson and Maronpot, 1971; Wilson *et al.*, 1973) and South African horses (Kellerman *et al.*, 1972; Kriek *et al.*, 1981a,b; Marasas *et al.*, 1976; Pienaar *et al.*, 1981). According to Marshall (1983) corn contaminated in the field of Arizona with *F. moniliforme* caused the death of many chickens, which led to serious financial losses for chicken farmers and feed suppliers.

There are few documented cases of mycotic ulcerative keratitis (Anderson and Chick, 1963; Anderson *et al.*, 1959) and disseminated infection in humans (Young *et al.*, 1978) caused by *F. moniliforme* (*G. fujikuroi*).

The role that *F. moniliforme* plays in the potential mycotoxicoses of farm animals consuming agricultural commodities that have been produced in semitropical and tropical countries has not been determined. However, it is suspected that mycotoxicoses do occur in these regions. Most investigations on fusariotoxicoses outbreaks have centered on crops produced in the temperate zone.

A comprehensive survey of *F. moniliforme* and its varieties, which belong to the Liseola section, would provide valuable information about their role in mycotoxicoses in semitropical and tropical climates (Joffe, 1984).

## Gibbosum Section

*F. equiseti.* Degnala disease is a seasonal disease of cattle occurring during the winter months. The symptoms usually seen are lameness, edema, necrosis, and gangrene of tail, ear, and legs. Fescue-foot in the United States is a somewhat similar syndrome. Buffalo and cattle were affected by this syndrome in 1972 in India, the outbreaks occurring in areas where rice was grown. *F. equiseti* (*G. intricans*) isolated from the rice straw fed to the animals produced a hemorrhagic reaction when applied to rabbit skin. The authors assumed that butenolide was responsible for the condition (Kalra *et al.*, 1973).

Calves fed rice straw, suspected of being the reason for the Degnala outbreak, revealed lameness, edema of the forelegs, and dry gangrene on the tips of ears and tail (Kalra *et al.*, 1977). Rice straw, inoculated with *F. equiseti* and incubated at 15°C for 1 month, was fed to three buffalo calves. One calf developed edema of the fetlock region of the forelegs, and the other two developed lameness. One calf also developed dry gangrene at the tip of the tail.

## Arachnites Section

*F. nivale.* Numerous toxicoses have been associated with *Fusarium*-infected cereals (the so-called "red mold" disease or Akakabibyo) (Saito and Tatsuno, 1971). Only a few outbreaks, however, have been directly connected with *F. nivale*.

*F. nivale* and *F. roseum* (*F. graminearum*) were isolated from moldy rice associated with an outbreak of nausea, vomiting, drowsiness, vertigo, and headache among 25 young men in the Tokyo region of Japan in the summer of 1954 (Tsunoda *et al.*, 1957, 1958, 1968).

Udagawa *et al.* (1970) reported an occurrence of diarrhea among four families in the Kanagawa and Kagoshima areas of Japan that was associated with rice heavily infected with *F. nivale* and *F. graminearum*.

_F. nivale_ has been reported as causing keratomycosis in humans by Inokawa (1972).

## Eupionnotes Section

_F. dimerum_ **and** _F. episphaeria._   Cases of keratomycosis caused by _F. dimerum_ were reported from Argentina by Garcia _et al._ (1972); Zapater (1971); Zapater and Arrechea (1975); and Zapater _et al._ (1972). A case caused by _F. episphaeria_ (_F. aquaeductuum_) was reported in Colombia by Greer _et al._ (1973).

## Discolor Section

_F. culmorum._   There are very few reports in the literature implicating _F. culmorum_ in naturally occurring outbreaks of mycotoxicoses. Fisher _et al._ (1967) reported on a condition in dairy cattle in Southeast Australia characterized by loss of appetite and decreased milk production as well as staggering. _F. graminearum_ and _F. culmorum_ were both isolated from the suspect maize. These two strains were then grown on sterilized maize for 2 weeks at 21°C and extracted with ether. The ether extract (suspended in olive oil) of _F. culmorum_ caused a very severe, eventually lethal reaction when applied to rabbits' skin.

Fertility disturbances were noticed by Roine _et al._ (1971) in a herd of dairy cows (21 cows). _F. graminearum_ and _F. culmorum_ were isolated from the moldy grain. The authors stated that the fertility problems of the herd were probably due to the estrogenic mycotoxins present.

A herd of dairy cows showing decreased milk production, diarrhea, and premature calving was described by Moreau (1972). Rye grass (_Lolium perenne_) in the Normandy meadow where the cows were grazing was shown to be infected with _F. roseum_ var. _culmorum_ (_F. culmorum_). The author stated that such symptoms are usually associated with DAS and have been noticed in animals fed moldy grain, but rarely in those fed grass.

Mirocha _et al._ (1976b) mentioned a case of hyperestrogenism in swine in Washington. The causative organism was found to be _F. culmorum_ in the animal feed. Analyses showed the presence of deoxynivalenol (DON) and zearalenone (ZEN) when the isolate was grown on a rice substrate for 1 week at 24–27°C and 4 weeks at 14°C.

_F. graminearum._   Nakamura _et al._ (1951a,b) reported on abortions occurring in goats as a result of _F. graminearum_-infected wheat in Hokkaido in 1949.

McErlean (1952) described an outbreak of hyperestrogenism in swine in Ireland. Samples of the moldy barley, when fed to immature female pigs, reproduced the symptoms. The barley was found to contain _F. graminearum_ (_G. zeae_).

A young herd of swine in Minnesota developed estrogenic symptoms in 1963 when fed a pelleted feed. Guinea pigs and virgin weanling white rats, when given this feed, developed enlarged uteri. In 1964 another herd of swine in Minnesota developed the symptoms, this time while being fed corn from the 1962 harvest which had been stored in an open crib until late 1963. Virgin weanling white rats developed enlarged uteri when fed this corn, and an immature gilt developed typical estrogenic symptoms. Most of the isolates of fusaria included *F. graminearum* or *F. culmorum* (Christensen *et al.*, 1965).

Bugeac and Berbinschi (1967) reported an outbreak of hyperestrogenism affecting 70–90% of 300 gilts in July and August 1963. The condition could be reproduced by the isolate of *F. graminearum* which was found in the implicated feed.

Danko and Toth (1969) have also observed many cases of vulvar enlargement, prolapse of vagina and rectum, and edematous infiltration of the mammae in female swine in Hungary. Uterus expansion and prolapse were also seen. Reddening and expansion of mammae and enlargement of prepuce were observed in the males. Along with these symptoms, udder inflammation, decrease in milk production, disturbances of appetite, irregular heat, and uncertain pregnancy were noted in cattle and sheep. *F. graminearum* was isolated from corn used as food. Rats given this food inoculated with the isolate showed the same symptoms as those seen in swine, cattle, and sheep.

Mycotoxicoses causing hyperestrogenism and estrogenism in swine in Yugoslavia since 1963 have been shown to be caused by zearalenone, a toxin produced by *F. graminearum* (Ozegovic and Vukovic, 1972).

In the winter of 1972–1973 hyperestrogenism of pigs (principally between the ages of 3–5 months) was observed in Moldavia (Kurasova *et al.*, 1973). The maize or barley implicated was shown to contain *F. graminearum* (*G. zeae*). Pathological examination demonstrated degeneration of the liver and kidneys, enlarged suprarenal glands, splenic microinfarcts, and an edematous vaginal membrane with necroses and hemorrhages.

Incidences of vomiting and food refusal in swine were reported during 1972–1973 in the corn belt of the United States. Much of the corn was infected in the field with *F. graminearum* (*G. zeae*) (Tuite *et al.*, 1974).

Miller and Smith (1975) reported on food refusal by swine of corn from the 1972 crop and wheat from the 1973 crop. Both were infected with *F. graminearum*.

Collet *et al.* (1977) suggested that zearalenone from *F. graminearum*-contaminated maize was responsible for abortions in swine in France.

Maize is generally infected in storage by *F. roseum* "Graminearum" (*F. graminearum*), while wheat, barley, and oats are usually contaminated in the field under certain environmental conditions characterized by higher moisture and alternate warm and cold temperatures. The cereal grains when fed to animals, particularly swine, caused the estrogenic syndrome.

Zearalenone as an estrogenic mycotoxin produced by *F. graminearum* and other *Fusarium* species, occurred in cereal grains and also in mixed feeds (Mirocha *et al.*, 1974, 1976b; Shotwell *et al.*, 1980; Cohen and Lapointe, 1980; Prior, 1981).

*G. zeae.*   Ito (1912a,b) mentioned the refusal by horses of oats infected with *G. saubinetii* (*G. zeae*).

A case of refusal by swine of barley infected with *G. saubinetii* (*G. zeae*) was observed by Beller and Wedemann (1929). The barley involved was imported into Germany from America and caused many problems.

Swine consuming barley from the 1928 crop infected with *G. saubinetii* displayed nausea, vomiting, and weight loss (Christensen and Kernkamp, 1936).

In Japan in 1932 scabby wheat contaminated by *G. saubinetii* caused diarrhea in horses (Urakura, 1933).

Stob *et al.* (1962) observed hypertrophy of female genitalia during 1957–1958 in swine fed corn contaminated with *G. zeae.*

Sheep poisoning in Hokkaido in 1956 was caused by *G. zeae*-infected wheat (Ogasawara, 1965).

Farmers in Northern Indiana in 1965 complained about refusal, and also some vomiting, in relation to corn fed to swine. The corn was infected with *G. zeae*. Extracts of the infected corn, when administered orally, i.v., or i.p., caused emesis in pigs (Curtin and Tuite, 1966).

Futrell *et al.* (1976) mentioned that the epidemics of *G. zeae* on corn in Indiana with subsequent effects of emesis, vomiting, and refusal in swine, occurred in the years 1928, 1957, 1958, 1965, and 1972–1973.

*G. zeae* was identified in barley implicated in an outbreak of T-2 poisoning in British Columbia, Canada (Greenway and Puls, 1976). Many geese died, horses were depressed and febrile, pigs vomited and refused their food, and ducks were ill.

Tsunoda *et al.* (1968) and Tsunoda (1970) reported cases of poisoning due to molds in Japan. Among these they mentioned oats from Hokkaido, infected with *G. zeae*, which affected horses, and corn from the Tokachi area, infected with *G. zeae* also, which affected 30–40 head of sheep.

## INVESTIGATIONS UNDER LABORATORY CONDITIONS

The mycological and mycotoxicological investigations conducted in our laboratory conditions showed that the toxicity of trichothecene-producing *Fusarium* species is influenced by sharp temperature fluctuations. We also proved that the degree of contamination of cereal grains by naturally occurring trichothecenes, especially T-2 toxin, is high. This *Fusarium* metabolite has received considerable attention since it is extremely toxic to both humans and animals (Joffe, 1971, 1974c, 1978b, 1983b,c).

We have recently undertaken a comprehensive investigation to determine the toxicity of hundreds of single-spore strains of *Fusarium* species from our collection isolated from overwintered cereals in the Orenburg district of the USSR and those which were originally involved in ATA disease in the 1940s, in order to determine if they had not lost their ability for toxin production under laboratory conditions.

Cultivation and identification of the toxic activity of the *Fusarium* strains were performed by procedures described elsewhere (Joffe, 1974b; Joffe and Yagen, 1977; Yagen and Joffe, 1976; Yagen *et al.*, 1977).

In addition extracts of all *Fusarium* strains were also assayed and screened by rabbit skin test and tested for toxicity on brine shrimp larvae. The bioassay for skin test was performed by applying 5–10 μl of *Fusarium* crude extracts on the shaved rabbit skin. The brine shrimp test was confirmed by the method of Eppley (1974) using 10 μl of extract.

The toxin absorbed by the skin after applications caused local inflammation, edema, and necrotic reactions. After 12–24 hr there was loss of appetite and weakness; and after 48–72 hr there was a marked decrease of body weight and, as usual, death. Necropsy of the rabbits revealed multiple hemorrhages of the gastric and intestinal mucosa and hyperemia of the organs and tissues.

The strains of the Sporotrichiella section (*F. poae, F. sporotrichioides, F. sporotrichioides* var. *tricinctum*) produced large amounts of T-2 toxin and some quantities of neosolaniol and HT-2 toxin in smaller levels. The results are presented in Table 2.1.

The author (during sabbatical leave in FDA, Washington, D.C.) conducted a mycological and mycotoxicological investigation (in cooperation with Dr. R. M. Eppley) on *Fusarium* species isolated from cereal grains (wheat, rye, barley) and maize collected in the 1977 and 1978 crop years in the United States.

In 1976 at the Conference on Mycotoxins in Human and Animal Health the panel (in which the author took part) convened on Trichothecene Toxins made several research recommendations directed toward clarification of the problem. One of these recommendations was to "develop worldwide mycological surveys of foods for trichothecene-producing fungi" (Smalley *et al.*, 1977). In pursuance of this goal a survey of U.S. cereals was initiated for the 1977 harvest. Four major cereal crops, corn, wheat, barley, and rye, were included in the survey. The survey was continued through part of the 1978 harvest. The grain samples were obtained through the USDAs regional grader stations (Federal

**TABLE 2.1 Yields of Trichothecene Compounds from Strains of Sporotrichiella Section and Their Toxicity to Rabbit Skin and Brine Shrimp Larvae**

| Original Host | No. of Strains | Rabbit Skin Bioassay | Toxicity to Brine Shrimp Larvae Mortality (%) | TLC[a] Assay | Yields of T-2 Toxin (mg per 10 g Millet Grain) | Yields of Other Toxins (mg per 10 mg Millet Grain) |
|---|---|---|---|---|---|---|
| *F. sporotrichioides* | | | | | | |
| Millet, USSR | 11 | + + + + | 100 | Very strong | 12.5–28.7 | NS[b] 1.7–4.2 |
| Millet, USSR | 10 | + + + | 95–100 | Strong | 7.0–17.5 | NS 1.5–3.1 HT-2[c] 1.1–1.5 |
| Millet, USSR | 8 | + + | 70–78 | Medium | 2.5–4.0 | NS 0.9–1.8 |
| Millet, USSR | 12 | + | 55–67 | Slight | 0.7–1.6 | NS 0.4–0.7 |
| Millet, USSR | 21 | ± | 20–23 | Very slight | 0.1–0.3 | NS 0.1–0.3 |
| Wheat, USSR | 23 | + + + + | 100 | Very strong | 9–32.2 | NS 1.4–3.5 HT-2 0.2–0.8 |
| Wheat, USSR | 10 | + + + | 95–100 | Strong | 6.8–16 | NS 0.9–2.8 |

| | | | | | | |
|---|---|---|---|---|---|---|
| Wheat, USSR | 4 | ++ | 65–80 | Medium | 2.6–5.5 | NS 0.5–1.6 |
| Wheat, USSR | 4 | + | 45–70 | Slight | 1.2–2.4 | NS 0.3–0.5 |
| Wheat, USSR | 5 | ± | 30–42 | Very slight | 0.3–0.6 | NS 0.1 |
| Rye, USSR | 38 | ++++ | 100 | Very strong | 16.5–38.5 | NS 2.4–4.2 HT-2 0.6–1.8 |
| Rye, USSR | 13 | +++ | 98–100 | Strong | 11.7–14.2 | NS 1.7–3.3 HT 0.3–0.06 |
| Rye, USSR | 6 | ++ | 75–85 | Medium | 6.5–10.5 | NS 1.4–2.0 |
| Rye, USSR | 7 | + | 60–75 | Slight | 1.1–3.6 | NS 0.7–1.0 |
| Rye, USSR | 1 | ± | 30–45 | Very slight | 0.6–1.8 | NS 0.2–0.5 |
| Barley, USSR | 6 | ++++ | 100 | Very strong | 8.2–14.5 | NS 1.8 HT-2 0.4 |
| Barley, USSR | 4 | +++ | 92 | Strong | 6.5–10.8 | NS 1.1 |
| Barley, USSR | 5 | ++ | 70 | Medium | 2.0–6.6 | NS 0.6 |
| Barley, USSR | 3 | + | 46 | Slight | 1.5–4.8 | NS 0.3 |
| Barley, USSR | 7 | ± | 25–38 | Very slight | 0.2–0.5 | NS 0.1 |

**TABLE 2.1** Continued

| Original Host | No. of Strains | Rabbit Skin Bioassay | Toxicity to Brine Shrimp Larvae Mortality (%) | TLC[a] Assay | Yields of T-2 Toxin (mg per 10 g Millet Grain) | Yields of Other Toxins (mg per 10 mg Millet Grain) |
|---|---|---|---|---|---|---|
| | | | *F. poae* | | | |
| Millet, USSR | 4 | + + + + | 100 | Very strong | 14.8–22.6 | NS 1.2–3.4 HT-2 0.3–1.4 |
| Millet, USSR | 4 | + + + | 90–98 | Strong | 10.6–18.0 | NS 0.8–2.2 |
| Millet, USSR | 4 | + + | 76 | Medium | 5.5–8.4 | NS 0.5–1.0 |
| Millet, USSR | 4 | + | 50 | Slight | 26–51 | NS 0.1–0.4 |
| Millet, USSR | 12 | ± | 25–35 | Very slight | 0.7–2.4 | d |
| Wheat, USSR | 8 | + + + + | 100 | Very strong | 10–17.6 | NS 1.6–3.9 HT-2 0.3–0.7 |
| Wheat, USSR | 9 | + + + | 90–100 | Strong | 8.0–12.2 | NS 0.9–1.4 |
| Wheat, USSR | 10 | + + | 65–80 | Medium | 4.5–6.6 | NS 0.5–0.8 |
| Wheat, USSR | 11 | + | 35–55 | Slight | 1.8–3.0 | NS 0.2–0.5 |

| Origin | n | | % | | | |
|---|---|---|---|---|---|---|
| Wheat, USSR | 9 | ± | 20–30 | Very slight | 0.4–12 | NS 0.1 |
| Barley, USSR | 2 | ++++ | 100 | Very strong | 12.4 | NS 0.8–2.4 |
| Barley, USSR | 2 | +++ | 96 | Strong | 8.2 | NS 0.4–1.5 |
| Barley, USSR | 5 | ++ | 75–85 | Medium | 4.6 | NS 0.2–0.7 |
| Barley, USSR | 9 | + | 55–68 | Slight | 1.5–3.2 | NS 0.1 |
| Barley, USSR | 11 | ± | 30–40 | Very slight | 0.4–1.3 | |
| *F. sporotrichioides* var. *tricinctum* | | | | | | |
| Millet, USSR | 2 | ++++ | 100 | Very strong | 5.5 | NS 1.4 HT-2 0.2 |
| Millet, USSR | 2 | ++ | 68 | Medium | 3.1 | NS 0.6 |
| Millet, USSR | 1 | + | 38 | Slight | 1.7 | NS 0.2 |
| Millet, USSR | 1 | ± | 25 | Very slight | 0.4 | d |

[a]Thin layer chromatography.
[b]Neosolaniol.
[c]HT-2 toxin.
[d]Not detected.

Inspection Service) with the objective of collecting samples representative of U.S. agricultural production.

The survey consisted of two parts: (1) mycological isolation and identification of *Fusarium* species from the grain samples and (2) cultivation of representative isolates on wheat substrate followed by analysis for the production of trichothecenes and zearalenone. This work will be published by the author in cooperation with Drs. R. Eppley, P. Mislivec, M. Trucksess, and A. Pohland.

The investigations on toxic strains of *Fusarium* species involved in ATA disease and on *Fusarium* strains isolated from cereal grains in the United States revealed that *F. poae* and *F. sporotrichioides* strains are the most important producers of T-2 toxin. This compound was also strongly toxic to brine shrimp larvae and rabbit skin.

Production of T-2 toxin by *Fusarium* strains was correlated with toxicity to the brine shrimp larvae and irritation responses to the skin of rabbits.

There are many strains of the Sporotrichiella section in our collection that produced gram quantities of T-2 toxin when grown on 1 kg of wheat or millet grains at 12 or 5°C for 21 and 45 days, respectively (Joffe, 1983b).

The physicochemical analyses (thin layer chromatography (TLC), gas liquid chromatography (GLC), nuclear magnetic resonance, infrared, and mass spectrometry (MS)) have shown that the strains isolated in the Orenburg district, USSR produced greater amounts of T-2 toxin that those from other areas of the world. These strains did not decrease their toxic activity up to the present, indicating a remarkable persistence of toxigenic capability to produce high levels of T-2 toxin under our special laboratory conditions.

In addition to our investigations on the toxic *Fusarium* strains we include the results obtained by researchers on the effects of temperatures and duration of growth, substrate, and strains of *Fusarium* species that favor toxin production in laboratory conditions. The yield of mycotoxin concentrations detected in extracts of *Fusarium* strains from various sources are also demonstrated in Tables 2.2 through 2.10.

According to Eppley (1979); Smalley and Strong (1974); Tamm (1977); Ueno (1977a,b, 1980a,b,c, 1983b,c); there is information on 23 trichothecene metabolites produced by the *Fusarium* species under laboratory conditions.

Trichothecene toxins are tetracyclic sesquiterpenes, containing a 12,13-spiro-epoxy-ring, a double bond at C-9–C-10, and one or more hydroxyls, esterified with acetic or more complex acids.

The basic chemical structure of trichothecenes related tetracyclic sesquiterpenoid ring system is presented in Figure 2.1.

The toxicity of a particular trichothecene derivative depends on the presence of the 12,13-epoxy-ring, on the C-9 and C-10 double bond, on

**TABLE 2.2  The Concentrations of Mycotoxins from Strains of Sporotrichiella Section Incubated on Substrates at Various Temperatures and Lengths of Time**

| Strain/Source | Substrate | Time/T °C | Myco-toxins | Concentrations | References |
|---|---|---|---|---|---|
| | | *F. poae* | | | |
| HPB 071178-7 Wheat, Canada | Corn | 2–6 wk/19–25° + 1–2 wk/4° | DAS[a] | 17 µg/g | Scott et al., 1980 |
| HPB 071178-11 Corn, Canada | Corn | 2–6 wk/19–25° + 1–2 wk/4° | DAS | 5 µg/g | Scott et al., 1980 |
| NRRL 3287 Unknown, USA | PSC[b] | 12 d/26° | T-2[c] NS HT-2 | 0.30 g/liter crude toxin | Ueno et al., 1972b |
| 5253 Millet, USSR | Millet + 2% malt extract | 2 wk/23–27° + 1 wk/5–8° + 1 wk/23–27° | T-2 HT-2 NS T-2 | 200–520 µg/g 200–1600 µg/g 400–540 µg/g 50 µg/g | Szathmary et al., 1972b |
| 60/9 Millet, USSR | Wheat | 3 wk/12° | T-2 | 20.0 mg/10 g | Joffe and Yagen, 1977 |
| 396 Millet, USSR | Wheat | 3 wk/12° | T-2 | 21.0 mg/10 g | Joffe and Yagen, 1977 |
| 958 Wheat, USSR | Wheat | 3 wk/12° | T-2 | 8.0 mg/10 g | Joffe and Yagen, 1977 |
| 792 Barley, USSR | Wheat | 3 wk/12° | T-2 | 6.2 mg/10 g | Joffe and Yagen, 1977 |
| T-2 Corn, France | Wheat | 3 wk/12° | T-2 | 3.4 mg/10 g | Joffe and Yagen, 1977 |

**TABLE 2.2** Continued

| Strain/Source | Substrate | Time/$T$°C | Myco-toxins | Concentrations | References |
|---|---|---|---|---|---|
| | | *F. poae* (Cont.) | | | |
| NRRL 3287 Unknown, USA | Wheat | 3 wk/12° | T-2 | 1.0 mg/10 g | Joffe and Yagen, 1977 |
| NRRL 3299 Corn, USA | Wheat | 3 wk/12° | T-2 | 5.8 mg/10 g | Joffe and Yagen, 1977 |
| NRRL 3299 Corn, USA | WCG[d] | 3 wk/15° | T-2 | 9.96 g/12 kg | Burmeister, 1971 |
| NRRL 3299 Corn, USA | Rice | 3 wk/20° | T-2 | 0.199/1.2 kg | Burmeister, 1971 |
| T-2 Corn, France | PDA[e] | 3 wk/25° | T-2 | 103 mg/liter | Marasas et al., 1971 |
| NRRL 3287 USA | cracked corn | 13 d/28° | T-2 | 30 µg/g | Vesonder et al., 1981a |
| 3 strains Wheat, Czechoslovakia | wheat | 20 d/25° | T-2 | 600 mg/kg | Vesela et al., 1982 |
| | | *F. sporotrichioides* | | | |
| No. 38 Corn, Hungary | Millet + 20% malt extract | 1 wk/23–27° + 10 wk/5–8° + 1 wk/23–27° | ZEN[f] T-2 HT-2 T-2 tetraol NS | 200 µg/g 400 µg/g 900 µg/g 300 µg/g 300 µg/g | Szathmary et al., 1976 |

| | | | | | |
|---|---|---|---|---|---|
| F. 14 Millet, Hungary | Millet + 20% malt extract | 1 wk/23–27° + 10 wk/5–8° + 1 wk/23–27° | T-2<br>HT-2<br>NS<br>T-2 tetraol | 1070–1510 µg/g<br>270–400 µg/g<br>600–1480 µg/g<br>50 µg/g | Szathmary et al., 1976 |
| F. 38 Millet, Hungary | Millet + 20% malt extract | 1 wk/23–27° + 10 wk/5–8° + 1 wk/23–27° | ZEN<br>T-2<br>HT-2<br>NS | 990–1500 µg/g<br>1300–2300 µg/g<br>460–1000 µg/g<br>100–500 µg/g | Szathmary et al., 1976 |
| 5328a Millet, USSR | Millet + 20% malt extract | 1 wk/23–27° + 10 wk/5–8° + 1 wk/23–27° | ZEN<br>T-2<br>HT-2<br>NS | 340–1300 µg/g<br>960–3000 µg/g<br>420–650 µg/g<br>430–1750 µg/g | Szathmary et al., 1976 |
| M-1-1 Bean-hulls, Japan | PSC | 2 wk/27° | T-2, NS | 378 mg/liter crude toxin | Ueno et al., 1975 |
| M-1-1 Bean-hulls, Japan | PSC | 12 wk/25° | Fraction of crude toxin:<br>II T-2<br>III DAS<br>IV NS | 0.40 g/20 liter<br>0.22 g/20 liter<br>0.60 g/20 liter | Ueno et al., 1972 |
| NRRL 3510 Millet, USSR | PSC | 12 wk/25° | Fraction of crude toxin:<br>II T-2<br>III NS<br>IV Butenolide<br>NS, HT-2 | 0.488 g/20 liter<br>1.283 g/20 liter | Ueno et al., 1972 |
| NRRL 3510 Millet, USSR | Medium B[g] | 5 d/27° | T-2, NS | 10.5 g/liter<br>131 mg/liter crude toxin | Ueno et al., 1975 |

**25**

**TABLE 2.2** Continued

| Strain/Source | Substrate | Time/$T$ °C | Myco-toxins | Concentrations | References |
|---|---|---|---|---|---|
| | | *F. sporotrichioides* (Cont.) | | | |
| M-1-4 Bean-hulls, Japan | Medium C[h] (CDSP) | 2 wk/27° | T-2, NS | 299 mg/liter crude toxin | Ueno et al., 1975 |
| M-1-5 Bean-hulls, Japan | Medium C (CDSP) | 2 wk/27° | T-2, NS | 331 mg/liter crude toxin | Ueno et al., 1975 |
| 60/10 Millet, USSR | Wheat | 3 wk/12° | T-2 | 10.3 mg/10 g | Joffe and Yagen, 1977 |
| 347 Millet, USSR | Wheat | 3 wk/12° | T-2 | 23.2 mg/10 g | Joffe and Yagen, 1977 |
| 351 Millet, USSR | Wheat | 3 wk/12° | T-2 | 15.2 mg/10 g | Joffe and Yagen, 1977 |
| 738 Millet, USSR | Wheat | 3 wk/12° | T-2 | 7.8 mg/10 g | Joffe and Yagen, 1977 |
| 921 Rye, USSR | Wheat | 3 wk/12° | T-2 | 24.0 mg/10 g | Joffe and Yagen, 1977 |
| 1182 Wheat, USSR | Wheat | 3 wk/12° | T-2 | 10.2 mg/10 g | Joffe and Yagen, 1977 |
| 1823 Barley, USSR | Wheat | 3 wk/12° | T-2 | 4.6 mg/10 g | Joffe and Yagen, 1977 |
| NRRL 3249 Tall fescue, USA | Wheat | 3 wk/12° | T-2 | 4.0 mg/10 g | Joffe and Yagen, 1977 |

| Source | Substrate | Conditions | Toxin | Amount | Reference |
|---|---|---|---|---|---|
| NRRL 5908 Tall fescue, USA | Wheat | 3 wk/12° | T-2 | 0.6 mg/10 g | Joffe and Yagen, 1977 |
| 2061-C Corn, USA | Wheat | 3 wk/12° | T-2 | 1.5 mg/10 g | Joffe and Yagen, 1977 |
| YN-13 Corn, USA | Wheat | 3 wk/12° | T-2 | 2.0 mg/10 g | Joffe and Yagen, 1977 |
| HPB 071178-10 Wheat, Canada | Corn | 2–6 wk/19–25° +1–2 wk/4° | T-2 | 87 µg/g | Scott et al., 1980 |
| HPB 071178-18 Unharvested corn, Canada | Corn | 2–6 wk/19–25° +1–2 wk/4° | T-2 NS | 162 µg/g 20 µg/g | Scott et al., 1980 |
| HPB 071178-20 Cribbed corn | Corn | 2–6 wk/19–25° +1–2 wk/4° | T-2 | 320 µg/g | Scott et al., 1980 |
| HPB 071178-23 Unharvested corn, Canada | Corn | 2–6 wk/19–25° +1–2 wk/4° | T-2 | 280 µg/g | Scott et al., 1980 |
| HPB 071178-17 Corn | Corn | 2–6 wk/19–25° +1–2 wk/4° | T-2 | 34 µg/g | Scott et al., 1980 |
| HPB 071178-12 Cribbed corn, Canada | Corn | 2–6 wk/19–25° +1–2 wk/4° | T-2 | 380 µg/g | Scott et al., 1980 |
| HPB 071178-13 Cribbed corn, Canada | Corn | 2–6 wk/19–25° +1–2 wk/4° | T-2 HT-2 NS | 1350 µg/g 34 µg/g 140 µg/g | Scott et al., 1980 |
| HPB 071178-21 Cribbed corn, Canada | Corn | 2–6 wk/19–25° +1–2 wk/4° | T-2 NS | 260 µg/g 29 µg/g | Scott et al., 1980 |

**TABLE 2.2** Continued

| Strain/Source | Substrate | Time/$T$ °C | Myco-toxins | Concentrations | References |
|---|---|---|---|---|---|
| | | *F. sporotrichioides* (Cont.) | | | |
| HPB 071178-24 Unharvested corn, Canada | Corn | 2–6 wk/19–25° + 1–2 wk/4° | T-2 HT-2 NS | 51 µg/g 17 µg/g 17 µg/g | Scott et al., 1980 |
| 1 strain Wheat, Czechoslovakia | Wheat | 20 d/25° | T-2 | 600 mg/g | Vesela et al., 1982 |
| | | *F. tricinctum (F. sporotrichioides var. tricinctum)* | | | |
| FT-2 Soil, USA | Corn | 10 wk/16° or 2 wk/24° + 8 wk/12° | ZEN | 9.4 µg/g | Caldwell et al., 1970 |
| FT-3 Blue grass, USA | Corn | 10 wk/16° or 2 wk/24° + 8 wk/12° | ZEN | 14.4 µg/g | Caldwell et al., 1970 |
| FT-12 Corn, Kernels USA | Corn | 10 wk/16° or 2 wk/24° + 8 wk/12° | ZEN | 13.7 µg/g | Caldwell et al., 1970 |
| No. 223 Corn, Wisconsin, USA | PDA | 3 wk/25° | T-2 | 106 mg/liter | Marasas et al., 1971 |
| Unknown, USA | Corn-rice mixture | 21 d/27° + 24 d/8° | T-2 | 920 mg/kg | Ikediobi et al., 1971 |
| Unknown, USA | Corn culture | 30 d/8° | T-2 | 1515 mg/kg | Ikediobi et al., 1971 |

| Source | Substrate | Time/Temp | Toxin | Amount | Reference |
|---|---|---|---|---|---|
| Unknown, USA | Corn-rice mixture | 21 d/27° +24 d/8° | HT-2 | 750 mg/kg | Ikediobi et al., 1971 |
| Corn, USA | Corn grits | 3 wk/25° | T-2 | 0.67 g/1.2 kg | Lillehoj, 1973 |
| Corn, USA | Corn grits | 3 wk/20° | T-2 | 5.4 g/1.2 kg | Lillehoj, 1973 |
| Corn, USA | Corn grits | 3 wk/15° | T-2 | 9.9 g/1.2 kg | Lillehoj, 1973 |
| A-R-5 Blackened rice grains, Japan | Medium A[i] | 5 d/27° | T-2, NS | 67 mg/liter crude toxin | Ueno et al., 1975 |
| A-R-6 Blackened rice grains, Japan | Medium A | 5 d/27° | T-2, NS | 63 mg/liter crude toxin | Ueno et al., 1975 |
| A-R-7 Blackened rice grains, Japan | Medium A | 5 d/27° | T-2, NS | 56 mg/liter crude toxin | Ueno et al., 1975 |
| A-R-8 Blackened rice grains, Japan | Medium A | 5 d/27° | T-2, NS | 80 mg/liter | Ueno et al., 1975 |
| NRRL 3509 Unknown, USA | Wheat | 3 wk/12° | T-2 | 0.5 mg/10 g | Joffe and Yagen, 1977 |
| NRRL A-23377 Peanuts, USA | Liquid[j] culture | 32 d/27° | NSMA[k] | 250 mg/13 liter | Lansden et al., 1978 |
| NRRL 3197, USA | Cracked corn | 13 d/28° | VOM[l] | 33 mg/g | Vesonder et al., 1981a |
| NRRL 3197, USA | Cracked corn | 13 d/28° | T-2 | 1.5 mg/g | Vesonder et al., 1981a |

**TABLE 2.2** Continued

| Strain/Source | Substrate | Time/T °C | Myco-toxins | Concentrations | References |
|---|---|---|---|---|---|
| *F. tricinctum (F. sporotrichioides var. tricinctum)* (Cont.) | | | | | |
| 1 strain, wheat, Czechoslovakia | Wheat | 20 d/25° | ZEN | 50 mg/kg | Vesela *et al.*, 1982 |
| Fn-2B Barley, Japan | Rice | 2 wk/22–26° + 2 wk/10–12° | NIV[m] | 1.5 g/2 kg | Lee and Mirocha, 1984 |
| Fn-B Barley, Japan | Rice | 2 wk/22–26° + 2 wk/10–12° | FX[n] | 80 mg/2 kg | Lee and Mirocha, 1984 |
| *F. fusarioides (F. sporotrichioides var. chlamydosporum)* | | | | | |
| MRC 5 Peanuts | Corn | 21 d/25° | MON[o] | 0.42 g/kg | Rabie *et al.*, 1978 |
| MRC 7 Dried Fish, Mozambique | Corn | 21 d/25° | MON | 0.84 g/kg | Rabie *et al.*, 1978 |
| MRC 484 Soil, Pakistan | Corn | 21 d/25° | MON | 0.42 g/kg | Rabie *et al.*, 1978 |

| MRC 35 Millet, Southwest Africa | Corn | 21 d/25° | MON | 0.20 g/kg | Rabie et al., 1978 |
| 6 strains Millet, Namibia | Corn | 21 d/25° | MON | 0.32–1.30 g/kg | Rabie et al., 1982 |
| 1 strain Sorghum, Republic of South Africa (RSA) | Corn | 21 d/25° | MON | 1.47 g/kg | Rabie et al., 1982 |
| 1 strain Peanuts, Mozambique | Corn | 21 d/25° | MON | 0.80 g/kg | Rabie et al., 1982 |

[a] Diacetoxyscirpenol.
[b] Peptone supplemented Czapek-Dox medium.
[c] T-2 toxin.
[d] White corn grits.
[e] Potato dextrose agar.
[f] Zearalenone.
[g] Glucose, 40 g; corn steep liquor, 1 g; yeast extract, 2 g; water, 1 liter.

[h] Czapek-Dox supplemented with 10 g peptone per liter (CDSP).
[i] Glucose, 10 g; peptone, 1 g; yeast extract, 1 g; water, 1 liter.
[j] 15% sucrose; 5% mycological broth; 2% yeast extract.
[k] Neosolaniol monoacetate.
[l] Vomitoxin.
[m] Nivalenol.
[n] Fusarenon-X.
[o] Moniliformin.

**TABLE 2.3** The Concentrations of Mycotoxins from Strains of *F. oxysporum*, *F. solani* var. *coeruleum*, *F. semitectum*, and *F. concolor* Incubated on Substrates at Various Temperatures and Lengths of Time

| Strain/Source | Substrate | Time/$T$ °C | Myco-toxins | Concentrations | References |
|---|---|---|---|---|---|
| | | *F. oxysporum* | | | |
| T-M-1 Japan | Medium B (CDSP) | 5 d/27° | FX | 147 mg/liter | Ueno et al., 1975 |
| T-M-2 Japan | Medium B (CDSP) | 5 d/27° | FX | 131 mg/liter | Ueno et al., 1975 |
| Sp. 1028 Barley, RSA | Corn | 2 wk/25° | MON | 1150 ppm/g | Marasas et al., 1979b |
| 2 strains Sorghum malt, RSA | Corn | 21 d/25° | MON | 0.672–0.795 g/kg | Rabie et al., 1982 |
| 1 strain Barley malt, RSA | Corn | 21 d/25° | MON | 0.007 g/kg | Rabie et al., 1982 |
| 10 strains Peanuts, RSA | Corn | 21 d/25° | MON | 0.070–1.030 g/kg | Rabie et al., 1982 |

*F. solani* var. *coeruleum*

| | | | | | |
|---|---|---|---|---|---|
| Potatoes, Egypt | Potato tubers | HT-2 | 60 d/4° +4 d/23° | 0.24 µg/g | El-Banna et al., 1984 |
| Potatoes, Egypt | Potato tubers | DON[a] | 18 d/15° | 0.005–23.8 µg/g | El-Banna et al., 1984 |
| Potatoes, Egypt | Potato tubers | HT-2 | 18 d/15° | 0.29 µg/g | El-Banna et al., 1984 |
| Potatoes, Egypt | Potato tubers | ADON[b] | 18 d/15° | 1.6 µg/g | El-Banna et al., 1984 |
| *F. semitectum* | | | | | |
| Millet, Namibia | Corn | MON | 21 d/25° | 0.028 g/kg | Rabie et al., 1982 |
| *F. concolor* | | | | | |
| Millet, Namibia | Corn | MON | 21 d/25° | 9.55 g/kg | Rabie et al., 1982 |

[a] Deoxynivalenol.
[b] Acetyldeoxynivalenol.

**TABLE 2.4  The Concentrations of Mycotoxins from Strains of *F. equiseti*, *F. equiseti*, *F. equiseti* var. *acuminatum* Incubated on Substrates at Various Temperatures and Lengths of Time**

| Strain/Source | Substrate | Time/$T$ °C | Myco-toxins | Concentrations | References |
|---|---|---|---|---|---|
| | | *F. equiseti* | | | |
| 2 strains Sorghum malt, RSA | Corn | 21 d/25° | MON | 0.012–0.026 g/kg | Rabie *et al.*, 1982 |
| 1 strain Millet, RSA | Corn | 21 d/25° | MON | 0.017 g/kg | Rabie *et al.*, 1982 |
| | | *F. roseum* "Equiseti" (*F. equiseti*) | | | |
| 3 strains ATCC,[a] USA | Corn | 10 wk/16° or 2 wk/24° +8 wk/12° | ZEN | 0.6–2.0 µg/g | Caldwell *et al.*, 1970 |
| 9 Corn, USA | Corn | 4 wk/room +4 wk/12° | ZEN | 7.2–2.4 µg/g | Kotsonis *et al.*, 1975b |
| | | *F. roseum* "Scirpi" (*F. equiseti*) | | | |
| M-8-1 Bean-hulls, Japan | Medium C (CDSP) | 2 wk/27° | T-2; NS | 398 mg/liter (crude toxin) | Ueno *et al.*, 1975 |

| | | | | | |
|---|---|---|---|---|---|
| 3 strains ATCC, USA | Corn | 10 wk/15° or 2 wk/24° + 8 wk/12° | ZEN | 115–175 µg/g | Caldwell, 1970 |
| A-0-2 Feed | Rice | 14 d/25° + 14 d/12–15° | ZEN | 100 mg/kg | Ishii et al., 1974 |
| Corn, USA | Rice | 7 d/25° + 30 d/10–12° | ZEN | 563 µg/g | Hagler et al., 1979 |
| Corn, USA | Rice | 7 d/25° + 30 d/10–12° | ZEN | 8000 µg/g | Hagler et al., 1979 |
| Corn, USA | Rice | 7 d/25° + 30 d/10–12° | 8'HZEN[b] | 240 µg/g | Hagler et al., 1979 |
| Corn, USA | Rice | 7 d/25° + 30 d/10–12° | MAS[c] | 1000 ppm/kg | Pathre et al., 1976 |
| Corn, USA | Rice | 7 d/25° + 21 d/12° | ZEN | 12000 ppm/kg | Pathre et al., 1976 |

*F. acuminatum (F. equiseti var. acuminatum)*

| | | | | | |
|---|---|---|---|---|---|
| Dried bean leaves, Mozambique | Corn | 21 d/25° | MON | 0.012 g/kg | Rabie et al., 1982 |
| Millet, Namibia | Corn | 21 d/25° | MON | 3.40 g/kg | Rabie et al., 1982 |
| Sorghum, RSA | Corn | 21 d/25° | MON | 0.015 g/kg | Rabie et al., 1982 |

[a] American Type Culture Collection.
[b] 8'-Hydroxyzearalenone.
[c] Monoacetoxyscirpenol.

**TABLE 2.5  The Concentrations of Mycotoxins from Strains of *F. moniliforme* and *F. moniliforme* var. *subglutinans* Incubated on Substrates at Various Temperatures and Lengths of Time**

| Strain/Source | Substrate | Time/T °C | Myco-toxins | Concentrations | References |
|---|---|---|---|---|---|
| | | *F. moniliforme* | | | |
| 7 Corn, USA | Corn | 1 wk/room + 4 wk/12° | ZEN | 4.0 + 1.6 µg/g | Kotsonis et al., 1975b |
| Silos, Italy | Corn | 2 wk/25°, 8 wk/12° | ZEN | 2 ppm/g | Bottalico, 1976 |
| Maize, Italy | Corn | 2 wk/25°, 8 wk/12° | ZEN | 4.5–7.7 ppm/g | Bottalico, 1977 |
| 6 strains Corn, Mozambique | Corn | 21 d/25° | MON | 0.10–0.460 g/kg | Rabie et al., 1982 |
| 3 strains Sorghum, RSA | Corn | 21 d/25° | MON | 0.120–2.100 g/kg | Rabie et al., 1982 |
| 1 strain Sorghum, Mozambique | Corn | 21 d/25° | MON | 7.300 g/kg | Rabie et al., 1982 |
| 12 strains Millet, Namibia | Corn | 21 d/25° | MON | 0.010–33.70 g/kg | Rabie et al., 1982 |
| 3 strains Millet, Mozambique | Corn | 21 d/25° | MON | 0.100–1.40 g/kg | Rabie et al., 1982 |
| MRC 10 Corn, RSA | Corn | 21 d/25° | MON | 1.950 mg/kg[a] | Rabie et al., 1982 |

| Strain/Source | Substrate | Conditions | Toxin | Concentration | Reference |
|---|---|---|---|---|---|
| MRC 78 Corn, RSA | Corn | 21 d/25° | MON | 7.30 mg/kg[a] | Rabie et al., 1982 |
| MRC 1240 Corn, RSA | Corn | 21 d/25° | MON | 13.700 mg/kg[a] | Rabie et al., 1982 |

*F. moniliforme* var. *subglutinans*

| Strain/Source | Substrate | Conditions | Toxin | Concentration | Reference |
|---|---|---|---|---|---|
| MRC 115/1 Maize cobs, Transkei | Maize meal | 21 d/25° | MON | 9.7 g/kg | Kriek et al., 1977 |
| MRC 115/2 Maize cobs, Transkei | Maize meal | 21 d/25° | MON | 11.3 g/kg | Kriek et al., 1977 |
| MRC 115/3 Maize cobs, Transkei | Maize meal | 21 d/25° | MON | 6.1 g/kg | Kriek et al., 1977 |
| MRC 714 Corn, Northern Transvaal | Corn | 21 d/25° | MON | 533 mg/kg | Marasas et al., 1979a |
| MRC 750 Corn, Northern Transvaal | Corn | 21 d/25° | MON | 200 mg/kg | Marasas et al., 1979a |
| MRC 610 Corn, Western Transvaal | Corn | 21 d/25° | MON | 210 mg/kg | Marasas et al., 1979a |
| MRC 717 Corn, Western Transvaal | Corn | 21 d/25° | MON | 570 mg/kg | Marasas et al., 1979a |

**TABLE 2.5** Continued

| Strain/Source | Substrate | Time/$T$ °C | Myco-toxins | Concentrations | References |
|---|---|---|---|---|---|
| *F. moniliforme var. subglutinans* (Cont.) | | | | | |
| MRC 783 Corn, Western Transvaal | Corn | 21 d/25° | MON | 300 mg/kg | Marasas et al., 1979a |
| MRC 784 Corn, Western Transvaal | Corn | 21 d/25° | MON | 222 mg/kg | Marasas et al., 1979a |
| MRC 787 Corn, Western Transvaal | Corn | 21 d/25° | MON | 364 mg/kg | Marasas et al., 1979a |
| MRC 546 Corn, Eastern Transvaal | Corn | 21 d/25° | MON | 120 mg/kg | Marasas et al., 1979a |
| MRC 596 Corn, Eastern Transvaal | Corn | 21 d/25° | MON | 400 mg/kg | Marasas et al., 1979a |
| MRC 620 Corn, Eastern Transvaal | Corn | 21 d/25° | MON | 650 mg/kg | Marasas et al., 1979a |
| MRC 623 Corn, Eastern Transvaal | Corn | 21 d/25° | MON | 450 mg/kg | Marasas et al., 1979a |
| MRC 710 Corn, Eastern Transvaal | Corn | 21 d/25° | MON | 160 mg/kg | Marasas et al., 1979a |

| Source | Substrate | Conditions | Toxin | Yield | Reference |
|---|---|---|---|---|---|
| MRC 756 Corn, Eastern Transvaal | Corn | 21 d/25° | MON | 1.170 mg/kg | Marasas et al., 1979a |
| MRC 757 Corn, Eastern Transvaal | Corn | 21 d/25° | MON | 140 mg/kg | Marasas et al., 1979a |
| MRC 780 Corn, Eastern Transvaal | Corn | 21 d/25° | MON | 370 mg/kg | Marasas et al., 1979a |
| MRC 781 Corn, Eastern Transvaal | Corn | 21 d/25° | MON | 180 mg/kg | Marasas et al., 1979a |
| 1 strain Corn, Mozambique | Corn | 21 d/25° | MON | 0.900 g/kg | Rabie et al., 1982 |
| 3 strains Corn Sorghum, RSA | Corn | 21 d/25° | MON | 0.005–1.730 g/kg | Rabie et al., 1982 |
| 2 strains Corn Sorghum malt, RSA | Corn | 21 d/25° | MON | 0.020–0.948 g/kg | Rabie et al., 1982 |
| MRC 115 Corn Maize cobs, Transkei | Corn | 21 d/25° | MON | 11.300 mg/kg[a] | Rabie et al., 1982 |
| MRC 838 Corn, RSA | Corn | 21 d/25° | MON | 740 mg/kg[a] | Rabie et al., 1982 |

[a]Moniliformin yield in corn culture.

**TABLE 2.6  The Concentrations of Mycotoxins from Strains of *F. avenaceum* Incubated on Substrates at Various Temperatures and Lengths of Time**

| Strain/Source | Substrate | Time/*T* °C | Myco-toxins | Concentrations | References |
|---|---|---|---|---|---|
| | | *F. avenaceum* | | | |
| Sp. 803 Maize stems, Germany | Yellow maize kernels | 2 wk/25° +4 wk/10° | MON | 250 ppm/g | Marasas *et al.*, 1979b |
| Sp. 804 Barley grains, Germany | Yellow maize kernels | 2 wk/25° +4 wk/10° | MON | 2 ppm/g | Marasas *et al.*, 1979b |
| Sp. 805 Maize tassels, Germany | Yellow maize kernels | 2 wk/25° +4 wk/10° | MON | 7 ppm/g | Marasas *et al.*, 1979b |
| Sp. 881 Barley grains, Germany | Yellow maize kernels | 2 wk/25° +4 wk/10° | MON | 20 ppm/g | Marasas *et al.*, 1979b |
| Sp. 882 Barley grains, Germany | Yellow maize kernels | 2 wk/25° +4 wk/10° | MON | 80 ppm/g | Marasas *et al.*, 1979b |

| | | | | | |
|---|---|---|---|---|---|
| Sp. 888 Maize tassels, Germany | Yellow maize kernels | 2 wk/25° +4 wk/10° | MON | 10 ppm/g | Marasas et al., 1979b |
| Sp. 889 Maize tassels, Germany | Yellow maize kernels | 2 wk/25° +4 wk/10° | MON | 760 ppm/g | Marasas et al., 1979b |
| Sp. 894 Barley grains, Germany | Yellow maize kernels | 2 wk/25° +4 wk/10° | MON | 8 ppm/g | Marasas et al., 1979b |
| Sp. 896 Barley grains, Germany | Yellow maize kernels | 2 wk/25° +4 wk/10° | MON | 7 ppm/g | Marasas et al., 1979b |
| Barley malt, RSA | Corn | 3 wk/25° | MON | 0.032–1.20 g/kg | Rabie et al., 1982 |
| *F. roseum* "Avenaceum" (*F. avenaceum*) | | | | | |
| M-11-1 Bean-hulls, Japan | Medium C (CDSP) | 2 wk/27° | T-2, NS | 470 mg/liter | Ueno et al., 1975 |

**TABLE 2.7  The Concentrations of Mycotoxins from Strains of *F. nivale* and *F. epiphaeria* Incubated on Substrates at Various Temperatures and Lengths of Time**

| Strain/Source | Substrate | Time/$T$ °C | Myco-toxins | Concentrations | References |
|---|---|---|---|---|---|
| | | *F. nivale* | | | |
| Fn-2B Barley, Japan | Rice | 2 wk/27° | FX, NIV | 5.1 g/400 g (crude toxin) | Ueno et al., 1971a |
| Fn-2B Barley, Japan | PSC medium | 2 wk/27° | FX, NIV | 0.18 g/liter (crude toxin) | Ueno et al., 1971a |
| Fn-2-L-A Stock culture, Japan | Rice | 2 wk/25–27° | FX, NIV | 3.8 g/400 g (crude toxin) | Ueno et al., 1971a |
| Fn-2-L-A Stock culture, Japan | PSC medium | 2 wk/27° | FX, NIV | 0.18 g/liter (crude toxin) | Ueno et al., 1971a |
| Fn-2-L-B Stock culture, Japan | Rice | 2 wk/25–27° | FX, NIV | 3.7 g/400 g (crude toxin) | Ueno et al., 1971a |
| Fn-2-L-B Stock culture, Japan | PSC medium | 2 wk/27° | FX, NIV | 0.53 g/liter (crude toxin) | Ueno et al., 1971a |

| Strain/Source | Medium | Time/Temp | Toxins | Yield | Reference |
|---|---|---|---|---|---|
| Fn-2B Barley, Japan | Medium C (CDSP) | 5 d/27° | FX | 12.3/liter (crude toxin) | Ueno et al., 1975 |
| NRRL 3289 USA | Cracked corn | 13 d/28° | VOM | 1 mg/g | Vesonder et al., 1981a |
| *F. episphaeria (F. aquaeductuum)* | | | | | |
| Fn-M Vinyl plates, Japan | Rice | 2 wk/25°–27° | FX, NIV | 3.0 g/400 g (crude extract) | Ueno et al., 1971a |
| Fn-M Vinyl plates, Japan | PSC medium | 2 wk/27° | FX, NIV | 0.13 g/liter (crude extract) | Ueno et al., 1971a |
| Fn-M-L Stock culture, Japan | Rice | 2 wk/25°–27° | NIV | 4.4 g/400 g (crude extract) | Ueno et al., 1971a |
| Fn-M-L Stock culture Japan | PSC medium | 2 wk/27° | FX, NIV | 0.11 g/liter (crude extract) | Ueno et al., 1971a |
| Fn-M Vinyl plates, Japan | Medium C (CDSP) | 5 d/27° | FX | 7.1 g/liter | Ueno et al., 1975 |

**TABLE 2.8  The Concentrations of Mycotoxins from Strains of *F. culmorum*, *F. graminearum*, and *F. sambucinum* Incubated on Substrates at Various Temperatures and Lengths of Time**

| Strain/Source | Substrate | Time/T °C | Myco-toxins | Concentrations | References |
|---|---|---|---|---|---|
| | | *F. culmorum* | | | |
| 7289 USA | Liquid medium (CDP)[a] | 21 d/21° | DON | 20 mg/liter | Pathre and Mirocha, 1978 |
| 7289 USA | Rice | 21 d/21° | DON | 125 µg/g | Pathre and Mirocha, 1978 |
| Barley, England | Rice | 4 wk/25° + 6 wk/12° | ZEN | 2.2 mg/100 g | Hacking et al., 1976, 1977 |
| F.79 Millet, Hungary | Millet + 2% malt extract | 1 wk/23–27° + 2 wk/5–8° | ZEN | 540–3700 µg/ml | Szathmary et al., 1976 |
| F.247 Millet, Hungary | Millet + 2% malt extract | 1 wk/23–27° + 2 wk/5–8° + 1 wk/23–27° | ZEN | 200–3000 µg/ml | Szathmary et al, 1976 |
| FR22 Corn, USA | Starch-glutamine | 2 wk/16° | ZEN | 25 mg/liter | Bacon et al., 1977 |
| FR22 Corn, USA | Corn | 8 wk/10° | ZEN | 0.020 mg/g | Bacon et al., 1977 |
| 182 Poultry feed, USA | Starch-glutamine | 2 wk/16° | ZEN | 4.2 mg/liter | Bacon et al., 1977 |

| Strain/origin | Substrate | Conditions | Toxin | Amount | Reference |
|---|---|---|---|---|---|
| 182 Poultry feed, USA | Corn | 8 wk/10° | ZEN | 0.095 mg/g | Bacon et al., 1977 |
| Sp.878 Maize, Germany | Yellow maize kernels | 2 wk/25° +4 wk/10° | DON | 1 ppm/g | Marasas et al., 1979b |
| Sp.885 Barley, Germany | Yellow maize kernels | 2 wk/25° +4 wk/10° | ZEN DON ADON | 1400 ppm/g 7 ppm/g 1 ppm/g | Marasas et al., 1979b |
| Sp.886 Barley, Germany | Yellow maize kernels | 2 wk/25° +4 wk/10° | ZEN ADON | 600 ppm/g 1 ppm/g | Marasas et al., 1979b |
| Sp.887 Barley, Germany | Yellow maize kernels | 2 wk/25° +4 wk/10° | ZEN DON ADON | 320 ppm/g 15 ppm/g 2 ppm/g | Marasas et al., 1979b |
| Sp.892 Barley, Germany | Yellow maize kernels | 2 wk/25° +4 wk/10° | ZEN | 400 ppm/g | Marasas et al., 1979b |
| F 104 Germany | Maize | 21 d/25° +14 d/14° | ZEN | 325 mg/kg | Moller et al., 1978 |
| 13 strains Wheat, Czechoslovakia | Wheat | 20 d/25° | ZEN | 600–1400 mg/kg | Vesela et al., 1982 |

*F. roseum* "Culmorum" (*F. culmorum*)

| Strain/origin | Substrate | Conditions | Toxin | Amount | Reference |
|---|---|---|---|---|---|
| 3 strains ATCC, USA | Corn | 10 wk/16° or 2 wk/24° +8 wk/12° | ZEN | 1–20 µg/g | Caldwell et al., 1970 |

**TABLE 2.8** Continued

### F. graminearum

| Strain/Source | Substrate | Time/$T$ °C | Myco-toxins | Concentrations | References |
|---|---|---|---|---|---|
| F-59 Millet, Hungary | Millet + 2% malt extract | 1 wk/23–27° + 14 wk/5–8° + 14 wk/23–27° | ZEN | 200–730 µg/g | Szathmary et al., 1976 |
| F-184a Millet, Hungary | Millet + 2% malt extract | 1 wk/23–27° + 14 wk/5–8° + 14 wk/23–27° | ZEN | 350–1900 µg/g | Szathmary et al., 1976 |
| NRRL 5883 Corn, USA | Rice | 13 d/28° | VOM | 60 mg/kg | Vesonder et al., 1976 |
| 1 strain Corn, USA | Corn | 2 wk/25° + 8 wk/12° | ZEN | 24 µg/g | Bottalico, 1977 |
| MRC 460 Hand-selected corn, RSA | Yellow maize | 2 wk/25° + 6 wk/12° | ZEN DON | 1140.0 ppm/kg 1.3 ppm/kg | Marasas et al., 1977 |
| MRC 121 Corn | Yellow maize | 2 wk/25° + 6 wk/12° | ZEN DON | 1270.0 ppm/kg 6.0 ppm/kg | Marasas et al., 1977 |
| F-59 Millet, Hungary | Liquid medium (CDP) | 7 d/25° + 21 d/12° | DON | 15 mg/liter | Pathre and Mirocha, 1978 |
| F-59 Millet, Hungary | Rice | 7 d/25° + 21 d/12° | DON | 320 µg/g | Pathre and Mirocha, 1978 |

| Strain/Source | Substrate | Conditions | Toxin | Amount | Reference |
|---|---|---|---|---|---|
| 7137 Feedstuff, USA | Rice | 7 d/25° +21 d/12° | DON | 125 µg/g | Pathre and Mirocha, 1978 |
| 716-46 Feedstuff, USA | Rice | 7 d/25° +21 d/12° | DON | 256 µg/g | Pathre and Mirocha, 1978 |
| NRRL 5864 ARSCC,[e] USA | RWR[b] | 15 d/25°[c] 19 d/25° 38 d/25° | ZEN ZEN ZEN | 170 µg/g 200 µg/g 220 µg/g | Lindenfelser et al., 1978 |
| NRRL 5864 ARSCC, USA | Grain Sorghum | 15 d/25°[cd] 19 d/25° 38 d/25° | ZEN ZEN ZEN | 280 µg/g 400 µg/g 1500 µg/g | Lindenfelser et al., 1978 |
| NRRL 5883, USA | RWR | 2–3 wk/28° | VOM | 5 µg/g | Lindenfelser et al., 1978 |
| F 107 Gordon Collection, Canada | Maize | 21 d/25° +14 d/14° | ZEN | 248 mg/kg | Moller et al., 1978 |
| F 110 Gordon Collection, Canada | Maize | 21 d/25° +14 d/14° | ZEN | 324 mg/kg | Moller et al., 1978 |
| NRRL 6449 Corn, Australia | Cracked corn | 13 d/28° | VOM | 16 µg/g | Vesonder and Ciegler, 1979 |
| NRRL 6450 Corn, Australia | Cracked corn | 13 d/28° | VOM | 11 µg/g | Vesonder and Ciegler, 1979 |

**TABLE 2.8** Continued

| Strain/Source | Substrate | Time/T °C | Myco-toxins | Concentrations | References |
|---|---|---|---|---|---|
| | | *F. graminearum* (Cont.) | | | |
| NRRL 8451 Corn, Australia | Cracked corn | 13 d/28° | VOM | 16.7 µg/g | Vesonder and Ciegler, 1979 |
| NRRL 6452 Corn, Australia | Cracked corn | 13 d/28° | VOM | 3.7 µg/g | Vesonder and Ciegler, 1979 |
| NRRL 5883 Corn, USA | Cracked corn | 40 d/30° | VOM | 362 µg/g | Vesonder et al., 1982 |
| NRRL 5883 Corn, USA | Cracked corn | 30 d/28° | VOM | 250 µg/g | Vesonder et al., 1982 |
| NRRL 5883 Corn, USA | Cracked corn | 60 d/28° | VOM | 625 µg/g | Vesonder et al., 1982 |
| 3 strains Wheat, Czechoslovakia | Wheat | 3 wk/25° | VOM | 70–600 mg/kg | Vesela et al., 1982 Vesela and Vesely, 1983 |
| 3 strains Wheat, Czechoslovakia | Wheat | 3 wk/25° | ZEN | 250–600 mg/kg | Vesela and Vesely, 1983 |

| Source | Substrate | Conditions | Toxin | Amount | Reference |
| --- | --- | --- | --- | --- | --- |
| Rice Barley field, Japan | Rice + 3 ml of 10% peptone | 7 d/25° | NIV | 1.1–340 µg/g | Ichinoe et al., 1983 |
| Rice Barley field, Japan | Rice + 3 ml of 10% peptone | 7 d/25° | FX | 1.1–7.0 µg/g | Ichinoe et al., 1983 |
| Rice Barley field, Japan | Rice + 3 ml of 10% peptone | 7 d/25° | ZEN | 1.1–30 µg/g | Ichinoe et al., 1983 |

*F. roseum* "Graminearum" (*F. graminearum*)

| Source | Substrate | Conditions | Toxin | Amount | Reference |
| --- | --- | --- | --- | --- | --- |
| 21 strains USA | Corn | 10 wk/16° or 2 wk/24° +8 wk/12° | ZEN | 0.2–230 µg/g | Caldwell et al., 1970 |
| 2 Corn, USA | Corn | 1 wk/room +4 wk/12° | ZEN | 22.8–6.7 µg/g | Kotsonis et al., 1975b |
| 3 Corn, USA | Corn | 1 wk/room +4 wk/12° | ZEN | 47.8–10.6 µg/g | Kotsonis et al., 1975b |
| 5 Corn, USA | Corn | 1 wk/room +4 wk/12° | ZEN | 17.8–6.0 µg/g | Kotsonis et al., 1975b |
| 11 Corn, USA | Corn | 1 wk/room +4 wk/12° | ZEN | 13.8–5.4 µg/g | Kotsonis et al., 1975b |

**TABLE 2.8 Continued**

| Strain/Source | Substrate | Time/$T$ °C | Myco-toxins | Concentrations | References |
|---|---|---|---|---|---|
| | | *F. sulphureum (F. sambucinum)* | | | |
| Tomato, Canada | Corn | 2–6 wk/15° | HT-2 <br> T-2 <br> NS | 1.9–37.8 µg/g <br> 0.38–37.5 µg/g <br> 5.6 µg/g | Harwig *et al.*, 1979 |
| HPB110178-19 Tomato, Canada | Corn | 2–6 wk/19–25° <br> + 1–2 wk/4° | T-2 | 620 µg/g | Scott *et al.*, 1980 |
| | | *F. sambucinum* | | | |
| Potatoes, Egypt | Potato tubers | 60 d/4° <br> + 4 d/23° | DON | 0.017 µg/g | El-Banna *et al.*, 1984 |
| Potatoes, Egypt | Potato tubers | 60 d/4° <br> + 4 d/23° | NIV | 0.22 µg/g | El-Banna *et al.*, 1984 |
| Potatoes, Egypt | Potato tubers | 67 d/4° <br> + 4 d/23° | NIV | 0.046 µg/g | El-Banna *et al.*, 1984 |
| Potatoes, Egypt | Potato tubers | 67 d/4° <br> + 4 d/23° | HT-2 | 0.33 µg/g | El-Banna *et al.*, 1984 |
| Potatoes, Egypt | Potato tubers | 71 d/4° <br> + 4 d/23° | DON | 0.14 µg/g | El-Banna *et al.*, 1984 |

[a] Liquid medium Czapek-Dox peptone.
[b] Raw wild rice.
[c] Initial moisture content, 35.5%.
[d] Initial moisture content, 28%.
[e] ARSCC: Agricultural Research Service Culture Collection.

**TABLE 2.9  The Concentrations of Mycotoxins from Strains of G. *zeae* Incubated on Substrates at Various Temperatures and Lengths of Time**

| Strain/Source | Substrate | Time/$T$ °C | Myco-toxins | Concentrations | References |
|---|---|---|---|---|---|
| | | *G. zeae* | | | |
| FR 26 Corn, USA | Starch-glutamate | 2 wk/16° | ZEN | 35 mg/liter | Bacon, et al., 1977 |
| FR 26 Corn, USA | Corn | 8 wk/10° | ZEN | 0.320 mg/g | Bacon, et al., 1977 |
| FR 29 Corn, USA | Starch-glutamate | 2 wk/16° | ZEN | 40 mg/liter | Bacon, et al., 1977 |
| FR 29 Corn, USA | Corn | 8 wk/10° | ZEN | 0.420 mg/g | Bacon, et al., 1977 |
| 163 Corn, USA | Starch-glutamate | 2 wk/16° | ZEN | 86 mg/liter | Bacon, et al., 1977 |
| 163 Corn, USA | Corn | 8 wk/10° | ZEN | 0.596 mg/g | Bacon, et al., 1977 |

**TABLE 2.9** Continued

| Strain/Source | Substrate | Time/$T$ °C | Myco-toxins | Concentrations | References |
|---|---|---|---|---|---|
| | | *G. zeae* | | | |
| 165 Corn, USA | Starch-glutamate | 2 wk/16° | ZEN | 30 mg/liter | Bacon, et al., 1977 |
| 165 Corn, USA | Corn | 8 wk/10° | ZEN | 0.410 mg/g | Bacon, et al., 1977 |
| 265 Grain sorghum, USA | Starch-glutamate | 2 wk/16° | ZEN | 22 mg/liter | Bacon, et al., 1977 |
| 265 Grain sorghum, USA | Corn | 8 wk/10° | ZEN | 0.090 mg/g | Bacon, et al., 1977 |
| 266 Grain sorghum, USA | Starch-glutamate | 2 wk/16° | ZEN | 76 mg/liter | Bacon, et al., 1977 |
| 266 Grain sorghum, USA | Corn | 8 wk/10° | ZEN | 0.520 mg/g | Bacon, et al., 1977 |
| Ohoita-II Barley, Japan | Rice | 2 wk/25–27° | NIV | 4.2 g/400 g | Ueno et al., 1971a |

| | | | | | |
|---|---|---|---|---|---|
| Rice<br>Barley field,<br>Japan | Rice +<br>3 ml of<br>10% peptone | 7 d/25° | NIV | 1.1–67 µg/g | Ichinoe<br>et al., 1983 |
| Rice<br>Malting barley,<br>Japan | Rice +<br>3 ml of<br>10% peptone | 7 d/25° | DON | 3–134 µg/g | Ichinoe<br>et al., 1983 |
| Rice<br>Barley field,<br>Japan | Rice +<br>3 ml of<br>10% peptone | 7 d/25° | FX | 1.1–134 µg/g | Ichinoe<br>et al., 1983 |
| Rice<br>Malting barley,<br>Japan | Rice +<br>3 ml of<br>10% peptone | 7 d/25° | ADON | 1–34 µg/g | Ichinoe<br>et al., 1983 |
| Rice<br>Malting barley,<br>Japan | Rice +<br>3 ml of<br>10% peptone | 7 d/25° | ZEN | 1.1–3.0 µg/g | Ichinoe<br>et al., 1983 |
| 24 samples<br>Preharvest<br>corn,<br>USA | Rice | 13 d/28° | VOM | 0.5–10.7 µg/g | Vesonder<br>et al., 1978 |

**TABLE 2.10  The Concentration of Mycotoxins from Strains of *F. roseum* Incubated on Substrates at Various Temperatures and Lengths of Time**

| Strain/Source | Substrate | Time/$T$ °C | Myco-toxins | Concentrations | References |
|---|---|---|---|---|---|
| | | *F. roseum* | | | |
| 3 strains ATCC USA | Corn | 10 wk/16° or 2 wk/24° + 8 wk/12° | ZEN | 0.6–110 µg/g | Caldwell et al., 1970 |
| Mapleton 10, USA | Corn | 2–3 wk/22–25° | ZEN | 500–600 µg/g | Nelson, 1973 |
| M-3-2 Feed, Japan | Rice | 14 d/25° + 14 d/12–15° | ZEN | 40 mg/kg | Ishii et al., 1974 |
| M-3-2 Feed, Japan | Rice | 14 d/25° + 14 d/12–15° + 14 d/25° | ZEN | 250 mg/kg | Ishii et al., 1974 |
| M-3-2 Feed, Japan | Rice | The same + 1 percent peptone in rice grains | ZEN | 407 mg | Ishii et al., 1974 |

| | | | | | |
|---|---|---|---|---|---|
| My 73-45 Feedstuff, USA | Liquid medium (CDP) | 3 wk/21° | DON | 10 mg/liter | Pathre and Mirocha, 1978 |
| My 73-45 Feedstuff, USA | Rice | 7 d/25° +3 wk/12° | DON | 55 µg/g | Pathre and Mirocha, 1978 |
| India-3 Feedstuff, USA | Rice | 14 d/25° +14 d/12–15° | DON | 40 µg/g | Pathre and Mirocha, 1978 |
| India-3 Feedstuff, USA | Liquid medium (CDP) | 3 wk/21° | DON | 2–5 mg/liter | Pathre and Mirocha, 1978 |
| FS-R-3 Feedstuff, USA | Rice | 7 d/25° +3 wk/12° | DON | 70 µg/g | Pathre and Mirocha, 1978 |
| NRRL 6101 Barley, USA | Cracked corn | 41 d/26° | VOM | 189 µg/g | Vesonder et al., 1982 |
| NRRL 6101 Barley, USA | Cracked corn | 60 d/28° | VOM | 170 mg/g | Vesonder et al., 1982 |

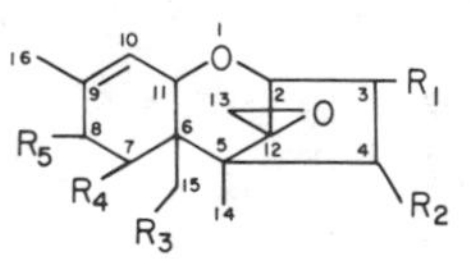

**Figure 2.1** The basic chemical structure of trichothecene mycotoxins and related tetracyclic sesquiterpenoid ring system.

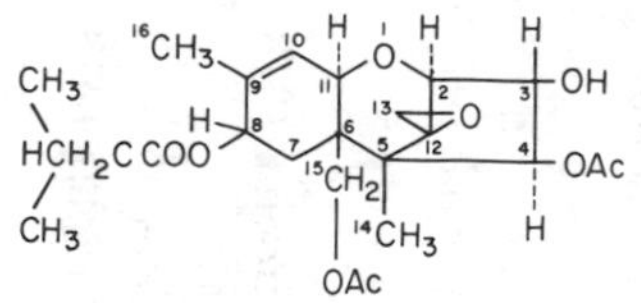

**Figure 2.2** Chemical structure of T-2 toxin ($R_1$ = OH, $R_2$ = OAc, $R_3$ = OAc, $R_4$ = H, $R_5$ = $(CH_3)_2CH_2OCO$).

the number and position of the hydroxy-group, and on the type of the esterifying acids.

Chemical modification of the hydroxyl group at C-3 (R1), C-4 (R2), C-15 (R3), C-7 (R4), and C-8 (R5) and reduction of the double bond of C-9, C-10 increased the toxicity of the trichothecene compounds.

The best known trichothecene compounds produced in significant quantities by various strains of the *Fusarium* are mainly T-2 toxin (4β, 15-diacetoxy-3α-hydroxy-8α-(3-methylbutyryloxy)-12,13-epoxytrichothec-9-ene) (Fig. 2.2) and some others such as HT-2 toxin (15-acetoxy-3α, 4β-dihydroxy-8α-(3-methylbutyryloxy)-12, 13-epoxytrichothec-9-ene) (Fig. 2.3); neosolaniol (4β, 15-diacetoxy-3α,8α-dihydroxy-12,13-epoxy-trichothec-9-ene) (Fig. 2.4); diacetoxyscirpenol (4β, 15-diacetoxy-3α-hydroxy-12, 13-epoxytrichothec-9-ene) (Fig. 2.5); nivalenol (3α, 4β, 7α, 15-tetrahydroxy-12, 13-epoxytrichothec-9-en-8-one) (Fig. 2.6); fusarenon-X (4β-acetoxy-3α, 7α, 15-trihydroxy-12, 13-epoxytrichothec-9-en-8-one) (Fig. 2.7); and deoxynivalenol (3α,7α,15-trihydroxy-12,13-epoxy-trichothec-9-en-8-one) (Fig. 2.8).

Trichothecene toxins are crystalline, colorless, and chemically stable and can be stored for a long time without significant deterioration. Structure elucidation and determination of relative and absolute configuration was achieved first for trichotecolone, trichodermol, and verrucarol. Then, by the use of chemical and spectroscopic methods, mainly NMR and X ray spectroscopy, the majority of *Fusarium* toxins were related to these three trichothecenes (Bamburg, 1969; Bamburg and Strong, 1971).

The chemistry of certain trichothecene toxins has been studied thoroughly, including degradation, partial and total synthesis, and ring open-

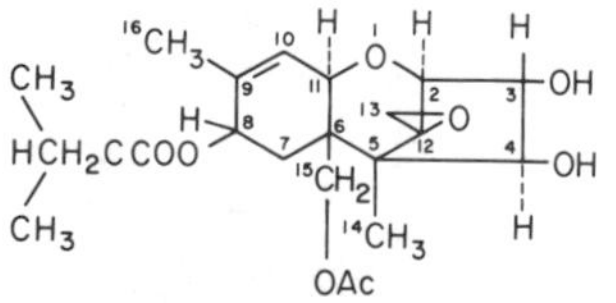

**Figure 2.3** Chemical structure of HT-2 toxin ($R_1$ = OH, $R_2$ = OH, $R_3$ = OAc, $R_4$ = H, $R_5$ = $(CH_3)_2CH_2OCO$).

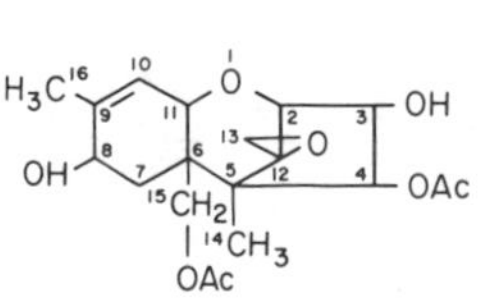

**Figure 2.4** Chemical structure of neosolaniol ($R_1$ = OH, $R_2$ = OAc, $R_3$ = OAc, $R_4$ = H, $R_5$ = OH).

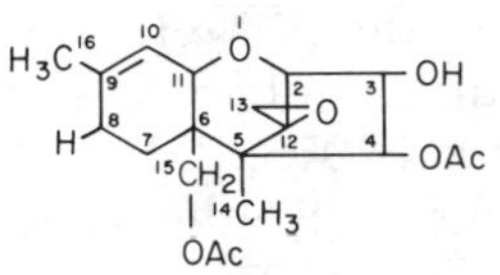

**Figure 2.5** Chemical structure of diacet-oxyscirpenol ($R_1$ = OH, $R_2$ = OAc, $R_3$ = OAc, $R_4$ = H, $R_5$ = H).

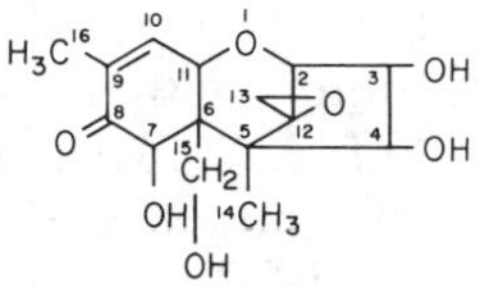

**Figure 2.6** Chemical structure of niva-lenol ($R_1$ = OH, $R_2$ = OH, $R_3$ = OH, $R_4$ = OH, $R_5$ = O).

ings, while with most toxins few chemical studies were reported apart from isolation and simple acetylation and hydrolysis.

The main and general reactions performed and those possible with *Fusarium* trichothecenes will be summarized here. Detailed descriptions are given by Bamburg (1969, 1972, 1976); Bamburg and Strong (1971); Bamburg *et al.* (1968a, 1969); Kurata (1978); Saito and Tatsuno (1971); Ueno (1977a,b, 1980a,b,c); Ueno and Ueno (1978); and Yamazaki (1978).

The hydroxyl groups at the various positions are readily acetylated and the esters saponified easily. Thus a dilute solution of potassium carbonate, sodium hydroxide, or ammonium hydroxide hydrolyzes the T-2 toxin and neosolaniol to T-2 tetraol and mono- and diaceto-xyscirpenol to scirpentriol (Mirocha *et al.*, 1977a,b). Some selective acetylation and deacetylation is achieved enzymatically.

Catalytic hydrogenation reduces the C-9–C-10 olefinic bond, while LiAlH4 opens the 12,13-epoxy ring giving tertiary alcohol at C-12.

Oxidation of alcohols at C3, C4, or C8 positions with various $CrO_3$ reagents yields a ketone at the corresponding position.

Allylic oxidation with $SeO_2$ yields C-8-keto and C-16 aldehyde from the methylene at C-8 and methyl at C-16. Epoxidation with perbenzoic acid gives oxirane ring at the olefinic C-9–C-10 bond.

The total synthesis of simpler derivatives of trichothecenes has been accomplished (Colvin *et al.*, 1973; Fujimoto *et al.*, 1972, 1974).

The epoxide ring at C-12–C-13 is relatively shielded to nucleophilic attack and quite resistant to alkali treatment, while mineral acids or prolonged boiling in water open the epoxide ring and lead to rearranged tricyclic skeleton (Bamburg and Strong, 1971; Mirocha *et al.*, 1977a).

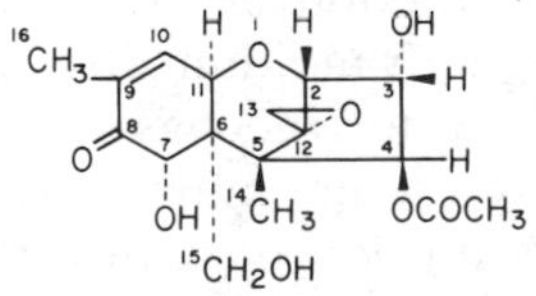

**Figure 2.7** Chemical structure of fusare-non-X ($R_1$ = OH, $R_2$ = OAc, $R_3$ = OH, $R_4$ = OH, $R_5$ = O).

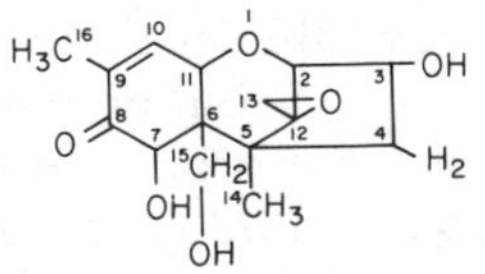

**Figure 2.8** Chemical structure of deoxy-nivalenol ($R_1$ = OH, $R_2$ = H, $R_3$ = OH, $R_4$ = OH, $R_5$ = O).

Most of the analytical methods and procedures were developed mainly to detect either T-2 toxin or deoxynivalenol in feed samples (Forsyth *et al.*, 1977; Hsu *et al.*, 1972; Ishii *et al.*, 1975; Puls and Greenway, 1976; Vesonder *et al.*, 1976). Other methods (Eppley *et al.*, 1974; Nakano *et al.*, 1973; Naoi *et al.*, 1974; Mirocha *et al.*, 1976b; Romer *et al.*, 1978; Tatsuno *et al.*, 1973; Yoshizawa *et al.*, 1976) were developed for detection of diacetoxyscirpenol, T-2 toxin, fusarenon-X, and nivalenol. However, none of these methods have been sufficiently evaluated for recovery and sensitivity (Eppley *et al.*, 1979).

At present, because of the importance of the trichothecene mycotoxins, it is necessary to work out a simple analytical method for their rapid detection in foodstuffs and feeds contaminated by toxigenic *Fusarium* species.

The toxic *Fusarium* strains grown on suitable sterile substrates at different temperatures according to their taxonomy are also briefly described in this chapter.

## Sporotrichiella Section

*Fusarium poae.*   According to Bamburg *et al.* (1968a) and Marasas *et al.* (1969), *F. tricinctum* T-2 (*F. poae*) was grown at 8°C for 30 days on Gregory's medium (Gregory *et al.*, 1952) producing T-2 toxin.

This fungus was shown to produce another toxin (apart from T-2 toxin and diacetoxyscirpenol), HT-2 toxin, when grown at 24°C for 14 days (Bamburg and Strong, 1969; Bamburg *et al.*, 1968b).

According to Wei *et al.* (1971), T-2 toxin was obtained from *F. tricinctum* NRRL 3299 (*F. poae*) grown at 8°C on corn. The authors developed a procedure for the interconversion of T-2 toxin and HT-2 toxin. *F. tricinctum* NRRL 3299 (*F. poae*) produced T-2 toxin and butenolide when incubated at 15°C on Sabouraud agar for 10 days (Yates *et al.*, 1970; Grove *et al.*, 1970). T-2 toxin was also extracted from this strain when grown on white corn grits (Wyatt *et al.*, 1972b, 1973a, 1975a,b).

Ueno *et al.* (1971a, 1973d) stated that *F. tricinctum* NRRL 3299 produced neosolaniol (NS), T-2 toxin, HT-2 toxin, and diacetoxyscirpenol (DAS) when grown on a peptone-supplemented Czapek medium (PSC) at 25–27°C for 2 weeks.

*F. tricinctum* NRRL 3299, isolated from toxic corn, was grown 3 weeks at 15°C on white corn grits. The production of T-2 toxin was highest on white corn grits and at 8°C on Gregory's medium, less on rice, and was nonexistent on wheat grown at 20°C. A decline was seen in the amount of toxin produced when the incubation temperature was raised to 20°, 25°, and 32°C; 4 µg T-2 toxin, when applied on a disc of filter paper, was sufficient to inhibit the growth of *Rhodotorula rubra* NRRL Y-7222 (a fungus which is sensitive to T-2 toxin) (Burmeister, 1971). Detection of T-2 toxin-producing fusaria was performed by utilizing sensitive microorganisms (*Rhodotorula rubra* and *Penicillium digitatum*). The *Fusarium* strains inhibited both indicator microorganisms, but T-2 toxin was not detected by TLC in 3 cases (Burmeister *et al.*, 1972).

T-2 toxin was produced from *F. tricinctum* NRRL 3299 (*F. poae*), also on rice incubated at room temperature for 10 days, and at 11°C for 3 months (Weaver *et al.*, 1978a,c). *F. tricinctum* NRRL 3287 (*F. poae*) when grown on PSC at 25–27°C for 12 days, produced T-2 toxin, HT-2 toxin, and neosolaniol (Ueno *et al.*, 1972b, 1973d), and on corn and Richard's solution at 25°C produced T-2 toxin (Ellison and Kotsonis, 1973). Acetyl T-2 toxin (3α, 4β, 15-triacetoxy-8α-isovaleroxy-12, 13-epoxy-$\Delta^9$-trichothecene) was isolated from this strain which was also grown in Richard's solution for 4 weeks. This metabolite was suspected of being the immediate biogenetic precursor to T-2 toxin (Kotsonis *et al.*, 1975a).

*Fusarium trincinctum* NRRL 3299 (*F. poae*) was found to produce T-2 toxin and some other trichothecene compounds and used in experiments on animals by Boonchuvit *et al.* (1975); Chi and Mirocha (1978); Chi *et al.* (1977a,b,c, 1978b, 1981); Coulter *et al.* (1977); Cullen and Smalley (1981); Doer *et al.* (1974, 1981); Hagler *et al.* (1981); Joffe and Yagen (1977); Patterson *et al.* (1979, 1980); Weaver *et al.* (1978a,b,d, 1980); Witlock *et al.* (1977); Wyatt *et al.* (1972b, 1973a,b,c, 1975a,b).

According to Vesonder *et al.* (1977a, 1981a) *F. poae* NRRL 3287, when grown on cracked corn at 28°C for 13 days, produced T-2, HT-2, acetyl T-2 toxins, fusarenon-X, and vomitoxin, and *F. tricinctum* NRRL 3197 produced T-2 toxin and vomitoxin.

*F. tricinctum* T-2, NRRL 3287 and NRRL 3299 were reidentified by Joffe and Palti (1975) as *F. poae* and have been shown to produce T-2 toxin at levels of 1.0, 3.4, and 5.9 mg/10 g wheat grain, respectively (Joffe and Yagen, 1977).

*F. poae* isolates (from actual cases of ATA in Orenburg, USSR) were grown on wheat for 21 days at 12°C. A total of 25 isolates of *F. poae* produced 0.1 to 21 mg/ml T-2 toxin (Joffe, 1960a; Yagen and Joffe, 1976).

An extract of *F. poae* obtained by Mirocha from the USSR was found to produce T-2 toxin, neosolaniol, T-2 tetraol, and zearalenone (Mirocha and Pathre, 1973). Bukharbayeva and Piotrovski (1972) reported that *F. sporotrichiella* var. *poae* (Pk) Bilai 365 (*F. poae*) isolated from toxic wheat harvested in Alma Ata, Kazakhstan in 1963 and grown on Bilai synthetic medium, produced a steroid similar to poaefusarin. They named this new steroid poaefusariol.

**F. sporotrichioides.** *F. sporotrichioides* M-1-1 (originally classified as *F. solani*) was reidentified by Ichinoie (1978). It was isolated from bean-hulls used as feed and bedding for horses in Hokkaido, produced neosolaniol, diacetoxyscirpenol (DAS), and mainly T-2 toxin when grown on PSC for 12 days at 27°C (Ishii and Ueno, 1981; Ishii *et al.*,

1971; Kato *et al.*, 1979; Masuko *et al.*, 1977; Matsumoto *et al.*, 1978; Ohta *et al.*, 1977; Sato and Amano, 1976; Sato *et al.*, 1975, 1978; Tatsuno *et al.*, 1973; Ueno *et al.*, 1972a, 1975; Yoshizawa and Morooka, 1975a,b; Yoshizawa *et al.*, 1980a).

*F. sporotrichioides* M-1-1 was found to produce T-2 toxin, HT-2 toxin, NS, DAS, and butenolide (Ueno *et al.*, 1972b, 1973b,d).

*F. sporotrichioides* M-1-4, M-1-5, as well as NRRL 3510 (*F. sporotrichioides* NRRL 3510 (=738) isolated from overwintered millet) (Joffe, 1960a) produced the largest quantities of T-2 toxin and NS when grown on the PSC medium in shake culture for 5 days at 27°C (Ueno *et al.*, 1975).

T-2 toxin was isolated from *F. sporotrichioides* NRRL 3510 (=738) by Burmeister (1972, 1981); Joffe and Yagen (1977); Lafarge-Frayssinet *et al.* (1979); Lafont and Lafont (1980); Ueno *et al.* (1972b,d).

We have isolated T-2 toxin (Joffe and Yagen, 1977, Joffe 1978b) from the following strains: NRRL 3249, 5908, 2061-C; YN-13 from the United States (which were widely used in a variety of studies) which were reidentified as *F. sporotrichioides* by Joffe and Palti (1975).

The following researchers have isolated T-2 toxin from strains of *F. sporotrichioides* from overwintered cereal grains in Russia: Ermakov *et al.* (1978); Joffe (1960a,b, 1962a, 1965, 1969a, 1971, 1973b, 1974b,c, 1978b, 1983b,c); Joffe and Yagen (1977, 1978); Kvashnina (1979); Kotik *et al.* (1979); Yagen and Joffe (1976); and Yagen *et al.* (1977, 1980). In Canada it was isolated by Mills and Frydman (1980) and Scott *et al.* (1980), and in Hungary by Szathmary *et al.* (1976, 1977).

From a strain of *F. sporotrichioides* 72187 (from cereal grain in Finland) T-2 toxin, NS, and HT-2 toxin were also isolated by Ilus *et al.* (1977, 1981), Niku-Paavola and Nummi (1977), and Niku-Paavola *et al.* (1977).

A total of 71 very toxic isolates of *F. sporotrichioides* were grown for 21 days at 12°C on autoclaved wheat producing 4–21 mg/ml T-2 toxin in crude extract obtained from 10 g infected wheat (Yagen and Joffe, 1976).

Poaefusarin and sporofusarin (as reported by Olifson, 1965a,b and Olifson *et al.*, 1969, 1972, 1975) were not identified in the toxic cultures of *F. sporotrichioides* and *F. poae* (Joffe and Yagen, 1977; Szathmary *et al.*, 1976; Yagen *et al.*, 1977).

**F. sporotrichioides var. tricinctum.** *F. tricinctum* NRRL 3509 (*F. sporotrichioides* var. *tricinctum*) of unknown origin, grown at 12°C for 21 days, produced 0.5 mg/10 g of wheat grain of T-2 toxin (Joffe and Yagen, 1977).

*F. sporotrichioides* var. *tricinctum* cultures isolated from overwintered wheat, barley, and millet produced various amounts of T-2 toxin (1.7–2.9 g/kg wheat grain) when grown at 12°C for 21 days (Joffe, unpublished data). These strains isolated from wheat, rye, and barley in the

1977 harvest year and from wheat in 1978 in the United States produced T-2 toxin: 32 out of 55 isolates were from wheat, 7 out of 10 from rye, and 16 out of 18 from barley, along with 7 zearalenone and 1 DAS from wheat and 1 DAS from rye (Eppley and Joffe, unpublished results).

**F. sporotrichioides** var. **chlamydosporum.** *F. sporotrichioides* var. *chlamydosporum* was isolated from overwintered wheat, barley, and millet and also from soil. Some isolates when grown at 12°C for 21 days produced T-2 toxin (Joffe, unpublished results).

Isolates of *F. fusarioides* (*F. sporotrichioides* var. *chlamydosporum*) from beans, sorghum, peaches, soil, dried fish, peanuts, and quite frequently from millet grown on corn at 25°C for 21 days, produced moniliformin. *F. fusarioides* MRC 35 incubated at 31°C for 21 days produced the largest amounts of moniliformin (Rabie *et al.*, 1978).

Strains of *F. sporotrichioides* var. *chlamydosporum* were isolated from wheat, rye, and barley in the 1977 harvest in the United States. Only two strains out of four from rye produced T-2 toxin and one DAS (Eppley and Joffe, unpublished results).

Abbas *et al.* (1984) studied the toxicity of extracts from 34 various *Fusarium* species (including *F. poae, F. sporotrichioides*, and *F. tricinctum*) isolated from widely separated geographic areas of the world. The authors have isolated from *F. sporotrichioides, F. poae*, and *F. tricinctum* the following mycotoxins: T-2, NS, HT-2, 4-deacetylneosolaniol, 8-acetyl T-2 tetraol, T-2 tetraol, 8-acetylneosolaniol, deacetyl HT-2, and acetyl T-2 and several unknown trichothecenes.

**F. tricinctum.** *F. tricinctum* B-24 when grown at 7°C produced diacetoxyscirpenol (Gilgan, 1966; Bamburg *et al.*, 1968a). A toxin (related to diacetoxyscirpenol) was extracted from this fungus grown at 8°C for 30 days (Marasas *et al.*, 1967).

Smalley *et al.* (1970) found that the most toxic isolates from moldy corn were of *F. tricinctum*, and that toxin production in liquid culture by *F. tricinctum* is enhanced by a low incubation temperature. At higher temperatures less toxin is produced, and after a long incubation period the toxin is degraded.

T-2 toxin and butenolide from *F. tricinctum* were tested on fungi, bacteria, and plants (pea seeds). Fifty μg of T-2 toxin was not bacteriostatic but was phytotoxic and fungistatic. Butenolide was not fungistatic, phytotoxic, or bacteriostatic (Burmeister and Hesseltine, 1970).

Kosuri *et al.* (1971) isolated T-2 toxin from *F. tricinctum* which caused stalk or ear rot of maize.

Three *F. tricinctum* strains were isolated from wheat and 10 from barley. All were cultured on a 1 : 1 : 1 mixture of sterile wheat-barley-oats at 20°C for 2 weeks, 4°C for 5 days, and 20°C for a further 2 weeks. The authors noted a great variation in the toxic effect of different strains (Korpinen and Uoti, 1974).

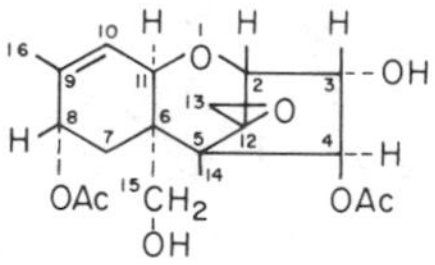

**Figure 2.9** Chemical structure of T-1 toxin ($R_1$ = OH, $R_2$ = OAc, $R_3$ = OH, $R_4$ = H, $R_5$ = OAc).

*F. roseum* as well as *F. tricinctum* contamination of corn was indicated in a survey of corn of the 1972 United States harvest. Seventeen percent of the samples were found to contain zearalenone, and 54% of the samples tested had a skin irritating factor when tested on rabbits. The authors assumed that the skin irritating factor was T-2 toxin (Eppley *et al.*, 1974).

Four isolates of *F. tricinctum*, isolated in 1973 from blackened rice grains in Hokkaido (AR-5, AR-6, AR-7 and AR-8), produced the largest amount of crude toxin in PSC incubated for 5 days at 27°C in shake culture. However, thin layer chromatography was negative for T-2 and neosolaniol. A medium containing glucose, peptone, and yeast extract, incubated as above, yielded a very small amount of crude toxin, but the TLC analyses were positive for T-2 toxin and neosolaniol (Ueno *et al.*, 1975).

*F. tricinctum* No. 72187 was cultured on a 1:1:1 mixture of sterile wheat-barley-oats at 20°C for 2 weeks, 4°C for 5 days, and again at 20°C for a total of 4 months. T-2, HT-2, neosolaniol, and a new toxin named T-1 (Fig. 2.9) (4β, 8α-diacetoxy-3α-15-dihydroxy-12,13-epoxytrichotec-9-ene) (isomeric with neosolaniol) were isolated (Ilus *et al.*, 1977).

According to Eugenio *et al.*, 1970a; Ishii *et al.*, 1974; and Mirocha *et al.*, 1969 *F. tricinctum, F. oxysporum, F. moniliforme, F. solani, F. decemcellulare*, and *F. lateritium* were not capable of producing zearalenone (ZEN) when grown on autoclaved rice grain for 2 weeks at 22–24°C, and then 5 weeks at 10–12°C, or for 2 weeks at 22–25°C, followed by 8 weeks at 8–11°C.

Only one of 4 isolates of *F. tricinctum* was able to produce ZEN on rice grains with incubation of 14 days at 25°C and then 14 days at 12–15°C (Ishii *et al.*, 1974).

*F. tricinctum* produced zearalenone on autoclaved maize after 16°C incubation for 3 weeks, as shown by TLC analysis (Lasztity *et al.*, 1977).

## Martiella Section

*F. solani (F. javanicum).*   In Israel *F. solani* and *F. javanicum*, isolated from various substrates, were phytotoxic to plants and toxic on rabbit skin when incubated at 24°C for 10–12 days (*see* Chapter 5).

A highly toxic *F. solani* was isolated from cabbage bought in a supermarket. The fungus was grown on nutrient-amended shredded wheat and incubated 14–21 days at 25°C (Davis *et al.*, 1975).

T-2 toxin and neosolaniol were obtained from *F. solani*, cultivated on PSC medium and incubated at 27°C for 5 days on a rotary shaker, according to Ueno *et al.* (1973d, 1975) and Sato *et al.* (1975).

Six strains of *F. solani* were isolated from river sediments by Ueno *et al.* (1977b). They were incubated on PSC medium at 25–27°C for 2 weeks and analyzed for trichothecenes. Then they were incubated on rice at 24°C for 2 weeks and 10–15°C for one week and then analyzed for zearalenone. Chemical analyses did not show the presence of trichothecenes, and one of the isolates produced zearalenone.

*F. solani* produces zearalenone when grown on maize for 3 weeks at 16°C, as seen by analyses by TLC (Lasztity *et al.*, 1977).

El-Banna *et al.* (1984) isolated *F. solani* var. *coeruleum* and *F. sambucinum* from potato tubers in Egypt. From the tubers contaminated with *F. solani* var. *coeruleum*, the authors obtained deoxynivalenol (DON), acetyldeoxynivalenol (ADON), nivalenol (NIV), and HT-2 toxin (HT-2), and from those contaminated with *F. sambucinum*, they obtained NIV, DON and HT-2 toxin.

The results proved that mycotoxins of Type B trichothecenes may occur naturally in potatoes contaminated with *Fusarium* strains and could be dangerous to livestock.

## Elegans Section

*F. oxysporum.*   In Israel *F. oxysporum* from various substrates was phytotoxic and toxic on rabbit skin when cultivated at 24°C for 10–12 days (*see* Chapter 5).

*F. oxysporum*, isolated from corn suspected of causing toxicosis in cattle, was inoculated onto corn and incubated 2 weeks at 22–25°C followed by 3–11 weeks at 10–12°C (Meronuck *et al.*, 1970).

Thirteen isolates of *F. oxysporum* were tested for their capability to produce trichothecene mycotoxins. They were grown on PSC medium and incubated 2 weeks at 25–27°C. Two strains of *F. oxysporum* var. *niveum*, both from melon, showed the presence of fusarenon-X and diacetylnivalenol when analyzed with thin layer chromatography (TLC) (Ueno *et al.*, 1973d).

*F. oxysporum* No. 597 was isolated from a carrot purchased in a supermarket. The crude extract was found to be moderately toxic to brine shrimp and lethal to all chick embryos when the fungus was grown on nutrient-amended shredded wheat 14–21 days at 25°C (Davis *et al.*, 1975).

Nelson *et al.* (1973), reported that certain isolates of *F. oxysporum* were found to produce large amounts of zearalenone.

El-Kady and El-Maraghy (1982) examined by TLC analysis zearalenone, isolated from 36 strains of *F. oxysporum*, 8 *F. moniliforme* and 1 *F. equiseti*. The authors stated that the *F. oxysporum* strains produced the maximum quantities of zearalenone at pH 7 after 12 days of incubation at 20°C on synthetic medium containing mainly glutamine and riboflavin.

## Liseola Section

*F. moniliforme.* One of the toxins produced by *F. moniliforme* is zearalenone (F-2).

F-3, a compound related, but not identical, to F-2, was produced by an isolate of *F. moniliforme*. This compound appeared in small amounts in cultures and has been found in feed samples suspected of causing abortion in dairy cattle in Minnesota (Mirocha *et al.*, 1968a,b).

F-2 was produced by two isolates of *F. moniliforme* (from shelled corn and feed) inoculated on autoclaved corn, and incubated for 2 weeks at 22–26°C and then for 6 weeks at 10–12°C. Identification of F-2 was accomplished by TLC, gas liquid chromatography (GLC), and infrared spectrophotometry. GLC also disclosed the presence of F-3 (Mirocha *et al.*, 1969).

Nelson *et al.* (1973) found some isolates of *Fusarium moniliforme* (cultivated on autoclaved moist corn) able to produce zearalenone but only in small amounts.

*F. moniliforme*, incubated on autoclaved maize for 3 weeks at 16°C, was found to produce zearalenone (detectable by TLC). The authors stated that the amount of zearalenone is increased with prolonged storage of the grain and that biosynthesis is activated by low temperatures (Lasztity *et al.*, 1977).

Fourteen strains of *F. moniliforme*, isolated from corn samples, produced F-2 toxin (Stankushev *et al.*, 1977).

In a survey made in Lesotho and Swaziland, Martin and Keen (1978) found *F. moniliforme* in raw and fermented food products. This isolate, when incubated on sterile maize at 25°C for 7 days, produced zearalenone.

Moniliformin has also been shown to be produced by strains of *F. moniliforme* (Burmeister *et al.*, 1979; Cole *et al.*, 1973; Hayes and Hood, 1977; Rabie *et al.*, 1982; Springer *et al.*, 1974; Steyn *et al.*, 1978a).

Some other compounds in addition to zearalenone and moniliformin are produced by *F. moniliforme*, and histopathological changes in animals were noted by Prentice *et al.* (1959) and Prentice and Dickson (1964, 1968), who examined various species of *Fusarium* for the production of emetic substances on artificial media. They found that *F. moniliforme, F. culmorum, F. nivale,* and *F. poae* produced emetic activity when grown on Richard's solution. One substance produced by *F. moniliforme* was found to be active at doses 100 μg when given intravenously to pigeons but was not lethal; the second also produced emesis but killed the pigeons in less than 12 hr. These chemical compounds were not investigated.

Scott (1965) studied the toxicity of *F. moniliforme* strains on 10-day-old white Pekin ducklings and found that two of them caused acute toxicoses.

Casas-Campillo and Bautista (1965) studied the factors involved in the hydroxylation of estrogens produced by four strains of *F. moniliforme* and *G. fujikuroi*. The authors proved that only two strains of *F. moniliforme* (1H 4 and 9851) were the most effective in the accumulation of significant amounts of 15α hydroxyestrone.

Arai and Ito (1970) described the biological activity of fusariocin A, a new toxin isolated from a strain of *F. moniliforme*.

Ueno *et al.* (1977a) observed radiomimetic injury and karyorrhexis in the thymus and lymph nodes of mice administered with the crude toxins isolated from *F. moniliforme* Nos. 5003 and 5012.

Van Rensburg *et al.* (1971) described the histopathological changes occurring in mice fed 10 strains of *F. moniliforme*. Only one mouse died, and the survivors exhibited severe loss of weight. The authors observed hyperemia in the renal glomeruli and vascular degenerative changes in the proximal tubules associated with necrosis in all mice.

Birone *et al.* (1972) stated that *F. moniliforme* inhibited the spermiogenesis in 20% of the white rats.

According to Korpinen and Ylimaki (1972) *F. moniliforme* strains isolated from oats and homemade feed caused colic and gastric symptoms in horses and commercial hen feed caused retarded growth in chickens.

Marasas and Smalley (1972) noted that one-day-old White Leghorn cockerels died after 14 days of feeding on a pure culture of *F. moniliforme* incubated at 10°C and isolated from maize.

Fritz *et al.* (1973) stated that *F. moniliforme* No. 13 paralyzed four White Plymouth Rock chicks with signs of typical thiamine deficiency. After intraperitoneal injection of 2.5 mg thiamine hydrochloride the chicks appeared to be normal. When again the chicks were subsequently fed cultures of *F. moniliforme* they were paralyzed with the typical syndrome of polyneuritis. *F. moniliforme* was also administered with and without additional thiamine in the diet, but chicks that received the *F. moniliforme* without thiamine developed polyneuritis and all died within 11–12 days.

According to Sharby *et al.* (1973) various levels of corn diet infected separately with two isolates of *F. moniliforme* caused significant reduction in weight gain and feed efficiency of chicks (when incubated for 4, 6, and 8 weeks at room temperature). The 12-day-old chicks exhibited severe leg deformity and had caeca distended with brown material, mild enteritis, and lesions of the sciatic nerves.

Vanyi *et al.* (1973a) also stated that rats died after feeding on maize contaminated with *F. moniliforme* and showed intestinal catarrh and enteritis without changes in the uterus (white rats fed grain contaminated with strains of *F. graminearum* or *F. culmorum* developed uterine hypertrophy).

Archer (1974) noted that all of the *Fusarium* species (*F. equiseti*, *F. graminearum* and *F.* sp.) proved nontoxic except some extracts of *F. moniliforme*, which caused teratogenic effects in chick embryos.

Marasas *et al.* (1978, 1979a,c) reported that corn samples with diseased kernels were found to be infected mainly with *F. moniliforme* which caused marked loss of weight in rats. The authors stated that none of the 14 toxic isolates of *F. moniliforme* was found to produce moniliformin.

Marasas *et al.* (1978) noted that *F. moniliforme* MRC 137 isolated from Zambian maize caused the death of two ducklings and one rat of a group of 4 ducklings and 4 rats. The mycotoxin produced by this fungus has not yet been characterized chemically.

Steyn *et al.* (1979) reported on four dark red pigments which were isolated and identified from *F. moniliforme* Sheld. (MRC 602), isolated from moldy maize ears in the Transkei and incubated on autoclaved maize at 25°C for 21 days. The main pigment was 6.8-dimenethoxy-5-hydroxy-3-methyl-2-azaanthra-quinone and three chemically related naphthaquinones. It is important to state that the culture of *F. moniliforme* (MRC 602) did not produce moniliformin.

Robb *et al.* (1982) isolated diacetoxyscirpenol from an extract of *F. moniliforme* by TLC and GLC.

All these reports (communications) reveal that *F. moniliforme* is a very important mycotoxin producer.

Five strains of *F. moniliforme* out of 37 isolated from overwintered cereals (wheat and barley) and soil were toxic (Joffe, 1960a, 1971).

Out of 145 isolates of *F. moniliforme* (from Israel and foreign sources), 89 produced toxic skin reactions when tested on rabbit skin after incubation on wheat grains for 10–11 days at 24°C (*see* Table 5.13).

***F. moniliforme* var. *subglutinans*.**   A very toxic strain of *F. moniliforme* var. *subglutinans* (MRC 115), isolated from maize in the Transkei, South Africa, was grown on autoclaved yellow maize at 25 ± 2°C in the dark for 21 days. The culture yielded 10 g/kg moniliformin (Steyn *et al.*, 1978a).

*F. moniliforme* var. *subglutinans* was the most prevalent *Fusarium* isolated from corn in the Eastern Transvaal region of South Africa (the area with the warmest climate), and the majority of toxic strains were isolated there (Kriek *et al.*, 1977).

The moniliformin was isolated from *F. moniliforme* var. *subglutinans* grown on maize (Thiel, 1978, 1981).

Out of 15 isolates of *F. moniliforme* var. *subglutinans* (from Israel and foreign sources), 11 produced toxic skin reactions on rabbits (*see* Table 5.13).

## Spicarioides Section

**F. decemcellulare.**   Four isolates of *F. rigidiusculum* (*F. decemcellulare*) were incubated on PSC for 5 days at 27°C with shaking. Isolate M-1-3 produced neosolaniol, T-2 toxin, and diacetoxyscirpenol under conditions as seen with TLC (Ueno *et al.*, 1973d).

Out of four isolates of *F. decemcellulare* (isolated from foreign sources) three produced toxic reactions on rabbit skin (*see* Table 5.8).

## Gibbosum Section

**F. equiseti.**   Brian *et al.* (1961) examined *F. equiseti* and *F. scirpi* (*F. equiseti*) and found that these strains produced a highly phytotoxic compound diacetoxyscirpenol (DAS). Dawkins (1966); Dawkins *et al.* (1965); Flory *et al.* (1965); and Sigg *et al.* (1965) isolated a substance from *F. equiseti* which was also identified as DAS. This compound is a potent toxin, both for plants and animals.

Six strains of *F. equiseti* (isolated from fescue hay) were grown on white corn grits at 15°C for 21 days. They were then screened for the production of T-2 toxin by using TLC in conjunction with inhibition of T-2 sensitive microorganisms (*Rhodotorula rubra* and *Penicillium digitatum*). Five isolates produced T-2 toxin, as seen by inhibition of both microorganisms and confirmed by TLC, but the sixth only inhibited *P. digitatum*, and the TLC analysis was negative for T-2 toxin (Burmeister *et al.*, 1972).

*F. roseum* "Scirpi" (*F. equiseti*) M-8-1 grown on PSC for 2 weeks at 25–27°C was found to produce NS, T-2, and DAS. Mice died after i.p. injections of 10 and 40 mg/10 g, and 14 C-leucine uptake in rabbit reticulocytes was inhibited. None of three isolates of *F. gibbosum* (*F. equiseti*) produced trichothecene toxins under the same conditions when examined with TLC, but one isolate did inhibit protein synthesis (Ueno *et al.*, 1973b).

A strain of *F. roseum* "Gibbosum" (*F. equiseti*) was isolated from feed suspected of causing illness and death in dairy cattle and abortion in swine. It was grown on rice for 7 days at 25°C and then 21 days at 12°C, and a toxic metabolite, monoacetoxyscirpenol (MAS), and zearalenone were isolated (Mirocha and Christensen, 1974b).

On solid corn grits *F. roseum* "Gibbosum" also produced monoacetoxyscirpenol, a major toxic compound (15-acetoxy-3α, 4β-dihydroxy-12, 13-epoxytrichothec-9-ene) and scirpentriol (3α, 4β, 15-trihydroxy, 12, 13-epoxytrichothec-9-ene) and a small amount of ZEN. (Diacetoxyscirpenol was produced as well but only in a liquid Czapek's medium and not in rice or corn medium) (Pathre *et al.*, 1976).

Autoclaved corn (40% moisture content) was inoculated with _F. roseum_ "Gibbosum" (_F. equiseti_) and incubated at 25°C for 2 weeks and then at 12°C for 4 weeks. Analyses of the infected corn showed 2.500 ppm zearalenone, 50 ppm monoacetoxyscirpenol, and no T-2 toxin (Speers _et al._, 1977).

Ishii _et al._ (1978) isolated _F. roseum_ "Gibbosum" (_F. equiseti_) from corn harvested in Minnesota and, after growing it on Czapek-peptone medium, obtained a new trichothecene toxin, 4-acetoxyscirpenediol (4β-acetoxy-3α,15-dihydroxy-12,13-epoxytrichothec-9-ene). From _F. roseum_ V-18 the authors isolated another compound which proved to be (a monoacetate of neosolaniol) 8-acetylneosolaniol or monoacetyl-neosolaniol (4β, 8α, 15-triacetoxy-3α-hydroxy-12,13-epoxytrichothec-9-ene) and which was also isolated from Czapek-peptone medium.

Gareis _et al._ (1984) noted that 0.05–0.125% sorbic acid stimulated T-2 toxin production of _F. acuminatum_ (_F. equiseti_ var. _acuminatum_) grown on maize meal. The maximum production of T-2 toxin was detected in 14-day-old cultures containing 0.025% sorbic acid. The identity of T-2 toxin was carried out by GLC and mass spectroscopy. The stimulation of toxin production may be due to the special antimicrobial action of sorbic acid.

Out of 162 isolates of _F. equiseti_ (isolated from Israel and foreign sources) 115 produced toxic skin reactions on rabbits (_see_ Table 5.14).

Schroeder and Hein (1975), using _F. roseum_ "Gibbosum," demonstrated that higher yields of zearalenone were produced on grain sorghum when grown at 25°C for 9 days than at 10°C. Stipanovic and Schroeder (1975) isolated strain _F. roseum_ "Gibbosum" S-74-1C from sorghum grains. This strain was grown on autoclaved grain sorghum or cracked yellow corn for 20 days at 25°C and the authors obtained zearalenone and two compounds: one was zearalenone and the second compound was B-8-hydroxyzearalenone.

Hagler and Mirocha (1980) found that 14 C-labeled acetate incorporation into zearalenone by _F. roseum_ "Gibbosum" incubated at 10 and 25°C on rice varied between 1.63 and 46.5 μCi-mmol depending on the substrate, temperature of incubation, and time of addition of the isotope to the medium.

## Lateritium Section

_F. lateritium._ When incubated on corn steep liquor medium for 5 days at 26°C on a shaker, _F. lateritium_ produced diacetoxyscirpenol (Cole and Rolinson, 1972).

Eleven strains of _F. lateritium,_ isolated from fescue hay, were grown on white corn grits and incubated at 15°C for 21 days. All isolates were found to produce trichothecene T-2 toxin when tested by TLC, and to inhibit the two indicator microorganisms, _R. rubra_ and _P. digitatum_ (Burmeister _et al._, 1972).

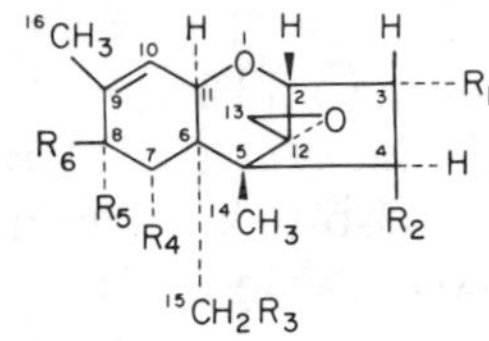

**Figure 2.10** Trichothecene mycotoxins produced by *F. lateritium* No. 5036.

|  |  | $R_1$ | $R_2$ | $R_3$ | $R_4$ | $R_5$ | $R_6$ |
|---|---|---|---|---|---|---|---|
| I. | diacetoxyscirpenol | OH | OAc | OAc | H | H | H |
| II. | 7α-hydroxydiacetoxyscirpenol | OH | OAc | OAc | OH | H | H |
| III. | 7,8α-dihydroxydiacetoxyscirpenol | OH | OAc | OAc | OH | OH | H |
| IV. | neosolaniol | OH | OAc | OAc | H | OH | H |
| V. | diacetylnivalenol | OH | OAc | OAc | OH | └─O─┘ |  |

*F. lateritium* M produced no toxic metabolites (as seen by TLC and the use of bioassays) when grown on PSC for 2 weeks at 25–27°C (Ueno *et al.*, 1973d).

Four strains of *F. lateritium* were isolated from river sediment by Ueno *et al.*, 1977b. One isolate was not lethal, while the other three produced radiomimetic injury and karyorrhexis in mice. Rice incubated at 24°C for 2 weeks followed by 10–15°C for 1 week with strain 5013 produced zearalenone. On PSC strain 5036 was seen to produce five fractions: I: diacetoxyscirpenol; II: 7α-hydroxydiacetoxyscirpenol (4β, 15-diacetoxy-3α,7α-dihydroxy-12, 13-epoxytrichothec-9-ene); III:7α, 8α-dihydroxydiacetoxyscirpenol (4β, 15-diacetoxy-3α, 7α, 8α-trihydroxy-12, 13-epoxytrichothec-9-ene) (Ishii, 1975; Ueno *et al.*, 1977b); IV: neosolaniol; and V: diacetylnivalenol (Fig. 2.10). The $LD_{50}$ for compound II is 3.5 mg/kg and III is 6 mg/kg for male mice i.p. Bleeding in the intestines was marked. The fungus 5036 was seen to produce both type A and type B trichothecene (Ueno *et al.*, 1977b).

Out of 43 isolates of *F. lateritium* (from Israel and foreign sources) 23 produced toxic reactions on rabbit skin (*see* Table 5.12).

An isolate of *F. lateritium* No. 553 (from orange) was grown on nutrient-amended shredded wheat for 14–21 days at 25°C. Extract of the culture was highly toxic to brine shrimp and fatal to 50% of chicken embryos (Davis *et al.*, 1975).

*F. lateritium* isolate was found to produce zearalenone when grown for 14 days at 25°C and for a further 14 days at 12–15°C on autoclaved rice grain. The presence of zearalenone was seen with TLC (Ishii *et al.*, 1974).

## Arthrosporiella Section

*F. semitectum.* Isolation of zearalenone and zearalenol from cultures of *F. roseum* "Semitectum" (*F. semitectum*) grown on moist autoclaved grain sorghum for 20 days at 20°C was reported by Stipanovic and Schroeder (1975) and Schroeder and Hein (1975).

Out of 45 isolates of *F. semitectum* (from Israel and foreign sources) 24 produced toxic reactions on rabbit skin when inoculated on autoclaved wheat at 24°C for 10 days (*see* Table 5.11).

*F. semitectum* var. *majus.* Isolated from moldy sorghum, *F. incarnatum* (*F. semitectum* var. *majus*) was grown on rice grains for 1 week at 30°C. Ether and ethyl acetate extracts of both the *Fusarium*-infected rice and the moldy sorghum were toxic when injected i.p. into rats. There were marked skin reactions to all extracts tested on guinea pigs. Protein synthesis in rabbit reticulocytes was inhibited. Rats treated with extracts of both cultures did not exhibit increase in uterine weights. T-2 toxin was detected by means of TLC (Rukmini and Bhat, 1978).

Out of 12 isolates (11 from Israel and 1 from a foreign source) 9 produced toxic reactions on rabbit skin (*see* Table 5.11).

### Roseum Section

*F. avenaceum.* One of two isolates of *F. roseum* "Avenaceum" (*F. avenaceum*) produced zearalenone when grown on rice grains for 14 days at 25°C followed by 14 days at 12–15°C (Ishii *et al.*, 1974), and one strain of *F. avenaceum* isolated from barley seed when grown 10 days at 25°C also produced ZEN (Hacking *et al.*, 1976).

*F. roseum* "Avenaceum" produced neosolaniol, T-2 toxin, and diacetoxyscirpenol when incubated on PSC for 2 weeks at 25–27°C. The crude toxin inhibited protein synthesis in rabbit reticulocytes and was lethal to mice (Ueno *et al.*, 1973d).

Lappe and Barz (1978) stated that *F. anguioides* (*F. avenaceum*) and *F. avenaceum* caused degradative reaction for pisatin isolated from infected pea seedlings.

Out of 32 isolates (from Israel and foreign sources) 23 produced toxic reactions on rabbit skin (*see* Table 5.10).

### Arachnites Section

*F. nivale.* Joffe (1960a) isolated *F. nivale* from millet in the USSR. Crude extract of millet on which this isolate was grown was toxic to rabbit skin.

*F. nivale* was isolated from moldy wheat (Ide *et al.*, 1967; Okubo *et al.*, 1966a,b, 1969; Tatsuno, 1968, 1969; Tatsuno *et al.*, 1968a,b, 1971, 1973; Ueno *et al.*, 1969a,b,c, 1970a,b, 1971b) and from barley (Ueno *et al.*, 1971a). This fungus was inoculated on sterilized rice, wheat, and PSC at 25–27°C for 6–14 days from which some trichothecene mycotoxins (nivalenol, fusarenon-X, *et al.*) were isolated (Morooka and Tatsuno, 1970; Morooka *et al.*, 1971; Ohtsubo *et al.*, 1972; Saito *et al.*, 1969;

Ueno *et al.*, 1969c, 1970a,c, 1971b,c, 1973b,d). On rice the incubation temperatures were 24–27°C, and production of fusarenon-X was higher that at temperatures below 20°C. *F. tricinctum* produces diacetoxyscirpenol at lower temperatures, in contrast to fusarenon-X (Ueno *et al.*, 1970c).

Vesonder *et al.* (1973) isolated *F. nivale*, as well as *F. graminearum* (which was the predominant species), *F. moniliforme*, and others from samples of 1972 corn from a northwest Ohio farm. This corn had caused vomiting in swine. Each fraction during the toxin isolation procedure was assayed for its emetic activity by intubation into pigs. The toxic metabolite isolated was given the name vomitoxin.

*F. nivale* NRRL 3289, cultured on corn (35–40% mc) and incubated at 28°C for 13 days, was then offered to swine. Forty-two percent of the corn was refused by the swine (Vesonder *et al.*, 1977a). Only one out of 21 isolates of *F. nivale* isolated from barley grain in England was shown to produce zearalenone (Hacking *et al.*, 1976) when grown on rice for 4 weeks at 25°C and then 6 weeks at 12°C.

## Eupionnotes Section

*F. episphaeria.*  Isolated from vinyl plates on the southern island of Japan, *F. episphaeria* (*F. aquaeductuum*) FnM was incubated on PSC at 25° and 27°C for 2 weeks. TLC examination showed the presence of fusarenon-X and nivalenol. Another strain of *F. episphaeria* Fn-M-L grown on rice for 2 weeks at 27°C produced only nivalenol. The third strain—M—was negative (Ueno *et al.*, 1973d).

## Discolor Section

*F. culmorum.*  When incubated on PSC medium at 25–27°C for 2 weeks, *F. roseum* "Culmorum" (*F. culmorum*) was found, by Ueno *et al.* (1973d), to produce neosolaniol, T-2 toxin, HT-2 toxin, and DAS.

Mirocha *et al.* (1976b) isolated deoxynivalenol and zearalenone from *F. culmorum* when the isolate was grown on a rice substrate for 1 week at 24–27°C and 4 weeks at 14°C.

According to Vesonder *et al.* (1977a) corn of 35–40% m.c. incubated for 13 days at 25°C with *F. culmorum* produced vomitoxin.

When an isolate of *F. culmorum*, originally from barley harvested in Northeast England with 18–20% moisture content, was grown on autoclaved rice for 6 weeks at room temperature and then 1 week at 15°C, it produced F-2 (zearalenone) (Miller *et al.*, 1973).

*F. culmorum* and *F. graminearum* (from grain and fodder samples in Hungary) grown on maize under optimal conditions produced 900 ppm of F-2 toxin (Vanyi *et al.*, 1973a).

Gross and Robb (1975) artificially infected a barley field with *F. culmorum* (W. C. Smith) Sacc., a species occurring naturally in Scotland.

The isolate was taken from barley associated with a case of infertility in pigs. The infected field was harvested at 18.4% moisture content, the control field at 19.8%. Samples taken from both fields showed the presence of *F. culmorum* but no zearalenone. After 20 weeks of storage, both the control and the artificially infected barley contained zearalenone.

*F. culmorum* No. 53, from an Irish dairy pasture, produced culmorin and 4β,15-diacetoxy-3α,7α,dihydroxy-12,13-epoxytrichothec-9-en-8-one (acetyl fusarenon). Another strain (No. 34) produced culmorin and two new fungal metabolic products, 3α-acetoxy-7α,15-dihydroxy-12,13-epoxytrichothec-9-en-8-one (acetyldeoxynivalenol) and 2-acetylquinazolin-4(3H)-one. Strain 53 was incubated in glucose-ammonium nitrate medium at 25°C for 14 days. The resulting extract produced a necrotic skin reaction on albino rat skin. Strain 34 was incubated on Czapek-Dox medium at 25°C for 10 days (Grove, 1970a; Blight and Grove, 1974).

Cyclonerotriol [6-(3-hydroxy-2,3-dimethylcyclopentyl)-2-methylhept-2-ene-1,6-diol], a sesquiterpenoid metabolite, was isolated from a culture of *F. culmorum*. The culture was grown on Raulin Thom medium for 21 days stationary or 10 days shake culture (Hanson *et al.*, 1975).

**F. graminearum.**   Ishii *et al.* (1975) isolated deoxynivalenol from a sample of *F. graminearum* (*G. zeae*)-infected corn from the 1972 crop in the United States Midwest.

Corn of 35–40% mc was fermented by *F. graminearum* NRRL 5883 for 13 days at 28°C. TLC and infrared analysis showed that the corn contained vomitoxin, but analyses for T-2 toxin, HT-2 toxin, acetyl T-2, and fusarenon-X were negative (Vesonder *et al.*, 1977a).

Vesonder *et al.* (1982) isolated *F. graminearum* NRRL 5883 from corn and *F. roseum* NRRL 6101 from barley. Both these *Fusarium* species produced vomitoxin. The *Fusarium* strain isolated from corn produced twice as much vomitoxin on cracked corn as the *Fusarium* strain from barley under the same conditions of growth. Maximum production of vomitoxin by *F. graminearum* occurred at 30°C for 40 days and that by *F. roseum* at 26°C for 41 days.

The authors isolated, from 1g of infected cracked corn with *F. graminearum* NRRL 5883, 133 µg/g of vomitoxin.

Vomitoxin was not isolated from rice (35–40% mc) which was incubated with *F. graminearum* NRRL 5883 for 13 days at 25°C, but resulted in the identification of a new metabolite, 4-acetamido-2 butenoic acid (Vesonder *et al.*, 1977b).

Vesonder *et al.* (1981) isolated vomitoxin and zearalenone from 16 *F. graminearum* and *F. culmorum* strains cultivated on cracked corn at 28°C. Vomitoxin was produced in quantities of 5–236 µg/g of infected corn. This toxin showed weak antibiotic activity against *P. digitatum*, *Mucor ramannianus*, and *Saccharomyces bayanus* but did not inhibit gram positive and gram negative bacteria.

Both the fungi and yeast were inhibited by 5 μg of T-2 toxin while DAS inhibited the fungi at 5 μg and the yeast at 50 μg.

These two toxins (vomitoxin and zearalenone) are frequently found in cereal grains and corn in the field (Vesonder and Ciegler, 1979; Vesonder et al., 1973).

*F. graminearum* (106 isolates) isolated from barley and wheat in southern Japan were cultured on a 0.5% peptone-enriched Czapek-Dox medium at 25°C for 2 weeks. These strains produced various trichothecenes (mainly deoxynivalenol or nivalenol derivatives), butenolide, and zearalenone (Yoshizawa et al., 1979).

*F. graminearum* (*G. zeae*), isolated from moldy Indiana corn in 1957–1958, was inoculated onto sterile ground corn (35% mc) and, when inoculated at 24°C for 2–3 weeks and then at a low temperature for a few weeks, produced zearalenone (Stob et al., 1962).

Low amounts of vomitoxin were obtained with *F. graminearum* NRRL 5883 incubated on wild rice (35.5% mc) at 28°C for 2–3 weeks. *F. graminearum* NRRL 5864 incubated on wild rice (mc 35.5, 37, and 40%) at 25°C for 19 and 38 days showed a yield of 220 μg/g zearalenone. Wild rice was seen to be an inferior substrate for zearalenone and vomitoxin production (Lindenfelser et al., 1978).

Mirocha et al. (1967) produced F-2 by growing *F. graminearum* on moist autoclaved corn for 2 weeks at 25–28°C, followed by different periods of time at 12°C.

Mirocha et al. (1968b) showed that yields of zearalenone by *F. graminearum* were highest at 27°C when inoculated for 3, 5, 7, and 9 days on sterile corn, after the culture was first subjected for a sufficient amount of time at 12°C.

Christensen and Kaufman (1969) produced F-2 by inoculating *F. graminearum* onto corn and incubating at 20–25°C for 2 weeks and then several months at 12°C. They also used other grains, including rice.

Kurtz et al. (1969) prepared crystalline F-2 by growing *F. graminearum* on autoclaved corn and incubating at 20–25°C for 3 weeks, followed by 2 weeks at 12°C. In Hungary Palyusik (1973) obtained F-2 from maize infected with *F. graminearum* and incubated 1 week at room temperature and 1 month at 10–12°C.

Zearalenone was extracted from rice infected with *F. graminearum* by Ueno et al. (1974c) in Japan.

*F. graminearum* (isolated from cereal grains in Finland) was cultured on a mixture of sterile wheat, barley, and oats (1:1:1) for 2 weeks at 20°C. The overall toxicity of *F. graminearum* was found to be intermediate, with that of *F. tricinctum* most toxic and *F. poae* least toxic (Korpinen and Uoti, 1974).

Extracts of two isolates of *F. graminearum*, 152102 and 152103, from hay (shown by Mirocha et al., 1968a to be producers of zearalenone), were found to be nonlethal to chick embryos (Archer, 1974). The isolates had been grown on cereals (rye, maize, and wheat) and incubated at 25°C for 2–3 weeks, and then at 12°C for a further 2–3 weeks. Only some extracts of *F. moniliforme* were toxic to chick embryos.

Lasztity and Wöller (1975a,b) stated that the optimum temperature for toxin production by *F. graminearum* on sterile maize is 12°C. They also stated that an increase in the incubation period produces an increase in the production of the toxin. After incubation at 16°C for three weeks on autoclaved maize, the authors found that zearalenone was produced by *F. roseum* "Culmorum," *F. roseum* "Equiseti," *F. roseum* "Gibbosum," *F. roseum* "Graminearum," and *F. tricinctum*. No production of zearalenone was observed for *F. moniliforme, F. nivale, F. oxysporum,* and *F. solani*. Mirocha *et al.* (1968b, 1969), however, found *F. moniliforme* to produce zearalenone.

Berisford and Ayres (1976a,b) incubated *F. graminearum* on potato dextrose broth at 27°C for 2 weeks, which provided them with enough zearalenone ($2 \times 10^3$ µg) for use in their study of the inhibition of zearalenone production by naled. They found that naled, at levels of 30 and 100 µg/liter, completely inhibited toxin production both in potato dextrose broth and on corn. Naled must be applied before the initial growth of the fungus.

There are many reports in the literature on the effects of moisture and temperature on the production of zearalenone. Christensen *et al.* (1965) commented that large amounts of zearalenone can be produced in corn, both in the field or in corn crib as a result of a spring in which the temperature is above freezing and there is an increase in moisture.

Moisture content of the grain is also important for the growth of *F. graminearum* in the field and in storage.

Mirocha *et al.* (1971), describing the conditions of harvest and storage of corn in the corn belt of the United States, stated that corn is, in most cases, harvested dry and stored on the cob in open cribs of wire mesh. Fall rains can add 20–22% to the moisture content of the harvested ears, thereby aiding the proliferation of *F. graminearum*. Favorable conditions for zearalenone synthesis are created by autumnal weather, warmer during the day and cold at night, often with a cold period following a warm one. The estrogenic syndrome in swine can often be traced to corn stored in conditions such as these.

Sherwood and Peberdy (1972a,b, 1974a) experimentally examined the production of zearalenone by *F. graminearum* in barley, wheat, oats, and corn, in relation to storage conditions. The greatest toxin production was on wheat with a 37% moisture content, incubated at 25°C for 4 weeks and then at 12°C for 6 weeks. Cultures transferred to 12°C after a period of incubation at 25°C produced the greatest amount of zearalenone (500–2000 µg/g grain). Toxicity tests were not performed on animals. At moisture contents above 18%, not even grains stored at low temperatures (7°C) were free from *Fusarium* growth. A difference was seen in the various grains in their tolerance to increases in storage temperatures at moisture contents of 18 and 24%, which are optimal for the establishment of *F. graminearum* at low temperatures. Maize and wheat were less resistant than oats and barley to the establishment of *F. graminearum* at low temperatures. Maximum zearalenone concentrations in all four grains were at 12 and 18°C, especially when coupled with high moisture content. Reduced temperatures during the period of growth of *F. graminearum* were shown to increase zearalenone production.

Caldwell and Tuite (1974) and Tuite *et al.* (1974) described the weather conditions conducive to *G. zeae* infection of corn. In Indiana in 1972 planting was delayed owing to a very wet spring. Summer was unusually cool and fall mild. Harvest was delayed because of wet weather in some areas until January 1973. That year's corn caused much trouble among hog farmers because it contained zearalenone.

Mirocha *et al.* (1974) stated that a high moisture environment is needed for the production of zearalenone in barley, oats, and wheat, with alternate warm days followed by cold nights or cold spells.

Shotwell *et al.* (1975) reported that zearalenone was discovered, along with *Aspergillus flavus,* in a bin of corn of crop year 1972, harvested in January 1973 because of unfavorable weather conditions and dried artificially to 15% moisture content. The window of the bin had been open the entire winter and sleet, rain, or snow could have entered and dampened the corn. With enough moisture zearalenone could have been synthesized at the low temperatures.

In 83.5% of corn visibly contaminated with *Fusarium,* especially *F. graminearum,* Collet *et al.* (1977) found zearalenone. They also stated that the warm, wet winter of southwest France favors development of the fusaria.

Lasztity *et al.* (1977) reported on the 1972 maize crop in Hungary, which was contaminated with *F. graminearum* and exhibited high zearalenone activity in swine and other domestic animals. They mentioned the autumn of 1972 as being cool and rainy, and stated that toxin production is greatly influenced by temperature. They also mentioned that after a long period of storage of corn, it is possible to find an increased amount of zearalenone toxin, which they said agreed with observations that the biosynthesis of zearalenone is activated by low temperatures.

Lovelace and Nyathi (1977) found zearalenone in samples from Zambia of home-brewed maize beer. The beer examined had been brewed with corn from the 1974 season which had been stored for 8–12 months prior to use. During poor harvest years in particular, maize highly infected with mold is used for animal food or brewing. *F. graminearum* was the most common fungus isolated from the moldy maize. During 1973–1974 the rainy season extended into the cold season. The moisture content of the corn in the field was very high and periods of low temperature (12°C) occurred, favoring the production of zearalenone. On the other hand, the rainy season in 1975 was shorter, ending with a warm dry period before the cold season began, and maize from the 1975 harvest had a low zearalenone content.

Christensen (1979) described the environmental conditions for the production of zearalenone in cereal grains. He noted that high moisture content and moderate, followed by low, temperature permitted vigorous growth of *F. graminearum* and also the production of great levels of the toxin.

Sutton *et al.* (1980a,b) noted that climatic, environmental conditions influenced the growth in the field of corn contaminated with *F. graminearum* which showed a tendency to produce zearalenone and vomitoxin.

**G. zeae.**　Bacon *et al.* (1977) described a method for producing zearlenone on a semisynthetic liquid medium. *G. zeae* FR26 and FR29 and *F. culmorum* FR22 were inoculated into a starch-glutamate medium, incubated as a static culture in the dark for 2 weeks at 24–28°C.

Vesonder *et al.* (1978) observed that *G. zeae* infection of corn was favored by low temperature and high moisture conditions. These conditions often led to a delayed harvest, allowing the fungus to grow for a longer period with more chance for mycotoxin production.

**F. roseum.**　When grown on corn for 2 weeks at 22–24°C and then for 5 weeks at 10–12°C, *F. roseum* was able to produce F-2 (3600 µg per ml) (Mirocha *et al.*, 1969).

Eugenio *et al.* (1970a) incubated cultures of *F. roseum* (isolated from corn stored in cribs or from feeds containing corn on Minnesota and South Dakota farms) for 2 weeks at 22–25°C and then lowered the temperature to 15°C. This produced the highest yield of F-2. The largest amount of F-2 was found to be produced on polished rice, followed by corn and wheat, very small amounts were produced on oats and barley, and none at all on soybeans and peas. The optimal moisture content of polished rice was found to be 60–65%, and that of corn, 45%.

Nelson *et al.* (1973), using *F. roseum* and *F. graminearum*, produced zearalenone on a substrate of autoclaved rice of 40–60% mc, with an incubation of 2–3 weeks at 22–25°C and 2 or more weeks at 14°C and tested the toxin produced on swine. They mentioned that these temperatures approximate the weather conditions common in the late fall and winter in the American corn belt.

Zearalenone was isolated from *F. roseum* M-3-2 grown on rice (Ueno *et al.*, 1977a).

*F. roseum* 70-K-11 produced the greatest amount of T-2 toxin and neosolaniol on a medium containing glucose, peptone, and yeast extract, incubated at 22°C for 5 days (Ueno *et al.*, 1975).

*F. roseum* V-18 was found to produce neosolaniol monoacetate, diacetoxyscirpenol, neosolaniol, and T-2 toxin (Ishii and Mirocha, 1975).

**F. sambucinum.**   Strains of *F. sambucinum* were isolated in the 1977 and 1978 harvests of wheat and rye in the United States. Of 13 strains isolated from wheat, 1 produced zearalenone and 2 produced diacetoxy-scirpenol, while from rye only one isolate produced zearalenone (Eppley and Joffe, unpublished data).

In Israel 96 *F. sambucinum* isolates from various agricultural crops and soil were moderately to strongly phytotoxic and also toxic to rabbit skin when cultivated at 24 and 30°C for 10–12 days (*see* Table 5.15).

## CONCLUSION

In this chapter we reviewed the natural occurrence of *Fusarium* species, their distribution and origin, and their ability to produce trichothecene toxins in the field as well as in storage.

*Fusarium* strains producing the largest quantities of trichothecenes (mainly T-2 toxin) were *F. sporotrichioides, F. sporotrichioides* var. *tricinctum,* and *F. poae.*

A detailed description of the environmental conditions favoring *Fusarium* trichothecene toxin production in the field and in the laboratory is present.

The extent of toxin formation by strains of *Fusarium* species, their form of growth, and their degree of toxicity may be modified by the substrates (media) used. The nutritional-physiological aspect of *Fusarium* species in effective production of trichothecene metabolites and their yield and concentration after growing on various substrates at various temperatures and lengths of time is therefore of paramount importance.

The author paid special attention to the yields and intensity of toxicity of various *Fusarium* species and varieties as illustrated in Tables 2.2 through 2.10.

*Fusarium* strains appear especially on millet, wheat, barley, oats, and rye in temperate zones (USSR), on maize and sorghum in temperate and warm regions, and on rice in warm zones.

The species and varieties of *Fusarium* that commonly contaminate cereal grains and fodder in temperate zones belong to the Sporotrichiella section.

Mycotoxin research has been concentrated primarily in temperate agricultural zones with relatively little emphasis on the problem in tropical and semitropical areas. In the subtropical regions, such as Israel, the widespread fungi are *F. oxysporum, F. equiseti, F. solani,* and *F. moniliforme.* The part that these *Fusarium* species and mainly *F.*

*moniliforme* play in potential mycotoxicoses in farm animals, consuming agricultural commodities produced in tropical and semitropical countries, has not been determined to any extent. These species are therefore of great importance and their role needs to be elucidated.

Mycotoxicoses associated with trichothecene compounds produced by toxic *Fusarium* species are a worldwide problem affecting both humans and domestic animals.

T-2 toxin is the best known among the toxic trichothecene compounds and has been studied extensively by the author and his colleagues for its chemistry and toxicology.

This toxin is widespread in naturally contaminated cereal grains in temperate climatic regions. T-2 toxin and some other trichothecenes (DAS, NS, HT-2 toxin) were isolated from various strains of *F. sporotrichioides, F. poae,* and *F. sporotrichioides* var. *tricinctum* by Eppley *et al.* (in press); Joffe and Yagen (1977); Yagen and Joffe (1976); and Yagen *et al.* (1977).

The author has undertaken, over many years, a detailed study to elucidate the environmental and ecological conditions conducive to the formation of *Fusarium* toxins in temperate regions such as the USSR, the United States, and subtropical areas such as Israel. This study was devoted to better understanding the role of toxic *Fusarium* strains involved in the trichothecene toxin production and to the development of arrangement for prevention of serious outbreaks in animals and humans. Further comprehensive studies in this direction are highly recommended.

# CHAPTER 3

## PRINCIPAL TOXINS PRODUCED BY *FUSARIUM* SPECIES

A variety of *Fusarium* species associated with toxicoses occur naturally in tropical, subtropical, and temperate areas of the world. During their growth on natural substrates in the field (cereal grain, feed, fodder), under preharvest, postharvest, and storage conditions, these species produce metabolites chemically identified as fusariotoxins which cause outbreaks of diseases when ingested by humans and animals.

Brook and White (1966) listed 96 species of fungi belonging to 27 genera which caused toxic problems in domestic animals. The majority of these species were cited by Joffe (1960a,b) and were associated with various species of *Cladosporium*, *Penicillium*, and *Fusarium*.

Major fusariotoxins which have caused a variety of symptoms in humans and animals in different parts of the world are presented in Table 3.1.

Numerous outbreaks of human and animal diseases have been traced to *Fusarium* species (*see* Chapters 6, 8, and 9), and considerable work has been done on the isolation and identification of the toxins responsible. However, toxins have also been produced by *Fusarium* species isolated from substrates that were not associated with any toxicoses. A toxic strain of *F. oxysporum* was isolated from carrots, a toxic *F. solani* from cabbage, and a toxic *F. lateritium* from oranges (Davis *et al.*, 1975). Christensen and Kaufman (1969) isolated a strain of *F. roseum* from muskmelon which produced toxins when cultured at 12°C on corn medium over a long period (after initial incubation for 2 weeks at 24°C).

During an investigation into the causes of ATA, numerous strains of *F. sporotrichioides* and *F. poae*, isolated from nontoxic wheat, barley,

**TABLE 3.1  Toxic Effects of Major Fusariotoxins in Man and Animals**

| Disease | Fusarium Species | Substrate (Source) | Toxin | Animal Species | Symptoms | Location | References |
|---|---|---|---|---|---|---|---|
| Drunken bread toxicosis | G. saubinetti (G. zeae) F. roseum F. graminearum | Cereal crops, mainly rye | Alkaloids Zearalenone | Humans, horse, pig, poultry | Headache, nausea, vomiting, weakness, exhaustion, dizziness | Siberia | Gabrilovich, 1906; Naumov, 1916; Woronin, 1891; Palchevski, 1891 |
| Moldy corn toxicosis | F. tricinictum | Maize | T-2 toxin | Cattle, pig, poultry | Vomiting, emesis, feed refusal, hemorrhages, death | USA Europe | Bamburg, 1969; Greenway and Puls, 1976; Hsu et al., 1972; Smalley and Strong, 1974; Smalley et al., 1970 |
| Red mold toxicosis (Akakabibyo) | F. graminearum F. nivale F. episphaeria F. poae F. oxysporum | Wheat, barley, oats, rye, rice | Fusarenon-X Nivalenol Deoxynivalenol | Humans, horse, pig, cow | Vomiting, diarrhea, hemorrhages in various organs, feed refusal, nausea, abortion | Japan | Ishii et al., 1975; Tsunoda, 1970; Tsunoda et al., 1957, 1958, 1968; Ueno et al., 1971a, 1974a |
| Bean-hull toxicosis | F. sporotrichioides[a] | Bean-hull fodder | T-2 toxin Neosolaniol | Horse | Disturbances of nervous and circulatory systems | Japan | Ueno et al., 1972a |
| Alimentary toxic aleukia | F. poae F. sporotrichioides | Overwintered cereals | T-2 toxin | Humans, cattle, pig, chick, lab. animals | Leukopenia, hemorrhagic diathesis, anemia, necrotic angina, oral and pharyngeal lesions | USSR | Joffe, 1971, 1974c, 1978b; Sarkisov, 1954; Yagen et al., 1977 |

| Disease | Fusarium sp. | Substrate | Toxin | Affected | Symptoms | Country | References |
|---|---|---|---|---|---|---|---|
| Kashin-Beck's or Urov disease | *F. sporotrichioides* var. *poae* | Cereal grains | Not defined | Humans | Skeletal deformities mainly in children, osteo-dystrophy | USSR Taiwan North China Korea Japan | Perkel, 1957, 1960; Rubinstein, 1949, 1950a,b, 1951a,b, 1953, 1960b; Sergiyevski, 1952; Takizawa et al., 1956, 1957, 1958, 1959, 1960, 1961, 1963 |
| Degnala disease | *F. equiseti* (*G. intricans*) | Rice straw | Butenolide | Buffalo, cattle | Lameness, edema, necrosis and gangrene of tail, ears, and legs | India | Kalra et al., 1973, 1977 |
| Hyperestrogenism (estrogenic toxicosis) | *F. graminearum* *F. moniliforme* | Maize, barley, wheat | Zearalenone | Pig | Necrosis and inflammation of genitalia, vulvovaginitis, abortion, genital hypertrophy, swollen vulva and enlarged mammae, splayleg piglets | USA | Mirocha and Christensen, 1974b; Mirocha et al., 1967 |

**TABLE 3.1** Continued

| Disease | Fusarium Species | Substrate (Source) | Toxin | Animal Species | Symptoms | Location | References |
|---|---|---|---|---|---|---|---|
| Leukoen-cephalo-malacia | *F. moniliforme* | Maize | Not defined | Horse, donkey | Necrotic lesions of brain, hepatic congestion, hemorrhagic enteritis and cystitis | Egypt USA | Badiali et al., 1968; Marasas et al., 1976; Haliburton et al., 1979; Wilson 1971; Wilson and Maronopot, 1971; Wilson et al., 1973 |
| Moniliformin toxicosis | *F. moniliforme F. fusarioides* | Maize, peanuts, wheat, barley, sorghum, millet, soil | Moniliformin | Chicken, duckling, rat, mouse | Muscular weakness, respiratory distress, cyanosis, coma, death | USA RSA | Burmeister et al., 1979, 1980a; Cole et al., 1973; Kriek et al., 1977; Rabie et al., 1978; Springer et al., 1974; Thiel, 1978 |

| Disease | Fungus | Substrate | Toxin | Animal | Symptoms | Country | References |
|---|---|---|---|---|---|---|---|
| Fescue toxicosis (fescue-foot) | *F. tricinctum* (*F. sporotrichioides*) *F. equiseti* | Tall fescue | T-2 toxin Diacetoxyscripenol Butenolide | Cattle | Lameness, gangrene, arched back, swelling of the hind legs | USA | Ellis and Yates, 1971; Garner and Cornell, 1978; Grove et al., 1970; Kosuri et al., 1970; Tookey et al., 1972; Yates, 1971; Yates et al., 1968, 1969 |
| Pulmonary toxicosis | *F. javanicum* *F. solani* | Sweet potato | Ipomeamarone Ipomeanine | Cattle | Pulmonary edema and congestiom | USA | Boyd et al., 1979; Doster et al., 1978; Peckham et al., 1972; Wilson, 1973a; Wilson et al., 1970, 1971 |

*Source:* Drawn from various sources.

[a] According to Ueno, 1980. Originally classified incorrectly as *F. solani* M-1-1.

and millet, proved to be toxic when grown in the laboratory (Joffe, 1960a). Similarly no zearalenone was found in barley infected by _Fusarium_ species, whereas several of the same species produced zearalenone when isolated and grown in pure culture (Hacking _et al._, 1977).

_F. solani_ isolated from bananas gave toxic extracts when cultured on wheat kernels at 18, 24, and 30°C; no toxicity was detected in cultures grown at 4 or 35°C (Joffe and Palti, 1972).

Many workers have shown toxin production to be temperature dependent both in the amount and type of toxin produced (Bamburg and Strong, 1969; Bamburg _et al._, 1968b; Burmeister, 1971; Joffe, 1974b; Marasas and Smalley, 1972; Ueno _et al._, 1970a).

Not all isolates of a particular species are toxic. Thus Ueno _et al._ (1977b) found that just over half the _Fusarium_ species they isolated from river sediment were toxic. Prentice and Dickson (1968) found that only one out of five isolates of _F. moniliforme_ produced an emetic toxin. Marasas _et al._ (1979b), in a recent survey of _Fusarium_ in maize and barley in Germany, found that three out of five strains of _F. culmorum_ produced deoxynivalenol and four of the five produced zearalenone. In a recent survey of _Fusarium_ species isolated from cereals in the United States the author observed the same phenomenon (_see_ Chapter 2).

The age, origin, and substrate of a particular isolate can also influence toxin production. Many authors have found that after repeated transfers isolates lose their ability to produce toxic properties. Thus although Prentice and Dickson (1968) found several toxic strains, only one (_F. poae_ NRRL 3287) proved to be toxic when tested a few years later (Ellison and Kotsonis, 1973). Tatsuno (1968) found that the original strain of _F. nivale_ has lost its toxicity. However, _F. sporotrichioides_ No. 921 and _F. poae_ No. 958, isolated from overwintered rye and wheat, respectively, by Joffe in 1943 through 1947 (Joffe, 1960a), were still highly toxic many years later (Joffe, 1983b,c; Joffe and Yagen, 1977; Yagen _et al._, 1977).

Many fusariotoxins are highly stable and can persist in food a long time after all traces of the live _Fusarium_ have disappeared. Thus, for example, zearalenone has been found in animal feed (Mirocha _et al._, 1976b), corn products (Martin and Keen, 1978; Stoloff, 1973, 1976; Stoloff and Dalrymple 1977; Stoloff _et al._, 1976), wheat, and soybeans (Shotwell _et al._, 1975), and T-2 toxin has been found in animal feed (Stahr _et al._, 1978).

## CHEMICAL CHARACTERISTICS

Fusariotoxins have been the subject of several recent reviews, among them Chu (1977); Chu _et al._ (1979, 1984); Ciegler (1978, 1979); Kurata (1978); Mirocha and Christensen (1982); Mirocha _et al._ (1977a, 1980);

Scott (1978); Siegfried (1978); Siegfried and Frank (1978); Ueno (1977a,b, 1980a,b,c, 1983b,c); Ueno and Ueno (1978); Wilson (1978).

Ciegler and Vesonder (1983) demonstrated the structures and chemicophysical characteristics of fungal toxins in tables. They also included the sources and biological effects of mycotoxins and mycotoxin-producing fungi and their natural occurrence in foods and feeds.

The major animal fusariotoxins may be divided into four general groups: (1) trichothecenes, (2) zearalenone-type toxins, (3) moniliformin, and (4) butenolides.

Each group has its own chemical and toxicological characteristics which will be outlined in this chapter. Many of the toxins are produced by more than one species of *Fusarium* and, depending on cultural conditions, a single *Fusarium* species can produce more than one type of toxin. Thus, for example, T-2 toxin can be produced by more than 10 different *Fusarium* species (Table 3.2) and *F. equiseti* has been shown to produce diacetoxyscirpenol (Brian *et al.*, 1961), butenolide (Kalra *et al.*, 1973), and zearalenone (Scott *et al.*, 1972). More than one species of *Fusarium* can often invade a single host producing its own toxin. Thus several species were found in maize stalks affected by stalk rot (Mirocha *et al.*, 1979a), and the toxins detected were zearalenone (2.8 μg/g), T-2 toxin (110 μg/g). and deoxynivalenol (vomitoxin) (1.5 μg/g). These were in sufficient quantities to cause food refusal or hyperestrogenism in swine.

## Trichothecenes

Trichothecenes are biologically and chemically active secondary fungal metabolites produced mainly by various *Fusarium* species (Joffe and Yagen, 1977; Mirocha *et al.*, 1977b; Yagen and Joffe, 1976; Ueno, 1977a,b, 1980a,b, 1983b,c). Some trichothecene compounds are also produced by other fungi including *Trichoderma, Myrothecium, Cephalosporium, Stachybotrys, Cladosporium, Trichothecium, Verticimonosporium*, and possibly others (Davis and Diener, 1979; Smalley and Strong, 1974; Smalley *et al.*, 1970; Ueno, 1977a,b).

Ichinoe and Kurata (1983) described the natural occurrence of trichothecene-producing (T-2 toxin, diacetoxyscirpenol, deoxynivalenol, and nivalenol) *Fusarium* species and also some fungi other than Fusaria.

Tseng (1983) has reported the natural occurrence of trichothecene mycotoxins (mainly T-2 toxin and deoxynivalenol) in imported corn from South Africa and the United States. From 428 polished rice samples in Taiwan, no *Fusarium* trichothecene toxins (T-2 toxin, deoxynivalenol, or diacetoxyscirpenol) were found, although zearalenone was present in some animal feedstuffs.

Hald and Krogh (1983) concluded that the frequency of *Fusarium*

**TABLE 3.2  Major Trichothecene Toxins Produced by *Fusarium* Species on Original Sources**

| *Fusarium* Species | Source | Toxin | References |
|---|---|---|---|
| *F. nivale* | Wheat | Nivalenol | Morooka and Tatsuno, 1970; Saito and Okuba, 1970; Saito et al., 1969; Scott et al., 1970; Tatsuno, 1968; Tatsuno et al., 1966, 1968a,b, 1969, 1970, 1971; Ueno et al., 1971b,c, 1973a,d |
| *F. nivale* | Wheat | Fusarenon-X | Saito and Okuba, 1970; Saito and Tatsuno, 1971; Scott et al., 1970; Ueno, 1970; Ueno et al., 1969c, 1971a,c, 1973d |
| *F. episphaeria* | Vinyl plates | Nivalenol | Ueno et al., 1971a, 1972a, 1973d |
| *F. episphaeria* | Vinyl plates | Fusarenon-X | |
| *F. poae* | Scabby barley, maize, millet, mixed feed, overwintered cereals, Fescue hay | T-2 toxin, HT-2 toxin, diacetoxyscirpenol, neosolaniol, acetyl T-2 toxin | Ueno et al., 1971a,c, 1973d Bamburg and Strong, 1969, 1971; Bamburg et al., 1968b; Burmeister et al., 1972; Grove et al., 1970; Joffe and Yagen, 1977; Kosuri et al., 1971; Kotsonis et al., 1975a; Mirocha and Pathre, 1973; Mirocha et al., 1976b; Prentice et al., 1959; Scott et al., 1980; |

| Species | Substrate | Toxins | References |
|---|---|---|---|
| *F. sporotrichioides* | Overwintered cereals, scabby barley, bean-hulls, millet, maize, wheat, oats, barley | T-2 toxin, diacetoxyscirpenol, neosolaniol, T-1 | Strong, 1969, 1971; Szathmary et al., 1976; Ueno et al., 1971a, 1972b, 1973d; Yagen, 1977; Yagen and Joffe, 1976; Yagen et al., 1977; Joffe and Yagen, 1977; Scott et al., 1980; Szathmary et al., 1976; Ueno et al., 1972a,b, 1973d; Yagen and Joffe, 1976; Yagen et al., 1977; Ylimaki et al., 1979 |
| *F. tricinctum* | Maize | Diacetoxyscirpenol | Gilgan et al., 1966; Marasas et al., 1967 |
| *F. tricinctum* | Maize, fescue hay | T-2 toxin | Bamburg and Strong, 1971; Hsu et al., 1972; Kosuri et al., 1970; Smalley et al., 1970; Ueno et al., 1972a; Yates, 1971 |
| *F. tricinctum* | Peanuts | Neosolaniol monoacetate | Lansden et al., 1978 |
| *F. decemcellulare* (*F. rigidiusculum*) | Bean-hulls, maize | T-2 toxin, diacetoxyscirpenol, neosolaniol | Bamburg and Strong, 1971; Ueno et al., 1972a, 1973d |
| *F. roseum* "Equiseti" "Gibbosum" "Culmorum" "Graminearum" "Sambucinum" | Maize, fescue hay, bean-hulls, barley, wheat | T-2 toxin, HT-2 toxin, diacetoxyscirpenol, nivalenol, fusarenon-X, neosolaniol, dehydronivalenol, deoxynivalenol (Rd. toxin, vomitoxin), deoxynivalenol-monoace- | Bamburg and Strong, 1971; Brian et al., 1961; Brook and White, 1966; Burmeister et al., 1972; Dawkins, 1966; Dawkins and Grove, 1970; Dawkins et al., 1965; Flury et al., 1965; Grove, |

**TABLE 3.2** Continued

| *Fusarium* Species | Source | Toxin | References |
|---|---|---|---|
| | | tate, triacetoxyscirpenol, triacetylnivalenol, diacetyl-deoxynivalenol, acetoxy-scirpendiol, T-2 tetraol, scirpentriol, monoace-toxyscirpenol | 1970a,b; Lindenfelser *et al.*, 1974; Pathre et Mirocha, 1978; Pathre *et al.*, 1976; Sigg *et al.*, 1965; Smalley and Strong, 1974; Tidd, 1967; Ueno *et al.*, 1971b, 1972a, 1973b,d; Vesonder *et al.*, 1977a; Yoshizawa, 1978; Yoshizawa and Morooka, 1973, 1974; Yoshizawa *et al.*, 1976, 1978, 1979, 1980a,b |
| *G. zeae* | Fescue hay, barley | T-2 toxin, nivalenol, fusarenon-X | Burmeister *et al.*, 1972; Ueno *et al.*, 1971a,b, 1973d |
| *F. sulphureum* | Tomato | T-2 toxin, HT-2 toxin | Harwig *et al.*, 1979 |
| *F. solani* | Soy beans | T-2 toxin, diacetoxyscir-penol | Hitokoto *et al.*, 1977 |
| *F. lateritium* | River sediment | Diacetoxyscirpenol | Ueno *et al.*, 1977b |
| *F. lateritium* | Fescue hay | T-2 toxin | Burmeister *et al.*, 1972 |
| *F. semitectum var. majus* | Sorghum | T-2 toxin | Rukmini and Bhat, 1978 |
| *Calonectria nivalis* | Cereals | Calonectrin, 15-diacetyl-calonectrin | Gardner *et al.*, 1972 |

*Source:* Drawn from various sources.

toxins (mainly trichothecenes and zearalenone) in contaminated cereals was found to be rather low in Denmark.

The chemical, biochemical, toxicological, and phytotoxicological properties of trichothecene fusariotoxins have been reviewed by several excellent works: Bamburg (1969, 1972, 1976); Bamburg and Strong (1971); Borker et al. (1966); Christensen (1971); Ciegler (1975, 1978, 1979); Ciegler and Lillehoj (1968); Ciegler et al. (1970); Hesseltine (1976a,b); Herzburg (1970); Jemmali (1979); Joffe (1983a,b,c); Kraybill and Shapiro (1969); Marth and Calanog (1976); Mirocha and Christensen (1974a); Mirocha et al. (1977b); Newberne (1974a,b); Ohtsubo and Saito (1977); Pathre and Mirocha (1977); Perlman and Peruzzotti (1970); Purchase (1974); Saito and Ohtsubo (1974); Sato and Ueno (1977); Sato et al. (1978); Smalley and Strong (1974); Tamm (1974, 1977); Ueno (1977a,b, 1980a,b,c, 1983b,c); Uraguchi and Yamazaki (1978); Wilson (1978); Wilson and Hayes (1973); Wogan (1975); Wyllie and Morehouse (1977, 1978a,b); Yoshizawa (1978, 1983a,b); Yoshizawa and Hosokawa (1983); Yoshizawa and Morooka (1973, 1974, 1975a,b); Yoshizawa and Sakamoto (1982); Yoshizawa et al. (1976, 1978, 1979, 1980a,b,c, 1981, 1982a,b, 1984).

Vesonder and Hesseltine (1981) briefly described the biological activity and physicochemical data and quoted reference sources of all *Fusarium* trichothecenes, zearalenone derivatives, and other *Fusarium* mycotoxins and their structures. In addition the authors included specific groups of various *Fusarium* metabolites such as pigments, antibiotics, phytotoxins, and some phytoalexins.

The trichothecenes are sesquiterpenes. They contain carbon, hydrogen, and oxygen and have a tetracyclic trichothecene skeleton (Godtfredsen et al., 1967). All naturally occurring trichothecenes contain an olefinic bond at C-9, C-10 and an epoxy group at C-12, C-13 and usually have a hydroxyl or ester group at C-3, C-7, C-8, and C-15. The common name "scirpene," which designates 12, 13-epoxy $\Delta^9$-trichothecenes, is therefore used for many toxins of the trichothecene family. To date 50 naturally occurring trichothecene compounds were isolated, the majority of which are associated with *Fusarium* producing trichothecenes.

Ueno et al. (1973d) and Ueno (1977a) worked out a chemical procedure for screening and determination of the toxic *Fusarium* strains isolated from cereal grains and feeds. The identification of trichothecenes and their toxic principles was carried out by TLC, using various solvent systems and examined under longwave ultraviolet light (360 nm) (after spraying the plate with 20% $H_2SO_4$ and then heating at 110°C for 10–20 min) for determination of typical Rf values and observed colors.

On the basis of this procedure Ueno et al. (1973d) and Ueno (1977a) divided trichothecene mycotoxins into four types: A, B, C, and D. The type A trichothecene toxins (produced by *F. sporotrichioides, F. poae, F. sporotrichioides* var. *tricinctum*, and others) exhibited sky-blue

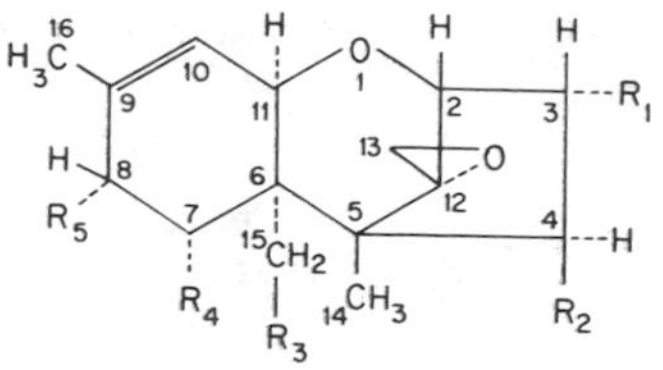

**Figure 3.1**  The chemical structure for trichothecene mycotoxins of the type A derivatives of the *Fusarium* species.

fluorescence and type B (produced by *F. nivale, F. episphaeria,* and others) produced a brown spot. The first type A included T-2 toxin, diacetoxyscirpenol, neosolaniol, HT-2 toxin, and others (Fig. 3.1). The second type B of trichothecenes included nivalenol, fusarenon-X, deoxynivalenol, and diacetylnivalenol (Fig. 3.2). The difference between type A and type B lies in the functional groups that are connected with C-7 and C-8, but the main difference is at C-8. The type A toxins lack the ketone function at C-8 and contain an H,OH, or ester at C-8. Type B of trichothecenes is characterized by ketone function at position C-8.

The type A and type B include naturally occurring trichothecenes produced by various species of *Fusarium* (Tables 3.2 and 3.3).

Ishii (1983) described the isolation, structure, production, and chemistry of nonmacrocyclic trichothecenes.

Type C contains oxirane ring at C-7–C-8 characterized by an epoxide function at C-7-8 position and includes one compound: crotocin (Fig. 3.3), which is produced by *Cephalosporium crotocinigenum*.

The fourth type, type D, includes the macrocyclic derivatives of verrucarins and verrucarol (Fig. 3.4). This type contains a macrocyclic ring at C-4 and C-15 and includes various compounds isolated from several fungi such as *Myrothecium* (Tamm, 1977), *Stachybotrys* (Eppley and Bailey, 1973; Eppley *et al.*, 1977, 1980; Rodricks and Eppley, 1974; Szathmary *et al.*, 1976; El-Kady and Moubasher 1982), *Cylindrocarpon* (Matsumoto *et al.*, 1977), and *Verticimonosporium* (Minato *et al.*, 1975).

Breitenstein and Tamm (1975, 1977, 1978) and Tamm and Breitenstein (1980) studied the trichothecene derivatives of the verrucarol, verrucarins, and their biosynthesis. Chi (1982) described the pharmacolog-

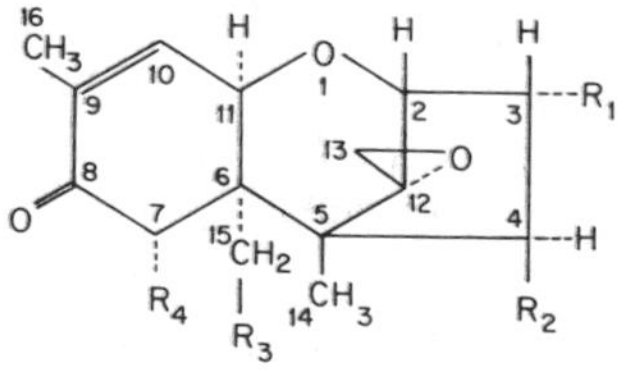

**Figure 3.2**  The chemical structure for trichothecene mycotoxins of the type B derivatives of the *Fusarium* species.

**TABLE 3.3 Trichothecene Toxins Isolated from *Fusarium* Species (Only Major Toxins Given)**

| Toxin | *Fusarium* Species | Culture Medium | References |
|---|---|---|---|
| | *Type A* | | |
| T-2 toxin | F. tricinctum | Gregory's[a], Sabouraud's, GPY[b], WCG[c] | Bamburg et al., 1968b; Burmeister et al., 1972; Grove et al., 1970; Ueno et al., 1975 |
| T-2 toxin | F. poae | Richard's solution | Kotsonis and Ellison 1975 |
| T-2 toxin | F. poae | Wheat, millet | Joffe and Yagen, 1977; Yagen and Joffe, 1976; Yagen et al., 1977 |
| T-2 toxin | F. poae | Cracked corn | Vesonder et al., 1981 |
| T-2 toxin | F. poae | CzD-P[d] | Ueno et al., 1972b |
| T-2 toxin | F. sporotrichioides[e] | CzD-P | Ueno et al., 1972b |
| T-2 toxin | F. sporotrichioides | PSC[f] | Ueno et al., 1972a |
| T-2 toxin | F. sporotrichioides | GPY | Ueno et al., 1975 |
| T-2 toxin | F. sporotrichioides | Wheat, millet | Joffe and Yagen, 1977; Yagen and Joffe, 1976; Yagen et al., 1977 |
| T-2 toxin | F. lateritium | WCG | Burmeister et al., 1972 |
| T-2 toxin | F. decemcellulare (F. rigidiusculum) | CzD[g] | Ueno et al., 1973d |
| T-2 toxin | F. solani | PDA[h] | Hitokoto et al., 1977 |
| T-2 toxin | F. roseum "Scirpi" "Avenaceum" "Culmorum" | CzD, GPY | Ueno et al., 1973d, 1975; Ueno et al., 1973d, 1975; Ueno et al., 1973d |

**TABLE 3.3** Continued

| Toxin | *Fusarium* Species | Culture Medium | References |
|---|---|---|---|
| T-2 toxin | *G. zeae* | WCG | Burmeister *et al.*, 1972 |
| T-2 toxin | *F. moniliforme* | Maize | Ghosal *et al.*, 1978a |
| T-2 toxin | *F. sulphureum* | Tomato | Harwig *et al.*, 1979 |
| T-2 toxin | *F. oxysporum* | Richard's solution | Chakrabarti *et al.*, 1976 |
| T-2 toxin | *F. oxysporum* | CzD | Ueno *et al.*, 1973d |
| T-2 toxin | *F. incarnatum* (*F. semitectum* var. *majus*) | Rice | Rukmini and Bhat, 1978 |
| HT-2 toxin | *F. roseum* | CzD | Ueno *et al.*, 1973d |
| HT-2 toxin | *F. poae* | Richard's solution | Kotsonis and Ellison, 1975a |
| HT-2 toxin | *F. poae* | CzD | Ueno *et al.*, 1973d |
| HT-2 toxin | *F. sporotrichioides* | Millet seeds with 2% malt extract | Szathmary *et al.*, 1976 |
| HT-2 toxin | *F. sporotrichioides* | CzD | Ueno *et al.*, 1973d |
| HT-2 toxin | *F. tricinctum* | Gregory's | Bamburg and Strong, 1969 |
| HT-2 toxin | *F. sulphureum* | Tomato | Harwig *et al.*, 1979 |
| Diacetoxyscirpenol | *F. lateritium* | Synthetic | Cole and Rolinson, 1972 |
| Diacetoxyscirpenol | *F. tricinctum* | Gregory's | Gilgan *et al.*, 1966 |
| Diacetoxyscirpenol | *F. sporotrichioides* | Sabouraud's | Yates *et al.*, 1969 |
| Diacetoxyscirpenol | *F. sporotrichioides* | Gregory's | Bamburg *et al.*, 1968b |
| Diacetoxyscirpenol | *F. equiseti* | CzD, GAN[i] | Brian *et al.*, 1961 |
| Diacetoxyscirpenol | *F. equiseti* | CzD-Yeast extract | Dawkins, 1966; Dawkins *et al.*, 1965 |

| Compound | Fusarium species | Substrate | Reference |
|---|---|---|---|
| Diacetoxyscirpenol | G. intricans | GAN, CzD-Yeast extract | Brian et al., 1961 |
| Diacetoxyscirpenol | F. sulphureum | Corn meal | Steyn et al., 1978b |
| Diacetoxyscirpenol | F. solani | PDA | Hitokoto et al., 1977 |
| Diacetoxyscirpenol | F. solani var. coeruleum | CzD | Ripperger et al., 1975 |
| Diacetoxyscirpenol | F. moniliforme | Maize | Ghosal et al., 1978a |
| Diacetoxyscirpenol | Fusarium species | Not quoted | Grove, 1969 |
| Monoacetoxyscirpenol | F. roseum "Gibbosum" | Rice | Pathre et al., 1974, 1976 |
| 7-Hydroxy-diacetoxyscirpenol | F. lateritium | PSC | Ishii 1975; Ueno et al., 1977b |
| 7,8-Dihydroxydiacetoxyscirpenol | F. lateritium | PSC | Ueno et al., 1977b |
| 4-Acetoxyscirpendiol | F. roseum "Gibbosum" | PSC | Ishii et al., 1978 |
| Neosolaniol | F. decemcellulare | CzD | Ueno et al., 1973d |
| Neosolaniol | F. roseum "Culmorum" | CzD | Ueno et al., 1973d |
| Neosolaniol | F. roseum "Scirpi" | CzD | Ueno et al., 1973d |
| Neosolaniol | F. poae | Millet + 2% malt extract | Szathmary et al., 1976 |
| Neosolaniol | F. sporotrichioides | Millet + 2% malt extract | Szathmary et al., 1976 |
| Neosolaniol | F. sporotrichioides | CzD-P | Ishii et al., 1971; Ueno et al., 1972b |
| Neosolaniol | F. sporotrichioides | PSC | Ueno et al., 1972a |
| Neosolaniol | F. sulphureum | Tomato | Harwig et al., 1979 |
| Neosolaniol monoacetate | F. tricinctum | Artificial | Lansden et al., 1978 |
| Neosolaniol monoacetate | F. nivale | | Tatsuno et al., 1969 |
| Neosolaniol monoacetate | F. nivale | | Ishii and Mirocha, 1975 |

**TABLE 3.3** Continued

| Toxin | *Fusarium* Species | Culture Medium | References |
|---|---|---|---|
| 8-Acetylneosolaniol | *F. roseum* | Cz-P | Ishii *et al.*, 1978 |
| | | *Type B* | |
| Nivalenol | *F. nivale* | Rice | Tatsuno, 1968 |
| Nivalenol | *F. nivale* | CzD | Ueno *et al.*, 1969d, 1970a |
| Nivalenol | *F. nivale* | PDA | Ueno *et al.*, 1971a |
| Nivalenol | *F. nivale* | Rice, Cz + 10% P | Tatsuno *et al.*, 1970 |
| Nivalenol | *F. episphaeria* | PDA | Tatsuno *et al.*, 1970 |
| Deoxynivalenol | *F. graminearum* | Rice | Vesonder *et al.*, 1976, 1979 |
| Deoxynivalenol | *G. zeae* | Maize | Ishii *et al.*, 1975 |
| Deoxynivalenol | *F. nivale* | Maize | Vesonder *et al.*, 1981 |
| Deoxynivalenol | *F. tricinctum* | Maize | Vesonder *et al.*, 1981 |
| Deoxynivalenol | *F. roseum* | Maize | Mirocha *et al.*, 1976b |
| Deoxynivalenol | *F. graminearum* | Maize | Ueno *et al.*, 1974a |
| Deoxynivalenol | *F. graminearum* | Maize | Ishii *et al.*, 1975 |
| Deoxynivalenol | *F. culmorum* | Rice, maize | Pathre and Mirocha, 1978 |
| Deoxynivalenol | *F. culmorum* | Rice | Vesonder *et al.*, 1977a |
| Deoxynivalenol | *F. culmorum* | MEA[j] | Marasas *et al.*, 1979b |

| Compound | Species | Medium | Reference |
|---|---|---|---|
| Deoxynivalenol monoacetate | *F. roseum* | Barley | Yoshizawa and Morooka, 1973 |
| Deoxynivalenol monoacetate | *F. culmorum* | Barley | Blight and Grove, 1974 |
| Acetyldeoxynivalenol | *F. culmorum* | MEA | Marasas et al., 1979b |
| Fusarenon-X | *F. nivale* | PSC | Ueno et al., 1969c, 1971b |
| Fusarenon-X | *F. nivale* | Rice, Cz[k] | Tatsuno et al., 1970 |
| Fusarenon-X | *F. nivale* | Rice | Morooka et al., 1971 |
| Fusarenon-X | *F. episphaeria* | CzD-P | Ueno et al., 1971a,c |
| Fusarenon-X | *F. oxysporum* "niveum" | CzD | Ueno et al., 1973d |
| Fusarenon-X | *G. zeae* | PSC | Ueno et al., 1971a,c |

*Source:* Drawn from various sources.

[a] Gregory et al. (1952).

[b] Glucose-Peptone-Yeast.

[c] White corn grits.

[d] Czapek Dox-Peptone.

[e] *F. solani* M-1-1 (Ishii et al., 1971) was reclassified as *F. sporotrichioides* by Ueno (1980).

[f] Peptone supplemented Czapek.

[g] Czapek Dox.

[h] Potato dextrose agar.

[i] Glucose ammonium nitrate.

[j] Malt extract agar.

[k] Czapek.

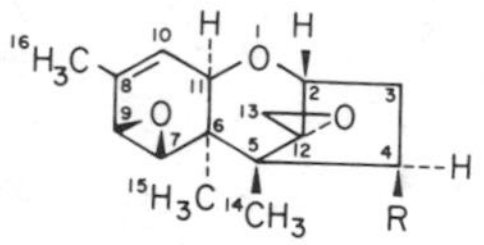

**Figure 3.3**  The chemical structure of the type C trichothecene—crotocin.

ical properties of some macrocyclic derivatives and synthesis of trichothecene compounds. Jarvis *et al.* (1980, 1981, 1982, 1983) studied the chemistry and bioproduction and biosynthesis of macrocyclic trichothecenes. They also described procedures for isolation of the trichoverroids, new roridins, and verrucarins and production of verrucarol. However, since these macrocyclic trichothecenes are not produced by *Fusarium* species they will not be discussed further here.

In addition to differences in chemical structure of types A and B (*see* Figs. 3.1 and 3.2) Ueno *et al.* (1973d) pointed out a geographical distinction between these two types in Japan; *Fusarium* species producing type A toxins were more common in the colder, northern areas, whereas those producing type B toxins were found mainly in the south of Japan. In general, toxicoses caused by type A toxin producers are associated with harsh temperate climates.

The *Fusarium* species that are known to produce trichothecenes, together with both original and experimental substrates and the cultural conditions used for toxin production, are listed in Tables 3.2 and 3.3.

Most of the work on *Fusarium*-produced trichothecene toxins was carried out under laboratory conditions. However, fusariotoxins isolated from natural substrates (cereals or animal feed) and associated with human and animal toxicoses, are limited to the following mycotoxins: T-2 toxin, diacetoxyscirpenol, nivalenol, and deoxynivalenol (vomitoxin) (*see* Table 3.4).

Structures of T-2 toxin and related trichothecenes belonging to type A and structures of naturally occurring trichothecenes of type B are presented in Figures 3.1 and 3.2.

All natural trichothecenes have the same stereochemistry (Sigg *et al.*, 1965; Grove 1969, 1970a,b; Wilson, 1973b): $\alpha$ at C-3 (R1), $\beta$ at C-4 (R2), C-15 (R3), $\alpha$ at C-7 (R4), and $\alpha$ at C-8 (R5) for group A and B (*see* Fig. 2.1). The full systematic chemical name can therefore be easily built up by designating the appropriate substituent attached to the scirpene or 12,13-epoxy-$\Delta^9$-trichothecene structure. For example, T-2 toxin (Ta-

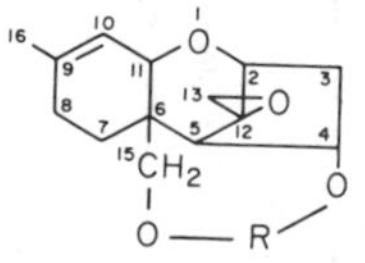

**Figure 3.4**  The chemical structure of the type D macrocyclic trichothecenes.

ble 3.5) is 3α-hydroxy-4β, 15-diacetoxy-8α-(3 methylbutyryloxy)-12,13-epoxy-$\Delta^9$-trichothecene (*see* Fig. 2.2) and fusarenon-X (Table 3.6) is 3α, 7α, 15-trihydroxy-4β-acetoxy-12,13-epoxy-$\Delta^9$-trichothecene-8-one (*see* Fig. 2.7).

The trichothecenes belonging to types A and B are listed in Tables 3.5 and 3.6, and the data of *Fusarium* produced toxins, arranged in order of increasing chemical complexity, are given. That is, the parent trialcohol scirpentriol, lacking substituents at C-7 and C-8, is given first, followed by its mono-, di- and triacetate derivatives and then by the parent alcohol T-2 tetraol and its acetates, and so on.

*F. roseum* "Gibbosum" (*F. equiseti*) produced two new toxins, 4β-acetoxy 12,13-epoxytrichothec-9-ene-3α, 15-diol (4-acetoxyscirpen-3, 15-diol), and 4β, 8α, 15-triacetoxy 12,13-epoxytrichothec-9-ene-3 α-ol (neosolaniol monoacetate), as well as diacetoxyscirpenol, monoacetoxy-scirpenol, scirpentriol, and neosolaniol and T-2 toxin (Ishii and Mirocha, 1975).

*F. roseum* V-18 (from corn) was grown on PSC. A new trichothecene compound, 8-acetylneosolaniol, was isolated from the culture filtrate (Ishii *et al.*, 1978).

Deoxynivalenol (DON) and butenolide were obtained from *F. roseum* 117 (Morooka *et al.*, 1972) as well as deoxynivalenol monoacetate (Yoshizawa and Morooka, 1973, 1974, 1977).

Deoxynivalenol, 3-acetyldeoxynivalenol, and a new trichothecene, 3α,15-diacetoxy-7α-hydroxy-12,13-epoxy-trichothec-9-en-8-one, were also obtained from *F. roseum* 117 after incubation for 15 days at 25°C on rice (Yoshizawa *et al.*, 1978).

Ishii and Ueno (1981) also isolated two new trichothecene toxins from the culture filtrate of *F. sporotrichioides* M-1-1. These toxins were characterized as 4β,8α-diacetoxy-12,13-epoxytrichothec-9-one-3α diol (named NT-1) and 4β-acetoxy-12,13-epoxytrichothec-9-one-3α,8α,15-triol (named NT-2). In addition the authors also isolated T-2 toxin, neosolaniol, and HT-2 toxin. The authors noted that these toxins inhibited protein synthesis similar to that of neosolaniol.

Ishii *et al.* (1971) and Ueno *et al.* (1972) stated that T-2 toxin and neosolaniol were isolated from *F. solani* M-1-1, which was later reidentified by Ichinoe (1978b) as *F. sporotrichioides.*

*F. tricinctum*, grown on corn, caused estrogenic symptoms when fed to rats, swine, and poultry. Certain isolates of *F. tricinctum* were also found to produce the F-2 toxin in large amounts, in association with toxic trichothecenes such as T-2 toxin and neosolaniol (Nelson *et al.*, 1973).

Ikediobi *et al.* (1971) developed a GLC procedure for detecting and determining mycotoxins in foods and feeds. Using their method the authors were able to identify T-2 toxin in corn artificially infected with *F. tricinctum* and T-2 and HT-2 toxins in a corn-rice mixture also infected with *F. tricinctum.*

**TABLE 3.4  Natural Occurrence of Trichothecene Toxins**

| Trichothecene | Source | Symptoms | Levels | Location | References |
|---|---|---|---|---|---|
| T-2 toxin | Maize | Lethal toxicosis in dairy cattle | 2 mg/kg | USA | Hsu et al., 1972 |
| T-2 toxin | Maize | Severe skin reactions | 0.2–1.0 mg/kg | USA | Eppley et al., 1974 |
| T-2 toxin | Barley | Lethal toxicosis in geese | 25 mg/kg | Canada | Greenway and Puls, 1976; Puls and Greenway, 1976 |
| T-2 toxin | Maize | Skin reaction | 5 mg/kg | Yugoslavia | Balzer et al., 1977b |
| T-2 toxin | Mixed feed | Bloody stools, bovine | 0.076 mg/kg | USA | Mirocha et al., 1976b |
| Deoxynivalenol | Maize | Vomiting, feed refusal in swine | 3 mg/kg | USA | Vesonder et al., 1976, 1977a |
| Deoxynivalenol | Preharvest maize | Vomiting | 0.005–0.017 mg/kg | USA | Vesonder et al., 1978 |
| Deoxynivalenol | Maize | Vomiting, feed refusal | 7.5 mg/kg | USA[a] | Ishii et al., 1975 |
| Deoxynivalenol | Barley | Feed refusal, emesis | 5 mg/kg | Japan | Yoshizawa et al., 1976, 1978 |
| Deoxynivalenol | Maize | Feed refusal, infertility | 0.14–0.6 mg/kg | France | Jemmali et al., 1978 |

| Toxin | Substrate | Effect | Concentration | Country | Reference |
|---|---|---|---|---|---|
| Deoxynivalenol | Maize kernels | Feed refusal in swine | 0.1–1.8 mg/kg | USA | Mirocha et al., 1976b |
| Deoxynivalenol | Mixed feed | Feed refusal in swine, vomiting in dogs | 1 mg/kg | USA | Mirocha et al., 1976b |
| Deoxynivalenol | Commercial mixed feed | Feed refusal and bloody stools in swine | 0.04–0.06 mg/kg | USA | Mirocha et al., 1976b |
| Deoxynivalenol | Barley | Feed refusal | 7.3 mg/kg | Japan | Morooka et al., 1972 |
| Nivalenol | Maize | Feed refusal, infertility | 1.18–4.28 mg/kg | France | Jemmali et al., 1978 |
| Nivalenol | Barley | Death of mice | Not quoted | Japan | Morooka et al., 1972 |
| Diacetoxyscirpenol | Mixed feed | Hemorrhagic bowel syndrome in swine | 0.380–0.500 mg/kg | USA | Mirocha et al., 1976b |
| Diacetoxyscirpenol | Safflower seed | Yellowish discoloration of leaves and chlorosis | Not quoted | India | Chakrabarti et al., 1976; Ghosal et al., 1976, 1977 |

Source: Drawn from various sources.
[a]Moldy corn obtained from C. W. Hesseltine (U.S.).

99

**TABLE 3.5  Trichothecene Fusariotoxins of Type A**

| Trichothecenes (Compound Name) | R1 | R2 | R3 | R4 | R5 | References |
|---|---|---|---|---|---|---|
| Scirpentriol | OH | OH | OH | H | H | Bamburg and Strong, 1971; Christopher et al., 1977; Ishii et al., 1978; Pathre et al., 1974; Sigg et al., 1965 |
| 4-Acetoxyscirpendiol | OH | OAc[a] | OH | H | H | Ishii et al., 1978; Claridge and Schmitz, 1979; Sigg et al., 1965 |
| Monoacetoxyscirpenol | OH | OH | OAc | H | H | Bamburg and Strong, 1971; Ishii et al., 1978; Pathre et al., 1976; Sigg et al., 1965 |
| Pentahydroxyscirpenol | OH | OH | OH | OH | OH | Stahr et al., 1978 |
| Diacetoxyscirpenol | OH | OAc | OAc | H | H | Bamburg and Strong, 1971; Dawkins, 1966; Dawkins et al., 1965; Flury et al., 1965; Liao et al., 1976; Mirocha et al., 1976b; Sigg et al., 1965 |
| Triacetoxyscirpene | OAc | OAc | OAc | H | H | Claridge and Schmitz, 1979; Ohta et al., 1978; Sigg et al., 1965 |
| Tetraacetoxyscirpene | OAc | OAc | OAc | H | OAc | Ohta et al., 1978 |

| Compound | | | | | | Reference |
|---|---|---|---|---|---|---|
| Neosolaniol | OH | OAc | OAc | H | OH | Bamburg and Strong, 1971; Ishii et al., 1971; Mirocha and Pathre, 1973; Ueno et al., 1972a |
| Monoacetylneosolaniol | OH | OAc | OAc | H | OAc | Ishii et al., 1978 |
| 7α-Hydroxydiacetoxyscirpenol | OH | OAc | OAc | OH | H | Ishii, 1975; Ueno et al., 1977b |
| 7,8α-Dihydroxydiacetoxyscirpenol | OH | OAc | OAc | OH | OH | Ishii, 1975; Ueno et al., 1977b |
| T-2 tetraol | OH | OH | OH | H | OH | Bamburg 1976; Bamburg et al., 1968b; Ishii et al., 1978; Wei et al., 1971 |
| T-1 | OH | OAc | OH | H | OAc | Ilus et al., 1977; Ylimäki et al., 1979 |
| T-2 triol | OH | OH | OH | H | OIV[b] | Wei et al., 1971 |
| T-2 toxin | OH | OAc | OAc | H | OIV | Bamburg et al., 1968b; Bamburg and Strong, 1971; Hesseltine, 1976b; Mirocha and Pathre, 1973; Saito et al., 1969; Wilson, 1973b; Yagen et al., 1977 |
| HT-2 toxin | OH | OH | OAc | H | OIV | Bamburg and Strong, 1969, 1971 |
| Acetyl T-2 toxin | OAc | OAc | OAc | H | OIV | Kotsonis et al., 1975a; Ohta et al., 1978 |
| Calonectrin | OAc | H | OAc | H | H | Gardner et al., 1972 |
| 15-Desacetylcalonectrin | OAc | OAc | OH | H | H | Gardner et al., 1972 |

*Source:* Drawn from various sources.

[a] Acetate (OCOCH₃)

[b] Isovaleryl (OCO-CH₂-CH(CH₃)₂)

**TABLE 3.6  Trichothecene Fusariotoxins of Type B**

| Trichothecenes (Compound Name) | R1 | R2 | R3 | R4 | R5 | References |
|---|---|---|---|---|---|---|
| Nivalenol | OH | OH | OH | OH | =O | Bamburg and Strong, 1971; Stahr et al., 1978; Tatsuno et al., 1968a, 1969 |
| Fusarenon-X (Monoacetylnivalenol) | OH | OAc | OH | OH | =O | Bamburg and Strong, 1971; Ueno et al., 1969c, 1971b |
| Diacetylnivalenol | OH | OAc | OAc | OH | =O | Bamburg and Strong, 1971; Grove, 1970a; Tatsuno et al., 1970; Tidd, 1967; Ueno et al., 1972a, 1977b |
| Deoxynivalenol (Rd toxin) Vomitoxin | OH | H | OH | OH | =O | Forsyth et al., 1977; Mirocha et al., 1976b; Romer, 1977; Ueno et al., 1973b; Vesonder et al., 1973, 1976; Yoshizawa and Morooka, 1973, 1977 |
| Deoxynivalenol monoacetate (Rc toxin) | OAc | H | OH | OH | =O | Ueno et al., 1973b; Yoshizawa and Morooka, 1973, 1977 |

*Source:*  Drawn from various sources.

## Biosynthesis

Biogenetically the trichothecene is formed by cyclization of farnesyl pyrophosphate followed by two consecutive 1,2-methyl group shifts. [$^{14}$C] mevalonate, when supplied to growing *Trichothecium roseum* fungus, resulted in trichothecin incorporating 3 molecules of labeled mevalonate, originating from C-2 of mevalonate, at C-4, C-10, and C-14 (Dawkins, 1966; Godtfredsen and Vangedal, 1965; Hanson and Achilladelis, 1967; Jones and Lowe, 1960).

Studies with [$^3$H] and [$^{14}$C] mevalonate and farnesyl pyrophosphate gave more detailed insight into biosynthetic routes (Achilladelis and Hanson, 1968; Achilladelis *et al.*, 1972; Machida and Nozoe, 1972a,b).

Hagler *et al.* (1981) used *F. tricinctum* NRRL 3299 (*F. poae*) for T-2 toxin production experiments. This *Fusarium* species was incubated on solid rice medium in the presence of [1-$^{14}$C] sodium acetate and [2-$^3$H] mevalonic acid of high specific activity. This investigation demonstrated that biosynthesis of radiolabeled T-2 toxin can be obtained biologically in the presence of [$^{14}$C] acetate or [2-$^3$H] mevalonic acid.

Tritium-labeled T-2 toxin was synthesized by a two-step process of oxidation of T-2 toxin in the C-3 position with dimethylsulfide-N-chlorosuccimide, followed by reduction with sodium [$^3$H] borohydride, as described by Wallace *et al.*, 1977.

A summary of up-to-date knowledge of chemical synthesis and biosynthetic pathways is given by Ciegler (1979), Mirocha *et al.*, (1977b, 1980), and Tatsuno (1983).

Yoshizawa and Morooka (1975a,b) reported on the microbial and chemical modifications of nivalenol, T-2 toxin, and deoxynivalenol and their derivatives by *F. nivale*, *F. solani*, and *F. roseum*.

## Toxicology

More than half of the trichothecenes which have been isolated to date are elaborated by various *Fusarium* species. The most important trichothecene metabolites (T-2 toxin, DAS, NIV, DON, and others) have been found in cereal grains and feedstuffs contaminated by *Fusarium* cultures in several zones of the world and caused fusariotoxicoses in domestic animals and humans.

The LD$_{50}$ values of various trichothecene compounds are presented in Table 3.7. Unfortunately not all LD$_{50}$ have been measured, and no systematic comparative toxicological studies with standard animals and route of administration are available. However, the general conclusion is that derivatives with OH or OAC at C-3 and C-4 have about the same toxicity (slightly reduced with OH), and reducing the C-9–C-10 double bond only slightly decreases toxicity. Opening the C-12–C-13 epoxide abolishes toxicity (Bamburg, 1969; Bamburg and Strong, 1971; Grove, 1969, 1970a).

**TABLE 3.7  Lethal Doses of Trichothecene Fusariotoxins According to Mode of Administration and Animal**

| Fusariotoxin | Mode of Administration | Animals | LD$_{50}$ (mg/kg b.w.) | References |
|---|---|---|---|---|
| | | | *Type A* | |
| T-2 toxin | i.p. | Mice | 3.0 | Bamburg, 1972 |
| T-2 toxin | i.p. | Mice | 3.04 | Hesseltine, 1976b; Yates *et al.*, 1968 |
| T-2 toxin | i.p. | Mice | 3.3 | Ylimäki *et al.*, 1979 |
| T-2 toxin | i.p. | Mice | 5.2 | Ueno, 1977b; Ueno and Ueno, 1978; Ueno *et al.*, 1973b |
| T-2 toxin | p.o. | Mice | 4.8 | Ylimäki *et al.*, 1979 |
| T-2 toxin | p.o. | Mice | 7.0 | Ueno, 1977b; Ueno and Ueno, 1978; Ueno *et al.*, 1973b |
| T-2 toxin | p.o. | Mice | 10.5 | Sato and Ueno, 1977 |
| T-2 toxin | s.c. | Newborn mice | 0.15 | Sato and Ueno, 1977 |
| T-2 toxin | i.p. | Rats | 2.0 | Ylimäki *et al.*, 1979 |
| T-2 toxin | i.p. | Rats | 3.0 | Wilson, 1973b |
| T-2 toxin | p.o. | Rats | 3.8 | Bamburg, 1972; Kosuri *et al.*, 1971; Ueno and Ueno, 1978 |
| T-2 toxin | p.o. | Rats | 5.2 | Sato and Ueno, 1977 |
| T-2 toxin | p.o. | Rats | 4.0 | Smalley, 1973 |
| T-2 toxin | p.o. | Swine | 4.0 | Smalley, 1973 |
| T-2 toxin | p.o. | Chicks | 3.6 | Sato and Ueno, 1977 |
| T-2 toxin | p.o. | Guinea pigs | 2.0 | Sato and Ueno, 1977 |
| T-2 toxin | s.c. | Peking duckling | 2.0 | Ellison and Kotsonis, 1973 |
| T-2 toxin | Crop intubation | Cockerels | 1.84 | Lansden *et al.*, 1978 |

| Compound | Route | Animal | Value | Reference |
| --- | --- | --- | --- | --- |
| T-2 toxin | p.o. | Trout | 6.1 | Marasas et al., 1967 |
| T-2 toxin | s.c. | Cats | <0.5 | Sato and Ueno, 1977 |
| T.1 toxin | i.p. | Mice | 15.0 | Ylimäki et al., 1979 |
| T.1 toxin | p.o. | Mice | 10.0 | Ylimäki et al., 1979 |
| HT-2 toxin | i.p. | Mice | 9.0 | Bamburg, 1976; Ueno, 1973a; Ueno et al., 1973b |
| HT-2 toxin | i.p. | Mice | 14.0 | Ylimäki et al., 1979 |
| HT-2 toxin | p.o. | Mice | 20.0 | Ylimäki et al., 1979 |
| HT-2 toxin | i.p. | Rats | 1.7 | Ylimäki et al., 1979 |
| Diacetoxy-scirpenol | i.v. | Mice | 10.0 | Bamburg, 1972; Ueno and Ueno, 1978 |
| | i.v. | Mice | 12.0 | Sato and Ueno, 1977 |
| | i.p. | Mice | 23.0 | Sato and Ueno, 1977; Ueno, 1973a; Ueno and Ueno, 1978 |
| | s.c. | Newborn mice | 0.17 | Sato and Ueno, 1977 |
| | i.v. | Rats | 1.3 | Sato and Ueno, 1977 |
| | i.p. | Rats | 0.75 | Bamburg, 1972; Brian et al., 1961; Sato and Ueno, 1977; Ueno and Ueno, 1978; Wilson, 1973b |
| | p.o. | Rats | 7.3 | Bamburg, 1972; Sato and Ueno, 1977; Ueno and Ueno, 1978 |
| | i.v. | Dogs | 1.1 | Sato and Ueno, 1977 |
| | i.v. | Rabbits | 1.0 | Sato and Ueno, 1977 |
| Scirpentriol | i.p. | Rats | 0.81 | Brian et al., 1961 |
| Neosolaniol | i.p. | Mice | 14.5 | Bamburg, 1976; Ishii et al., 1971; Sato and Ueno, 1977; Ueno, 1973a, 1977b; Ueno and Ueno, 1978; Ueno et al., 1972a, 1973b |
| Neosolaniol | i.p. | Mice | 22.1 | Ylimäki et al., 1979 |
| Neosolaniol | p.o. | Mice | 27.1 | Ylimäki et al., 1979 |
| Neosolaniol monoacetate | Crop intubation | Cockerels | 0.789 | Lansden et al., 1978 |

**TABLE 3.7** Continued

| Fusariotoxin | Mode of Administration | Animals | LD$_{50}$ (mg/kg b.w.) | References |
|---|---|---|---|---|
| 7-Hydroxy-diacetoxy-scirpenol | i.p | Mice | 3.5 | Ueno et al., 1977b |
| | i.p. | Mice | 4.0 | Ueno and Ueno, 1978 |
| 7,8-Dihy-droxy-acetoxy-scirpenol | i.p. | Mice | 6.0 | Ueno and Ueno, 1978; Ueno et al., 1977b |
| | | | *Type B* | |
| Nivalenol | i.p. | Mice | 4.0 | Tatsuno, 1968; Wilson, 1973b |
| Nivalenol | i.p. | Mice | 4.1 | Sato and Ueno, 1977; Ueno, 1973a; Ueno and Ueno, 1978; Ueno et al., 1973b |
| Nivalenol | s.c. | Newborn mice | 0.14 | Sato and Ueno, 1977 |
| Nivalenol | i.p. | Adult male mice | 6.9 | Yoshizawa and Morooka, 1974 |
| Nivalenol | i.p. | Female mice | 6.2 | Yoshizawa and Morooka, 1974 |
| Fusarenon-X | i.v. | Adult mice | 3.4 | Sato and Ueno, 1977; Ueno and Ueno, 1978; Ueno et al., 1971c |
| Fusarenon-X | s.c. | Mice | 4.2 | Sato and Ueno, 1977; Ueno and Ueno, 1978; Ueno et al., 1971c |
| Fusarenon-X | i.p. | Mice | 3.3 | Ueno, 1973a, 1977b; Ueno et al., 1973b |
| Fusarenon-X | i.p. | Mice | 3.4 | Sato and Ueno, 1977; Ueno and Ueno, 1978 |
| Fusarenon-X | p.o. | Mice | 4.5 | Bamburg, 1972; Sato and Ueno, 1977; Ueno and Ueno, 1978 |

| Fusarenon-X | i.p. | Adult male mice | 5.6 | Yoshizawa and Morooka, 1974 |
| Fusarenon-X | p.o. | Adult male mice | 5.5 | Yoshizawa and Morooka, 1974 |
| Fusarenon-X | i.p. | Adult female mice | 5.2 | Yoshizawa and Morooka, 1974 |
| Fusarenon-X | s.c. | Newborn mice | 0.23 | Sato and Ueno, 1977 |
| Fusarenon-X | s.c. | Newborn mice | 0.3 | Ueno, 1973a |
| Fusarenon-X | p.o. | 11-month-old mice | 7.2 | Saito et al., 1980 |
| Fusarenon-X | p.o. | 11-month-old mice | 7.4 | Saito et al., 1980 |
| Fusarenon-X | p.o. | 5-week-old mice | 3.4 | Saito et al., 1980 |
| Fusarenon-X | s.c. | Newborn mice | 0.1 | Saito et al., 1980 |
| Fusarenon-X | p.o. | Adult male rats | 4.4 | Saito et al., 1980; Sato and Ueno, 1977; Ueno, 1973a; Ueno and Ueno, 1978; Ueno et al., 1971c |
| Fusarenon-X | p.o. | Female rats (adult) | 4.0 | Ueno et al., 1971c |
| Fusarenon-X | i.p. | Guinea pigs (adult) | <0.5 | Sato and Ueno, 1977; Ueno, 1973a |
| Fusarenon-X | s.c. | Guinea pigs (adult) | <0.1 | Sato and Ueno, 1977; Ueno, 1973a |
| Fusarenon-X | s.c. | Adult cats | <5.0 | Sato and Ueno, 1977 |
| Fusarenon-X | s.c. | 1-week-old cats | <1.0 | Ueno, 1973a |
| Fusarenon-X | s.c. | Ducklings | 2.0 | Sato and Ueno, 1977 |

TABLE 3.7 Continued

| Fusariotoxin | Mode of Administration | LD$_{50}$ (mg/kg b.w.) | Animals | References |
|---|---|---|---|---|
| Diacetyl-nivalenol | i.p. | 9.6 | Mice | Sato and Ueno, 1977; Ueno, 1973a, 1977b; Ueno and Ueno, 1978; Ueno et al., 1973b |
| Dihydro-nivalenol | i.p. | 3.5 | Male mice | Saito and Ohtsubo, 1974 |
|  | i.p. | 15.0 | Mice | Ueno and Ueno, 1978 |
| Deoxy-nivalenol (Rd-toxin) | i.p. | 70.0 | Adult male mice | Sato and Ueno, 1977; Ueno and Ueno, 1978 |
|  | p.o. | 46.0 | Adult male mice | Sato and Ueno, 1977; Yoshizawa and Morooka, 1974 |
|  | i.p. | 77.0 | Adult female mice | Yoshizawa and Morooka, 1974 |
|  | s.c. | 3.8 | Dogs | Sato and Ueno, 1977 |
|  | s.c. | 27.0 | Peking duck-lings | Sato and Ueno, 1977; Ueno and Ueno, 1978; Yoshizawa and Morooka, 1974 |
| Deoxy-nivalenol-monoacetate (Rc-toxin) | i.p. | 46.9 | Mice | Ueno and Ueno, 1978 |
|  | i.p. | 49.0 | Mice | Sato and Ueno, 1977 |
|  | p.o. | 34.0 | Mice | Sato and Ueno, 1977 |
|  | s.c. | >1.0 | Dogs | Sato and Ueno, 1977 |
|  | s.c. | 37.0 | Ducklings | Sato and Ueno, 1977; Ueno and Ueno, 1978; Yoshizawa and Morooka, 1974 |
| Dihydro-nivalenol | i.p. | ~18.0 | Mice | Saito and Ohtsubo, 1974 |

Source: Drawn from various sources.

The toxic biological activity of trichothecene toxins shows a similar pattern. Thus dermal toxicity of T-2 tetraol, the parent alcohol, was lower than of T-2 toxin, and reducing the double bond of T-2 toxin decreased toxicity only slightly.

Ueno (1983c) summarized the lethal toxicity ($LD_{50}$ values) of various trichothecene metabolites to newborn and adult mice, rat, guinea pig, rabbit, dog, swine, duckling, chick, and trout by i.v., i.p., s.c., and p.c. administration. The author also summarized the toxic, cytotoxic, and biochemical properties of trichothecene derivatives.

The various toxic effects of the trichothecene fusariotoxins may be summarized as follows:

1. Necrosis of the epidermis and oral lesions in mice and rats (Bamburg and Strong, 1971; Bamburg et al., 1968a,b; Chung et al., 1974; Gilgan et al., 1966; Hayes and Schiefer, 1979b; Hayes et al., 1980; Joffe, 1960a,b, 1973b, 1974c; Lansden et al., 1978; Marasas et al., 1969; Sato et al., 1978; Ueno et al., 1970a), guinea pigs (De Nicola et al., 1978), poultry (Chi and Mirocha, 1978; Chi et al., 1977b,c; Christensen et al., 1972a; Coffin and Combs, 1980; Hamilton et al., 1971; Joffe, 1978a; Joffe and Yagen, 1978; Palyusik and Koplic-Kovacs, 1975; Palyusik et al., 1968; Richard et al., 1978; Speers et al., 1972, 1977), pigs (Weaver et al., 1981), cattle (Kosuri et al., 1970; Patterson et al., 1979), and sheep (Kurmanov, 1978b).

2. Vomiting (Ellison and Kotsonis, 1973; Joffe, 1978a; Joffe and Yagen, 1978; Vesonder et al., 1973, 1976, 1981b, 1982; Weaver et al., 1978a,b,c).

3. Food refusal may be a result of one and two according to Vesonder et al. (1973, 1976, 1979a, 1981a).

4. Hemorrhaging in mice and rats (Hayes and Schiefer, 1980; Hayes et al., 1980; Joffe 1960a; Korpinen and Ylimäki, 1972; Korpinen and Uoti, 1974; Kosuri et al., 1971), guinea pigs (De Nicola et al., 1978), cats (Joffe, 1960a, 1971, 1974c; 1978b, 1983b; Lutsky et al., 1978, 1981a,b; Sato et al., 1975; Yagen et al., 1977), monkeys (Rukmini et al., 1980; Stahelin et al., 1968), poultry (Joffe, 1978a; Joffe and Yagen, 1978; Kurmanov, 1978a; Palyusik and Koplic-Kovacs, 1975; Pearson, 1978), pigs (Weaver et al., 1978b,d), cattle (Grove et al., 1970; Kosuri et al., 1970; Kurmanov et al., 1978b; Pier et al., 1976; Ribelin, 1978; Tookey et al., 1972), and sheep (Kurmanov, 1978b).

5. Hematological changes and immunosuppression in mice and rats (Lafarge-Frayssinet et al., 1979; Lafont et al., 1977; Masuko et al., 1977; Otokawa et al., 1979; Rosenstein et al., 1978, 1979, 1981; Salazar et al., 1980; Schoental and Joffe, 1974; Schoental et al.,

1979; Ueno *et al.*, 1971c), rabbits (Gentry and Cooper, 1981), cats (Joffe, 1960a, 1971, 1974c, 1983b; Lutsky and Mor, 1981a,b; Sato *et al.*, 1975; Yagen *et al.*, 1977), poultry (Broonchuvit *et al.*, 1975; Doerr *et al.*, 1974, 1981; Hoerr *et al.*, 1981a,b; Joffe, 1978a, Joffe and Yagen, 1978; Otokawa *et al.*, 1979; Richard *et al.*, 1978), pigs (Weaver *et al.*, 1978a,d), and sheep (Kurmanov, 1978b).

6. Neurologic disturbances (Ishii *et al.*, 1971; Joffe, 1978a; Joffe and Yagen, 1978; Wyatt *et al.*, 1973a)

7. Ulcerations in some organs in mice and rats (Hayes *et al.*, 1980; Ohtsubo and Saito, 1977; Schoental and Joffe, 1974; Schoental *et al.*, 1979), cats (Lutsky and Mor, 1981a,b; Lutsky *et al.*, 1978), poultry (Greenway and Puls, 1976; Palyusik and Koplic-Kovacs, 1975; Palyusik *et al.*, 1968; Puls and Greenway, 1976), and pigs (Weaver *et al.*, 1978c).

Type A and type B mycotoxins have very similar lethal toxicities (*see* Table 3.7). T-2 toxin is one of the most important trichothecene mycotoxins of the type A occurring naturally in various agricultural commodities and products (Mirocha *et al.*, 1976b; Puls and Greenway, 1976; Rukmini and Bhat, 1978; Szathmary, 1983; Joffe, 1983b,c). These contaminated products, after consumption, caused severe disorders in humans and farm animals and usually lethal toxicoses (Cirilli, 1983; Szathmary, 1983; Joffe, 1983b). Type A toxins produced by various *Fusarium* species caused stronger vomiting, refusal and dermal necrotizing effects, hemorrhages in the gastrointestinal tract, leukopenia, reduction in weight gains and feed efficiency, infertility, abortion, and inhibition of protein synthesis in various animal and biological systems (Joffe, 1978b, 1983b; Ueno, 1980a,b,c, 1983b,c).

Type B is held to be particularly responsible for emetic effects and refusal by swine after consumption of corn contaminated by *F. graminearum* (Vesonder *et al.*, 1973, 1976, 1979b, 1981, 1982; Vesonder and Hesseltine, 1980/81).

Some derivatives belonging to type A and type B (T-2 toxin, fusarenon-X, and related trichothecenes) caused severe radiomimetic cellular injuries and karyorrhexis in the actively dividing cells of the bone marrow, thymus, small intestine, testes, and ovary in laboratory animals (mice, rats, and cats) (Ueno, 1971a,c, 1973d; Saito and Ohtsubo, 1970; Saito *et al.*, 1969; Saito and Ohtsubo, 1974; Sato *et al.*, 1980; Saito, 1983).

The mode of administration can affect the degree of toxicity. Working with pure fusarenon-X, Ueno *et al.* (1971c), found that the $LD_{50}$ for mice was 3.4 mg/kg i.v., 4.2–4.6 mg/kg s.c., 3.4 mg/kg i.p., and 4.5 mg/kg when administered orally. Diarrhea was also often noted (Ueno *et al.*, 1971c) and degenerative changes, atypical mitoses, and nuclear deformation were found in the intestinal system. The mice often seemed to have died from intestinal bleeding.

Similar histopathological findings were reported in rats. Fusarenon-X seemingly affects cell proliferation in both animals. In guinea pigs degradation of bone marrow cells was noted, in addition to the intestinal tract damage. Cytotoxic damage was also seen in cat intestine, in addition to karyorrhexis of intrasinosoidal cells of the liver. In ducks fusarenon-X caused cytotoxic changes in bone marrow as well as degenerative changes in the intestine. More precise clinical details are given in the chapters on animal and human toxicoses (Chapters 6 and 8).

The individual toxins in any group have differing toxicities. These differences can be linked to the chemistry of the toxins concerned. In type A trichothecenes, HT-2 toxin, which is less toxic than T-2 toxin, has been shown to be a metabolite of T-2 toxin (Bamburg and Strong, 1969; Ellison and Kotsonis, 1974). Similarly in the case of type B toxins, fusarenon-X is more toxic than nivalenol (Ueno et al., 1970c). It has been hypothesized that the C-4 hydroxyl group may play an important role in type B toxicity since deoxynivalenol and its monoacetate (which both lack the C-4 hydroxy group) are far less toxic than fusarenon-X or nivalenol (Yoshizawa and Morooka, 1974).

Studies of various aspects of the inhibition of protein synthesis in cell free systems by trichothecenes show marked selectivity. The intact epoxy ring of C-12–C-13 is essential for activity. Esterification of certain alcohols makes them more or less specific and potent inhibitors of protein and DNA synthesis, reticulocytes, Ehrlich ascites tumor cells, rat liver and microbial cells depending on the position of the hydroxyl and enzyme system studied (Ueno et al., 1973b; Wei and McLaughlin, 1974).

The effect of a lethal dose of trichothecene toxins is similar to lethal whole-body irradiation (Saito and Ohtsubo, 1974).

## Metabolism

Quite a bit is known on the metabolic fate of *Fusarium* toxins in plants and animals. Ueno et al. (1971c) studied the distribution of [³H]-labeled fusarenon-X in mice and found it in urine and feces after 24 hr. It seems that the toxin is eliminated rapidly after 30 min through the kidneys, intestines, and bile, but no products of metabolism were identified.

More recent publications have shown that fusarenon-X and T-2 toxin were biotransformed by the microsomal enzyme system of the liver (Ueno and Ohta, 1977; Ueno, 1977b, 1980a). Patterson (1973) also found that duck and rat liver microsomal preparations rapidly metabolize scirpene triacetate.

According to Ohta et al. (1977, 1978) T-2 toxin, fusarenon-X, and other trichothecenes administered to rats and rabbits were metabolized by C-4 deacetylation to HT-2 toxin, nivalenol, and other metabolites. Matsumoto et al. (1978) found [³H]-labeled T-2 toxin in feces and urine.

Deacetylation mainly of C-4 and also of C-3 and C-8 acetates was found in liver preparation in vitro by Ellison and Kotsonis (1974) and Yoshizawa *et al.* (1980b).

Yoshizawa *et al.* (1980c) worked out a method for the detection of T-2 toxin metabolites in the excreta of broiler chickens after administration of [$^3$H]-labeled T-2 toxin.

After oral administration of [$^3$H]-labeled T-2 toxin the authors have extracted from the excreta of broiler chickens, aside from T-2 toxin, neosolaniol and T-2 tetraol, eight unknown derivatives named TB-1 to TB-8. One of these metabolites (TB-6) was identified as 4-deacetylneosolaniol    (15-Acetyl-3α,4β,8α-trihydroxy-12,13-epoxytrichothec-9-ene). Yoshizawa *et al.* (1981) studied the metabolism of T-2 toxin in lactating Jersey cows. The chromatographic analysis of the tritium residues in cow tissues and excreta revealed that T-2 toxin was metabolized to at least eight derivatives, TC-1 to TC-8, and also detected HT-2 toxin, neosolaniol, and 4-deacetylneosolaniol. Three major metabolites (TC-1, TC-3, and TC-6) and TC-8 were unidentified.

Yoshizawa *et al.* (1982a) described the structure of metabolites named TC-1 and TC-3 found in the plasma, milk, and excreta of lactating cow. These metabolites may correspond to unidentified metabolites found in the excreta of chicks and rats (Matsumoto *et al.*, 1978; Yoshizawa *et al.*, 1980c).

The isolated compounds were new metabolites TC-1 and TC-3 of T-2 toxin, such as 3′-hydroxy T-2 toxin, 4β,15-diacetoxy-3α-hydroxy-8α-(3-hydroxy-3-methylbutyryloxy)-12,13-epoxy-trichothec-9-ene   and   3′hydroxy HT-2 toxin, 15-acetoxy-3α-4β-dihydroxy-8α-(3-hydroxy-3-methyl-butyryloxy)-12,13-epoxytrichothec-9-ene (Yoshizawa *et al.*, 1982a,b).

The LD$_{50}$ values of male mice of the dds strain, administered i.v. for acute toxicity of 3′-hydroxy T-2 toxin, was 4.63 ± 0.63 mg/kg and for 3′-hydroxy HT-2 toxin was 22.8 ± 2.0 mg/kg.

Yoshizawa *et al.* (1984) studied in vitro metabolism of T-2 toxin in the hepatic homogenates of mice and monkeys for elucidation of the formation of the hydroxylated products (as 3′-hydroxy T-2 toxin and 3′hydroxy HT-2 toxin) including metabolites of HT-2 toxin, neosolaniol, 4-deacetylneosolaniol, 15-deacetylneosolaniol, and T-2 tetraol. The formation of these hydroxylated metabolites was detected in the microsomes in the presence of NADPH, and the hydroxylation reaction was increased by treating mice with phenobarbital.

The hydroxylated metabolites of T-2 toxin were also observed in the liver homogenates of rabbits, swine, and cows, but HT-2 toxin was the sole metabolite in homogenates of chickens (Yoshizawa and Sakamoto, 1982). According to Yoshizawa *et al.* (1980b) T-2 toxin was rapidly metabolized to HT-2 toxin, which was then converted into T-2 tetraol via 4-deacetylneosolaniol in liver homogenates of rats.

Robison *et al.* (1979a) studied the distribution and excretion of tritium-labeled T-2 toxin in two weanling pigs. The percentage of administered radioactivity found 18 hr after dosing in one pig intubated with 0.1 mg of [$^3$H] T-2/kg body weight in 50% aqueous ethanol was 0.7% (3.1 ppm) in muscle, 0.43% (13.8 ppm) in liver, 0.08% (15.9 ppm) in kidney, 0.06% in bile, 21.6% in urine, and 25.0% in feces. The percentage of administered radioactivity in the second pig intubated with 0.4 mg/kg of labeled T-2 toxin was 0.7% (11.5 ppm) in muscle, 0.29% (37.7 ppm) in liver, 0.08% (61.4 ppm) in kidney, 0.14% in bile, 17.6% in urine, and 0.84% in feces.

Robison *et al.* (1979b) investigated the occurrence of T-2 toxin in milk of a pregnant Holstein cow and of a crossbred sow. The T-2 toxin had been administered to the cow by intubation at a rate equivalent to 50 ppm in the feed (182 mg for 15 consecutive days). Milk samples were analyzed for T-2 toxin on the day of toxin intubation (on the second, fourth, fifth, eighth, tenth, and twelfth days). All samples (except those on the fourth and eighth day of intubation) contained T-2 toxin from 10–165 ppb.

T-2 toxin was also detected in the milk of a sow fed a diet containing 12 ppm for 220 days. In a milk sample of 58 g, six days after parturition, 76 ppm of T-2 toxin was isolated.

However, the authors point out that the level of T-2 toxin occurring naturally in feeds (2 ppm) would produce levels of the toxin in milk too low to be detected.

The role of naturally occurring trichothecenes in feedstuffs (T-2 toxin, diacetoxyscirpenol, deoxynivalenol, and nivalenol) and in causing mycotoxicoses and other potential hazards and disorders in animals and humans was described by Pathre and Mirocha (1979). The authors suggest that the residue from administered T-2 toxin and deoxynivalenol in domestic animals can be transmitted into humans; however, the concentrations were extremely small.

Fontelo *et al.* (1983) analyzed blood and urine for detection of T-2 toxin by modified radioimmunoassay procedure directly in assay tubes without previous extraction. The reaction between antibody and ligands was optimal at 1 hr intervals. Albumin-coated charcoal was used to separate bound from free ligand. This method was simpler than the one associated with ammonium sulphate precipitation used by Chu *et al.* (1979). The sensitivity was 1 nanogram per assay or 10 ng/ml.

A sensitive method for production of antibody against T-2 toxin was developed using radioimmunoassay and enzyme-linked immunosorbent assay for T-2 toxin (Chu *et al.*, 1979; Lee and Chu, 1981a,b; Pestka *et al.*, 1981). Because antibody against T-2 toxin did not cross-react with diacetoxyscirpenol (DAS) (Chu *et al.*, 1979; Fontelo *et al.*, 1983), Chu *et al.* (1984) recently attempted to work out a method for production of antiserum in rabbits which could specifically be used against DAS. The authors used hemisuccinate or hemiglutarate derivatives of DAS for immunizing of rabbits for subsequent conjugation to bovine serum albumin (BSA). The last (DAS-hemiglutarate-BSA) was found to be a much better immunogen than the former (DAS-hemisuccinate-BSA) because the antibody titers were higher after using DAS-hemiglutarate-BSA (DAS-HG-BSA). The antibodies against DAS were tested by radioimmunoassay procedure with antiserum from rabbits that were immunized with DAS-HG-BSA.

The authors state that the antibody was most specific for DAS. The concentration causing 50% inhibition of binding of [$^3$H] DAS to antisera by unlabeled DAS and 4-monoacetoxyscirpenol (MAS) and 15-MAS were found to be 1.5, 130, and 300 ng per assay, respectively. The authors concluded that the antibodies elicited have a high binding efficiency against DAS.

The recent interest in DAS _Fusarium_ compound is associated with its high toxicity in humans and animals (Ueno, 1983c) and with the fact that it naturally occurs in cereal grains and chiefly because of its antibiotic and antitumor effects (Bamburg and Strong, 1971; Kaneko _et al._, 1982). Diacetoxyscirpenol is clinically called anguidine (Claridge _et al._, 1978a,b, 1979). Murphy _et al._ (1976), Haas _et al._ (1977); Goodwin _et al._ (1978); Belt _et al._ (1979); Diggs _et al._ (1978); DeSimone _et al._ (1979); and Yap _et al._ (1979), noted that anguidine was administered to terminal patients who had undergone extensive chemotherapy.

Gyongyossy-Issa _et al._ (1984) worked out a method using a charge-shift electrophoresis to define the behavior of T-2 toxin in agarose gels. This method can be adapted with small hydrophobic molecules such as T-2 toxin and cholesterol.

Ueno (1980a,b,c) summarized the results on the trichothecene-producing _Fusarium_ species in Japan and described the toxicological and pharmacological characteristics and the metabolism of trichothecenes in animal organs, as well as an epoxide-specific detection method, known as the "blue-spot test."

Ichinoe _et al._ (1983) tested trichothecene compounds and zearalenone (ZEN) from 113 single-spore cultures of _Gibberella zeae_ isolated from rice stubbles in barley and wheat fields in two geographically different regions in Japan. The isolates were divided into two chemotaxonomic types: nivalenol (NOV) and fusarenon-X (FX) producers and deoxynivalenol (DON) and 3-acetyldeoxynivalenol (ADON) producers. No cross production of these two types of trichothecenes was observed in these isolates. Zearalenone was found in 68% of the isolates, but no clear relationship could be seen regarding its position with respect to the chemotaxonomic types. The authors revealed that the chemotaxonomy of _G. zeae_ could explain the geographic differences of trichothecene contamination in the world.

## Zearalenone

Zearalenone [6-(10-hydroxy-6-oxo-*trans*-1-undecenyl)-β-resorcylic acid-μ-lactone] (Urry *et al.*, 1966) (Fig. 3.5) has caused estrogenic disorders and various diseases in swine ingesting *Fusarium*-contaminated maize, wheat, and barley grains as well as in other farm animals in Midwestern and Southeastern United States and in laboratory animals, mainly rats and mice.

A derivative of zearalenone, zearalanol [6-(6,10-dihydroxy-*trans*-1-undecenyl)-β-resorcylic acid-μ-lactone], has been used commercially to enhance the weight gain of cattle and sheep (Brown, 1970; Dixon, 1983; Perry *et al.*, 1970; Sharp and Dyer, 1968, 1971; Willemart and Bouffault, 1983) and to serve as a drug for alleviation of postmenopausal distress in women as an estrogen substitute (Utian, 1973).

Zearalenone and zearalenol are estrogenic secondary metabolites produced by some *Fusarium* strains especially by *F. roseum* "Graminearum" (*F. graminearum*) or *F. roseum* "Gibbosum" (*F. equiseti*) (Hagler and Mirocha, 1980; Palyusik *et al.*, 1980; Stipanovic and Schroeder, 1975; Steele *et al.*, 1976). Zearalenone was associated with hyperestrogenism and other reproductive disorders in swine (Miller *et al.*, 1973), cattle (Mirocha *et al.*, 1968c), and poultry (Allen, 1981a,c; Meronuck *et al.*, 1970), and zearalenol was evaluated under laboratory conditions (Hagler *et al.*, 1979; Mirocha *et al.*, 1978, 1979b; Peters, 1972).

According to Mirocha *et al.* (1980) zearalenone and zearalenol are the only naturally occurring derivatives.

The derivatives of zearalenone isolated from strains of *F. roseum* "Gibbosum" and *F. roseum* "Graminearum" are shown in Figure 3.6.

Several early researchers associated food refusal and estrogenic symptoms in pigs with moldy corn (McNutt *et al.*, 1928) or with barley infected by *F. graminearum* (Christensen and Kernkamp, 1936). In Iowa Koen and Smith (1945) observed hyperestrogenism in pigs and some other symptoms. Similar cases of vulvovaginitis in swine were noted in Australia by Pullar and Lerew (1937). In Ireland McErlean (1952) found that barley infected by *F. graminearum* (*G. zeae*) produced a toxin which caused hyperestrogenism. The same symptoms were noted by Lelievre *et al.* (1962) in France, Paita (1962) in Italy, Stamatovic *et al.* (1963) in Yugoslavia, Bugeau and Berbinschi (1967) in Romania, Danko and Aldasy (1968) in Hungary, Eriksen (1968) in Denmark, and Bristol and Djurickovic (1971) in Canada.

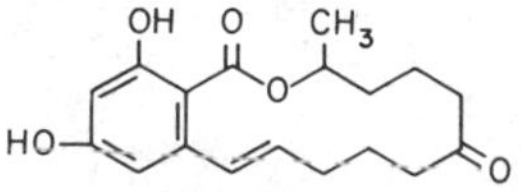

**Figure 3.5**  The chemical structure of zearalenone.

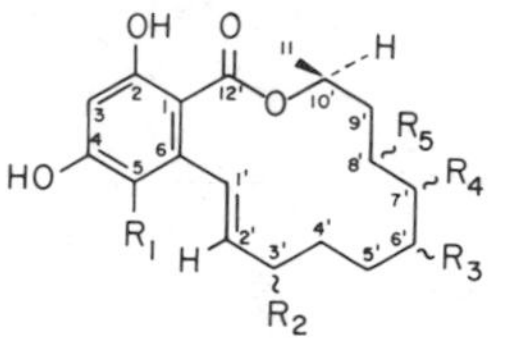

**Figure 3.6**   The chemical structure of zearalenone's naturally occurring derivatives.

|                                    | $R_1$ | $R_2$  | $R_3$ | $R_4$  | $R_5$  |
|------------------------------------|-------|--------|-------|--------|--------|
| 1. Zearalenol                      | H     | $H_2$  | OH    | $H_2$  | $H_2$  |
| 2. Zearalenone                     | H     | $H_2$  | O     | $H_2$  | $H_2$  |
| 8'-hydroxyzearalenone              | H     | $H_2$  | O     | $H_2$  | OH     |
| 6',8'-dihydroxyzearalene           | H     | $H_2$  | OH    | $H_2$  | OH     |
| 3'-hydroxyzearalenone              | H     | OH     | O     | $H_2$  | $H_2$  |
| 5-formylzearalenone                | CHO   | $H_2$  | O     | $H_2$  | $H_2$  |
| 7'-dehydrozearalenone              | H     | $H_2$  | O     | H      | H      |

Fennel (1979) described the incidence in Ontario, Canada of zearalenone associated with corn during six years, 1972–1977. He found that the concentrations in 266 samples ranged from 0.01 to 141 ppm. Those samples containing levels of 1 ppm zearalenone or more were significant enough to affect swine.

Neish *et al.* (1982) reported results of field study in Eastern Ontario associated with corn ears inoculated with three isolates of *F. graminearum* for their ability to produce zearalenone and some trichothecene compounds. The authors noted that the average level of deoxynivalenol was 860 ppm/g, but no ZEN was detected. However, rice inoculated with *Fusarium* isolates gave maximum level of 542 ppm/ g of DON and 1.255 ppm of ZEN.

Greenhalgh *et al.* (1983) screened three Canadian isolates of *F. graminearum* (DAOM 18077, 180378, and 180379) for their ability to produce mycotoxins on corn and rice at various temperatures, moisture content, and various times. They all produced DON and ZEN on corn, but on rice only DAOM 18078 and 18079 produced significant levels of these mycotoxins (with levels of DON higher than those of ZEN). The authors also studied the effects of the initial moisture contents (IMC) before autoclaving, incubation temperature, and time only with isolate DAOM 18078. This isolate produced ZEN 333 ppm/g at 19.5°C after incubation of 40 days at the highest moisture level, and the concentration of DON was low, only 20 ppm. At 28°C DON levels were higher than ZEN levels (515 ppm/g) after 24 days incubation at 40% IMC; at the same temperature 399 ppm/g ZEN was produced but at an IMC of 35%. Other factors such as pH, oxygen, and carbon dioxide concentrations and size of the culture flasks influenced the production of mycotoxins.

The authors state that rice was a better substrate than corn for DON formation.

Scott *et al.* (1984) examined samples of winter wheat contaminated with *F. graminearum*, from 5 fields in Southwestern Ontario, for occurrence of deoxynivalenol. The authors note that the concentration of DON in the grain declined significantly from 1.56 μg/g on July 7th to 0.21 μg/g on July 14th and 0.11 μg/g on July 18th. The decline in DON levels occurred as the grain matured.

Martin and Johnston (1982) note that a lack of correlation between severity of head blight symptoms and DON levels was observed in the Atlantic regions. According to Seaman (1982) symptoms of head blight in wheat were noted in eastern Ontario when DON was not detected. Miller *et al.* (1983) proved that the decline in DON levels may be a result of reaction with plant components or of metabolism by host plant enzymes.

Hagler *et al.* (1984) studied the occurrence of deoxynivalenol, zearalenol, T-2 toxin, and aflatoxin in samples of wheat from the 1982 harvest year from 4 states (Kansas, Nebraska, Illinois, and Texas). Deoxynivalenol was isolated in 31 out of 33 samples, zearalenone in 3 out of 33 samples and aflatoxin [$B_1$] in 23 out of 31 samples. None of the samples contained T-2 toxin. The mean concentration of DON was $1.782 \pm 262$ mg per g and the mean concentration of zearalenone in the 3 samples were 35, 90, and 115 ng per g, while the mean concentration of aflatoxin [$B_1$] was from 0.8–17 0 ng per g. This is the first report on the simultaneous occurrence of DON, ZEN, and aflatoxin [$B_1$] in wheat from the 1082 harvest crop.

The occurrence of DON and ZEN together has been reported previously by Mirocha *et al.* (1976).

Kallela and Saastamoinen (1981) note that the preservative "Luprosil®" is a propionic acid preparation of BASF, and "Gasol" is one of the Farmline Grain Preservatives produced by Farmos-Yhtyma Oy. "Gasol" contains organic acids and some additional compounds. These preservatives inhibited the growth of *F. graminearum* in oat grains. "Gasol" decreases the amount of zearalenone, but "Luprosil" has no effect on the toxin content of oats.

## Fusarium Species

Zearalenone and its natural derivatives have been produced by the following *Fusarium* species: *F. roseum*, *F. roseum* "Avenaceum," *F. roseum* "Culmorum," *F. roseum* "Equiseti," *F. roseum* "Gibbosum," *F. lateritium*, *F. moniliforme*, *F. oxysporum*, *F. sambucinum*, *F. solani*, *F. sporotrichioides*, *F. tricinctum*, and others (Bennett and Shotwell, 1979; Bennett *et al.*, 1976, 1978; Caldwell and Tuite, 1970; Caldwell *et al.*, 1970; Curtin and Tuite, 1966; Davis and Diener, 1979; Hesseltine, 1974,

1976; Ishii *et al.*, 1974; Lasztity and Woller, 1975a,b, 1977; Lew, 1978; Marasas *et al.*, 1979b,c; Martin and Keen, 1978; Mirocha and Christensen, 1974a, 1982; Mirocha *et al.*, 1974, 1967, 1968a,b,c, 1969, 1976b, 1977a,b, 1980; Moller *et al.*, 1978; Muller, 1978; Scott, 1978; Shannon *et al.*, 1980; Steele *et al.*, 1976; Stipanovic and Schroeder, 1975; Ueno, 1973b; Ueno *et al.*, 1974c, 1977a,b; Wilson and Hayes, 1973; Wimmer, 1978).

### Environmental Conditions in the Production of Zearalenone

Low temperatures or alternating high and low temperatures favor zearalenone production, and this temperature regime has been used by most workers (*see* Table 3.8). Thus when *F. roseum* (M-3-2) was grown on rice (supplemented by 10% of various nutrients) for 14 days at 25°C followed by 14 days at 14–15°C, 32.0–560 mg/kg zearalenone was produced. An additional 2 weeks of incubation at 12–15°C increased the yield of zearalenone (Ishii *et al.*, 1974).

Mirocha *et al.* (1967) found that after initial incubation at 25–28°C for 2 weeks, the yield of zearalenone obtained from *F. graminearum* could be increased by prolonging the incubation period at 12°C. Thus after 8 weeks at 12°C the yield was 3500 ppm of zearalenone. In contrast to these findings, an isolate of *F. roseum* cultured on sorghum by Schroeder and Hein (1975) produced more zearalenone at 25°C than at 10°C, and Naik *et al.* (1979) noted that an initial incubation period at 10°C only enhanced zearalenone production in one out of the five isolates of *F. graminearum* isolated from maize in Ontario. Table 3.8 shows that several researchers have in fact elected to use higher incubation temperatures (Bacon *et al.*, 1977; Martin and Keen, 1978; Stipanovic and Schroeder, 1975; Ueno *et al.*, 1977).

Mirocha *et al.* (1967) and Ishii *et al.* (1974) found that liquid media gave low yields of zearalenone and most workers used maize or rice media. Hidy *et al.* (1977) developed an artificial medium with vermiculite for the large-scale production of zearalenone from strains of *G. zeae*. Flasks containing the medium were incubated in water-cooled trays at 17–18°C. Bacon *et al.* (1977) used a medium containing glutamic acid, starch, and yeast extract incubated at 24°C for 2 weeks. Ueno *et al.* (1977b) used Czapek-Dox Peptone medium, incubating the *Fusarium* species at 24°C for 2 weeks followed by 10–15°C for one week.

Zearalenone-producing *Fusarium* species have been isolated from numerous natural substrates. Although usually associated with grain used for animal feed, zearalenone producers have also been isolated from bean-hulls, rice-straw, barley, oats, and even river sediment (Ishii *et al.*, 1974).

The occurrence of zearalenone in foods and feeds and its chemistry and biological activity have been described by Hidy *et al.* (1977); Hobson *et al.* (1977b); Jemmali *et al.* (1978); Martin and Keen (1978); Mirocha and Christensen (1974b); Mirocha *et al.* (1974); Pathre and Mirocha (1976); and Patterson *et al.* (1977). Zearalenone has been detected in a wide range of natural products (*see* Table 3.9) including human food such as maize beer (Lovelace and Nyathi, 1977) and cereals which might be used for human consumption, such as corn, wheat, barley, and oats (Sherwood and Peberdy, 1972a,b,c,d, 1973b; Shotwell *et al.*, 1970, 1971, 1976) and maize, barley, sorghum, and silage (Mirocha and Christensen, 1974b; Mirocha *et al.*, 1974, 1980). However, the toxin is usually associated with *Fusarium*-infected maize stored under moist conditions in open cribs (Collet *et al.*, 1977; Jemmali, 1973; Mirocha *et al.*, 1974).

Wide variations have been noted in the yearly occurrence of zearalenone, usually linked to environmental climatic conditions. Thus Palyusik (1973, 1978) noted that in Hungary vulvovaginitis in pigs (usually associated with zearalenone) was most prevalent in cold, wet years. In Italy Bottalico (1977) found low levels of zearalenone (0.24 µg/g) in 1975, a dry harvest year, but in 1974 and 1976, when heavy October rain delayed harvest, 24 µg/g zearalenone were detected in newly harvested maize.

Bottalico *et al.* (1980) noted that 3 ppm of zearalenone was isolated from Indonesian cassava meal which had been imported to Italy. Aflatoxin was also isolated from the cassava meal.

In Yugoslavia rainy weather in 1974 encouraged *G. zeae* contamination, and zearalenone was detected in 50% of the corn samples with concentrations reaching 200 ppm (Balzer *et al.*, 1977a,b). Numerous cases of estrogenism in swine were also noted that year. However, in 1975, when the weather was dry, little zearalenone was detected and levels were low (less than 10 ppm). Bennett and Anderson (1978) noted that in 1972 an unusually wet season may have been responsible for the high incidence of zearalenone contamination in maize. Caldwell and Tuite (1974) stated that zearalenone contamination was not a problem in freshly harvested corn in Indiana, and Stob (1973) suggested that contamination and toxin production occurred during storage. Gross and Robb (1975) detected no zearalenone in freshly harvested barley but found that zearalenone could be detected after 7–10 weeks storage. The maximum amount produced was 208 µg/g at 15°C and 45% of moisture content. Sherwood and Peberdy (1974a,b) noted that in grain held at 15–18% mc zearalenone production increased with temperatures up to 18°C (maximum 2000 µg/g grain) but that at 25°C zearalenone production rarely exceeded 100 µg/g grain. They suggested that grain was only safe if stored at moisture content of 14% or less, although at 15–18% mc mycelial growth of *F. graminearum* could be prevented for up to 6 months by keeping the storage temperature at or below 7°C.

**TABLE 3.8  Zearalenone Production by *Fusarium* Species**

| *Fusarium* species | Culture Medium | Temperature/Time (°C) | References |
|---|---|---|---|
| *F. lateritium* | Rice | 25° for 14 d + 12–15° for 14 d | Ishii *et al.*, 1974 |
| *F. lateritium* | CDSP[a] | 25° for 2 wk | Ueno *et al.*, 1977b |
| *F. equiseti* | Wheat | Not stated | Scott *et al.*, 1972 |
| *F. avenaceum* | Rice | 25° for 4 wk + 12° for 6 wk | Hacking *et al.*, 1977 |
| *F. graminearum* | Rice | 25° for 4 wk + 12° for 6 wk | |
| *F. graminearum* | Malt extract agar | 25° for 5–7 d | Marasas *et al.*, 1978 |
| *F. graminearum* | Maize | 25–28° for 2 wk | Mirocha *et al.*, 1966, 1967, 1968c |
| *F. graminearum* | Wild rice and grain sorghum | Room temperature on a shaker for 4 d | Lindenfelser *et al.*, 1978 |
| *F. graminearum* *F. culmorum* | Maize | Optimal conditions | Vanyi *et al.*, 1973a |
| *F. culmorum* | Rice | 24–27° for 1 wk + 14° for 4 wk | Mirocha *et al.*, 1976b |
| *F. culmorum* | Rice | 25° for 4 wk + 12° for 6 wk | Hacking *et al.*, 1976, 1977 |
| *F. culmorum* | Maize | Room temperature for 1 wk + 10–12° for 1 wk | Kovacs *et al.*, 1975a |

| Species | Substrate | Conditions | Reference |
|---|---|---|---|
| *F. sambucinum* var. *coeruleum* | Rice | 25° for 4 wk + 12° for 6 wk | Hacking *et al.*, 1977 |
| *F. moniliforme* | Maize | 26° for 2 wk + 10–12° for 6 wk | Mirocha *et al.*, 1969 |
| *F. moniliforme* | Rice | 25° for 4 wk + 12° for 6 wk | Hacking *et al.*, 1976, 1977 |
| *F. moniliforme* | Maize | 25° for 7 d | Martin and Keen, 1978 |
| *F. nivale* | Rice | 25° for 4 wk + 12° for 6 wk | Hacking *et al.*, 1976, 1977 |
| *F. roseum* | Maize | 25° for 2 wk | Stipanovic and Schroeder, 1975 |
| *F. roseum* | Maize | 22–25° for 2 wk + 10–12° for 3 wk | Meronuck *et al.*, 1970 |
| *F. roseum* "Gibbosum" | Rice | 25° for 7 d + 10–12° for 30 d | Hagler and Mirocha, 1980; Hagler *et al.*, 1979 |
| *F. roseum* "Culmorum"<br>*F. roseum* "Equiseti"<br>*F. roseum* "Gibbosum"<br>*F. roseum* "Graminearum" | Maize | 16° for 3 wk<br>or<br>24° for 2 wk + 12° for 8 wk | Caldwell *et al.*, 1970 |
| *F. tricinctum* | Maize | 16° for 3 wk | Caldwell *et al.*, 1970 |
| *G. zeae* | Maize | 16° for 3 wk | Adams and Tuite, 1976 |
| *G. zeae* | Carboxymethyl-cellulose | Room temperature on shaker for 5 d | Bacon and Marks, 1976 |
| *G. zeae* | Starch-yeast | 24° for 2 wk | Bacon *et al.*, 1977 |
| *G. zeae* | Vermiculite | 17–18° for 4–5 wk | Hidy *et al.*, 1977 |
| *Fusarium* species | Rice | 25° for 2 wk | Scott *et al.*, 1970 |

*Source:* Drawn from various sources.

[a]Czapek-Dox medium supplemented with peptone.

**TABLE 3.9  Natural Occurrence of Zearalenone**

| *Fusarium* Species | Substrate | Zearalenone Levels | Country | References |
|---|---|---|---|---|
| G. zeae | Maize | 1.5–10 µg/g | USA | Caldwell and Tuite, 1974 |
| G. zeae | Wheat | 0.36–11.05 ppm | USA | Shotwell et al., 1977 |
| G. zeae | Wheat | 3.64–11.054 ppb | USA | Hesseltine et al., 1978 |
| G. zeae (F. graminearum) | Maize | ≈5.1–200 ppm | Yugoslavia | Balzer, 1977b |
| G. zeae (F. graminearum) | Maize | 0.24–24.0 µg/g | Italy | Bottalico, 1977 |
| F. graminearum | Maize | 170 ppm | France | Collet et al., 1977 |
| F. graminearum | Feed | 25 ppm | Finland | Korpinen, 1972 |
| F. graminearum | Maize | 3.6 ppm | Yugoslavia | Loncarevic et al., 1977 |
| F. graminearum | Maize | 12.8 ppm 6.4 ppm | Zambia Transvaal | Marasas et al., 1977 |
| F. graminearum | Maize | 20–3500 ppm | USA | Mirocha et al., 1967 |
| F. graminearum | Hay | 14.0 ppm | England | Mirocha et al., 1968c |
| F. graminearum | Maize | 2.7 ppm | USA | Mirocha et al., 1974 |
| F. graminearum | Maize | 32.0 ppm | USA | Mirocha et al., 1974 |
| F. graminearum | Silage | 87.0 ppm | USA | Mirocha et al., 1974 |
| F. graminearum | Maize | 2.5 ppm | USA | Mirocha et al., 1974 |
| F. graminearum | Pig feed | 0.1 ppm | USA | Mirocha et al., 1974 |
| F. graminearum | Dairy ration | 1.0 ppm | USA | Mirocha et al., 1974 |
| F. graminearum | Pig feed | 0.01 ppm | USA | Mirocha et al., 1974 |
| F. graminearum | Pig feed | 50.0 ppm | USA | Mirocha et al., 1974 |

| Organism | Substrate | Concentration | Country | Reference |
|---|---|---|---|---|
| *F. graminearum* | Sorghum | 12.0 ppm | USA | Mirocha *et al.*, 1974 |
| *F. graminearum* | Maize | 36.6 ppm | Yugoslavia | Mirocha *et al.*, 1974 |
| *F. graminearum* | Pig feed | 0.5 ppm | Yugoslavia | Mirocha *et al.*, 1974 |
| *F. graminearum* | Maize | 306.0 ppm | England | Mirocha *et al.*, 1974 |
| *F. graminearum* | Maize | 2.5–35.6 ppm | Yugoslavia | Ozegovic, 1970 |
| *F. culmorum* | Maize | 61–270 µg/g; 25–256 µg/g | Italy | Bottalico, 1975 |
| *F. graminearum* *F. culmorum* | Maize | 70–80 ppm | Hungary | Lasztity and Wöller, 1975a |
| *F. graminearum* *F. moniliforme* | Maize | Not stated | Yugoslavia | Ozegovic and Vukovic, 1972 |
| *F. graminearum* *F. equiseti* *F. moniliforme* | Maize | 4.0–42.8 µg/g | USA | Kotsonis *et al.*, 1975b |
| *F. graminearum* (*G. zeae*) | Maize | 100–210 µg/kg | USA | Stoloff *et al.*, 1976 |
| *F. roseum* | Feedstuff | 0.1–290 ppm | USA | Mirocha and Christensen, 1974b |
| *F. roseum* | Feed | 175–3600 µg/kg | USA | Mirocha *et al.*, 1976b |
| *F. roseum* | Maize | 6.400 µg/kg | USA | Mirocha *et al.*, 1976b |
| *F. roseum* | Mixed feed | 120 µg/kg | USA | Mirocha *et al.*, 1976b |
| *F. roseum* | Milo | 2000–5600 µg/kg | USA | Mirocha *et al.*, 1976b |
| *F. roseum* | Animal rations | 66–1000 µg/kg | Canada | Mirocha *et al.*, 1976b |
| *F. roseum* | Oats | 25–235 µg/g and 1.5–4.0 µg/g (zearalenol) | Finland | Mirocha *et al.*, 1979b |

**TABLE 3.9**  Continued

| *Fusarium* Species | Substrate | Zearalenone Levels | Country | References |
|---|---|---|---|---|
| *F. moniliforme* | Maize | 2.0–7.7 ppm | Italy | Bottalico, 1976 |
| *F. moniliforme* | Maize | Not stated | India | Ghosal et al., 1978 |
| *F. culmorum* | Barley | Not stated | USA | Gross and Robb, 1975 |
| *F.* species | Maize | 2350 µg/kg | France | Jemmali, 1973 |
| *F.* species | Barley | Traces (not quantified) | Scotland | Shreeve et al., 1975 |
| *F.* species | Maize | 2.5–10 ppm | France | Jemmali et al., 1978 |
| Not stated | Maize | 800 ppb | USA | Shotwell et al., 1970 |
| Not stated | Maize | 450–750 ppb | USA | Shotwell et al., 1971 |
| Not stated | Cereal grain (barley, rye, oats, maize) | 50–200 µg/kg | Poland | Juszkiewicz and Piskorska-Pliszczynska, 1976, 1977 |
| Not stated | Animal feedstuffs | 1–2 ppm | England | Patterson et al., 1977 |
| Not stated | Maize | Up to 12.500 µg/kg | USA | Shotwell et al., 1976 |
| Not stated | Maize beer | <4.6 mg/l | Zambia | Lovelace and Nyathi, 1977 |
| Not stated | Maize beer | 0.3–2 µg/g | Lesotho | Martin and Keen, 1978 |
| Not stated | Cereal or oil seed crops | Not stated | England | Patterson, 1978 |
| Not stated | Cornmeal | 11.0–69 ng/g | USA | Ware and Thorpe, 1978 |

*Source:*  Drawn from various sources.

## *Estrogenic Symptoms Associated with Zearalenone*

Originally both estrogenic symptoms and feed refusals were associated mainly with zearalenone. It now seems that some other toxin may be responsible (or partially responsible) for the latter syndrome (Kotsonis *et al.*, 1975b; Kurtz and Mirocha, 1978), namely T-2 toxin or deoxynivalenol. These mycotoxins have often been found together with zearalenone in cases of feed refusal (Jemmali *et al.*, 1978; Mirocha *et al.*, 1976b).

In India Ghosal *et al.* (1978a) isolated zearalenone (16 µg/kg) from sweet corn and also DAS and T-2 toxin which occurred together with zearalenone.

In general most of the toxicoses associated with zearalenone have involved hyperestrogenism (Mirocha and Christensen, 1974b; Mirocha *et al.*, 1971, 1974), abortion (Collet *et al.*, 1977; Jemmali *et al.*, 1978; Mirocha and Christensen, 1974b; Mirocha *et al.*, 1976b), and vulvovaginitis (Ozegovic and Vukovic, 1972) in pigs. Vulvovaginitis and loss of appetite and milk production in cows were also traced to zearalenone in the fodder (Vanyi *et al.*, 1974b). Some cases of infertility in cows (Mirocha *et al.*, 1968c) were thought to have been due to zearalenone, and estrogenic symptoms caused by the toxin have been observed in sheep (Vanyi *et al.*, 1973a,b), pigs, and even in deer in the agricultural regions (Lasztity and Woller, 1975a, 1977). Experimental work with zearalenone has usually supported these observations on natural estrogenic toxicoses (Miller *et al.*, 1973; Mirocha *et al.*, 1967; Stob *et al.*, 1962; Stojanovic *et al.*, 1978; Vanyi and Romvaryne, 1974; Vanyi *et al.*, 1973a, 1974a,b; and others). However, some reports give conflicting results. Thus according to Bristol and Djurickovic (1971) the morbidity of swine feeding corn contaminated by *F. graminearum* was usually light. Ruhr *et al.* (1978) noted that boars fed 60 ppm zearalenone for 8 weeks showed no impairment of reproductive potential. Shreeve *et al.* (1978) noted that moldy wheat containing 2.2 µg/g of zearalenone caused no clinical abnormalities or significant effect on the reproductive performance of seven pregnant sows. No zearalenone residues were detected in the sows' livers or milk 12 hr after farrowing. According to Shreeve *et al.* (1979) two adult cows also remained clinically normal after being fed a diet containing 385–1925 µg/kg zearalenone. During a period of 7 weeks there was no obvious effect on milk yields, and no zearalenone was found in the kidney, liver, milk, muscle, serum, or urine.

Splaylegged and weakened offspring, mummified fetuses, reduced litter size, stillbirth, agalacia, and reduced immunological response have also been reported as a result of feeding zearalenone to pregnant sows (Miller *et al.*, 1973; Sharma *et al.*, 1974; Wilson *et al.*, 1967), although not all these symptoms have been proved to be due to zearalenone (Kurtz and Mirocha, 1978). However, Ruzsas *et al.* (1978), working with rats, showed that injecting 30 μg zearalenone into 2-day-old rats reduced the pregnancy rate from 99 to 33%. A 50% reduction in fertility was noted in the offspring of rats fed corn infected with *F. graminearum* (a known zearalenone producer) during pregnancy.

Poultry are little affected by zearalenone. It reduced both number and fertility of eggs layed by 1-year-old rhineland geese (Palyusik and Koplik-Kovacs, 1975). Feed containing 900 ppm zearalenone inhibited spermatogenesis in adult ganders after one week of feeding (Vanyi *et al.*, 1974a). Adams and Tuite (1976) found that corn infected with *G. zeae* and containing 0.2 and 0.5 ppm zearalenone reduced feed consumption and egg production in Single Comb White Leghorn (SCWL) hens. However, there was no effect on egg weight or mortality. In contrast Speers *et al.* (1971) found that zearalenone had no significant effect on egg production although low levels of toxin increased the weight of the bursa of Fabricius. Increased comb and testes weights were observed in 4-day-old cockerels (Sherwood and Peberdy, 1972b).

Zearalenone has been shown to have estrogenic effects on several laboratory animals, including rats, mice, and guinea pigs (Mirocha *et al.*, 1968a).

Katzenellenbogen *et al.* (1979) noted that estrogenic metabolites such as zearalenone and zearalenol [2,4-dihydroxy-6-(6',10'-dihydroxy-1-undecenyl)-benzoic acid-μ-lactone], isolated from *F. graminearum* and other species, acted like other estrogenic agents. They became bound to the same cytoplasmic and nuclear receptors as do estrogenic hormones which induced tumors in organs (Schoental, 1981c).

Kiang *et al.* (1978), Kiessling and Petterson (1978), Obsen *et al.* (1981), and Thouvenot *et al.* (1980) studied the zearalenone and zearalanol properties on their biochemical mode of action associated with cytosol estrogenic and nuclear receptors in calf uteri, metabolism in rat liver, and the steroid-metabolizing enzymes of the human prostate gland, respectively.

According to Kiang *et al.* (1978) zearalenone and zearalanol can compete with 17β estradiol for binding with cytosol estrogen receptors. The binding of zearalenone to estrogen receptors is structurally specific. Six zearalenone derivatives were tested for their potency to compete with estradiol for binding at the receptor sites. Four of the six derivatives, mainly *trans*- and *cis* zearalenone, zearalenol, and zearalanol showed significant competitive binding compared to 17β estradiol. The other two derivatives (8'-hydroxyzearalenone and b'-aminozearalenone) could not compete with 17β estradiol at the receptor sites, and did not exhibit estrogenic activity in the rats. The *cis* zearalenone has the highest binding affinity to receptors among the tested derivatives. The authors concluded that zearalenone derivatives are capable of binding to the estrogen receptors.

[³H] and [¹⁴C]-labeled zearalenone, produced by *F. roseum*

"Graminearum" on Czapek liquid medium at 25°C on a rotary shaker for 4 days, was seen to bind preferentially to one of the two peaks containing uncharacterized proteins from the cytosol of young mycelium (from an isolate of *F. roseum*) (Inaba and Mirocha, 1979). Zearalenone has also been found to bind to the receptor protein found in the cytosol of the rat uterus (Kiang *et al.*, 1978).

Richardson *et al.* (1984a, pp. 1206–1209) reported on the conversion of $\alpha[^{14}C]$ zearalenol and $\beta$-$[^{14}C]$ zearalenol to zearalenone in cultures of *F. roseum* "Gibbosum" grown on rice in order to determine if a precursor-product relationship exists between these mycotoxins. The authors stated that $\beta$-$[^{14}C]$ zearalenol and -$\alpha[^{14}C]$ zearalenol added to cultures were converted to zearalenone within 7–14 days. But when $[^{14}C]$ zearalenone was added to cultures no label was detected in the zearalenol fraction, and zearalenone was not converted to zearalenol. This observation supports the conclusion of Steele *et al.* (1976) that catabolism of zearalenone did not include zearalenol formation.

The mechanism of zearalenone action in rat liver was studied by Kiessling and Petterson (1978) and Olsen *et al.* (1981). They noted that the conversion of zearalenone to zearalenol in the liver was catalyzed by a hydroxysteroid dehydrogenase, which is normally involved in the metabolism of steroids.

Thouvenot and Morfin (1980) noted that both zearalenone and zearalenol competitively inhibited the 3α and 3β-hydroxysteroid dehydrogenases of human prostate tissue, whereas estradiol-17β competitively inhibited only the 5α-reduction of testosterone. Kallela and Vasenius (1982) noted that the amount of zearalenone was decreased by ruminal digestion and they concluded that the estrogenic effect therefore was also most reduced.

Kiessling *et al.* (1984) also studied the effect of the bacteria and protozoa in rumen fluid from sheep and cattle in the presence or absence of millet feed on six mycotoxins (aflatoxin B, ochratoxin A, zearalenone, T-2 toxin, diacetoxyscirpenol and deoxynivalenol). The authors noted that rumen fluid had no effect on aflatoxin A and deoxynivalenol. The other four mycotoxins were all metabolized, and protozoa were more active than bacteria. Metabolism of zearalenone and DAS was moderately or slightly (respectively) inhibited by addition of millet feed in vitro. The rumen fluid reduced zearalenone to α-zearalenol and degraded to a lesser degree to β-zearalenol; DAS and T-2 toxin were deacetylated to monoacetoxyscirpenol and HT-2 toxin, respectively.

Richardson *et al.* (1984b, pp. 643–646) reported on three methods for detection of toxicogenic *Fusarium* species (*F. roseum* "Gibbosum," *F. tricinctum* NRRL 3299 (*F. poae*) and an unidentified *Fusarium* sp.) using three media (autoclaved rice, vermiculite moistured with nutrient broth in tubes, and liquid medium of Bacon *et al.* 1977). All three *Fusarium* species produced the specific mycotoxins in the three media (zearalenone, zearalenol by _F. roseum_ "Gibbosum"; T-2 toxin by _F. tricinctum_; and zearalenone, zearalenol, and deoxynivalenol by _Fusarium_ sp.). The authors noted that the vermiculite cultures appeared to be the best medium for the production and detection of the tested mycotoxins.

Ueno and Ueno (1978) mention the danger of secondary toxicoses in humans due to zearalenone in milk or meat, although no zearalenone residues were found in the muscle, liver, kidney, serum, or milk of cows fed zearalenone (385–1925 µg/kg) for 7 weeks prior to slaughter (Shreeve _et al._, 1979). However, zearalenone has been found in fermented maize products in areas of South Africa where cervical cancer is prevalent (Martin and Keen, 1978), and Marasas _et al._ (1976c) found that the level of zearalenone (and deoxynivalenol) was higher in maize from Southwest Transkei (a high esophageal cancer area) than in Northeast Transkei (a low cancer area), although the number of samples was too small to make any statistical connection between the mycotoxin and cancer. However, this work supports the hypothesis put forward by Purchase _et al._ (1975) that the Transkei diet contained an unidentified factor that might cause cancer.

Christensen _et al._ (1965) isolated an estrogenic metabolite named F-2 produced from corn contaminated with _Fusarium_ isolates and from commercially prepared pelleted feeds from farms in Minnesota where estrogenic symptoms occurred in swine herds.

In studies associated with biosynthesis of F-2 Mirocha _et al._ (1968a,b, 1969) detected another naturally occurring unidentified compound related to F-2 that was named F-3. This compound was found in small amounts produced by _F. graminearum_. F-3 was also isolated from feed samples suspected of causing abortion or infertility in dairy cattle in Minnesota. Another fungus was isolated from these feed samples, _F. moniliforme_, which was also suspected of causing infertility in dairy cattle (Mirocha _et al._, 1968b). F-3 proved to be a highly labile compound for oxidation and was isolated using the same solvent as for F-2. Its absorption spectrum is identical with F-2, but F-3 lacks an absorption at 314 nm.

F-3, like F-2, reacts with silating agents [N, O-bis (trimethylsil)-acetamide] to form the trimethylsilyl ether which has a retention time about 2 min less than that of F-2 at 260°C.

Mirocha _et al._ (1971) reported on naturally occurring derivatives of zearalenone, F-5-3 and F-5-4, which were determined by Jackson _et al._ (1974) as isomers of 8'-hydroxyzearalenone. A new component named F-4 (with Rf value of 0.41) was also isolated from _F. graminearum_ and _F. culmorum_ by Lasztity and Woller (1975a,b, 1977) and Lasztity _et al._ (1977) and was shown to have estrogenic effects on rats.

In addition to zearalenone itself there are several naturally occurring derivatives and isomers noted by Hagler _et al._ (1979, 1980); Mirocha

(1977); Mirocha and Christensen (1974b, 1982); Mirocha and Pathre (1979); Mirocha *et al.* (1974, 1977a,b); Taylor and Watson (1976).

The naturally occurring derivatives of zearalenone were isolated by the following researchers: 5-formylzearalenone and 7'-dehydro-zearalenone (Bolliger and Tamm, 1972); α- and β-8'-hydroxyzearalenone (Bolliger and Tamm, 1972; Jackson *et al.*, 1974; Stipanovic and Schroeder, 1975; Taylor and Watson 1976; 6',8'-dihydroxyzearalenone (Steele *et al.*, 1976); α- and β-zearalenols (Hidy *et al.*, 1977; Hagler *et al.*, 1979; Peters, 1972; Shipchandler, 1975); 4',5'-dihydroxyzearalenone (Robison *et al.*, 1979); α- and β-3'-hydroxyzearalenone (Pathre *et al.*, 1980).

Mirocha *et al.* (1978) found that in rats the *cis* isomer of zearalenone was significantly more active uterotropically than *trans* zearalenone. This contrasts with the findings of Peters (1972) in mice where the *cis* isomer of zearalenone was slightly less active than the *trans* isomer but with the zearalenols the reverse was true and the *cis* isomers were substantially more active than the *trans* isomers. Hagler and Mirocha (1980) isolated a diastereomer of *trans* zearalenol from a culture of *F. roseum* "Gibbosum." This was identical with the α isomer observed by synthesis which was three times more estrogenic than zearalenone. The isomer formed 7% of the total zearalenone found; two other diastereomers of 8'-hydroxyzearalenone each formed 3% of the total zearalenone.

It is interesting to note that in addition to its effects on animals' reproductive systems zearalenone also has a regulatory effect on the fungus that produces it.

Nelson (1971) noted that zearalenone (F-2) affected reproduction in animals and also the formation of the sexual stage of *G. zeae*. A close association was seen between zearalenone production and perithecial formation (Eugenio, 1968a,b).

Addition of dichlorvos to cultures of *F. roseum* "Graminearum" (*F. graminearum*) inhibited perithecia production as well as zearalenone production. The authors stated that the production of perithecia may depend upon the production of zearalenone in those strains which normally produce the sexual stage (Lieberman *et al.*, 1971).

Wolf *et al.* (1972) and Wolf and Mirocha (1973) showed that dichlorvos inhibits both zearalenone biosynthesis by *F. roseum* "Graminearum" (*F. graminearum*) and also perithecia development. Zearalenone prevents or reverses the inhibition by dichlorvos. Zearalenone was seen by Wolf and Mirocha (1971, 1973) to enhance the production of perithecia by *F. roseum* "Graminearum" (*F. graminearum*). Thus low amounts of zearalenone (1–10 ng/g culture) enhance sexual reproduction in *G. zeae* (the perfect stage of *F. graminearum*) whereas higher levels (10–100 ng/g) inhibit sexual reproduction (Mirocha and Swanson, 1983; Wolf and Mirocha, 1973, 1977).

Zearalenone, tested against *Bacillus thuringiensis*, has an antibacterial activity causing cellular changes (decreased growth rate and formation of atypical cells). After 3 days incubation at 21°C, 5 μg/ml zearalenone blocked the sporulation of _B. thuringiensis_ and the biosynthesis of phosphamidase and α-glucosidase (Boutibonnes 1979a,b; Boutibonnes and Loquet, 1979).

However, although presumptive screening of zearalenone for mutagenic and carcinogenic characteristics was positive (DNA attacking ability test), no mutagenic activity was observed during the routine test combining _Salmonella typhimurum_ strains and liver homogenate (Boutibonnes and Loquet, 1979).

Because of the importance of toxic effects in commercial agricultural feeds several attempts have been made to prevent zearalenone production (by suitable storage conditions) or to detoxify contaminated corn. Corn grits containing 60 ppm zearalenone were subjected to 50 min treatment with Cl gas. After suitable aeration it was fed to rats and did not produce any deleterious effects (Sarudi _et al._, 1979). Several chemical and physical treatments destroyed zearalenone (Bennett _et al._, 1979, 1980, 1981; and Moerk _et al._, 1980).

## Moniliformin

The toxin moniliformin has been isolated from natural substrates infected with cultures of _F. moniliforme_ (Cole _et al._, 1973, Steyn _et al._, 1978a), _G. fujikuroi_ (perfect state of _F. moniliforme_) (Springer _et al._, 1974) and _F. fusarioides_ (Rabie _et al._, 1978).

Rabie _et al._ (1978) extracted moniliformin from a culture of _F. fusarioides_ (_F. sporotrichioides_ var. _chlamydosporum_) isolated from millet obtained from the homes of South-West African patients suffering from the hemorrhagic disease Onyalai. Maximum toxin production was reached at 34°C (or possibly 36–37°C). Isolates obtained from peanuts, sorghum, peaches, soil, and dried fish were all toxic to ducklings. The moniliformin produced was shown to be highly toxic to ducklings and rats by i.v. injection. _F. fusarioides_ was also grown on maize meal and incubated at 31°C for 7 days. Moniliformin has also been found in cultures of _F. avenaceum_ and _F. oxysporum_ (Marasas _et al._, 1979b). Springer _et al._ (1974) reported on the structure and synthesis of moniliformin and it was shown to be the sodium or potassium salt of 1-hydroxycyclobut-1-ene-3,4-dione (Fig. 3.7).

Moniliformin synthesis and structure was studied by Bellus _et al._ (1978) and its isolation and purification by Burmeister _et al._ (1979) and Steyn _et al._ (1978). Spectroanalysis of _F. moniliforme_ metabolites was carried out by Lansden _et al._ (1974).

Fusaria known to produce moniliformin have been isolated from maize (Cole _et al._, 1973; Marasas _et al._, 1979b) and from peanuts, sorghum, millet, dried fish, and soil (Rabie _et al._, 1978).

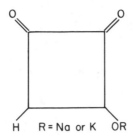

H      R = Na or K      OR

**Figure 3.7**   The chemical structure of moniliformin.

Moniliformin has been produced in the laboratory on maize at 25°C (Kriek *et al.*, 1977) and on corn grits at 28°C (Burmeister *et al.*, 1979). Rabie *et al.* (1978) found that the highest yield of moniliformin from maize meal culture incubated with *F. fusarioides* was obtained after 7 days at 31°C. Longer incubation and/or lower temperatures reduced the yield. The authors described the methods for analysis and isolation of moniliformin from maize infected by *F. fusarioides*. They obtained 200– 840 mg/kg of moniliformin by the TLC method.

Rabie *et al.* (1982) isolated the following species of *Fusarium* which produced moniliformin from sorghum and millet in southern Africa: *F. acuminatum, F. concolor, F. equiseti,* and *F. semitectum.*

*F. acuminatum* and *F. concolor* both produced large amounts of moniliformin (3.4 and 9.5 g/kg, respectively). Also *F. oxysporum, F. avenaceum, F. fusarioides, F. moniliforme,* and *F. moniliforme* var. *sub- glutinans* produced moniliformin.

A number of strains of *F. moniliforme* produced more than 10 g/kg of moniliformin, with even one strain, which was isolated from corn cul- tivated for 5 weeks at 25°C, producing 33.7 g/kg.

Some isolates that did not produce moniliformin, such as *F. acuminatum, F. equiseti, F. fusarioides,* and *F. moniliforme* were highly toxic to ducklings.

According to Marasas *et al.* (1979c) a number of *F. moniliforme* iso- lates also did not produce moniliformin.

Steyn *et al.* (1978a) also described a successful procedure for isolating pure moniliformin from *F. moniliforme* and *F. moniliforme* var. *sub- glutinans* grown on yellow maize for 21 days at 25 ± 1°C.

Although *F. moniliforme* has been found in feed in several outbreaks of animal toxicoses, no naturally occurring toxicosis has been attributed to moniliformin. However, moniliformin has been shown to be ex- tremely toxic to rats, ducklings, mice, and chicks, and it is quite possible that it has been responsible for some fusariotoxicoses (Kriek *et al.*, 1977; Burmeister *et al.*, 1979). It should be noted, however, that the toxin did not produce a skin reaction in mice (Burmeister *et al.*, 1980a), although Joffe (1960a) and Joffe *et al.* (1973) showed that extracts of *F.*

*moniliforme* and *F. moniliforme* var. *subglutinans* cultures gave toxic reactions on rabbit skin.

Moniliformin together with T-2 toxin and zearalenone were ruled out as the mutagenic factor in *F. moniliforme* (Bjeldanes and Thomson, 1979).

According to Cole *et al.* (1973) the oral lethal dose for 1-day-old cockerels was 4 mg/kg, and Kriek *et al.* (1977) found that although the $LD_{50}$ for the pure toxin in male rats was higher than that of the female rats (50 mg/kg as opposed to 41.57 mg/kg), the female rats were more resistant than the males when the toxin was fed in low doses. These workers found that moniliformin caused progressive muscular weakness, respiratory problems, cyanosis, coma, and eventual death in rats and ducklings. Pathological investigation of the rats showed acute congestive heart failure, myocardial degeneration and necrosis and scattered single cell necrosis in the liver, kidney, pancreas, adrenal glands, gastric mucosal glands, and crypt of the small intestine.

Kriek *et al.* (1981a) isolated 21 *F. moniliforme* strains from South African and Transkeian maize and examined their toxicity to ducklings and rats. None of the strains produced moniliformin. Only one strain (MRC 602) was studied in detail and compared to the other isolates. This strain (MRC 602) caused hepatotoxicity, lesions in heart, nephrotoxicity, and other disorders in organs and tissues in rats, which led to the eventual death of the animals.

All ducklings fed *F. moniliforme* isolates died without producing symptoms of hepatic and cardiac lesions. Therefore, according to the authors, ducklings are unsuitable as a bioassay model for screening of hepato- and cardiotoxic strains of *F. moniliforme.*

Thiel (1978) showed that low concentrations of moniliformin (<5 µM), isolated from *F. moniliforme* var. *subglutinans* grown on maize, selectively inhibited mitochondrial pyruvate and α-ketoglutarate oxidations. He suggested that these inhibitory effects could play a major role in the clinical symptoms of moniliformin toxicity.

Recently, in regions of southern Africa, where maize was contaminated with *F. verticillioides* (*F. moniliforme*), Thiel (1981) isolated moniliformin by the high pressure liquid chromatography (HPLC) procedure. The moniliformin was monitored by ultraviolet absorption at 229 nm and by comparing the chromatogram with that of authentic pure moniliformin.

*F. moniliforme* NRRL 6322 was grown on white corn grits for 16–20 days at 28°C. Pure moniliformin was injected into mice ($LD_{50}$ 20.9 mg/kg for female and 29.1 mg/kg for male), intubated into day-old chicks ($LD_{50}$ 5.4 mg/kg), and injected into chick embryos ($LD_{50}$ 2.8 mg) with no teratogenic effects observed. There was no toxic reaction when purified moniliformin was applied to mouse skin (Burmeister *et al.*, 1979).

According to Burmeister *et al.* (1980b) moniliformin and butenolide have no effect on young white mice after 21 days consumption of water containing both toxins.

Now that the toxicity of moniliformin has been established and assay methods developed (Kriek *et al.*, 1977; Rabie *et al.*, 1978, 1982; Steyn *et al.*, 1978a,b; Thiel, 1978, 1981) it is necessary to determine if this mycotoxin can become a hazard in human and animal nutrition.

## Butenolide

Fescue foot is a seasonal disease of cattle, occurring in the cold months of the year. This syndrome was first described in New Zealand by Cunningham (1949), then in South Australia by Pulsford (1950), and later in Italy by Rossi (1959). In the United States the area of Missouri was connected with the highest incidence of fescue-foot. The clinical signs showed that the disease was not caused by ergot (Jacobson and Miller, 1961; Jacobson *et al.*, 1963). The latter researchers suggested initially that alkaloids isolated from tall fescue grass were suspected of causing the disease; but after administration to a cow for 11 days the workers did not obtain the significant clinical symptoms of fescue foot. Yates and Tookey (1965) isolated an alkaloid festucine from tall fescue which was mildly toxic to mice. Only later did Keyl *et al.* (1967) isolate, from tall fescue grass, a culture (initially identified by these authors as *F. nivale*) of *F. tricinctum* NRRL 3249 (*F. sporotrichioides*, according to the classification of Joffe and Palti, 1975) that had been associated with fescue foot syndrome in cattle.

Yates *et al.* (1969) isolated 13 of 29 *Fusarium* strains from tall fescue (*Festuca arundinacea*). The strains produced a toxin which killed test mice. *Fusarium* strains isolated from fescue hay and orchard grass grown near the tall fescue fields were also most toxic to mice.

A toxic metabolite named butenolide was isolated from *F. tricinctum* NRRL 3249 and characterized as 4-acetamide-4-hydroxy-2-butenoic acid γ-lactone (Ellis and Yates, 1971; Grove *et al.*, 1970; Tookey *et al.*, 1972; Yates *et al.*, 1967, 1968, 1969, 1970).

Butenolide is a water-soluble mycotoxin and may be the cause of fescue foot in cattle (Grove *et al.*, 1970; Tookey *et al.*, 1972). Ueno *et al.* (1972b) described the chemical structure of butenolide (Fig. 3.8).

White (1967) isolated butenolide from two strains of *F. equiseti*. *F. equiseti* isolated from moldy paddy rice straw was associated with a gangrenous syndrome in buffaloes (Kwatra and Singh, 1973). The authors noted that lameness and ear and tail necrosis are similar to those observed in tall fescue toxicosis, but *F. equiseti* was not isolated from a tall fescue.

According to Qin *et al.* (1981) another similar disease of cattle in China, known as "sore foot disease," was also caused by *F. equiseti* and *F. semitectum*.

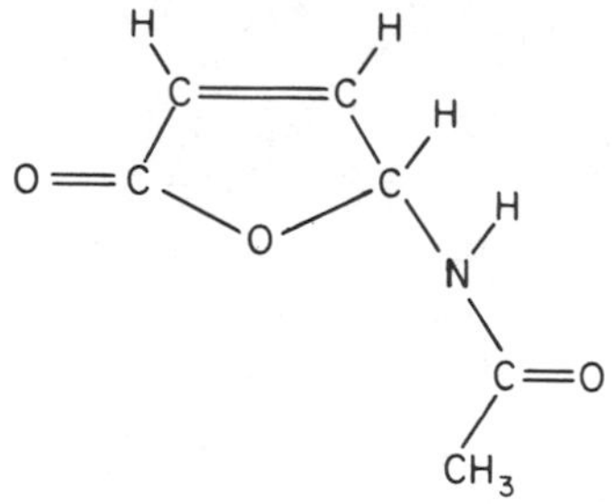

**Figure 3.8**   The chemical structure of butenolide.

The following species have been reported as producers of butenolide: *F. tricinctum, F. equiseti, F. sporotrichioides, F. semitectum, F. roseum, F. graminearum,* and *F. lateritium* (Burmeister *et al.,* 1971; Morooka *et al.,* 1972; Ueno *et al.,* 1972b; Yoshizawa *et al.,* 1979, 1983a).

Butenolide has a $LD_{50}$ in mice of 43.6 mg/kg body weight by i.p. injection and an oral toxicity of 275 mg/kg and causes a very weak skin reaction on rabbits. According to Ueno and Kubota (1976) butenolide has negative Rec-assay with *B. subtilis.*

Yates *et al.* (1967, 1968, 1969, 1970) isolated three metabolites from the same *F. tricinctum* NRRL 3249 grown on Sabouraud's agar at 15°C for 4 weeks or at temperatures alternating between 7 and 20°C. They were butenolide, T-2 toxin, and an unknown metabolite. In fact these researchers noted that no toxin was obtained when the culture was kept at room temperature and that several strains lost their ability to produce toxin when transfers were cultured at room temperature. However, butenolide was isolated from cultures of *F. sporotrichioides* NRRL 3510 grown on Czapek-Dox Peptone medium for 12 days at 26°C (Ueno *et al.,* 1972b). In a later work Yates (1971) described in detail the chemical properties, synthesis, and physiological effects of *F. tricinctum* NRRL 3249 toxins.

A possible metabolic intermediate (4-acetamido-2 butenoic acid) in the biosynthesis of butenolide was isolated from a culture of *F. graminearum* NRRL 5883 grown on rice for 13 days at 28°C. This metabolite was not toxic (Vesonder *et al.,* 1977b).

Although several workers tried to implicate butenolide as the cause of fescue-foot in cattle, administration of the pure toxin failed to produce all the symptoms of the syndrome (Grove *et al.,* 1970; Kosuri *et al.,* 1970; Tookey *et al.,* 1972). Only two of the symptoms, gangrene of the tail and an arched back, were caused by the toxin. Grove *et al.* (1970) pointed out that butenolide had not been isolated from tall fescue and, since this is a common winter pasture grass, any connection between it and the syndrome may be accidental. On the other hand there may be some factor present in the grass that predisposes the cattle to the toxicity of butenolide. Williams *et al.* (1975) stated that the ethanolic extract of toxic tall fescue hay was the causative agent of the fescue foot syndrome, and Futrell *et al.* (1974) found some fungi in the rumen of fistulated steers grazing tall fescue, mainly *Aspergillus terreus* and *F. tricinctum,* which were associated with fescue toxicosis.

Apart from the work on cattle and on the toxicity of butenolide to laboratory animals, little interest has been shown in this toxin despite its potentially lethal toxicity to cattle (Tookey *et al.*, 1972). According to Garner and Cornell (1978) the causal agent of fescue foot has not been proved. It is not clear if the symptoms are caused by a microbial toxin or a plant metabolite made toxic by biological activity. *F. tricinctum* has been associated with fescue foot, but neither butenolide nor T-2 toxin have been isolated from the suspect tall fescue grass. Garner and Cornell (1978) described in detail the clinical symptoms and histopathological findings of the fescue-foot disease in cattle.

## DETECTION OF FUSARIOTOXINS

Detection of fusariotoxins in animal and human food is of extreme importance, in particular because the toxins are often stable and remain in food long after all traces of *Fusarium* fungal infection have disappeared. Thus, for example, T-2 toxin was detected in feedstuffs held to be responsible for the death of beef cattle (Stahr *et al.*, 1978) and zearalenone has been detected in corn and corn products (Eppley *et al.*, 1974; Shotwell, 1977a,b).

Both biological and chemical methods have been devised for the detection and quantification of the fusariotoxins.

Biological and physicochemical methods for detection and determination of *Fusarium* species toxins have been carried out by Hagen and Tientjen (1975); Mirocha *et al.* (1977); Pareles *et al.* (1976); Ueno (1980b); and a sensitive and specific radioimmunoassay for T-2 toxin in corn and wheat was recently described by Lee and Chu (1981a,b).

The bioassay methods generally have the advantage that they do not require expensive equipment or highly purified toxin and are in fact often carried out with crude extracts of *Fusarium* cultures or contaminated feed. However, they often lack the specificity and quantitative accuracy of chemical methods. Nevertheless they are extremely useful as rapid screening methods. In addition biological methods often give an indication of the toxic ability of a fungus even when no known toxin can be detected by chemical assays. Thus Köhler *et al.* (1978), working on a strain of *F. moniliforme* known to cause rachitis in commercial broiler chickens, resulting in demineralization of the spongiosa and production of fibrous bone marrow accompanied by some degree of granulomatous proliferation of osteoclasts, attempted to detect the metabolites by physicochemical methods but failed, whereas biological assays (rabbit skin test and chicken embryo) demonstrated the toxic nature of these products. The chemical methods, while requiring careful clean-up and purification of samples, give precise identification and quantitative data.

## Bioassay Methods

The bioassay methods developed for the detection of fusariotoxins are based on their specific toxicity to various biological systems (_see_ Tables 3.10 and 3.11). Some of these methods (e.g. the toxic skin reaction in rabbits, rats, and guinea pigs) can be used to screen a wide range of toxins while others are specific (e.g. the rat uterus test for zearalenone). Many of the tests have been widely used (toxic skin test, rabbit reticulocyte assay, _T. pyriformis_ assay, HeLa cells, etc.) whereas others have only been used once or twice (pea seed germination, wheat coleoptile, etc.)

Bioassays can be divided into the following basic types:

1. Phytotoxicity
2. Skin toxicity
3. Lethal toxicity (multicellular systems)
4. Cytotoxicity
5. Inhibition of protein synthesis
6. Enzyme and antibody reactions
7. Specific toxic responses

### _Phytotoxic Assays_

The phytotoxic properties of the fusariotoxin diacetoxyscirpenol were demonstrated by Brian _et al._ (1961). They found that 1 µg/ml of the toxin sprayed on pea plants caused a just detectable inhibition of stem elongation and a noticeable scorching of the leaves; 10 µg/ml was often lethal to the plants. However, they detected no translocation of the toxin within the plant (although, according to Sutton _et al._, 1976, zearalenone was shown to be translocated from stem to ears of maize). Mustard, wheat (var. Victor), beetroot, and carrots were unaffected by 10 µg/ml. T-2 toxin has also been shown to inhibit pea seedling growth (Bamburg, 1972; Marasas _et al._, 1971), but the method has not been used extensively to detect this toxin. Burmeister and Hesseltine (1970) found that 0.3 µg/ml T-2 toxin caused 50% inhibition of pea seed germination, although butenolide had no effect even at concentrations as high as 200 µg/ml. In contrast to the findings of Brian _et al._ (1961) with diacetoxyscirpenol, another trichothecene toxin, neosolaniol monoacetate, inhibited wheat coleoptile elongation (Lansden _et al._, 1978). T-2 toxin has been shown to inhibit the elongation of excised soybean hypocotyls, the inhibition being more noticeable in auxin promoted elongation (Stahl *et al.*, 1973). It was suggested that T-2 toxin may inhibit the protein synthesis necessary for elongation (cf. protein inhibition test). T-2 toxin also retarded the germination of barley (Nummi *et al.*, 1975) and the logarithmic growth rates of tobacco callus tissue (Helgeson *et al.*, 1973), but other researchers have not used these screening methods. The growth of *Chlorella* is also inhibited by fusariotoxins but as yet this inhibition has only been used to screen the toxicity of extracts of various *Fusarium* species rather than specific toxins. This method was used to test extracts of *F. tricinctum* (Nummi, 1977; Main and Hamilton, 1972) and to test a wide range of mycotoxins, including zearalenone and diacetoxyscirpenol (Sullivan and Ikawa, 1972). The latter researchers used several strains of *Chlorella* and found wide variations in the response of the different strains to various mycotoxins. Main and Hamilton (1972) found that, when cultured on synthetic media, *F. tricinctum* was toxic to *Chlorella* but not to mice. However, when cultured on Main's medium, which provided a source of plant protein, *F. tricinctum* lost its phytotoxic ability but became toxic to mice. They suggested that *F. tricinctum* needs a plant substrate to produce the animal toxin and that plant and animal toxins have different requirements for their metabolic synthesis. These workers also found that *F. oxysporum, F. lycopersici,* and *F. oxysporum* f. *niveum* were all toxic to *Chlorella* but not to mice. Ueno (1971) note that fusarenon-X was much more toxic to the animal system than to plants. Thus the germination of *B. oleracea* was inhibited only at relatively high levels, and the yeast, fungi, and bacteria systems tested were resistant. Ueno (1971) suggested that fusarenon-X has a high affinity only for animal cells and that it is a cytotoxic zootoxin. In contrast to these results, Joffe (1973b, 1983b) found a direct correlation between the percentage of seedling mortality in some plants and the skin necrotizing effect of extracts of various isolates of *F. poae, F. sporotrichioides, F. sporotrichioides* var. *tricinctum*, and *F. sporotrichioides* var. *chlamydosporum* among others (*see* Chapter 5).

Yeast cells have also been used to detect fusariotoxins. Mishustin *et al.* (1946) studied the effect of ethanol extracts of cereal grains infected with toxic *F. poae* and *F. sporotrichioides* on yeast, *Saccharomyces cerevisiae*. The authors established that the extracts depressed the fermentation action of the yeasts, as well as their reproduction. They also suggested that a fermentation test might be used to detect the toxicity of grain or toxic *Fusarium* species. Schapper and Khachatourians (1983, 1984) reported on the growth inhibitory action of T-2 toxin on *S. cerevisiae* and *S. carlsbergensis*. The authors noted that the inhibition was dependent on the carbon source, concentration of T-2 toxin and length of the exposure to T-2 toxin. They stated that T-2 toxin at 10 μg/ml caused nearly 50% reduction in growth. The authors also studied the interaction

**TABLE 3.10  Minimum Lethal Doses of Fusariotoxins in Major Bioassay Tests**

| Bioassay Test | Toxin/Fusarium Species | Value | References |
|---|---|---|---|
| Brine shrimp | Acetone extract of F. tricinctum | 1–20 µg/ml extract | Brown, 1969 |
| Brine shrimp | T-2 toxin<br>Diacetoxyscirpenol<br>Zearalenone | 2 µg/disc<br>0.2 µg/disc<br>10 µg/disc | Harwig and Scott, 1971 |
| Brine shrimp | T-2 toxin<br>Diacetoxyscirpenol | 200 ng/0.5 ml | Eppley, 1974 |
| Brine shrimp | T-2 toxin<br>Diacetoxyscirpenol | 5 µg/ml<br>10 µg/ml | Tsubouchi et al., 1976 |
| Brine shrimp | Methanol extracts of:<br>F. oxysporum<br>F. solani<br>F. lateritium | 20–59% mortality<br>0–19% mortality<br>60–100% mortality | Davis et al., 1975 |
| Brine shrimp | Diacetoxyscirpenol | 0.5–1.0 µg/ml | Reiss, 1972 |
| Chicken embryo | Methanol extracts of:<br>F. oxysporum<br>F. solani<br>F. lateritium | embryo death/eggs:<br>10 out of 10<br>10 out of 10<br>5 out of 10 | Davis et al., 1975 |
| Chicken embryo | Toxic extract of:<br>F. sporotrichiella var. poae | Died within 24 hr. from doses of 0.1, 0.25, 0.5 ml | Ivanov, 1968 |
| Chicken embryo | Fusarenon-X | 6 µg/egg (LD$_{50}$) | Ueno, 1971 |
| Chicken embryo | Moniliformin | 2.8 µg/egg (LD$_{50}$) | Burmeister et al., 1979 |

| Test system | Toxin | Dose/concentration | Reference |
|---|---|---|---|
| Yeast (*Rhodotorula rubra*) | T-2 toxin | 4 μg/disc | Burmeister, 1971; Burmeister and Hesseltine, 1970; Burmeister *et al.*, 1972 |
| Pea seed germination | T-2 toxin | 0.5 μg/ml (48% inhibition) | Burmeister and Hesseltine, 1970 |
| Larvicidal test *Aedes aegypti* | T-2 toxin (Scirpentriol and its acetate; Nivalenol and its diacetate) | 25 μg/ml | Grove and Hosken, 1975 |
| *Lucilia sericata* | Diacetoxyscirpenol | 7.5 μg/ml | Cole and Rolinson, 1972 |
| *Culex pipieus pallens* | Fusarenon-X | 0.195 μg/ml | Yoshizawa and Morooka, 1974 |
| Skin test: | | | |
| Rabbit (topical) | T-2 toxin | 0.1 μg | Yagen and Joffe, 1976 |
| Rabbit (topical) | T-2 toxin and neosolaniol monoacetate | 0.02–0.04 μg | Lansden *et al.*, 1978 |
| Rabbit (topical) | Butenolide | 30 mg in 5 ml | Yates *et al.*, 1968 |
| Rabbit (topical) | T-2 toxin | 0.01 μg/test (50 ppb) | Chung *et al.*, 1974 |
| Rabbit (topical) | Diacetoxyscirpenol / Fusarenon-X / Neosolaniol | 1.0 μg / 10 μg / 10 μg | Ueno *et al.*, 1971a |
| Rabbit (i.d.) | Zearalenone | 0.02–1.0 μg/side | Chung *et al.*, 1974 |
| Rat | T-2 toxin | 0.05–0.1 μg | Wei *et al.*, 1972 |
| Rat | T-2 toxin | 0.02–0.04 μg | Lansden *et al.*, 1978 |
| Rat | T-2 toxin, HT-2 toxin | 0.1–0.17 mg | Bamburg and Strong, 1969 |
| Rat | Diacetoxyscirpenol | 0.5 mg | Gilgan *et al.*, 1966 |
| Rat | Zearalenone | 0.3–0.6 mg | Mirocha *et al.*, 1977a |

**TABLE 3.10  Continued**

| Bioassay Test | Toxin/Fusarium Species | Value | References |
|---|---|---|---|
| Rat | Neosolaniol monoacetate | 0.02 µg | Lansden et al., 1978 |
| Mouse | Moniliformin | 0.5 mg/ml | Burmeister et al., 1980a |
| Mouse | Diacetoxyscirpenol | 1.0 µg | Ueno et al., 1970a |
| | Fusarenon-X | 0.2–1.0 µg | |
| | Nivalenol | 100 µg | |
| Guinea pig | T-2 toxin | 0.04 µg | Lansden et al., 1978 |
| | Neosolaniol monoacetate | 0.04 µg | |
| Guinea pig | Diacetoxyscirpenol | 0.2 µg | Ueno et al., 1970a |
| | Fusarenon-X | 0.2 µg | |
| | Nivalenol | 100 µg | |
| Guinea pig | T-2 toxin | 0.02–0.1 µg | Chung et al., 1974 |
| | Zearalenone | 0.05–0.1 µg | |
| Wheat coleoptile growth | Neosolaniol monoacetate | $10^{-6}$–$10^{-5}$M | Lansden et al., 1978 |
| Pea seedling growth | Diacetoxyscirpenol | $2.73 \times 10^{-6}$M | Brian et al., 1961 |
| | | $2.73 \times 10^{-5}$M | |
| | T-2 toxin | $10^{-5}$M | Bamburg, 1972 |
| | T-2 toxin | ~1.5 ppm | Marasas et al., 1971 |
| Cauliflower seed germination (Brassia oleracea) | Fusarenon-X | 10–100 µg/ml | Ueno, 1971 |
| Chlorella | Extracts of various Fusarium species | — | Main and Hamilton, 1972 |
| Chlorella | Extracts of F. tricinctum | — | Nummi, 1977 |

| | | | |
|---|---|---|---|
| *Chlorella* | Diacetoxyscirpenol | Inhibition zone from 4 mm to 40 mm for 1 mg/ml | Sullivan and Ikawa, 1972 |
| *Chlorella* | Zearalenone | 0–15 mm for 1 mg/ml | Sullivan and Ikawa, 1972 |
| Rat uterus | Zearalenone: oral | 1.5–1.0 mg | Mirocha et al., 1977a |
| | topical | 0.3–0.6 mg | |
| Rat uterus | Zearalenone | 0.02–0.04 mg (lowest concentration) | Mirocha et al., 1966 |
| Rat uterus | Zearalenone and deoxynivalenol | 81.8–99.4 mg (Control 17.0 mg) | Marasas et al., 1977 |
| Enzyme assay | For trichothecenes | 1 µg/g | Foster et al., 1975 |
| Antibody assay | T-2 toxin | 3.5 ng | Chu et al., 1979 |
| | HT-2 toxin | 20.0 ng | |
| | T-2 triol | 168.0 ng | |
| | Neosolaniol | 1588.0 ng | |
| | T-2 tetraol | 5000.0 ng | |
| Antiprotozoal | Diacetoxyscirpenol | 0.05 µg/ml | Ueno and Yamakawa, 1970 |
| *Tetrahymena pyriformis* | Fusarenon-X | 5 µg/ml | |
| | Nivalenol | 25 µg/ml | |
| *T. pyriformis* | Deoxynivalenol | 4.6 µg/ml | Yoshizawa and Morooka, 1974 |
| | Deoxynivalenol monoacetate | 29.0 µg/ml | |
| | Nivalenol | 8.8 µg/ml | |
| | Fusarenon-X | 5.0 µg/ml | |

*Source:* Drawn from various sources.

**TABLE 3.11  Effect of Fusariotoxins in Four Bioassay Methods**

| Bioassay | Toxin | Level | Reference |
|---|---|---|---|
| Rabbit reticulocyte | T-2 toxin | 0.02 µg/ml | Ueno, 1971 |
| Rabbit reticulocyte | Diacetoxyscirpenol | 0.05 µg/ml | |
| Rabbit reticulocyte | Fusarenon-X | 0.25 µg/ml | |
| Rabbit reticulocyte | Zearalenone | 2.5 µg/ml 10–15 µg/ml | |
| Rabbit reticulocyte | Deoxynivalenol | 2.0 µg/ml | Yoshizawa and |
| | Deoxynivalenol monoacetate | 10.0 µg/ml | Morooka, 1974 |
| | Nivalenol | 3.0 µg/ml | |
| | Fusarenon-X | 0.25 µg/ml | |
| Rabbit reticulocyte | T-2 toxin | 0.03 µg/ml | Ueno et al., 1973b |
| | Diacetoxyscirpenol | 0.03 µg/ml | |
| | Neosolaniol | 0.03 µg/ml | |
| | HT-2 toxin | 0.03 µg/ml | |
| | Nivalenol | 3.0 µg/ml | |
| | Fusarenon-X | 0.25 µg/ml | |
| | Diacetylnivalenol | 0.10 µg/ml | |
| | Deoxynivalenol (Rd-toxin) | 2.0 µg/ml | |
| | Deoxynivalenol monoacetate (Rc-toxin) | 10 µg/ml | |
| Rabbit reticulocyte | Zearalenone | 100 µg/ml | Ueno et al., 1977a |
| Ehrlich ascite | Diacetoxyscirpenol | 0.30 µg/ml | Ueno et al., 1973b |
| | Nivalenol | 6.0 µg/ml | |
| | Fusarenon-X | 0.35 µg/ml | |
| Ehrlich ascite | Diacetoxyscirpenol | 0.035 µg/ml | Ueno and Yamakawa, 1970 |
| Inhibition of multi-plication of HeLa cells | Fusarenon | 0.5–1.5 µg/ml | Ohtsubo, 1973; Ohtsubo and Saito, 1970; Saito et al., 1974 |
| | Nivalenol | 1.0 µg/ml | Yoshizawa and |
| | Deoxynivalenol | 5.0 µg/ml | Morooka, 1974 |
| | Deoxynivalenol monoacetate | 2.0 µg/ml | |
| | Nivalenol | 0.5 µg/ml | Ohtsubo et al., 1968 |
| | Fusarenon-X | 0.13 µg/ml | Ohtsubo et al., 1972 |
| | Diacetoxyscirpenol (anguidine) | 0.0024 µg/ml | Claridge et al., 1978 |
| Protein inhibition in cell-free rat liver | T-2 toxin | 5 µg/ml | Ueno et al., 1973b |
| | Diacetoxyscirpenol | 50 µg/ml | |
| | Neosolaniol | 20 µg/ml | |
| | Nivalenol | 8.0 µg/ml | |
| | Fusarenon-X | 8.0 µg/ml | |

*Source*:  Drawn from various sources.

**142**

of T-2 toxin with yeasts and demonstrated the effects of membrane-modulating agents on the growth reduction caused by T-2 toxin in *S. carlsbergensis* and membrane mutant of *S. cerevisiae*. They concluded that the toxic action of T-2 toxin is dependent on membrane integrity, and therefore the membrane is important in T-2 toxin sensitivity.

Sukroongreung *et al.* (1984) described the results of the survey that was conducted on the sensitivity of 75 yeasts (belonging to 12 yeast genera) to T-2 toxin. The authors found that one of the yeasts studied, *Kluyveromyces fragilis*, showed the greatest sensitivity (which ranged between 0.5 and 2.5 µg) of T-2 toxin per ml of culture medium in comparison with other yeasts.

*F. acuminatum* NRR 6227 (*F. equiseti* var. *acuminatum*), when grown on white corn grits and incubated at 16–20°C for 2 weeks with periodic shaking, produced an antifungal metabolite which inhibited the germination of *P. digitatum* and caused swelling of its conidia (Burmeister *et al.*, 1977, 1981). The antibiotic has been identified as a cyclic peptide composed of alanine, glumatic acid, leucine, threonine, and tyrosine.

*F. equiseti* (Corda) Sacc. NRRL 5537 produced equisetin, an antibiotic active against some Gram-positive bacteria, when grown on corn grits at room temperature. The $LD_{50}$ in mice was found to be 63 mg/kg. Its structure and synthesis were described by Vesonder *et al.* (1979c).

Jayaraman and Parihar (1975) reported *F. moniliforme* as producing a growth-promoting pigment in rice culture under laboratory conditions. Four dark red pigments were isolated and identified from *F. moniliforme* Sheld. (MRC 602) isolated from moldy maize ears in the Transkei and incubated on autoclaved maize at 25°C for 21 days. The main pigment was 6, 8-dimethoxy-5-hydroxy-3-methyl-2-azaanthraquinone, and 3 chemically related naphthaquinones (Steyn *et al.*, 1979).

Burmeister *et al.* (1972) found a reasonably good correlation between the inhibition of *R. rubra* and TLC detection of T-2 toxin, although in some cases yeast growth was inhibited though no toxin was detected on the TLC plates. The authors suggested that this may indicate either that the yeast test is more sensitive than TLC for detecting toxin in crude extracts or that some other factor may be involved. In the same series of experiments they found that *P. digitatum* was sometimes inhibited while *R. rubra* was not, suggesting that the *Fusarium* strain tested may have produced a second antibiotic which was only effective against *P. digitatum*.

Phytotoxic bioassays have not, on the whole, been extensively developed, and in general most bioassay methods have been based on the zootoxic effects of fusariotoxins.

### Animal Skin Test

This test, which involves the reaction of shaved animal skin to drops of toxin, has generally been used to detect the trichothecene group of fusariotoxins but also zearalenone (Chung *et al.*, 1974), butenolide (Yates *et al.*, 1968), and toxins of various *Fusarium* species and samples of overwintered and normal cereal grains and soils (Joffe, 1960a,b, 1963a, 1965; Joffe and Yagen, 1977; Mironov *et al.*, 1947b; Yagen and Joffe, 1976; Yagen *et al.*, 1977).

Joffe (1960a) used rabbits to determine the toxicity of numerous strains of *Fusarium*. The toxic properties of *Fusarium* strains were assessed by skin tests performed on male and female rabbits, colored grey or brown, weighing 1.5–2 kg. Only rabbits with nonpigmented skin were used, because their skin is thinner and they are more sensitive to toxins. Rabbits were kept in separate cages and maintained on laboratory feed. On the back and sides of each rabbit squares of skin measuring 3 × 3 cm were carefully cleared of hair so that five rows of 7–10 squares each were obtained (chessboard pattern) (Figs. 3.9 and 3.10). Extracts made from one of the strains (or pure toxin) were applied by micropipette or by Hamilton syringe and tested on two rabbits, the test being repeated where results were not identical. If the second test still did not produce uniform reactions, which was rare, the results were disregarded. Two squares of each rabbit served for control and were treated with alcohol extract of autoclaved wheat or millet grains not infected by any fungus. Twenty-four hr after the first application a similar dose of the same extract was applied and results were recorded after a further 24 hr. The extent and intensity of reactions has been described in detail elsewhere

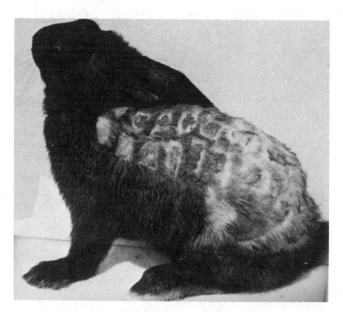

**Figure 3.9** General view of a rabbit treated with toxic extracts of *Fusarium* species.

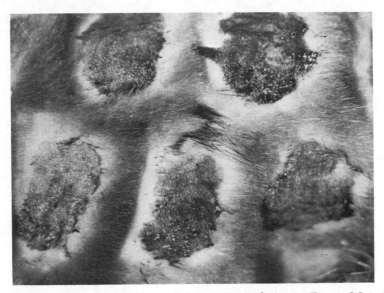

**Figure 3.10** Detailed view of some of the reactions shown in Figure 3.9 in higher magnification.

(Joffe, 1960a,b, 1971, 1973b, 1974b). Rats, mice, and guinea pigs have also been used. Guinea pigs were found to be more sensitive than either rabbits or mice in detecting diacetoxyscirpenol, fusarenon-X, and nivalenol (Ueno *et al.*, 1970a). However, the rat was more sensitive to neosolaniol monoacetate than the rabbit or guinea pig, and, in this case, the guinea pig was the least sensitive of the three animals to T-2 toxin. Thus the sensitivity of the skin test varies both according to the toxin and animals used. Ueno *et al.* (1970a) found that the necrotizing effects of toxins tested were in the order diacetoxyscirpenol > fusarenon-X > nivalenol > butenolide.

Wei *et al.* (1972) managed to detect 0.1 μg T-2 toxin with a modified form of the rat skin test, and Lansden *et al.* (1978) detected 0.02–0.04 μg T-2 toxin, also using the rat. Joffe and Yagen (1977) found an excellent correlation between the intensity of rabbit skin reaction and the T-2 toxin TLC assay, as can be seen in Table 3.12.

Hayes and Schiefer (1979b) described a severe irritation response to rat skin bioassay which was caused by an extract isolated from *F. sporotrichioides* #7452 (cultured on sterile rye grain). This extract contained a concentration of 4500 μg/ml of T-2 toxin.

### Lethal Toxicity

The highly toxic nature of the fusariotoxins led several scientists to devise screening tests using the lethal effects of these toxins on small animals.

**TABLE 3.12   The Toxic Strains Used, Their Origin, and T-2 Toxin Yield**

| Strain | Origin | Rabbit Skin | T-2 toxin[a] |
|--------|--------|-------------|--------------|
| | *F. poae* | | |
| 60/9 | Millet, USSR | + + + + | 20.0 |
| 396 | Millet, USSR | + + + + | 21.0 |
| 958 | Wheat, USSR | + + + | 8.0 |
| NRRL 3287 | Unknown, USA | + | 1.0 |
| NRRL 3299 | Corn, USA | + + + | 5.8 |
| T-2 | Corn, France | + + | 3.4 |
| | *F. sporotrichioides* | | |
| 60/10 | Millet, USSR | + + + | 10.3 |
| 347 | Millet, USSR | + + + + | 23.2 |
| 351 | Millet, USSR | + + + + | 15.2 |
| 921 | Rye, USSR | + + + + | 24.0 |
| NRRL 3249 | Fescue, USA | + + | 4.0 |
| NRRL 5908 | Fescue, USA | + | 0.6 |
| 2061-C | Corn cobs, USA | + | 1.5 |
| YN-13 | Corn, USA | + | 2.0 |
| | *F. sporotrichioides* var. *tricinctum* | | |
| NRRL 3509 | Unknown, USA | + | 0.5 |

*Source*:   Adapted from Joffe and Yagen (1977).
[a]As determined by gas liquid chromatography (GLC) in mg/10 g wheat grain.

**Mice.**   The lethal effect of i.p. injections of 10 and 40 mg/10 g body weight of mice was used to screen the toxicity of *Fusarium* isolates (Ueno *et al.*, 1973d), and a reasonably good correlation was found between this bioassay and the rabbit reticulocyte test. Other workers have also employed this mouse test, mainly for screening trichothecenes (Bamburg, 1972; Morooka *et al.*, 1972; Terao and Ueno, 1978; Ueno, 1971; Ueno and Ueno, 1978; Yoshizawa and Morooka, 1974).

**Chicken Embryo Test.**   Ivanov (1968) injected toxic extracts from *F. sporotrichiella* var. *poae* (*F. poae*) into 9–12-day-old embryos and found that the 9-day-old embryos gave the most sensitive results and that this test was more sensitive than the rabbit skin test. Archer (1974) used this method to screen mycotoxins and counted an extract as toxic if the egg hatch was less than 50%. This method was used to screen the toxicity of crude extracts of *Fusarium* (Davis *et al.*, 1975) and has also been used to detect pure toxins such as T-2 toxin (Burmeister and Hesseltine, 1970), fusarenon-X (Ueno, 1971) and moniliformin (Burmeister *et al.*, 1979).

*Day-Old Chicks and Ducklings.* The lethal toxicity of fungal extracts to day-old chicks was used by Wells and Payne (1976) to screen fungi (*Penicillium, Fusarium,* and *Aspergillus*) from pecans. Day-old chicks have also been used to detect trichothecenes (Lansden *et al.,* 1978; Joffe and Yagen, 1978), and Rabie *et al.* (1978) used the lethal reaction in ducklings (and rats) to detect moniliformin. Ducklings were also used by Yoshizawa and Morooka (1974) to detect deoxynivalenol and its mono-acetate, and they were used by Ueno (1971) to detect fusarenon-X.

*Brine Shrimp Assay.* Brown (1969) investigated the toxic effect of several mycotoxins on the hatching of brine shrimps (*Artemia salina*). He found that 1–20 µg/ml of an acetone extract of *F. tricinctum* caused 100% mortality. Harwig and Scott (1971) used the brine shrimp method to detect mycotoxins in fungal extracts, and they also tested known toxins. They found that the larvae were sensitive to diacetoxyscirpenol (0.47 µg/ml) dropped into test suspension of larvae (assay method). They also devised a disc method for screening toxins and found that this method could detect 0.2 µg/disc diacetoxyscirpenol and 2 µg/disc T-2 toxin. The test was less sensitive to zearalenone, and 10 µg/disc caused only 18% mortality. Butenolide had no inhibitory effect at 10 µg/disc. These researchers point out the advantages of the simplicity of this test and the fact that no live cultures have to be maintained. Reiss (1972) also suggested that the brine shrimp test would be useful for screening mycotoxins. Eppley (1974) used this method to screen several trichothecenes, and Davis *et al.* (1975) used it to screen the toxicity of fungi isolated from food. However, these researchers point out that when testing crude extracts this test should be used in conjunction with some other test since some fungi produce fatty acids that are toxic to brine shrimp (Curtis *et al.,* 1974). Tsubouchi *et al.* (1976) found the brine shrimp sensitive to T-2 toxin (5 µg/ml) and diacetoxyscirpenol (10 µg/ml) but not to fusarenon-X which had no effect at 20 µg/ml. Recently this method has been used by Eppley and Joffe (unpublished data) to screen the toxicity of *Fusarium* species isolated from cereals in the United States and from overwintered cereals in USSR by Joffe (*see* Table 2.1).

According to Eppley (1974); Tanaka *et al.* (1975); Durachova *et al.* (1977); and Scott *et al.* (1980) brine shrimp larvae are very sensitive to T-2 toxin and DAS.

*Insect Larvicidal Assays.* Cole and Rolinson (1972) found that 7.5 µg/ml diacetoxyscirpenol from *F. lateritium* was larvicidal to *L. sericata.* Grove and Hosken (1975) used *A. aegypti* larvae and found reasonable correlation between this test and the toxicity test using tissue cultures of mammalian cells. These larvicidal bioassays have not been as widely used or developed as other bioassay methods, and in general more interest has been shown in the actual development of insecticides from *Fusarium* toxins than in the use of larvae as tools for the bioassay of fusariotoxins.

### Cytotoxic Bioassays

*T. pyriformis.* Nakano (1968) found that fusarenon isolated from *F. nivale* completely inhibited cell division in *T. pyriformis*, but in experimental mass culture and in temperature-induced synchronous cultures he found that protein synthesis was severely affected. He suggested that the inhibition of cell division was probably due to the blockage of cellular protein synthesis. Later it was shown that lipid synthesis and phosphate uptake are also affected (Chiba *et al.*, 1972). Ueno and Yamakawa (1970) found that 0.05 µg/ml diacetoxyscirpenol, 5 µg/ml fusarenon-X, and 25 µg/ml nivalenol gave 100% inhibition of synchronized division in *Tetrahymena*. As little as 2 µg/ml gave an $LD_{50}$ (Ueno, 1971). Yoshizawa and Morooka (1974) found that in addition to nivalenol ($LD_{50}$ 5 µg/ml) and fusarenon-X (8.8 µg/ml), deoxynivalenol (4.6 µg/ml) and its monoacetate (23 µg/ml) were also toxic to *Tetrahymena*.

*Paramaecium caudatum.* Drabkin (1950) and Drabkin and Joffe (1950, 1952) concluded that paramecia (*P. caudatum*) might be utilized as a test for determining the toxicity of *Fusarium* species and extracts of overwintered cereal crops or grain infected by fusaria. These extracts completely deformed the paramecia. The toxic extracts which destroyed the paramecia also produced a strong reaction on rabbit skin and stopped the fermentation process of yeasts.

*Tissue Cultures (Cell and Organ Cultures).* Tissue culture plates have been used to screen the cytotoxic effects of mycotoxins causing a zone of inhibition around the point of application, usually a filter paper disc (Bamburg, 1972). Various cell lines have been used. Grove and Mortimer (1969) used $HE_{p2}$ and baby hamster kidney cultures to test diacetoxyscirpenol and found the hamster cells (minimum lethal dose 1.5 ng/ml) more sensitive than the $HE_{p2}$ (5 mg/ml). Human KB cells were used by Bamburg (1972) who compared the inhibition produced by 1 mg/ml of various trichothecenes. He found T-2 toxin (zone 49 mm) slightly more toxic than diacetoxyscirpenol (48.5 mm).

Oldham *et al.* (1980) attempted to study the toxicological evaluation of the mycotoxins T-2 and T-2 tetraol using normal human fibroblasts *in vitro* in order to elucidate the immunological mechanism of action of these toxins.

Bodon and Zöldac (1974) found the tissue culture assay (permanent pig kidney epithelial cell line PK.15) more sensitive to T-2 toxin (0.005 mg/assay tube) than the rabbit skin test (0.5 mg). Nivalenol completely inhibited the multiplication of HeLa cells at doses of 0.5 µg/ml (Ohtsubo *et al.*, 1968). Protein and DNA synthesis in HeLa cells were almost entirely suppressed by 5 µg/ml nivalenol although RNA synthesis was unaffected. Zearalenone and fusarenon-X had no significant effect on RNA synthesis by RNA polymerases from rat liver or *E. coli* (Tashiro *et al.*, 1979). Fusarenon-X also inhibited DNA synthesis in HeLa cells (Umeda *et al.*, 1972). Arai and Ito (1970) isolated a toxin from *F. moniliforme* (Fusariocin A) which was toxic to HeLa cells and Ehrlich tumor cells. Ohtsubo and Saito (1970) and Saito *et al.* (1974) showed complete inhibition of HeLa cells by fusarenon-X (0.5 and 1.5 µg/ml), dihydronivalenol, and dihydrofusarenon (both at 3.0 µg/ml). Mouse fibroblast L cells were used to test the toxicity of fusarenon-X but were found to be less sensitive ($LD_{50}$ 0.2 µg/ml) than HeLa cells ($LD_{50}$ 0.13 µg/ml) (Ohtsubo *et al.*, 1972). The effect of fusarenon-X on protein synthesis ($LD_{50}$ 0.2 µg/ml) and DNA synthesis ($LD_{50}$ 0.3 µg/ml) were also tested. The inhibition of protein synthesis in HeLa cells has also been used for diacetoxyscirpenol with a $TD_{50}$ of 0.0024 µg/ml (Claridge *et al.*, 1978). The tissue culture technique has also been used to detect zearalenone. In this case special cell lines were developed. Primary cell lines from calf, turkey, cock, and swine testicles and chicken fibroblasts were used. Only swine and turkey testicle cultures were adversely affected; 250–500 µg per assay tube caused complete degeneration. Even doses as low as 31–63 µg/tube caused noticeable degeneration (Vanyi and Szailer, 1974; Vanyi and Romvaryne, 1974). It was pointed out that there is a good correlation between the sensitivity of the monolayer testicle tissue culture and that of the species from which the cell line was prepared, and it was suggested that this may be used as a rapid screening test of the sensitivity of different animal species and breeds to zearalenone.

Terao (1983) described serial changes of microorganelles in various cells after the administration of trichothecene metabolites, and summarized the effect of those trichothecenes on the various cell types such as free cells and fixed cells.

### Inhibition of Protein Synthesis

As mentioned previously the trichothecenes inhibit protein synthesis, and this inhibition has been used to detect fusariotoxins using both whole cell systems such as HeLa cells and yeast spheroblasts (Cundliffe *et al.*, 1974), rabbit reticulocytes (Ueno *et al.*, 1968; Ueno, 1971; Tatsuno, 1968; Rukmini and Bhat, 1978), and Ehrlich tumor cells (Ueno and Yamakawa, 1970) as well as cell-free systems (Ueno, 1971, 1977a,b, 1980a; Smith *et al.*, 1975).

The inhibition of rabbit reticulocyte protein synthesis by nivalenol was demonstrated by Tatsuno (1968). He found that 10 µg/ml of nivalenol inhibits 90% of the incorporation of $[C^{14}]$-leucine in rabbit reticulocyte protein. Ueno *et al.* (1968) showed that the cell-free system was much more sensitive to nivalenol ($LD_{50}$ 0.5 µg/ml) than the whole-cell system ($LD_{50}$ 2.5 µg/ml). Ueno and Yamakawa (1970) showed that diacetoxyscirpenol ($LD_{50}$ rabbit reticulocytes 0.05 µg/ml; Ehrlich tumor

0.035 μg/ml) was more toxic than either fusarenon-X ($LD_{50}$ 0.25 and 0.35 μg/ml) or nivalenol ($LD_{50}$ 2.3 and 6 μg/ml). T-2 toxin was an even stronger inhibitor with an $LD_{50}$ 0.02 μg/ml in the rabbit reticulocyte system and similar results were found in the cell-free rat liver system (Ueno, 1971). Ueno *et al.* (1969a) found that fusarenon-X was ten times more potent than nivalenol in inhibiting protein synthesis in rabbit reticulocytes.

Ueno *et al.* (1969c) reported that fusarenon-X was a potent inhibitor of protein synthesis in animal cells; 0.5 μg/ml fusarenon-X inhibited the uptake of [$C^{14}$]-leucine and [$C^{14}$]-thymidine in Ehrlich tumor cells and of [$C^{14}$]-leucine in rabbit reticulocytes and interfered with the polyU-directed synthesis of polyphenylalanine in both systems. According to Ueno *et al.* (1971a) all the *F. nivale* strains freshly isolated from paddies showed inhibition of protein synthesis in the rabbit reticulocyte assay. These authors revealed that 100 μg/ml of crude toxin of *G. zeae* Ishii and *G. zeae* Ohoita-II caused complete inhibition of [$C^{14}$]-leucine when tested with rabbit reticulocytes. Ueno *et al.* (1973d) later showed the toxins of *G. zeae* Ohoita-II to be fusarenon-X and nivalenol.

The rabbit reticulocyte test has been used to screen toxic *Fusarium* species (Ueno *et al.*, 1972a,b, 1973d). A comparison of several reticulocyte systems showed that guinea pig reticulocytes were more sensitive to T-2 toxin ($LD_{50}$ 0.007 μg/ml) than either hen or rabbit reticulocytes (both $LD_{50}$ 0.03 μg/ml). With neosolaniol the guinea pig was again the most sensitive (0.01 μg/ml as compared with 0.18 and 0.25 μg/ml), but for fusarenon-X the hen was the most sensitive ($LD_{50}$ 0.18 as compared with 0.20 guinea pig and 0.25 μg/ml rabbit). These researchers also found nivalenol to be six times more active in the cell-free system than in the whole-cell system. However diacetoxyscirpenol was at least 20 times more active in the whole-cell system while fusarenon-X and neosolaniol showed similar activity in both systems. It is suggested that trichothecenes have some effect on cell membrane permeability and that the different toxins have different modes of action (Ueno, 1977a,b). Ueno and Shimada (1974) stated that the rabbit reticulocyte test is specific to the trichothecenes and that neither moniliformin nor fusaric acid cause inhibition of protein synthesis in this system.

Fractionation of *F. graminearum* (*G. zeae*)-infected corn (from the 1972 harvest in the U.S. Midwest) produced a fraction causing vomiting in ducklings and cats and death in mice. This fraction markedly inhibited the uptake of [$^{14}$C]-leucine in the rabbit reticulocyte assay (Ueno *et al.*, 1974a,b). Ishii *et al.* (1975) also studied fractions of *F. graminearum* (*G. zeae*)-infected corn for their ability to produce vomiting. The rabbit reticulocyte assay was positive, and TLC showed the presence of deoxynivalenol in one of the fractions. Ueno *et al.* (1977b) detected zearalenone in cultures of *F. lateritium* #5013 and other *Fusarium* species originally selected for their effect in the reticulocyte assay.

Ohtsubo (1973) stated that a 1 hr treatment with nivalenol (1–2 µg/ml) severely inhibited DNA and protein synthesis, but RNA synthesis was unaffected at these concentrations.

Eppley (1975) described the various biological (dermal toxicity, cytotoxicity, and inhibition of protein synthesis) and physicochemical (TLC, GLC) methods then in use for the detection and determination of trichothecene toxins.

Grollman and Huang (1976) noted that inhibitors of protein synthesis are determined as (1) inhibitors of initiation, (2) inhibitors of peptide-chain elongation, and (3) inhibitors of peptide-chain termination. Carter *et al.* (1976) noted that high concentrations of trichodermin (25 µg/ml) inhibited the elongation phase of protein synthesis and that low concentrations (0.25 µg/ml) inhibited peptide-bond formation at both initiation and elongation.

According to Carter *et al.* (1978) fusarenon-X, depending on the concentration, had a strong effect on protein synthesis.

Cannon *et al.* (1976a) studied the interaction of T-2 toxin with its receptor site on eukaryotic ribosomes and stated that T-2 toxin inhibited peptide-bond formation on ribosomes. Cannon *et al.* (1976b) noted that trichothecene mycotoxins inhibit protein and DNA synthesis.

The mechanism of protein inhibition has been investigated by several researchers (Cundliffe *et al.*, 1974; Smith *et al.*, 1975; Bamburg, 1976; Liao *et al.*, 1976; Christopher *et al.*, 1977; Cundliffe and Davis, 1977; McLaughlin *et al.*, 1977; and Ciegler and Bennett, 1980) and it is generally accepted that the fusariotrichothecene toxins inhibit protein initiation (unlike trichodermin which blocks elongation).

The biological tests were based on necrotic skin reaction in laboratory animals and on investigations of inhibition of protein synthesis of trichothecene mycotoxins in HeLa cells and cultured rabbit reticulocytes and mouse fibroblasts (L-cells) (Bamburg *et al.*, 1968, 1969; Gilgan *et al.*, 1966; Joffe, 1960a,b, 1971; Ohtsubo and Saito, 1970; Ohtsubo *et al.*, 1972; Saito *et al.*, 1971; Ueno, 1971; Ueno and Shimada, 1974; Ueno *et al.*, 1969, 1973b). These biological tests are very sensitive but not specific, whereas physicochemical analyses are specific but not sensitive. Therefore a method that is sensitive and also has the requisite chemical specificity would be of considerable value. Such a method is based on the reaction between the epoxide moiety of the trichothecene mycotoxins with GSH in the presence of glutathione-S-epoxide-transferase. The residue GSH indicates the extent of the reaction. A sensitive and specific method may be achieved by using physicochemical procedures (TLC, GLC) in conjunction with enzyme and antibody assays.

### Enzyme and Antibody Assays

Several attempts have been made to establish rapid, specific methods for detecting fusariotoxins. Foster *et al.* (1975) devised an enzymatic method based on the reaction between the epoxy moiety of the trichothecene toxins and GSH in the presence of glutathione-S-epoxide-transferase. However, although this method could be used to detect as little as 1 μg/ml total trichothecene, it could not distinguish between the various toxins. Ueno and Matsumoto (1975b) stated that some trichothecenes have a capacity to bind with the thiol residues of SH-enzyme protein. According to Ueno and Matsumoto (1975a) fusarenon-X and T-2 toxin are associated with the protein moiety of ribosomes.

Nakamura *et al.* (1977) studied the effect of trichothecenes (T-2 toxin and fusarenon-X) on epoxide hydrolase and GSH-S-transferase (glutathione-S-transferase) and stated that the epoxide of these trichothecenes is inert to epoxide hydrolase, GHS-S-transferase and GSH.

An antibody has been produced against T-2 toxin (Chu *et al.*, 1979). This gave the greatest binding effect with T-2 toxin followed by HT-2 toxin and T-2 triol. Neosolaniol, T-2 tetraol, and 8-acetyl-neosolaniol gave weak cross-reactions, and diacetoxyscirpenol, trichodermin, vomitoxin, and verrucarin gave virtually no reaction. The authors suggest that this antibody method could be used for a sensitive immunological detection of T-2 toxin.

Otokawa (1983) studied immunological disorders caused by T-2 toxin. The author described the mode of action of trichothecenes on antibody responses and various types of cell-mediated immune reactions in mice.

## Specific Toxic Responses

Certain fusariotoxins produce specific toxic reactions. Zearalenone causes increased uterus weight in many animals, and this factor has been used in a bioassay test for this toxin (Christensen *et al.*, 1965; Marasas *et al.*, 1977; Mirocha *et al.*, 1967). Uteri from previously ovariectomized mice treated with varying doses of zearalenone were incubated in vitro at 37°C in Eagle's HeLa medium supplemented with glutamine (1.0 mM). An increase in RNA and protein synthesis and an accelerating effect on the cellular permeability to $[^{14}C]$-α-aminoisobutyric acid, $[^{14}C]$-(3-0-methyl)-glucose, and $[^3H]$-uridine were demonstrated. The alteration of uterine permeability is presumed to be one of zearalenone's primary effects (Ueno and Yagasaki, 1975).

Vomiting in ducklings has also been used to detect fusarenon-X (Ueno, 1971) as well as several other trichothecenes. The doses required were, for T-2 toxin, neosolaniol, and HT-2 toxin: 0.1 mg/kg; for diacetoxyscirpenol: 0.2 mg/kg; for nivalenol: 1.0 mg/kg; fusarenon-X: 0.4–0.5 mg/kg; and for diacetylnivalenol: 0.4 mg/kg (Ueno *et al.*, 1974a).

Ueno (1983b) described several biological assay methods for the detection of trichothecenes, including skin toxicity and cytotoxicity tests, biochemical assay, and radioimmunoassay.

## Chemical Methods

Although the bioassay methods described above are useful for detecting mycotoxins, chemical methods are required for the unequivocable identification and quantification of fusariotoxins. These methods have been developed in two main directions for trichothecenes and zearalenone:

1. Multitoxin screening methods to determine the safety of foods and feedstuffs.
2. Refined methods for the accurate detection and quantification of individual fusariotoxins.

### *Multitoxin Screening Test for Trichothecenes*

Trichothecene mycotoxins have received less attention than zearalenone in multiscreening assays and are usually dealt with as a separate group. Thus nivalenol, deoxynivalenol, fusarenon-X, diacetoxyscirpenol, and T-2 toxin in cereals were detected by combining (Amberlite-florisil) column and gas chromatography (Kamimura *et al.*, 1978). Gas liquid chromatography was used by Romer (1976) and Romer *et al.* (1978) to detect T-2 toxin and diacetoxyscirpenol in corn and mixed feed. T-2 toxin, diacetoxyscirpenol, and deoxynivalenol were detected in food by gas chromatography (sensitive to 10 ng) and combined TLC and mass spectroscopy (100 ng) (Stahr *et al.*, 1979). A more refined gas chromatographic method using electron capture detection could detect 0.02–5 ppm trichothecenes in barley and wheat (Kuroda *et al.*, 1979). The limits for multitoxin and the individual toxins are given in Table 3.13 and are as low as 0.002 ng. Sphon *et al.* (1977) used desorption mass spectrometry to detect several mycotoxins, including T-2 toxin and zearalenone (10–100 ng) but did not recommend this method for general use since several factors, such as sensitivity, control, and the amount of clean-up required, had not yet been finalized. Kato *et al.* (1979) worked out a spectrophotometric-fluorometric method for detecting the following trichothecenes: neosolaniol, nivalenol, tetraacetylnivalenol, diacetoxyscirpenol, and HT-2 toxin.

Kato *et al.* (1976) described a fluorometric method of fusarenon-X by zirconyl nitrate–ethylenediamine reagent. This method can also be used in determining nivalenol and dehydroxynivalenol but not for T-2 toxin, HT-2 toxin, neosolaniol, and trichothecin.

Naoi *et al.* (1974) detected T-2 toxin and diacetoxyscirpenol in wheat by two-dimensional TLC under ultraviolet after spraying the plate with 20% $H_2SO_4$ and then heating at 105°C for 20 min. Pareles *et al.* (1976) worked out a method for rapid screening of T-2 toxin and HT-2 toxin in

**TABLE 3.13  Chemical Assays of Major Fusariotoxins**

| Toxin | Assay | Limit | References |
|---|---|---|---|
| T-2 toxin<br>HT-2 toxin | GLC[a] | 300 ppb | Pareles et al., 1976 |
| T-2 toxin<br>Diacetoxyscirpenol | GLC | 100 ng<br>25 ng | Romer et al., 1978 |
| T-2 toxin<br>HT-2 toxin | Partition chromatography; TLC[b] | 0.6 mg/kg (from cereal) | Niku-Paavola et al., 1976 |
| Deoxynivalenol | GC[c], MS[d] identified GCL quantitative | 0.250 µg/g | Pathre and Mirocha, 1978 |
| T-2 toxin<br>Diacetoxyscirpenol | GC | 10 ng | Stahr et al., 1979 |
| Deoxynivalenol | TLC, MS | 100 ng | |
| Nivalenol<br>Deoxynivalenol<br>Fusarenon-X<br>Diacetoxyscirpenol<br>T-2 toxin | Amberlite florisil column chromatography and GC | 0.1 ppm (in cereals) | Kamimura et al., 1978 |
| Nivalenol<br>Deoxynivalenol<br>Fusarenon-X<br>T-2 toxin<br>Diacetoxyscirpenol<br>Neosolaniol | (1) GC with flame ionization;<br>(2) electron capture detection | (1)    (2)<br>7 ng  0.002 ng<br>5 ng  0.002 ng<br>10 ng  0.004 ng<br>80 ng  0.4  ng<br>30 ng  0.4  ng<br>30 ng  0.2  ng | Kuroda et al., 1979 |
| T-2 toxin | TLC | 1 mg/kg corn | Eppley et al., 1974 |
| Zearalenone | TLC/UV[e] | 50–100 µg/kg corn | Eppley et al., 1974<br>Thomas et al., 1975 |
| Zearalenone | TLC | 2 ppm | Vanyi et al., 1974a,b |
| Zearalenone | TLC | 200–500 µg/kg grain | Stoloff et al., 1971 |
| Zearalenone | TLC | 0.1 µg (sensitivity 50 ppb) | Mirocha et al., 1974 |
| Zearalenone | GLC | 0.05 ppm | Sugimoto et al., 1975 |
| Zearalenone | TLC, GLC/MS | 0.36–11.05 ppm | Shotwell et al., 1976 |

**TABLE 3.13 Continued**

| Toxin | Assay | Limit | References |
|-------|-------|-------|------------|
| Zearalenone | TLC | 200 ppb | Seitz and Mohr, 1976 |
| Zearalenone | HPLC$^f$/GLC | 10 ppb | Holder *et al.*, 1977 |
| Zearalenone | TLC | 35 µg/kg cereals | Josefsson and Möller, 1977 |
| Zearalenone | TLC/UV GC | 0.1 mg/kg grain 0.01 mg/liter beer | Lovelace and Nyathi, 1977 |
| Zearalenone | TLC HPLC (confirmed GLC/MS) | 5 ng (20 µg/kg in cornflakes) (5 µg/kg in cornflakes | Scott *et al.*, 1978 |
| Zearalenone | TLC | 200 ppb | Balzer *et al.*, 1978 |
| Zearalenone | HPLC and fluorescence | 11–69 ng/g | Ware and Thorpe, 1978 |
| Zearalenone | HPLC | 0.01 ppm | Frischkorn *et al.*, 1978 |
| Zearalenone | HPLC and laser fluorimetry | 5 ppb | Diebold *et al.*, 1979 |
| Zearalenone | HPLC and UV detector | 1 ng | Engstrom *et al.*, 1977 |

*Source:* Drawn from various sources.
[a] Gas liquid chromatography
[b] Thin layer chromatography
[c] Gas chromatography
[d] Mass spectrometry
[e] Ultra violet
[f] High pressure liquid chromatography

milk by mass fragmentography. This method can be used for the confirmation and semiquantification of these toxins. Recently Steyn (1981) described some multimycotoxin methods involving extraction and clean-up, separation of extract compounds by TLC, and some other methods for detection of mycotoxins in feedstuff extracts.

Howell *et al.* (1981) described a sensitive and economical method (using TLC and HPLC) for the determination of six mycotoxins in mixed feeds, including laboratory animal diets and raw materials.

Scott (1982) described quantitative methods for the precise determination of trichothecenes in agricultural commodities using TLC and GLC with flame ionization detection (GLC-FID) and GLC with electron capture or mass spectrometric detection (GLC-MS).

Kamimura _et al._ (1981) described a systematic method for the simultaneous determination of the following _Fusarium_ mycotoxins in cereal grains and foodstuffs: nivalenol, deoxynivalenol, fusarenon-X, diacetoxyscirpenol, neosolaniol, T-2 toxin, HT-2 toxin, butenolide, moniliformin, and zearalenone.

Ellison and Kotsonis (1976) assigned [$^{13}$C]-NMR spectra to 8 trichothecenes, including T-2 toxin, HT-2 toxin, diacetoxyscirpenol, T-2 triol, T-2 tetraol, and acetyl T-2 toxin.

Cox and Cole (1983) discussed the general aspects of nuclear magnetic resonance ($^{13}$CNMR) for structural determination, stereochemical analysis, and biosynthetic studies for the identification of trichothecenes.

The authors stated that there are now more than 50 naturally occurring compounds having the 12,13-epoxytrichothec-9-ene ring system.

As multiscreening methods have been refined they have become more sensitive and smaller amounts of mycotoxins can be detected. However, for really accurate, sensitive estimations the methods devised for individual toxins give the best results.

## _Chemical Methods for Individual Trichothecene Toxins_

The assay methods for these toxins have been recently reviewed by Pathre and Mirocha (1977) and Eppley (1979). They described analytical procedures for the following naturally occurring trichothecenes: T-2 toxin, diacetoxyscirpenol (DAS), deoxynivalenol (DON), and nivalenol (NV). Eppley (1982) also recommended useful methods for the detection and quantification of trichothecenes.

The following procedures were used for the analysis of T-2 toxin in cereal grains, foods, and feeds: gas chromatography-mass spectrometry (Mirocha _et al._, 1976b); TLC and flame ionization gas chromatography (Kamimura _et al._, 1981; Stahr _et al._, 1979); electron capture gas chromatography (Kamimura _et al._, 1981; Romer _et al._, 1978; Scott, 1982); and radioimmunoassay (Lee and Chu, 1981a,b).

Thin layer chromatography has been used extensively to detect trichothecene toxins both in crude extracts and as pure toxins. Detection of these toxins by TLC is difficult because they do not significantly absorb ultraviolet and show no fluorescence.

It was found that T-2 toxin could be detected by spraying the plates with $H_2SO_4$ and p-anisaldehyde reagent followed by heating (Scott _et al._, 1970; Bamburg and Strong, 1971), which gives characteristic colors to different trichothecenes. Later it was shown that after the $H_2SO_4$ heat treatment, T-2 toxin, HT-toxin, neosolaniol, and diacetoxyscirpenol fluoresce under long wave (356 nm) ultraviolet radiation (Ueno _et al._, 1973d). Kotsonis and Ellison (1975) used a combination of TLC and fluorodensitometric method to quantify T-2 toxin and HT-2 toxin.

Since the $R_f$ values of each of the toxins depends on the solvent used (Eppley, 1975; Pathre and Mirocha, 1977), the various toxins can be separated out on the TLC. Kotsonis and Ellison (1975) found that by using suitable systems the toxins in the crude extracts can be separated from any interfering substances, thus making any elaborate clean-up unnecessary. In general the limit of the TLC method is 0.1 μg for group A trichothecenes and 1 μg for group B (Pathre and Mirocha, 1977). Gas liquid chromatography (GLC) (Ikediobi et al., 1971) and a combination of gas chromatography (GC) and mass spectrometry (MS) have also been used (Pathre and Mirocha, 1977) to detect trichothecenes. Quantitative analysis has usually employed some form of GLC or GC, and recent refinements have allowed the detection of as little as 0.002–0.004 ng of group B toxins (Kuroda et al., 1979). In this case the analysis of group B toxins was more sensitive than group A (minimum detection 0.2–0.4 ng). Table 3.13 shows some of the detection limits obtained for the trichothecenes by the various analytical methods.

Details for separation and identification of the common trichothecenes by GLC were published as their acetates or trimethylsilyl ether derivatives. Detection and quantification are very good but, in order to avoid misinterpretation, separation by preparative TLC or GLC should be performed first (Bamburg et al., 1969; Ikediobi et al., 1971; Tanaka et al., 1974; Pathre et al., 1976; Pathre and Mirocha, 1977, 1978).

Collins and Rosen (1979) found that gas-liquid chromatographic-mass spectral (GLC-MS) analysis of T-2 toxin in milk can be examined by using 6 ppb electron impact and 3 ppb chemical ionization.

Milama and Lelievre (1979) described a GLC method for T-2 toxin and diacetoxyscirpenol. The limit of sensitivity was 250 and 150 ppb respectively by using flame ionization detection and, when analysed by electron capture, 0.9 and 0.3 ppb respectively.

Cohen and Lapointe (1982) described a GLC method which uses an electron capture detector for determining vomitoxin in wheat, oats, barley, and corn grains at levels of 0.05 ppm.

Swanson et al. (1983) described a GLC method for the determination of T-2 toxin in plasma. The limit of detection is 25 ng/ml. The method used is a modification of a heptafluorobutyryl derivatization technique reported by Romer et al. (1978) and is associated with GLC and electron capture detection.

The best procedure is combined GLC-MS of the TMS-ester, which gives unequivocal identification of the trichothecene toxin by its mass spectrum (Mirocha et al., 1977b; Pathre and Mirocha, 1977).

Kato et al. (1977) noted that fusarenon-X and T-2 toxin gave a colored product by the reaction with chromotropic acid. This reaction can also be used for neosolaniol, nivalenol, tetraacetylnivalenol, and HT-2 toxin but not for trichothecin and dihydronivalenol. Kato et al. (1978) also developed a colorimetric reaction for the trichothecene metabolites by using 4-p-nitrobenzyl-pyridine.

Table 3.14 lists the various clean-up procedures used on samples of trichothecenes and zearalenone prior to the final assay.

**TABLE 3.14  Some Examples of Clean-Up and Assay Steps Used in Estimating Fusariotoxins**

| Toxin | Initial Clean-Up | Final Assay | Detection | References |
|---|---|---|---|---|
| T-2 toxin DAS[a] | Silica gel column | TLC | Spray | Eppley et al., 1974 |
| T-2 toxin HT-2 toxin | — | TLC | Fluorodensitometry | Kotsonis and Ellison, 1975 |
| T-2 toxin DAS Zearalenone | Membrane clean-up | TLC | Recoveries of mycotoxins were detected by TLC | Roberts and Patterson, 1975 |
| T-2 toxin DAS Zearalenone | Ferris gel precipitation | TLC | — | Stahr, 1975 |
| T-2 toxin DAS | (1) Ammonium sulphate was used to denature and precipitate proteinaceous compound (2) Silica gel column chromatography | GLC | $^{63}$Ni electron capture | Romer et al., 1978 |
| Zearalenone | Liquid/liquid separation | TLC | Fluorescence | Stoloff et al., 1971 |

| Zearalenone | Bis-diazotized benzidine spray reagent for treatment; silica gel column chromatography | TLC | UV | Malaiyandi et al., 1976 |
| Zearalenone | Aqueous $NH_4$ sulphate/Freon-113 liquid/liquid separation | TLC | Fluorescence | Seitz and Mohr, 1976 |
| Zearalenone | Column chromatography; liquid/liquid separation | TLC | Fluorescence | Shotwell et al., 1976 |
| Zearalenone | Silica gel column chromatography | TLC | GLC-MS | Shotwell et al., 1977 |
| Zearalenone | Liquid/liquid partition into benzene; Sephadex LH-20 column; silica gel column chromatography | HPLC | UV | Holder et al., 1977 |
| Zearalenone | Hexane wash, pH adjusted liquid/liquid separation, silica gel minicolumn chromatography | TLC HPLC GLC-MS | Fluorescence | Scott et al., 1978 |
| Zearalenone | Liquid/liquid partition into benzene | HPLC | Fluorescence | Ware and Thorpe, 1978 |

*Source:* Drawn from various sources.

[a] Diacetoxyscirpenol.

Naoi (1983) described clean-up procedures and GLC analysis for trichothecene mycotoxins. For the chemical analysis of these trichothecenes, the authors based their methods on those of Kamimura *et al.* (1981). Also Takitani and Asabe (1983) described some clean-up methods and detecting reagents associated with TLC analysis of trichothecenes as well as structures of trichothecene mycotoxins and their Rf values.

Recently several analytical methods have been developed for isolation and purification of deoxynivalenol isolated from cereal grain samples or *Fusarium* strain extracts (Bennet *et al.*, 1981b; Ehrlich and Lillehoj, 1984; Ehrlich *et al.*, 1983; Pathre and Mirocha, 1978; Scott, 1982; Scott *et al.*, 1984; Trucksess *et al.*, 1984; and Vesonder *et al.*, 1982).

Ehrlich and Lillehoj (1984) worked out a new procedure for deoxynivalenol and 3-acetyldeoxynivalenol isolation and purification based on the method of Yoshizawa (1975a).

### Chemical Procedures for Zearalenone Identification

Zearalenone was first isolated by Stob *et al.* (1962), the chemical structure being identified by Urry *et al.* (1966) and later synthesized by Taub *et al.* (1968). The chemical structure of zearalenone (Fig. 3.5) is 6-(10-hydroxy-6-oxo-*trans*-1-undecenyl)-β-resorcylic acid lactone.

The chemistry, biosynthesis, metabolism, and the physical and chemical properties of zearalenone and its derivatives have been covered in the following publications and reviews: Bennett and Shotwell (1979); Bollinger and Tamm (1972); Brooks *et al.* (1971); Bullerman (1974); Ellestad *et al.* (1978); Girotra and Wendler (1967); Hagler and Mirocha (1980); Hagler *et al.* (1979); Hidy *et al.* (1977); Hobson *et al.* (1977a,b); Hurd and Shah (1973); Inaba and Mirocha (1979); Jackson (1973); Jackson *et al.* (1974); Jensen *et al.* (1972); Johnston *et al.* (1970); Kuo *et al.* (1967); Kurtz and Mirocha (1978); Lieberman and Mirocha (1970); Mirocha (1977); Mirocha and Christensen (1974a,b); Mirocha and Pathre (1979); Mirocha *et al.* (1978a, 1979b, 1980, 1981); Pathre and Mirocha (1976, 1977); Pathre *et al.* (1978, 1980); Peters (1972); Shannon *et al.* (1980); Shipchandler (1975); Shotwell *et al.* (1980); Steele *et al.* (1974, 1976, 1977a,b); Taylor and Watson (1976); Ueno (1973b); Ueno and Yagasaki (1975); Ueno *et al.* (1974c, 1977a); Vandenheuvel (1968); Wehrmeister and Robertson (1968); Windholz and Brown (1972).

Yoshizawa (1983a) described the chemistry, biosynthesis, and toxicity of zearalenone and its naturally occurring derivatives as well as fescue-foot butenolide, leukoencephalomalacia disease, and moniliformin toxicosis.

Most of the early work was carried out to detect aflatoxin but other toxins were soon added, including zearalenone (Stoloff *et al.*, 1968). In fact most of the multitoxin screening tests include zearalenone and other mycotoxins (Balzer *et al.*, 1978; Eppley, 1968; Hagan and Tietjen, 1975; Jemmali, 1974; Roberts and Patterson, 1975; Seitz and Mohr, 1976; Shotwell *et al.*, 1975; Stahr, 1975; Stoloff *et al.*, 1971; Thomas *et al.*, 1975; Wilson *et al.*, 1976). A screening method that included zearalenone, butenolide, diacetoxyscirpenol, and nivalenol was devised by Scott *et al.* (1970); the screening method of Eppley *et al.* (1974) included zearalenone and T-2 toxin, and that of Niku-Paavola *et al.* (1976) T-2 toxin, HT-2 toxin, and zearalenone. Marasas *et al.* (1979b) used a modification of the method devised by Nakano *et al.* (1973) for screening deoxynivalenol, acetyldeoxynivalenol, and zearalenone. Most of the early multitoxin assays were based on TLC. More recently GLC combined with mass spectrometry were added to the TLC technique (Shotwell *et al.*, 1976, 1977) and also a novel membrane clean-up procedure (Roberts and Patterson, 1975). Patterson and Roberts (1979) modified the multimycotoxin method associated with membrane clean-up. They used two-dimensional TLC and an appropriate solvent system. The limit of detection for T-2 toxin and zearalenone was found to be 20–200 µg/kg. HPLC has been used in a multitoxin assay that included zearalenone and other mycotoxins by Hunt *et al.* (1978) and Frischkorn *et al.* (1978).

Stoloff *et al.* (1971) and Stoloff (1972) detailed the basic chemical procedure and emphasized the importance of proper sample preparation. Five major stages are included in the screening procedure: extraction, lipid removal (defatting), clean-up, separation, and quantitation.

The economic significance of the presence of zearalenone in agricultural feeds, cereal grains, and maize led to the development of numerous detection methods. TLC and GLC are covered by Stoloff (1971, 1972, 1975–1978); Mirocha *et al.* (1974); Patterson and Roberts (1979); Scott *et al.* (1978); Seitz and Mor (1976); Shotwell (1977b); Shotwell *et al.* (1976, 1977); Stoloff *et al.* (1976); Sugimoto *et al.* (1975); Thouvenot and Morfin (1979), and Wilson *et al.* (1976), and gas chromatography is covered by Suzuki *et al.* (1978).

Trenholm *et al.* (1980) described a sensitive gas chromatographic method for the quantitative analysis of zearalenone in blood serum.

Trenholm and Warner (1981) described a sensitive, HPLC method for the quantitative determination of zearalenone and α-zearalenol in blood serum, which can be used to detect levels as low as 100 ng/ml.

The fluorescent properties of the toxin have made it possible to use a fluorodensitometry technique for quantification (Jemmali, 1974; Sarudi *et al.*, 1976), and a combined laser fluorometry with high pressure liquid chromatography (Diebold *et al.*, 1979).

Gillespie and Schenk (1977) noted that maximum fluorescence of zearalenone in ethyl alcohol occurs at short wavelength radiation of 275 nm with emission at 465 nm.

Several recent publications have demonstrated the use of HPLC for the determination of zearalenone in maize, various cereals, and feeds as follows: Cohen and Lapointe (1980); Engstrom *et al.* (1977); Holder *et al.* (1977); Hunt *et al.* (1978); Kallela and Saastamoinen (1979); Kovacs *et al.* (1975a,b); Moller and Josefsson (1978); Scott *et al.* (1978). HPLC was described either alone or in combination with fluorescence detection by Ware and Thorpe (1978), and gas-liquid chromatography and high resolution mass spectrometry was demonstrated by Scott *et al.* (1978) and Pathre *et al.* (1979).

James *et al.* (1982) developed a HPLC technique for determining zearalenone and its metabolites in rat urine and liver.

Turner *et al.* (1983) described a method for the extraction and analysis of zearalenone in chicken fat and in heart, muscle, and kidney tissues by HPLC with ultraviolet absorption detection.

For detecting zearalenone in amounts of 2.0 ng, *bis*-diazotized benzidine has been used as a highly sensitive spray reagent on TLC plates by Malaiyandi *et al.* (1976). Zearalenone at 0.5 mg/ml developed a blue color after the application of copper-Folin reagent solution enabling it to be studied without the interference of various other mycotoxins in protein (Siraj *et al.*, 1978). Stoloff and Francis (1980) offered an analytical method for the detection of aflatoxin and zearalenone in canned and frozen sweet corn.

A rapid minicolumn procedure was described for zearalenone in corn, wheat, and sorghum by Holiday (1980).

Dixon (1980) described the results of the radioimmunoassay of zearalenol, a synthetic dihydro-derivative of zearalenol, and also preparation and properties of a specific antibody to this zearalenol metabolic agent.

Thouvenot and Morfin (1983) described the preparation of specific porcine antibodies for resorcylic acid lactones, the 6'-carboxymethyl-oxime derivatives of zearalenone and zearalanol. These porcine antibodies can be used for a radioimmunoassay for zearalenone, zearalenol, and for their transformation products, which may be found in both human and animal sera. The authors noted that the limit of detection in human serum was 5 ppb.

## CONCLUSION

The role of *Fusarium* isolates and their toxins as causative agents in human and animal health disorders has acquired considerable significance and importance. Therefore rapid, simple, sensitive, and specific methods are necessary for the detection and determination of *Fusarium* tricothecene toxins (mainly T-2 toxin, DAS, and other compounds) from various cereal grains, corn, feeds, and foods as well as from animal sources (milk, eggs, meat, urine, and blood). The complicated physicochemical procedures which are known at present (GLC, capillary

GC-MS) and many others, used for identification and quantitative assessment (yield, concentrations) of mycotoxins, require special skilled personnel and also expensive and sophisticated apparatuses. Only well equipped laboratories are able to carry out the physicochemical analyses. The majority of peripheral laboratories, however, have no suitable conditions and less complicated methods are needed.

The biological activity of *Fusarium* strains and their extracts containing trichothecene toxins can be assayed by skin tests on rabbits and by toxicity on brine shrimp larvae. Skin reactions criteria were widely used by the author in comprehensive studies of the etiology of ATA in the USSR, which was associated with toxic *Fusarium* isolates from overwintered cereal grain samples and soil. Also skin reactions criteria were used in a study of aflatoxin problems during the years 1963–1968 in Israel. There is an excellent agreement, for example, between amounts of T-2 toxin present in crude extracts and their toxicity as measured by the rabbit skin test and brine shrimp larvae. Therefore I recommend the use of both these bioassays.

The intensity of the skin reactions caused by different strains of *Fusarium* are assessed in a scale of grades determined by the author (Joffe, 1960a,b, 1974b, 1983b; Joffe and Palti, 1974b; Joffe and Yagen, 1977). Fusaria of the Sporotrichiella section with intensive pigments are shown to be stronger producers of trichothecene metabolites, mainly T-2 toxin, than the light pigment-producing strains.

# CHAPTER 4

## ANTIBACTERIAL, ANTIPROTOZOAL, AND INSECTICIDAL EFFECTS OF TOXIC *FUSARIUM* SPECIES

### ANTIBACTERIAL EFFECT

The *Fusarium* species are capable of producing a number of antibiotics, and therefore these fungi attracted our attention.

A study was carried out on the effect of some *Fusarium* species on the tubercle bacilli, on several intestinal bacilli, and on *Staphylococcus aureus*. *Penicillium, Cladosporium, Alternaria, Mucor, Aspergillus,* and *Thamnidium* were also tested previously on the same microorganisms (Joffe, 1955, 1956b,c; Joffe and Yeshmantaite, 1955, 1958; Khaikina *et al.*, 1955; Joffe *et al.*, 1958).

A study of *Fusarium* species was undertaken to determine antibiotic activity with respect to the intestinal, dysenteric, and typho-paratyphoid bacilli. The *Fusarium* isolates were incubated at 24°C on liquid media in Erlenmayer flasks of 250 ml containing 50–70 ml of Czapek (I) and carbohydrate-peptone (III) media (*see* Chapter 6). The liquid was removed on the seventh, eleventh to twelfth, and fourteenth to fifteenth (occasionally on the twentieth to twenty-first) day of growth and the pH determined. The liquids were passed through a Seitz filter and then the sterile filtrates tested against strains of *Salmonella typhi, S. paratyphi* B, *Shigella flexneri,* and *Sh. dysenteriae* on solid medium (meat-peptone agar) in Petri dishes and on a liquid medium (meat-peptone broth with

**164**

**TABLE 4.1**  Antibiotic Effect of *Fusarium* Filtrates on Several Intestinal
Bacilli

| Bacillus Tested | Medium I | % Active Cultures | Medium III | % Active Cultures |
|---|---|---|---|---|
| S. typhi | 5 | 5.9 | 4 | 4.8 |
| S. paratyphi B | 3 | 3.7 | 1 | 1.2 |
| Sh. flexneri | 5 | 5.9 | 4 | 4.8 |
| Sh. dysenteriae | — | — | 4 | 4.8 |

*Source:*  From Joffe (1960a).

1% glucose) by serial solution method (Joffe, 1955, 1960a; Joffe and
Yeshmantaite, 1955). The results are given in Table 4.1. From Table 4.1
it may be seen that the effect of the tested fusaria grown on Media I and
III on *S. typhi*, *S. paratyphi* B, and *Sh. flexneri* was not significantly
different. However, on Czapek's medium I the *Fusarium* cultures were
completely inactive with respect to *Sh. dysenteriae*, whereas their positive effect on carbohydrate-peptone medium III was noted in four cases.
The most active *Fusarium* isolates on the bacilli were: *F. lateritium, F.
moniliforme, F. equiseti, F. avenaceum, F. sambucinum, F. semitectum,
F. solani,* and, mainly, *F. poae* and *F. sporotrichioides.* One hundred and
five cultures of *Fusarium* isolates were also tested for their effect on
*Staph. aureus. Fusarium* cultures grown on carbohydrate-peptone
medium (III) had only an insignificant antibiotic effect on *Staph. aureus*,
2.8% of cultures as compared with 6.7% with Czapek's medium (I).
*Fusarium* isolates active against *Staph. aureus* were *F. sambucinum, F.
poae, F. equiseti, F. sporotrichioides* var. *tricinctum*, and *F. javanicum.*
    We also studied the action of 63 *Fusarium* isolates against tuber-
culosis bacilli: *Mycobacterium tuberculosis* var. *avium, bovis,* and
*hominis.* The *Fusarium* isolates were grown on a liquid carbohydrate-
peptone medium (III) and on a synthetic Czapek medium (I) in flasks.
On the seventh, twelfth, fifteenth, and twenty-third days filtrates were
removed and also passed through a Seitz filter. The filtrates were then
tested for their effect on the human type (*M. tuberculosis* var. *hominis*),
on the bovine type (*M. tuberculosis* var. *bovis*), and on the avian type (*M.
tuberculosis* var. *avium*) in solid media (egg type medium, serum,
potato, or glycerol agar) in tubes. The sterile *Fusarium* filtrates were
introduced into the tubes in the amount of 0.5–1.0 ml. The tubes were
sealed with paraffin and placed in an incubator at 37°C. The control
series was prepared by inoculating the corresponding tubercle bacilli to
which had been added only filtrate of media used for growing fusaria
isolates (but without the fusaria).
    These experiments involved over 1100 tubes; every filtrate was exam-
ined twice in four to five tubes. At the same time 5–7 ml of sterile

TABLE 4.2   *Fusarium* Isolates Active Against *M. Tuberculosis* var. *Avium*, *Bovis*, and *Hominis*

| Culture Medium | Avium | % Active Cultures | Bovis | % Active Cultures | Hominis | % Active Cultures |
|---|---|---|---|---|---|---|
| Czapek (I) | 4 | 6.3 | 3 | 4.7 | 5 | 7.9 |
| Carbohydrate peptone (III) | 4 | 6.3 | — | — | — | — |

*Source:*   From Joffe (1960a).

*Fusarium* filtrates were added to the flasks containing 50–70 ml of 5% glycerine veal broth. Cultures of the tubercle bacilli were then introduced on cork plates which floated on the surface of the broth. As a control inoculated flasks containing 5% glycerine veal broth without *Fusarium* isolates were used.

*Fusarium* isolates grown on Czapek medium inhibited the growth of the tuberculosis culture in four instances for the avian type (6.3%), in three instances for the bovine type (4.7%), and in five instances for the human type (7.9%). When the carbohydrate-peptone medium was used, the *Fusarium* isolates only inhibited the growth of the avian type of tubercle bacillus (Table 4.2). The following *Fusarium* species inhibited the growth of *M. tuberculosis* var. *avium*, *bovis*, and *hominis*: *F. orthoceras* (*F. oxysporum*), *F. moniliforme*, *F. sambucinum*, *F. avenaceum*, *F. javanicum*, *F. solani*, *F. semitectum*, and *F. sporotrichioides* (Joffe, 1960a).

The filtrates of *Fusarium* isolates used the tuberculosis bacilli entirely leaving only a drop of oil on the bottom of the test tube and prevented the growth of the bacilli both on solid and in liquid media. The tubercle bacilli to which filtrates of *Fusarium* isolates were added were in the majority of cases thick and polymorphic with numerous granules, while in the control cultures the bacilli were slender and long and generally contained one to two granules.

Our observations indicated that *Fusarium* isolates under unfavorable environmental conditions of overwintering (low temperatures and sharp fluctuations of temperature accompanied by freezing and thawing and other factors) undoubtedly favored the formation of antibiotic substances.

## ANTIPROTOZOAL EFFECT

The inhibitory effect of fusarenon-X isolated from *F. nivale* on the synchronously dividing cells of the protozoa *Tetrahymena pyriformis* was induced by heat treatment. Fusarenon-X toxin was added to a culture of

*T. pyriformis* and 5 μg/ml of the toxin completely inhibited cell division of the protozoa (Nakano, 1968; Ueno *et al.*, 1968, 1971c; Ueno, 1971).

Ueno and Yamakawa (1970) studied the antiprotozoal activity of diacetoxyscirpenol, fusarenon-X, and nivalenol on *T. pyriformis* and found that the inhibitory effect of diacetoxyscirpenol at a dosage of 0.05 μg/ml, fusarenon-X at 5.0 μg/ml, and nivalenol at 25 μg/ml also completely inhibited cell division in *T. pyriformis*.

The protozoa *T. pyriformis* was grown in culture medium containing 2% protease-peptone, 0.5% yeast extract, and 87% dextrose (Nakano, 1968).

Chiba *et al.* (1972) studied the effect of fusarenon-X on lipid synthesis and phosphate uptake in *T. pyriformis*. They found that lipid synthesis and phosphate uptake were inhibited by fusarenon-X in the protozoa cells. According to Ueno and Yamakawa (1970) *T. pyriformis* was a useful biological assay test for some trichothecene metabolites isolated from contaminated cereal grains and foodstuffs.

We studied the effect of toxic extracts obtained from *F. poae* and *F. sporotrichioides* on protozoa (*Paramaecium caudatum*) and extracts isolated from overwintered millet grains contaminated by the above fusaria. The procedure was described by Drabkin and Joffe (1950, 1952). The extract from overwintered millet grains, in dilutions of 1 : 50, 1 : 100, 1 : 200, and 1 : 400, caused destruction of the paramecia whereas an extract from normal millet grain exhibited no effect. The toxic extracts which destroyed the paramecia produced a strong reaction on rabbit skin and also stopped the alcoholic fermentation process of yeast (*Saccharomyces cerevisiae*) according to Mishustin *et al.* (1946) and Drabkin (1950).

Mironov (1944) and Mironov and Alisova (1947) revealed that the toxic substance in overwintered millet grains contaminated with *F. poae* and *F. sporotrichioides* was largely concentrated in the surface layers of the grain, particularly in the husks. We therefore divided the husks from the flour and made extracts separately. Table 4.3 shows that the toxic action of the husk extracts against the paramecia was stronger (1 : 800) than that of the flour (1 : 200) obtained from overwintered millet.

We also examined extracts of 41 samples of overwintered toxic millet contaminated by *F. poae* and *F. sporotrichioides*, which were associated with outbreaks of ATA disease. Five extracts from normal millet grains were used as control. All extracts from overwintered millet grains showed a strong toxic effect on paramecia. After some minutes, depending on the concentration of the extract and its degree of toxicity, the motion of the paramecia was halted completely, they became deformed and bent and exhibited granular disintegration. Extracts from normal millet grains had no effect (Drabkin and Joffe, 1950).

We carried out a series of tests with toxic *Fusarium* species. Normal, sterilized millet grains were infected with *F. poae* No. 60/9 and *F.*

**TABLE 4.3  The Effect of Crude Extract from Normal and Overwintered Millet on Paramecia**

| Type of Sample | Action on Paramecia at Concentrations of | | | | | Rabbit Skin Test | Fermentation Test— Gas Evolved in 48 hr (ml) |
|---|---|---|---|---|---|---|---|
| | 1:50 | 1:100 | 1:200 | 1:400 | 1:800 | | |
| Lemma of normal millet | 0 | 0 | 0 | 0 | 0 | Non-toxic | 2.5 |
| Flour of normal millet | 0 | 0 | 0 | 0 | 0 | Non-toxic | 6.0 |
| Lemma of toxic millet | + + | + + | + + | + + | + + | Strongly toxic | 0 |
| Flour of toxic millet | + + | + + | + + | 0 | 0 | Strongly toxic | 0 |

*Source:* From Drabkin and Joffe (1950).

*sporotrichioides* No. 60/10 which exhibited strong toxicity on rabbit skin and when fed to laboratory animals (frogs, mice, rabbits, guinea pigs, and rats) (Joffe, 1960a,b, 1971; Schoental and Joffe, 1974). The extracts obtained were strongly toxic to paramecia in dilutions of 1 : 100 to 1 : 1000 (Table 4.4).

We were interested to find out whether *Fusarium* extracts have an effect only on paramecia or whether their antiprotozoal action-spectrum is wider. We therefore prepared alcoholic extracts of toxic *F. poae* Nos. 60/9, 396, 444, and 975 and *F. sporotrichioides* Nos. 60/10, 341, 738, 921, and 1823 isolated from overwintered cereal grains, and tested them on four different protozoa: *P. caudatum, Stylonychia mylitus, Opalina ranarum*, and *Nyctotherus cordiformis*. The results obtained proved that *F. poae* and *F. sporotrichioides* had a destructive effect on all protozoa tested in concentrations of 1 : 200 to 1 : 1200 after 2–25 min, depending on the protozoa tested. It was thus established that cultures of *F. poae* and *F. sporotrichioides* have a lethal effect on living cells.

## INSECTICIDAL EFFECT

Numerous works have been published on the effect of aflatoxin B on various insects but there is very little material on the action of *Fusarium* mycotoxins (zearalenone or trichothecenes) on insect larvae or on egg production.

Kishaba *et al.* (1962) found that a trichothecene (verrucarin) inhibited feed consumption by larvae and adult Mexican bean beetle (*Epilachna varivestis*).

Eugenio *et al.* (1970b) fed zearalenone contaminated rice flour to the confused flour beetle (*Tribolium confusum* J. duVal) and to the lesser meal worm (*Alphitobius diaperinus* Panzer) and found that zearalenone persisted in the insect throughout metamorphosis. After 6 days of being fed meal containing 10,000 ppm of zearalenone, after death 8 and 14 ppm was detected in the two insects, respectively. The amount of zearalenone obtained from the insects increased with rising zearalenone concentration in the rice flour.

Rao *et al.* (1971) studied various fungus metabolites to determine their effect on survival and reproduction of the confused flour beetle. They found that *Fusarium* species (*F. roseum* (Link) Snyder and Hansen, *F. tricinctum* (Corda) Snyd. and Hans., *F. moniliforme* (Sheld.) Snyd. and Hans.) had little toxic effect on mortality (5–17.5% of parents) but enhanced larval production (all except one strain of *F. roseum* which reduced larval numbers by half).

Harein *et al.* (1971) studied the effects of pure zearalenone (F-2) and F-2 in mixture with rice flour on the reproduction of the confused flour beetle. The latter were fed rice with F-2 produced by *F. roseum* var.

TABLE 4.4  The Effect of Extract from Millet Infected by *F. Poae* and *F. Sporotrichioides* on Paramecia

| Type of Sample | Action on Paramecia at Concentrations of | | | | | Rabbit Skin Test | Fermentation Test— Gas Evolved in 48 hr (ml) |
|---|---|---|---|---|---|---|---|
| | 1:100 | 1:200 | 1:400 | 1:800 | 1:1000 | | |
| Normal millet grain | 0 | 0 | 0 | 0 | 0 | Non-toxic | 6.5 |
| Millet grain infected by *F. poae* | ++ | ++ | ++ | ++ | 0 | Strongly toxic | 0 |
| Millet grain infected by *F. sporotri-chioides* | ++ | ++ | ++ | ++ | ++ | Strongly toxic | 0 |

*Source:*  From Drabkin and Joffe (1950).

*graminearum.* Five culture media were prepared for the examination of the effect of the F-2 mixture and the pure F-2, administered in rice flour with and without brewer's yeast, on the fecundity and fertility of the confused flour beetle. The authors stated that zearalenone (F-2) in the rice flour mixture did not affect fecundity or fertility, and that pure F-2 had no effect on the production which remained the same as in the untreated controls. In all cases the emerging larvae were smaller and more sluggish than the controls.

According to Sinha (1966), confused flour beetles developed successfully when fed food contaminated with many fungi. However, *F. moniliforme*, after a 3–5-day larval period, resulted in only 8% pupated larvae (control larval period: 12 days and 90% pupation).

Cole and Rolinson (1972) studied the effect of various fungi, including *F. lateritium,* on insect activity. The culture of *F. lateritium* IMI 140879 produced an insecticidal substance, diacetoxyscirpenol, which resulted in 100% mortality in the tested larvae of *Lucilia sericata.*

Reiss (1972) stated that diacetoxyscirpenol (in a dose of 0.5 and 1 $\mu$g/ 0.5 ml) caused mortality of more than 80% of brine shrimp larvae (*Artemia salina* L.). Reiss proposed the brine shrimp as an effective bioassay method for mycotoxins. Reiss (1975), in larvicide tests with some mycotoxins, also examined diacetoxyscirpenol in various concentrations and found that it was less active against *Drosophila melanogaster* than aflatoxin B or rubratoxin B.

Grove and Hosken (1975) tested various naturally occurring 12,13-epoxytrichothec-9-enes and their derivatives against larvae of the mosquito *Aedes aegypti.* They found that T-2 toxin, scirpentriol, its triacetate, nivalenol, and its diacetate had a larvicidal activity on the mosquito and only moderate mammalian cytotoxicity.

Wright *et al.* (1976) fed the confused flour beetle wheat flour containing various concentrations of T-2 toxin and zearalenone (F-2). T-2 toxin at 100 ppm increased the egg production and fecundity of the *T. confusum* for 60–90 days but reduced their fertility (larval hatch) by 10–12% below the controls. T-2 toxin inhibits protein synthesis (Ellison and Kotsonis, 1974), and this may be the cause of the reduced fertility, decreased egg production, and slow development noted in larvae on medium with concentration of 100 ppm T-2 toxin. However, the authors pointed out that T-2 toxin is liable to make food unpalatable, and that the young beetles may become hyperactive in their search for wholesome food. The loss of fertility and slow development in treated larvae may be due to inadequate food consumption.

According to Vea and Wright (1973) larvae fed T-2 toxin developed more slowly than controls. Zearalenone (F-2) had little effect over the 60 days of treatment, but older beetles on medium containing 1000 and 10,000 ppm F-2 showed increased egg production, although this was not statistically significant. Possibly F-2 can be broken down during insect metabolic processes into some form that prolongs reproductive life; this could explain the longterm fecundity of adults exposed to F-2 for the first time and of their progeny.

Recently Wright *et al.* (1982) described some *Fusarium* species and their metabolites which were toxic to insects. The authors noted that *F. sporotrichiella* (*F. sporotrichioides*) was toxic to all development stages of the spider mite (*Tetranychus telarius* L.), and *F. solani* produced secondary metabolites with insecticidal properties.

Davis *et al.* (1975) used the yellow mealworm larvae (*Tenebrio molitor*) for test screening of mycotoxins. Davis and Smith (1977) fed the yellow mealworm larvae sterile rye grains infected with three strains of *F. sporotrichioides*, all of which were isolated from the Canadian prairies and were harmful to this kind of larvae. According to Davis and Schiefer (1982a) the growth of larvae was depressed by dietary protein and by 64 and 128 ppm of T-2 toxin in the diet.

Davis *et al.* (1982b) found that *F. sporotrichioides* No. 7452, isolated from hay in Saskatchewan, produced T-2 toxin which was toxic to yellow mealworm larvae and rats. The rats died within 48 hr, and radiomimetic lesions were observed in follicles of the lymphatic system and acute necrosis and hyperkeratosis of the gastric mucosa were also observed.

The same lesions were found by Hayes *et al.* (1980) in mice and by Hayes and Schiefer (1980) in rats, both being fed on pure T-2 toxin isolated from *F. sporotrichioides* No. 921 by Joffe (1960a) and prepared by Makor Chemicals, Jerusalem, Israel.

Ciegler (1977) briefly quoted some references connected with feeding insects zearalenone (F-2), T-2 toxin, and diacetoxyscirpenol. He concluded that "the use of mycotoxins as insecticides appears impractical."

# CHAPTER 5

## PHYTOTOXIC EFFECT OF *FUSARIUM* SPECIES

Species of the genus *Fusarium* occur in great numbers in soil and find a favorable substrate in a variety of plants.

Various *Fusarium* isolates produce substances in the infected plants known as phytotoxins. They penetrate into host tissues under auspicious environmental conditions causing wilt disease or death.

The phytotoxic potential of the Sporotrichiella section of the *Fusarium* species has been studied to a limited extent (Seemüller, 1968; Marasas *et al.*, 1971), but *Fusarium* species belonging to other sections have not, generally, been studied sufficiently. The author therefore carried out a comprehensive, 13-year study (1964–1977) of the phytotoxic effect of various species and varieties of the genus *Fusarium* on selected test seedlings of field crops. They were compared with the action of crude extracts of the same fusaria on rabbits using the skin test as a criterion.

This comparison was made with a view to determining whether a positive relationship exists between toxicity to animals (rabbit skin) and mortality of seedling plants. The histological changes induced by application of *Fusarium* crude extracts to rabbit skin were then studied in detail (stained with Hematoxylin and Eosin, HE).

The strong toxicity of many isolates is shown by the extent to which inflammation, edema, and necrosis are induced.

For this reason the author studied the distribution and phytotoxic and toxigenic properties of other fusaria isolated from various sources (soil and plants) of subtropical (semiarid) zones in Israel (Joffe, 1962a,b, 1963c, 1967c, 1968a,b, 1969b,c, 1972, 1975; Joffe and Lisker, 1968–1970; Joffe and Nadel-Schiffmann, 1967; Joffe and Palti, 1962, 1963, 1964a,b,

1965, 1967, 1970–1972, 1977; Joffe and Schiffmann-Nadel, 1972; Joffe *et al.*, 1964, 1965, 1967, 1972–75; Palti and Joffe, 1971) and other countries, chiefly from temperate zones in the USSR, the United States, Canada, Finland, and others (Joffe, 1960a, 1963a, 1965, 1969a, 1973b, 1974b; Joffe and Palti, 1974b, 1975).

## ISOLATION, IDENTIFICATION, CULTIVATION, AND INOCULATION TESTS WITH SEEDLING PLANTS AND TOXICITY ON RABBIT SKIN

The procedure for isolation and identification of *Fusarium* strains derived from soil and plant material and inoculation of seedlings has been described in detail elsewhere (Borut and Joffe, 1966; Joffe, 1963b,d, 1966, 1967a,b, 1973a, 1974a, 1977, 1983a; Joffe and Borut, 1966; Joffe and Palti, 1967, 1972, 1974a, 1975; Joffe *et al.*, 1973, 1974). The cultivation of *Fusarium* isolates and preparation of crude extract were carried out in the following way: natural media of wheat or millet grain (containing 10 g grain and 20 ml tap water) were sterilized for 30 min twice at 1 atm, then inoculated with suspensions of monospore cultures of *Fusarium* species and incubated for 40, 21, 15, 12, 10, and 8 days at temperatures ranging from 6–35°C (6°, 12°, 18°, 24°, 30°, and 35°C, respectively). After incubation infected natural media were heated at 100°C for 20 min and extracted with ethyl alcohol (96%) in a vacuum rotary evaporator, yielding biologically active oillike residue.

The bioassay was performed by applying 10 μg of *Fusarium* crude extract to the shaved rabbit skin (*see* Fig. 3.6). The intensity of the skin reaction produced by each isolate was assessed on the following scale:

N = no skin reaction;
M = slightly (±) or moderately (+) toxigenic—slight inflammation, reddening, edema, and formation of slight leukocytic reaction;
t = toxic (+ +)—edematous, hemorrhagic, necrotic, and leukocytic reaction;
S = strongly or very strongly toxic (+ + +, + + + +) reaction—acute edema and severe necrotic reaction.

Phytotoxic action on crop seedlings treated with suspensions was studied with 1022 strains of *Fusarium* isolated from Israel and 250 from other countries in inoculation tests with 10 field and vegetable plants.

In order to determine the phytotoxic range and potential of various *Fusarium* species and varieties, 200 seedlings for each *Fusarium* isolate were used for screening (i.e., 20 per test crop) taken from the following test plants: bean, cucumber, watermelon, cotton, tomato, onion, eggplant, pepper, wheat, and maize. Wheat and maize were rarely affected by *Fusarium* extracts, and the results relating to these crops have therefore been omitted from the tables.

Reisolation of *Fusarium* isolated from wilted (dead) plants was also carried out. These reisolates were tested as before, this time using 400–500 instead of 200 seedlings per *Fusarium* strain. The results showed that strains became more phytotoxic to seedlings upon reisolation and showed their strongest effect upon their own source plants. This is seen with onion, eggplant, tomato, watermelon, and others.

## Toxicity of Fusarium Isolates Grown at Temperatures from 6–35°C

Phytotoxic action to plant seedlings and toxicity on rabbit skin treated with suspensions and crude extracts of *Fusarium* strains isolated from various sources in Israel and from other countries are presented in Tables 5.1 to 5.4.

None of the cultures grown at 6°C and collected in subtropical areas in Israel produced a toxic response. *Fusarium* isolates cultivated at 24 and 30°C were more frequently toxic than those incubated at lower (12 or 18°C) or higher (35°C) temperatures. In contrast isolates derived from temperate regions were most toxic when inoculated at low temperatures (6–12°C). This phenomenon is clearly demonstrated in Tables 5.1 through 5.4 and in the tables separately summarized for each section of the genus *Fusarium*.

The experimental results obtained from rabbit skin tests after application of toxic crude extracts which had been isolated from various *Fusarium* species in subtropical regions of Israel and from countries other than Israel, are presented in Appendices 1 and 2.

Though the toxicity of fusaria exhibited in this study by isolates from Israel was moderately high (Joffe, 1983a), it was nevertheless far below that of species of the Sporotrichiella section (*F. sporotrichioides, F. poae*) (Joffe, 1973b, 1974c) isolated only in temperate environmental conditions.

The results of the rabbit skin reaction are given for the optimum temperature at which toxic effects are produced by each *Fusarium* species or variety. The strength of these effects depends on the *Fusarium* strain, isolate, the source, temperature, and incubation time.

# RELATION BETWEEN PHYTOTOXICITY TO PLANTS AND TOXICITY TO RABBIT SKIN

The determination of the phytotoxic properties of *Fusarium* fungi isolated from various soil types and host plants in field and storage under subtropical and temperate climatic conditions was first carried out

**TABLE 5.1  Phytotoxicity of *Fusarium* Strains from Field and Storage Crops in Israel to Seedlings of Eight Test Plants**

| | No. of Isolates | Seedlings (% of 20 per Strain) Killed After Treating with *Fusarium* Suspension | | | | | | | |
|---|---|---|---|---|---|---|---|---|---|
| | | Bean | Cucumber | Water-melon | Cotton | Tomato | Onion | Egg-plant | Pepper |
| *F. dimerum* | 6 | 17.5 | 15.0 | 19.1 | 29.2 | 32.5 | 50.0 | 45.0 | 15.0 |
| *F. merismoides* | 5 | 8.0 | 10.0 | 9.0 | 16.0 | 18.0 | 24.0 | 6.0 | 11.0 |
| *F. coccidicola* | 3 | 1.7 | 13.3 | 8.3 | 11.7 | 6.7 | 13.3 | 11.7 | 8.3 |
| *F. avenaceum* | 27 | 3.5 | 10.2 | 5.0 | 10.6 | 7.8 | 17.2 | 15.3 | 14.0 |
| *F. arthro-sporioides* | 6 | 1.7 | 11.7 | 0 | 16.7 | 13.3 | 21.7 | 10.0 | 8.3 |
| *F. semitectum* | 50 | 4.6 | 15.4 | 12.8 | 18.3 | 15.0 | 10.3 | 18.1 | 7.0 |
| *F. semitectum var. majus* | 5 | 18.0 | 8.0 | 21.0 | 23.0 | 13.0 | 19.0 | 22.0 | 5.0 |
| *F. lateritium* | 26 | 12.0 | 16.5 | 9.0 | 20.7 | 28.5 | 10.8 | 31.3 | 11.3 |
| *F. moniliforme* | 133 | 4.4 | 9.3 | 10.0 | 13.7 | 5.8 | 13.8 | 9.9 | 9.3 |

| Species | | | | | | | | |
|---|---|---|---|---|---|---|---|---|
| F. moniliforme var. subglutinans | 12 | 0 | 11.7 | 5.8 | 0 | 11.7 | 11.2 | 5.0 | 0 |
| F. equiseti | 156 | 6.0 | 18.6 | 13.2 | 9.0 | 11.4 | 17.4 | 15.8 | 5.8 |
| F. equiseti var. acuminatum | 4 | 7.5 | 21.2 | 26.3 | 30.0 | 31.3 | 41.3 | 25.0 | 25.0 |
| F. equiseti var. compactum | 2 | 0 | 10.0 | 10.0 | 0 | 20.0 | 25.0 | 15.0 | 5.0 |
| F. equiseti var. caudatum | 2 | 0 | 2.0 | 1.0 | 6.0 | 8.0 | 2.0 | 4.0 | 10.0 |
| F. sambucinum | 66 | 8.6 | 14.7 | 10.0 | 9.5 | 11.4 | 12.5 | 13.6 | 9.5 |
| F. culmorum | 40 | 6.9 | 16.9 | 5.3 | 13.2 | 5.5 | 19.4 | 12.3 | 7.8 |
| F. graminearum | 3 | 0 | 0 | 13.3 | 15.0 | 6.6 | 26.6 | 10.0 | 16.6 |
| F. oxysporum | 236 | 10.9 | 28.8 | 29.8 | 14.3 | 26.7 | 33.8 | 21.1 | 10.3 |
| F. oxysporum var. redolens | 6 | 19.2 | 36.7 | 37.5 | 37.5 | 36.7 | 40.0 | 13.3 | 20.0 |
| F. solani | 207 | 3.5 | 9.2 | 7.4 | 11.8 | 17.0 | 15.3 | 11.0 | 5.8 |
| F. javanicum | 27 | 6.7 | 33.3 | 36.3 | 12.4 | 20.2 | 21.7 | 12.2 | 6.8 |

TABLE 5.2  Toxic Reactions on Rabbit Skin to Isolates of *Fusarium* Species and Varieties from Israel after Culturing at Various Temperatures

| | No. of Isolates | 12°C | | | | 18°C | | | | 24°C | | | | 30°C | | | | 35°C | | | |
|---|---|---|---|---|---|---|---|---|---|---|---|---|---|---|---|---|---|---|---|---|---|
| | | N | M | T | S[a] | N | M | T | S | N | M | T | S | N | M | T | S | N | M | T | S |
| F. dimerum | 6 | 6 | 0 | 0 | 0 | 6 | 0 | 0 | 0 | 3 | 3 | 0 | 0 | 4 | 2 | 0 | 0 | 6 | 0 | 0 | 0 |
| F. merismoides | 5 | 5 | 0 | 0 | 0 | 5 | 0 | 0 | 0 | 2 | 3 | 0 | 0 | 3 | 2 | 0 | 0 | 5 | 0 | 0 | 0 |
| F. coccidicola | 3 | 3 | 0 | 0 | 0 | 3 | 0 | 0 | 0 | 1 | 2 | 0 | 0 | 0 | 3 | 0 | 0 | 1 | 2 | 0 | 0 |
| F. avenaceum | 27 | 27 | 0 | 0 | 0 | 22 | 5 | 0 | 0 | 8 | 19 | 0 | 0 | 21 | 6 | 0 | 0 | 27 | 0 | 0 | 0 |
| F. arthrosporioides | 6 | 6 | 0 | 0 | 0 | 4 | 2 | 0 | 0 | 3 | 3 | 0 | 0 | 6 | 0 | 0 | 0 | 6 | 0 | 0 | 0 |
| F. semitectum | 50 | 50 | 0 | 0 | 0 | 41 | 9 | 0 | 0 | 12 | 32 | 6 | 0 | 19 | 27 | 4 | 0 | 46 | 4 | 0 | 0 |
| F. semitectum var. majus | 5 | 5 | 0 | 0 | 0 | 5 | 0 | 0 | 0 | 2 | 0 | 3 | 0 | 2 | 3 | 0 | 0 | 5 | 0 | 0 | 0 |
| F. lateritium | 26 | 26 | 0 | 0 | 0 | 23 | 3 | 0 | 0 | 14 | 12 | 0 | 0 | 3 | 22 | 1 | 0 | 17 | 7 | 2 | 0 |
| F. moniliforme | 133 | 131 | 2 | 0 | 0 | 112 | 21 | 2 | 0 | 34 | 87 | 11 | 1 | 62 | 61 | 10 | 0 | 107 | 24 | 2 | 0 |
| F. moniliforme var. subglutinans | 12 | 12 | 0 | 0 | 0 | 10 | 2 | 0 | 0 | 3 | 9 | 0 | 0 | 6 | 6 | 0 | 0 | 12 | 0 | 0 | 0 |

| Species | N | M | T | S | N | M | T | S | N | M | T | S | N | M | T | S | N | M | T | S |
|---|---|---|---|---|---|---|---|---|---|---|---|---|---|---|---|---|---|---|---|---|
| *F. equiseti* | 156 | 151 | 5 | 0 | 93 | 6 | 2 | 0 | 18 | 108 | 30 | 0 | 92 | 54 | 9 | 1 | 150 | 6 | 0 | 0 |
| *F. equiseti* var. *acuminatum* | 4 | 4 | 0 | 0 | 0 | 4 | 0 | 0 | 0 | 3 | 1 | 0 | 4 | 0 | 0 | 0 | 4 | 0 | 0 | 0 |
| *F. equiseti* var. *compactum* | 2 | 2 | 0 | 0 | 2 | 0 | 0 | 0 | 1 | 1 | 0 | 0 | 2 | 2 | 0 | 0 | 2 | 0 | 0 | 0 |
| *F. equiseti* var. *caudatum* | 2 | 2 | 0 | 0 | 2 | 0 | 0 | 0 | 0 | 2 | 2 | 0 | 0 | 2 | 0 | 0 | 2 | 0 | 0 | 0 |
| *F. sambucinum* | 66 | 63 | 2 | 1 | 53 | 10 | 2 | 1 | 21 | 36 | 8 | 1 | 36 | 25 | 4 | 1 | 64 | 2 | 2 | 0 |
| *F. culmorum* | 40 | 39 | 1 | 0 | 34 | 5 | 1 | 0 | 14 | 23 | 3 | 0 | 30 | 10 | 0 | 0 | 38 | 2 | 2 | 0 |
| *F. graminearum* | 3 | 3 | 0 | 0 | 1 | 2 | 0 | 0 | 2 | 1 | 0 | 0 | 3 | 0 | 0 | 0 | 3 | 0 | 0 | 0 |
| *F. oxysporum* | 236 | 227 | 9 | 0 | 145 | 82 | 8 | 1 | 32 | 143 | 54 | 7 | 81 | 114 | 35 | 6 | 214 | 22 | 0 | 0 |
| *F. oxysporum* var. *redolens* | 6 | 6 | 0 | 0 | 3 | 3 | 0 | 0 | 0 | 4 | 2 | 0 | 3 | 3 | 0 | 0 | 6 | 0 | 0 | 0 |
| *F. solani* | 207 | 207 | 0 | 0 | 169 | 38 | 0 | 0 | 65 | 114 | 28 | 0 | 132 | 66 | 9 | 0 | 199 | 8 | 0 | 0 |
| *F. javanicum* | 27 | 27 | 0 | 0 | 22 | 5 | 0 | 0 | 9 | 13 | 5 | 0 | 16 | 9 | 2 | 0 | 25 | 2 | 0 | 0 |

[a]N = nontoxic; M = mildly toxic; T = toxic; S = strongly toxic.

**TABLE 5.3  Phytotoxicity of *Fusarium* Strains from Various Sources (Other than Israel) to Seedlings of Eight Test Plants**

| | No. of Isolates | Seedlings (1% of 20 per strain) Killed After Treating with *Fusarium* Suspension | | | | | | | |
|---|---|---|---|---|---|---|---|---|---|
| | | Bean | Cucumber | Water-melon | Cotton | Tomato | Onion | Egg-plant | Pepper |
| *F. nivale* | 14 | 9.5 | 3.0 | 4.5 | 1.5 | 17.0 | 12.5 | 21.5 | 2.5 |
| *F. dimerum* | 1 | 0 | 15 | 10 | 0 | 0 | 25 | 30 | 5 |
| *F. merismoides* | 6 | 10 | 7 | 1 | 5 | 18 | 11 | 20 | 8 |
| *F. melanochlorum* | 2 | 7.5 | 7.5 | 10 | 12.5 | 0 | 17.5 | 7.5 | 0 |
| *F. aquaeductuum* | 3 | 1 | 13 | 7 | 15 | 29 | 18 | 8 | 6 |
| *F. tabacinum* | 1 | 0 | 0 | 0 | 0 | 0 | 20 | 20 | 15 |
| *F. sporotrichioides*[a] | 8 | 52 | 27.5 | 31.5 | 59.5 | 67 | 53 | 45 | 26.5 |
| *F. sporotrichioides*[b] | 4 | 6.5 | 7.5 | 11.5 | 24 | 32.5 | 22.5 | 29 | 24 |
| *F. sporotrichioides*[c] | 12 | 21 | 11.5 | 9 | 21.5 | 34.5 | 27 | 21.5 | 11.5 |
| *F. poae*[a] | 5 | 57 | 41 | 34 | 23 | 55 | 81 | 58 | 38 |
| *F. poae*[b] | 5 | 6 | 13 | 14 | 14 | 21 | 20 | 15 | 0 |
| *F. poae*[c] | 6 | 8.5 | 25 | 21.5 | 21.5 | 21.5 | 25.5 | 33.5 | 14.5 |
| *F. sporotrichioides*[a] var. *tricinctum* | 2 | 17.5 | 37.5 | 17.5 | 12.5 | 50 | 30 | 40 | 10 |
| *F. sporotrichioides*[b] var. *tricinctum* | 2 | 55 | 50 | 30 | 35 | 30 | 52.5 | 32.5 | 12.5 |
| *F. sporotrichioides*[c] var. *tricinctum* | 7 | 3 | 5.5 | 8.5 | 3 | 5 | 33 | 10 | 13 |
| *F. sporotrichioides*[a] var. *chlamydosporum* | 2 | 65 | 45 | 20 | 10 | 55 | 70 | 20 | 0 |
| *F. sporotrichioides*[c] | 5 | 21 | 6 | 5 | 4 | 11 | 15 | 8 | 1 |

| | | | | | | | | | |
|---|---|---|---|---|---|---|---|---|---|
| var. chlamydosporum | | | | | | | | | |
| F. sporotrichioides[d] var. chlamydosporum | 4 | 0 | 2.5 | 10 | 5 | 0 | 4 | 12.5 | 0 |
| F. decemcellulare | 4 | 5 | 6 | 7.5 | 0 | 7.5 | 16 | 2.5 | 1 |
| F. coccidicola | 3 | 2 | 1.5 | 7 | 10 | 13 | 18 | 12 | 7 |
| F. coccophilum | 4 | 6 | 7.5 | 12.5 | 0 | 7.5 | 2.5 | 0 | 0 |
| F. avenaceum | 5 | 4 | 10 | 10 | 8 | 17 | 18 | 18 | 21 |
| F. arthrosporioides | 4 | 4 | 0 | 6.5 | 20 | 15 | 20 | 10 | 5 |
| F. semitectum | 4 | 2.5 | 2.5 | 0 | 16.5 | 0 | 5 | 12.5 | 2.5 |
| F. semitectum var. majus | 1 | 10 | 10 | 0 | 0 | 35 | 20 | 20 | 0 |
| F. camptoceras | 2 | 2.5 | 20 | 22.5 | 7.5 | 20 | 7.5 | 30 | 10 |
| F. concolor | 3 | 2 | 11.5 | 5 | 20 | 22 | 11.5 | 22 | 8.5 |
| F. lateritium | 17 | 4.5 | 11 | 4.5 | 9 | 11.5 | 14.5 | 9 | 0.5 |
| F. stilboides | 5 | 8 | 13 | 5 | 11 | 16 | 5 | 9 | 0 |
| F. xylarioides | 4 | 6 | 17.5 | 5 | 12.5 | 6 | 10 | 14 | 2.5 |
| F. moniliforme | 10 | 0 | 3 | 3 | 6 | 24.5 | 17.5 | 14 | 24.5 |
| F. moniliforme var. subglutinans | 3 | 6.5 | 8.5 | 0 | 10 | 10 | 13.5 | 5 | 16.5 |
| F. moniliforme var. anthophilum | 2 | 7.5 | 2.5 | 0 | 5 | 25 | 15 | 15 | 0 |
| F. equiseti | 6 | 0 | 21 | 18.5 | 14 | 8.5 | 27.5 | 15 | 8.5 |
| F. equiseti var. compactum | 6 | 0 | 9 | 17.5 | 18.5 | 29 | 31 | 21 | 11 |
| F. equiseti var. acuminatum | 8 | 2.5 | 11.5 | 13 | 22.5 | 30 | 37.5 | 25.5 | 12 |

**TABLE 5.3 Continued**

| | No. of Isolates | Seedlings (1% of 20 per strain) Killed After Treating with *Fusarium* Suspension | | | | | | | |
|---|---|---|---|---|---|---|---|---|---|
| | | Bean | Cucumber | Water-melon | Cotton | Tomato | Onion | Egg-plant | Pepper |
| *F. heterosporum* | 7 | 7 | 8.5 | 4.5 | 6.5 | 18.5 | 19.5 | 15.5 | 8 |
| *F. graminearum* | 5 | 9 | 4 | 4 | 10 | 22 | 22 | 14 | 10 |
| *Gibberella zeae* | 7 | 10 | 1.5 | 6.5 | 0 | 18 | 17 | 21.5 | 5.5 |
| *F. sambucinum* | 7 | 14.5 | 0 | 13 | 19.5 | 22 | 27 | 20 | 13.5 |
| *F. sambucinum* var. *coeruleum* | 4 | 7.5 | 4 | 1.5 | 2.5 | 7.5 | 10.5 | 5 | 7.5 |
| *F. sambucinum* var. *trichothecioides* | 1 | 20 | 10 | 20 | 30 | 50 | 80 | 55 | 20 |
| *F. culmorum* | 4 | 2.5 | 16.5 | 12.5 | 11.5 | 9 | 36.5 | 25 | 10 |
| *F. tumidum* | 3 | 0 | 0 | 0 | 0 | 1.5 | 6.5 | 3 | 0 |
| *F. oxysporum* | 13 | 3 | 17 | 17.5 | 14.5 | 17 | 9.5 | 28 | 14.5 |
| *F. oxysporum* var. *redolens* | 1 | 0 | 20 | 10 | 0 | 60 | 25 | 25 | 0 |
| *F. solani* | 9 | 4.5 | 9.5 | 9.5 | 16 | 12 | 18 | 13 | 4 |
| *F. solani* var. *coeruleum* | 2 | 0 | 0 | 0 | 12.5 | 17.5 | 2.5 | 10 | 10 |
| *F. solani* var. *ventricosum* | 4 | 0 | 12.5 | 10 | 11.5 | 22.5 | 4 | 25 | 9 |
| *F. javanicum* | 3 | 0 | 10 | 5 | 10 | 25 | 10 | 17.5 | 6.5 |

[a] USSR.
[b] USA.
[c] From other sources.
[d] India, Thailand, Pakistan.

**TABLE 5.4  Toxic Reactions on Rabbit Skin by Extracts of *Fusarium* Species and Varieties from Various Sources (Other than Israel) Grown at Various Temperatures**

| | No. of Isolates | 6°C | | | | 12°C | | | | 18°C | | | | 24°C | | | | 30°C | | | | 35°C | | | |
|---|---|---|---|---|---|---|---|---|---|---|---|---|---|---|---|---|---|---|---|---|---|---|---|---|---|
| | | N | M | T | S | N | M | T | S | N | M | T | S | N | M | T | S | N | M | T | S | N | M | T | S |
| *F. nivale* | 14 | 14 | — | — | — | 13 | 1 | — | — | 3 | 8 | 3 | — | 6 | 7 | 1 | — | 13 | 1 | — | — | 14 | — | — | — |
| *F. dimerum* | 1 | 1 | — | — | — | 1 | — | — | — | 1 | — | — | — | — | 1 | — | — | — | 1 | — | — | 1 | — | — | — |
| *F. merismoides* | 6 | 6 | — | — | — | 6 | — | — | — | 6 | — | — | — | 4 | 2 | — | — | 1 | 4 | 1 | — | 4 | 2 | — | — |
| *F. melanochlorum* | 2 | 2 | — | — | — | 2 | — | — | — | 2 | — | — | — | 2 | — | — | — | — | 2 | — | — | 1 | 1 | — | — |
| *F. aquaeductuum* | 3 | 3 | — | — | — | 3 | — | — | — | 3 | — | — | — | 2 | — | 1 | — | — | 3 | — | — | — | 3 | — | — |
| *F. tabacinum* | 1 | 1 | — | — | — | 1 | — | — | — | 1 | — | — | — | — | 1 | — | — | — | 1 | — | — | 1 | — | — | — |
| *F. sporotrichioides*[a] | 8 | — | — | — | 8 | — | — | 1 | 7 | — | 1 | 2 | 5 | — | 3 | — | 5 | 3 | 5 | — | — | 8 | — | — | — |
| *F. sporotrichioides*[b] | 4 | — | 2 | 2 | — | — | 3 | 1 | — | — | 4 | — | — | 2 | 2 | — | — | 3 | 1 | — | — | 4 | — | — | — |
| *F. sporotrichioides*[c] | 12 | 8 | — | 1 | 3 | 9 | — | 1 | 2 | 6 | 3 | 3 | — | 6 | 5 | 1 | — | 8 | 4 | — | — | 12 | — | — | — |
| *F. poae*[a] | 5 | — | — | — | 5 | — | — | — | 5 | — | — | 2 | 3 | 3 | 2 | — | — | 4 | 1 | — | — | 5 | — | — | — |
| *F. poae*[b] | 5 | 2 | 1 | — | 2 | 2 | 1 | — | 2 | 1 | 4 | — | — | 4 | 1 | — | — | 4 | 1 | — | — | 5 | — | — | — |
| *F. poae*[c] | 6 | 2 | 1 | 1 | 2 | 2 | 1 | 3 | — | 2 | 3 | 1 | — | 5 | 1 | — | — | 5 | 1 | — | — | 5 | 1 | — | — |
| *F. sporotrichioides*[a] var. *tricinctum* | 2 | — | 1 | 1 | — | — | 1 | 1 | — | — | 2 | — | — | 2 | — | — | — | 2 | — | — | — | 2 | — | — | — |
| *F. sporotrichioides*[b] var. *tricinctum* | 2 | — | 1 | — | 1 | — | 1 | — | 1 | — | 1 | 1 | — | 1 | 1 | — | — | 1 | 1 | — | — | 1 | 1 | — | — |
| *F. sporotrichioides*[c] var. *tricinctum* | 7 | 6 | 1 | — | — | 6 | 1 | — | — | 3 | 4 | — | — | 3 | 4 | — | — | 5 | 2 | — | — | 7 | — | — | — |
| *F. sporotrichioides*[a] var. *chlamydosporum* | 2 | — | 1 | — | 1 | — | 1 | — | 1 | — | 1 | 1 | — | 2 | — | — | — | 2 | — | — | — | 2 | — | — | — |
| *F. sporotrichioides*[c] var. *chlamydosporum* | 5 | 4 | — | 1 | — | 4 | — | 1 | — | 4 | — | 1 | — | 4 | 1 | — | — | 4 | 1 | — | — | 4 | 1 | — | — |

**183**

TABLE 5.4 Continued

| | No. of Isolates | 6°C | | | | 12°C | | | | 18°C | | | | 24°C | | | | 30°C | | | | 35°C | | | |
|---|---|---|---|---|---|---|---|---|---|---|---|---|---|---|---|---|---|---|---|---|---|---|---|---|---|
| | | N | M | T | S | N | M | T | S | N | M | T | S | N | M | T | S | N | M | T | S | N | M | T | S |
| *F. sporotrichioides*[d] var. *chlamydosporum* | 4 | 4 | — | — | — | 4 | — | — | — | 3 | 1 | — | — | 1 | 3 | — | — | 4 | — | — | — | 4 | — | — | — |
| *F. decemcellulare* | 4 | 4 | — | — | — | 4 | — | — | — | 4 | — | — | — | 2 | 2 | — | — | 2 | 2 | — | — | 4 | — | — | — |
| *F. coccidicola* | 3 | 3 | — | — | — | 3 | — | — | — | 3 | — | — | — | — | 3 | — | — | 1 | 2 | — | — | 2 | 1 | — | — |
| *F. coccophilum* | 4 | 4 | — | — | — | 4 | — | — | — | 4 | — | — | — | 3 | 1 | — | — | 2 | 1 | — | — | 3 | 1 | — | — |
| *F. avenaceum* | 5 | 5 | — | — | — | 4 | 1 | — | — | 4 | — | 1 | — | 1 | 4 | — | — | 3 | 2 | — | — | 5 | — | — | — |
| *F. arthrosporioides* | 4 | 4 | — | — | — | 3 | 1 | — | — | 3 | 1 | — | — | 2 | 1 | 1 | — | — | 4 | — | — | 4 | — | — | — |
| *F. semitectum* | 4 | 4 | — | — | — | 4 | — | — | — | 3 | 1 | — | — | 2 | 2 | — | — | 3 | 1 | — | — | 4 | — | — | — |
| *F. semitectum* var. *majus* | 1 | 1 | — | — | — | 1 | — | — | — | 1 | — | — | — | 1 | — | — | — | — | 1 | — | — | 1 | — | — | — |
| *F. camptoceras* | 2 | 2 | — | — | — | 1 | 1 | — | — | 1 | 1 | — | — | — | 1 | 1 | — | 1 | 1 | — | — | 1 | 1 | — | — |
| *F. concolor* | 3 | 3 | — | — | — | 3 | — | — | — | 3 | — | — | — | 3 | — | — | — | 1 | 2 | — | 8 | 1 | 1 | 1 | — |
| *F. lateritium* | 17 | 17 | — | — | — | 16 | 1 | — | — | 14 | 3 | — | — | 4 | 12 | 1 | — | 8 | 9 | — | — | 16 | 1 | — | — |
| *F. stilboides* | 5 | 5 | — | — | — | 4 | 1 | — | — | 3 | 2 | — | — | 1 | 4 | — | — | 3 | 2 | — | — | 5 | — | — | — |
| *F. xylarioides* | 4 | 4 | — | — | — | 4 | — | — | — | 4 | — | — | — | 2 | 2 | — | — | 1 | 3 | — | — | 2 | 2 | — | — |
| *F. moniliforme* | 10 | 10 | — | — | — | 10 | — | — | — | 9 | 1 | — | — | 3 | 6 | 1 | — | 5 | 5 | — | — | 8 | 2 | — | — |
| *F. moniliforme* var. *subglutinans* | 3 | 3 | — | — | — | 3 | — | — | — | 1 | 1 | 1 | — | 2 | 1 | — | — | 3 | — | — | — | 3 | — | — | — |
| *F. moniliforme* var. *anthophilum* | 2 | 2 | — | — | — | 2 | — | — | — | 1 | 1 | — | — | 1 | 1 | — | — | 1 | 1 | — | — | 2 | — | — | — |
| *F. equiseti* | 6 | 5 | 1 | — | — | 5 | — | 1 | — | 4 | 1 | — | — | 4 | — | — | 1 | 3 | 1 | 2 | — | 4 | 2 | — | — |

184

| | | | | | | | | | | | | | | | | | | | |
|---|---|---|---|---|---|---|---|---|---|---|---|---|---|---|---|---|---|---|---|
| F. equiseti var. compactum | 6 | 6 | — | 6 | — | — | — | 6 | — | — | 1 | 2 | 3 | — | 5 | — | 1 | 6 | — | — |
| F. equiseti var. acuminatum | 8 | 8 | — | 8 | — | — | — | 4 | 4 | — | 3 | 3 | 1 | 1 | 4 | 2 | 2 | 7 | — | 1 |
| F. heterosporum | 7 | 7 | — | 6 | 1 | — | — | 6 | 1 | 5 | 2 | 2 | — | 3 | 3 | 1 | — | 5 | 2 | — |
| F. graminearum | 5 | 5 | — | 5 | — | — | — | 2 | 3 | 1 | 4 | 4 | — | 2 | 1 | 2 | — | 5 | — | — |
| G. zeae | 7 | 7 | — | 7 | — | — | — | 3 | 4 | — | 5 | 5 | 1 | 7 | — | — | — | 7 | — | — |
| F. sambucinum | 7 | 7 | — | 7 | — | — | — | 4 | 2 | 1 | 3 | 3 | — | — | 5 | 2 | — | 7 | — | — |
| F. sambucinum var. coeruleum | 4 | 4 | — | 4 | — | — | — | 3 | 1 | — | 3 | 3 | 1 | 3 | 3 | 1 | — | 3 | 1 | — |
| F. sambucinum var. trichothecioides | 1 | 1 | — | 1 | — | 1 | 1 | — | — | — | — | — | — | — | — | 1 | — | 1 | — | — |
| F. culmorum | 4 | 4 | — | 4 | — | — | — | 2 | 2 | — | 2 | 2 | — | 3 | 3 | 1 | — | 3 | 1 | — |
| F. tumidum | 3 | 3 | — | 3 | — | — | — | 3 | — | — | 3 | 3 | — | 2 | 2 | 1 | — | 3 | — | — |
| F. oxysporum | 13 | 13 | — | 13 | — | — | — | 9 | 4 | — | 11 | 11 | 2 | 9 | 9 | 4 | — | 13 | — | — |
| F. oxysporum var. redolens | 1 | 1 | — | 1 | — | — | — | 1 | — | — | 1 | — | 1 | — | 1 | — | 1 | 1 | — | — |
| F. solani | 9 | 9 | — | 9 | — | — | — | 7 | 2 | — | 2 | 6 | 1 | 5 | 5 | 4 | — | 9 | — | — |
| F. solani var. coeruleum | 2 | 2 | — | 2 | — | — | — | 1 | 1 | — | 1 | 1 | — | 1 | 1 | 1 | — | 2 | — | — |
| F. solani var. ventricosum | 4 | 4 | — | 4 | — | — | — | 4 | — | — | 1 | 3 | — | 2 | 2 | 1 | 1 | 4 | — | — |
| F. javanicum | 3 | 3 | — | 3 | — | — | — | 2 | 1 | — | 1 | 1 | 1 | 3 | 3 | — | — | 3 | — | — |

[a] USSR.
[b] USA.
[c] From other sources.
[d] India, Thailand, Pakistan.

widely, as far as we are aware, in the author's laboratory in Israel. A total of 1272 *Fusarium* species and varieties from over 130 sources in subtropical and temperate zones were studied for their phytotoxic effects and toxicity on rabbit skin. The results in the following tables indicate a surprisingly strong correlation between the toxicity of *Fusarium* strains to rabbit skin and their phytotoxicity to seedling plants.

The histologic findings in rabbit skin which showed dermal lesions produced by topical application of crude extracts of *F. sporotrichioides* and *F. poae* and their effects on test plant seedlings are illustrated in Figures 5.1 and 5.2.

The effect on plants and rabbit skin is demonstrated for each section of *Fusarium* in detail.

### Arachnites Section

*F. nivale.*   This is a well-known plant pathogen causing disease in cereals. Fourteen strains of *F. nivale* were studied: three from *Secale cereale*, one from *Agrostis*, one from grass, one from *Poa annua*, three from *Triticum aestivum*, and five from *Lolium perenne* (one strain from Germany, two strains from Finland, and two other strains from turf from CMI in Kew, England) (Table 5.5).

*F. nivale* grown at 18 and 24°C gave moderate to strong reactions on rabbit skin and were phytotoxic to onion, tomato, and eggplant.

### Eupionnotes Section

*F. dimerum, F. merismoides, F. melanochlorum, F. aquaeductuum, F. tabacinum.*   These *Fusarium* strains, grown at 24 and 30°C, were moderately phytotoxic and mildly toxic on rabbit skin.

Thirteen strains of *F. dimerum*, nine strains of *F. merismoides*, and one strain of *F. melanochlorum* from Israel were studied (Table 5.6).

One strain of *F. dimerum* from *Matthiola incana* in Germany and six *F. merismoides* strains (one from goldfish bowl water in Holland; one from CBS Baarn in Holland; one from river water in England; one from sugar beet in England; one from *T. aestivum* in Germany; and one from pear decline in the United States) were examined. Two strains of *F. melanochlorum* (from *Fagus silvatica* and from beech tree in Austria); three strains of *F. aquaeductuum* (from rubber tube in Holland; from *F. silvatica* in Germany; and from potato tuber in Australia); and one strain of *F. tabacinum* (from CBS Baarn in Holland) were also examined (Table 5.6).

### Sporotrichiella Section

*Fusarium* species and varieties (*F. sporotrichioides, F. poae, F. sporotrichioides* var. *tricinctum,* and *F. sporotrichioides* var. *chlamydo-*

**Figure 5.1** Wilting plant seedlings treated with suspensions of *F. sporotrichioides* and *F. poae* strains and healthy control plants. Upper row: *F. sporotrichioides* Nos. 347 and 921 on bean and watermelon. Second row: *F. sporotrichioides* Nos. 921 and 1182 on cotton and tomato, respectively. Third row: *F. poae* No. 958 on cucumber and pepper. Fourth row: *F. poae* Nos. 24 and 60/9 on tomato and onion, respectively.

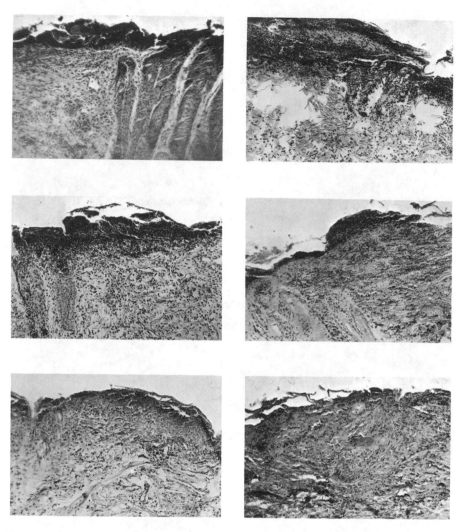

**Figure 5.2** Histologic findings in rabbit skin with the same strains of *F. sporotrichioides* and *F. poae* (stained with Hematoxylin and Eosin). Upper row: *F. sporotrichioides* No. 347—necrosis of epidermis; deeper layers of dermis are edematous including hair follicles infiltrated by leukocytes. *F. sporotrichioides* No. 921—necrosis of the dermis with seropurulent exudate and heavily infiltrated by inflammatory cells; the deep layers infiltrated by leukocytes. Second row: *F. sporotrichioides* No. 1182—acute vesicular dermatitis; vesicles filled with seropurulent exudate are seen in the outer layer of dermis; edematous dermis infiltrated with eosinophils and polymorphs. *F. poae* No. 958—necrosis of epidermis; destruction of hair follicles; edema in dermis with large number of infiltrated leukocytes. Third row: *F. poae* No. 24—necrosis of the epithelium with vesicle formation; the deep layers of the dermis infiltrated with eosinophils. *F. poae* No. 60/9—necrosis of the epidermis with infiltrated cells and vesicle formation.

**TABLE 5.5  Relation between Phytotoxicity[a] and Rabbit Skin Test[b] in Countries Other Than Israel—Arachnites Section (F. nivale)**

| No. of Isolates | Bean | Cucumber | Water-melon | Cotton | Tomato | Onion | Egg-plant | Pepper | Rabbit Skin Test | | | Incu-bation Temper-ature (°C) |
| | | | | | | | | | O | M | S | |
|---|---|---|---|---|---|---|---|---|---|---|---|---|
| 14 | 9.5 | 3.0 | 4.5 | 1.5 | 17.0 | 12.5 | 21.5 | 2.5 | 2 | 8 | 4 | 18°–24° |

Source:  Joffe, unpublished results.
[a] Average % of plants killed.
[b] O = nontoxic; M = mildly toxic; S = strongly toxic.

**TABLE 5.6  Relation Between Phytotoxicity[a] and Rabbit Skin Test[b]—Eupionnotes Section**

| Species/ Substrate | No. of Isolates | Bean | Cucumber | Water- melon | Cotton | Tomato | Onion | Egg- plant | Pepper | Rabbit Skin Test | | | Incubation Temper- ature (°C) |
|---|---|---|---|---|---|---|---|---|---|---|---|---|---|
| | | | | | | | | | | O | M | S | |
| *From Israel* | | | | | | | | | | | | | |
| *F. dimerum* | | | | | | | | | | | | | |
| Cucumber | 3 | 10 | 5 | 5 | 20 | 25 | 20 | 10 | 10 | 1 | 2 | 0 | 30° |
| Tomato | 2 | 10 | 25 | 20 | 15 | 45 | 25 | 20 | 17.5 | 1 | 1 | 0 | 30° |
| Soil | 8 | 4 | 11 | 10 | 12 | 31 | 29 | 24 | 16 | 2 | 6 | 0 | 24°–30° |
| *F. merismoides* | | | | | | | | | | | | | |
| Potato | 3 | 17 | 0 | 6 | 12 | 14 | 13 | 6 | 0 | 1 | 2 | 0 | 24°–30° |
| Maize | 2 | 0 | 0 | 10 | 20 | 30 | 40 | 5 | 0 | 0 | 2 | 0 | 24° |
| Soil | 4 | 5 | 12.5 | 9 | 17.5 | 19 | 22.5 | 7.5 | 11 | 1 | 3 | 0 | 24°–30° |
| *F. melanochlorum* | | | | | | | | | | | | | |
| Mango | 1 | 0 | 0 | 0 | 0 | 0 | 0 | 0 | 0 | 1 | 0 | 0 | 30° |
| *From Other Countries* | | | | | | | | | | | | | |
| *F. dimerum* | 1 | 0 | 15 | 10 | 0 | 0 | 25 | 30 | 5 | 0 | 1 | 0 | 24° |
| *F. merismoides* | 6 | 10 | 7 | 1 | 5 | 18 | 11 | 20 | 8 | 0 | 5 | 1 | 24°–30° |
| *F. melanochlorum* | 2 | 7.5 | 7.5 | 10 | 12.5 | 0 | 17.5 | 7.5 | 0 | 0 | 2 | 0 | 30° |
| *F. aquaeductuum* | 3 | 1 | 13 | 7 | 15 | 29 | 18 | 8 | 6 | 0 | 3 | 0 | 24°–30° |
| *F. tabacinum* | 1 | 0 | 0 | 0 | 0 | 0 | 20 | 20 | 15 | 0 | 1 | 0 | 24° |

*Source:* Joffe, unpublished results.

[a] Average % of plants killed.

[b] O = nontoxic; M = mildly toxic; ʂ = strongly toxic.

*sporum*) were isolated from overwintered cereal grain in the field and from soil in Russia and both caused ATA disease in humans (Joffe, 1960a,b, 1971, 1974c, 1978b, 1983b,c). The phytotoxic properties and toxicity on rabbit skin of 63 strains (including 46 obtained from other countries) were studied (Table 5.7).

No strains of *Fusarium* Sporotrichiella section have been isolated in Israel.

**F. sporotrichioides.** This was isolated from overwintered millet, wheat, rye, and barley in the Orenburg district, USSR (Joffe, 1960a,b). Eight of these strains were studied: five from millet (23, 60/10, 347, 351, and 738), one from rye (921), one from wheat (1182), and one from barley (1823). The strains from millet, rye, and wheat were severely phytotoxic to bean, cucumber, watermelon, cotton, and tomato (Fig. 5.1) and also to onion and eggplant. In histological examination all these strains grown at 6–12°C gave very strong skin reactions when extracts were treated on rabbit skin (Fig. 5.2).

Four strains from the United States were tested (NRRL 3249 and NRRL 5908 from tall fescue, and 2061-C and YN-13 from Minnesota corn): NRRL 3249 and 2061-C were strongly phytotoxic and others moderately phytotoxic.

Twelve strains from other sources were studied: strain 22-26 from Hungary (source unknown); strain 2416 from oats in Canada; two strains (5054 and 5050) from wheat in Australia; strain 15661 from *Scirpus* in Scotland; and seven strains from Germany (10329 from *Malus sylvestris*, 10339 from *Avena sativa*, 10360 from *Picea abies*, 10362 from *Pinus nigra*, 63421 from *P. sylvestris*, 62054 from maritime forshore sediment, and 62424 from *Solanum tuberosum*). Strain 10339 was the most phytotoxic to bean, cotton, tomato, and eggplant, and strongly toxic to rabbit skin.

**F. poae.** This was isolated from overwintered millet, wheat, barley, sunflower, and soil in the USSR (Joffe, 1960a,b).

Five of these strains were studied: three from millet (24, 60/9, and 396), one from barley (792), and one from wheat (958). All these strains were strongly phytotoxic to bean, cucumber, tomato, onion, and pepper (Fig. 5.1) and their toxicity is illustrated in Figure 5.2.

Six strains from the United States were tested: NRRL 3287 (source unknown), NRRL 3299 (from corn), M309 (AMTCC), M484 and M618 (from FDA), and 15654 (from barley seeds). None of these were phytotoxic to pepper. These strains were generally phytotoxic to tomato and onion. The most phytotoxic strains were NRRL 3287 and NRRL 3299. Among the strains from other sources tested, T-2 (from corn in France) was the most phytotoxic to eggplant. Strain 10317 from *A. sativa* in Germany was most phytotoxic to eggplant and cucumber; 10426 from

**TABLE 5.7  Relation between Phytotoxicity[a] and Rabbit Skin Test[b]—Sporotrichiella Section in Countries Other Than Israel**

| Species/Source | No. of Isolates | Bean | Cucumber | Water-melon | Cotton | Tomato | Onion | Egg-plant | Pepper | Rabbit Skin Test | | | Incubation Temperature (°C) |
|---|---|---|---|---|---|---|---|---|---|---|---|---|---|
| | | | | | | | | | | O | M | S | |
| *F. sporotrichioides* | | | | | | | | | | | | | |
| USSR | 8 | 52 | 27.5 | 31.5 | 59.5 | 67 | 53 | 45 | 26.5 | 0 | 0 | 8 | 6° 12° |
| USA | 4 | 6.5 | 7.5 | 11.5 | 24 | 32.5 | 22.5 | 29 | 24 | 0 | 2 | 2 | 6–18° |
| Others | 12 | 21 | 11.5 | 9 | 21.5 | 34.5 | 27 | 21.5 | 11.5 | 5 | 3 | 4 | 18° 24° |
| *F. poae* | | | | | | | | | | | | | |
| USSR | 5 | 57 | 41 | 34 | 23 | 55 | 81 | 58 | 38 | 0 | 0 | 5 | 6–12° |
| USA | 6 | 6 | 13 | 14 | 14 | 21 | 20 | 15 | 0 | 0 | 4 | 2 | 6–24° |
| Others | 6 | 8.5 | 25 | 21.5 | 21.5 | 21.5 | 25.5 | 33.5 | 14.5 | 1 | 3 | 2 | 12° |
| *F. sporotrichioides* var. *tricinctum* | | | | | | | | | | | | | |
| USSR | 2 | 17.5 | 37.5 | 17.5 | 12.5 | 50 | 30 | 40 | 10 | 0 | 1 | 1 | 6–12° |
| USA | 2 | 55 | 50 | 30 | 35 | 30 | 52.5 | 32.5 | 12.5 | 0 | 1 | 1 | 6° 12° |
| Others | 7 | 3 | 5.5 | 8.5 | 3 | 5 | 33 | 10 | 13 | 3 | 4 | 0 | 6° 18° 24° |
| *F. sporotrichioides* var. *chlamydosporum* | | | | | | | | | | | | | |
| USSR | 2 | 65 | 45 | 20 | 10 | 55 | 70 | 20 | 0 | 0 | 1 | 1 | 6° 12° |
| India, Thailand, Pakistan | 4 | 0 | 2.5 | 10 | 5 | 0 | 4 | 12.5 | 0 | 2 | 2 | 0 | 24° |
| Others | 5 | 21 | 5 | 5 | 4 | 11 | 15 | 8 | 1 | 3 | 1 | 1 | 24° |

*Source:*  Joffe, unpublished results.

[a] Average % of plants killed.

[b] O = nontoxic; M = mildly toxic; S = strongly toxic.

*Anthoxanthum odoratum* in Germany was nonphytotoxic; 16557 from barley in Scotland, and 22-205 from Hungary (source unknown) were weakly phytotoxic. Strain 3918 from *Sambucus* in Canada, grown at 12°C, was very phytotoxic to cucumber, tomato, onion, watermelon, cotton, eggplant, and pepper, nonphytotoxic to bean, and strongly toxic to rabbit skin (Table 5.7).

*Fusarium sporotrichioides* var. *tricinctum.* Two strains from the USSR were examined: 1227 from barley and 2457 from millet. Strain 1227, grown at 6–12°C, was strongly phytotoxic to tomato, cucumber, and eggplant and toxic to rabbit, and 2457, grown at 6–18°C, was moderately toxic on rabbit skin.

From the United States two strains were studied: M304 and NRRL 3509 (sources unknown). M304 was very phytotoxic to bean, cucumber, onion, and cotton (much more phytotoxic than NRRL 3509) and was strongly toxic on rabbit skin at 6–12°C.

In addition seven strains from other sources were tested: strain 10390 (from *Trifolium pratense*) and 10458 (from *Calamagrostis epigaeos*), both from Germany, were nontoxic, while 62448 (from *T. aestivum* in Germany) showed strongest phytotoxicity to onion. The strain from Hungary, 22-144 (source unknown) was nontoxic, and three strains from Finland (703011, 7047S11-5, and 711081-22—all from unknown sources) were weakly phytotoxic, and moderately toxic when grown at 6–24°C (Table 5.7).

*Fusarium sporotrichioides* var. *chlamydosporum.* Two strains from the USSR were examined: 1174 from millet was strongly phytotoxic to bean and onion and strongly toxic on rabbit skin, and strain 2388 from barley was moderately phytotoxic when grown at 24°C.

The author recently screened 315 strains of *F. sporotrichioides, F. sporotrichioides* var. *trincinctum,* and *F. poae* isolated from overwintered cereal grain and soil which caused the fatal disease of ATA in order to study their ability to produce a variety of trichothecene compounds. These strains have strong phytotoxic action and severe effects on rabbit skin. The majority of the strains produced T-2 toxin in large quantities and neosolaniol and HT-2 toxin in smaller quantities. The results are presented in Table 2.1.

Four strains from South Asia were studied: 4693 (from grassland soil in India) and 62049 (from *Crotalaria juncea* in Thailand) were nonphytotoxic, but 15615 (from soil in India) and 62171 (from soil in Pakistan) were weakly phytotoxic and moderately toxic when tested on rabbit skin grown at 24°C (Table 5.7).

Five other strains were studied: strain 4347 (from wheat in Canada) was strongly phytotoxic to bean and strongly toxic on rabbit skin; strain

10356 (from *Populus candicans* in England), 10354 (from *T. aestivum* in Canada) and 62170 (from *Lycopersicum esculentum* in Turkey) were nonphytotoxic, and strain 157755 (from bovine feed in Australia) was weakly phytotoxic and moderately toxic when grown at 24°C.

## Spicarioides Section

**F. decemcellulare (Calonectria rigidiuscula).** This is widely distributed in tropical and subtropical areas of different continents on cocoa and other host plants (Booth, 1971b).

In Israel only two strains have been examined, both from cropped heavy soil (Joffe, 1963c); they were nontoxic and nonphytotoxic.

From other countries only four strains were studied: one from Baarn in Holland; one from the FDA in the United States; one from AMTCC in the United States; and one from *Miconia,* branch gall in Costa Rica (Table 5.8).

Three strains of *F. decemcellulare* grown at 24 and 30° were moderately phytotoxic and toxic on rabbit skin (Table 5.8).

## Macroconia Section

The tested *Fusarium* strains were weakly phytotoxic and moderately toxigenic when grown at 30°C.

*F. coccidicola* (*C. diploa*) and *F. coccophilum* (*Nectria coccophila*) occur as parasites on insects (Booth, 1971b).

**F. coccidicola.** In Israel three isolates from Valencia oranges were examined (Table 5.9).

Three strains from other countries were examined: two from orange in Canada and one from *Prunus vulgaris* in Iran.

**F. coccophilum.** Four strains were tested: one from AMTCC in the United States, one from *Gleditschia caspica* in Iran, one strain from *Coffea arabica* in New Guinea, and one strain from *Miconia* sp. in Costa Rica (Table 5.9).

## Roseum Section

Most strains grown at 24°C showed moderate toxicity when tested on rabbit skin and were most phytotoxic on onion, eggplant, pepper, and tomato (Table 5.10).

**F. avenaceum and F. arthrosporioides.** Twenty-seven strains of *F. avenaceum* from Israel were studied from various host plants and soil (Table 5.10).

TABLE 5.8 Relation between Phytotoxicity[a] and Rabbit Skin Test[b]—Spicarioides Section in Countries Other Than Israel

| Species | No. of Isolates | Bean | Cucumber | Water-melon | Cotton | Tomato | Onion | Egg-plant | Pepper | Rabbit Skin Test | | | Incubation Temperature (°C) |
|---|---|---|---|---|---|---|---|---|---|---|---|---|---|
| | | | | | | | | | | O | M | S | |
| *F. decemcellulare* | 4 | 5 | 6 | 2.5 | 0 | 7.5 | 16 | 2.5 | 1 | 1 | 3 | 0 | 24° 30° |

*Source:* Joffe, unpublished results.

[a] Average % of plants killed.

[b] O = nontoxic; M = mildly toxic; S = strongly toxic.

**TABLE 5.9  Relation between Phytotoxicity[a] and Rabbit Skin Test[b]—Macroconia Section**

| Species/Substrate | No. of Isolates | Bean | Cucumber | Water-melon | Cotton | Tomato | Onion | Egg-plant | Pepper | Rabbit Skin Test | | | Incubation Temperature (°C) |
|---|---|---|---|---|---|---|---|---|---|---|---|---|---|
| | | | | | | | | | | O | M | S | |
| *From Israel* | | | | | | | | | | | | | |
| *F. coccidicola* | | | | | | | | | | | | | |
| Valencia orange | 3 | 2 | 13.5 | 8.3 | 11.6 | 7 | 13 | 12 | 8 | 0 | 3 | 0 | 30° |
| *From Other Countries* | | | | | | | | | | | | | |
| *F. coccidicola* | | | | | | | | | | | | | |
| Valencia orange | 3 | 2 | 15 | 7 | 10 | 13 | 18 | 12 | 7 | 0 | 3 | 0 | 24° |
| *F. coccophilum* | 4 | 6 | 7.5 | 12.5 | 0 | 7.5 | 2.5 | 0 | 0 | 2 | 2 | 0 | 30° |

*Source:* Joffe, unpublished results.
[a] Average % of plants killed.
[b] O = nontoxic; M = mildly toxic; S = strongly toxic.

TABLE 5.10  Relation between Phytotoxicity[a] and Rabbit Skin Test[b]—Roseum Section

| Species/Substrate | No. of Isolates | Bean | Cucumber | Water-melon | Cotton | Tomato | Onion | Egg-plant | Pepper | Rabbit Skin Test | | | Incubation Temperature (°C) |
|---|---|---|---|---|---|---|---|---|---|---|---|---|---|
| | | | | | | | | | | O | M | S | |
| *F. avenaceum* | | | | | | | | | | | | | |
| *From Israel* | | | | | | | | | | | | | |
| Beet | 1 | 10 | 0 | 0 | 15 | 5 | 20 | 20 | 10 | 0 | 1 | 0 | 24° |
| Medicago | 2 | 12.5 | 0 | 0 | 17.5 | 25 | 30 | 22.5 | 20 | 0 | 2 | 0 | 24° |
| Leek | 1 | 0 | 10 | 10 | 30 | 10 | 5 | 35 | 10 | 0 | 1 | 0 | 24° |
| Onion | 1 | 0 | 20 | 5 | 10 | 0 | 40 | 15 | 30 | 0 | 1 | 0 | 24° |
| Tomato | 3 | 0 | 10 | 5 | 0 | 25 | 10 | 18.5 | 11.5 | 1 | 2 | 0 | 24° |
| Watermelon | 3 | 3.5 | 0 | 0 | 10 | 11.5 | 21.5 | 18.5 | 11.5 | 0 | 3 | 0 | 24° |
| Avocado | 1 | 0 | 0 | 0 | 10 | 10 | 0 | 20 | 0 | 0 | 1 | 0 | 24° |
| Carnation | 2 | 5 | 17.5 | 7.5 | 15 | 12.5 | 7.5 | 0 | 2.5 | 1 | 1 | 0 | 24° |
| Cotton | 1 | 0 | 0 | 0 | 10 | 5 | 0 | 10 | 15 | 0 | 1 | 0 | 24° |
| Soil | 12 | 3 | 15 | 7.5 | 10 | 12.5 | 19 | 14 | 17.5 | 5 | 7 | 0 | 24° |
| *From Other Countries* | | | | | | | | | | | | | |
| *F. avenaceum* | 5 | 4 | 10 | 10 | 8 | 17 | 18 | 18 | 21 | 0 | 4 | 1 | 24° |
| *F. arthrosporioides* | 4 | 4 | 0 | 6.5 | 20 | 15 | 20 | 10 | 5 | 0 | 4 | 0 | 30° |

*Source:* Joffe, unpublished results.

[a] Average % of plants killed.

[b] O = nontoxic; M = mildly toxic; S = strongly toxic.

Five strains of *F. avenaceum* from other countries were examined: one from red clover in Canada; one from Finland (source unknown); one from *Betula verrucosa* in Germany; and two from *S. cereale* in Germany.

Four strains of *F. arthrosporioides* were tested: one from Finland (source unknown); one from wheat seed in the United States; one from Gramineae in Finland; and one from *Azalea* in New Zealand (Table 5.10).

## Arthrosporiella Section

*Fusarium* strains of the Arthrosporiella section grown at 24°C were moderately and strongly toxic to rabbit skin and moderately or highly phytotoxic to watermelon, eggplant, and cotton (Table 5.11).

**F. semitectum, F. semitectum var. majus, F. camptoceras, F. concolor.** In Israel 57 strains of *F. semitectum* from various sources and 11 strains of *F. semitectum* var. *majus* from cotton, sorghum, and avocado were examined (Table 5.11).

Four strains of *F. semitectum* from other countries were examined: one strain from CBS, Baarn in Holland, one from tomatoes in Egypt, one from soybean seed in Tanganyika, and one from *Citrus* in British Guiana.

One strain of *F. semitectum* var. *majus*, from coccidia on *Prunus* from Iran, was examined.

Two strains of *F. camptoceras* from cocoa bean in Costa Rica and three strains of *F. concolor*; two strains from CBS, Baarn in Holland; and one strain from Hungary (source unknown) were also examined.

## Lateritium Section

Most strains belonging to Lateritium section showed moderately toxic or toxic reactions on rabbit skin and moderately phytotoxic action on cotton, tomato and eggplant when grown at 24 and 30°C.

**F. lateritium.** Twenty-six strains of *F. lateritium* from Israel were studied (Table 5.12). From other countries, three strains of *F. lateritium* (one from *Cajamus indicus* in Tanganyika and two from *Coffea excelsa* in French Equatorial Africa) were tested.

In addition to these, 12 strains of *F. lateritium* were tested: one from CBS, Baarn in Holland; four from *Ficus elastica, Carpinus betulus, Juglans regia,* and *Tillandsia latifolia* from Germany; one from dead tree shoot in Italy; one from *Eriosoma* leaf gall on *Ulmus glabra* in Austria; one from *Hibiscus syriacus* in Canada; one from AMTCC in the United States; two from *Passiflora edulis* in New Zealand; and two strains from *Citrus* in Iran (Table 5.12).

*F. stilboides.*   Five strains were examined: one from AMTCC in the United States and four others from *Coffea* in Tanganyika, Nyasaland, Southern Rhodesia, and Brazil.

*F. xylarioides.*   Four strains were examined: one from AMTCC in the United States and one from CMI in Kew, England; and two from *Coffea robusta* in Guinea (Table 5.12).

## Liseola Section

*F. moniliforme* and their varieties grown at 24°C showed moderate and strong toxicity on rabbit skin and moderately or strongly phytotoxic action on cucumber, watermelon, onion, and eggplant.

*F. moniliforme.*   One hundred and thirty-three strains from Israel were studied from various sources; also twelve strains of *F. moniliforme* var. *subglutinans* from mango and one strain of *F. moniliforme* var. *anthophilum* from maize were studied (Table 5.13).

Three strains of *F. moniliforme* from tropics in other countries were also studied: one from *Musa* in Libya and two strains from peanuts in Mozambique.

In addition seven strains from the following sources were also examined: four from the FDA in the United States; two from corn in Canada and the United States; and one from *Freesia* seed in England.

*F. moniliforme* var. *subglutinans.*   Three strains were tested: one from corn in Canada; one from *Haemanthus* in Germany; and one from *Saccharum officinarum* in India.

*F. moniliforme* var. *anthophilum.*   Two strains were tested: one from *S. officinarum* in Nigeria and one strain from *Hippeastrum* in Germany (Table 5.13).

## Gibbosum Section

*Fusarium* strains of Gibbosum section grown at 24°C caused moderately and strongly phytotoxic and toxic action on various crop seedlings and rabbit skin, respectively.

*F. equiseti, F. equiseti* var. *compactum, F. equiseti* var. *caudatum, F. equiseti* var. *acuminatum.*   In Israel 156 strains of *F. equiseti* were studied from various sources. Also one strain of *F. equiseti* var. *compactum*, two strains of *F. equiseti* var. *caudatum*, and four strains of *F. equiseti* var. *acuminatum* were tested (Table 5.14).

Six strains of *F. equiseti* from other countries were examined: one from CBS, Baarn in Holland; one from the FDA in the United States;

TABLE 5.11  Relation between Phytotoxicity[a] and Rabbit Skin Test[b]—Arthrosporiella Section

| Species/Substrate | No. of Isolates | Bean | Cucumber | Water-melon | Cotton | Tomato | Onion | Egg-plant | Pepper | Rabbit Skin Test | | | Incu-bation Temper-ature (°C) |
|---|---|---|---|---|---|---|---|---|---|---|---|---|---|
| | | | | | | | | | | O | M | S | |
| *F. semitectum* | | | | | *From Israel* | | | | | | | | |
| Trifolium | 1 | 0 | 15 | 5 | 0 | 0 | 5 | 20 | 0 | 0 | 1 | 0 | 24° |
| Cotton | 18 | 3 | 19.5 | 13 | 18 | 0 | 18 | 13 | 4.5 | 1 | 16 | 1 | 24° |
| Cucumber | 1 | 0 | 35 | 15 | 0 | 0 | 10 | 5 | 0 | 0 | 1 | 0 | 24° |
| Sorghum | 4 | 0 | 0 | 0 | 20 | 10 | 0 | 20 | 17.5 | 1 | 3 | 0 | 24° |
| Pepper | 1 | 0 | 0 | 0 | 0 | 0 | 0 | 0 | 0 | 1 | 0 | 0 | 24° |
| Tomato | 2 | 10 | 57.5 | 10 | 27.5 | 20 | 0 | 12.5 | 5 | 0 | 1 | 1 | 24° |
| Orange | 1 | 0 | 0 | 0 | 0 | 0 | 0 | 0 | 0 | 1 | 0 | 0 | 24° |
| Peanuts | 2 | 5 | 20 | 17.5 | 25 | 0 | 0 | 12.5 | 2.5 | 0 | 2 | 0 | 24° |
| Avocado | 5 | 0 | 5 | 20 | 35 | 15 | 10 | 25 | 0 | 0 | 3 | 2 | 24° |

| | | | | | | | | | O[b] | M[b] | S[b] | Temp. |
|---|---|---|---|---|---|---|---|---|---|---|---|---|
| Melon | 1 | 25 | 10 | 60 | 0 | 20 | 45 | 20 | 0 | 1 | 0 | 24° |
| Grapefruit | 1 | 10 | 5 | 5 | 15 | 5 | 10 | 10 | 0 | 1 | 0 | 24° |
| Soil | 20 | 6 | 12 | 13 | 21 | 9 | 26.5 | 11 | 10 | 8 | 2 | 24° |
| *F. semitectum* var. *majus* | | | | | | | | | | | | |
| Cotton | 5 | 11.5 | 19 | 6.5 | 11.5 | 19 | 24 | 5 | 2 | 3 | 0 | 24° |
| Sorghum | 1 | 0 | 0 | 0 | 0 | 0 | 15 | 0 | 0 | 1 | 0 | 24° |
| Avocado | 5 | 4 | 14 | 5 | 30 | 5 | 19 | 10 | 1 | 3 | 1 | 24° |
| *From Other Countries* | | | | | | | | | | | | |
| *F. semitectum* | 4 | 2.5 | 2.5 | 0 | — | 5 | 12.5 | 2.5 | 2 | 2 | 0 | 24° |
| *F. semitectum* var. *majus* | 1 | 10 | 10 | 0 | 35 | 20 | 20 | 0 | 0 | 1 | 0 | 30° |
| *F. camptoceras* | 2 | 2.5 | 20 | 22.5 | 20 | 7.5 | 30 | 10 | 0 | 2 | 0 | 24° |
| *F. concolor* | 3 | 2 | 11.5 | 5 | 22 | 11.5 | 22 | 8.5 | 1 | 2 | 0 | 30–35° |

*Source:* Joffe, unpublished results.

[a] Average % of plants killed.

[b] O = nontoxic; M = mildly toxic; S = strongly toxic.

**TABLE 5.12  Relation between Phytotoxicity[a] and Rabbit Skin Test[b]—Lateritium Section**

| Species/ Substrate/ Source | No. of Isolates | Bean | Cucumber | Water- melon | Cotton | Tomato | Onion | Egg- plant | Pepper | Rabbit Skin Test | | | Incu- bation Temper- ature (°C) |
|---|---|---|---|---|---|---|---|---|---|---|---|---|---|
| | | | | | | | | | | O | M | S | |
| *F. lateritium* | | | | | *From Israel* | | | | | | | | |
| Tomato | 3 | 13.5 | 15 | 1.5 | 18.5 | 16.5 | 0 | 20 | 6.5 | 1 | 2 | 0 | 30° |
| Grapefruit | 5 | 14 | 29 | 22 | 29 | 31 | 11 | 30 | 12 | 0 | 5 | 0 | 30° |
| Lemon | 1 | 10 | 0 | 0 | 20 | 30 | 15 | 35 | 20 | 0 | 1 | 0 | 30° |
| Soil | 17 | 11.5 | 14 | 7 | 19 | 30 | 12 | 33.5 | 11.5 | 2 | 15 | 0 | 30° |
| *F. lateritium* | | | | | *From Other Countries* | | | | | | | | |
| Africa | 5 | 2 | 13 | 2 | 6 | 7 | 18 | 4 | 0 | 1 | 4 | 0 | 24° |
| Others | 12 | 7 | 9 | 7 | 13 | 16 | 11 | 14 | 1 | 3 | 7 | 2 | 24° |
| *F. stilboides* | 5 | 8 | 13 | 5 | 11 | 16 | 5 | 9 | 0 | 1 | 4 | 0 | 24° |
| *F. xylarioides* | 4 | 6 | 17.5 | 5 | 12.5 | 6 | 10 | 14 | 2.5 | 1 | 3 | 0 | 30° |

*Source:*  Joffe, unpublished results.

[a] Average % of plants killed.

[b] O = nontoxic; M = mildly toxic; S = strongly toxic.

two from cereals in the United States; one from *Hordeum vulgare* in Germany; and one from *Cassava* sp. in Nigeria. In addition six strains of *F. equiseti* var. *compactum* were studied: one from CBS, Baarn in Holland; one from IMB in Germany; one from AMTCC in the United States; two from peanut root in Tanganyika; and one from *Pinus elliotii* seedling in Australia.

The following eight strains of *F. equiseti* var. *acuminatum* were studied: one from *P. ponderosa* in Canada; one from *Lupinus angustifolius* in Germany; one from Finland (source unknown); one from CMI in Kew, England; one from CBS Baarn in Holland; one from clover rhizosphere in Australia; one from potato tuber in Australia; and one from *Citrus* in Iran (Table 5.14).

## Discolor Section

The *Fusarium* strains of this section grown at 24 and 30°C gave moderately and strongly toxic reactions on rabbit skin and showed moderate and strong phytotoxicity to bean, cucumber, watermelon, tomato, onion, eggplant, and pepper.

**F. graminearum, G. zeae, F. sambucinum, F. sambucinum var. coeruleum, F. sambucinum var. trichothecioides, F. culmorum, F. heterosporum, F. tumidum.** In Israel one strain of *F. graminearum* from pepper and two from maize were examined (Table 5.15).

Sixty-six strains of *F. sambucinum* were studied in Israel and also 40 strains of *F. culmorum* isolated from various sources (Table 5.15).

In Israel three strains of *F. heterosporum* from soil were examined. They were not phytotoxic or toxic (Joffe, 1963c).

Seven strains of *F. heterosporum* from other countries were studied: one from rose twig in Canada; one from white pine in British Columbia; one from raspberry in Scotland; one strain from *Lolium perenne* in Germany; one from *Claviceps purpurea* in Germany; one from corn in Iran; and one from *Pterocarya fraxinifolia* in Iran.

Five strains of *F. graminearum* were studied: one from corn in Canada; one from the FDA in the United States; one from CBS Baarn in Holland; one from Hungary (source unknown); and one from wheat in Australia.

Seven strains of *G. zeae* were studied: three of them from the FDA in the United States; two from *Dianthus caryophyllus* in the United States; and two from *Avena sativa* and *L. perenne* in Germany.

The following seven strains of *F. sambucinum* were examined: one from potato dry rot in Canada; two from carrot storage rot in Canada; one from *Polygonum siebaldia* in England; one from CBS Baarn in Holland; one from hops in Tasmania; and one from poultry feed in Australia.

Four strains of *F. sambucinum* var. *coeruleum* were tested: one from Hungary (source unknown); one from pasture soil in New Zealand; one from Sitka spruce in England; and one from strawberry in Canada.

TABLE 5.13 Relation between Phytotoxicity[a] and Rabbit Skin Test[b]—Liseola Section

| Species/ Substrate Source | No. of Isolates | Bean | Cucumber | Water-melon | Cotton | Tomato | Onion | Egg-plant | Pepper | Rabbit Skin Test O | M | S | Incubation Temperature (°C) |
|---|---|---|---|---|---|---|---|---|---|---|---|---|---|
| *F. moniliforme* | | | | | | | | | | | | | |
| *From Israel* | | | | | | | | | | | | | |
| Maize | 31 | 8.5 | 13 | 13.5 | 14.5 | 5.5 | 16.5 | 10 | 9.5 | 7 | 20 | 4 | 24° |
| Wheat | 5 | 7 | 17 | 11 | 9 | 7 | 23 | 17 | 11 | 1 | 4 | 0 | 24° |
| Onion | 13 | 14.5 | 6.5 | 11 | 13 | 4.5 | 16 | 22.5 | 14.5 | 3 | 10 | 0 | 24° |
| Cotton | 3 | 0 | 15 | 16.5 | 28.5 | 0 | 11.5 | 16.5 | 0 | 0 | 3 | 0 | 24° |
| Banana | 1 | 0 | 0 | 10 | 20 | 20 | 20 | 10 | 0 | 0 | 1 | 0 | 24° |
| Pepper | 2 | 5 | 0 | 0 | 0 | 0 | 15 | 17.5 | 0 | 0 | 2 | 0 | 24° |
| Millet | 2 | 7.5 | 27.5 | 17.5 | 45 | 17.5 | 27.5 | 7.5 | 2.5 | 1 | 2 | 2 | 24° |
| Sorghum | 5 | 7 | 27 | 16 | 25 | 10 | 32 | 15 | 8 | 1 | 2 | 0 | 24° |
| Barley | 3 | 2 | 3 | 17 | 20 | 0 | 15 | 22 | 3 | 1 | 2 | 0 | 24° |
| Grapefruit | 20 | 2 | 8.5 | 8 | 13 | 6 | 7 | 7 | 10 | 8 | 12 | 0 | 24° |
| Mango | 10 | 0 | 16 | 12.5 | 18 | 10.5 | 22 | 13.5 | 14 | 3 | 6 | 1 | 24° |
| Anona | 3 | 0 | 0 | 0 | 3 | 3.3 | 1.7 | 1.3 | 0 | 1 | 2 | 0 | 24° |

| | | | | | | | | | | | | | |
|---|---|---|---|---|---|---|---|---|---|---|---|---|---|
| Castor-oil plant | 1 | 10 | 45 | 35 | 0 | 20 | 24 | 10 | 0 | 0 | 1 | 0 | 24° |
| Orange | 3 | 6.5 | 6.5 | 0 | 0 | 8.5 | 8.5 | 0 | 11.5 | 1 | 2 | 0 | 24° |
| Avocado | 5 | 0 | 5 | 24 | 44 | 1 | 20 | 0 | 16 | 1 | 4 | 0 | 24° |
| Soil | 26 | 3 | 5.5 | 7 | 10 | 6 | 13 | 9 | 10 | 9 | 13 | 4 | 24° |
| *F. moniliforme* var. *subglutinans* | | | | | | | | | | | | | |
| Mango | 12 | 8.5 | 13.5 | 12 | 14.5 | 16.5 | 13.5 | 10.5 | 10 | 3 | 9 | 0 | 24° |
| *F. moniliforme* var. *anthophilum* | | | | | | | | | | | | | |
| Maize | 1 | 0 | 0 | 12 | 0 | 25 | 26 | 0 | 12 | 0 | 0 | 1 | 24° |
| *From Other Countries* | | | | | | | | | | | | | |
| *F. moniliforme* | | | | | | | | | | | | | |
| Tropics | 3 | 0 | 3 | 0 | 6.5 | 15 | 21.5 | 13.5 | 6.5 | 0 | 3 | 0 | 24° |
| Others | 7 | 0 | 5.5 | 5.5 | 5.5 | 10.5 | 15.5 | 15 | 18.5 | 3 | 3 | 1 | 24° |
| *F. moniliforme* var. *subglutinans* | 3 | 6.5 | 8.5 | 0 | 10 | 10 | 13.5 | 5 | 16.5 | 1 | 1 | 1 | 18° |
| *F. moniliforme* var. *anthophilum* | 2 | 0 | 7.5 | 2.5 | 0 | 5 | 25 | 15 | 15 | 0 | 2 | 0 | 18° |

*Source:* Joffe, unpublished results.

[a] Average % of plants killed.

[b] O = nontoxic; M = mildly toxic; S = strongly toxic.

**TABLE 5.14  Relation between Phytotoxicity[a] and Rabbit Skin Test[b]—Gibbosum Section**

| Species/Substrate | No. of Isolates | Bean | Cucumber | Water-melon | Cotton | Tomato | Onion | Egg-plant | Pepper | Rabbit Skin Test | | | Incu-bation Temper-ature (°C) |
|---|---|---|---|---|---|---|---|---|---|---|---|---|---|
| | | | | | | | | | | O | M | S | |
| *F. equiseti* | | | | | | | | | | | | | |
| | | | | | | *From Israel* | | | | | | | |
| Onion | 14 | 5 | 6 | 3 | 11 | 14 | 29.5 | 20 | 11 | 0 | 11 | 3 | 24° |
| Carrot | 1 | 10 | 20 | 20 | 15 | 20 | 81 | 45 | 20 | 0 | 0 | 1 | 24° |
| Eggplant | 1 | 15 | 15 | 0 | 0 | 0 | 0 | 30 | 10 | 0 | 1 | 0 | 24° |
| Tomato | 3 | 3.5 | 33 | 10 | 0 | 0 | 15 | 25 | 3.5 | 1 | 2 | 0 | 24° |
| Potato | 5 | 5 | 29 | 4 | 18 | 17 | 29 | 17 | 12 | 1 | 4 | 0 | 24° |
| Pepper | 4 | 0 | 15 | 15 | 24 | 10 | 6.5 | 0 | 44 | 0 | 3 | 1 | 24° |
| Wheat | 5 | 1 | 13 | 5 | 13 | 8 | 18 | 20 | 5 | 1 | 4 | 0 | 24° |
| Maize | 10 | 19 | 36.5 | 11.5 | 14.5 | 20 | 30 | 6 | 5.5 | 1 | 7 | 2 | 24° |
| Barley | 3 | 8.5 | 23 | 5 | 53 | 3 | 13 | 18 | 5 | 0 | 3 | 0 | 24° |
| Lemon | 3 | 3.5 | 5 | 0 | 3.5 | 28.5 | 6.5 | 25 | 11.5 | 1 | 2 | 0 | 24° |
| Clementine | 3 | 0 | 31.5 | 11.5 | 15 | 0 | 0 | 30 | 0 | 0 | 3 | 0 | 24° |
| Banana | 1 | 0 | 30 | 60 | 0 | 5 | 75 | 0 | 0 | 0 | 1 | 0 | 24° |
| Watermelon | 10 | 4.5 | 23 | 43 | 12 | 9.5 | 39.5 | 25.5 | 6.5 | 0 | 5 | 5 | 24° |
| Melon | 6 | 7.5 | 19 | 11.5 | 11 | 8.5 | 16 | 7.5 | 1.5 | 1 | 2 | 3 | 24° |
| Cucumber | 31 | 10 | 32 | 29 | 15.5 | 14.5 | 37 | 20 | 3 | 0 | 21 | 10 | 24° |
| Strawberry | 1 | 0 | 40 | 0 | 0 | 0 | 0 | 0 | 5 | 0 | 1 | 0 | 24° |
| Peanuts | 7 | 5.5 | 10.5 | 15 | 12 | 15.5 | 7 | 36.5 | 5.5 | 0 | 4 | 3 | 24° |

| | | | | | | | | | | | | | |
|---|---|---|---|---|---|---|---|---|---|---|---|---|---|
| Cotton | 5 | 2 | 7 | 4 | 5 | 0 | 7 | 4 | 0 | 0 | 5 | 0 | 24° |
| Carnation | 7 | 7 | 33.3 | 6.5 | 5.5 | 32.5 | 8 | 27 | 3 | 0 | 6 | 1 | 24° |
| Sorghum | 1 | 0 | 10 | 0 | 20 | 15 | 0 | 25 | 0 | 0 | 1 | 0 | 24° |
| Millet | 3 | 0 | 5 | 0 | 15 | 8 | 8 | 21 | 0 | 1 | 2 | 0 | 24° |
| Avocado | 5 | 0 | 13 | 13 | 15 | 15 | 4 | 23 | 8 | 2 | 3 | 0 | 24° |
| Sesame | 1 | 0 | 0 | 20 | 20 | 15 | 0 | 0 | 25 | 0 | 1 | 0 | 24° |
| Soil | 26 | 4.5 | 9 | 1.5 | 1 | 8.5 | 7 | 22 | 6 | 8 | 17 | 1 | 24° |
| *F. equiseti* var. *compactum* Cotton | 1 | 0 | 10 | 10 | 0 | 20 | 25 | 15 | 5 | 0 | 1 | 0 | 24° |
| *F. equiseti* var. *caudatum* Cotton | 2 | 0 | 5 | 2.5 | 15 | 20 | 5 | 10 | 2.5 | 1 | 1 | 0 | 24° |
| *F. equiseti* var. *acuminatum* Cotton | 4 | 7.5 | 21.5 | 26.5 | 30 | 31.5 | 41.5 | 25 | 25 | 0 | 3 | 1 | 24° |
| *From Other Countries* | | | | | | | | | | | | | |
| *F. equiseti* | 6 | 0 | 21 | 18.5 | 14 | 8.5 | 27.5 | 15 | 8.5 | 1 | 4 | 1 | 24° |
| *F. equiseti* | 6 | 0 | 9 | 17.5 | 18.5 | 29 | 31 | 21 | 11 | 1 | 2 | 3 | 24° |
| *F. equiseti* var. *compactum* | 8 | 2.5 | 11.5 | 13 | 22.5 | 30 | 37.5 | 25.5 | 12 | 3 | 3 | 2 | 24° |
| *F. equiseti* var. *acuminatum* | | | | | | | | | | | | | |

*Source:* Joffe, unpublished results.

[a] Average % of plants killed.

[b] O = nontoxic; M = mildly toxic; S = strongly toxic.

**TABLE 5.15** Relation between Phytotoxicity[a] and Rabbit Skin Test[b]—Discolor Section

| Species/Substrate | No. of Isolates | Bean | Cucumber | Watermelon | Cotton | Tomato | Onion | Eggplant | Pepper | Rabbit Skin Test | | | Incubation Temperature (°C) |
|---|---|---|---|---|---|---|---|---|---|---|---|---|---|
| | | | | | | | | | | O | M | S | |
| *F. graminearum* | | | | | | | | | | | | | |
| Pepper | 1 | 10 | 0 | 0 | 15 | 20 | 30 | 20 | 20 | 0 | 1 | 0 | 18° 24° |
| Maize | 2 | 8 | 0 | 0 | 21 | 18 | 34 | 22 | 24 | 0 | 2 | 0 | 24° |
| *F. sambucinum* | | | | | | | | | | | | | |
| | | | | *From Israel* | | | | | | | | | |
| Beet | 2 | 7.5 | 0 | 27.5 | 0 | 20 | 7.5 | 17.5 | 5 | 0 | 2 | 0 | 24° 30° |
| Tomato | 2 | 0 | 0 | 10 | 0 | 0 | 0 | 5 | 0 | 1 | 1 | 0 | 24° 30° |
| Pink | 5 | 6 | 3 | 18 | 1 | 11 | 7 | 15 | 2 | 1 | 4 | 0 | 24° 30° |
| Banana | 2 | 10 | 22.5 | 15 | 0 | 5 | 7.5 | 15 | 0 | 0 | 2 | 0 | 24° 30° |
| Carnation | 1 | 20 | 15 | 20 | 20 | 0 | 30 | 30 | 10 | 0 | 1 | 0 | 24° 30° |
| Castor-oil plant | 1 | 90 | 90 | 90 | 90 | 90 | 90 | 90 | 90 | 0 | 0 | 1 | 24° 30° |
| Onion | 3 | 10 | 15 | 8.5 | 21.5 | 10 | 35 | 18.5 | 13.5 | 0 | 3 | 0 | 24° 30° |
| Gladiolus | 3 | 16.5 | 50 | 36.5 | 20 | 20 | 50 | 23.5 | 11.5 | 0 | 0 | 3 | 24° 30° |
| Grapefruit | 2 | 0 | 5 | 2.5 | 0 | 0 | 7.5 | 7.5 | 0 | 2 | 0 | 0 | 24° 30° |
| Orange | 2 | 5 | 12.5 | 7.5 | 17.5 | 2.5 | 2.5 | 10 | 0 | 1 | 1 | 0 | 24° 30° |
| Peanuts | 1 | 0 | 45 | 0 | 15 | 15 | 0 | 20 | 0 | 0 | 1 | 0 | 24° 30° |
| Cotton | 4 | 5 | 17.5 | 2.5 | 20 | 17.5 | 0 | 14 | 12.5 | 0 | 4 | 0 | 24° 30° |
| Strawberry | 4 | 0 | 4 | 1.5 | 2.5 | 0 | 0 | 0 | 2.5 | 4 | 0 | 0 | 24° 30° |
| Potato | 11 | 11.5 | 14 | 0 | 0 | 17.5 | 0 | 13 | 11 | 3 | 5 | 3 | 24° 30° |
| Soil | 23 | 8 | 15 | 22 | 11 | 10 | 27 | 17 | 14 | 8 | 11 | 4 | 24° 30° |

*F. culmorum*

| | | | | | | | | | | | | | |
|---|---|---|---|---|---|---|---|---|---|---|---|---|---|
| Cotton | 4 | 0 | 17.5 | 1.5 | 10 | 2.5 | 2.5 | 9 | 4 | 2 | 2 | 0 | 24° |
| Banana | 1 | 20 | 65 | 0 | 15 | 10 | 50 | 0 | 0 | 0 | 1 | 0 | 24° |
| Cucumber | 3 | 5 | 38.5 | 8.5 | 25 | 13.5 | 28.5 | 6.5 | 6.5 | 0 | 2 | 1 | 24° |
| Tomato | 1 | 0 | 0 | 0 | 0 | 0 | 0 | 0 | 0 | 1 | 0 | 0 | 24° |
| Grapefruit | 1 | 0 | 0 | 0 | 0 | 0 | 15 | 15 | 5 | 1 | 0 | 0 | 24° |
| Peanuts | 1 | 15 | 45 | 35 | 70 | 20 | 65 | 10 | 10 | 0 | 0 | 1 | 24° |
| Melon | 1 | 0 | 0 | 0 | 0 | 0 | 15 | 10 | 10 | 0 | 1 | 0 | 24° |
| Gladiolus | 1 | 10 | 20 | 10 | 0 | 0 | 20 | 10 | 15 | 1 | 1 | 0 | 24° |
| Potato | 1 | 0 | 0 | 0 | 0 | 0 | 0 | 0 | 0 | 0 | 0 | 0 | 24° |
| Carnation | 5 | 16 | 18 | 5 | 0 | 0 | 24 | 16 | 13 | 0 | 5 | 0 | 24° |
| Maize | 2 | 5 | 12.5 | 10 | 0 | 0 | 15 | 10 | 5 | 1 | 1 | 0 | 24° |
| Soil | 19 | 6.5 | 15.5 | 5 | 17 | 7 | 29 | 15 | 9 | 8 | 9 | 2 | 24° |

*From Other Countries*

| | | | | | | | | | | | | | |
|---|---|---|---|---|---|---|---|---|---|---|---|---|---|
| *F. heterosporum* | 7 | 7 | 8.5 | 4.5 | 6.5 | 18.5 | 19.5 | 15 | 8 | 3 | 3 | 1 | 30° |
| *F. graminearum* | 5 | 9 | 4 | 1 | 10 | 22 | 22 | 14 | 10 | 1 | 3 | 1 | 30° |
| *G. zeae* | 7 | 10 | 1.5 | 6.5 | 0 | 18 | 17 | 21.5 | 5.5 | 1 | 5 | 1 | 24° |
| *F. sambucinum* | 7 | 14.5 | 0 | 13 | 19.5 | 22 | 27 | 20 | 13.5 | 0 | 5 | 2 | 18–30° |
| *F. sambucinum* var. *coeruleum* | 4 | 7.5 | 4 | 1.5 | 2.5 | 7.5 | 10.5 | 5 | 7.5 | 3 | 1 | 0 | 18–35° |
| *F. sambucinum* var. *trichothecioides* | 1 | 20 | 10 | 20 | 30 | 50 | 80 | 55 | 20 | 0 | 0 | 1 | 18° |
| *F. culmorum* | 4 | 2.5 | 16.5 | 12.5 | 11.5 | 9 | 36.5 | 25 | 10 | 0 | 2 | 2 | 24° |
| *F. tumidum* | 3 | 0 | 0 | 0 | 0 | 1.5 | 6.5 | 3 | 0 | 2 | 1 | 0 | 30° |

*Source:* Joffe, unpublished results.

[a] Average % of plants killed.

[b] O = nontoxic; M = mildly toxic; S = strongly toxic.

**209**

Only one strain of *F. sambucinum* var. *trichothecioides* (from *Solanum tuberosum* in Iran) was tested.

Four strains of *F. culmorum* were examined: one from soil in the United States; one from AMTCC in the United States; one from barley seed in England; and one from CBS, Baarn in Holland.

Two strains of *F. tumidum* (one from CBS Baarn in Holland and one from CMI in Kew, England) were tested (Table 5.15).

### *Elegans* Section

Strains of *Fusarium oxysporum* and *F. oxysporum* var. *redolens* were moderately toxic or toxic to rabbit skin and phytotoxic to almost all plants tested (except maize and wheat) after growing at 24°C.

**F. oxysporum and F. oxysporum var. redolens.**   From Israel, a total of 236 strains and 6 strains of *F. oxysporum* var. *redolens* from various plants and soil were studied (Table 5.16).

From other countries, the following 13 strains of *F. oxysporum* were studied: nine strains from unknown sources in France; one strain from the FDA in the United States; one strain from carnation in England; and two strains from *Gladiolus* in Australia.

Only one strain of *F. oxysporum* var. *redolens* (from *Convallaria majalis* in Germany) was tested (Table 5.16).

### *Martiella* Section

The *Fusarium* strains in this section, when grown at 18 and 24°C, were most phytotoxic to tomato, watermelon, onion, and cucumber and moderately toxic or toxic to rabbit skin.

**F. solani, F. solani var. coeruleum, F. solani var. ventricosum, and F. javanicum.**   In Israel, *F. solani* was studied from various host plants and soil. Two hundred and seven strains of *F. solani* were studied (Table 5.17).

Twenty-seven strains of *F. javanicum* were studied from melon, vegetable marrow, and soil.

Nine strains of *F. solani* from other countries were studied: one from the FDA in the United States; one from *Gossypium purpurascens* in Canada; five from CBS, Baarn in Holland; one from *Pisum sativum* in Germany; and one from *Theobroma cacao* in Costa Rica.

Two strains of *F. solani* var. *coeruleum* were tested: one from CBS Baarn in Holland and one from potato tuber in Canada.

Four strains of *F. solani* var. *ventricosum* were studied: one from soil in Germany; one from CBS Baarn in Holland; one from peanuts in Mozambique; and one from soil in the Sudan.

Three strains of *F. javanicum* from CBS Baarn in Holland were tested.

# THE PHYTOTOXIC ACTION OF THE SPOROTRICHIELLA SECTION AND SOME OTHER *FUSARIUM* SPECIES

## The Phytotoxic Effect of Fusarium Culture Filtrates on Seed Germination

The phytotoxic effect on seed germination of media on which fusaria of the Sporotrichiella section had been grown was studied in experiments on pea, bean, wheat, and barley seed. Cultures of six isolates each of *F. poae* (24, 60/9, 344, 396, 792, and 958) and *F. sporotrichioides* (60/10, 347, 351, 738, 921, and 1182), two isolates of *F. sporotrichioides* var. *tricinctum* (M304 and 1227), and two culture filtrates of *F. sporotrichioides* var. *chlamydosporum* (1174 and 4337), were inoculated for 5–7 days on media I and II (*see* chapter 6) in Erlenmeyer flasks at 25°C, then for 40 days at 10°C, followed by 75 days at 4°C. Cultures were then killed at 100°C for 15 min. Equal amounts of filtrates were applied to paper filters which were placed in large Petri dishes and the seeds placed on the filters for 4–5 days (5–10 seeds for each of 10 Petri dishes for every plant/*Fusarium* filtrate/temperature for both media). Control seeds were placed on filter paper wetted only with sterilized tap water or with media without fusaria.

Results showed that all filtrates substantially reduced seed germination in all four plants, reduction increasing as the temperature, at which cultures were grown, decreased (Table 5.18).

## The Effect of Fusarium Culture Filtrates and their Extracts on Leaf Blades of Young Plants

The following *Fusarium* culture filtrates and their crude ethanol extracts were studied on young pea, bean, and tomato plants:

| | |
|---|---|
| *F. oxysporum* | 10 isolates |
| *F. solani* | 10 isolates |
| *F. moniliforme* | 6 isolates |
| *F. equiseti* | 9 isolates |
| *F. culmorum* | 4 isolates |
| *F. sambucinum* | 4 isolates |
| *F. lateritium* | 2 isolates |
| *F. semitectum* | 3 isolates |
| *F. sporotrichioides* | 10 isolates |
| *F. poae* | 8 isolates |
| *F. sporotrichioides* var. *tricinctum* | 3 isolates |
| *F. sporotrichioides* var. *chlamydosporum* | 3 isolates |

**TABLE 5.16  Relation between Phytotoxicity[a] and Rabbit Skin Test[b]—Elegans Section**

| Species/ Substrate | No. of Isolates | Bean | Cucumber | Water-melon | Cotton | Tomato | Onion | Egg-plant | Pepper | Rabbit Skin Test | | | Incu-bation Temper-ature (°C) |
|---|---|---|---|---|---|---|---|---|---|---|---|---|---|
| | | | | | | | | | | O | M | S | |
| *F. oxysporum* | | | | | | | | | | | | | |
| *From Israel* | | | | | | | | | | | | | |
| Onion | 44 | 12 | 31 | 27 | 15 | 26.5 | 47 | 21 | 12 | 1 | 22 | 21 | 24° |
| Carrot | 2 | 0 | 0 | 5 | 7.5 | 7.5 | 0 | 7.5 | 0 | 2 | 0 | 0 | 24° |
| Eggplant | 4 | 11.5 | 24 | 45 | 20 | 10 | 35 | 66 | 15 | 0 | 2 | 2 | 24° |
| Tomato | 38 | 11 | 42 | 27.5 | 21.5 | 56 | 45.5 | 28 | 10 | 1 | 20 | 17 | 24° |
| Potato | 5 | 9 | 33 | 17 | 32 | 30 | 17 | 7 | 17 | 0 | 5 | 0 | 24° |
| Pepper | 4 | 9 | 11.5 | 4 | 5 | 35 | 20 | 7.5 | 29 | 1 | 3 | 0 | 24° |
| Garlic | 2 | 12.5 | 65 | 42.5 | 17.5 | 12.5 | 57.5 | 45 | 12.5 | 0 | 1 | 1 | 24° |
| Broad bean | 3 | 12 | 28 | 12 | 20 | 30 | 12 | 3 | 2 | 1 | 2 | 0 | 24° |
| Maize | 1 | 20 | 0 | 10 | 5 | 25 | 35 | 10 | 5 | 0 | 1 | 0 | 24° |
| Barley | 3 | 0 | 27 | 10 | 20 | 5 | 42 | 15 | 35 | 0 | 3 | 0 | 24° |
| Leek | 1 | 0 | 0 | 30 | 30 | 70 | 30 | 10 | 20 | 0 | 1 | 0 | 24° |
| Grapefruit | 11 | 5.5 | 23 | 12.5 | 32 | 28 | 21.5 | 30 | 18 | 1 | 8 | 2 | 24° |
| Orange | 6 | 18.5 | 23 | 29 | 36.5 | 18.5 | 41.5 | 19 | 15 | 0 | 4 | 2 | 24° |
| Clementine | 1 | 10 | 70 | 45 | 0 | 5 | 20 | 40 | 10 | 0 | 1 | 0 | 24° |

| | | | | | | | | | | O[b] | M | S | |
|---|---|---|---|---|---|---|---|---|---|---|---|---|---|
| Avocado | 5 | 8 | 17 | 17 | 16 | 13 | 18 | 8 | 14 | 1 | 4 | 0 | 24° |
| Watermelon | 7 | 7 | 35 | 89.5 | 12 | 34.5 | 43 | 28 | 28.5 | 0 | 3 | 4 | 24° |
| Melon | 11 | 20 | 33 | 30 | 16.5 | 25 | 37 | 40 | 8 | 1 | 7 | 3 | 24° |
| Cucumber | 25 | 20 | 42 | 52 | 12 | 28 | 48 | 34 | 17 | 2 | 11 | 12 | 24° |
| Peanuts | 10 | 11 | 15 | 15.5 | 1 | 26 | 12 | 8.5 | 3.5 | 3 | 6 | 1 | 24° |
| Cotton | 9 | 10 | 48 | 48 | 14 | 11.5 | 28 | 9.5 | 10.5 | 1 | 7 | 1 | 24° |
| Sesame | 1 | 0 | 0 | 0 | 0 | 0 | 0 | 0 | 0 | 1 | 0 | 0 | 24° |
| Safflower | 6 | 25 | 12.5 | 17 | 28 | 53 | 58 | 37 | 15 | 1 | 5 | 0 | 24° |
| Medicago | 3 | 16.5 | 16.5 | 5 | 0 | 33.5 | 25 | 6.5 | 1.5 | 0 | 3 | 0 | 24° |
| Dianthus | 10 | 16.5 | 32 | 36.5 | 26 | 25 | 35 | 29.5 | 5.5 | 1 | 4 | 5 | 24° |
| Amaranthus | 1 | 5 | 15 | 50 | 5 | 5 | 65 | 15 | 5 | 0 | 1 | 0 | 24° |
| Malva | 1 | 10 | 65 | 15 | 5 | 15 | 70 | 20 | 0 | 0 | 0 | 1 | 24° |
| Soil | 22 | 4 | 13.5 | 29 | 4.5 | 11.5 | 8.5 | 2.5 | 2 | 9 | 12 | 1 | 24° |
| *F. oxysporum* var. *redolens* Carnation | 6 | 19 | 36.5 | 37.5 | 27.5 | 36.5 | 40 | 13.5 | 20 | 0 | 4 | 2 | 24° |
| *From Other Countries* | | | | | | | | | | | | | |
| *F. oxysporum* | 13 | 3 | 17 | 17.5 | 14.5 | 17 | 9.5 | 28 | 14.5 | 0 | 11 | 2 | 24° |
| *F. oxysporum* var. *redolens* | 1 | 0 | 20 | 10 | 0 | 60 | 25 | 25 | 0 | 0 | 0 | 1 | 24° |

*Source:* Joffe, unpublished results.

[a] Average % of plants killed.

[b] O = nontoxic; M = mildly toxic; S = strongly toxic.

**TABLE 5.17  Relation between Phytotoxicity[a] and Rabbit Skin Test[b]—Martiella Section**

| Species/Substrate | No. of Isolates | Bean | Cucumber | Watermelon | Cotton | Tomato | Onion | Eggplant | Pepper | Rabbit Skin Test | | | Incubation Temperature (°C) |
|---|---|---|---|---|---|---|---|---|---|---|---|---|---|
| | | | | | | | | | | O | M | S | |
| *F. solani* | | | | | | | | | | | | | |
| *From Israel* | | | | | | | | | | | | | |
| Onion | 15 | 4.5 | 14 | 13 | 18.5 | 31.5 | 22.5 | 13.5 | 10 | 4 | 9 | 2 | 24° |
| Carrot | 3 | 0 | 0 | 6.5 | 15 | 23.5 | 5 | 30 | 10 | 2 | 0 | 1 | 24° |
| Tomato | 7 | 7 | 8.5 | 12 | 27 | 28 | 24.5 | 15.5 | 10 | 3 | 3 | 1 | 24° |
| Potato | 5 | 17 | 3 | 9 | 24 | 22 | 4 | 12 | 24 | 0 | 5 | 0 | 24° |
| Pepper | 7 | 5.5 | 13 | 17 | 23 | 40 | 32 | 23 | 15.5 | 2 | 4 | 1 | 24° |
| Lupinus | 2 | 22.5 | 2.5 | 5 | 20 | 27.5 | 22.5 | 25 | 13 | 0 | 2 | 0 | 24° |
| Bean | 4 | 14 | 19 | 11.5 | 5 | 27.5 | 19 | 6.5 | 10 | 2 | 1 | 1 | 24° |
| Broad bean | 2 | 17.5 | 17.5 | 15 | 12.5 | 20 | 32.5 | 12.5 | 2.5 | 0 | 2 | 0 | 24° |
| Maize | 6 | 3.5 | 17.5 | 5 | 17.5 | 9 | 12.5 | 6 | 11.5 | 2 | 3 | 1 | 24° |
| Wheat | 6 | 5 | 6 | 15 | 4 | 20 | 14 | 7.5 | 20 | 2 | 4 | 0 | 24° |
| Grapefruit | 10 | 1.5 | 4.5 | 5 | 6.5 | 10 | 13.5 | 5.5 | 6 | 6 | 2 | 2 | 24° |
| Orange | 11 | 1.5 | 5 | 8 | 10 | 13 | 5.5 | 2.5 | 2.5 | 6 | 4 | 1 | 24° |
| Lemon | 2 | 5 | 10 | 15 | 0 | 17.5 | 20 | 0 | 7.5 | 1 | 1 | 0 | 24° |
| Avocado | 3 | 0 | 0 | 0 | 0 | 0 | 0 | 0 | 0 | 3 | 0 | 0 | 24° |
| Anona | 1 | 0 | 20 | 30 | 0 | 20 | 20 | 30 | 5 | 0 | 1 | 0 | 24° |
| Banana | 2 | 12.5 | 17.5 | 2.5 | 0 | 45 | 72.5 | 15 | 2.5 | 0 | 1 | 1 | 24° |

|  |  |  |  |  |  |  |  |  |  |  |  |  | Temp. |
|---|---|---|---|---|---|---|---|---|---|---|---|---|---|
| Watermelon | 8 | 2 | 19.5 | 9.5 | 19 | 29.5 | 27 | 12.5 | 12 | 1 | 5 | 2 | 24° |
| Melon | 17 | 3 | 9 | 13 | 22.5 | 21 | 23 | 7 | 4.5 | 6 | 10 | 1 | 24° |
| Cucumber | 29 | 4.5 | 13 | 14 | 9.5 | 17.5 | 17.5 | 18.5 | 3.5 | 10 | 13 | 6 | 24° |
| Peanut | 15 | 8.5 | 9 | 5.5 | 16.5 | 21.5 | 17.5 | 16.5 | 3 | 3 | 9 | 3 | 24° |
| Cotton | 2 | 0 | 0 | 0 | 0 | 0 | 0 | 0 | 0 | 2 | 0 | 0 | 24° |
| Sesame | 4 | 0 | 4 | 9 | 10 | 5 | 0 | 12.5 | 43 | 2 | 2 | 0 | 24° |
| Trifolium | 1 | 0 | 0 | 0 | 0 | 0 | 0 | 0 | 0 | 1 | 0 | 0 | 24° |
| Gladiolus | 1 | 0 | 0 | 0 | 20 | 50 | 0 | 40 | 0 | 0 | 1 | 0 | 24° |
| Carnation | 1 | 10 | 0 | 15 | 0 | 20 | 10 | 0 | 75 | 0 | 1 | 0 | 24° |
| Chrysanthemum | 1 | 0 | 0 | 0 | 0 | 0 | 0 | 5 | 0 | 1 | 0 | 0 | 24° |
| Malva | 1 | 0 | 0 | 0 | 0 | 20 | 40 | 20 | 0 | 0 | 1 | 0 | 24° |
| Soil | 41 | 2 | 9 | 2 | 1 | 10 | 11 | 9.5 | 27 | 7 | 29 | 5 | 24° |
| *F. javanicum* |  |  |  |  |  |  |  |  |  |  |  |  |  |
| Melon | 5 | 4 | 30 | 26 | 14 | 20 | 23 | 11 | 7 | 0 | 4 | 1 | 24° |
| Vegetable marrow | 6 | 9 | 39 | 43.5 | 11.5 | 19 | 10 | 6 | 3.5 | 2 | 3 | 1 | 24° |
| Soil | 16 | 6.5 | 39 | 37 | 12 | 21 | 26 | 15 | 8 | 5 | 9 | 2 | 24° |
| *From Other Countries* |  |  |  |  |  |  |  |  |  |  |  |  |  |
| *F. solani* | 9 | 4.5 | 9.5 | 9.5 | 16 | 12 | 28 | 13 | 4 | 2 | 6 | 1 | 18–24° |
| *F. solani* var. *coeruleum* | 2 | 0 | 0 | 0 | 12.5 | 17.5 | 2.5 | 10 | 10 | 1 | 1 | 0 | 18° |
| *F. solani* var. *ventricosum* | 4 | 0 | 12.5 | 10 | 11.5 | 22.5 | 4 | 25 | 9 | 1 | 3 | 0 | 24–30° |
| *F. javanicum* | 3 | 0 | 10 | 5 | 10 | 25 | 20 | 17.5 | 6.5 | 1 | 1 | 1 | 24° |

*Source:* Joffe, unpublished results.

[a] Average % of plants killed.

[b] O = nontoxic; M = mildly toxic; S = strongly toxic.

**TABLE 5.18**   Effect of Sporotrichiella Section Fusaria Isolates on Plant Germination

| Species | Conditions of Incubation and Range of Germination (%)[a] | | |
| --- | --- | --- | --- |
| | 5 Days at 25°C | 40 Days at 10°C | 75 Days at 4°C |
| *F. sporotrichioides* | 28–49% | 11–21% | Total suppression |
| *F. poae* | 32–47% | 17–24% | 3–7% |
| *F. sporotrichioides* var. *tricinctum* | 51–57% | 40–51% | 21–33% |
| *F. sporotrichioides* var. *chlamydosporum* | 53–59% | 42–53% | 23–30% |

*Source:*   Adapted from Joffe (1973b).
[a]Germination in the control groups ranged from 89–91%.

All fusaria were cultivated on medium I. *Fusarium* culture filtrates and their crude extracts were applied to leaf blades or leaf axils of young plants. At the site of application various necrotized spots appeared after 6–72 hr. The action of crude extracts on plants was stronger than that of filtrates (Table 5.19).

### Effect of Media on Phytotoxicity of Fusarium Species

Four *Fusarium* species were incubated at 24°C for 10 days in four different substrates (media): I, II, III, and IV (*see* chapter 6). Application to pea, bean, and tomato branches caused wilting within 15.0–29.5 hr as shown in Table 5.20.

### Effect of Fusarium and Pure Filtrates on Plant Branches

Seven *Fusarium* species and two varieties were tested by the author for their phytotoxic effect on a variety of plants. The *Fusarium* cultures were grown on liquid Potato-Dextrose medium with pH 5.6 in Erlenmeyer flasks containing 70 ml of substrate. After incubation at 12°C for 21 days the cultures were killed by 1 atm for 20 min. After cooling the cotton corks were removed and the Erlenmeyer flasks covered with aluminum foil. Branches of test plants were introduced into each flask, and the rate at which they wilted was recorded.

Each *Fusarium* strain was tested in five separate flasks on branches of the same plant. Twenty-two experiments were carried out using 220

**TABLE 5.19**  Necrotic Effect of *Fusarium* Isolates on Young Pea, Bean, and Tomato Plants

| *Fusarium* | Average Time Required for Effect to Appear (Hr) | Color of Spots on Leaf Blade or Axil |
|---|---|---|
| *F. oxysporum* | | |
| Filtrate | 32 | White |
| Crude extract | 14 | Yellow |
| *F. solani* | | |
| Filtrate | 42 | Light cream |
| Crude extract | 17 | Light brown |
| *F. moniliforme* | | |
| Filtrate | 62 | Rose |
| Crude extract | 19 | Light violet to red |
| *F. equiseti* | | |
| Filtrate | 70 | Light yellow |
| Crude extract | 21 | Deep yellow |
| *F. culmorum* | | |
| Filtrate | 46 | Light brown |
| Crude extract | 18 | Brown |
| *F. sambucinum* | | |
| Filtrate | 31 | Yellow-brown |
| Crude extract | 21 | Brown |
| *F. lateritium* | | |
| Filtrate | 48 | Cream-light green |
| Crude extract | 15 | Light brown |
| *F. semitectum* | | |
| Filtrate | 42 | Yellow to light brown |
| Crude extract | 17 | Brown to dark brown |
| *F. sporotrichioides* | | |
| Filtrate | 18 | Brown |
| Crude extract | 6 | Black |
| *F. poae* | | |
| Filtrate | 24 | Red to light brown |
| Crude extract | 9 | Brown |
| *F. sporotrichioides* var. *tricinctum* | | |
| Filtrate | 31 | Rose red |
| Crude extract | 16 | Purple red |
| *F. sporotrichioides* var. *chlamydosporum* | | |
| Filtrate | 38 | Light carmine |
| Crude extract | 19 | Carmine red to brown |

*Source:*  Joffe, unpublished results.

**TABLE 5.20 Effect of Growth Conditions on Phytotoxicity of Four *Fusarium* Species (Wilting Time in Hr)**

| Substrate | *F. moniliforme* No. 34 | | | *F. equiseti* No. 91/1 | | | *F. oxysporum* No. 106 | | | *F. solani* No. 54 | | |
|---|---|---|---|---|---|---|---|---|---|---|---|---|
| | Pea | Bean | Tomato | Pea | Bean | Tomato | Pea | Bean | Tomato | Pea | Bean | Tomato |
| I | 23 | 29 | 21.5 | 28 | 24.5 | 21 | 18.5 | 22 | 16 | 19 | 24 | 19.5 |
| II | 27 | 22.5 | 26 | 23 | 29 | 26.5 | 17 | 24 | 17.5 | 24 | 27.5 | 15.5 |
| III | 22 | 24 | 29.5 | 25 | 29.5 | 28 | 20 | 21 | 18.5 | 26.5 | 28.5 | 18.5 |
| IV | 21 | 20 | 24 | 21 | 27.5 | 25 | 18 | 22 | 15 | 23.5 | 26 | 19 |

*Source:* Joffe, unpublished results.

TABLE 5.21   Phytotoxic Action of *Fusarium* Species on Plant Branches

| Plant | *Fusarium* Species | Wilted Within (Hr) |
|-------|--------------------|--------------------|
| *Tagetes patula* | *F. poae* No. 958 | 12 |
| *Zinnia elegans* | *F. poae* No. 958 | 9 |
| *Melilotus albus* | *F. poae* No. 958 | 14 |
| *Xanthium strumarium* | *F. poae* No. 958 | 4 |
| *Allium cepa* | *F. poae* No. 958 | 10 |
| *Phaseolus vulgaris* | *F. poae* No. 958 | 6 |
| *Solanum lycopersicum* | *F. poae* No. 958 | 18 |
| *Eschscholtzia californica* | *F. sporotrichioides* No. 921 | 6 |
| *Melilotus albus* | *F. sporotrichioides* No. 921 | 13 |
| *Vicia faba* | *F. sporotrichioides* No. 921 | 8 |
| *Solanum lycopersicum* | *F. sporotrichioides* No. 921 | 16 |
| *Pisum sativum* | *F. sporotrichioides* No. 921 | 12 |
| *Sorghum* | *F. sporotrichioides* No. 921 | 14 |
| *Vicia faba* | *F. sporotrichioides* No. 347 | 8 |
| *Solanum lycopersicum* | *F. sporotrichioides* No. 738 | 18 |
| *Zea mays* | *F. sporotrichioides* NRRL5908 | 12 |
| *Hordeum sativum* | *F. sporotrichioides* var. *tricinctum* No. 1227 | 14 |
| *Phaseolus vulgaris* | *F. sporotrichioides* var. *chlamydosporum* No. 1174 | 15 |
| *Pisum sativum* | *F. solani* No. 251 | 26 |
| *Zea mays* | *F. moniliforme* No. 63 | 30 |
| *Hordeum sativum* | *F. culmorum* No. 543/3 | 17 |
| *Hordeum sativum* | *G. zeae* No. 280 | 15 |

*Source:*   Joffe, unpublished results.

flasks: 110 for experimental and 110 for control purposes. All branches treated with fusaria wilted completely within 4–26 hr, depending on the species of *Fusarium* tested (Table 5.21 and Figs. 5.3 to 5.8).

Experiments on phytotoxicity were carried out in my greenhouse using *Fusarium* species isolated from various sources in subtropical and temperate regions.

We have prepared crude extracts from strains of *F. sporotrichioides, F. sporotrichioides* var. *tricinctum,* and *F. poae* (involved in ATA disease

**Figure 5.3** Plant branches treated with *F. poae* No. 958 and control plants. (*a*) *Tagetes patula* wilted after 12 hr; (*b*) *Zinnia elegans* wilted after 9 hr; (*c*) *Melilotus albus* wilted after 14 hr; (*d*) *Xanthium strumarium* wilted after 4 hr. (Adapted from Joffe, 1983b.)

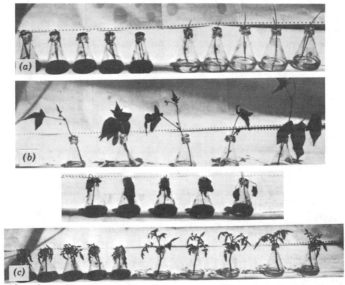

**Figure 5.4** Plant branches treated with *F. poae* No. 958 and control plants. (*a*) *Allium cepa* wilted after 10 hr; (*b*) *Phaseolus vulgaris* wilted after 6 hr; (*c*) *Solanum lycopersicum* wilted after 18 hr. (Joffe, unpublished results.)

**Figure 5.5** Plant branches treated with *F. sporotrichioides* No. 921 and control plants. (*a*) *Eschscholtzia californica* wilted after 6 hr; (*b*) *Melilotus albus* wilted after 13 hr; (*c*) *Vicia faba* wilted after 8 hr. (Adapted from Joffe, 1983b.)

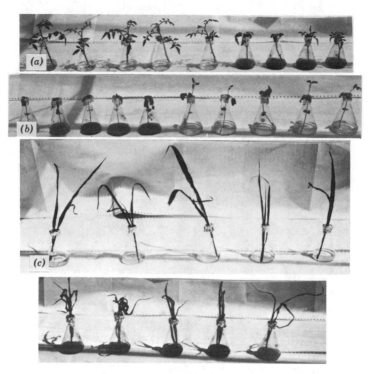

**Figure 5.6** Plant branches treated with *F. sporotrichioides* No. 921 and control plants. (*a*) *Solanum lycopersicum* wilted after 16 hr; (*b*) *Pisum sativum* wilted after 12 hr; (*c*) *Sorghum* wilted after 14 hr. (Joffe, unpublished results.)

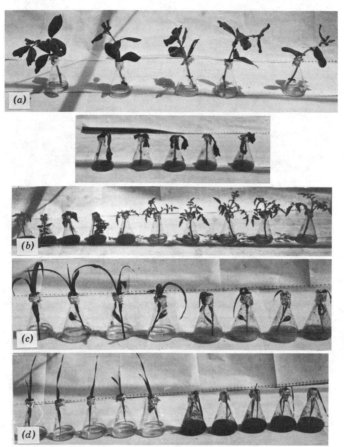

**Figure 5.7** Plant branches treated with *Fusarium* species and control plants. (*a*) *Vicia faba* wilted after 8 hr with *F. sporotrichioides* No. 347; (*b*) *Solanum lycopersicum* wilted after 18 hr with *F. sporotrichioides* No. 738; (*c*) *Zea mays* wilted after 12 hr with *F. sporotrichioides* NRRL 5908; (*d*) *Hordeum sativum* wilted after 14 hr with *F. sporotrichioides* var. *tricinctum* No. 1227. (Joffe, unpublished results.)

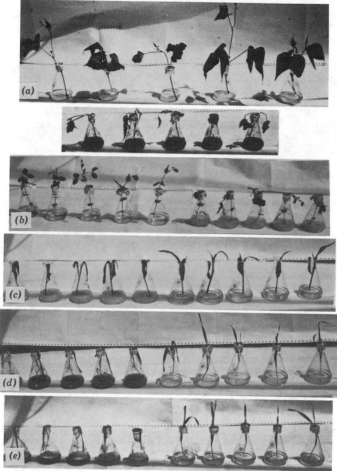

**Figure 5.8** Plant branches treated with _Fusarium_ species and control plants. (_a_) _Phaseolus vulgaris_ wilted after 15 hr with _F. sporotrichioides_ var. _chlamydosporum_ No. 1174; (_b_) _Pisum sativum_ wilted after 26 hr with _F. solani_ No. 251; (_c_) _Zea mays_ wilted after 30 hr with _F. moniliforme_ No. 63; (_d_) _Hordeum sativum_ wilted after 17 hr with _F. culmorum_ No. 543/3; (_e_) _Hordeum sativum_ wilted after 15 hr with _G. zeae_ No. 280. (Joffe, unpublished results.)

in humans) for testing their capability to produce trichothecene compounds after inoculating experiments with seedling hosts. The filtrates and extracts of these fusaria produced mainly great amounts of T-2 toxin and caused deaths of all eight rabbits 18–48 hr after application on their skin and killed all (100%) brine shrimp larvae. Necropsy of the rabbits revealed multiple hemorrhages in the gastrointestinal tract and hyperemia in various organs and tissues.

# CHAPTER **6**

# EFFECTS OF FUSARIOTOXINS IN HUMANS

*Fusarium* species are known to have caused a variety of diseases in humans in different regions of the world.

## ALIMENTARY TOXIC ALEUKIA

A very severe disease known as alimentary toxic aleukia (ATA) was widespread in the USSR during World War II and up to 1947.

Between 1942 and 1947 ATA occurred widely as a very serious, in most cases lethal, disease accompanied by a variety of symptoms that have been fully described in Russian scientific literature (Chilikin, 1944, 1945, 1947; Gajdusek, 1953a; Grinberg, 1943; Gromashevski, 1945; Malkin and Odelevskaya, 1945; Manburg, 1944; Manburg and Rachalski, 1944, 1947; Mayer, 1953a,b; Romanova, 1947; Yefremov, 1948; and by the author (Joffe, 1971, 1974c, 1978b, 1983b).

The conditions of near-famine prevailing in parts of the USSR during the period in question may be related to the severity of the outbreaks of ATA. The population was forced to collect grain that had been left in the field under snow during the winter. Outbreaks of ATA were due to the consumption of such cereals by the population of the Orenburg district and other districts of the USSR. The disease was called "alimentary toxic aleukia" since it was characterized by a progressive leukopenia and often led to a subsequent stage known as "septic angina."

The main signs of the ATA disease were necrotic lesions of the oral mucosa. The necrotic changes also developed in the pharyngeal tonsils and later in the throat and even in the esophagus inducing necrotic

**225**

angina. This clinical picture resembles some of the signs of sepsis and therefore was termed septic angina.

Many different names have been used to describe the clinical aspects of ATA: septic angina (Chernikov, 1944; Chilikin, 1944, 1945, 1947; Geminov, 1945b; Genkin, 1944; Germanov, 1945; Gorodijskaja, 1945; Gromashevski, 1945; Karlik, 1945; Kasirski and Alekseyev, 1948; Kozin and Yershova, 1945; Kudryakov, 1946; Kurbatova, 1948; Lando, 1935, 1939; Lopatin, 1946; Lovla, 1944; Lozanov and Tsareva, 1944; Malkin and Odelevskaja, 1945; Nakhapetov, 1944; Nesterov, 1948; Popova, 1948; Radkevich, 1952; Reisler, 1943; Smirnova, 1945; Yefremov, 1944a,b, 1948), alimentary hemorrhagic aleukia (Myasnikov, 1935; Perstneva, 1949; Talayev et al., 1936), hemorrhagic aleukia and agranulocytosis (Koza, 1945; Lyass, 1940), cereal agranulocytosis (Grinberg, 1943), acute myelotoxicosis (Manburg, 1944), alimentary panhematophy (Zodzishki, 1933), aplastic mesenchymopathy (Koza, 1945; Koza et al., 1944), and endemic panmyelotoxicosis (Mayer, 1953a,b).

In 1943 a committee of the Soviet Ministry of Health selected "alimentary toxic aleukia" as the correct official name for the disease. This name emphasizes the progressive leukopenia and the fact that ingestion of grain (alimentary) and the secretion of toxin are necessary for an outbreak of the disease.

To establish a diagnosis of ATA the recent dietary history of the patient had to be carefully investigated. The quantity and duration of feeding on overwintered toxic grains or their products had to be determined.

When environmental conditions during the autumn-winter-spring periods are unfavorable cereal grains may become moldy. The causal agents of ATA were found to be certain *Fusarium* species which develop on grain overwintered in the field and exposed to the extreme climatic conditions of autumn-winter-spring seasons prevailing in some parts of the USSR.

Mycological investigations have been carried out on the grains and vegetative parts of cereals overwintered in the field as well as on the soil in which they were grown. In addition material was collected from special trial fields laid out for the purpose in various parts of the Orenburg district (Joffe, 1963a).

Most frequently found among the fungi isolated in the course of these studies were species of *F. poae* and *F. sporotrichioides* (Joffe, 1947a,b,c,d, 1950, 1960a, 1963a, 1965, 1971, 1974c, 1978b, 1983b; Joffe and Mironov, 1944; Mironov and Joffe, 1947a,b; Mironov et al., 1944, 1947a), which belong to the Sporotrichiella section of the genus *Fusarium*.

The isolation and identification of the toxic agents produced by authentic strains of *F. poae* and *F. sporotrichioides* isolated from overwintered grain in the USSR is established as being trichothecenes, chiefly

T-2 toxin, which is the cause of ATA. The chemistry of the toxins from the above species involved in outbreaks of ATA is discussed in this chapter.

## Distribution of ATA in the USSR

Outbreaks of ATA have been recorded in the USSR from time to time, probably since the nineteenth century. It was reported in Eastern Siberia and in the Amur region as a food intoxication in 1913 (Beletski, 1945; Joffe, 1971; Yefremov, 1944a,b, 1945, 1948). In the spring of 1932 the area of outbreaks widened, and ATA suddenly appeared in endemic form in several districts of Western Siberia (Chilikin, 1945; Friedman, 1945b; Geminov, 1945b, 1948; Karlik, 1945; Rubinstein, 1960b), in Bash-kiria (Zodzishki, 1933), and in Kazakhstan (Beletski, 1945; Rubinstein, 1960b). In May and June of 1934 the disease was recorded again in Western Siberia (Talayev *et al.*, 1936). After that year there were new outbreaks of ATA in Ryazan, Molotov (Perm), Sverdlovsk, Omsk, Novosibirsk (Myasnikov, 1935), Altai territory (Grinberg, 1943; Sarkisov, 1950, 1954; Yefremov, 1944a,b), and some counties of Kazakhstan and the Kirgiz Republic (Lando, 1935, 1939). The disease became wide-spread in 1942 and, according to a report by Beletski (1945), appeared again at the beginning of World War II in several republics and districts, including Molotov (Perm), Kirov (Onegov and Naumov, 1943), Saratov (Sirotinina, 1945), Gorkov, Yaroslavl, Kuibyshev, Chelyabinsk, Oren-burg (Alisova, 1947a,b; Alisova and Mironov, 1944, 1947; Chilikin, 1944, 1945, 1947; Davydova, 1947; Joffe and Mironov, 1944, 1947a,b; Lovla, 1944; Manburg, 1944; Manburg and Rachalski, 1944, 1947; Manoilova, 1947; Mironov, 1944, 1945a,b; Mironov and Alisova, 1947; Mironov and Davydova, 1947; Mironov and Fok, 1944, 1947a,b; Mironov and Joffe, 1947a,b; Mironov *et al.*, 1947a,b; Romanova, 1947; Yudenitch *et al.*, 1944), and also in the Udmurt Republic (Slonevski, 1945) as well as in the Tatar Republic (Elpidina, 1945a,b, 1946b; Lozanov and Tsareva, 1944), and again in Bashkiria (Gorodijskaya, 1945; Shklovskaya and Brodskaya, 1944; Smirnova, 1945; Tatarinov, 1945; Teregulov, 1945; Zhukhin, 1945). In 1943 the disease occurred in the districts of Lenin-grad (Nakhapetov, 1944), Ulyanovsk (Teregulov, 1945), and Stalingrad, and also in the Moldavian Republic and Mari ASSR (Beletski, 1945). The disease spread considerably during 1944, appearing in the north-western and southeastern sections of the Ural (Beletski, 1945; Chilikin, 1944, 1945; Geminov, 1945b; Sarkisov, 1948; Joffe, 1950) and approach-ing regions of the Volga River (Beletski, 1945). ATA also broke out in the Ukraine, in the central parts of European Russia, and also in Central Asia and Far Eastern Siberia (Ryazanov, 1947).

To sum up there were outbreaks of ATA in 34 districts and counties of the USSR during 1944. In the same year the food situation deteriorated even further, and a considerable portion of the population was forced to collect grain that had been left in the fields throughout the winter.

In 1945 ATA occurred in the Voronezh district (Nesterov, 1948), in the Komi Republic, and in 12 other districts and regions (Ryazanov, 1947). In 1946 ATA was reported for the first time in the Dostromsk district, in the Kabardinian Republic, and in other areas; in 19 districts and counties in all (Ryazanov, 1948). In 1947 ATA appeared in 23 regions, among them Tomsk, Omsk, Arkhangelsk, and Novosibirsk (Nesterov, 1948).

The disease was found to occur most frequently in longitudes of approximately 40 and 140° east and latitudes 50 and 60° north (Mayer, 1953b). ATA recurred several times in this zone. In the Altai territory, for example, outbreaks were reported every year for 14 years (Yefremov, 1944a, 1948); in the Molotov (Perm) district for 8 consecutive years; and in Bashkiria for 7 years (Tatarinov, 1945; Teregulov, 1945; Zhukhin, 1945). In 6 of the 31 regions affected by the disease ATA appeared only once, namely in the Komi Republic (Ryazanov, 1947).

In 1952, 1953, and 1955 severe cases of ATA appeared again in a number of regions of the USSR (Bart, 1960; Rubinstein, 1956a, 1960b).

## Epidemiology

Epidemiological investigations showed that ATA occurred in families which had gathered grain from fields after the snow had thawed in the spring. In a few, sporadic cases the disease occurred in other seasons, when people bought the overwintered toxic grain, meal, or products prepared therefrom in the local markets, villages, or towns. In 1945–1947 and in 1948–1949 ATA occurred comparatively seldom and was caused chiefly by the sale of overwintered toxic cereal crops (Joffe, 1962a, 1971, 1978b).

Mortality in the period 1942–1944 was high. Whole families, sometimes whole villages, were affected, mainly in agricultural areas (Joffe, 1963a, 1965).

The fact that the typical clinical syndrome of ATA could not be reproduced in laboratory animals hindered research. In 1932 the nature of the disease was still unknown, with outbreaks being wrongly labeled as diphtheria or cholera.

Because of its sudden outbreaks and high mortality rate ATA was at first thought to be an epidemic disease of infectious origin. This hypothesis, however, was not confirmed by epidemiological and bacteriological studies, and, furthermore, the fact that none of the medical staff who cared for ATA patients was ever affected by the disease led to the rejection of the contagious hypothesis.

For a considerable time ATA was thought to be caused by a dietary deficiency of vitamins $B_1$, C, and riboflavin (Reisler, 1943, 1952). This hypothesis, like others relating ATA to bacterial infection, had to be rejected. Although a riboflavin deficiency was found in experimental animals suffering from severe aleukia, agranulocytosis, and anemia, the diet of ATA patients was not riboflavin deficient (Gordon and Levitskij, 1945; Rubinstein, 1956a).

Recognition of the real nature of ATA was delayed by these various hypotheses, but eventually it was realized that the disease was caused by the consumption of fungal contaminated overwintered grain or other agricultural products forming the staple diet of the peasant population in the areas of Russia (Joffe, 1960a, 1963a, 1965, 1971, 1974c, 1978b). The identification of the specific fungi responsible for ATA eventually led to the introduction of proper prophylactic measures.

The first outbreaks in the Orenburg district were in 1924 and again in 1934 (Geminov, 1945a,b). In 1942 and 1943 ATA reappeared on a large scale (with 19 and 30 counties, respectively, being affected), and in the spring of 1944 it was extremely widespread, affecting 47 out of 50 counties. The only unaffected counties were 3 in the easterly part of the district (Fig. 6.1).

ATA outbreaks were not uniform throughout the Orenburg district: they were limited to specific counties, sometimes even to one small village within a county (Joffe, 1974c, 1978b).

It can be seen from Table 6.1 and Figure 6.1 that in 1944, the peak year, over 10% of the population was affected and there was a high

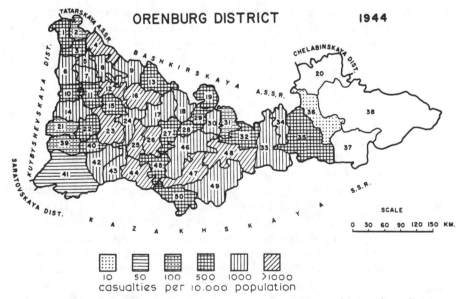

**Figure 6.1** The outbreaks of ATA disease in various counties in the Orenburg district in 1944 (Joffe, 1963a).

TABLE 6.1  Incidence of ATA in 50 Counties[a] of the Orenburg District

| | Years | | | | | | |
|---|---|---|---|---|---|---|---|
| | 1941–1942 | 1942–1943 | 1943–1944 | 1944–1945 | 1945–1946 | 1946–1947 | 1947–1948 |
| No disease | 31 | 20 | 3 | 36 | 42 | 38 | 0 |
| 0–50 cases | 12 | 17 | 3 | 14 | 8 | 11 | 0 |
| 50–500 cases | 7 | 13 | 19 | 0 | 0 | 0 | 0 |
| 500–1000 cases | 0 | 0 | 16 | 0 | 0 | 0 | 0 |
| Over 1000 cases | 0 | 0 | 9 | 0 | 0 | 0 | 0 |

*Source:* Adapted from Joffe, 1978b.

[a] Number of counties in which the population was affected by ATA (cases per 10,000 of population).

mortality rate in 9 of the 50 counties of the Orenburg district (Joffe, 1950, 1963a, 1971, 1974c, 1978b). In 1945, 1946, and 1947 the disease decreased in intensity, affecting 14, 8, and 12 counties, respectively, but by 1948 and 1949 it was almost nonexistent.

The spread of ATA was particularly marked in districts where it occurred early in the spring. It was clearly a seasonal disease, the main outbreaks occurring from April and May to mid-June. The rate of spread declined in July and August, and in autumn and winter there were no signs of the disease. In 1942 ATA appeared in Orenburg later than usual, that is, in July and August, but in most of the cases it appeared where spring starts early, with the melting of the snow. A few cases of ATA were reported at other seasons, but these were urban patients who had received overwintered grain later than peasants in agricultural areas.

### Etiology

The seasonal occurrence of ATA, its endemicity, and the composition of the affected population pointed to the importance of climatic and ecological factors in outbreaks of the disease. Surveys made in the affected areas showed the major nutrients of patients to be cereal grain (wheat, prosomillet, barley, oats, buckwheat) and leguminous plants. It was assumed, therefore, that there was a correlation between these foods and the development of the disease. Since peak outbreaks occurred in the spring after the snow had melted, when the rural population consumed overwintered grain from the fields, investigators searched for changes in the grains as a result of having been under snow cover throughout the winter. In fact considerable changes were found in the mycoflora of the overwintered cereals, and this suggested that ATA was probably caused by fungi which developed well under cover of snow.

The occurrence of ATA in man was found to be influenced by the following etiological factors:

1. *The Quantity of Overwintered Grain Ingested.* The disease usually appeared after at least 2 kg of food prepared from toxic overwintered grain had been consumed.

2. *The Duration of Feeding on Toxic Grain.* Lesions in the hematopoietic system were the result of toxic substance accumulating in the body and usually appeared 2–3 weeks after the consumption of toxic grain. If, for example, a patient ate toxic grain in sufficient quantity at the end of April, he or she would fall ill with ATA in mid-May. Death occurred 6–9 weeks after the initial grain consumption (Chilikin, 1947; Manburg, 1944; Manburg and Rachalski, 1947; Mironov and Myasnikov, 1947; Nesterov, 1948; Romanova, 1947; Yefremov, 1948).

3. *Concentration of Toxin in the Food.* Considerable variations were found in the toxicity of overwintered grain, even within samples from the same field. Families consuming nontoxic overwintered grain were unaffected, whereas entire families who had ingested highly toxic grain were all affected by ATA.

4. *Type of Cereal Consumed.* Prosomillet and wheat were found to be most toxic (Joffe, 1947a,b, 1950, 1963a).

5. *Individual Sensitivity to Toxin.* People having a balanced diet were less sensitive to the toxin than undernourished persons, while the most seriously affected were those whose diet consisted almost entirely of overwintered cereals (Joffe, 1971, 1978b).

6. *Season of Harvesting.* Grain harvested during spring thaws was toxic while grain harvested during autumn or winter before the snow melted was either nontoxic or slightly toxic. Grain harvested after a less severe winter, with abundant snow followed by frequent alternate freezing and thawing in spring, showed such climatic conditions to be conducive to the growth of toxic fungi (Joffe, 1963a).

7. *Age.* Breast-fed babies of less than 1-year-old were not affected by ATA since the toxic substance was not secreted in the milk of the sick mother (Chilikin, 1947; Manburg, 1944; Sergiev, 1946, 1948). The disease occurred in infants over 1 year if they had eaten products made of toxic overwintered cereals (Lozanov and Tsareva, 1944; Sergiev, 1945, 1946, 1948; Yefremov, 1948). The most frequent occurrence and the highest mortality rates were between the ages of 8 and 50 (Chilikin, 1947; Romanova, 1947).

8.  *Gender.* The literature did not report any sex difference in ATA mortality rates, a view also suggested by the author, although Talayev *et al.* (1936) noticed a susceptibility to the disease in middle-aged women.

## Mycological and Mycotoxicological Studies

Prosomillet was originally considered to be the most dangerous source of ATA since many people fell ill after eating this grain. Prosomillet was indeed an extensive source of the disease: it was widely grown in the Orenburg district, as well as in other parts of the Soviet Union, and ripened late, thus producing an abundant harvest so that large amounts were left to overwinter in the field. Later on, however, it was shown that wheat and barley were the main causes of ATA, followed by oats, rye, and buckwheat. All these cereals were left unharvested during the winter, under snow, and the climatic, meteorologic and ecologic conditions were favorable to the growth of toxic fungi.

Researchers working on ATA suggested that the disease was associated with a toxic origin, more particularly with overwintered grain contaminated by toxic fungi. This hypothesis was indeed correct and was supported by many investigators (Bilai, 1953; Elpidina, 1945a, 1946b; Forgacs and Carll, 1962; Joffe, 1950, 1960a,b, 1962a, 1963a, 1965, 1971, 1973b, 1974c, 1978b, 1983b; Karatygin and Rozhnova, 1947; Kopytkova, 1948a,b; Kvashnina, 1948; Mironov and Joffe, 1947a,b; Rubinstein, 1948, 1950b, 1951a,b, 1956a,b, 1960a,b; Rubinstein and Lyass, 1948; Sarkisov, 1945, 1948, 1950, 1954, 1960; Tsvetkov, 1946).

The author's studies described in this chapter deal with the Orenburg district in particular (*see* Fig. 6.1) and were performed during 8 years at the Mycological Division of the Institute of Epidemiology and Microbiology in the USSR Ministry of Health. The Institute set up a special ATA (septic angina) laboratory to investigate every aspect of the disease. The following factors were studied: the role of overwintered cereal crops, the mycoflora of these grains, the toxic properties of cryophilic fungi developing at particularly low temperatures, climatic and ecological conditions for toxin production in grain, characteristics of such toxins in animals, the chemistry of the toxins, and, mainly, clinical symptoms and pathological findings in humans. These studies were subsequently continued at the Department of Botany of the Hebrew University of Jerusalem in the Laboratory of Mycology and Mycotoxicology. Samples were taken from the most dangerous loci for mycological and toxicological analysis. The fungi most frequently associated with the toxic grain causing ATA belonged to the Sporotrichiella section of the genus *Fusarium* and included the following species: *F. poae, F. sporotrichioides, F. sporotrichioides* var. *tricinctum,* and *F. sporotrichioides*

**Figure 6.2**   Sample of overwintered prosomillet infected by *F. sporotrichioides* consumed by two members of a family who died (Joffe, 1978b).

var. *chlamydosporum* (Joffe, 1950, 1960a,b, 1962a, 1963a, 1965, 1971, 1974a,c, 1977, 1978b, 1983b).

Identification and determination of ATA cases were established by the following steps: (a) regional offices of the USSR Ministry of Health gave notification of death caused by ATA; (2) samples of grain and food products consumed by the deceased were taken from their homes; and (3) these samples were carefully investigated with the help of mycological analysis (Figs. 6.2 and 6.3) and the fungal extracts obtained examined by the skin test method on rabbits, while being fed at the same time to animals.

The toxins produced by *F. poae* and *F. sporotrichioides*, which cause ATA, were investigated in detail on various animals by the author (Joffe, 1950, 1960a,b, 1965, 1971, 1974c, 1978b, 1983b). When these animals exhibited symptoms resembling those observed in humans, the toxic principle was considered to be inherent in the samples.

The toxins of *F. poae* and *F. sporotrichioides* were also isolated from various overwintered grains and soil samples collected from lethal fields in the Orenburg district.

In addition to these investigations experiments were conducted between 1943 and 1949 on 39 variously treated trial plots in the Orenburg district (Joffe, 1963a), the results of which contributed much to deter-

**Figure 6.3**  Wheat ear overwintered in the field infected by *F. poae*, which caused the death of three persons after consumption (Joffe, 1978b).

mining the conditions for toxin production in overwintered grain (Joffe, 1950). For this study 39 experimental plots were set up under normal field conditions in different counties of the Orenburg district. The plots varied in size from 100 sq m to 0.5 ha. The experimental crops included the most commonly grown cereals, namely millet, wheat, and barley.

During harvesting the crop from half of each plot was cut and arranged in stacks, while the other half was left uncut. Samples of the stacked cereals were taken from the upper layer of the stack as well as from the layer at soil level. Plant and soil samples were usually collected twice a month from these plots, from August or September of each year to the following May. Cereal samples were first threshed and then dried at 45°C; soil samples were dried similarly. Samples of dried and threshed grain weighing 30–50 g, samples of vegetative parts (stems, leaves, ears, panicles, husks, etc.) weighing 15–30 g, and soil samples of not less than 200 g were soaked in ether or alcohol. After 3–5 days soaking with repeated shaking the ether or alcohol was driven off in a distilling apparatus. The residue obtained after evaporation was assayed for toxicity by application to the shaved skin of a rabbit. The testing method has been fully described elsewhere (Joffe and Mironov, 1947a,b; Mironov and Joffe, 1947a,b; Joffe, 1960a,b, 1962a, and in Chapter 3 of this book). A total of 4702 tests with cereal and soil samples were performed on 528 rabbits.

During the 6-year investigation from 39 experimental plots in the field 1336 samples of grain and vegetative parts and 327 soil samples were tested, as well as 300 samples of cereals and 122 soil samples collected at harvest times for controls. In all 2085 cereal and soil samples were assayed for toxicity during the experimental period.

Concurrent investigations were also conducted during the same period on 1931 samples of overwintered cereals and 230 soil samples from all counties of the Orenburg district. Controls consisted of 296 cereal samples and 160 soil samples collected during the normal summer harvesting period. Thus a total of 2617 cereal and soil samples were studied for toxicity. Material investigated consisted of millet, wheat, barley, oats, and buckwheat, as well as occasional samples of sunflower seeds, acorns, legumes, and so on gathered in the fields.

## The Ecological Conditions in Field Outbreaks

For many years we have been conducting a comprehensive study to elucidate the environmental conditions conducive to the formation of *Fusarium* toxins in plants and soil in the field in temperate and subtropical regions. This study has been devoted to an understanding of the role of environmental, ecological conditions in the mechanism of formation of *Fusarium* toxins in food and feed.

Outbreaks of fusariotoxicoses in farm animals have been reported by a number of researchers, the majority of whom, however, have concentrated their attention on toxin production under laboratory conditions and not environmental, ecological conditions in the field during the period of the outbreak.

Laboratory studies make it possible to isolate the metabolite(s) but cannot determine the conditions conducive to *Fusarium* toxin formation in the field leading to outbreaks among animals and humans.

The ecology of trichothecene (chiefly T-2 toxin) production in the field by *Fusarium* must be given increased attention. An example illustrating the paramount importance of environmental factors—including meteorological conditions—in the formation of *Fusarium* toxins is associated with the etiology of the fatal ATA disease.

Outbreaks of acute or chronic fusariotoxicosis in animals and humans in temperate regions are dependent on environmental factors. In the warm regions, for example, *F. moniliforme* occurred (Brodnik, 1975; Joffe *et al.*, 1973; Palti, 1978; Marasas *et al.*, 1979a). In the cooler, temperate areas the most toxic species and varieties have been *F. sporotrichioides*, *F. poae*, and *F. sporotrichioides* var. *tricinctum* (Joffe, 1960a, 1965, 1983b,c), which have been implicated as toxin producers. The development of the latter three isolates in overwintered cereal grains correlated with toxic outbreaks in animals and humans.

The discussion here will focus on ATA, an outbreak in which *F. poae* and *F. sporotrichioides* produced mycotoxins (especially T-2 toxin) in special ecological conditions (Joffe, 1960a,b, 1963a, 1965, 1971, 1974c, 1978b, 1983b,c; Joffe and Yagen, 1977; Yagen and Joffe, 1976; Yagen *et al.*, 1977).

Most of the toxic strains of *F. sporotrichioides* and *F. poae* isolated from overwintered cereal grains were characterized by cryophilic properties and satisfactory growth at low temperatures. Temperature fluctuations, with freezing and thawing, intensified toxin production both in pure and mixed cultures. Cultures of *F. poae* and *F. sporotrichioides* were most toxic during the period of abundant sporulation when grown at $-2--7°C$ (Joffe, 1971). No toxin was produced at higher temperatures of 25–30°C. Mixtures of toxic cultures grown together appeared to be more toxic than cultures grown singly, indicating a synergistic relationship (Joffe, 1960a,b).

Various ecological conditions affected growth and toxin production by *F. poae* and *F. sporotrichioides*. The most important factors in field outbreaks were: moisture (relative humidity), temperature, and biological factors such as the presence of toxigenic *Fusarium* strains. For many outbreaks specific ecological factors influencing toxin production under field conditions were not recorded. In a review of the literature it was seen that few authors published information on environmental conditions in their articles. In the case of ATA we studied the environmental conditions involved in toxin production, the distribution of *F. poae* and *F. sporotrichioides* and their biological and chemical characteristics, and methods of prevention and control.

The strains of *F. poae* and *F. sporotrichioides*, which produce T-2 toxin, have occurred widely in the field in low temperature conditions. Both incidence and level of *Fusarium* toxins were found to vary with environmental conditions, changing from one geographical region to another and even within the same field. When isolated from varying ecological conditions the same strain was able to synthesize several different toxic metabolites.

For the determination of T-2 toxin produced by *F. poae* and *F. sporotrichioides* from different geographical regions it was important to have accurate information concerning weather, the climatic conditions in which the formation of toxins occurred, the amount and degree of activity of the toxins, and the syndromes caused in animals (Joffe, 1960a,b, 1963a, 1971, 1974c, 1978b, 1983b; Joffe and Yagen, 1977, 1978; Lutsky and Mor, 1980a,b; Lutsky *et al.*, 1978; Yagen and Joffe, 1976; Yagen *et al.*, 1977, 1980).

### Environmental Factors Favoring Toxin Formation in ATA Disease

What factors influenced the high degree of toxin production in overwintered cereals, particularly in 1943–1944 when ATA was widespread throughout almost the entire Orenburg district? The year of highest incidence of ATA, 1943–1944, was characterized by specific environmental conditions. A comparison with the years 1941–1948 can be seen in Table 6.1.

A comprehensive analysis of the connection between various environmental conditions prevailing in autumn and winter and the incidence of ATA in the 50 counties of the Orenburg district was made during the 7-year period 1941–1942 to 1947–1948 (see Tables 6.1 and 6.2). The data show that the year of heaviest incidence of ATA, 1943–1944, was characterized by higher temperatures in January and February.

The mean temperature during January and February 1944 was only −8.3°C, as compared with the much lower 10 year average of −14.7°C for these months. The importance of temperature in toxin formation has been emphasized by many researchers, for example, Asami (1932); Eugenio et al. (1970a); Hesseltine (1976a); Kalra et al. (1977); Keyl et al. (1967); Martin (1974); Oya and Morimoto (1942); Sutton et al. (1976); Ueno et al. (1972a); Yates et al. (1968).

During the 30 years preceding the 1943–1944 ATA outbreaks similar combinations of low September and October rainfall and relatively mild January and February temperatures were recorded in the Orenburg district twice: in 1924–1925 and again in 1934–1935 (Joffe, 1950). It is interesting to note that, according to Geminov (1945a,b), cases of poisoning occurred in considerable numbers in the rural population of this district in those 2 years.

Other factors were also of great importance: depth of snow, moisture, and altitude. The average depth of snow in January through March 1944 reached 80–120 cm, far exceeding that of all other years, and in February and March it was 108 cm (Table 6.2) compared with a 10-year average of only 41 cm.

The depth to which the soil was frozen in January through March 1944 was only 22 cm, as compared with 77 cm for a 6-year average.

Total rainfall in September and October 1943 was only 15.2 mm as compared with a 10-year average of 57.0 mm (Table 6.3). The rain, and the relatively cold autumn nights together with ample dew, favored development of toxin-producing *Fusarium* in vegetative cereal parts from where they proceeded to attack the grain under the favorable conditions of the following spring (Joffe, 1963a, 1983c).

The presence of moisture, whether in the form of rainfall, dew, or even the moisture content of the grain, is necessary for the growth of *Fusarium* fungi, as demonstrated in the literature by the following workers: Balzer et al. (1977a,b); Bristol and Djurickovic (1971); Caldwell and Tuite (1974); Christensen (1979); Christensen and Kernkamp (1936); Christensen et al. (1965); Ciegler (1978); Collet et al. (1977); Gross and Robb (1975); Lovelace and Nyathi (1977); McErlean (1952); Mirocha

**TABLE 6.2** Relationship between Autumn and Winter, and the Incidence of ATA in the Orenburg District, 1941–1942 to 1947–1948

| Weather Conditions | Month | 1941–1942 | 1942–1943 | 1943–1944 | 1944–1945 | 1945–1946 | 1946–1947 | 1947–1948 |
|---|---|---|---|---|---|---|---|---|
| Temperature (°C) | January | −17.7 | −18.8 | −8.3 | −16.2 | −12.4 | 16.6 | −7.0 |
| | February | −18.7 | −15.7 | −8.0 | −20.8 | −15.4 | −15.1 | 12.6 |
| Snow cover (in cm) | December | 14 | 40 | 29 | 5 | 11 | 4 | 8 |
| | January | 22 | 53 | 50 | 5 | 29 | 6 | 15 |
| | February | 23 | 50 | 49 | 6 | 50 | 25 | 21 |
| | March | 25 | 43 | 108 | 17 | 70 | 18 | 28 |
| Depth to which soil was frozen (in cm) | November | 52 | 21 | 12 | 25 | 40 | 33 | — |
| | December | — | 36 | 14 | 30 | 50 | 75 | 41 |
| | January | — | — | 27 | — | 50 | 106 | 68 |
| | February | 100 | — | 29 | 80 | — | 123 | 118 |
| | March | — | — | 10 | 80 | 91 | — | 124 |

*Source*: Adapted from Joffe, 1978b.
— = depth not recorded.

**TABLE 6.3** Rainfall in the Orenburg District in September and October, 1939–1948 (in mm)

| Month | 1939 | 1940 | 1941 | 1942 | 1943 | 1944 | 1945 | 1946 | 1947 | 1948 | Average |
|---|---|---|---|---|---|---|---|---|---|---|---|
| September | 30.3 | 8.0 | 19.5 | 11.3 | 7.1 | 7.0 | 92.8 | 83.3 | 31.8 | 42.1 | 26.0 |
| October | 17.1 | 74.8 | 49.5 | 29.2 | 8.1 | 38.2 | 31.3 | 34.8 | 83.4 | 72.1 | 31.0 |
| Total | 47.4 | 82.8 | 69.0 | 40.5 | 15.2 | 45.2 | 124.1 | 118.1 | 62.2 | 114.8 | 57.0 |

*Source*: From Joffe, 1950.

and Christensen (1974a,b); Mirocha *et al.* (1971, 1974); Nishikado (1957); Pullar and Lerew (1937); Sherwood and Paberdy (1972a,b,c,d, 1974a,b); Shotwell *et al.* (1975); Tuite *et al.* (1974); Urakura (1933); and Vesonder *et al.* (1978).

An additional factor affecting the degree of toxicity of overwintered cereals in 1943–1944 was altitude. ATA was not found in counties with altitudes above 350–400 m because of their lower winter temperatures. The altitude of the three eastern counties generally exceeds 400 m above sea level (Fig. 6.1), and all the cereal and soil samples taken from these counties were nontoxic. However, in various counties of the Orenburg district which were only 70–200 m above sea level, the population was strongly affected by ATA in the same period (Table 6.1).

An example of the possible importance of altitude in toxin formation is found in the literature: Martin and Keen (1978) isolated *F. monili- forme* from samples of food and fermented material in Swaziland. They suggested that altitude might influence the incidence of zearalenone, judging from their findings in sour drinks, sour porridges, and beer con- sumed by humans.

There were years in which overwintered grain was consumed but no outbreaks of ATA resulted because the climatic and ecological condi- tions did not favor toxin production. Conditions prevailing in spring 1944 seem to have favored toxin formation: relatively low temperatures and, in particular, the dense snow cover which prevented the soil from freezing to its usual depth. In the spring of that year thawing and freez- ing alternated more often than in other years and this favored develop- ment of toxin-producing fungi on overwintering grain.

The weather prevailing in 1943/1944 thus greatly favored the accumu- lation of toxin in overwintering cereals under snow and caused the disas- trous ATA outbreaks in 1944.

## Annual Comparison of Toxicity of Cereals between 1943–1949[*]

A comparison of the toxicity of cereals from field samples and from the experimental plots in different years (Figs. 6.4 and 6.5) shows that sam- ples collected in 1945, 1948, and 1949 were mostly nontoxic. The years 1946 and 1947 hold an intermediate position since the toxicity of sam- ples investigated in those years, though not high, was quite significant. The years 1943 and 1944 were characterized by exceptionally potent toxicity of the grains. Samples obtained from the 1943–1944 period al- most invariably produced a positive skin reaction. Samples collected from different counties in the spring of 1944 were particularly toxic (Fig. 6.4). The number of toxic samples from experimental plots in the spring

[*]In cases where data on years of observation are given, for example, 1944–1949, they should be read as from autumn 1944 to spring 1949.

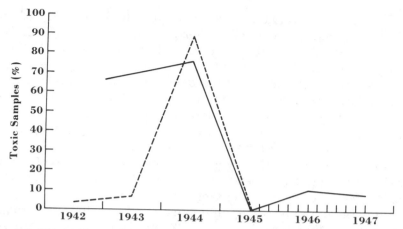

**Figure 6.4**  The incidence of ATA compared with the incidence of toxic samples collected from overwintered cereal grains in the Orenburg district from 1942–1947. Solid line, percentage of toxic samples of overwintered cereals; broken line, percentage of consumers of overwintered cereals affected by ATA disease (Joffe, 1963a).

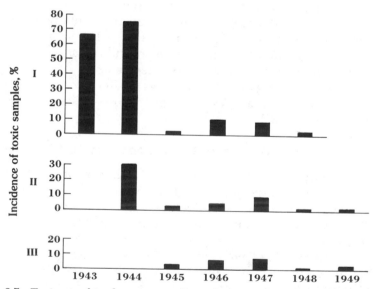

**Figure 6.5**  Toxic samples of overwintered cereal crops collected from: I. counties of the Orenburg district; II. 39 experimental plots in the Orenburg district; III. the experimental plot of the Orenburg district Experimental Station, as percent of total samples. (Adapted from Joffe, 1963a).

**240**

**TABLE 6.4** Toxicity of Cereal and Soil Samples from Experimental Plots for the Periods 1943–1944 and 1944–1949

| Years | Samples | No. Examined | Toxic | | Slightly Toxic | | Nontoxic | |
|---|---|---|---|---|---|---|---|---|
| | | | No. | % | No. | % | No. | % |
| 1943–1944 | Cereal | 287 | 89 | 31.0 | 101 | 35.2 | 97 | 33.8 |
| | Soil | 38 | 14 | 38.9 | 4 | 11.1 | 18 | 50.0 |
| 1944–1949 | Cereal | 1049 | 35 | 3.5 | 55 | 5.2 | 959 | 91.4 |
| | Soil | 291 | 8 | 2.8 | 24 | 8.2 | 259 | 89.0 |

*Source:* Adapted from Joffe (1963a).

of 1944 exceeded the total number of toxic samples collected during the entire period from autumn 1944 to spring 1949. This is clearly indicated in Table 6.4.

A comparison was made of the toxicity of cereals gathered in different years from the same experimental plot. A plot on the grounds of the District Experimental Station of Orenburg County, which was under observation for 5 consecutive years from 1945, may serve as an example (Fig. 6.5). It is clear that the toxicity of cereals on the same plot varied from year to year. Both the collated material for the period 1943–1949 and the results obtained for the same plot in different years indicate that the degree of toxicity of overwintered cereals varied according to the year. On the basis of the data presented it may be assumed that only in certain years are environmental conditions favorable to the formation and accumulation of toxic substances in cereals overwintering under snow.

*Toxicity of Cereals and Incidence of ATA.* The correlation between the incidence of ATA and the toxicity of overwintered cereals over a period of years is presented in Figure 6.4. In 1943 the percentage of toxic samples was high but was not paralleled by high incidence of the disease due to the fact that in 1943 food was not yet so scarce in Russia as to force large numbers of people to collect overwintered grain. In 1944, however, the high prevalence of toxicity in the samples coincided with an extreme food shortage, and large sections of the population could only subsist by searching for overwintered grain. This resulted in a very high incidence of disease in that year. In the following years the toxicity of samples decreased and the food situation improved with a consequent reduction in the incidence of the disease.

*Seasonal Effects.* Toxin formation in overwintered cereals generally took place during the autumn-winter-spring period. Investigations showed the largest number of toxic samples to be present in spring: in winter and in autumn the number of toxic samples recorded was less

TABLE 6.5   Toxicity of Cereal and Soil Samples Taken from Various Counties of the Orenburg District during Autumn, Winter, and Spring Seasons, 1943–1944 and 1944–1949

| Years | Toxicity | Autumn | | Winter | | Spring | |
|---|---|---|---|---|---|---|---|
| | | No. of Samples | % | No. of Samples | % | No. of Samples | % |
| 1943–1944 | Toxic | 27 | 24.0 | 43 | 27.7 | 121 | 75.9 |
| | Nontoxic | 81 | | 112 | | 37 | |
| 1944–1949 | Toxic | 41 | 7.9 | 54 | 10.6 | 84 | 11.7 |
| | Nontoxic | 473 | | 455 | | 633 | |

*Source:*   Joffe, unpublished results.

frequent than that of the spring season (Joffe, 1960a, 1971, 1974c, 1978b) (*see* Tables 6.5 and 6.6).

The contrast between numbers of toxic samples collected in the different seasons is well illustrated by an analysis of the material relating to 1943–1944. Since that period was characterized by a very high degree of toxicity in cereals, the data from that year may be regarded as conclusive. The results obtained from one experimental plot in 1944, for example, were as follows: During the autumn-winter period investigations were carried out on vegetative parts and grain of millet and on soil samples. Out of 102 samples of vegetative parts tested, 48 were toxic or slightly toxic and 54 were nontoxic. Of the 36 samples of millet grain, only one sample gathered in January was toxic: slight toxicity was found in 10 cases and 25 grain samples were unaffected. Of six soil samples, three were found to be toxic.

The first spring samples were collected in April, from beneath the snow, from moist places free of snow, and from dry places. The investi-

TABLE 6.6   Toxicity of Cereal and Soil Samples Taken from Experimental Fields during Autumn, Winter, and Spring Seasons, 1943–1944 and 1944–1949

| Years | Toxicity | Autumn | | Winter | | Spring | |
|---|---|---|---|---|---|---|---|
| | | No. of Samples | % | No. of Samples | % | No. of Samples | % |
| 1943–1944 | Toxic | 34 | 28.8 | 46 | 30.8 | 56 | 100.0 |
| | Nontoxic | 84 | | 103 | | — | |
| 1944–1949 | Toxic | 27 | 6.3 | 51 | 9.6 | 44 | 11.2 |
| | Nontoxic | 395 | | 484 | | 339 | |

*Source:*   Joffe, unpublished results.

**Figure 6.6** (*a*) Overwintered prosomillet grain taken from an experimental field which caused severe cases of ATA. Growth of *F. sporotrichioides* was heavy. (*b*) Grains of non-toxic normal prosomillet gathered in autumn 1943 from the experimental field. (Adapted from Joffe, 1978b.)

gation covered eight samples of grain separated from the vegetative parts, 16 extracts in all being prepared. Application of the extracts to rabbit skin resulted in edema, hemorrhage, and necrosis. Two weeks later, after a full thaw, 11 samples were collected from different parts of the plot. Strong toxicity was displayed in 22 extracts prepared from vegetative parts and from grain. Figure 6.6 shows that all samples collected in the spring were highly toxic and those gathered in the summer were nontoxic. By the end of April 1944 46 kg of millet were harvested from the same experimental plot, and the grain was shown to be strongly toxic.

The findings from samples collected from one half hectare experimental field plot during the different seasons of the 1943–1944 millet crop are given in Table 6.7.

**TABLE 6.7   Toxicity of Overwintered Millet Grain from One Experimental Plot Sampled in Autumn, Winter, and Spring of 1943–1944**

| Toxicity | Autumn | | Winter | | Spring | |
|---|---|---|---|---|---|---|
| | No. of Samples | % | No. of Samples | % | No. of Samples | % |
| Toxic | 16 | 20.5 | 7 | 11.2 | 38 | 100.0 |
| Slightly toxic | 18 | 23.1 | 19 | 30.1 | — | — |
| Nontoxic | 44 | 56.4 | 37 | 58.7 | — | — |
| Total no. of samples examined: | 78 | | 63 | | 38 | |

*Source:*   Adapted from Joffe (1963a).

***Toxic Properties of Stored Grain.***   Strongly toxic samples of millet from the 1943–1944 season were reinvestigated in 1949 and 1950 in order to determine whether the toxicity of overwintered cereals is modified by prolonged storage. Results showed that the toxic ingredient in the grain was unaffected by 6–7 years of storage, and toxicity remained unchanged (Joffe, 1962a, 1971, 1974c, 1978b). However, concurrent mycological examination of these samples failed to isolate *F. poae* and *F. sporotrichioides*, which had clearly perished. This may explain the difficulties met in trying to isolate toxic *Fusarium* fungi from cereal grains after prolonged storage.

## Toxin Formation in Cereals and Soil

Contamination of cereal crops by *Fusarium*-produced toxins is a worldwide problem which appears in many geographical areas in a variety of ecological conditions. The most significant substrates for toxin production have been the small grains (millet, rice, wheat, barley, rye, oats, and sorghum) and also maize.

Large-scale investigations were undertaken between 1943 and 1949 on samples from cereal grain, vegetative parts, and soil, all collected from all counties and experimental plots in the Orenburg district in which ATA outbreaks were heaviest (Joffe, 1963a, 1971, 1974c, 1978b, 1983b,c).

Comparisons of the toxicity of soil and grain samples and vegetative cereal parts taken from counties of the Orenburg district and from experimental fields, made in 1943–1944 and 1944–1949, are shown in Tables 6.8 and 6.9. These tables separately demonstrate results for 1943–1944, when the outbreaks of ATA associated mainly with *Fusarium* contaminated millet were very severe, and for the other years, when outbreaks were lighter. The data show the enormous rise in the

**TABLE 6.8** Toxicity of Vegetative Parts, Cereal Grains, and Soil Samples from Various Counties of the Orenburg District in 1943–1944 and 1944–1949

| Years | Toxicity | Vegetative Parts No. of Samples | % | Cereal Grains No. of Samples | % | Soil No. of Samples | % |
|---|---|---|---|---|---|---|---|
| 1943–1944 | Toxic | 36 | 33.0 | 171 | 80.5 | 49 | 51.0 |
| | Nontoxic | 75 | | 43 | | 47 | |
| 1944–1949 | Toxic | 23 | 5.0 | 83 | 8.2 | 17 | 12.6 |
| | Nontoxic | 460 | | 1040 | | 117 | |

*Source:* Joffe, unpublished results.

percentage of toxic samples in the year which saw the most serious outbreaks.

In subsequent years (1944–1949) samples of wheat were more often affected than millet and barley (Table 6.10).

The role of the soil studied in specially designated experimental field plots in various regions of the Orenburg district and the toxicity determined in samples of soil, grain and vegetative parts during the period 1944–1949 are given in Table 6.11.

The tabulated data show that toxicity occurred most frequently in soil (11.2%) and rather less in grain (10.7%), while the vegetative parts of cereals were the least toxic (6.2%).

The detection of 11.2% toxic soil samples points to the presence of a toxic principle in the soil. This soil factor apparently spreads to the cereal plant, affecting first the vegetative parts and then the grain. The conditions prevailing in the latter are evidently particularly favorable for toxin accumulation, and toxicity was therefore more frequent in grain

**TABLE 6.9** Toxicity of Vegetative Parts, Cereal Grains, and Soil Samples from Experimental Fields in 1943–1944 and 1944–1949

| Years | Toxicity | Vegetative Parts No. of Samples | % | Cereal Grains No. of Samples | % | Soil No. of Samples | % |
|---|---|---|---|---|---|---|---|
| 1943–1944 | Toxic | 17 | 18.4 | 98 | 50.3 | 18 | 50.0 |
| | Nontoxic | 75 | | 97 | | 18 | |
| 1944–1949 | Toxic | 32 | 6.2 | 58 | 10.7 | 32 | 11.2 |
| | Nontoxic | 478 | | 481 | | 259 | |

*Source:* Joffe, unpublished results.

**TABLE 6.10** Toxicity of Cereal Samples from Experimental Plots for the Period 1944–1949

| Source of Samples | No. of Samples Examined | Percentage of Samples | | |
|---|---|---|---|---|
| | | Toxic | Slightly Toxic | Nontoxic |
| Millet | 420 | 2.6 | 5.2 | 92.2 |
| Wheat | 415 | 4.6 | 4.6 | 90.8 |
| Barley | 255 | 1.9 | 5.5 | 92.6 |

*Source:* Adapted from Joffe (1963a).

than in the vegetative parts. In light of this reasoning the findings described below, and their practical value, may be understood.

Out of a total of 92 samples of vegetative parts gathered between September and November 1943, 17 (18.4%) were toxic. Out of a total of 136 assayed samples collected during a corresponding period in 1944, only 1 (0.7%) was toxic. That is to say in autumn 1943—preceding the severe outbreak of ATA—toxicity was detected in a large number of vegetative samples, whereas in autumn 1944 there were practically no toxic samples. The accumulation of toxic substances in the vegetative parts during the autumn (October and November) may, therefore, serve to predict the probable appearance of ATA the following spring. Infection and formation of toxin in the grain are apparently secondary processes greatly influenced by environmental conditions.

The fungi most commonly encountered in the soil samples and cereal grains were representatives of the genera *Penicillium* and *Fusarium*, with fewer examples of *Mucor, Cladosporium*, and *Alternaria* (Table 6.12). These fungi varied greatly in their degree of toxicity, with those of the genus *Fusarium*, particularly *F. poae* and *F. sporotrichioides*, being the most highly toxic.

We revealed a relationship between the toxicity of these *Fusarium* species isolated in 1943–1944 from overwintered cereals and from soil. The resemblance between the toxicity of the *Fusarium* isolated from

**TABLE 6.11** Toxicity of Samples of Cereal Grains, Plant Organs, and Soil from Experimental Plots for the Period 1944–1949

| Source of Samples | No. of Samples Examined | Percentage of Samples | | |
|---|---|---|---|---|
| | | Toxic | Slightly Toxic | Nontoxic |
| Vegetative Parts | 510 | 1.9 | 4.3 | 93.8 |
| Grains | 539 | 4.6 | 6.1 | 89.2 |
| Soil | 291 | 2.8 | 8.4 | 88.8 |

*Source:* Adapted from Joffe (1963a).

TABLE 6.12 Genera of Fungi Associated with Toxin Production in Overwintered Grain

| Genus | Total No. of Isolates | Highly Toxic (%) | Mildly Toxic (%) |
|---|---|---|---|
| *Fusarium* | 546 | 20.7 | 19.8 |
| *Cladosporium* | 480 | 5.4 | 8.5 |
| *Alternaria* | 506 | 2.8 | 5.3 |
| *Penicillium* | 830 | 1.6 | 3.8 |
| *Mucor* | 335 | 3.0 | 7.2 |

*Source:* Adapted from Joffe, 1971.

cereal grains and those isolated from soil pointed primarily to the role of the soil *Fusarium* species in the infection of cereals and in the process of toxin formation (Joffe, 1950, 1960a,b, 1965, 1978b, 1983b).

A relationship was also established between the fungal flora of over-wintered cereals and the soil microflora in 1944 and 1945 (Geimberg and Babusenko, 1949; Joffe, 1950, 1960a,b, 1965, 1971, 1974c, 1978b; Mironov and Joffe, 1947a,b). The resemblance between microflora of overwintered cereals and of soil indicates the part played by the soil in fungal infection of cereals. The microflora of the experimental fields, as determined in the spring of the years 1947–1949, did not differ materially in generic composition from the fungal population of samples received from other counties (Joffe, 1950).

## Laboratory Analysis of Toxic *Fusarium* Species

### Effect of Temperature on Growth and Toxin Production

In investigating the cryophilic properties of the isolated toxic fungi we mainly used toxic cultures of *F. poae* and *F. sporotrichioides*. For purposes of comparison investigations were conducted concurrently on nontoxic cultures isolated from normal, high-quality, nonoverwintered cereals. These cultures included *F. sporotrichioides, C. epiphyllum,* and others (Joffe, 1962a). Growth under specified conditions was determined daily. Observations indicated that for the most part toxic fungi grew well at temperatures of 0–1°C and that satisfactory growth also took place at 23–25°C. However, cultures isolated from normal cereal samples produced luxuriant growth at 23–25°C, whereas at 0–1°C if any growth occurred, it was generally very scanty.

Additional experiments were carried out in which toxic cultures of *F. poae, F. sporotrichioides, C. epiphyllum, C. fagi, A. tenuis, P. brevicompactum, M. hiemalis,* and *M. racemosus* were grown on the same nutrient media at temperatures of $-2--7°C$. Nontoxic cultures of the same species (with the exception of *F. poae*), derived from normal nonover-wintered cereals, were grown for comparison. Toxic cultures of these fungi developed in 19–47 days at −2−−7°C. It should be stressed here that the development of these fungi took place on media which were in a frozen state. Nontoxic cultures showed no growth during the entire ex-perimental period which extended over 72 days.

Results showed that toxic cultures of *F. poae, F. sporotrichioides, C. epiphyllum, C. fagi, M. hiemalis, A. tenuis,* and so on are cryophilic, while nontoxic cultures of these species are not.

There is ample evidence confirming the tolerance to low tempera-tures of fungi belonging to the genera studied.

In view of the fact that in most cases *Fusarium* and *Cladosporium* were found in association on overwintered cereals, it was considered of interest to investigate the effect of the temperature factor on the growth associations of these fungi. Experiments were therefore carried out on *F. poae, F. sporotrichioides, C. epiphyllum,* and *C. fagi.*

A study of the effect of temperature regimes on the growth of *F. poae* and *C. epiphyllum* in mixed culture showed that active growth of *F. poae* takes place at temperatures of +25°C down to −7°C. At temperatures lower than −7°C the growth of *Fusarium* was arrested. *C. epiphyllum* developed actively at low temperatures from −2−−10°C.

Experiments conducted with *F. sporotrichioides* showed that growth of these cultures took place at temperatures down to −2°C, while cul-tures of *C. epiphyllum* also grew between −2 and −10°C. *F. sporo-trichioides* and *C. fagi* yielded analogous results. It appears from this experiment that *F. poae* in mixed cultures grows at temperatures from 25°C down to −7°C, but not from −7−−10°C. *Cladosporium* grows at temperatures from 25−−10°C but its growth was weaker than *Fusarium* in all tests.

The results of our experiments were in agreement with observations conducted under natural conditions. Various forms of growth associa-tions of fungi were observed at different times of the year, both in sam-ples supplied from counties of the Orenburg district and from experi-mental plots. During the autumn-winter period copious growth of *Cladosporium* was in evidence on the ears of cereals, while the amount of *Fusarium* was insignificant. During the spring, however, *Fusarium* was predominant on contaminated ears (Figs. 6.2 and 6.3).

In order to determine the effect of temperature factors on the forma-tion of toxin, a number of experiments were set up with *F. poae* and *C. epiphyllum.* Pure and mixed cultures of these fungi were sown on potato, potato-acid agar and carbohydrate-peptone media and also on sterile moistened millet grain. Growth took place at temperatures listed in Table 6.13. Within a month of initiation of the experiment, ether extracts were prepared from dried experimental cultures and transferred onto rabbit skin.

**TABLE 6.13  Accumulation of Toxin in Fungi Due to Sharp Temperature Fluctuations**

| | Toxin Accumulation | | | | | | | | |
| | *F. poae* | | | *C. epiphyllum* | | | *Fusarium Cladosporium* | | |
| Conditions | L | E | N[a] | L | E | N | L | E | N |
|---|---|---|---|---|---|---|---|---|---|
| Room temperature (18°C) | − | +[b] | − | + | + | − | − | − | − |
| In snow | + | + | ++ | + | ++ | − | − | +++ | + |
| Room temperature—in snow | − | − | − | − | − | − | +++ | + | − |
| In snow—room temperature | − | − | − | − | − | − | − | + | ++ |
| Room temperature—freezing—room temperature | +++ | +++ | +++ | + | +++ | − | − | − | − |
| In snow—freezing (−2°C) (10°C) | +++ | +++ | +++ | ++ | + | − | − | − | − |
| Room temperature—freezing—in snow | − | − | − | − | − | − | − | +++ | − |
| In snow—freezing—in snow | − | − | − | − | − | − | − | +++ | − |
| In snow—freezing—room temperature | − | − | − | − | − | − | − | +++ | ++ |
| Alternating room temperature—freezing | − | − | − | − | − | − | − | +++ | + |
| Alternating in snow—freezing | − | − | − | − | − | − | − | ++ | − |

*Source:*  From Joffe, 1962.

[a]Reactions: L = leukocytic; E = edematous; N = necrotic.

[b]Degree of toxicity: + = mildly toxic; ++ = toxic; +++ = strongly toxic.

Table 6.13 shows that cultures of *Fusarium* and *Cladosporium* grown at low temperatures were more toxic than cultures maintained at room temperature. Sharp fluctuations of temperature greatly increased the toxicity of extracts, which produced acute edema and necrosis when applied to rabbit skin. Pure cultures of *F. poae* and mixed cultures of *F. poae* and *C. epiphyllum* led to a more intense skin reaction than *C. epiphyllum* alone. *C. epiphyllum* is characterized by an edemo-leukocytic reaction, *F. poae* by an edemo-necrotic reaction. Mixed cultures were characterized as a rule by strong edema or edematous necrotic reaction. Three repetitions of the tests produced analogous results.

In order to clarify the dependence of toxin formation on temperature and the developmental stage of the fungus many experiments were carried out in 1946 and again in 1952 (Joffe, 1971, 1974c). Pure cultures of *F. poae* and *F. sporotrichioides* sown on liquid medium (synthetic with starch and carbohydrate-peptone) were grown at different temperatures. At each of the three stages of development (namely prior to sporulation), at the stage of abundant sporulation, and at senescence, toxicity assays were carried out with native filtrate, evaporated filtrate, extracts of the fungal mass, and ether extracts of the liquid media by means of skin tests on rabbits. The liquid substrates were killed, passed through a Seitz filter, and the sterile filtrates thus obtained were tested for toxicity on white mice by s.c. injections of 0.2, 0.5, and 1.0 ml. Results obtained with *F. poae* and *F. sporotrichioides* are given in Table 6.14 and show clearly that injection of filtrates of cultures obtained at different stages of development at a temperature of 23–25°C was not lethal to white mice.

Extracts of cultures maintained at low temperatures, as well as those grown at 6–12°C, with intervening freezing caused death of mice within 12–48 hr, depending on dosage and fungal species, due to systemic toxemia. Postmortem examination disclosed necroses in the digestive system and other organs in all cases. Highest toxicity was displayed by filtrates of *Fusarium* obtained during the stage of abundant sporulation from cultures grown at 1–2°C, while extracts obtained at an advanced stage of senescence were considerably less toxic or nontoxic.

Heating of filtrates at 100°C for 30 min and passing through an asbestos filter did not result in any reduction in toxicity. In some cases the presence of toxin in the filtrate could be detected at a very early stage of development. This was apparently due to the fact that spores were occasionally present during the first stage, even though in far smaller numbers than in the second stage.

Results of investigations on the toxicity of native liquid and of ether extracts thus proved that active accumulation of toxin takes place under low temperature conditions, particularly in the abundant sporulation stage.

The tabulated data bring out another interesting characteristic: the application of ether extracts of liquid substrate to rabbit skin produced a

**TABLE 6.14  Effect of Temperature and Development Stage of *Fusarium* Species on Toxin Formation**

| Temperature (°C) | Development Stage of *Fusarium* | *F. poae* No. 958 Effect on Mice[a] | | | *F. poae* No. 958 Toxicity to Rabbits[b] | | | *F. sporotrichioides* No. 921 Effect on Mice | | | *F. sporotrichioides* No. 921 Toxicity to Rabbits | | |
|---|---|---|---|---|---|---|---|---|---|---|---|---|---|
| | | 0.2 | 0.5 | 1.0 | L | E | N | 0.2 | 0.5 | 1.0 | L | E | N |
| 24–30° for 10–40 days | Before sporulation | A[c] | A | A | 0 | 0 | 0 | A | A | A | 0 | 0 | 0 |
| | Sporulation | A | A | A | 0 | +[d] | 0 | A | A | A | ± | 0 | 0 |
| | Senescence | A | A | A | ± | 0 | + | A | A | A | 0 | 0 | 0 |
| 6–12° for 21–76 days | Before sporulation | A | A | A | 0 | 0 | 0 | A | A | A | 0 | 0 | 0 |
| | Sporulation | D | D | D | 0 | ++++ | +++ | D | D | D | 0 | ++++ | +++ |
| | Senescence | A | A | A | 0 | + | 0 | A | A | A | 0 | 0 | 0 |
| 1–-2° for 40–104 days | Before sporulation | A | D | D | 0 | + | + | D | D | D | 0 | 0 | 0 |
| | Sporulation | D | D | D | 0 | ++++ | ++++ | D | D | D | ++ | ++++ | ++++ |
| | Senescence | A | A | A | 0 | + | 0 | A | A | A | 0 | 0 | 0 |

*Source:*  Joffe, unpublished results.

[a] Subcutaneous application of ether crude extract (0.2, 0.5, and 1.0 ml) to mice.

[b] Skin reaction of rabbit to ether crude extract: L = leukocytic; E = edematous; N = necrotic.

[c] Condition of mice: A = alive; D = dead.

[d] Degree of toxicity: 0 = nontoxic; ± = very slightly toxic; + = slightly toxic; ++ = moderately toxic; +++ = toxic; ++++ = strongly toxic.

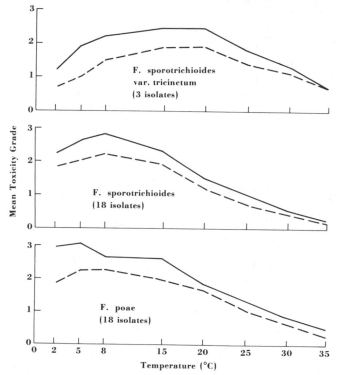

**Figure 6.7**   Effect of temperature and light on toxicity of the Sporotrichiella section. Solid line, dark; broken line, light (Joffe, 1974b).

stronger reaction than extracts of the fungal film of *F. poae* and *F. sporotrichioides* cultures, indicating that toxins of these two fusaria are excreted into the surrounding medium and thus act as exotoxins.

The overall toxicity produced by extracts from wheat or millet cultures grown in light and darkness at eight temperatures, ranging from 2–35°C, is shown in Figure 6.7. *F. poae* and *F. sporotrichioides* produced highest toxicity in both light and darkness at 5 and 8°C and *F. sporotrichioides* var. *tricinctum* at 15 and 20°C. Darkness clearly favored toxicity development in this, as in the other two, species and, in general, darkness obviously favored the development of toxicity in all three fungi at all temperatures.

Very similar results concerning temperature and light effects on production of toxicity by these species were obtained in additional test series in which fungi were grown on liquid substrate (Joffe, 1974b).

There is very little literature on the relationship between toxic properties of fungal species and the temperatures in which they were grown. Only Sarkisov and Kvashnina (1948) found that cultures of *F. sporotrichioides* grown at 1.5–4°C were more toxic than when grown at 22–25°C. Lacicowa (1963) found that development of *F. poae* on potato-dextrose agar and maltose agar media was best at 24–25°C, which is in agreement with our results. This author stated that the fungus developed equally well in light and darkness.

Seemüller (1968), working with two to six isolates of the following fungi, determined temperature relationships as follows: *F. poae* had its minimum growth at 2.5°C, optimum at 22.5–27.5°C, and maximum at 32.5°C, the latter being appreciably lower than in our work (35°C). *F. sporotrichioides* and, what Seemüller called *F. tricinctum* (*F. sporotrichioides* var. *tricinctum*), had lower minima at 0°C, optima at 27.5 and 22.5°C, respectively, and maxima at 35 and 32.5°C, respectively. Seemüller's *F. chlamydosporum* (*F. sporotrichioides* var. *chlamydosporum*) had a higher minimum at 5°C, an optimum at 27.5°C, and maximum at 35°C or slightly above. This agrees with our observation that *F. sporotrichioides* var. *chlamydosporum* grew better at 35°C than any of the other fungi.

There are some reports in the literature concerning the effect of temperature on laboratory production of trichothecenes and other toxins by *Fusarium* species. Thus, for example, *F. tricinctum* strain T-2 produced T-2 toxin and diacetoxyscirpenol when cultured at 8°C (Bamburg *et al.*, 1968b) and HT-2 toxin when cultured at 24°C (Bamburg and Strong, 1969). *F. poae* and *F. sporotrichioides* were much more toxic when grown at 8°C than at 25°C (Joffe, 1974b). Siegfried (1977), working with *F. tricinctum*, found that more total toxin was produced at 22° than at 8°C. However, at 22°C, 87% of the toxin produced was T-2 toxin and 13% diacetoxyscirpenol. At 8°C the relative amounts were 96 and 4%. The total toxin production could be greatly increased (from 14.76 to 46.71 mg) by alternating the temperatures and using a regime of 5 days at 22°C, 22 days at 8°C, and then again at 22°C.

In general the toxicity of the *Fusarium* species producing type A toxins is enhanced by low temperatures. However, the production of the principal toxin in the type B group, fusarenon-X, is also favored by higher temperatures (Ueno *et al.*, 1973d). Thus, the optimal temperature for fusarenon-X production was 27°C, and less toxin was produced at lower temperatures (Ueno *et al.*, 1970c). These temperature effects may explain, at least partially, the geographical distribution noted by Ueno *et al.* (1973d).

### Effects of Substrate and pH on the Growth of Toxic Fungi

As shown by our observations, the extent of toxin formation by cultures of *F. poae*, *F. sporotrichioides*, *C. epiphyllum*, and *A. tenuis* is not only determined by temperature. Both the character of growth and the toxin-forming propensities of certain fungi may be modified by the substrate used, and the type of substrate may even mask temperature effects.

Joffe (1974b) found that with *F. poae*, *F. sporotrichioides*, and *F.*

*sporotrichioides* var. *tricinctum* better yields of toxin were produced on starch medium than on Czapek medium or other carbohydrate medium. Barley substrates gave higher toxicity than wheat or prosomillet with *F. poae* or *F. sporotrichioides,* although prosomillet was the best substrate for *F. sporotrichioides* var. *tricinctum.* In addition more toxin was produced in darkness than in light at all temperatures, although here again lower temperatures favored T-2 toxin production irrespective of the medium used, even though these low temperatures are not optimal for fungal growth. Thus although 24°C was the optimum growth temperature for *F. poae* and *F. sporotrichioides,* the highest toxicity in the rabbit skin test was obtained from cultures grown at 6°C (Joffe, 1974b).

We used various natural media (millet, wheat, barley, oats, rice, potato, etc.), as well as solid and liquid synthetic media, for culturing toxic fungi (Joffe, 1950, 1960a,b) and investigated numerous sources of nitrogenous and carbohydrate nutrients and their effect on the formation of toxin by fungal cultures. The following sources of nitrogen and carbon were tested: peptone, casein, glycocol, cysteine, albumin, asparagine, alanine, arginine, histidine, tryptophan, tyrosine, glutamic acid, urea, ammonium sulfate, sodium nitrate, sodium nitrite, ammonium nitrate, L-arabinose, D-arabinose, dextrose, galactose, glucose, saccharose, maltose, lactose, D-mannose, D-fructose, mannitol, starch, amylose, pectin, cellulose, sodium citrate, sodium acetate, and sodium oxalate. These tests involved 274 cultures of toxic fungi.

The best nutrient sources among organic substances for *F. poae* and *F. sporotrichioides* proved to be carbohydrates (starch and glucose), while peptone and asparagine were the best nitrogen suppliers. Best results among inorganic substances were obtained with ammonium sulfate and sodium nitrate. Organic acids produced very meagre growth of *Fusarium.* It is worth noting that satisfactory development and production of toxic substances were obtained on filter paper with cultures of *F. poae* and *F. sporotrichioides* (Joffe, 1971).

Problems concerning the nutritional physiology of *F. sporotrichioides* were investigated by Sarkisov (1948, 1954) and Kvashnina (1948). Among the substances tested, the best sources of nitrogen and carbon were found to be peptone, casein, asparagine, glucose, starch, and mannitol. Bilai (1947, 1953) stated that aspartic acid, glutamic acid and their amides, alanine, glycol, and ammonium carbonate, as well as gaseous ammonia, all provided suitable sources of nitrogen for *Fusarium* species of the Sporotrichiella section.

The effect of substrate on overall toxicity production was first studied by growing *F. poae* and *F. sporotrichioides* at six different temperatures, ranging from 2–35°C, on three grain substrates (wheat, barley, and prosomillet) and on liquid substrates IV, V, and VII (*see* p. 265).

In the case of the liquid cultures, toxicity was determined separately for the mycelial mass grown on the liquid (thallus) and for the substrate after removal of that mass. Results are summarized in Table 6.15. These figures show that toxicity derived from the liquid substrate was always higher than that from the thallus grown on that substrate. Overall toxin production was therefore strongest on the starch substrate and, in the case of *F. poae*, it was somewhat stronger on Czapek's than on the carbohydrate peptone substrate.

Among the grain substrates barley yielded growth with higher overall toxicity of *F. poae* and *F. sporotrichioides* both in light and darkness. The favorable effect of darkness on toxin production was evident on all three substrates.

As regards the source of isolates, those of *F. poae* were most highly toxic when isolated from soil and those of *F. sporotrichioides* when isolated from barley. Here again darkness favored toxin production regardless of the source of the isolate.

Further studies with liquid substrates at three pH levels were carried out with *F. poae* and *F. sporotrichioides* at 8 and 25°C. All eight substrates listed in this chapter were used but, since substrates I and II gave very similar results, those obtained with substrate II are omitted. The purpose of these studies was to ascertain the effect of the various substrates on the toxicity rating of toxic isolates and the extent to which toxic and nontoxic isolates changed the pH level of substrates on which they were grown. Results obtained (Joffe, 1974b) are given in Table 6.16, and show that *F. poae* and *F. sporotrichioides* produced higher toxicity ratings at 8°C than at 25°C. This rating was highest on all substrates at pH 5.6. It may be assumed that toxin formation is also conditioned by the acidity of the medium. The most suitable pH values were found to be 4.6–5.4 for *Fusarium*.

*Fusarium Growth and Toxigenicity.*   Is there any relation between the growth of these fungi *in vitro* and the degree of toxicity of their culture extract to rabbit skin? Conclusions may be drawn from results obtained with five isolates of *F. poae* and six of *F. sporotrichioides* showing strong toxicity (Joffe, 1974b). Growth peak for both species was at 18–24°C where toxicity was slight. Toxicity was strongest at 6–12°C where growth was limited. At temperatures of 30–35°C growth was minimal and toxicity weak or absent. There would seem, therefore, to be no relation between vigor of growth at any of the above temperatures and the degree of toxicity of these fungi. This is clearly shown in Figure 6.8.

Repeated investigation has shown that low temperatures promote rapid accumulation of toxin in cultures of *F. poae* and *F. sporotrichioides*, despite the slow growth of the mycelia. Cultures grown at high temperatures, while producing luxuriant mycelial growth, were nontoxic or only slightly toxic. It should be noted that absence of sporulation was occasionally observed in cultures of *F. poae* and *F. sporotrichioides* grown at room temperature, 18–20°C. Such cultures were

TABLE 6.15 Effect of Substrate and Source of Isolate on Toxin Production (Mean of Results Obtained at Eight Temperatures)

| | F. poae | | F. sporotrichioides | | F. sporotrichioides var. tricinctum | |
|---|---|---|---|---|---|---|
| **Mean Grade of Toxicity per Isolate** | | | | | | |
| Liquid substrates | 9 isolates | | 9 isolates | | 3 isolates | |
| | Liquid | Thallus | Liquid | Thallus | Liquid | Thallus |
| No. IV Carbohydrate-peptone | 1.2[a] | 0.5[a] | 1.4[a] | 0.8[a] | 1.5[b] | 0.7[b] |
| No. V Czapek's | 1.4 | 0.7 | 1.5 | 0.9 | 1.5 | 0.7 |
| No. VII Starch | 1.6 | 0.8 | 1.8 | 1.0 | 1.7 | 1.0 |
| Mean | 1.4 | 0.7 | 1.6 | 0.9 | 1.6 | 0.8 |
| Grain substrates | 18 isolates | | 18 isolates | | 3 isolates | |
| | Light | Dark | Light | Dark | Light | Dark |
| Prosomillet | 1.3 | 2.2 | 1.5 | 2.0 | 1.6 | 2.0 |
| Wheat | 1.3 | 1.8 | 1.3 | 1.9 | 1.4 | 1.9 |
| Barley | 1.7 | 2.3 | 1.7 | 2.1 | 1.1 | 1.8 |
| Mean | 1.4 | 2.1 | 1.5 | 2.0 | 1.4 | 1.9 |
| Source of isolates | | | | | | |
| Prosomillet | 1.3 | 2.0 | 1.4 | 1.9 | 1.5 | 2.1 |
| Wheat | 1.7 | 2.2 | 1.2 | 1.7 | | |
| Barley | 1.2 | 1.8 | 2.2 | 2.8 | 1.1 | 1.4 |
| Rye | 1.4 | 2.1 | 1.7 | 1.9 | | |
| Soil | 2.1 | 2.6 | 1.1 | 1.5 | | |

*Source:* From Joffe, 1974b.

[a] Each figure in this column represents the mean of 144 cultures.
[b] Each figure in this column represents the mean of 48 cultures.

**TABLE 6.16** Effects of Substrates and Temperatures on Toxicity Produced by Isolates of *F. poae* and *F. sporotrichioides* and the Changes Induced by These Isolates in the pH Values of the Substrate (Means for All Isolates Tested)

| | | F. poae | | | | | | F. sporotrichioides | | | | | |
| | | 7 toxic isolates | | | | 2 nontoxic isolates | | 7 toxic isolates | | | | 2 nontoxic isolates | |
| | | 8°C | | 25°C | | 8°C | 25°C | 8°C | | 25°C | | 8°C | 25°C |
| Substrate No. [a] pH | | T [b] | pH | T | pH | T | pH | T | pH | T | pH | T | pH |
|---|---|---|---|---|---|---|---|---|---|---|---|---|---|
| I | 3.8 | 0.6 | 4.0 | 0.1 | 4.4 | 3.7 | 3.5 | 1.1 | 3.9 | 0.4 | 4.2 | 3.6 | 3.8 |
|   | 5.6 | 3.7 | 6.3 | 1.9 | 6.7 | 5.6 | 5.5 | 2.1 | 6.5 | 1.2 | 7.0 | 6.0 | 6.3 |
|   | 7.2 | 2.4 | 7.1 | 1.2 | 7.3 | 7.0 | 7.2 | 1.6 | 7.3 | 0.5 | 7.6 | 7.1 | 7.1 |
| III | 3.8 | 1.7 | 4.5 | 0.4 | 5.0 | 4.0 | 4.3 | 0.5 | 4.2 | 0.2 | 4.5 | 4.1 | 4.2 |
|   | 5.6 | 2.2 | 6.6 | 0.5 | 7.0 | 5.6 | 6.1 | 1.4 | 6.3 | 0.5 | 7.3 | 6.0 | 6.2 |
|   | 7.2 | 1.7 | 7.4 | 0.3 | 7.7 | 7.1 | 7.2 | 1.6 | 7.5 | 0.8 | 8.0 | 7.3 | 7.4 |
| IV | 3.8 | 1.1 | 3.8 | 0.3 | 4.3 | 3.3 | 3.4 | 0.6 | 3.7 | 0.3 | 4.1 | 3.6 | 3.2 |
|   | 5.6 | 2.6 | 6.1 | 0.6 | 6.7 | 5.5 | 6.1 | 1.7 | 6.1 | 0.6 | 6.9 | 5.7 | 5.8 |
|   | 7.2 | 1.3 | 7.2 | 0.3 | 7.3 | 6.6 | 7.2 | 1.6 | 7.1 | 1.1 | 7.4 | 6.8 | 7.1 |
| V | 3.8 | 0.8 | 3.5 | 0.5 | 4.1 | 3.5 | 3.4 | 0.9 | 3.6 | 0.5 | 3.9 | 3.7 | 4.0 |
|   | 5.6 | 2.2 | 5.8 | 1.6 | 6.3 | 5.7 | 5.7 | 1.9 | 5.8 | 1.2 | 6.5 | 6.0 | 6.0 |
|   | 7.2 | 1.6 | 7.0 | 0.8 | 7.4 | 6.8 | 6.9 | 1.3 | 7.2 | 0.8 | 7.5 | 7.4 | 7.5 |
| VI | 3.8 | 1.7 | 4.3 | 0.9 | 5.0 | 4.0 | 4.1 | 0.6 | 4.4 | 0.4 | 5.0 | 4.1 | 4.2 |
|   | 5.6 | 2.7 | 6.5 | 2.0 | 7.2 | 6.1 | 6.5 | 1.9 | 6.5 | 1.4 | 7.2 | 6.1 | 6.3 |
|   | 7.2 | 1.6 | 7.4 | 1.0 | 8.1 | 7.3 | 7.5 | 2.9 | 7.6 | 0.9 | 8.0 | 7.2 | 7.3 |
| VII | 3.8 | 2.1 | 4.4 | 1.7 | 4.9 | 4.1 | 4.5 | 1.1 | 4.6 | 0.7 | 5.2 | 4.1 | 4.6 |
|   | 5.6 | 3.3 | 7.1 | 2.4 | 7.3 | 6.1 | 6.4 | 3.1 | 6.6 | 1.8 | 7.5 | 6.2 | 6.7 |
|   | 7.2 | 1.7 | 7.9 | 1.5 | 8.3 | 7.3 | 7.4 | 2.2 | 7.9 | 1.6 | 8.7 | 7.5 | 7.8 |
| VIII | 3.8 | 1.0 | 4.0 | 0.4 | 4.4 | 3.7 | 3.8 | 0.9 | 3.9 | 0.2 | 4.4 | 4.3 | 4.5 |
|   | 5.6 | 1.9 | 6.0 | 1.2 | 6.5 | 5.7 | 6.1 | 1.4 | 6.2 | 0.7 | 6.7 | 6.5 | 6.5 |
|   | 7.2 | 1.5 | 7.3 | 1.4 | 7.5 | 6.7 | 6.6 | 1.1 | 7.2 | 0.9 | 7.9 | 7.2 | 7.5 |

*Source:* Adapted from Joffe, 1974b.

[a] Composition of substrates is detailed in the section on methods in this chapter.

[b] Toxicity rating (0–4).

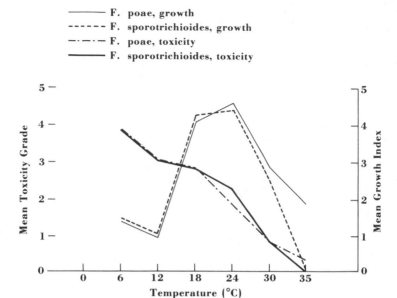

**Figure 6.8** Relation between growth *in vitro* and degree of toxicity of *F. poae* (five isolates) and *F. sporotrichioides* (six isolates) at various temperatures (Joffe, 1974b).

shown to be nontoxic or slightly toxic in rabbit skin tests. Cultures of these two fungi, grown at low temperatures or in conditions of alternate freezing and thawing, were characterized by prolific spore production and strong toxicity.

The presence of abundant nonsporulating aerial mycelium usually coincided with an absence of toxicity. However, scanty aerial mycelium with a large number of spores is normally associated with high toxicity. Toxicity in these cultures is apparently associated with intensive spore production.

The presence of pigmentation in the Sporotrichiella section was an indication of toxicity. But *Fusarium* cultures from various other sections with an insignificant amount of pigment infrequently proved to be highly toxic, while strongly pigmented cultures sometimes had low toxicity. The germination rate of spores of various toxic fungi was also found to be dependent on temperature (Joffe, 1962a, 1973b).

*Toxic Fusarium Species and their Variability.* In the course of studies on the toxicity of isolated cultures we met with instances of morphological and cultural variability of strains of *Fusarium* and also a weakening of their toxic properties. The variability within the genus *Fusarium* has been mentioned by many authors. According to Appel and Wollenweber (1910), Brown (1928), and Brown and Horn (1926) cultures of *Fusarium* mutate readily. Similar observations have been reported by Leonian (1929), who obtained some 50 varieties from a single species.

He noted considerable variability in *Fusarium* cultures as regards ability to produce spores and stated that "the only thing constant in *Fusarium* is its variability." The great variability of morphological characteristics within the genus *Fusarium* has also been indicated by Raillo (1935, 1936, 1950) and Bilai (1955, 1970a,b, 1977, 1978a,b).

Growth of fungal mycelia, type of sporulation, intensity of pigmentation, morphology of conidia and sclerotia, and other attributes vary according to environmental conditions and to the derivation and age of the culture. Snyder and Hansen (1940, 1945) and Snyder *et al.* (1975) observed a marked variability of *Fusarium* isolates. Prasad (1949) noted morphological and cultural variability in connection with his studies of 33 cultures of *F. solani* f. *cucurbitae*. Subramanian (1951) recorded differences in the character of growth and spore formation in *Fusarium* cultures belonging to the Elegans, Martiella, Gibbosum, and Arthrosporiella sections which had been isolated from different plants and soils.

We have established that the morphological and cultural properties of fungi isolated from cereals overwintered under snow display great variability under the influence of unfavorable ecological conditions. This variability is also associated with changes in physiological behavior (Joffe, 1956b,c). Changes in morphological and cultural properties of toxic cultures of *Fusarium* were often seen in connection with frequent transferences into liquid and solid media. These changes were linked with loss of toxicity, proven by application to rabbit skin and by animal feeding experiments. Pure *Fusarium* cultures, sown on liquid and carbohydrate-peptone agar medium and on sterile normal millet under room temperature conditions, underwent marked changes in their morphological characteristics and their toxic properties weakened. Degree of toxicity was determined by biological rabbit skin tests and s.c. injections of liquid substratum of cultures into mice (Joffe, 1960a).

Only in the toxic and strongly toxic strains isolated from overwintered cereals (e.g., *F. poae* and *F. sporotrichioides*) is an unimpaired capacity for high-level production of T-2 toxin retained. This phenomenon is certainly connected with the specific environmental conditions obtained in the Orenburg district in the years when these strains were originally isolated (and ATA outbreaks were widespread).

## Toxin Production in *Fusarium* Cultures and Cereal Samples in Subtropical and Temperate Areas

Over 17,000 strains of *Fusarium* from semitropical (Israel) and temperate (the United States and the USSR) continental climatic areas and varying environmental conditions have been isolated (*see* Table 11.4). Hundreds of these strains were investigated morphologically according to our modern taxonomy (Joffe, 1974a, 1977; Joffe and Palti, 1975), as well as for their ability to produce mycotoxins in laboratory conditions (Joffe, 1960a,b, 1965, 1971, 1974c, 1978b, 1983b; Joffe and Eppley, unpublished results; Joffe and Yagen, 1977; Yagen and Joffe, 1976; and Yagen *et al.*, 1977, 1980).

The effects of temperature, nutrient media, humidity, light, age, nature of sporulation, growth, pH, geographical distribution, origin, and other factors on toxin production *in vitro* under laboratory conditions were studied. With regard to temperature Table 5.2 shows that the majority of these *Fusarium* species, such as *F. solani, F. oxysporum, F. moniliforme,* and *F. equiseti,* isolated in semitropical regions (Israel), were weak to moderate toxin producers in the 18–30°C range. In temperate areas, however, these species appeared in small numbers and showed an insignificant degree of toxicity. There is a very important exception: the species of the Sporotrichiella section, *F. poae* and *F. sporotrichioides,* isolated in the Orenburg district, produced most of their toxins in the 6–12°C range (Table 6.17). Determination of the amount of T-2 toxin isolated from these various strains at different temperatures has given the same results. Much more T-2 toxin was produced at 6 and 12°C, while at higher temperatures (24–35°C) toxin production was minimal or absent.

It is of interest to note that, in addition to toxic strains of *F. poae, F. sporotrichioides,* and *F. sporotrichioides* var. *tricinctum* isolated in the United States, many moderately to strongly toxic isolates of *F. equiseti* var. *acuminatum* and *F. sambucinum* were also found (Table 6.18). In the USSR the last two strains were isolated infrequently and were less toxic (Table 6.17). In general, moderate and strongly toxic *F. sporotrichioides, F. poae,* and *F. sporotrichioides* var. *tricinctum* were isolated only from temperate regions and none from semitropical areas such as Israel (Table 5.2).

We feel that much more attention should be given to the geographical distribution and origin of strains and to the ecological field conditions favoring toxin production. This is probably the best way to predict where the danger of toxin production exists and then minimize it by adjusting field conditions. The importance of origin of toxigenic strains is clearly shown by a comparison of trichothecene T-2 toxin production by authentic ATA strains from the USSR and strains used in American studies. The Russian strains produced greater amounts of toxin than the American strains, as shown by rabbit skin (Joffe, 1960a) and brine shrimp tests (Joffe and Eppley, unpublished results) and by physical and chemical analysis including TLC, GLC, and mass spectroscopy (Joffe and Yagen, 1977; Yagen and Joffe, 1976; Yagen *et al.*, 1977).

The yield of T-2 toxin produced by strains from the USSR was of the order of 20–24 mg/10 g wheat grain for four samples, and between 4.6–15.2 mg/10 g wheat for the other seven strains (*see* Table 3.12). By contrast the strains from the United States in general produced T-2 toxin

**TABLE 6.17 Toxicity of *Fusarium* Strains Isolated from Many Substrates in Temperate Regions of the Orenburg District and Incubated at Various Temperatures (Rabbit Skin Test and Brine Shrimp Test)**

| Species | No. of Isolates | 6° | | | 12° | | | 18° | | | 24° | | | 30° | | | 35° | | |
|---|---|---|---|---|---|---|---|---|---|---|---|---|---|---|---|---|---|---|---|
| | | S | M | W[a] | S | M | W | S | M | W | S | M | W | S | M | W | S | M | W |
| *F. sporotrichioides* | 83 | 54 | 19 | 3 | 51 | 20 | 5 | 40 | 14 | 16 | 8 | 8 | 10 | 0 | 0 | 12 | 0 | 0 | 0 |
| *F. poae* | 81 | 53 | 15 | 8 | 44 | 15 | 11 | 24 | 20 | 19 | 0 | 14 | 14 | 0 | 0 | 3 | 0 | 0 | 1 |
| *F. sporotrichioides* var. *tricinctum* | 11 | 1 | 1 | 3 | 1 | 0 | 4 | 1 | 0 | 7 | 0 | 1 | 5 | 0 | 0 | 3 | 0 | 0 | 1 |
| *F. sporotrichioides* var. *chlamydosporum* | 10 | 2 | 0 | 0 | 2 | 0 | 0 | 1 | 1 | 1 | 0 | 1 | 4 | 0 | 0 | 1 | 0 | 0 | 1 |
| *F. solani* | 41 | 0 | 0 | 0 | 0 | 0 | 0 | 0 | 0 | 6 | 0 | 1 | 13 | 0 | 1 | 6 | 0 | 0 | 0 |
| *F. oxysporum* | 44 | 0 | 0 | 0 | 0 | 0 | 0 | 0 | 0 | 4 | 1 | 3 | 14 | 0 | 0 | 7 | 0 | 0 | 0 |
| *F. moniliforme* | 44 | 0 | 0 | 0 | 0 | 0 | 0 | 0 | 0 | 2 | 0 | 1 | 11 | 1 | 3 | 6 | 0 | 0 | 1 |
| *F. equiseti* | 88 | 0 | 0 | 1 | 1 | 1 | 3 | 2 | 2 | 4 | 7 | 5 | 7 | 2 | 1 | 2 | 1 | 0 | 1 |
| *F. equiseti* var. *acuminatum* | 9 | 0 | 0 | 0 | 0 | 0 | 0 | 0 | 0 | 4 | 1 | 1 | 3 | 0 | 2 | 2 | 0 | 0 | 0 |
| *F. lateritium* | 53 | 0 | 0 | 0 | 0 | 0 | 1 | 0 | 0 | 3 | 2 | 2 | 12 | 2 | 3 | 13 | 0 | 1 | 0 |
| *F. semitectum* | 27 | 0 | 0 | 0 | 0 | 0 | 0 | 0 | 0 | 1 | 2 | 2 | 2 | 0 | 1 | 1 | 0 | 0 | 0 |
| *F. avenaceum* | 43 | 0 | 0 | 0 | 0 | 0 | 1 | 1 | 1 | 3 | 3 | 3 | 5 | 2 | 2 | 3 | 0 | 0 | 1 |
| *F. arthrosporioides* | 19 | 0 | 0 | 0 | 0 | 0 | 1 | 0 | 0 | 3 | 0 | 1 | 2 | 0 | 0 | 4 | 0 | 0 | 0 |
| *F. nivale* | 33 | 0 | 0 | 0 | 0 | 2 | 1 | 0 | 4 | 12 | 0 | 0 | 8 | 0 | 0 | 0 | 0 | 0 | 0 |
| *F. culmorum* | 17 | 0 | 0 | 0 | 0 | 0 | 0 | 0 | 0 | 2 | 0 | 4 | 3 | 0 | 1 | 5 | 0 | 0 | 1 |
| *F. sambucinum* | 25 | 0 | 0 | 0 | 0 | 0 | 0 | 2 | 1 | 3 | 2 | 2 | 5 | 0 | 3 | 6 | 0 | 0 | 3 |
| *F. graminearum* | 14 | 0 | 0 | 0 | 0 | 0 | 0 | 0 | 0 | 6 | 0 | 1 | 9 | 0 | 2 | 3 | 0 | 0 | 0 |

*Source:* Joffe, unpublished results.

[a]S = strongly toxic; M = medium; W = weak.

**TABLE 6.18**  Toxicity of *Fusarium* Strains Isolated from Cereal Grains in the United States (1977) and Incubated at 12°C (Brine Shrimp Test)

| Species | No. of Strains | Strong | Medium | Weak | Nontoxic |
|---------|---------|--------|--------|------|----------|
| *F. sporotrichioides* | 76 | 54 | 7 | 1 | 14 |
| *F. poae* | 106 | 58 | 15 | 1 | 32 |
| *F. sporotrichioides* var. *tricinctum* | 54 | 42 | 2 | 2 | 8 |
| *F. sporotrichioides* var. *chlamydosporum* | 15 | 7 | 1 | 2 | 5 |
| *F. moniliforme* | 84 | 4 | 6 | 6 | 68 |
| *F. equiseti* | 11 | 3 | 1 | 2 | 5 |
| *F. equiseti* var. *acuminatum* | 70 | 24 | 12 | 2 | 32 |
| *F. lateritium* | 4 | 1 | 1 | 1 | 1 |
| *F. semitectum* | 18 | 4 | 2 | 0 | 12 |
| *F. avenaceum* | 8 | 1 | 2 | 0 | 5 |
| *F. arthrosporioides* | 4 | 2 | 0 | 1 | 1 |
| *F. dimerum* | 2 | 2 | 0 | 0 | 0 |
| *F. merismoides* | 7 | 3 | 0 | 0 | 4 |
| *F. culmorum* | 11 | 2 | 1 | 1 | 7 |
| *F. sambucinum* | 30 | 10 | 4 | 0 | 16 |
| *F. graminearum* | 35 | 5 | 2 | 1 | 27 |

*Source:*  Joffe, unpublished results.

in amounts between 0.6–5.8 mg/10 g wheat. The isolate from France gave a moderate reaction on rabbit skin and produced 3.4 mg/10 g T-2 toxin (Joffe and Yagen, 1977). Ecological background, environmental factors, and geographical distribution and origin must be considered as the most important reasons for greater T-2 toxin production by strains from the USSR as compared to those from the United States.

Ueno *et al.* (1973d) reported on the geographical differences of *Fusarium* isolates in Japan, based on the type of toxin they produce. Type A toxin producers are found in the northern areas of Japan where the temperatures are lower and include *F. tricinctum* and *F. sporotrichioides;* type B toxin producers are found in the southern region where it is warmer, and they include *F. nivale* and *F. episphaeria.*

The USSR strains were first isolated in the spring of 1944 when ATA occurred with particular severity in the Orenburg district. The special conditions under which overwintering grain was then affected by *Fusarium* species have been described in detail by Joffe (1963a, 1965, 1971, 1974c, 1978b). These conditions may have been conducive to the extraordinarily high levels of T-2 toxin formation in wheat, millet, barley, rye, and other cereals contaminated by fusaria, Sporotrichiella section. Russian strains of *F. sporotrichioides* and *F. poae,* isolated 40 years ago, have retained their high level of T-2 toxin production up to the present day. This indicates a remarkable persistence of toxigenic capability of strains from particular sources, which are associated with special environmental conditions.

Among the strains in our collection associated with the fatal ATA outbreaks, many produced gram quantities of T-2 toxin when grown on 1 kg of wheat or millet grains at 12 or 5°C for a period of 21 and 45 days, respectively. For example, from *F. sporotrichioides* No. 921 a total of 4.1 g of T-2 toxin and from *F. poae* No. 958 a total of 2.8 g T-2 toxin were isolated from 1 kg infected millet at 12°C for 21 days (Yagen *et al.,* 1977).

It should be noted that American and Japanese investigators have in most cases referred to the American strains as *F. tricinctum* according to the Snyder and Hansen (1945) taxonomy. Classification according to our taxonomy is given in Table 9.1 (Joffe, 1974a, 1977; Joffe and Palti, 1975).

### Comparison of Toxicity in Fungi and Cereals

Toxicity of cultures was compared with the toxicity of cereal samples from which they had been isolated. Strongly and mildly toxic fungi were detected on both toxic and nontoxic overwintered cereals in 1945–1949, whereas in 1944, when ATA was widespread in the Orenburg district, mildly and strongly toxic cultures were isolated almost exclusively from toxic samples of overwintered cereals.

A comparison of the type of reaction obtained from highly toxic fungi with that obtained from the respective cereal samples from which they had been isolated (samples gathered in spring 1944) showed that toxic *Cladosporium* cultures, producing a reaction of the leukocytic edematous type, and *Fusarium* cultures, giving an edematous-necrotic reaction, had been isolated only from strongly toxic cereal samples. It is most probable, therefore, that the decisive role in toxicity development in overwintered cereals was played by the *Fusarium* and *Cladosporium* species which produced these types of reaction.

Strukov and Mironov (1944, 1947) studied the histology of the skin to which several toxic fungi from our material had been applied. These authors indicated that the tissue changes which developed following an application of ether extracts of these fungi to the skin of rabbits were similar to reactions caused by extracts from toxic millet.

As we have stated elsewhere (Joffe, 1960a, 1965), toxic species of *Fusarium* were always strongly toxic when inoculated on sterilized grain.

Myasnikov (1948), who tested *F. poae* in our laboratory, found that its toxic ingredient was similar to that contained in overwintered cereals.

Similar results were obtained from our material by Olifson (1955a,b, 1956a,b,c, 1957a,b, 1960, 1962, 1965a,b), who studied the chemical composition of millet after it had been infected experimentally with pure cultures of *F. poae* and *F. sporotrichioides*.

## Mycoflora of Overwintered Cereals and Their Toxicity

We studied a large number of cereal samples collected from trial plots after being covered with snow all winter (Joffe, 1963a), samples of soil on which the plants being investigated had grown, and overwintered cereals from various areas of the Orenburg district. Summer harvested grain and soil collected in the same fields, as well as grain samples from large government storehouses, served as controls.

### *General*

The superficial flora of the grains was studied by culturing on various media. For study of the internal flora of the grains the latter were surface-sterilized and then rinsed and cultured. A variety of synthetic culture media were used as well as substrates prepared from sterile normal prosomillet, barley, wheat, and rice. Fungi which were found to produce toxins on sterilized millet and wheat were invariably found to produce their toxins also on agar or on a liquid medium. Throughout the years of investigation, therefore, toxic cultures were grown on both agar and natural substrates.

The number of genera found on samples of overwintered cereals was proportionately much larger than that found on summer-harvested samples. The genera represented by numerous isolates on overwintered cereals were *Penicillium, Fusarium, Cladosporium, Alternaria*, and *Mucor*. On normal (wholesome) samples only three genera, *Penicillium, Mucor*, and *Alternaria*, were present as a rule, while *Fusarium* and *Cladosporium* were only found in a few cases or were absent altogether.

We isolated 3549 cultures belonging to 42 genera and 192 species (Joffe, 1960a,b) from over 1000 selected samples of overwintered toxic and nontoxic cereals from various regions of the Orenburg district and from the experimental field plots.

***Isolation of Fusarium Strains.*** From each overwintered sample 100–200 grains were dipped in 0.1% solutions of mercuric chloride ($HgCl_2$) or 2–5% hypochlorite for 1–3 min, washed 4–5 times in sterile distilled water, and plated on potato-dextrose agar (PDA), generally five grains per Petri dish. The antibiotic added just before the plates were poured effectively inhibited bacterial growth. Plates were incubated at 24 or 18° for 11 to 15 days, respectively, and the pure cultures maintained as monoconidial isolates were stored on PDA slants in test tubes and on sterilized soil at 3°C. *Fusarium* strains were identified according to the taxonomic system published by Joffe (1974a, 1977, 1978b, 1983b) and Joffe and Palti (1975). Microcultures were also frequently used for identification of the fungi of *Fusarium* and other species.

The pure isolates were grown on different liquid and solid media at varying temperatures (1, 6, 12, 18, 24, 35, and 40°C) for a period of 6 to 76 days, and sometimes from +4 to −5, −7, and −10°C on natural media for a period of 30 to 90 days or longer.

When using natural media, prosomillet, wheat, or barley grains were prepared in 150, 250, 500 ml or 1 liter flat flasks containing 10, 30, 50, or 100 g grain and 15, 45, 75, or 140 ml tap water, respectively. The agar cultures were sometimes placed in a 3 liter flask on whose walls a layer of agar had been deposited by rolling. A variety of liquid media was used (Joffe 1973b, 1974b). The cultures were incubated in 1000 or 2000 ml Fernbach flasks containing 400–800 ml liquid medium, respectively.

The liquid media were as follows:

| | |
|---|---|
| Substrate I | 2 g $KNO_3$, 1 g $KH_2PO_4$, 0.5 g KCl, 0.5 g $MgSO_4$, $7H_2O$, traces of $FeSO_4$, 10 g sucrose, and 1000 cc distilled water. |
| Substrate II | First five compounds as in I, 5 g soluble starch, 2.5 g dextrose, 2.5 g sucrose, and 1000 cc distilled water. |
| Substrate III | 25 g malt extract and 1000 cc distilled water. |
| Substrate IV | 30 g sucrose, 10 g glucose, 1 g peptone, and 1000 cc distilled water (carbohydrate-peptone medium). |
| Substrate V | 2 g $NaNO_3$, 1 g $K_2HPO_4$, 0.5 g KCl, 0.5 g $MgSO_4$, $7H_2O$, traces of $FeSO_4$, 30 g sucrose, 1000 cc distilled water. |
| Substrate VI | 200 g potato, 20 g dextrose and 1000 cc tapwater. |
| Substrate VII | 2 g $NaNO_3$, 1 g $K_2HPO_4$, 0.5 g KCl, 0.5 g $MgSO_4$, $7H_2O$, 0.001 g $FeSO_4$, 25 g soluble starch and 1000 cc distilled water. |
| Substrate VIII | 0.5 g $K_2HPO_4$, 0.25 g $MgSO_4$, $7H_2O$, 0.5 g KCl, 5 g glucose, 3 g yeast extract, and 1000 cc distilled water. |

Media were autoclaved at 0.5–1 atm for 20–30 min or sometimes steam sterilized 2 or 3 times for 1 hr. The pH of the media varied according to the time of cultivation and the type of medium.

***Preparation of Crude Extract.*** After incubating different isolates of *F. poae* and *F. sporotrichioides* on natural media, wheat or millet, at 5 or 12°C for 46 or 21 days, respectively and heating at 100°C for 30 min, they were taken out and dried overnight at 45–50°C and extracted with ethyl alcohol 96% in a vacuum rotary evaporator. Evaporation of the solvents yielded a biologically active oil. To purify the oil from water-soluble compounds it was extracted five times with absolute ethyl alcohol and the combined alcoholic extracts concentrated to about 1 ml. The presence of T-2 toxin in the sample of crude extract was shown by TLC and GLC.

After cultivating and heating the liquid media, infected by strains of *Fusarium*, they were filtered through filter paper using a Buchner funnel or passed through a Seitz filter and then extracted with 96% ethanol or ether and concentrated in a vacuum rotary evaporator.

### Fusarium Species and Their Toxicity

To evaluate the role of *Fusarium* in toxin production, it was necessary to study the toxicity of isolated cultures.

The mycotoxic properties of the *Fusarium* crude extracts were assessed by skin test on rabbits. Details of this bioassay method and the intensity of skin reaction have been described by the author (Joffe, 1960a).

Among the isolates studied, 80 of *F. poae* (out of 82), 76 of *F. sporotrichioides* (out of 83), 8 of *F. sporotrichioides* var. *tricinctum* (out of 32), and 7 of *F. sporotrichioides* var. *chlamydosporum* (out of 15) were toxic. Toxicity results and characteristics of *Fusarium* strains according to substrate, place, and time of isolation are given in tables by the author (Joffe, 1983b).

During a 4 year period (1944–1948) we also isolated 49 toxic strains belonging to various *Fusarium* species from overwintered cereals and soil: 20 of them were strongly toxic and the remaining 29 were slightly toxic (Joffe, 1983b).

All species of *Fusarium* isolated from overwintered toxic and nontoxic grain, summer harvested cereals, or soil, whether collected in experimental fields or received from various parts of the Orenburg district, have been listed by Joffe (1983b).

Among 546 *Fusarium* cultures isolated from these substrates, 40.5% showed toxicity of varying degrees, 20.7% were strongly toxic, and 19.8% slightly toxic (*see* Table 6.12). Strongly and mildly toxic cultures were much less common in *Cladosporium*, and still less in *Mucor, Alternaria*, and *Penicillium*.

In view of the supposition that grain-borne fungi may cause disease, each fungus isolate was tested for toxicity. Joffe (1983b) lists the *Fusarium* species isolated and their respective degrees of toxicity. Results of tests on different types of fungi other than *Fusarium*, isolated from overwintered toxic and nontoxic grain of various cereals and capable of producing toxic compounds, are of great importance and were described by the author (Joffe, 1960b, 1978b, 1983b).

It is clearly shown in the data cited that 199 toxic and strongly toxic cultures and 309 mildly toxic cultures were isolated from overwintered cereals and their soils. No toxic cultures were found on grain or vegetative parts of summer-harvested plants (Joffe, 1960b).

Further studies were aimed at assessing the genera most likely to produce toxins. It was assumed that the toxicogenic properties of different genera of fungi might be estimated from the frequency of their occurrence on overwintered cereals and from the appearance of highly toxic strains among them. From toxic and strongly toxic cultures 13 genera were isolated, and 17 genera were isolated from mildly toxic cultures. The most frequently occurring toxic fungi belonged to the *Alternaria, Mucor,* and *Penicillium* genera and in particular to *Fusarium* and *Cladosporium,* each of which was represented by many species. *F. poae* and *F. sporotrichioides,* both very common on overwintered cereals in all those parts of the Orenburg district where samples had been collected, were present in most cultures. Also common were *C. epiphyllum* and *C. fagi.* Cultures of *A. tenuis, M. hiemalis, M. racemosus, P. brevicompactum, P. steckii,* and others showed considerable toxicity.

Other mycologists working on similar lines obtained varying amounts of toxic *Fusarium* species and other ATA-related fungi from overwintered cereal grains. Murashinski (1934) isolated several strains of *Alternaria* and *Fusarium* from toxic Siberian wheat, and Sirotinina (1945) isolated some *Alternaria* from toxic prosomillet from the Saratov district. Kvashnina (1948) isolated 1227 strains belonging to 83 species, among which she found *Aspergillus calyptratus, Phoma* sp., *Hymenopsis* sp., 14 cultures of *F. poae,* and 32 of *F. sporotrichioides,* all very toxic. These were taken from 107 samples of wheat, rye, oats, barley, peas, and sunflowers gathered in the Altai territory, Bashkiria, the Tatar Republic, the Belorussian Republic, and the Ivanovo, Saratov, Kuibyshev, Tambov, Orenburg, and Yaroslavl districts.

Pidoplichka and Bilai (1946, 1960) and Bilai (1947, 1953) examined some 1400 strains belonging to 23 genera and 160 species, isolated from 765 grain samples (of prosomillet, wheat, oats, and buckwheat) obtained in Bashkiria and the Ukraine. The most toxic isolates were *M. hiemalis* and *M. albo-ater,* and other toxic isolates including *Piptocephalis freseniana, Mortierella polycephala, M. candelabrum* var. *minor, F. lateritium, Gliocladium ammoniophilum, Trichoderma lignorum,* 34 isolates of *F. poae* (only 14 toxic), and 36 of *F. sporotrichioides* (only 12 toxic).

From cultures of *F. poae* and *F. sporotrichioides,* isolated from overwintered wheat, prosomillet, and oats, Bilai (1947, 1953) found 11.5, 7.5, and 4.7%, respectively, to be toxic. In our isolates the incidence of toxicity was generally higher.

The author (Joffe, 1983b) listed the different species, in particular the *Fusarium* fungi, isolated from overwintered and summer-harvested cereals and soil. The data given show that the group most frequently associated with overwintered cereal grains and with producing outbreaks of ATA was *Fusarium* of the Sporotrichiella section, principally *F. poae* and *F. sporotrichioides*. According to Bilai (1947, 1953, 1965), Joffe (1960a, 1965, 1971, 1974c, 1978b), Joffe and Palti (1974a, 1975), Pidoplichka and Bilai (1946, 1960), Rubinstein (1948, 1950a, 1951a,b, 1956a, 1960a,b), Rubinstein and Lyass (1948), and Sarkisov (1944, 1948, 1954), one of the characteristic biological properties of *F. poae* and *F. sporotrichioides* compounds was their inflammatory and irritative action on rabbit skin.

A relationship was established between the nature of the toxic *Fusarium* cultures and the toxicity of the samples from which they had been isolated. Some of the *Fusarium* cultures produced reactions on rabbit skin analogous to those produced by toxic, overwintered cereals (Joffe, 1960a, 1965).

A detailed description of morphological and cultural properties of the toxic fungi of the Sporotrichiella section may be found in the author's publications (Joffe, 1947b,c,d, 1962a, 1971, 1974a, 1977, 1983b) and in Chapter 10 of this volume.

### Antigenic Properties of Some Toxic Fungi

Following our proof of the role played by certain fungi in the etiology of ATA, our material was used by Mironov and Alisova (1947) in work on the immunization of rabbits with toxic cultures of *F. poae* and *C. epiphyllum*, with the aim of determining the antigenic properties of the latter fungi. For the immunization filtrates of cultures, and suspensions of mycelium and spores were used. On the basis of their results Mironov and Alisova concluded that it is possible to obtain specific immunization sera for toxic cultures of both the above fungi. They further found that in order to obtain these sera it is necessary to immunize rabbits with filtrates of cultures started at the stage of intensive sporulation and cultured at a temperature range of −2 to +2°C.

It should be noted that extracts derived from sterile millet grain infected with toxic fungal cultures displayed more pronounced toxic action than extracts from dry mycelial film grown on starch or carbohydrate-peptone medium.

### Distribution of Toxic Fusarium Strains

The distribution of *Fusarium* fungi of the Sporotrichiella section is fairly widespread in plants, soils, and other substrates. Remnants of vegetative parts of cereals in the field, as well as grain left after harvesting, constitute a good medium for fungal development, as do cereals mown, piled up in the field, and then wetted by rain or harvested late in autumn when rains have already begun.

Since toxin may already be present in vegetative parts in autumn (Joffe, 1947d, 1963a), there is a danger that spring-harvested cereal grains will also be toxic. The toxin is not distributed equally in the grain. In prosomillet grains it was observed that there are light and heavy grains. After separating them with 10–25% sodium chloride solution, it was found that the light grains, which floated in the solution, were toxic, whereas the heavy grains were either less toxic or nontoxic (Mironov and Fok, 1944; Mironov et al., 1944, 1947a). This observation led to the assumption that toxin is produced within the grain. If the light, toxic grains are pressed slightly, they turn into powder, in contrast to the heavy, nontoxic grain which cannot easily be ground. The powder thus derived from the fungus-contaminated grains contains the highest concentration of toxin.

The toxic fungus is believed to develop first in the embryo of the grain, and then later the mycelium spreading throughout the grain. This is why the percentage of germination of overwintered grain infected with toxic fungi is much lower than in normal grain (Joffe, 1973b).

### Bioassay Methods of Toxic Fusarium

In feeding experiments on animals we used different bioassay methods. The toxic grain responsible for outbreaks of ATA was investigated chiefly by skin tests on rabbits. This skin test is currently accepted as a laboratory test for toxicity of cereals and also for toxins produced by various fungi.

Culture filtrates, mycelium (dry fungi), infected grains, ethanol, or ether extract of toxic *F. poae*, *F. sporotrichioides*, or *F. sporotrichioides* var. *tricinctum* were administered p.o., and culture filtrate, crude extract, and pure T-2 toxin from *F. sporotrichioides* No. 921 were administered s.c. or i.p. in varying doses to laboratory animals (Joffe, 1960a,b, 1971, 1974c, 1978a, 1983b; Joffe and Yagen, 1978; Lutsky et al., 1978; Schoental and Joffe, 1974; Schoental et al., 1979). Gastric fistulas were used in experiments on dogs (Khrutski et al., 1953) and intragastric intubation for mice and rats (Schoental and Joffe, 1974; Schoental et al., 1978b, 1979).

The following animals were used in our studies for the biological assay test with toxic strains of *F. poae*, *F. sporotrichioides*, and *F. sporotrichioides* var. *tricinctum* (including observations on weight loss, mortality, and histopathological changes in organs and tissues): frogs, mice, rats, guinea pigs, rabbits, dogs, cats, horses, chickens, ducklings (Alisova, 1947b; Alisova and Mironov, 1944, 1947; Antonov et al., 1951; Joffe, 1960a,b, 1974c; Joffe and Yagen, 1978; Lutsky et al., 1978; Lutsky and Mor, 1981; Schoental and Joffe, 1974; Schoental et al., 1979), and recently on vervet monkeys in cooperation with Kriek and Marasas (1983). All animals that died were autopsied and the tissues were stained with hematoxylin and eosin for histological investigation, and other stains were used if required.

Sarkisov (1948, 1954) conducted experiments per os on mice, rats, guinea pigs, rabbits, dogs, and cats, as well as on horses, cattle, sheep, and pigs. He fed them toxic overwintered grain infected by *F. sporotrichioides*. Bilai (1947, 1948, 1953, 1965) and Pidoplichka and Bilai (1946) fed young rabbits and guinea pigs grain infected by *F. poae* and *F. sporotrichioides* and used aqueous extracts by s.c. injection for the assessment. Rubinstein (1950a, 1951a,b, 1956a,b) and Rubinstein and Lyass (1948) carried out studies on feeding mice, rats, cats, and monkeys p.o. with *F. sporotrichioides*-infected grain. Getsova (1960) used i.p. injections on mice with toxins of ATA.

The effect of *Fusarium* toxins in animals has been reviewed by Joffe (1960a,b, 1971, 1974c, 1978b, 1983b), and a report on fusariotoxicoses in animals is given in Chapters 8 and 9.

## Chemistry of Toxins from Authentic Strains of *F. Poae* and *F. Sporotrichioides* Involved in Alimentary Toxic Aleukia

Numerous investigations have been published in the USSR concerning the structure and chemical properties and the toxicity of *Fusarium* belonging to the Sporotrichiella section associated with overwintered grain.

Gubarev and Gubareva (1945a,b) isolated two fractions from ether extracts of toxic grains. One fraction contained derivatives of fatty acids that caused local inflammatory skin reaction in rabbits, the second was of nonsaponifiable material that produced a necrotic skin reaction. Barer (1947) related toxins of overwintered grains to steroid type aromatic polycyclic compounds, and Okuniev (1948) isolated two separate fractions: a steroid resembling coumarol and vitamin E and associated with ATA, and a toxic product of hydroxy fatty acids which caused only local irritation of the mucous membranes.

Kretovitch (1945) and Kretovitch and Bundel (1945) found that toxins are present in overwintered grains as oxidation products of unsaturated fatty acids. Kretovitch and Sosedov (1946) found that toxic grains contain more nonprotein nitrogen and amino nitrogen and less starch than nontoxic grains. Toxic grains showed increased activity of dextrinogenic amylase and decreased activity of peroxidase and oxidase.

According to Kolosova (1949) the toxic materials were unsaturated acids, whereas Gabel (1947) concluded that they were sapogenins. Gabel (1947), Myasnikov (1947), and Svoyskaya (1947) found that these toxins had acidic properties, but they also believed that there were other toxins in the form of neutral lactones and compound ethers. Zavyalova (1946) isolated lipoproteins from wheat contaminated with *Fusarium* species causing ATA disease.

Mironov and Myasnikov (1947) and Yefremov (1944b, 1948) considered that the toxic extracts obtained from overwintered cereal grains infected with toxic *Fusarium* strains contained only one component for both rabbit skin reaction and ATA disease in man.

Kozin and Yershova (1945) considered that the lipid fraction isolated from overwintered grain serves only as a solvent for the toxic substance.

The studies of Pidoplichka and Bilai (1946, 1960) indicated that *F. poae* and *F. sporotrichioides* possess various enzymes which enable the fungi to utilize nutritional sources of nitrogen, carbon, and minerals in the grain. The fermentative system of fungi which operates in stored grains was studied in our laboratory, and we concluded that toxic fungi secrete enzymes which act on the components of the grain and thus render the grain toxic under suitable ecological conditions (Joffe, unpublished results).

It is obvious from the many theories put forward that there was no unanimous opinion on the part of Russian researchers on the chemical composition of *Fusarium* toxins isolated from overwintered toxic cereals as they have not yet succeeded in isolating a pure toxin and determining its structure correctly.

Many details have been added by the studies of Olifson (1957a,b), who found a method to purify and isolate toxins from overwintered grain and from normal grain previously autoclaved and then infected with toxic fungi (*F. poae* and *F. sporotrichioides*) of varying degrees of toxicity. Olifson's basic assumption was that the toxin is found in the lipid fraction of the grain. When such a lipid extract is applied to rabbit skin it causes a strong inflammatory reaction. Cats, mice, and guinea pigs fed on a diet mixed with this lipid fraction died. Olifson and Joffe (1954); Olifson (1955a,b, 1956a, 1962, 1965a,b, 1972); and Olifson et al. (1949, 1950, 1969, 1972, 1975, 1978) also determined the physicochemical constants for those lipids. The analysis indicated an increase of acid value from a normal 14.1–17.8 to 121.4 for toxic lipids and also an increase in peroxide value from a normal 1.3–1.8 to 6.8–8.4 in the nonsaponifiable residue. On the other hand, the refractive index decreased and the iodine value decreased from 132.4–66.6 for toxic lipids. If these constants for free and bound lipids are compared, it is evident that free lipids yield a higher acid value (121.4) and a lower peroxide value (6.8–8.4) than bound lipids whose acid value and peroxide values are 15.8 and 29.1, respectively. Thus it seems that a big difference exists in the chemical composition of free and bound lipids extracted from prosomillet grains contaminated with *F. poae* and *F. sporotrichioides*.

Under the influence of the *Cladosporium* fungi and *Mucor* the content of lipids was reduced; *Alternaria* and *Mucor* also reduced the proteins in prosomillet inoculated with these fungi by 35%.

A comparison was also made between the chemical composition of prosomillet grain intentionally infected with toxic cultures of *F. poae* and *F. sporotrichioides,* and overwintered grains which became toxic in the field. The grain infected with pure cultures had been grown for some days at − 10°C and then transferred to +3 and to +5°C for 52 to 36 days, respectively. The cultures were then autoclaved and, after the infected grain had completely dried, tests for ash content, protein, cellulose, acid value, and iodine value were performed. The infected grain showed an increased ash content and acid value and a decreased iodine value.

Olifson (1965b) isolated a neutral fraction from overwintered prosomillet and from normal prosomillet inoculated with *F. poae* and *F. sporotrichioides* and, infected with *C. epiphyllum* and *C. fagi,* an acid fraction. He found that the toxicity of the neutral fraction was more marked for *F. poae* and *F. sporotrichioides* than for some other toxigenic fungi, for example, species of *Cladosporium.*

Olifson (1957a,b, 1965b) isolated a steroidal glycoside which he named sporofusarin (Fig. 6.9) from prosomillet infected with *F. sporotrichioides.* Sporofusarin's empirical formula is $C_{65}H_{96}O_{25}$ and its melting point is 246–248°C. Hydrolysis of 4 hr with 5% $H_2SO_4$ yielded a saponin with an empirical formula of $C_{24}H_{31}O_4$; it was given the name sporofusariogenin (Fig. 6.10).

Olifson (1965b) also isolated a monoglycoside steroid from *F. poae* called poaefusarin (Fig. 6.11), with an empirical formula of $C_{35}H_{39}O_{12}$. This glycoside contains xylose and a steroidal aglycone. The aglycone was designated poaefusariogenin ($C_{24}H_{28}O_5$) and differed from sporofusariogenin by the presence of an aldehyde group instead of a methyl group (Fig. 6.12). These two derivatives were tested on cats and compared with the effect of a lipid material, called lipotoxol by Olifson (1957b), which was isolated from the lipid fraction of toxic overwintered prosomillet.

All these derivatives produced a similar syndrome in cats according to Olifson (1965b), characterized mainly by a constant leukopenia. The oral lethal dose was 0.5 mg, and all the animals died within 15–16 days. In cats and dogs lipotoxol inhibited the normal action of the heart. The lethal dose for mice was 0.06–0.07 mg of 1% lipotoxol extracted from the fatty fraction of prosomillet. On the skin of rabbits lipotoxol caused a typical hemorrhagic-edematous reaction. Lipotoxol resembles both

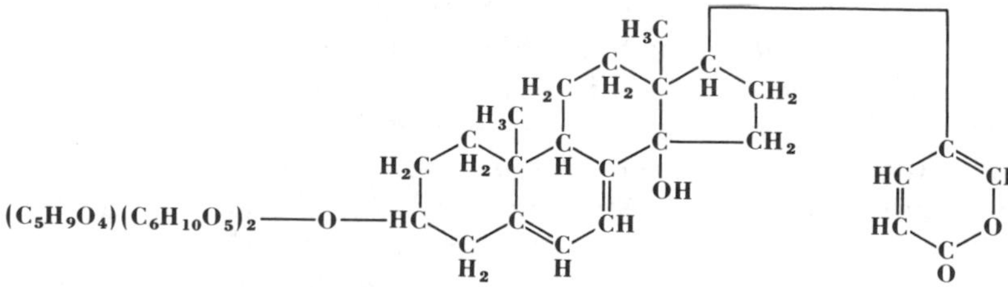

**Figure 6.9** Structure of sporofusarin (Bilai, 1977).

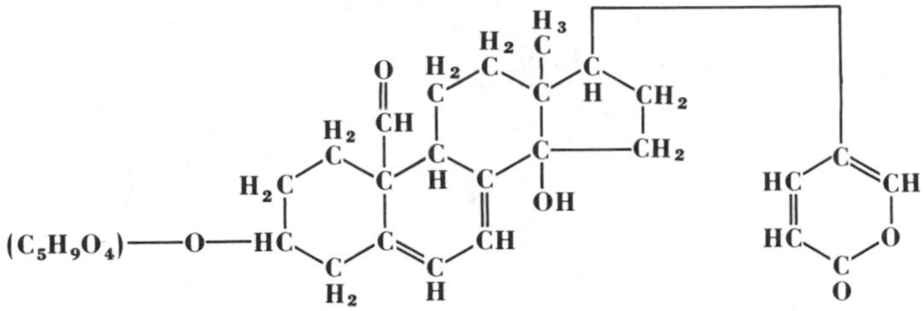

**Figure 6.10**  Structure of sporofusariogenin (Bilai, 1977).

sporofusariogenin and poaefusariogenin, which are both steroids, in structure and properties, and since the syndrome produced in cats was very much like ATA in humans, Olifson (1965b) concluded that the cause of the disease was the toxin secreted by the two fungi, *F. sporo-trichioides* and *F. poae*, in overwintered grain.

Olifson (1965b) also suggested that intermediate products of these final derivatives, such as unsaturated fatty acids, oxyacids, and steroidal lactones, are associated with the ATA syndrome. According to Olifson the two steroidal glycosides, poaefusarin and sporofusarin, and their aglycones are $C_{24}$ steroids having a doubly unsaturated six-membered lactone ring and carrying a 14β-hydroxyl group. He considered that these toxins contributed to the ATA disease.

Olifson (1960, 1972) and Misiurenko (1972) reported a method for the isolation and production of poaefusarin, further affirming its biological activity.

It should be noted that the structure of these compounds is reminiscent of the cardioactive steroidal lactones (Fieser and Fieser, 1959); they act specially on heart muscle and are not listed as skin irritants or compounds damaging to bone marrow. Only Ueno *et al.* (1972b), working with our authentic strain of *F. sporotrichioides* NRRL 3510, isolated a crude extract from this culture which contained the following compounds: neosolaniol, HT-2 toxin, butenolide, and chiefly T-2 toxin. (The strain in question was No. 738 in our collection; it was isolated from overwintered millet grain from the Orenburg district and, at the request of Dr. J. J. Ellis, was sent to the Fermentation Laboratory of the USDA in Peoria, Illinois on July 31, 1969 and delivered to Dr. Y. Ueno).

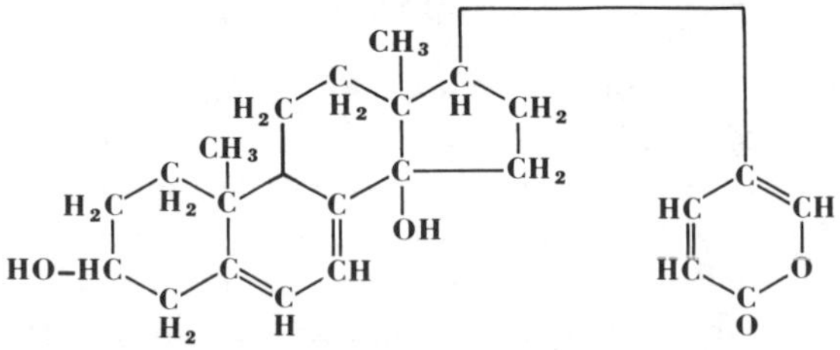

**Figure 6.11**  Structure of poaefusarin (Bilai, 1977).

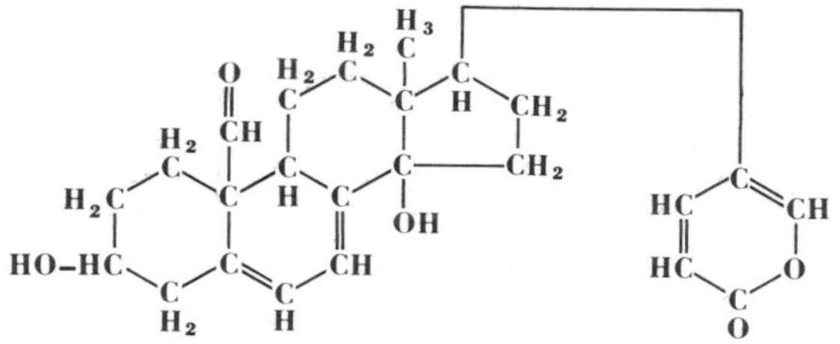

**Figure 6.12**   Structure of poaefusariogenin (Bilai, 1977).

Identification of Olifson's authentic sample of poaefusarin was performed by Mirocha and Pathre (1973), and they determined the following compounds: T-2 tetraol, neosolaniol, zearalenone, and mainly T-2
toxin. Later Szathmary *et al.* (1976, 1977) studied *F. poae* and *F. sporotrichioides* from Eastern Europe and isolated zearalenone, T-2 tetraol,
neosolaniol, HT-2 toxin, and again, mainly T-2 toxin from their extracts.

These scientists, together with Ueno *et al.* (1972b), stated that extracts
from *F. poae* and *F. sporotrichioides* did not contain any poaefusarin or
sporofusarin; they also could not confirm the presence of a steroid-type
compound in the toxic poaefusarin sample of Olifson.

Bamburg *et al.* (1968a,b, 1969) and Bamburg and Strong (1969, 1971)
have not been able to find poaefusarin or sporofusarin in their *Fusarium*
cultures either. These chemists found metabolites of trichothecenes and
zearalenone and indicated that T-2 toxin may be one of the principal
features of ATA in humans.

In view of the fact that Russian scientists have persisted in their stand-
point concerning the toxicity of the mycotoxins, poaefusarin and
sporofusarin, the author, together with Dr. Yagen, again undertook a
study covering all our original and authentic strains of *F. poae* and *F.
sporotrichioides,* that were isolated by the author from overwintered
grain collected at the time of the fatal ATA outbreaks in the Orenburg
district in the USSR. We examined the toxic metabolites from 131 isolates (106 *F. sporotrichioides* and 25 *F. poae*) cultivated at low temperatures in our laboratory (Yagen and Joffe, 1976). These toxic *Fusarium*
species were obtained from monoconidial cultures and maintained on
standard potato-dextrose agar medium at 3°C and mainly on sterile soil at
−8−−10°C. The cultures were inoculated and grown on sterilized
wheat or millet grain.

The first screening report on the distribution of T-2 toxin-producing
*Fusarium* fungi isolated in the USSR and associated with ATA showed
that over 95% of the *F. poae* and *F. sporotrichioides* isolates produced
T-2 toxin in varying quantities. Among the isolates in our collection
there were many which produced gram quantities of T-2 toxin when
grown on 1 kg wheat or millet at 12 or 5°C for a period of 21 and 45 days,
respectively (Yagen and Joffe, 1976; Joffe and Yagen, 1977). These par-

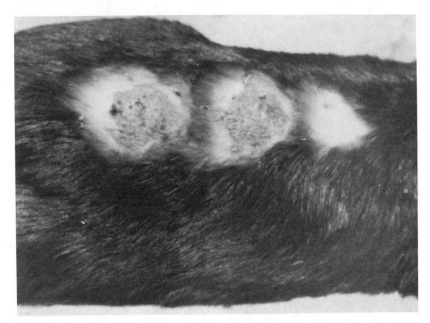

**Figure 6.13**   Edematous necrotic reaction on rabbit skin 24 hr after application of *F. poae* No. 958 (left), *F. sporotrichioides* No. 921 (center), and control (right) (Joffe, unpublished results).

ticular isolates were involved in the more severe ATA cases in the Orenburg district.

Identification of the isolated compounds, and chiefly of T-2 toxin, was determined by TLC, GLC, spectroscopic analyses, and by bioassay test on rabbit skin.

A good correlation was demonstrated between T-2 toxin detection by TLC and inflammatory skin reaction in rabbits, and the comparison of amounts of T-2 toxin determined by GLC also corresponded to the rabbit skin response.

The most toxic of all *Fusarium* isolates were *F. sporotrichioides* No. 921 and *F. poae* No. 958 (Joffe and Yagen, 1977) (*see* Fig. 6.13). *F. sporotrichioides* No. 921 was isolated in 1947 by the author from overwintered rye grain which had served as a general food source for two families of four and five persons. Three members of each family died within 6–8 weeks of consuming products prepared from the toxic grain.

Therefore we decided to analyze the metabolites produced by this culture. *F. sporotrichioides* No. 921 was inoculated on 1 kg millet grain at 12°C for 21 days. The following fractions were isolated from the crude extract obtained from 1 kg millet infected with this fungus: palmitic, oleic, and linoleic fatty acids, and six nontoxic sterols; β-sitosterol, camphesterol, stigmasterol, ergosterol (Fieser and Fieser, 1959; Itoh *et al.*, 1973), and a novel sterol metabolite of 12β-acetoxy-4,4-dimethyl-24-

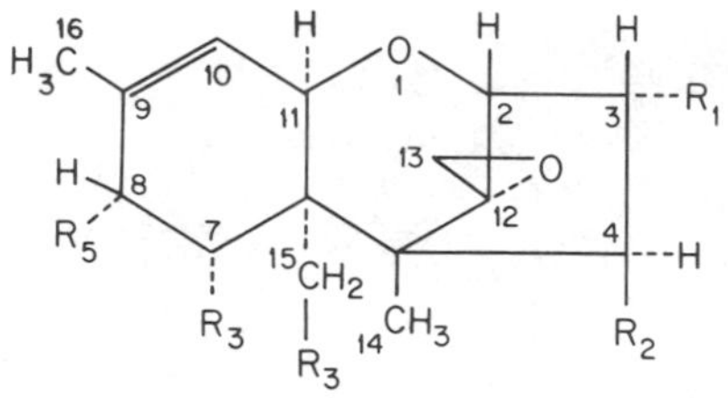

$$\text{H}_3\overset{16}{\text{C}} \quad \text{structure} \quad \text{R}_1$$

(Trichothecene structure with positions 1–16, epoxide at 12,13, and substituents R$_1$, R$_2$, R$_3$, R$_4$, R$_5$, CH$_2$ at 15, CH$_3$ at 14.)

|            | R$_1$ | R$_2$ | R$_3$ | R$_4$ | R$_5$ |
|------------|-------|-------|-------|-------|-------|
| T-2 toxin  | OH    | OAc   | OAc   | H     | OCOCH$_2$CH(CH$_3$)$_2$ |
| HT-2 toxin | OH    | OAc   | OAc   | H     | OCOCH$_2$CH(CH$_3$)$_2$ |
| Neosolaniol| OH    | OAc   | OAc   | H     | OH    |

**Figure 6.14** Trichothecenes from *F. sporotrichioides* No. 921. (Adapted from Ueno, 1973d.)

methylene-5α-cholesta-8,14-diene-3β,11α-diol (Yagen et al., 1980). This last fraction contained an oil which is in the process of being identified. Apart from these sterols, some trichothecenes—neosolaniol, HT-2 toxin, and chiefly T-2 toxin—were also isolated (Figs. 2.2 and 6.14).

The identification of the isolated trichothecene compounds and especially of T-2 toxin was carried out by GLC, combined gas chromatography-mass spectrometry, nuclear magnetic resonance, infrared, mass spectra, and optical rotation data (Yagen *et al.*, 1977).

A total of 4.1 g T-2 toxin was isolated from 1 kg infected millet. The six isolated steroids, a total of 0.15 g, were neither toxic nor skin irritants and did not have the structure of the poaefusarin or sporofusarin described by Olifson (1965a). Since they are not skin irritants and are present in such relatively small amounts, we assume that these steroids do not contribute in any significant way to intoxication. The fatty acid fractions isolated from the crude extract were neither skin irritants nor toxic (Dorell, 1971). The metabolites of HT-2 toxin and neosolaniol are present in the crude extract in very small amounts and therefore cannot cause any intoxication (Yagen *et al.*, 1977).

We also examined *F. poae* No. 958 using the same methods and analysis as for *F. sporotrichioides* No. 921, and a total of 2.8 g of T-2 toxin was isolated from 1 kg of infected millet (Yagen *et al.*, 1977).

We have shown in experiments on animals, mainly on cats and New Hampshire chicks (Joffe and Yagen, 1978; Lutsky *et al.*, 1978), that T-2 toxin (Fig. 6.14), a trichothecene (3α-hydroxy-4β,15-diacetoxy-8α-(3-methyl-butyryloxy)-12,13-epoxytrichothec-9-ene) from *F. sporotrichioides* No. 921, causes local inflammation, hemorrhage, and necrosis in skin, necrosis of the gastrointestinal tract, lymph nodes, and bone marrow, and also severe hematopoietic damage, mainly leukopenia (drastic decrease of leukocytes).

In view of the results of our analysis of 131 isolates of *F. sporotrichioides*, 25 isolates of *F. poae*, and of experiments on animals, we conclude that the intoxication seen in ATA is primarily a result of T-2 toxin poisoning.

We also undertook a comparative study of the amount of T-2 toxin produced by *F. poae, F. sporotrichioides,* and *F. sporotrichioides* var. *tricinctum* obtained from overwintered cereal grain involved in ATA disease in humans, with the yield of this toxin from different sources (Joffe and Yagen, 1977). The latter isolates were obtained from fescue hay, corn, and wheat from various parts of the United States and had been associated with outbreaks of severe toxicity in farm animals. These isolates have been referred to by American and Japanese investigators mostly as *F. tricinctum.* Closer study has shown them to belong to various species of the Sporotrichiella section, especially *F. poae* and *F. sporotrichioides* (Joffe, 1973b, 1974a, 1977; Joffe and Palti, 1975).

There were, in general, well-defined differences from overwintered grain sources in the USSR and strains isolated in the United States. The results are summarized in Table 3.12.

Thus the recent studies carried out in the United States (Mirocha and Pathre, 1973; Szathmary, 1976), in Japan (Ueno *et al.,* 1972b), and in Israel (Joffe and Yagen, 1977; Yagen and Joffe, 1976; Yagen *et al.,* 1977, 1980) could not confirm conclusions reached in the Russian studies concerning the role of steroidal lactones in causing ATA disease.

## Clinical Characteristics of Alimentary Toxic Aleukia

### Clinical Features

Clinical findings were described by Chilikin (1944, 1945, 1947); Lyass (1940); Manburg (1944); Manburg and Rachalski (1944, 1947); Myasnikov (1935); Nesterov (1948); Romanova (1947); and Yefremov (1948). The clinical features of ATA are usually divided into four stages. If the disease is diagnosed during the first stage, or even during the transition from the second to the third stages, early hospitalization may still enable the patient's life to be saved. However, if the disease is only detected during the third stage, the patient's condition is usually desperate, and in most cases death cannot be prevented. Very few patients in the third stage survive.

*First Stage.* The characteristic symptoms of this stage appear a short time after ingestion of the toxic grain. They may appear after a single meal of overwintered toxic grains and disappear completely even if the patient continues to consume the grain. The characteristics of the first stage include primary changes in the mouth cavity and gastrointestinal tract. Shortly after eating food prepared from toxic grain, the patient feels a burning sensation in the mouth, tongue, throat, palate, esophagus, and stomach as the toxin acts on the mucous membranes. The tongue may feel swollen and stiff and the mucosa of the oral cavity may be hyperemic. Inflammation of the gastric and intestinal mucosa results in vomiting, diarrhea, and abdominal pain. In most cases excessive salivation, headache, dizziness, weakness, fatigue, and tachycardia accompany this stage, and there may also be fever and sweating. The leukocyte count may decrease already at this stage to levels of 2000/mm$^3$ with relative lymphocytosis, and there may be an increased erythrocyte sedimentation rate (Friedman, 1945a; Yefremov, 1948; Yudenitch et al., 1944).

There is a danger of this stage not being detected since it appears and disappears relatively quickly; the patient may become accustomed to the toxin, so that a quiescent period follows while the toxin accumulates and the patient enters the second stage. The first stage may last from 3–9 days.

*Second Stage.* This is often called the latent stage (Chilikin, 1944, 1945, 1947) because the patient feels well and is capable of normal activity. It is sometimes also called the leukopenic stage (Manburg, 1944; Manburg and Rachalski, 1947; Romanova, 1947) because its main features are disturbances in the hematopoietic system characterized by a progressive leukopenia, a granulopenia and a relative lymphocytosis. In addition there is anemia and a decrease in the platelet count. The decrease in the number of leukocytes lowers the body's resistance to bacterial infection. Apart from changes in the hematopoietic system there are also disturbances in the central and autonomic nervous systems. Weakness, headache, palpitations, and mild asthmatic symptoms may occur. The skin and mucous membranes may be icteric, the pupils dilated, the pulse soft and labile, and the blood pressure decreased. Body temperature does not exceed 38°C and the patient may even be afebrile. There may be diarrhea or constipation.

The normal duration of this stage is usually from 3–4 weeks, but it may extend over a period of 2–8 weeks. If consumption of toxic grain continues the symptoms of the third stage develop rapidly.

*Third Stage.* The transition from the second to the third stage is sudden. By now the patient's resistance is low, and violent symptoms may be present especially under the influence of stress associated with physical exertion and fatigue. The first visible sign of this stage is the appearance of petechial hemorrhages on the skin of the trunk, in the axillary and inguinal areas, on the lateral surfaces of arms and thighs, on the chest (Figs. 6.15 and 6.16), and, in serious cases, on the face and head. The petechial hemorrhages vary from a few millimeters to larger areas a few centimeters in diameter (Chilikin, 1945, 1947). As a result of increased capillary fragility, any light trauma may cause hemorrhages to increase in size. Hemorrhages may also be found on the mucous membranes of mouth and tongue and on the soft palate and tonsils. Nasal,

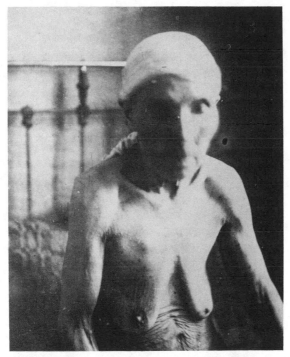

**Figure 6.15**  Petechial spots, first small and red, later blue or dark, caused by intradermal or submucous hemorrhage on chest and left arm (Joffe, 1978b).

gastric, and intestinal hemorrhages may occur (Chilikin, 1947; Yefremov, 1948).

Necrotic changes soon appear in the throat, causing difficulty and pain on swallowing. Necrotic lesions may extend to the vulva, gums, mouth, mucosa, larynx, and vocal cords, and are usually contaminated with a variety of avirulent bacteria. The necrotic areas are an excellent loci for bacterial infection, which can result from lowered body resistance due to the damage to the hematopoietic and reticuloendothelial systems. Bacterial infection causes an unpleasant odor from the mouth, due to the enzymatic activity of bacteria on proteins. Areas of necrosis may also appear on the lips and on the skin of the nose, jaws, and eyes (Fig. 6.17).

The regional lymph nodes are frequently enlarged, and the submandibular and cervical lymph nodes may become so large and the adjoining connective tissue so edematous that the patient experiences difficulty in opening his mouth. Esophageal lesions may occur, and involvement of the epiglottis may cause laryngeal edema and aphonia (loss of voice). In such cases death may occur by strangulation. Death in about 30% of patients was directly related to stenosis of the glottis (Peregud, 1947).

The blood abnormalities observed initially in the first and second

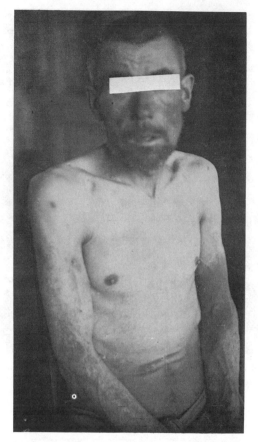

**Figure 6.16**  Hemorrhage on right and left arm and initially on chest (Joffe, 1978b).

stages intensify during the third stage. The leukopenia increases to counts of 100 or even fewer leukocytes per cubic millimeter. The lymphocytes may constitute 90% of the white cells present, the number of thrombocytes decreases to less than 5000 cells/mm$^3$, and the erythrocytes to below 1 million/mm$^3$.

Blood sedimentation rate is increased. The prothrombin time ranges from 20–56 sec, and clotting time is usually not very prolonged. There may be a deficiency in fibrinogen in severe cases (Germanov, 1945; Kurbatova, 1948; Veindrach and Fadeyeva, 1937). Some investigators found that patients suffer an acute parenchymatous hepatitis accompanied by jaundice. Bronchopneumonia, pulmonary hemorrhages, and lung abscesses are frequent complications.

*Fourth Stage.*  This is the stage of convalescence and its course and duration depend on the intensity of the toxicosis. Only 3–4 weeks of treatment, occasionally longer, are needed for the disappearance of ne-

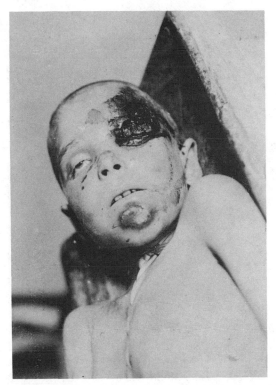

**Figure 6.17** Necrotic lesions around eye and face of a child who died from ATA disease after consuming overwintered prosomillet infected with *F. sporotrichioides*. (Adapted from Joffe, 1978b.)

crotic lesions and hemorrhagic diathesis and also the bacterial infections. Usually 2 mo or more elapse before the blood-forming capacity of the bone marrow returns to normal (Friedman, 1945a; Teregulov, 1945): as a general rule leukocytes come first, followed by granulocytes, the platelets, and finally the erythrocytes (Germanov, 1945; Gorodijskaja, 1945; Kasirski and Alekseyev, 1948; Shemshelevitch and Dubniakova, 1945).

### Pathological Findings in Humans

The toxic metabolite T-2 toxin derived from authentic strains of *F. poae* and *F. sporotrichioides* isolated from overwintered cereals caused fatal outbreaks of ATA in the USSR. Pathological findings in organs and tissues are therefore both useful and important.

Local action was manifested by clinical burning sensations in the mouth, soft palate, and tongue, which usually ended when the patient stopped consuming products made from toxic grain. If consumption continued, the first signs of toxicosis appeared in the pharynx, esophagus

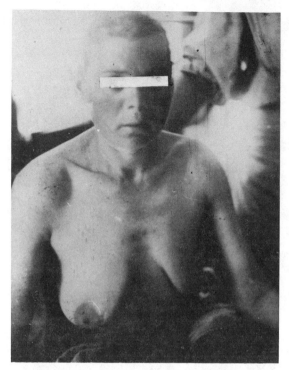

Figure 6.18    Hemorrhage rash on skin of chest and arms (Joffe, 1978b).

(Genkin, 1944), and stomach (Manburg, 1944), and later various hemorrhagic syndromes appeared, characterized by hemorrhagic rash on the
skin of the chest (Fig. 6.18), trunk, abdomen, legs, and arms (Figs. 6.15
and 6.16), and even on the face. Hemorrhagic petechiae appeared most
abundantly when necrotic angina began to develop (Chilikin, 1947;
Kudryakov, 1946; Manburg and Rachalski, 1947). Hemorrhagic and necrotic lesions also developed in stomach and intestines, including the
entire digestive tract, causing changes in intestinal function (Chilikin,
1947; Davydovski and Kestner, 1935; Manburg, 1944; Manburg and
Rachalski, 1947). Hemorrhage and necrosis of the appendix and cecum
and inflammation of the rectum also developed (Chilikin, 1947). Hemorrhages were also present in various visceral organs, in the adrenal and
thyroid glands, gonads, uterus, and chiefly in the pleura. Pulmonary
changes involving bronchopneumonia and severe hemorrhages in lung
tissue with pneumonic abscesses frequently developed (Chilikin, 1947).
Marked changes were observed in the heart accompanied by vascular
insufficiency and arterial hypotonia. Pressure was very low and throm-
bophlebitis and endocarditis were occasionally indicated (Nesterov,
1948). These changes brought about serious inflammation of the blood
vessels and affected the endocrine system.

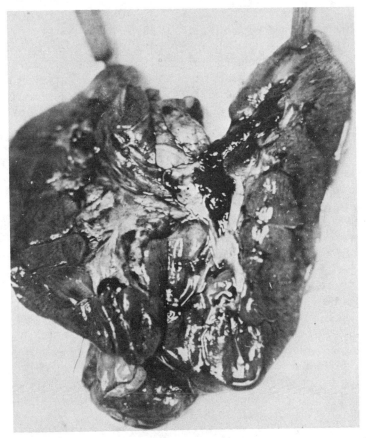

**Figure 6.19** Kidney from fatal case of ATA, showing marked hemorrhage in pelvic mucosa (Joffe, 1978b).

In the third stage of the disease hemorrhages in the liver were revealed with acute parenchymatous hepatitis accompanied by jaundice (Chilikin, 1947; Nesterov, 1948; Romanova, 1947; Yefremov, 1944a,b, 1948), and changes in the glucose, protein, mineral, and other metabolisms of the liver were observed (Chilikin, 1947; Manburg, 1944). Hemorrhages and necrosis in the kidneys also appeared (Fig. 6.19). In the necrotic angina stage severe changes were observed in the lymph nodes which became edematous and were characterized by the disappearance of the lymphoid elements. The entire lymphatic and reticuloendothelial systems were affected and showed proliferation of red blood cells and of endothelial capillaries and sinusoids. Some investigators have observed severe changes in the central and autonomic nervous systems resulting from ATA (Gurewitch, 1944; Kovalev, 1944; Poznanski, 1947) such as impaired nervous reflexes, meningitis, general depression and hyperesthesia, encephalitis, cerebral hemorrhages, and destructive lesions in nervous and sympathetic ganglia (Gurewitch, 1944; Kholodenko, 1947; Serafimov, 1945, 1946).

According to Tomina (1948) toxins from overwintered cereals which were absorbed in the stomach and intestines had a cumulative effect on various organs and tissues. The most severe effects were on the hematopoietic system, which correlated with the development of different stages of the disease and resulted in depression of leukopoiesis, erythropoiesis, and thrombopoiesis (Kasirski and Alekseyev, 1948; Koza *et al.*, 1944; Shemshelevitch and Dubniakova, 1945; and Teregulov, 1945). Progressive leukopenia appeared (the leukocyte count dropping to 100 mm$^3$ or even lower) as well as lymphocytosis (to 90%) and a decrease in erythrocytes (which dropped to 1 million mm$^3$). Hemoglobin content dropped to 8% and granulocytes disappeared completely. At the same time the sedimentation rate increased and showed a deficiency of prothrombin as a result of irritation of the bone marrow (Chilikin, 1944, 1945; Kasirski and Alekseyev, 1948). In severe cases destructive and hemorrhagic lesions frequently appeared in the blood circulation causing thrombosis in the blood vessels of different organs.

Destruction and sometimes atrophy of bone marrow (Chilikin, 1947; Davydovski and Kestner, 1935; Yefremov, 1944a,b, 1948) produced serious changes in the organism after long-term consumption of overwintered grain contaminated by toxic strains of *F. poae* and *F. sporotrichioides*. Hemorrhagic diathesis, necrotic angina, sepsis, and severe hematological changes also developed (Chilikin, 1947; Davydovski and Kestner, 1935; Gromashevski, 1945; Manburg, 1944; Manburg and Rachalski, 1947; Yefremov, 1948).

Important contributions to the pathogenesis of ATA were made by Strukov (1947) and Strukov and Tishchenko (1944, 1947), who showed that disturbances of the hematopoietic system were reversible and did not lead to bone marrow destruction. Strukov (1947), Aleshin *et al.* (1947), and Aleshin and Eyngorn (1944) thought that toxins of overwintered cereals did not act primarily on bone marrow but on an extramedullary apparatus which regulated the hematopoietic, autonomic nervous, and endocrine systems.

The necrotic changes in the final stage of ATA developed initially in the pharyngeal tonsils and later in so-called necrotic angina, in the throat, and even in the esophagus. Severe gangrenous pharyngitis also occurred (Fig. 6.20). Necrotic angina in the throat brought about an increase of fever and, in severe cases, signs of glottis edema with asphyxia were observed (Peregud, 1947; Smirnova, 1945). This caused weakness, apathy, and damage of the leukocytic, phagocytic, and reticuloendothelial functions. Necrotic lesions were present along the entire gastrointestinal tract as well as in other organs.

Strong and profuse menstrual and nose bleeding (Fig. 6.21) proved fatal in many cases.

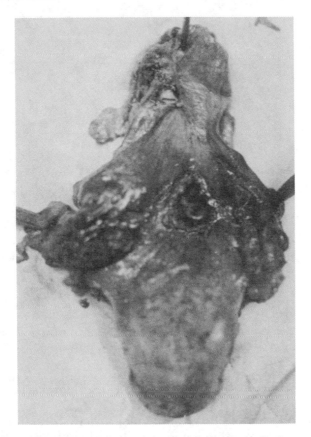

**Figure 6.20** Necrosis of pharyngeal mucosa caused by *F. poae,* which was isolated from prosomillet wintered under snow cover. The toxic prosomillet was eaten by this patient (Joffe, 1978b).

Recovery depended on the amount of toxic grain consumed, the type of therapy, and chiefly on the presence or absence of complications.

The disease sometimes recurred, with hemorrhagic and necrotic syndromes, after consumption of infected grain was resumed or after physical strain (Chilikin, 1945, 1947; Nesterov, 1948).

**Prophylaxis and Treatment**

The most important prophylactic measure is obviously to refrain from eating overwintered toxic grain, and the primary preventive measure, therefore, is to educate rural populations with reference to the etiology and clinical symptoms of ATA (Boldyrev and Shtenberg, 1950; Teregulov, 1945). These measures have reduced ATA outbreaks considerably. When outbreaks of the disease were first reported medical teams were sent to the affected areas, and the population was examined clini-

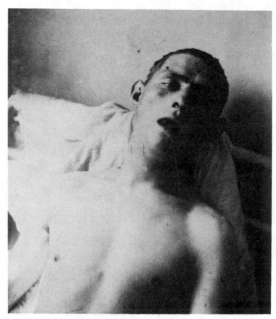

**Figure 6.21** General view of a patient with a severe form of ATA; nose bleed (epistaxis), respiratory distress, and hemorrhage on left arm (Joffe, 1978b).

cally and hematologically (Poliantseva, 1945; Shklovskaja and Brodskaja, 1944).

Grain samples should be examined for toxicity by skin test, and concurrently toxic overwintered cereals collected by the population should be replaced by wholesome grain. When ATA was detected in the second stage the treatment recommended at the time included blood transfusion (Kudryakov, 1946; Levin, 1946) and the administration of nucleic acid and calcium preparations, antibiotics (Levitski, 1948), Bogomolt's antireticular cytotoxic serum (Khabibullina, 1945; Mikhailovski, 1945), sulfonamides (Kavetski and Grinberg, 1945; Khabibullina, 1945) and vitamins C and K (Friedman, 1945b; Levin, 1946). When the number of leukocytes declined below 3000 mm$^3$, hospitalization was recommended (Karlik, 1945; Kasirski and Alekseyev, 1948).

The measures employed in the second stage were also used in the third stage but more intensively. Following recovery, a rich diet was given for one month, and the patient remained under periodic hematological checkup (Levin, 1945, 1946; Malkin and Odelevskaja, 1945).

The clinical and pathological findings in animal organs and tissues relating to toxins of *F. poae* and *F. sporotrichioides* are described in Chapters 8 and 9.

# UROV OR KASHIN-BECK DISEASE

Mycotoxicoses occur either in acute or chronic forms depending on the environmental, climatic, and meteorological conditions which favor the formation and accumulation of toxic metabolites by *Fusarium* fungi from cereal grains.

It is known that *Fusarium* species and varieties of the Sporotrichiella section, given favorable environmental conditions in the field and in storage, produce toxic compounds in cereal grains, foodstuffs, food products, and fodder. Depending on the degree of toxicity, these fungi cause diseases involving various clinical and histopathological changes in the organs and tissues of humans and animals.

Intoxication associated with toxic *Fusarium* isolates has usually occurred sporadically or spontaneously in various places and has produced acute diseases. One example is ATA, which has been studied in detail (Bilai, 1977; Joffe, 1971, 1974c, 1978a,b, 1983b; Joffe and Yagen, 1977, 1978; Lutsky *et al.*, 1978; Lutsky and Mor, 1981; Rubinstein, 1948, 1960a; Sarkisov, 1954; Yagen and Joffe, 1976; Yagen *et al.*, 1977, 1980).

Sometimes, however, the *Fusarium* fungi produce slightly toxic substances in grains and other substrates which cause only chronic diseases. In such cases the disease develops gradually and in its initial stage is frequently asymptomatic. Prolonged consumption of the slightly affected cereal grains or other food products can lead to harmful consequences for humans and animals. One such disease, connected in all probability with cereal grains in specific endemic regions and causing chronic toxicoses in humans, is Urov or Kashin-Beck disease, so called because it was detected for the first time along the Urov River in Transbaikal.

Sorokin (1890) described a case where the people of East Ussuri Land, after eating bread prepared from affected grain collected in the region, began to vomit and suffer from dizziness and other symptoms. The grain was contaminated by colored fungi which were not identified. Later Palchevski (1891) and Woronin (1891) identified these fungi as *F. graminearum, F. roseum, Gibberella saubinetii, Helminthorium* sp., and *C. herbarum,* which cause severe intoxication that affects humans and domestic animals. They described the syndromes associated with the grains, which they named "staggering grains" or "Taumelgetreide."

In the USSR Urov disease was reported in the Transbaikal and in the most easterly districts. In addition to endemic foci various cases of this disease were encountered near the Baikal and in the Irkutsk, Chita, Vologodsk, Pskovsk, Leningrad, and Kiev districts (Rubinstein, 1953). Cases of Urov disease have been described in Northern Sweden and Holland, but the disease is widespread chiefly in Taiwan, Northern China, Korea, and Japan (Aiiso, 1936; Aiiso and Nagashi, 1937; Hiyeda *et al.*, 1938; Iwatsu, 1959; Lin, 1959; Rubinstein, 1960b; Sergiyevski, 1952; Takizawa *et al.*, 1956, 1957, 1958, 1959, 1960, 1961, 1963, 1964; Zagrafski *et al.*, 1957). The Kashin-Beck disease has been known for over a hundred years and is named after two Russian scientists, Kashin and Beck, who studied the disease at the end of the nineteenth and the beginning of the twentieth century (Kashin, 1860; Beck, 1906).

Kashin-Beck disease is an endemic, bone-joint disease associated with an enchondral type of ossification (formation of bone substance) found chiefly in children of school age. The disease is manifested mainly as a shortening of the long bones, combined with a thickening and subsequent deformation of the joints, flexor contractures and muscular weakness, and atrophy. The disease develops gradually and has a chronic course. The early stage is characterized by marked pains in the joints (Babenkova *et al.*, 1955; Beck, 1906; Butko *et al.*, 1977; Chepurov and Cherkasova, 1954; Chepurov *et al.*, 1955; Chetvertakova, 1967; Damperov, 1939; Dobrovolski, 1925a; Kanshina, 1957; Kashin, 1860; Kravchenko, 1959, 1965, 1968; Muchkin, 1967, 1968; Nesterov, 1964; Perkel, 1957, 1960; Razumov and Rubinstein, 1951; Rokhlin, 1938; Roll, 1970; Rubinstein, 1949–1951a,b, 1953, 1960b; Rubinstein *et al.*, 1961; Sergiyevski, 1948, 1952).

The etiology of the disease has been linked to a variety of hypotheses and theories such as water sources, a deficiency of calcium and other minerals in water, as well as cereal crops, potatoes, and other food (Gamuzov, 1931; Goldstein and Nikiforov, 1931; Kashin, 1860; Oparin, 1939; Schwarzman, 1937; Shchipatchev, 1928; and Vinogradov, 1949). Another hypothesis was associated with a deficiency of vitamins in diet-avitaminosis and with changes or disturbances in the mineral metabolism (Reisler, 1952). Domaev (1976a) stated that the etiology of Urov disease was connected with hereditary factors and with clinicogenetic changes. He also studied bioelectrical muscular activity in relation to the severity of the pathological process and the electroencephalographic (EEG) data of 112 patients in different stages of the disease (Domaev, 1976b,c).

All these hypotheses, as it turned out, were groundless according to Georgiyevski (1952). Geller *et al.* (1954) sharply criticized Reisler's avitaminosis hypothesis, and Tikhonov (1977) strongly criticized the Domaev (1976a) theory of heredity.

Sergiyevski (1952) and Georgiyevski (1952) (and later with Geller *et al.*, 1954) simultaneously began an intensive investigation into the epidemiology and etiology of Kashin-Beck disease in its endemic foci, and they concluded, after a comprehensive chemical study, that the mineral content of water and cereal grains used by the local inhabitants in the endemic foci was no different from other, nonendemic regions.

After many years of investigation, Sergiyevski suggested that the etiology of the Urov disease was associated with specific environmental, climatic, meteorological conditions which favor toxin formation in the cereal grains of the endemic areas. Thus he considered the etiology of the Kashin-Beck disease to be connected only with local cereal grains in the endemic foci.

In the autumn of 1948 Rubinstein began to examine samples of cereal grains collected from fields in the endemic foci of the Chita district. She examined the fresh cereal grains from the 1948–1951 harvests and isolated 425 strains of *Fusarium* (Rubinstein, 1949, 1951a,b, 1953, 1960b). After identification of the *Fusarium* cultures and their determination by rabbit skin test, Rubinstein, in cooperation with Razumov, selected some strains of *F. sporotrichiella* var. *poae* for further investigation (Rubinstein, 1951b; Razumov and Rubinstein, 1951). Later Perkel (1957, 1960) isolated many *Fusarium* strains from wheat grains in the endemic regions of Eastern Transbaikal, in particular *F. sporotrichiella* var. *tricinctum* and *F. sporotrichiella* var. *poae*. According to Perkel (1957) the grain was contaminated by soil fusaria during harvesting; the longer the grain remained on threshing floors in the field the greater the contamination.

Razumov and Rubinstein (1951) and Rubinstein (1951a, 1953, 1960b) experimented on white rats and puppies, feeding them wheat grain infected with *F. sporotrichiella* var. *poae*. Results showed delay in the growth of bones lengthwise and shortening of the femur and tibial bones, with changes in the metaepiphyseal zone and fragility of the long bones. Experiments with the same strain of *F. sporotrichiella* var. *poae* over a period of 8 mon to 1 year increased the intensity of changes in the bone extremities. Control groups developed normally. Oparin (1939) and Kogan and Vasilyeva (1956) found greater moisture content and lesser calcium and phosphorus content in the bones of the experimental rats than in control animals. They suggested that the *Fusarium* strain had changed the mineral metabolism of the rats. Kogan and Yershova (1956) also observed a complete cessation of the enchondral growth of the long bones with pronounced contracture of the front extremity joints.

The phosphatase content of the bones was higher, particularly in the metaphyses, and thickening of the epiphyses and curvature of the humerus and femur were more pronounced in the experimental puppies than in the controls. In the experimental puppies the moisture content of the bones was higher and the total mineral, calcium, and phosphorus content lower than in the controls. The bone tissue of control puppies developed normally.

Experimental puppies had difficulty in movement. The clinical and histopathological signs in rats and puppies showed a delay in lengthwise bone growth, deformation of the joints, flexor contractures and, chiefly, muscular atrophy, similar to the symptoms of Urov disease in patients in the endemic foci. No changes in the internal organs or in hematopoiesis were noted in the experimental animals. Thus Rubinstein (1953, 1960b) and Perkel (1957, 1960), in experiments on growing white rats and puppies, succeeded in reproducing the characteristic symptoms of Kashin-Beck disease with the isolate of *F. sporotrichiella* var. *poae*. Together with this strain, other, nontoxic strains, which did not cause any pathological changes in the animals, were frequently isolated.

The experiments showed that the disease was connected with a toxic *Fusarium* fungus having a definite, specific effect on animals, and this strain was therefore classified as a special new physiological form, *F. sporotrichiella* var. *poae* f. *osteodystrophica* (Rubinstein, 1960b). It was shown to be capable of producing the specific substances which cause osteodystrophy in bone tissue. It is clear, therefore, that the Urov disease is a food fusariotoxicosis, like ATA, with a selective effect on bone development showing an enchondral type ossification.

On the basis of clinical and histopathological investigations on patients in the endemic regions and experiments on animals, Rubinstein (1953, 1960b) concluded that the etiology of Urov disease is associated with an alimentary-toxic hypothesis belonging to a group of mycotoxicoses.

Unfortunately not all investigators agree with this alimentary toxicosis theory or with a mycotoxic conception which seems to be most probable. For instance, Komissaruk (1957) criticized this mycotoxic alimentary conception but did not propose an alternative etiological hypothesis of the Urov disease. However, the disease requires further epidemiological and etiological studies, chiefly comprehensive investigations into the ecological and environmental conditions, including the soil conditions for toxin formation in the endemic regions, together with a determination of the chemical nature of the specific compounds produced by the *Fusarium* fungi.

Cases of animal bone disorders associated with other fusaria have also been described in Austria, Germany, Finland, France, and the United States.

According to Sharby *et al.* (1972) (from Arkansas) diets containing *F. roseum* f. *cereale* (ATCC 24090) caused leg bone abnormalities when fed to chicks. Two isolates of *F. moniliforme* (ATCC 24088 and 24089) isolated from poultry rations caused similar effects (together with reduced weight gain).

In Finland Korpinen and Ylimäki (1972) stated that *F. moniliforme* No. 19 was thought to be responsible for retarded growth in chickens; similarly, retarded chicken growth has also been linked to *F. tricinctum* No. 18 and *F. poae* Nos. 17 and 20.

*F. moniliforme* was found in barley fed to pigs and in prepared feed for young poultry in France. Bone dystrophy, osteomalacia, and paraplegia were observed in these animals (Moreau, 1974a,b, 1979).

It is interesting to note that other researchers (from Austria and Germany) have observed that *F. moniliforme* has been involved in bone disorders in chickens. This *Fusarium* species was isolated from pellets of chicken feed, which had been associated with an outbreak of a bone softening disease resulting in rapidly developing severe rickets in broiler chickens (Gedek *et al.*, 1978; Köhler *et al.*, 1978). No toxin was identified as the causal agent for this phenomenon, but it may be due to a lack of thiaminase in the observed vitamin $B_1$ deficiency (Fritz *et al.*, 1973; Moreau, 1974b) or lack of vitamin $D_3$, which was shown to relieve the ricket symptoms caused by *F. moniliforme*.

According to Stanford *et al.* (1975) (from Alabama) T-2 toxin has been shown to have a marked effect when applied prenatally to mice. Interperitoneal injections of 1.0 and 1.5 mg/kg into pregnant mice greatly increased maternal mortality and reduced prenatal survival. Fetus weight was reduced and 37% of the fetuses were badly malformed with bent, shortened, or missing tails and limb malformations. The authors suggested that T-2 toxin might have a teratogenic effect in prenatal mice. Similar prenatal effects were observed by Hood *et al.* (1978), who investigated the combined effects of prenatal exposure to Ochratoxin A and T-2 toxin. They found that teratological effects of the two toxins were additive.

Khera *et al.* (1982) studied the teratogenic action of vomitoxin on embryo toxicity in pregnant female Swiss-Webster mice using various doses (1, 2.5, 5, 10, and 15 mg/kg of body weight). The authors found a number of skeletal malformations in the mice.

Histological findings showed no changes in the fetal tissues after treatment with various doses of vomitoxin.

Zearalenone has also been shown to affect unborn mice and rats. Doses of 1–5 mg/kg administered to pregnant rats caused skeletal defects in the fetuses. The incidence of these defects increased from 12.8% at the lower dose to 26.1% at the higher. Defects included missing or malplaced sternal plates and various rib defects and, at the higher dose, some tarsus missing (Ruddick *et al.*, 1976). A pathological change in the bones of the treated animals was also noted by Gallo *et al.* (1977). Thirty mg/kg of zearalenone caused a decrease in body weight in both male and female rats.

## CONCLUSION

Strains of *F. sporotrichioides* and *F. poae* isolated from overwintered cereal grains associated with ATA disease were very strongly toxic to the following laboratory and domestic animals: mice, rats, guinea pigs, rabbits, pigeons, dogs, frogs, fish, poultry, sheep, cattle, and horses (Akhmedli *et al.*, 1972b, 1973; Bilai 1948, 1952b, 1970b; Bilai and Pidoplichko, 1970; Getsova, 1960; Joffe, 1960a,b, 1971, 1974c, 1978a,b, 1983b; Joffe and Yagen, 1978; Kurmanov, 1961, 1968, 1969, 1971, 1978a,b; Kvashnina and Gabrilova, 1956; Kvashnina, 1976, 1978, 1979;

Marchenko, 1963; Marianoshvili, 1964; Mayer 1953a,b; Sarkisov, 1960, 1961; Sarkisov *et al.*, 1972; Schoental and Joffe, 1974; Schoental *et al.*, 1976, 1978a, 1979). All attempts by these authors to produce an experimental model for inducing ATA in feeding studies with these animals have been unsuccessful.

The syndrome of ATA was induced in cats fed a mixture containing moldy grain or overwintered cereals contaminated with *F. sporotrichioides*, a pure culture of this fungi, grains of millet, or wheat infected by *F. sporotrichioides*, and these animals showed clinical and histopathological signs resembling ATA in humans (Alisova, 1947b; Alisova and Mironov, 1947; Joffe, 1960a,b, 1971, 1974c, 1978a,b, 1983a; Kvashnina 1976, 1978, 1979; Mironov, 1945a,b; Nesterov 1945; Rubinstein, 1956a, 1980b; Rubinstein and Lyass, 1948; Sarkisov, 1948, 1950; Vertinsky and Adutskevitch, 1948).

Recently ATA in humans has been reproduced in cats by administration of T-2 toxin, a naturally occurring trichothecene isolated from *F. sporotrichioides* (Joffe, 1971, 1978b, 1983b; Lutsky *et al.*, 1978; Lutsky and Mor, 1980, 1981; Sato *et al.*, 1975; Yagen *et al.*, 1977).

ATA was also reproduced in monkeys with toxic overwintered cereal grains contaminated by *F. sporotrichioides*, with pure cultures of this strain, and with pure T-2 toxin isolated from an authentic strain of *F. sporotrichioides* involved in the fatal ATA disease in humans (Kriek and Marasas, 1983; Kriek, Yagen, Joffe, and Marasas, unpublished data; Mayer, 1953b; Rubinstein and Lyass, 1948; Rubinstein, 1960a).

# CHAPTER 7

## HUMAN INFECTIONS ASSOCIATED WITH *FUSARIUM* SPECIES

*Fusarium* infections in humans occur in the nails, onychomycosis, and in the cornea, keratomycosis, or mycotic keratitis. They are also responsible for causing ulcers, necrosis, and other lesions of the skin in addition to infections in some organs and tissues.

## ONYCHOMYCOSIS

Diseases of the nail were described by Zaias (1966, 1971) in Florida. According to the author infections of the nail were mainly associated with *F. oxysporum*, which developed on the superficial layers of the toe nail plates and sometimes invaded the deeper layers.

Rush-Munro *et al.* (1971) reported cases of onychomycosis in New Zealand caused by *F. oxysporum*, which was isolated from the big toe nail in every case of milky lesions. These authors studied 53 *Fusarium* cultures isolated from toe nails, all of which were identified as *F. oxysporum*. Other *Fusarium* species from finger nails, toe clifts, and infected ulcers such as *F. solani*, *F. semitectum*, and *Cylindrocarpon* species were also isolated.

Walshe and English (1966) described two cases of onychomycosis of the toe nail associated with *Fusarium* species, and Suringa (1970) also described a case of onychomycosis due to *F. oxysporum*.

**293**

# KERATOMYCOSIS

Cases of keratomycosis, or mycotic keratitis, in the human cornea, associated with *Fusarium* species, have been reported by researchers in various parts of the world.

Ley and Sanders (1956) described three cases of fungal keratitis in the United States. They suggested that in the disease process of the cornea in these cases the fungi may have a triple role: they may occur as the primary pathogen or essential etiological factor, they may act as a secondary invader in a bacterial or viral infection, and they may be present only as a saprophyte.

Fazakas (1958) reported on the first cases of keratomycosis in which the pathogenicity of various fungi obtained from human eye lesions, including some *Fusarium* species such as *F. rozleri* (*F. merismoides*) and *F. moronei* (*F. equiseti*), was demonstrated primarily by animal experiments in Hungary.

The following researchers described cases of mycotic infections associated frequently with *F. oxysporum:*

Mikami and Stemmermann (1958) quoted a number of pertinent papers associated with mycotic corneal infection and described a case of keratomycosis caused by *F. oxysporum* in a 45-year-old man in Hawaii.

Barsky (1959) described a case of corneal ulcer with hypopyon in a 77-year-old woman in Michigan. Gingrich (1962) reported three cases of keratomycosis in Texas. Lynn (1964) described a severe case of keratitis accompanied by fever, headache, and sweating in the patient in Iowa. Francois (1968) made a comprehensive survey of 120 cases of keratomycosis in Europe and America due to *F. oxysporum.*

Jones *et al.* (1969b), in Southern Florida, and Inokawa (1972), in Japan, found one case each of keratitis. Greer *et al.* (1973) reported a case of mycotic infection of the cornea in a patient in Colombia, and Laverde *et al.* (1973), also in Colombia, described two cases.

Rowsey *et al.* (1979) described a case of *F. oxysporum* endophthalmitis (inflammation of the internal structures of the eye) in a patient, as did Lieberman *et al.* (1979) with *F. solani.*

*F. solani* is a worldwide distributed fungus and the most common agent associated with corneal keratitis. The following investigators reported incidences of *F. solani* keratitis in various regions of the United States:

Lynn (1964) described a number of cases, while Halde and Okumoto (1966) reported 35 cases of mycotic ulcers, four of which were linked with *F. solani.* Naumann *et al.* (1967) described a histopathologic study associated with mycotic keratitis due to *F. solani,* and Jones *et al.* (1969a,b,c, 1972) also described many cases of corneal mycotic infections by *Fusarium* sp., mainly *F. solani.*

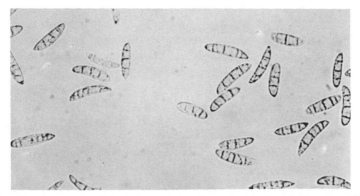

**Figure 7.1**   *F. solani* isolated from the cornea of a patient in Israel. Macroconidia. (Joffe, unpublished results.)

Polack *et al.* (1971) reported on 17 cases of keratomycosis associated with *F. solani*, while Wilson *et al.* (1971) described a case of keratitis caused by *F. solani* isolated from mascara that had been used by the patient.

Forster and Rebell (1975a,b,c) described cases of keratomycosis also due to *F. solani*, and Rebell (1981) described a variety of *Fusarium* infections in humans in addition to diagnosing a great number of mycotic keratitis conditions associated with *F. solani*.

In Israel *Fusarium* cultures isolated from two patients (Figs. 7.1 and 7.2) suffering from cases of keratitis, have recently been identified as *F. solani* (Joffe, personal communication, 1971, 1979).

Mycotic ulcer keratitis due to *F. solani* has been described by the following investigators from different parts of the world:

Balakrishnan (1962) described some cases of corneal ulcers in India, and Cordero-Moreno and Pifano (1970) also reported some incidences of mycotic keratitis from Venezuela.

Arrechea *et al.* (1971), Zapater (1971), Zapater and Arrechea (1975), and Zapater *et al.* (1976) from Argentina isolated many strains of *F. solani* in patients suffering from corneal mycotic infections. Laverde *et al.* (1972) and Greer *et al.* (1973) reported cases of keratitis from Colombia, and Matsumoto (1972), Matsumoto and Soejima (1976), and Suga (1972) reported cases from Japan.

Salceda *et al.* (1974) and Salceda (1976) studied a great amount of keratomycosis in the Philippines. Gugnani *et al.* (1976) described cases of mycotic keratitis in Nigeria, and Mitchell and Attleberger (1973) noted a case of *F. solani* keratomycosis in a horse.

Jones *et al.* (1969a) and Forster and Rebell (1975a) found an experimental model for the production of keratitis infections with *F. solani* in rabbits, and keratitis in owls, monkeys, and in rats was studied by Burda

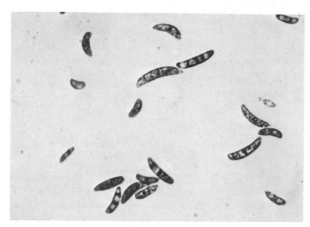

**Figure 7.2** A case of keratitis caused by *F. solani* in Israel. Micro- and macroconidia. (Joffe, unpublished results.)

and Fisher (1959) and Ishibishashi (1978). All these experiments were not very successful.

*Fusarium* species other than *F. solani* and *F. oxysporum* were also isolated from mycotic infections of the cornea by the following investigators:

Anderson *et al.* (1959) and Anderson and Chick (1963) described a case of mycotic ulcerative keratitis due to *F. moniliforme* in a 42-year-old male patient. A corn stalk had brushed against the eye of the patient a week before the symptoms began to appear. Kidd and Wold (1973) and Laverde *et al.* (1973) also described cases of mycotic keratitis connected with *F. moniliforme*.

Dudley and Chick (1964) produced corneal necrosis in rabbits using extracts isolated from *F. moniliforme,* and Perez (1966) and Inokawa (1972) isolated *F. nivale* from corneal mycotic keratitis.

Garcia *et al.* (1972) and Zapater *et al.* (1972) described two cases of keratomycosis associated with *F. dimerum*.

Zapater and Arrachea (1975) noted that in various parts of the world 112 cases of mycotic keratitis were connected with *F. episphaeria, F. nivale, F. oxysporum,* and mainly *F. solani.* Salcedo (1976) isolated *F. episphaeria* and *F. dimerum* from patients suffering from corneal ulcer mycotic infections.

Cases of mycotic ulcerative keratitis due to *Fusarium* sp. have been described by the following researchers without the identification of the exact species of *Fusarium* responsible: Singtenhorst and Gingrich (1957) from Texas; Gillespie (1963) from Alabama; Ming and Yu (1966) from China; Naumann *et al.* (1967) from the United States; Putanna (1967) from Peru; Newmark *et al.* (1970) from Florida; and Singh and Malik (1972) from India.

## ULCERS, NECROSES, AND LESIONS IN SKIN INFECTIONS AND IN ORGANS AND TISSUES

*Fusarium* infections resulting in skin irritation, wounds and other lesions, leg ulcers, facial granuloma, osteomyelitis, and other syndromes have all been associated with trauma and dissemination of chronic diseases.

In Japan *F. solani*, isolated from onion, when applied to human skin (the arm) caused hyperemia within 24 hr and hypertrophy and necrosis after 7 days (Kambayashi and Otake, 1936).

Peterson and Baker (1959), Holzegel and Kempf (1964), and Abramowsky *et al.* (1974) isolated *F. roseum, F. solani,* and *F. oxysporum,* respectively, from the wounds of severely burned children. Foley (1969) isolated various fungi from burn wounds including *Fusarium* sp., while Wheeler *et al.* (1981) described a case of disseminated burn wound invasion caused by *F. oxysporum* and three other cases of burn wounds associated with *F. moniliforme, F. solani,* and also *F. oxysporum.*

English (1968) isolated *F. solani* from necrotic skin of the toe webs in a 56-year-old diabetic woman and *F. redolens* from a necrotic ulcer on the foot of a 60-year-old diabetic patient. English (1972) and English *et al.* (1971) isolated *F. oxysporum* and *F. solani* from venous ulcers in the leg.

*Fusarium* fungi caused facial subcutaneous granuloma in a child suffering from chronic granulomatous disease (Benjamin *et al.,* 1970). Cho *et al.* (1973) described a case of skin lesions due to *F. solani* in an 18-month-old boy with acute leukemia. The skin lesions progressed into blackish necrotic centers surrounded by indurating erythema. Anderson and Chick (1963) described two cases of white grain mycetoma associated with *F. solani* and *F. moniliforme.*

Bourguignon *et al.* (1976) isolated *F. oxysporum* from ulcerous matter in a case of osteomyelitis of the tibia in a 7-year-old boy. Collins and Rinaldi (1977) isolated *F. moniliforme* from a patient with a severe cutaneous infection of the hand. A case of disseminated *F. moniliforme* infection was seen in a 32-year-old granulocytopenic man suffering from malignant lymphoma (Young *et al.,* 1978).

Guttman *et al.* (1975) described a case of a 66-year-old woman who suffered from gastrointestinal trouble and who had lost 15 lb. After a period of 9 months the proximal muscles were so weak that this patient was unable to climb and died several months later. Numerous ulcerative granulomas had developed in the esophagus, liver tissue, spleen, and cecum, from which *F. oxysporum* was isolated.

According to Parker and Klintworth (1971) local infections of the skin, nails, cornea, and urinary tract in men were commonly associated with *F. oxysporum,* while Lazarus and Schwarz (1948) noted that *Fusarium* species were isolated from the urine of 10 adult female patients and from a boy.

Young and Meyers (1979) reported the role of *F. oxysporum* as an opportunistic pathogen in a renal transplant patient, and Mutton *et al.* (1980) described disseminated infection caused by *Fusarium* species.

Steinberger *et al.* (1983) described a *Fusarium* brain abscess and diffuse meningitis in a 17-year-old white girl with chronic infectious mononucleosis syndrome and immunodeficiency.

# CHAPTER **8**

# FUSARIOTOXICOSES IN LABORATORY ANIMALS

The types of experiments which have been carried out by researchers to study the effect of fusariotoxins on laboratory animals are those associated mainly with feeding methods and skin tests. Skin reaction on rabbits caused by the action of different *Fusarium* extracts was distinguished by external appearance and histological changes. Thus the toxins of *Fusarium* cultures have both a localized and a general toxic effect. The localized effect is first apparent as an inflammatory reaction on rabbit skin and is accompanied by subsequent necrosis at the site of the toxin application. The general effect is apparent in defective blood production, acute degenerative processes in the internal organs, and extreme hyperemia, hemorrhages, and necrosis, especially in the digestive tract.

Pentman (1935) and later Kambayashi and Otake (1936) used the skin test on rabbits, but this method was not used again until large-scale skin tests were started in the author's laboratory in 1942. Today skin tests are considered to be a reliable method of determining the toxicity of various cereal crops, food, and feed, infected by *Fusarium* species and other fungi and the biological properties of a variety of isolated mycotoxins.

Symptoms in animals vary according to the potency and quantity of the toxin, route of administration, and sensitivity of the animal. T-2 toxin and some other trichothecene mycotoxins, isolated from foods and feeds contaminated with *Fusarium* species, have adverse effects on animals and humans (Agrelo and Schoental, 1980; Bamburg, 1972; Bamburg and Strong, 1971; Joffe, 1978a,b, 1983b). The general effects of toxins have been studied in a variety of animals, and the results of experiments carried out by the author and others will be briefly summarized.

## THE EFFECT OF TOXIC *FUSARIUM* SPECIES AND THEIR TOXINS ON RATS AND MICE

Rats and mice are commonly used to investigate the toxicity of drugs, chemicals, and so on. Fusariotoxins are no exception, and these small animals have been extensively used to investigate the potential toxic and pathological properties of the toxic *Fusarium* species and their products. The experimental work can be divided into two general categories:

1. Overall screening tests for toxicity of *Fusarium* species (or their metabolites).
2. Investigations into the specific effects of toxic *Fusarium* species, their clinical and histopathological effects, and the reproduction of symptoms (both clinical and histopathological) observed in domestic animals and humans.

### Mice and Rats as Test Animals for Screening Toxic *Fusarium* Species

Early screening tests involved the lethal effects of the *Fusarium* species tested on mice and/or rats. Scott (1965) screened a wide variety of fungi and found isolates of *F. moniliforme* and *F. roseum* that were toxic to mice and rats. Other animals, such as chicks or ducklings, were often used as additional test animals. Rats and mice have also been used as test animals in the "skin test" bioassay for trichothecenes and certain other fusariotoxins (Marasas *et al.*, 1969; Ueno *et al.*, 1970a,b; Wei *et al.*, 1972).

Several clinical or histopathological reactions in mice and rats have been used to detect specific toxins or groups of toxins. Examination of the uterus of rats or mice has been used for zearalenone or related compounds, and internal hemorrhaging and "radiomimetic effects" for trichothecenes. Thus, for example, Vanyi *et al.* (1973a) found that *F. sporotrichiella* (*F. sporotrichioides*) and *F. moniliforme* were lethal to rats, causing intestinal catarrh and enteritis but *no* enlarged uterus, whereas rats fed maize contaminated with *F. graminearum* and *F. culmorum* showed moderate uterine hypertrophy.

Birone *et al.* (1972) studied the toxicity of the *Fusarium* sp. on white rats resulting in an estrogenic syndrome accompanied by inhibition of spermiogenesis and follicle maturation in ovaries, increased follicle atresia, vascular degeneration, and pseudometaplasiatic changes of the endometrium. The most severe changes were caused by *F. sporotrichiella* (*F. sporotrichioides*). Animals fed a diet containing *F. semitectum* and *F. graminearum* (*Gibberella zeae*) showed regressive alteration in the gonads. *F. moniliforme* inhibited the spermiogenesis in 20% of the rats. Enteritis and general cachexy was caused only by *F. sporotrichiella*, which is known to produce T-2 toxin.

Christensen *et al.* (1968) noted that 65 out of 87 isolates of *Fusarium* tested resulted in the death of 21-day-old weanling rats within 4–5 days after being fed a diet contaminated by these fungi. Some isolates of *Fusarium* increased the uteri 8–10 fold in weight. These isolates presumably produced the estrogenic substance and other compounds which caused lethal toxicosis.

Van Rensburg *et al.* (1971) noted that two strains of *F. roseum* grown on maize meal and incorporated in the diet for mice caused death of all mice within 10–15 days after feeding. The histopathological findings revealed focal to massive necrosis of the proximal convoluted tubules, glomerule congestion and pyknosis of the medulla, and hepatic congestion and marked atrophy of the hepatocytes in the peripheral zones accompanied by cellular degeneration and necrosis. One mouse died after 3 days of feeding on the diet contaminated by *F. moniliforme*.

Ueno *et al.* (1973d) screened many *Fusarium* species for their ability to produce trichothecene mycotoxins.

*F. tricinctum*, grown on corn, caused estrogenic symptoms when fed to rats, swine, and poultry. Certain isolates of *F. tricinctum* were also found to produce the F-2 toxin (zearalenone) in large amounts, in association with toxic trichothecenes such as T-2 toxin and neosolaniol (Nelson *et al.*, 1973).

Korpinen and Uoti (1974) studied five *Fusarium* species isolated from cereal seeds (*F. avenaceum, F. culmorum, F. graminearum, F. poae*, and *F. tricinctum*) in a feeding experiment with rats during a period of two weeks. The authors noted that the effect of *F. tricinctum* was most toxic and caused the death of the animals. *F. poae* showed slight symptoms and most of the rats survived.

According to Ueno *et al.* (1977b) the crude toxins from four *F. oxysporum* cultures isolated from river sediments caused inhibition of protein synthesis in rabbit reticulocytes and radiomimetic cellular injury in some mice.

## Specific Effects of the Toxic *Fusarium* Species on Mice and Rats

### The Food Refusal Syndrome

Food refusal of moldy grain has been a common occurrence among domestic animals, especially pigs. Rats have also been found to be sensitive to food contaminated with toxic fungi. According to Greenway and Puls (1976) badly spoiled grain may be very unpalatable, and animals may refuse to eat it. The authors noted that prolonged ingestion of palatable grains contaminated with even low levels of mycotoxin may cause lethal mycotoxicosis. Ueno (1971) stated that mice, when administered with fusarenon-X isolated from *F. nivale* Fn-2B, became inactive, their fur became ruffled, and they refused food and water. Featherston (1973) found that rats refused corn infected with *G. zeae*, and he suggested that they could be used as an excellent screening test for detecting the "refusal factor" in corn. Washing corn has been found to increase the palatability of moldy corn to rats (deUriarte *et al.*, 1976; Forsyth, 1974). Specific toxins that have been shown to cause food refusal in rats include T-2 toxin. According to Kotsonis *et al.* (1975b) zearalenone was detected in *Fusarium* isolates and was shown to be partially responsible for refusal activity in rats which were fed diets consisting of the same corn cultures used in the pig study. The rats in this trial suffered from weight loss, feed refusal, and increased fresh weight of the uterus and ovaries. When both toxins (T-2 toxin and zearalenone) were present together the food intake was even more drastically reduced (50 μg/g T-2 toxin plus 50 μg/g zearalenone resulted in the consumption of only 2.86 g food) (Kotsonis *et al.*, 1975b). Deoxynivalenol and its three acetylated derivatives at doses of 100 μg/g and 150 μg/g, respectively, caused a 50% reduction in rat food consumption (Yoshizawa *et al.*, 1978).

Marasas *et al.* (1977) noted that *F. graminearum* produced two mycotoxins, zearalenone and deoxynivalenol, from hand-selected naturally contaminated Zambian and South African maize. The toxicity and feed refusal activity of *F. graminearum* is associated with the combined effects of zearalenone and deoxynivalenol and perhaps with other unidentified toxins under certain environmental conditions.

Burmeister *et al.* (1980b) proved that only diacetoxyscirpenol, T-2 toxin, and vomitoxin caused refusal responses in mice when added to drinking water.

Grove *et al.* (1984) prepared six derivatives of T-2 toxin for their feed refusal activity in the mouse drinking-water bioassay. Compounds were given to mice in their drinking water at concentrations of 50 mg/liter over a 20 hr period. The authors noted that T-2 toxin possesses the greatest refusal activity (71% drinking solution refused). The most striking results were obtained with the compound (Lithium aluminum hydrate—LAH, product hydrates, and ketone) in which the spiro epoxy group had been opened.

Fusarenon-X caused, in experimental animals (rats), various toxic signs such as vomiting, food refusal, hyperemia of the gastrointestinal tract, and diarrhea (Matsuoka and Kubota, 1981; Matsuoka *et al.* 1979; Ueno, 1977a; Ueno *et al.*, 1977b).

## The Effects of Trichothecene Toxins on Mice and Rats

Most researchers have consistently described the clinical and pathological findings in mice and rats resulting from the administration of these toxins. Getsova (1960) studied the effect of i.p. injections with toxins from ATA on mice. The early work done on rats by Marasas *et al.* (1969) showed that feeding 15 ppm T-2 toxin for 3 weeks caused stunted growth, inflammation around the mouth and nose, and small areas of cytoplasmic degeneration in the liver. T-2 toxin also caused reduction in body weight and feed consumption. Prolonged administration of 10 ppm over a period of 8 months appeared to have no adverse effect.

Hayes and Schiefer (1980) noted that 20 ppm T-2 toxin-contaminated diet (containing 8, 12, or 16% protein) caused reduction of growth and food consumption in young male weanling outbred Swiss mice after one to four weeks.

Mice fed T-2 toxin in a 16% protein diet had higher growth rates than those containing T-2 toxin in diets with a reduced protein content (8 or 12% protein), as described by Hayes *et al.* (1980).

Investigators in Japan discovered one of the characteristic effects of the trichothecenes on dividing cells, namely the so-called "radio-mimetic effect" (Saito *et al.*, 1969; Ueno et al., 1971c). Saito *et al.* described the clinical symptoms in mice after administration of toxic crude extracts of various trichothecenes (nivalenol and fusarenon-X) isolated from *F. nivale.* Histological examination of mice injected with the trichothecene metabolites of *F. nivale* showed characteristic cytotoxic changes in actively dividing cells such as the mucosa of the small intestine, germ center of the lymph follicles in the spleen, lymph node and other lymphatic tissues, the thymus and bone marrow, ileum, jejunum, stomach, testes, and hair follicles, and also hemorrhages in lungs and damage to liver and brain tissues. Fusarenon-X and nivalenol decreased thymus and spleen weights and caused typical radiomimetic cell necrosis. Single doses of nivalenol or nivalenol 4-0-acetate caused a drop in erythrocytes. The levels of circulating lymphocytes and leukocytes were initially increased but after 3 hr began to fall and, by 96 hr, fell well below normal (Saito *et al.*, 1969).

Ide *et al.* (1967), Ohtsubo *et al.* (1972), Saito and Okubo (1970), Tatsuno (1968), Tatsuno *et al.* (1966, 1968a,b), Terao and Ueno (1978) Tsunoda *et al.* (1968), Ueno (1971), and Ueno *et al.* (1969c, 1970a,b, 1971a,b,c, 1973b,d) carried out some biological assays with metabolites of *F. nivale* which were responsible for acute toxicity to mice and rats and other laboratory animals. According to these researchers the pathogenic findings in experiments on mice and rats given nivalenol, fusarenon-X, diacetylnivalenol, and T-2 toxin showed degeneration, necrosis and karyorrhexis of the mucosal epithels in the gastrointestinal tract, especially of the small intestine, as well as hemorrhages in the heart muscle. They also caused depression of leukopoiesis and nervous disorders.

According to Matsuoka *et al.* (1979) and Matsuoka and Kubota (1981, 1982) fusarenon-X induced diarrhea and caused a structural alteration of intestinal mucosal cells resulting in an abnormal increase in permeability of the mucosa. Therefore the authors stated that an increase in D-xylose absorption rate in rats, in everted intestinal sac and induced by fusarenon-X, is not connected with an absorption ability of the intestine but with an increase in the permeability of the intestinal mucosal membrane.

Cellular injury characterized by necrosis and karyorrhexis of proliferating cells in thymus and other organs of mice and rats have been shown to be caused by fusarenon-X (Ueno *et al.*, 1971a,b), T-2 toxin, neosolaniol, and HT-2 toxin (Ueno *et al.*, 1972b). Rats died after oral administration with less damage than in mice (Ueno *et al.*, 1971a).

Fusarenon-X was the most injurious, followed by nivalenol (Saito and Okubo, 1970; Saito *et al.*, 1971). Atypical mitosis, pyknosis, hyperchromatosis of the nuclear membrane, and fragmentation of the nuclei were recorded (Saito *et al.*, 1969).

Some strains of *F. nivale* and *F. episphaeria* (*F. aquaeductuum*) produced fusarenon-X and nivalenol and revealed high toxicity to mice and rabbit reticulocytes (inhibited $C^{14}$-leucine uptake) as well as toxicity on rabbit skin (Ueno *et al.*, 1971a, 1973b).

Mice and rats fed rice infected by *F. nivale* or *F. graminearum* in long-term experiments showed atrophy and hypoplasia of the hematopoietic tissues (Saito and Ohtsubo, 1974).

Ueno *et al.* (1973c, 1974a) concluded that crude extracts isolated from *F. nivale* invariably contained toxic trichothecenes (Saito and Okubo, 1970; Saito and Ohtsubo, 1974; Saito *et al.*, 1969; Ueno *et al.*, 1971b,c) and caused severe damage.

The crude extracts of *Fusarium* species gave strong reactions on rabbit skin (Ueno *et al.*, 1971a, 1972b, 1973d).

When the extracts of *F. poae* No. 958 and *F. sporotrichioides* No. 921 were applied to rabbit skin, the treated site became congested and edematous within 24–48 hr depending on the efficacy of the solution. Keratinization followed, with the scab falling off within 2–3 weeks. When the applications were repeated weekly or less often, some of the follicles became atrophic, ulceration was slower to heal, and the treated areas remained depilated for an increasingly longer time, possibly permanently.

Microscopically the treated skin exhibited disorganized architecture, and areas of desquamation and regeneration occurred next to each other. Changes were also present in the muscle, which sometimes became edematous, containing foci of infection and infiltration by inflammatory cells. When treatment was interrupted, the necrotic lesions tended to disappear, but the hyperplastic changes appeared to persist and may have become progressive (Schoental and Joffe, 1974).

Topical application of T-2 toxin to rat skin resulted in extensive inflammation and necrosis of the tissues but did not produce papillomas (Bamburg *et al.*, 1968a; Marasas *et al.*, 1969).

*F. moniliforme, F. poae, F. sporotrichioides,* and *F. tricinctum* all caused internal hemorrhaging in mice (Korpinen and Ylimäki, 1972).

Internal hemorrhaging in the intestinal tract, liver, and kidney is also a characteristic symptom of trichothecene poisoning.

According to Kosuri *et al.* (1971) T-2 toxin produced by *F. tricinctum* caused necrotic lesions and hemorrhages in the intestines, liver, and kidneys of rats. Yoshizawa and Morooka (1974) noted that deoxynivalenol and its monoacetate, nivalenol, and fusarenon-X administered intraperitoneally caused marked dilation with hemorrhage of the gastrointestinal tract and congestion of the testes in mice.

Ether and ethyl acetate extracts of *F. incarnatum* (*F. semitectum* var. *majus*) injected into rats did not exhibit an increase in uterine weight. Protein synthesis in rabbit reticulocytes was inhibited. T-2 toxin was detected by means of TLC (Rukmini and Bhat, 1978).

The trichothecenes also induced hematological effects in mice. A single administration (i.p. or orally) of T-2 neosolaniol or fusarenon-X caused temporary increases in leukocyte, lymphocyte, and neutrophil counts, although the erythrocytes and reduced platelet counts were unaffected. Neosolaniol increased the reticulocyte levels. Fusarenon-X and neosolaniol decreased albumin, urea, and gamma globulin levels but had no effect on alkaline phosphatase, chloride, or glutamate-pyruvate transaminase (Sato *et al.*, 1978).

In mice given 2–5 mg/kg of fusarenon-X and 5–20 mg/kg neosolaniol i.p. hematological changes occurred after 1–3 hr (Ueno and Ueno, 1978).

T-2 toxin incubated with both human and bovine liver homogenate supernatants resulted in its conversion to HT-2 toxin. The production of HT-2 toxin is probably due to a liver esterase.

Ohta *et al.* (1977) demonstrated that T-2 toxin is biotransformed into HT-2 toxin by nonspecific carboxylesterase mainly localized in the liver microsomes. When T-2 toxin was incubated with S-9 fraction from rat liver the T-2 toxin decreased and was shown to be replaced by HT-2 toxin and neosolaniol. After 60 min the sum of the residual T-2 toxin and HT-2 toxin present was almost equal to the original amount of T-2 toxin in the substrate. Tissue specificity of the microsomal diacetylation of T-2 toxin was examined in rats and rabbits. The microsomes derived from rat liver showed the highest activity, followed by those of the brain and kidney. In rabbits the hydrolyzing activity was high in liver, followed by the kidney and spleen. Rabbit capacity of liver microsomes to convert T-2 toxins into HT-2 toxin was 80 times higher than that in rats.

It was stated that a single dose of T-2 toxin and other trichothecene mycotoxins produced by *Fusarium* species have toxic action on the hematopoietic system and lymphoid tissue of mice (Fromentin *et al.*, 1980; Hayes, 1979; Hayes *et al.*, 1980; Lafarge-Frayssinet *et al.*, 1979, 1981; Sato *et al.*, 1978), cats (Joffe, 1971, 1978b, 1983b; Lutsky and Mor, 1981a,b; Lutsky *et al.*, 1978; Sato *et al.*, 1975; Ueno *et al.*, 1973b), swine (Weaver *et al.*, 1978a), and chickens (Bitay *et al.*, 1979). T-2 toxin and other derivatives caused immunosuppressive action. According to

Hayes and Schiefer (1979a) low-protein diets containing T-2 toxin at 10 ppm suppressed hematopoiesis in mice.

Hayes *et al.* (1980) described the hematological and morphological changes in subacute toxicity experiments in young Swiss mice fed a 16% protein diet containing 20 ppm T-2 toxin for 1–6 weeks. After 3 weeks of treating mice with T-2 toxin, lymphoid tissues, bone marrow, and splenic red pulp became hypoplastic causing anemia, lymphopenia, and eosinopenia. After 6 weeks exposure to T-2 toxin all lymphoid tissues remained atrophic.

In mice T-2 toxin also caused perioral dermatitis and hyperkeratosis with ulceration of the stomach mucosa. The authors concluded that young mice were susceptible to the irritation response and the hematopoietic-suppressive toxic action of dietary T-2 toxin.

Hayes and Schiefer (1982) compared the hematotoxicity of dietary T-2 toxin (10 or 20 ppm for 2 or 4 weeks) in young male Wistar rats and in young male and juvenile Swiss mice. T-2 toxin depressed the food consumption and weight gain in all rodents and caused hyperkeratosis of the squamous gastric mucosa, atrophy of the thymus, and lymphopenia. These disorders were most severe in juvenile mice and in rats.

After being fed 20 ppm of T-2 toxin, the juvenile mice developed erythroid hypoplasia and anemia and died. Rodents are generally relatively resistant to hematopoietic suppression.

The authors noted that subacute dietary T-2 toxicosis is different depending on amounts of toxin, kind of species, and age of rodents. The severe effects on the lymphoid tissues in mice and rats showed differences in their susceptibility to immunosuppression by the action of T-2 toxin.

Lafarge-Frayssinet *et al.* (1979) and Rosenstein *et al.* (1979) showed that T-2 toxin and diacetoxyscirpenol suppressed the *in vivo* antibody response to sheep red blood cells, and also delayed skin graft rejection in mice.

Masuko *et al.* (1977), Otokawa *et al.* (1979), and Rosenstein *et al.* (1981) indicated that suppressor T-cells or their precursors are inhibited by T-2 toxin in mice and that the immunosuppressive effect of the toxin is in its action against T-cells.

Masuda *et al.* (1982) found that fusarenon-X, when administered to mice, suppressed lymphocyte functions and inhibited antibody response *in vivo* and *in vitro*. The mitogenic response of mice splenic lymphocytes was suppressed by action with 0.05 μg/ml or higher concentrations of fusarenon-X for 24 hr. The authors stated that inhibition was more pronounced for T-cell mitogens than for B-cell mitogens.

Mann *et al.* (1982) noted that T-2 toxin also had a harmful effect on humoral immunity.

Friend *et al.* (1983a,c) studied the effect of 5, 10, or 20 ppm T-2 toxin on the immune system of young male white Swiss mice for 1–6 weeks and also the influence of T-2 toxin on reactivation of herpes simplex virus type I (HSV-I). The authors reported that T-2 toxin caused immunosuppression but this response did not result in reactivation of latent HSV-1.

Tenk *et al.* (1982) studied the effect of pure T-2 toxin, F-2 toxin, and diacetoxyscirpenol on the gut and the plasma glucocorticoid levels in rats and swine. High microflora counts in the gut are associated with an increase in plasma glucocorticoid levels.

The feeding of 5 µg/g in the diet of piglets and rats with T-2, T-2 and F-2, and diacetoxyscirpenol toxins for one week caused an increase of aerobic bacteria count in the gut.

The increase of the diacetoxyscirpenol dose caused no further enhancing in bacterial counts. T-2 toxin, when administered i.m., also increased the activity of the adrenal cortex and the gut microflora.

Carson and Smith (1983) stated that the feeding of alfalfa (a diet rich in dietary fibers) protects against T-2 toxicosis in rats by reducing intestinal absorption of the toxin and increasing fecal excretion, but not urinary excretion, of [$^3$H] T-2 toxin.

Fromentin *et al.* (1980) stated that a single dose of 4.5 mg/kg diacetoxyscirpenol caused weight decrease of spleen and thymus in mice. When candidiasis mice were treated with small doses of diacetoxyscirpenol they died more rapidly than untreated mice infected by *Candida albicans*. The authors concluded that diacetoxyscirpenol has an immunosuppressive action on mice.

Salazar *et al.* (1980) found that suspension of homogenized kidneys from experimental candidiasis mice, administered a single dose of 4.5 mg/kg diacetoxyscirpenol added to Sabouraud dextrose agar plates with *C. albicans*, increased the number of colonies of the latter.

The toxic properties of various *F. poae* and *F. sporotrichioides* strains (which produced T-2 toxin) were tested by the author on 288 experimental and 192 control white mice. These *Fusarium* species were separately incubated on solid and liquid Czapek medium and millet grain at low temperatures with alternate freezing and thawing. The mice died 2–7 days after having been fed p.o. various doses of the 12 *Fusarium* strains. The data are presented in Table 8.1.

Experiments with extracts and liquid medium of *F. poae* No. 60/9 and *F. sporotrichioides* No. 60/10 were conducted in 1946 (Joffe, 1971) and again in 1952 with *F. poae* No. 958 and *F. sporotrichioides* No. 921 (*see* Table 6.14). The *Fusarium* cultures were grown on liquid medium at different temperatures at each of the three stages of development, that is, before sporulation, during the period of abundant sporulation, and at senescence. Ethanol or ether extracts or filtrates displayed the highest toxicity of *F. poae* and *F. sporotrichioides* obtained during the stage of

**TABLE 8.1  Lethal Effect of Crude Toxin from *F. poae* and *F. sporotrichioides***

| Species | No. of Strain | Crude Toxin | | | | | | | |
| --- | --- | --- | --- | --- | --- | --- | --- | --- | --- |
| | | Czapek Agar | Millet Grain | | Millet Extract | | Filtrate of Czapek Medium | | Dry Mycelium |
| | | 20 mg | 50 mg | 150 mg | 5 mg | 8 mg | 0.2 ml | 0.3 ml | 10 mg |
| *F. poae* | 60/9 | 3/2[a] | 3/2 | 3/3 | 3/2 | 3/3 | 3/1 | 3/2 | 3/0 |
| | 24 | 3/2 | 3/3 | 3/3 | 3/1 | 3/3 | 3/2 | 3/3 | 3/0 |
| | 396 | 3/3 | 3/3 | 3/3 | 3/3 | 3/3 | 3/2 | 3/3 | 3/1 |
| | 792 | 3/3 | 3/3 | 3/3 | 3/3 | 3/3 | 3/2 | 3/3 | 3/1 |
| | 958 | 3/3 | 3/3 | 3/3 | 3/3 | 3/3 | 3/2 | 3/3 | 3/1 |
| *F. sporotrichioides* | 60/10 | 3/2 | 3/1 | 3/3 | 3/1 | 3/2 | 3/1 | 3/2 | 3/1 |
| | 347 | 3/3 | 3/3 | 3/3 | 3/2 | 3/3 | 3/2 | 3/3 | 3/1 |
| | 351 | 3/2 | 3/1 | 3/3 | 3/1 | 3/2 | 3/1 | 3/3 | 3/0 |
| | 738 | 3/3 | 3/3 | 3/3 | 3/2 | 3/3 | 3/2 | 3/3 | 3/1 |
| | 921 | 3/3 | 3/3 | 3/3 | 3/3 | 3/3 | 3/3 | 3/3 | 3/1 |
| | 1182 | 3/2 | 3/1 | 3/3 | 3/1 | 3/2 | 3/1 | 3/1 | 3/0 |
| | 1823 | 3/3 | 3/3 | 3/3 | 3/2 | 3/3 | 3/2 | 3/3 | 3/1 |

*Source:*  Joffe, unpublished results.
[a]Number of mice used/died.

abundant sporulation at low temperatures, causing strong reactions on rabbit skin and being lethal to mice, while those obtained at an advanced stage of senescence were not toxic.

## F. sporotrichioides and F. poae: Their Crude Extracts and Pure T-2 Toxin in Acute, Chronic, and Subchronic Intoxication in Rodents

Unfavorable climatic and ecological conditions favor the production of _Fusarium_ in overwintered cereal grains in the field causing intoxication in rodents. This intoxication may be acute, chronic, or subchronic depending on the size of the dose, the sensitivity of the animal, and the time elapsing between treatment and death of mice or rats.

Acute, chronic, and subchronic intoxication in mice and rats was studied by the author in cooperation with Drs. Regina Schoental and B. Yagen.

The results described in experiments on rodents are associated with crude extracts obtained from authentic species of _F. poae_ No. 958 and _F. sporotrichioides_ No. 921, which were responsible for the outbreak of the fatal ATA disease in Russia (Joffe, 1960a,b, 1965, 1971, 1974c, 1978b, 1983b), and with pure T-2 toxin isolated from these _Fusarium_ cultures by Joffe and Yagen (1977); Yagen and Joffe (1976); and Yagen _et al._ (1977).

Crude extracts and T-2 toxin have two types of action in rats and mice: local and systemic (Schoental, 1977b, 1979a,b,c, 1980a,b; Schoental and Joffe, 1974; Schoental _et al._, 1976, 1978a,b, 1979).

Doses of the crude extract (0.02–0.2 ml) or appropriately diluted solutions in 20% aqueous ethanol were administered to mice and rats of both sexes by local application, s.c. and i.p. injection, or i.g. intubation.

In rat experiments we also used pure crystalline T-2 toxin, isolated from extract of _F. sporotrichioides_ No. 921 by Yagen _et al._ (1977). Seventy weanling white male rats and one lactating female with her litter were used in the long-term experiments. The rats, of the Wistar-Porton strain (from the MRC Laboratory Animal Centre, Carshalton, Surrey, England), were separated by sex and kept in plastic or metal cages (not exceeding six rats per cage) and given a commercial pelleted diet and tap water _ad libitum_. T-2 toxin was dissolved in a few drops of ethanol before dosing and diluted with distilled water to the required volume. The final concentration of ethanol did not exceed 10%. The solution containing 1–4 mg/ml was given to the rats by stomach tube i.g. in doses of 0.2–4.0 mg/kg body weight at irregular, approximately monthly, intervals. Control rats were left untreated.

The acute and chronic or subchronic experiments with crude extract and with T-2 toxin were carried out on nearly 300 mice and rats. Animals which died or were killed were autopsied and their organs and tissues fixed in ethanol-formol and processed in the usual manner. Sections 5–6 µm thick were stained with haematoxylin for microscopic examination or occasionally with other stains when required. The intoxication caused by *F. sporotrichioides* and *F. poae* strains and their mycotoxins appears in hematopoietic organs (bone marrow, lymph nodes, spleen, and thymus) causing defective blood production. It produces degeneration of the internal organs and tissues (heart, lung, and kidney), especially of the digestive tract (mouth cavity, esophagus, duodenum, stomach, and intestines), accompanied by extreme hyperemia and hemorrhages of the digestive tract and of certain other organs with cardiovascular lesions of various degrees of severity. Intoxication also occurs in the reproductive organs (ovary, uterus, vulva, mammary glands, and testes) and in the nervous system (Joffe, 1971, 1974c, 1978b, 1983b; Schoental and Joffe, 1974; Schoental *et al.*, 1979).

*The Effects of Crude Extract.* Schoental and Joffe (1974) used the crude extracts from *F. poae* No. 958 and *F. sporotrichioides* No. 921 to test long-term effects in rodents when administered in various ways. Single doses, in the order of 0.1 ml, were toxic to young animals, causing them to die within a few days. Smaller doses allowed the animals to survive longer, and the dosage could be repeated at various time intervals. The authors drew attention to the chronic lesions that developed in animals that survived for several weeks or months, which suggested that the *Fusarium* metabolites may have been immunosuppressive. The larger doses of crude extract caused death within 1 day or longer regardless of the route of administration. The main lesions were present in the skin, lymphoid tissues, lung, heart, kidney, and digestive tract. After a subcutaneous dose of *F. sporotrichioides* No. 921 extract the lymph glands, particularly the retroperitoneal lymph nodes of mice, appeared enlarged and dark red, the spleen and the thymus appeared smaller than in the controls, the lungs were congested, the stomach and small intestine were congested and greatly distended with soft content which was sometimes dark when hemorrhage occurred from the congested mucosa.

Microscopically lymph glands were cystic and hemorrhagic (Fig. 8.1) and the central parts of the follicles often necrotic. The spleen was depleted of lymphoid elements and hemorrhagic. The walls of the digestive tract were edematous and the mucosa congested and/or desquamated.

When a drop of the extract was put into the mouths of baby rats they were unable to suckle, their tongues became very swollen, and they died within 2–3 days.

When weanling rats were given *Fusarium* extracts by short stomach tube, the lining of the esophagus and of the squamous part of the stomach became ulcerated, keratinized, and desquamated. In an extreme case the central part of the esophagus became distended and blocked with cellular debris (Fig. 8.2). Smaller doses of the extracts

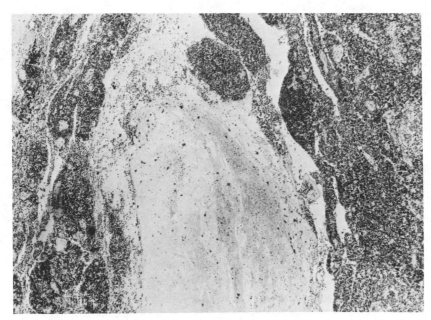

**Figure 8.1** Retroperitoneal lymph node. Mouse dying 1 day after s.c. dose of *F. sporo-trichioides* No. 921 extract; cystic distention and hemorrhage. Hematoxylin and eosin (HE) × 40. (Schoental and Joffe, 1974.)

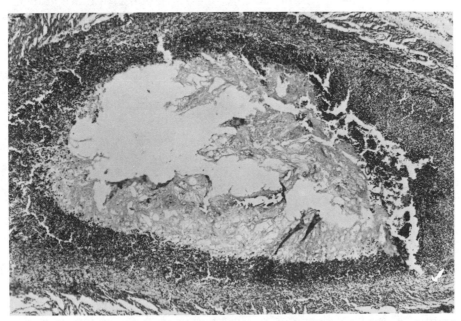

**Figure 8.2** Esophagus. Rat killed *in extremis* 15 days after first and 2 days after last of 3 doses of *F. poae* No. 958 given by stomach tube; local distension and blockage with cellular debris. HE × 40. (Schoental and Joffe, 1974.)

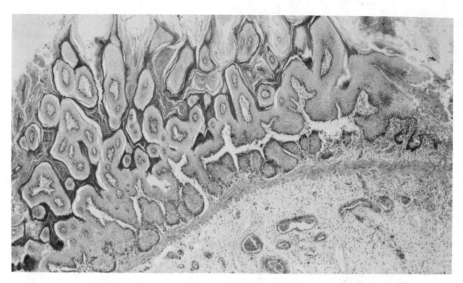

**Figure 8.3**  Forestomach. Female rat, dying 23 days after first, and 1 day after last (doubled), of 11 doses of *F. poae* No. 958 extract; squamous cell hyperplasia, edema of stomach wall, and ulceration (on right). HE × 60. (Schoental and Joffe, 1974.)

(0.02–0.05 ml or their solution in 20% aqueous ethanol) were tolerated well and could be given repeatedly, weekly, or daily. In these animals the acute lesions were superimposed on the more chronic ones: foci of ulceration and wall edema were found next to regenerative hyperplasia in both the squamous part of the stomach (Figs. 8.3 and 8.4) and in the esophagus. Animals killed for exploration after small doses did not appear to have gross lesions, though microscopically there were pockets of distention and invagination in the esophagus (Fig. 8.5) and thickened areas in the forestomach which were lined with multicellular squamous epithelium and showed hyperplasia of the basal cells.

Among the animals which died after several weeks (usually after more than 1 dose) autopsy showed widespread infections in various organs, appearing as whitish patches on the heart, lungs (Fig. 8.6), and kidneys (Fig. 8.7). Microscopically the foci of infection varied in size, and contained gram-positive organisms, some being accompanied by necrosis, fibrosis, and accumulation of inflammatory cells (Fig. 8.8). The spleens of these animals were enlarged and contained a variable number of megakaryocytes. No specific lesions were seen in livers of animals treated with the *Fusarium* extracts.

Among the animals which survived some continued to receive intermittent treatment while others were kept under observation without further dosing in order to establish whether progressive lesions might develop.

Some of the more acute effects induced by our crude *Fusarium* ex-

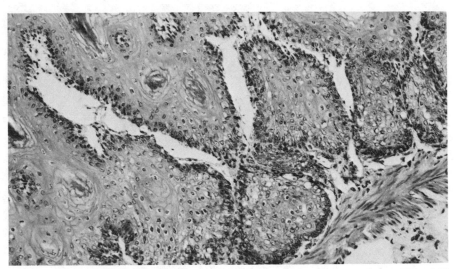

**Figure 8.4** Higher magnification of part of Figure 8.3; basal cell hyperplasia. HE × 240. (Schoental and Joffe, 1974.)

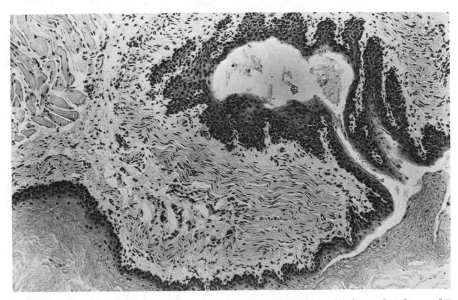

**Figure 8.5** Esophagus. Rat killed 9 weeks after first, and 7 days after last, of 10 doses of *F. poae* No. 958 extract given by stomach tube; invagination and basal cell hyperplasia of squamous epithelium. HE × 100. (Schoental and Joffe, 1974.)

**313**

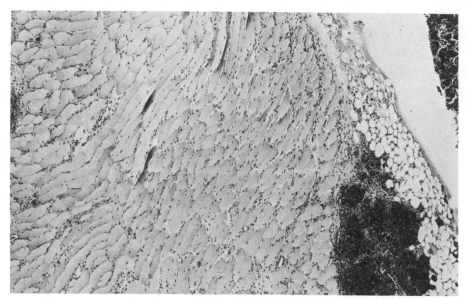

**Figure 8.6** Heart and (on right) lung. Mouse dying 10 weeks after last of weekly skin applications of *F. poae* No. 958 extract; foci of infection and infiltration with inflammatory cells. HE × 40. (Schoental and Joffe, 1974.)

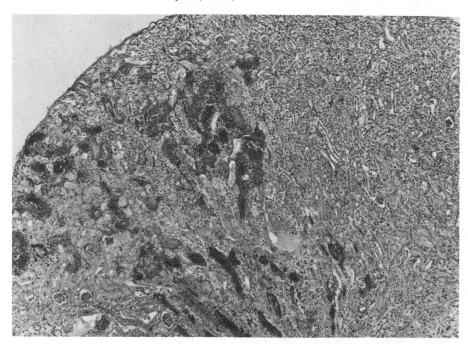

**Figure 8.7** Kidney. Male mouse dying 4 days after intraperitoneal dose of *F. poae* No. 958 extract; area of infection, necrosis, and infiltration by inflammatory cells. HE × 40. (Schoental and Joffe, 1974.)

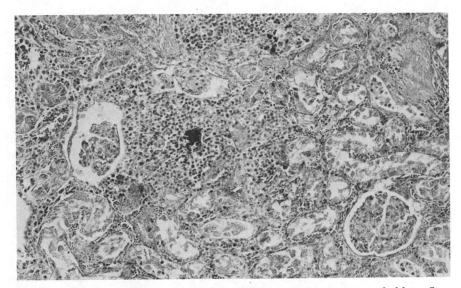

**Figure 8.8**  Kidney. Female rat from Figure 8.3; foci of infection surrounded by inflammatory cells. Gramstaining × 240. (Schoental and Joffe, 1974.)

tracts resembled those described in animals treated with certain 12,13-epoxy-trichothecenes (Bamburg and Strong, 1971; Marasas *et al.*, 1969; Ohtsubo and Saito, 1977; Saito and Tatsuno, 1971; Stähelin *et al.*, 1968).

We may conclude that large doses, which killed most of the animals, cause depletion of the lymphoid tissues, and the surviving rodents may develop subsequent widespread infections in various organs suggestive of an immunosuppressive action. Smaller doses applied to the skin, esophagus, and stomach have local cytotoxic effects followed by regeneration and basal hyperplasia of the squamous epithelium (Fig. 8.9).

The use of moldy cereals, contaminated by the inflammatory and irritant metabolites of *F. poae* No. 958 and *F. sporotrichioides* No. 921, may have played a part in the development of tumors of the digestive tract and other organs.

Rubinstein *et al.* (1967) found that *F. sporotrichiella* var. *poae* No. 1140/26 (*F. poae*), cultivated on wheat grains under environmental conditions with alternation of temperatures, had a pathological action on the alimentary tracts of albino rats. Extracts of infected wheat grain developed a pathological process in the mucosa of the preventriculus of rats and caused marked papillomatosis with hyperkeratosis.

*The Effects of Pure T-2 Toxin.*    The acute and some of the more chronic effects of T-2 toxin have been reported in rats, mice, trout, cats, and chickens (Bamburg and Strong, 1971; Joffe and Yagen, 1978; Lutsky and Mor, 1981a,b; Lutsky *et al.*, 1978; Marasas *et al.*, 1969; Ohtsubo and Saito, 1977; Schoental and Joffe, 1974; Wyatt *et al.*, 1973b).

**Figure 8.9**   Skin. Mouse dying 10 weeks after beginning of approximately weekly applications of *F. poae* No. 958 extract; areas of desquamation of keratinized epithelium, hyperplasia of basal cells, and foci of infection, some surrounded by inflammatory cells. HE × 70. (Schoental and Joffe, 1974.)

In a survey Bhat and Tulpule (1983) described the occurrence of *Fusarium* species producing T-2 toxin isolated from various plants in India which were toxic to chicks, mice, rats, and rhesus monkeys.

Schoental *et al.* (1979) carried out long-term experiments on rats with pure crystalline T-2 toxin isolated from crude extract of *F. sporotrichioides* No. 921. This mycotoxin is one of the most toxic and irritant among the trichothecene type metabolites of *Fusarium*. T-2 toxin caused severe cardiovascular lesions and high blood pressure in rats (Wilson *et al.*, 1982). Lesions and tumors were similar to those found among rats that survived when treated with crude ethanol extracts isolated from cultures of *F. poae* No. 958 and *F. sporotrichioides* No. 921 (Schoental, 1983c; Schoental and Joffe, 1974; Schoental *et al.*, 1978b).

After the first or after one of the subsequent treatments with T-2 toxin, about two-thirds of experimental rats died within a few days. At autopsy the stomach and small intestine were greatly distended with soft, often bloodstained content, hemorrhagic petechia, and erosions in the stomach; lymph glands and spleen appeared enlarged, the thymus appeared small, lungs were congested, the heart was engorged, and blood vessels were congested, this being particularly conspicuous in the brain.

Microscopically the stomach mucosa was usually denuded; there was glandular atrophy, cellular infiltration, and striking submucosal edema in the stomach and duodenum. The pancreas showed interlobular edema; there was depletion of the lymphoid elements in the hemorrhagic spleen and lymph glands. The thymus showed involution and foci of necrosis; lungs were often edematous. Necrotic changes were present in the gonads.

Details of the main degenerative or vascular and neoplastic lesions found in rats that survived 12–27.5 mon (mean: 22 mon) after the first dose and several weeks or months after the last of several doses of T-2 toxin are given in Table 8.2 (*also see* Schoental *et al.*, 1979).

Of the control rats, run parallel to the experimental ones, three died at 21.5–23 mon (from bronchopneumonia); the remaining 17 rats were killed at the age of 26–28 mon. No significant differences were seen between the control rats.

The acute effects which caused death within a few days after T-2 toxin administration included oliguria, hematuria, general congestion of blood vessels, and hemorrhages. T-2 toxin was known to produce hematological abnormalities, such as leukocytopenia and thrombocytopenia, and hemorrhages in several animals. Cats appeared to be singularly susceptible to the hematological changes and showed striking alterations in the morphology of the neutrophils. In animals which did not die the hematological changes appeared to be reversible (Lutsky *et al.*, 1978).

T-2 toxin has striking cytotoxic action in tissue cultures, which it shares with some of the other trichothecenes including diacetoxyscirpenol (Stähelin *et al.*, 1968; Ueno, 1977a). The cytotoxic effects manifest themselves at the site of application *in vivo*, either the skin or the digestive tract. From our results it appears that when T-2 toxin circulates in the blood stream after absorption from the stomach it can damage vascular endothelium, cause extravasation of the blood and hemorrhages in remote organs including the brain. If the animal survives, the arterial endothelium undergoes reparative processes, giving rise to thickening of the arterial wall and striking chronic cardiovascular lesions in rats. Various degrees of arteritis have been considered as "spontaneous" (Berg, 1967; Wilenz and Sproul, 1938a,b). We are inclined to interpret such spontaneous lesions as the result of occasional ingestion of food contaminated with *Fusarium* mycotoxins (Schoental *et al.*, 1979).

T-2 toxin appeared to be a versatile carcinogen to rats. When administered by i.g. intubation it can induce tumors of several organs, including the gastrointestinal tract, but not the liver. When a solution of T-2 toxin, containing 0.2–0.3 mg/ml in 10% aqueous ethanol, was applied to the clipped intrascapular regions of C57BL mice, it caused local irritation, hyperemia, edema, ulceration, and scab formation. When the scab fell off the healed area remained depilated and, in some animals, was surrounded by areas of depigmented hair.

T-2 toxin has been tested previously by several groups of workers for chronic toxicity in various animal species (Marasas *et al.*, 1969; Ohtsubo and Saito, 1977), and they concluded that T-2 toxin was not carcinogenic.

Lafont *et al.* (1977) and Rosenstein *et al.* (1979) studied the cytopatho-

**TABLE 8.2** Main Lesions, Degenerative or Vascular and Neoplastic, in Rats That Survived 12–27.5 Months After the First of 3–8 Doses of T-2 Toxin Given I.G. Alone or in Conjunction with I.P. Injections of Nicotinamide

| No. | Sex | Survival (Mon.)[a] | No. of Doses | Heart[b] | Arteries | Kidneys | Stomach — Squamous | Stomach — Glandular | Duodenum | Pancreas — Exocrine | Pancreas — Islet Cell | Brain | Pituitary | Others |
|---|---|---|---|---|---|---|---|---|---|---|---|---|---|---|
| 1 | M | 19.5 D | 3 + NA | *** | *** | * | + | | + | | | + | | |
| 2 | M | 23 K | 3 | * | | ** | | | | | + | | | |
| 3 | M | 21.5 K | 4 | * | | | | | | | | * | | |
| 4 | M | 23 K | 4 | * | ** | ** | | | | | ++ | **+ | | |
| 5 | M | 23.5 D | 4 | ** | *** | *** | ++ | | | | ++ | | | |
| 6 | M | 24 K | 4 | | * | ** | + | | | + | | * | | |
| 7 | M | 16 K | 5 | ** | | ** | ++ | +C | | ++ | | + | | |
| 8 | M | 16 K | 5 | ** | * | ** | + | ++C | +C | ++ | | | | |
| 9 | M | 22 K | 5 | | * | * | + | | ++ | ++ | | | | |
| 10 | M | 22 D | 5 | ** | | *C | | | | ++ | ++ | | + | Bladder + |
| 11 | M | 22 K | 5 | | | * | | | | ++ | + | | + | Parathyroid + |

| No. | Sex | Age | Fate[a] | Dose | | | | | | | | | Other organs[b] |
|---|---|---|---|---|---|---|---|---|---|---|---|---|---|
| 12 | M | 22 | D | 5 | *** | ** | | | C | | | | | Parathyroid + |
| 13 | M | 24 | D | 5 | ** | ** | | | C | | | | | |
| 14 | M | 24.5 | K | 5 | | ** | ** | + | | + | | | | |
| 15 | M | 24.5 | K | 5 | * | ** | ** | +++ | | +++ | + | + | ++ | |
| 16 | F | 21 | K | 6 | | * | * | +++ | | +++ | + | + | +++ | Mamma +++; Adrenal medullary +++ |
| 17 | F | 27.5 | D | 6 | * | * | ** | + | | +++ | * | + | ++ | |
| 18 | F | 27.5 | K | 6 | * | * | + | | | | * | + | +++ | Mamma +++ |
| 19 | M | 20 | K | 8 | *** | ** | **C | ++ | C | ++ | * | ++ | | Bladder |
| 20 | M | 21.5 | K | 8 | *** | *** | ** | + | | +++ | ++ | ++ | | |
| 21 | M | 22 | K | 8 | | | | | | | +++ | | | |
| 22 | M | 12.5 | K | 8 + NA | | ** | *C | + | *C | | +++ | +++ | | Bladder + Esophagus + |
| 23 | M | 23.5 | K | 8 + NA | ** | ** | | + | | +++ | + | | | |
| 24 | M | 23.5 | K | 8 + NA | | | | + | | +++ | + | ++ | | |
| 25 | M | 24.5 | K | 8 + NA | | ** | + | | | +++ | | | | |

*Source:* From Schoental *et al.* (1979).

[a] D = found dead; NA = nicotinamide; K = killed; C = foci of calcification.

[b] *** = thrombus; * = moderate; + = hyperplastic; ** = severe; ++ = neoplastic; +++ = malignant.

genic effect of *Fusarium* extracts and a trichothecene (T-2 toxin) on mice and rats. They stated that the T-2 toxin caused a blastogenic response in the spleen, thymus, and lymph nodes of mice and rats.

Among the rats that survived between 12 and 27.5 months after the first dose and several months after the last of 3–8 doses of T-2 toxin, cardiovascular lesions and various tumors were found. With the increase of the number of doses, the incidence, especially of the pancreatic tumors, increased (Schoental *et al.*, 1979).

The incidence of spontaneous pancreatic tumors in Wistar rats has been reported to be less than 1% (Rowlatt, 1967), and the incidence of spontaneous brain tumors has been reported to be even smaller (Garner *et al.*, 1967). However, the incidence of spontaneous tumors was known to show unexplained variations in different laboratories and even in the same laboratory at different times (Sass *et al.*, 1975). It is not unlikely that *Fusarium* mycotoxins may be involved in the variable occurrence of such spontaneous tumors when present in some of the animal diets as a result of occasional fungal growth (Schoental, 1979a).

Were the lesions and tumors seen in our rats caused by the administered T-2 toxin, or could the latter possibly have acted in conjunction with factors which may at times have contaminated the rats' diet? Our control rats were free of pancreatic, gastrointestinal, and brain tumors. The fact that administration of T-2 toxin caused acute lesions in the digestive tract and hemorrhagic lesions in the brain and other tissues supports the interpretation that T-2 toxin significantly contributed to the chronic lesions and tumors found in these organs.

Are the effects of T-2 toxin in rats relevant to human disorders? T-2 toxin, being a direct acting toxin, would be expected to act in a similar way in many animal species. People in the USSR who consumed bread made from moldy, toxic grain became ill and often died from ATA disease.

It would be of great interest if epidemiological studies could establish the incidence of tumors in various organs among people who survived the ATA disease 40 years ago.

T-2 toxin, having an epoxy ring, can be considered an alkylating agent; it is highly cytotoxic for several types of cells in tissue culture. It inhibits protein and DNA synthesis *in vitro* but is inactive in mutagenicity tests using *Salmonella typhimurium* or *Escherichia coli* (Ueno, 1977a). A related trichothecene, diacetoxyscirpenol (anguidine, NSC 141537), has been evaluated as an antitumor agent in cancer patients and has been found to have undesirable effects on the central nervous system, cardiovascular system, and the gastrointestinal tract, and also to have myelosuppressive action (Goodwin *et al.*, 1978).

Hartmann *et al.* (1978) reported that anguidine (diacetoxyscirpenol) inhibited DNA synthesis and caused inhibition of protein synthesis in spleen lymphocytes of mice.

A substance named poin, isolated from *F. sporotrichiella* var. *poae* (*F. poae*), was studied for its antitumor activity on Ehrlich's ascite mouse tumor and for other aspects of its effects (Elpidina 1945c, 1946a, 1956, 1958, 1959, 1960, 1961; Elpidina *et al.*, 1972a,b).

Detailed investigations are needed to evaluate the mechanisms of action of T-2 toxin leading to chronic cardiovascular lesions and tumors in rats and the role of dietary factors, especially the B vitamins therein, appear important. The low nutritional state of the people who consumed the bread made from moldy grain was suggested as a contributory factor, responsible for the severity of the ATA disease and for many deaths that followed.

Some experiments have been conducted on tumors in rodents.

Akhmeteli *et al.* (1972a, 1972b, 1973) fed mice 0.5 ml of an aqueous extract of barley infected with *F. sporotrichioides* No. 635 five times a week for 1 year. The animals were tested 16 months after the beginning of the experiment for the presence of carcinomas and lung adenomas, and serum alpha-feto-protein (AFP). There was a statistically significant increase in lung adenomas. A significant association of AFP-absence and hepatomas was seen, but the presence of AFP was not related to *F. sporotrichioides* treatment. One malignant lung adenoma and one malignant hepatoma were also found.

Sidorik *et al.* (1974) studied the effect of fusarin toxin isolated from *F. sporotrichiella* (*F. sporotrichioides*) on hematopoiesis in healthy rats and those with leukemia. Fusarin, in a dose of 20–40 mg/kg body weight, caused (by i.p. injection) a decrease in the number of leukocytes and erythrocytes in the peripheral blood and hypoplasia of the bone marrow in the healthy rats. Fusarin at the same dosage caused inhibition in the development of the leukosis process of rats, improved the hematological parameters, and decreased the foci of extramedullar hematopoiesis in comparison with the rats which were not given the fusarin.

In chronic toxicity experiments Saito *et al.* (1980) described tumors in Donryn male rats after being fed a diet containing fusarenon-X. The authors found the following tumors after long-term feeding of fusarenon-X to rats: adenocarcinoma of the stomach, two papillary carcinomas of the urinary bladder, one adrenocortical adenoma, and one leukemia. Animals of the experimental groups showed lower body weight than controls.

Saito (1983) described acute and chronic biological effects of fusarenon-X and tumor development in rats and mice due to long-term feeding with *F. nivale*, *F. graminearum*, and the toxic trichothecene fusarenon-X.

Ohtsubo (1983) described some cases of chronic toxicosis associated with neoplastic lesions in rats due to the treatment of the trichothecenes, T-2 toxin, and fusarenon-X.

## Fusarium Toxins in Tumors

Cancer of the digestive tract in humans and especially cancer of the esophagus has been considered to be linked to consumption of alcoholic beverages, mainly beer prepared from barley, and other foods, contaminated under unfavorable environmental conditions by *Fusarium* of the Sporotrichiella section (Doll, 1969; Etchevers *et al.*, 1977; Gjertsen *et al.*, 1964, 1965; Goff and Fine, 1979; Hacking *et al.*, 1976, 1977; Marshall, 1979; Sloey and Prentice, 1962; Trolle, 1969; Tuyns, 1979).

Many cases of esophageal cancer were detected in northern China and the Linshien county, where the cause was connected with corn bread and rice flour (contaminated with *F. moniliforme* and *Aspergillus flavus*) consumed by the local population (Li *et al.*, 1962).

Cook (1971) described cases of esophageal cancer in Kenya, possibly associated with consumption of alcoholic drinks made from maize. Warwick and Harington (1973) described the epidemiology and etiology of some cases of esophageal cancer in Transkei (South Africa) connected with maize flour consumed by the local people.

Many recent publications relate to the esophageal cancer of people in Africa, Asia, and other continents. In the area of Transkei there have been cases of esophageal cancer after consumption of maize or other foods infected with *Fusarium* species (Marasas *et al.*, 1979c, 1981). These authors correlated the incidences with the possible presence of zearalenone and deoxynivalenol.

*F. moniliforme* is most frequently associated with foods in the Henan region of Linshien county. This county is known for being one with the highest esophageal cancer rates in the world (Li *et al.*, 1979, 1980; Lin and Tang, 1980; Yang, 1980).

Li *et al.*, Lin and Tang, and Yong noted that corn meal contaminated with isolates of *F. moniliforme* from Linshien county caused tumors in several organs and also epithelial hyperplasia and papillomas of the esophagus and stomach in rats.

Lin and Tang (1980) showed that esophageal cancer among people and livestock in some regions of China was correlated with foods infected mainly by *Fusarium* species which produced carcinogenic nitrosamines (after addition of small amounts of sodium nitrite, 1 mg–0.1 mg/g of maize).

Schoental (1980a) suggested that *Fusarium* mycotoxins present in fermented beverages caused disorders and tumors in alcoholic drinkers.

Carcinogenic *Fusarium* mycotoxins in conjunction with nitrosamines were found in beer, whisky, and other fermented beverages by Goff and Fine (1979); Marshall (1979); and by Speigelhalter *et al.* (1979).

Magee and Barnes (1967) found that dimethylnitrosamine is carcinogenic to rats; many other nitrose compounds also proved to be carcinogenic (Druckery *et al.*, 1967).

The role of nitrosamines as a possible etiological factor in human esophageal cancer has been studied by certain researchers. The comparative mutagenicity of nitrosamines and their carcinogenic potential as a possible etiological agent in human esophageal cancer has been studied by the following: Bartsch *et al.* (1976, 1980); McCann *et al.* (1975); Montesano and Bartsch (1976); Sugimura *et al.* (1976); Tannenbaum *et al.* (1977); and Yahagi *et al.* (1977).

Some nitrosamines have been found in corn bread inoculated with isolates of *F. moniliforme* from Linshien county after including sodium nitrite (Li *et al.*, 1979, 1980; Lu *et al.*, 1980a,b). These researchers isolated the four following nitrosamines from corn bread in China: N-dimethylnitrosamine, N-diethylnitrosamine, N-methyl-N-benzylnitrosamine (MBNA), and a new N-3-methylbutyl N-1-methylacetonylnitrosamine (MAMBNA). One of these nitrosamines N-methyl-N-benzylnitrosamine (MBNA) produced tumors in the esophagus in rats. This nitrosome was also mutagenic in *S. typhimurium* (Lu *et al.*, 1980a).

Bjeldanes and Thomson (1979); Bjeldanes and Weib (1980); and Bjeldanes *et al.* (1978) also studied the mutagenic activity of various isolates of *F. moniliforme* (from the United States and Europe) in the *S. typhimurium* assay. They noted that these *F. moniliforme* isolates were shown to be mutagenic in *S. typhimurium*.

According to Kriek *et al.* (1981a) some strains of *F. moniliforme* isolated from corn in the esophageal cancer regions of the Transkei were toxic to ducklings, although they did not produce moniliformin. Two of these strains (MRC 826 and MRC 602) caused cirrhosis and nodular hyperplasia of the liver and intraventricular cardiac thrombosis in rats, leukoencephalomalacia in horses, pulmonary edema in pigs, nephrosis and hepatosis in sheep, and cardial congestion in baboons (Kriek *et al.*, 1981a,b).

The detection of mutagenic nitrosamines connected with esophageal carcinogens in bread inoculated with isolates of *F. moniliforme* (Li *et al.*, 1979, 1980; Lu *et al.*, 1979, 1980a,b; and Yang, 1980), the possible role in the cancer rates in the Transkei (Marasas *et al.*, 1981), and production of mutagenic compounds by *F. moniliforme* strains (Bjeldanes and Thomson, 1979; Bjeldanes and Weib, 1980) point out that this problem is very important and requests further comprehensive investigations for establishment of the role of *F. moniliforme* in human esophageal cancer.

Morel-Chany *et al.* (1980) studied the cytostatic effect of T-2 toxin isolates from *F. poae* and *F. sporotrichioides* on cultured human cells originating from rectal and colonic tumors. They concluded that T-2 toxin has a powerful cytostatic effect *in vitro*.

Sporadic episodes of unexplained indigestion are not unknown in Western countries, where the incidence of gastrointestinal tumors is particularly high (Bjelke, 1974). Some of these transient indispositions may be due to the presence of *Fusarium* mycotoxins in food and could possibly have cumulative effects and lead to chronic degenerative diseases and tumors (Schoental, 1979a).

Carcinogenic estrogenic compounds in certain foodstuffs infected with *Fusarium* strains which produce mycotoxins can also present a serious health hazard to animals and man (Lovelace and Nyathi, 1977; Marasas *et al.*, 1979a,c; Martin and Keen, 1978; Mirocha *et al.*, 1977a; Schoental, 1968, 1977a,b, 1978, 1979a,b,c,d, 1980a, 1981a,b,c,d, 1983a,b,c; Schoental and Joffe, 1974; Schoental *et al.*, 1976, 1978b, 1979; and Ueno *et al.*, 1977a,b).

According to Joffe (1978b, 1983b), Marasas *et al.* (1979b), Mirocha *et al.* (1977a), Möller *et al.* (1978), Patterson (1978), Schoental (1981a,b,d, 1983a,b), Schoental and Joffe (1974), Schoental *et al.* (1979), and Ueno *et al.* (1977a), people in zones with low temperatures, high humidity, and other unfavorable environmental conditions (e.g., Sweden, Finland, Germany, the United States, the USSR, or Japan) may be exposed to increased danger after consumption of cereal grains and other foodstuffs contaminated with *Fusarium* species which produce trichothecene mycotoxins such as T-2 toxin.

The role of T-2 toxin and of the other metabolites of *Fusarium* species in the etiology of certain idiopathic chronic disorders and tumors in humans and animals merits further study.

## The Effect of Zearalenone and Related Compounds on Mice and Rats

Stob *et al.* (1962) reported that corn molded with *G. zeae* caused a large increase in uterine weight when fed to ovariectomized mice. Injections of alcoholic extracts of the corn culture had similar effects. From the chemical data reported in that work this extract was later shown to have contained zearalenone, or F-2 toxin as it is known to many workers. Since then numerous researchers have used increased uterine weight in mice or rats as a bioassay method for zearalenone (Christensen *et al.*, 1965; Mirocha *et al.*, 1967, 1968a,b; Ueno *et al.*, 1974c; Vanyi *et al.*, 1973a). The rat bioassay method has been recommended as a simple bioassay method for determining the safety of animal feed suspected of containing zearalenone (Mirocha *et al.*, 1977a; Szathmary *et al.*, 1976). Uterotropic responses produced by zearalenone in mice, rats, swine, and dairy cattle feedstuffs were noticed by Funnel (1979); Nelson *et al.* (1973); and Ueno *et al.* (1977a).

In addition to the effect on uterine weight it was found that i.m. injections of *low* concentrations (20–40 µg) of zearalenone into rats caused significant increases in body weight. Higher doses failed to stimulate any significant changes in body weight although the uterus weight increase showed a linear response with increasing F-2 damage (Mirocha *et al.*, 1967). 30 mg/kg zearalenone caused a decrease in body weight in both male and female rats (Gallo *et al.*, 1977).

Oral administration of zearalenone was shown to be more effective in increasing uterine weight than injections (Mirocha *et al.*, 1968a; Ueno *et al.*, 1974c). Histological examination of the uterus of immature mice and rats given daily doses of 1–2 mg/kg for a week showed increased proliferation and mitosis in the uterine cells (Ueno *et al.*, 1974c). No changes were noted in the liver, kidneys, spleen, or small intestine. In the same series of experiments it was shown that no increase in uterine weight occurred after the fifth day of zearalenone administration, and that when treatment was stopped the uterine weight decreased and became "normal" by the twenty-fifth day. It was also shown that in mature mice zearalenone only caused increased uterine weight in ovariectomized animals.

Several derivatives and isomers of zearalenone have been shown to have uterotropic activity.

According to Brooks *et al.* (1971) zearalenone and its derivatives caused a considerable increase of uterotropic activity in immature rats and antifertility potency in mature rats. Uterotropic assays were conducted using 21-day-old female rats, and the uterotropic response was measured against diethylstilbesterol. For the determination of antifertility activity, adult virgin female rats were mated with fertile males, and vaginal smears were checked daily for presence of sperm. Uterotropic and antifertility properties of resorcylic acid lactone derivative, zearalane, and its 7'-formyl zearalane and 7'-carboxy zearalane were found to be estrogenic. Additions of the 7'-formyl zearalane markedly increases the estrogenicity and antiimplantation activities of zearalane. The more active isomer of 7'-carboxy zearalane is more than 100 times as estrogenic as zearalene (of the 7'-carboxy group) and one-tenth as active as diethylstilbesterol. Isomer B of 7'-formyl zearalene is more than 40 times as estrogenic as zearalane.

Mirocha *et al.* (1971) also isolated two derivatives of zearalenone (F-2): F-5-3 and F-5-4. Both were tested for their biological activity in the rat-uterus bioassay test. According to the authors F-5-3 has estrogenic activity but F-5-4 was inactive. In addition they found that 100 to 500 ppm gibberellic acid also increased the uterine weight in virgin weanling rats. Knaus (1978) reported that the mouse uterus test was not found to be suited to the demonstration of zearalenone in feed.

Mirocha *et al.* (1978) found that the uterotropic activity of *cis*-zearalenone in the white rat is much greater than that of the *trans* isomer. However, there was no significant difference between the activity of the *cis* and *trans* isomers of zearalenol. Kallela (1978), working with immature female rats, found that F-2 toxin increased the effects of natural estrogens. The determination of the estrogenic effect of zearalenone, isolated from different extracted samples, was carried out by the method of Kallela *et al.* (1978).

Zearalenone has also been shown to influence the fertility of the offspring of pregnant rats fed corn infected with *F. graminearum*. The offspring of the test rats showed a 50% reduction in fertility compared to the controls. Injection of 30 μg zearalenone into 2-day-old female rats reduced their normal pregnancy rate from 93 to 33%. Injected males sired 44% pregnancies in untreated females (Ruzsas *et al.*, 1978). Very high doses of zearalenone (1000 μg/kg/day) reduced the number of live litters in pregnant mice and halved the number of live births per litter. Lower doses (300 μg/kg/day) lowered the number of litters but not the number of live births per litter. Doses of 100 μg/kg/day of zearalenol had no effect on litter size or number of litters (Davis *et al.*, 1977).

Boyd and Wittliff (1978) demonstrated that dihydrozearalenol has higher affinity than zearalenone in the cytosol preparations of mammary glands from lactating rats.

Lasztity and Wöller (1977) investigated the fate of zearalenone within the test animal body. They fed rats with maize infected with *F. culmorum* and *F. graminearum* estimated to contain 50 μg/kg zearalenone. After 14 days the rats were sacrificed and the internal tissue examined for toxin residues. They found F-2 and the derivatives F-3 and F-4 in the liver, stomach, and inner genital tissues. These strains (*F. culmorum* and *F. graminearum*), when fed to rats, caused an enlargement of the uterus but did not produce dramatic responses on the rats' skin.

Ueno *et al.* (1974c) stated that administration of zearalenone to ovariectomized mice and rats induced an increase of uterine weight and stimulation of RNA synthesis.

Ueno *et al.* (1977a) found zearalenone in the adipose tissue, ovary, uterus, and other internal organs of treated rats. They showed that orally administered zearalenone is transferred into a secondary metabolite (metabolite I). This metabolite is still estrogenic to mice and was presumed to be a hydroxylated zearalenone. Zearalenone was excreted mainly in the feces as metabolite I, whereas another metabolite (metabolite II) and two glucuronides were excreted in the urine. Ueno and his co-workers found that, in the presence of NADH or NADPH, zearalenone was transformed into metabolite I by the 9000 g supernatant fraction of liver homogenates of several animals, including mice and rats. Similar findings were reported by Kiessling and Pettersson (1978), who suggested that there were two metabolic pathways: (1) via the conjugation with glucuronic acid, and (2) reduction to an isomer of zearalenone. However, they were unable to account for all the zearalenone metabolism. The addition of NADH or NADPH to rat liver homogenate doubled the rate of reduction of zearalenone to zearalenol. Hepatocytes could eliminate 100 μg/zearalenone/g liver/hr whereas the maximum for the liver homogenate was 82 μg/g/hr.

Tashiro *et al.* (1980) studied the mutual interaction between zearalenone, its derivatives and estrogen receptors, and RNA synthesis in rat uterus. The authors stated that the administration of zearalenone to rats enhanced the nuclear RNA polymerase activities when the uterine weight was increased.

According to Smith (1980a) diets rich in fiber have a dominant influence on zearalenone toxicoses in rats and swine. Alfalfa-fed rats and swine were shown to exhibit a change in their metabolism of zearalenone (Smith, 1980b; James and Smith, 1982).

## The Effects of Moniliformin on Mice and Rats

The toxin moniliformin has been shown to be toxic to several animals including rats (Kriek et al., 1977; Rabie et al., 1978) and mice (Burmeister et al., 1979). The $LD_{50}$ for rats was shown to be 50.0 and 41.57 mg/kg for male and female BD IX black rats (Kriek et al., 1977). Autopsy showed acute congestive heart failure and nonspecific lesions characterized by acute focal myocardial degeneration and necrosis. Severe cloudy swelling and scattered single cell necrosis occurred in the liver, kidney, pancreas, adrenal glands, gastric mucosal glands, and the crypts of the small intestine. Clinical symptoms included progressive muscular weakness, respiratory problems, cyanosis, and coma. Rats that recovered and survived moniliformin treatment were virtually normal apart from mild myocardial lesions.

Purified moniliformin was injected i.p. into mice. The $LD_{50}$ for mice was 29.1 mg/kg and 20.9 mg/kg for male and female mice, respectively (Burmeister et al., 1979). As with the rats, mice that survived toxin treatment showed no ill effects, and the toxin failed to produce any toxic reaction in the mouse skin test.

## THE EFFECT OF FUSARIOTOXINS ON GUINEA PIGS

In general guinea pigs show a response to fusariotoxins similar to that of mice and rats, although the intensity of the reaction may differ. Corn infected with G. saubinetii (G. zeae) was refused by guinea pigs (Mundkur and Cochran, 1930) just as it was by rats (Forsyth, 1974). Zearalenone causes uterotropic responses in guinea pigs as well as in rats and mice (Mirocha et al., 1968a,b). In a series of experiments carried out to screen toxic Fusarium species Stankushev et al. (1977) found that crude zearalenone (0.2 ml/day for 20 days) caused necrobiotic and degenerative changes in the reproductive systems of male and female guinea pigs. They also found that some Fusarium species produced toxins that were hepatotrophic or which caused hemodynamic problems or degenerative changes in the digestive tract, kidneys, heart, and nervous systems of guinea pigs. However, none of the toxins were identified. The guinea pig has been used for the skin test bioassay of fusariotoxins. Intradermal injections of zearalenone produced a toxic skin reaction in guinea pigs (but not in rabbits), and the guinea pigs reacted to intradermal injections of T-2 toxin, although they were not as sensitive as the rabbit to topical application of T-2 toxin (Chung et al., 1974). The guinea pig also gave toxic skin reactions to neosolaniol monoacetate, but this was not as strong as the reactions obtained with rabbits or rats (Lansden et al., 1978). According to Ueno et al. (1970a, 1971c) and Ueno (1971) the guinea pig skin reaction to diacetoxyscirpenol, fusarenon-X, and nivalenol was greater than that in mice or rabbits.

Forty-two guinea pigs in feeding tests were given liquid filtrates or dry powdered mycelium of the following toxic *Fusarium* species: *F. poae* No. 60/9, *F. sporotrichioides* No. 60/10, *F. sporotrichioides* var. *tricinctum* No. 1227, *F. equiseti* No. 2385, *F. lateritium* No. 1421, *F. sambucinum* No. 1904, and *F. semitectum* No. 2466, grown on liquid Czapek medium (Joffe, 1960a, 1971). The guinea pigs died within 5–21 days of receiving 50–70 mg of *Fusarium* mycelium or 1.0–2.0 ml of the liquid culture. Animals receiving the toxic *F. poae* and *F. sporotrichioides* in liquid form were more severely affected. The guinea pigs showed hemorrhages in organs and tissues and mainly hyperemia of the adrenal gland and necrosis in the gastrointestinal tract.

In guinea pigs the leukocyte count declined to $500/mm^3$ due to the consumption of overwintered cereal grains contaminated by *F. poae* of *F. sporotrichioides* (Alisova, 1947; Alisova and Mironov, 1947).

Pidoplichko and Bilai (1946) reported that administration of *F. poae* to guinea pigs and rabbits was lethal.

The acute and chronic effects on guinea pigs of T-2 toxin isolated from an authentic strain of *F. sporotrichioides* No. 921 associated with ATA disease (Joffe, 1960a; Joffe and Yagen, 1977; Yagen and Joffe, 1976; Yagen et al., 1977) were investigated by DeNicola et al. (1978). Single oral doses of 1.85 or 2.5 mg/kg body weight of T-2 toxin caused hemorrhage and necrosis in the duodenum (Fig. 8.10) and hyperemia and hemorrhage in the caecum and spleen (Fig. 8.11 and 8.12). The animals exhibited edematous intestinal lymphoidal tissue and hyperemia of the adrenal gland. Histological examination revealed typical radiomimetic symptoms: necrosis of lymphoidal tissue, bone marrow, and testis and necrosis and ulceration of the gastrointestinal tract (particularly the stomach and caecum). The oral $LD_{50}$ for guinea pigs was estimated to be 3.06 mg/kg body weight.

Ito et al. (1982) studied the immunosuppressive effects of T-2 toxin and fusarenon-X (isolated from *F. solani* and *F. nivale*) and cyclochlorotine (from *Penicillium islandicum*) by measuring the *in vivo* anti-2,4-dinitrophenyl (DNP) antibody response in guinea pigs immunized with DNP-bovine serum albumin (DNP-BSA).

These trichothecenes inhibited the *in vitro* blast transformation of guinea pig splenic cells stimulated with lipopolysaccharide of *E. coli* (B-

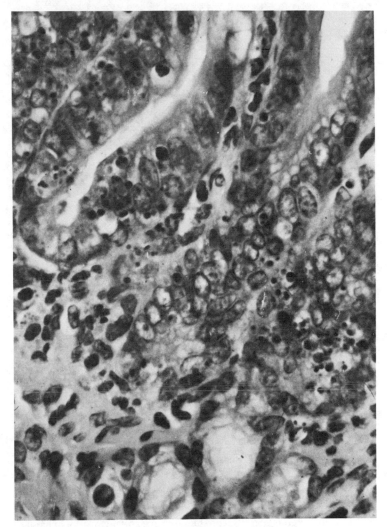

**Figure 8.10**  Duodenum. Severe necrosis of crypt epithelium in guinea pig. (Courtesy of Dr. DeNicola.)

cell mitogen) and concanavalin A (T-cell mitogen) using incorporation of [$^3$H]-thymidine into DNA.

The authors stated that T-2 toxin was most potent in reducing the DNA synthesis compared with fusarenon-X and cyclochlorotine.

T-2 toxin suppressed the antibody response in guinea pigs at a sublethal dose, but fusarenon-X did not exhibit any immunosuppressive activity *in vivo* when tested at the sublethal doses.

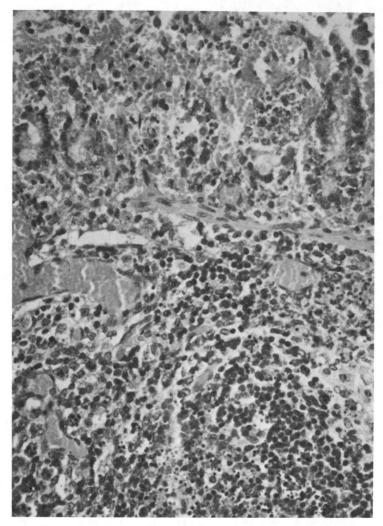

**Figure 8.11**  Caecum of guinea pig. Severe hemorrhage and necrosis of mucosa extending to muscular mucosa. Capillary hyperemia; necrosis of lymph follicle. (Courtesy of Dr. DeNicola.)

## THE EFFECT OF FUSARIOTOXINS ON RABBITS

Most of the work done with rabbits has involved the toxic skin test used for detecting trichothecenes and certain other toxins (*see* Chapter 3).

The author described experiments on 56 rabbits which received 100–150 mg of mycelium or 3–5 ml culture filtrate of the following *Fusarium* species: *F. poae* No. 60/9, *F. sporotrichioides* No. 60/10, *F. sporotrichioides* var. *tricinctum* No. 1227, *F. equiseti* No. 2385, *F. lateritium* No. 1421, *F. sambucinum* No. 1904, and *F. semitectum* No. 2466. Death

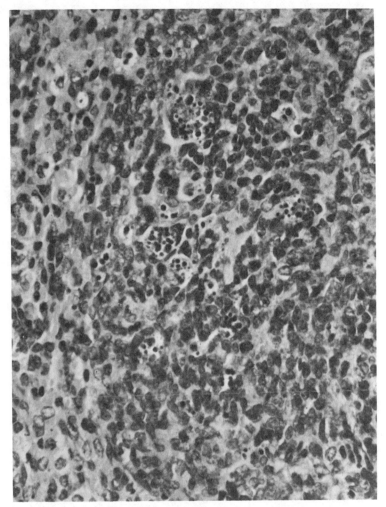

**Figure 8.12** Spleen of guinea pig. Necrosis in lymphoid follicles. (Courtesy of Dr. De-Nicola.)

occurred on the eighth to twenty-fourth day after being fed these toxic *Fusarium* cultures with a decreased number of leukocytes. The autopsy showed dilated blood vessels and hemorrhages in the intestinal walls, kidney, and other organs and tissues (Joffe, 1960a,b, 1971). After s.c. injecting two rabbits with ethanol extract of *F. poae* No. 792 marked infiltration of the dermis with disintegrated leukocytes (Fig. 8.13) and migration of leukocytes through the walls of small blood vessels were noted (Fig. 8.14). *F. sporotrichioides* No. 738 also showed marked leukocyte infiltration of muscle after application of toxic extract to rabbit skin (Fig. 8.15).

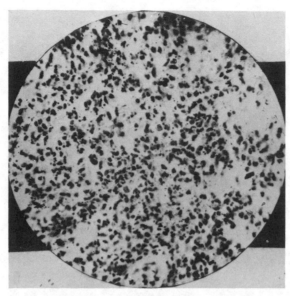

**Figure 8.13** Infiltration of the dermis by disintegrated leukocytes following application of toxic *F. poae* No. 792 to rabbit skin. (Adapted from Joffe, 1971.)

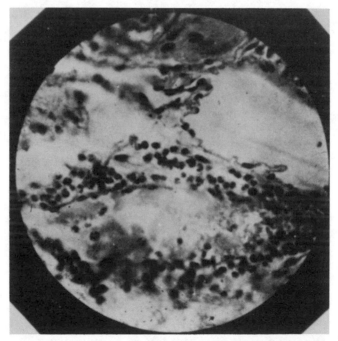

**Figure 8.14** Migration of leukocytes through walls of small blood vessels after subcutaneous injection of *F. poae* No. 792 into rabbits. (Adapted from Joffe, 1974c.)

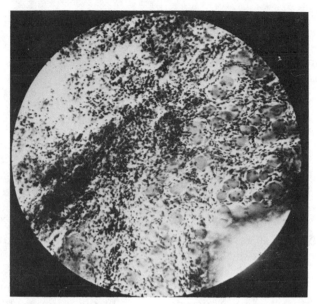

**Figure 8.15** Marked leukocyte infiltration of muscle following application of toxic *F. sporotrichioides* No. 738 to rabbit skin. (Adapted from Joffe, 1971.)

Bilai (1953) fed small rabbits weighing 600–900 g food infected with *F. poae* and *F. sporotrichioides* and found that the effect of the toxin depended on the weight of the animal and that rabbits were more sensitive than guinea pigs and mice. Postmortem examination of the rabbits revealed hyperemia of the subepidermal tissues and gastrointestinal tract as well as diffuse hemorrhages in the liver.

Maisuradge (1953, 1960) found that when rabbits were fed germinating oats infected with *F. sporotrichioides* they lost weight, body temperature rose, pulse became quicker, and breathing became heavy and slow. The leukocyte count rose to 37,000 per mm$^3$ in the middle of the experimental period and later fell to 600–200/mm$^3$. A decrease in the red cell count was also evident.

*F. roseum* as well as *F. tricinctum* contamination of corn was indicated in a survey of corn of the 1972 USA harvest. Seventeen percent of the samples were found to contain zearalenone, and 54% of the samples tested had a skin irritating factor when tested on rabbits. The authors assumed that the skin irritating factor was associated with trichothecene toxins (Eppley *et al.*, 1973, 1974).

Kurmanov (1971) reported that acute fusariotoxin poisoning in rabbits caused an increased level of nucleic acid in the blood. Leukocytes increased from 15,000–60,000 and the erythrocytes from 7.5–10 million.

Rabbits were also used to investigate the mycokeratitic action of *F. moniliforme*. Extracts taken from a strain of *F. moniliforme* isolated from a clinical case of corneal mycokeratitis were injected intracorneally into rabbits. Lesions were produced that were very similar to those found in human cases (Dudley and Chick, 1964). It appeared that the active factor was proteolytic and appeared to act on the collagen bundles within the corneal stroma.

## THE EFFECT OF FUSARIOTOXINS ON DOGS

The effect of various quantities of agar culture, inoculated by *F. poae* No. 396 isolated from overwintered millet grain, on the motor activity of dogs was studied by the author in cooperation with Khrutski *et al.* (1953). *F. poae* culture was fed orally and through a gastric fistula under various conditions. The fungus culture had been mixed with 200 g of meat, and gastric motility was registered on a kymograph. It was found that introduction of large doses of fungus (1 g and over) caused poisoning of the animal and cessation of stomach motor activity.

When dogs were fed oats infected with *F. poae* the following symptoms were detected by Maisuradge (1960): body temperature rose to 40°C and pulse to 120 beats/min; the leukocyte count decreased to 800/mm$^3$; stomatitis and gastrointestinal hemorrhages, degeneration of liver and kidneys, changes in the epicardium, and disturbances of the nervous system were evident.

## FUSARIOTOXICOSIS IN CATS

Cats have been of particular interest to scientists investigating the causes of ATA, since the clinical and histopathological syndromes induced in cats by the *Fusarium* species and their toxins assumed to be responsible for ATA closely resemble those occurring in man. The author studied the effect of *F. poae* No. 60/9 and *F. sporotrichioides* No. 60/10 (either millet grains or agar cultures, dried mycelium or culture filtrates and extract of inoculated wheat or millet grains by these *Fusarium* species), isolated from overwintered millet, on 26 cats. The daily dose of agar and millet cultures was 50–120 mg/kg, and the daily dose of the culture filtrate was 0.5–10 ml (Joffe, 1960a). In all the forms administered to cats both *Fusarium* species were lethal after periods of time depending on the daily dose of fungus and the individual properties of the respective organism. In the majority of cases the cats died on the sixth to twelfth day, death following a failure in blood production. A fall in leukocytes, neurophils, and hemoglobin red cells was observed with a related increase in lymphocytes. The lowest leukocyte count found was 100/mm$^3$. The *Fusarium* species, after being fed p.o. to the cats, caused symptoms that included vomiting, hemorrhagic diathesis, and neurological disturbances. Autopsy revealed marked hyperemia and hemorrhages of internal organs, in particular the digestive tract and kidneys, and extreme changes in the adrenal glands (Joffe, 1971). The cats that died revealed drastic changes, mainly in the blood-producing tissue, that were similar to those seen in men. These studies were carried out partially in cooperation with Alisova (1947b).

It was also interesting to study mixtures of toxic species of fungi which appear on overwintered cereal grains in the field (*Cladosporium epiphyllum, Mucor hiemalis, Alternaria tenuis,* and *P. steckii*) together with individual *Fusarium* species such as *F. poae* No. 60/9, *F. sporotrichioides* No. 60/10, *F. sporotrichioides* var. *tricinctum* No. 1227, *F. equiseti* No. 2385, *F. lateritium* No. 1421, *F. sambucinum* No. 1094, and *F. semitectum* No. 2466 (all of which had been shown previously to be toxic to guinea pigs and rabbits). The cultures were incubated on PDA medium and on millet grains which had been infected by a suspension from the above mixed cultures to which single *Fusarium* isolates had been added. These cultures were then fed to 28 cats. The mixtures, together with a single *Fusarium* isolate, were more toxic than the individual fungi alone, showing a synergistic relationship (Joffe, 1960a,b, 1965). In all the cats tested leukopenia was recorded, and all perished within 4–42 days. Autopsy revealed symptoms resembling ATA in humans.

Alisova (1947a,b) and Alisova and Mironov (1947) studied the effect of toxic isolates of *F. poae* and *F. sporotrichioides* on cats. They all died within 6–16 days.

Sarkisov (1954) fed seven cats with prosomillet infected by *F. sporotrichioides*. The lethal dose was 0.4–16.5 g, and death occurred within 2–34 days. Body temperature rose to 41°C, and a progressive leukopenia appeared, while the leukocyte count fell to 50–200 cells/mm$^3$. Rubinstein and Lyass (1948) fed prosomillet infected with *F. sporotrichioides* to cats and monkeys, and these also developed symptoms resembling ATA in humans.

Later Sato *et al.* (1975) reported that T-2 toxin caused emesis, vomiting, diarrhea, anorexia, ataxia of the hind legs, and discharge from the eyes. Both crude and pure T-2 toxin caused leukopenia after continuous administration. After each dose of T-2 toxin cats suffered temporary leukocytosis.

In addition to radiomimetic cellular damage in the rapidly dividing tissues of intestine, spleen, lymph nodes, and bone marrow (Ueno *et al.*, 1971a), cats also had meningeal hemorrhage in the brain, bleeding in the lung, and vacuolic degeneration of the renal tubes (Sato *et al.*, 1975).

After evaluating the specific pathogenicity of T-2 toxin on chicks (Joffe and Yagen, 1978) we decided to carry out a study on cats because they showed a clinical picture that, in most respects, resembled ATA in man.

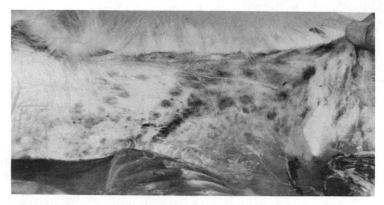

**Figure 8.16** Subcutaneous ecchymotic hemorrhages. Cat after 14 doses of crude extract (containing T-2 toxin 0.06 mg per kg). (Courtesy of I. Lutsky.)

In cooperation with Lutsky *et al.* (1978) we demonstrated and evaluated the specific pathogenicity to cats using crude extract and T-2 toxin isolated from a strain of *F. sporotrichioides* No. 921. Twenty-four mature male and female short-haired healthy cats with normal blood values were used in the study. The cats died 6–40 days after receiving various doses of T-2 toxin in crude extract or pure form. The gross pathological findings included severe emaciation, subcutaneous, ecchymotic hemorrhage, stomach and intestinal hemorrhages, enlarged lymph nodes, and depletion of hematopoietic tissue with replacement of fatty vacuoles of the bone marrow (Figs. 8.16 through 8.19).

Lutsky *et al.* (1978) noted that extracts of *F. sporotrichioides* No. 921, from which T-2 toxin had been removed, caused no harmful effects when fed to cats, suggesting that T-2 toxin was the agent responsible for

**Figure 8.17** Hemorrhagic blood-filled intestine of cat after 14 doses of T-2 toxin. (Courtesy of I. Lutsky.)

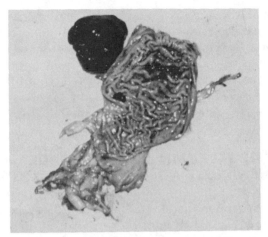

**Figure 8.18** Stomach of cat showing hemorrhagic erosions and necrosis and also clotted blood. (Courtesy of I. Lutsky.)

all the clinical symptoms and histopathological changes associated with ATA disease in humans.

Cats have been shown to be sensitive to other trichothecenes such as fusarenon-X. Subcutaneous injection of 5 mg/kg proved fatal to an adult male cat (1.58 kg). The cat vomited 30 min after administration of the toxin, became immobile, and died overnight (Ueno, 1971; Ueno *et al.*,

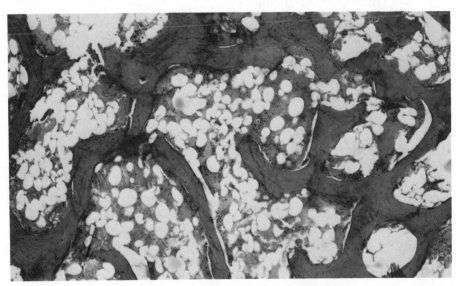

**Figure 8.19** Bone marrow section of femur from cat (after seven doses of T-2 toxin at 0.1 mg/kg) showing depletion of hematopoietic tissue with replacement by fatty vacuoles. (Courtesy of I. Lutsky.)

1971c). A dose of 1 mg/kg fusarenon-X was fatal to a 1-week-old kitten. Both these animals had marked karyorrhexis of the intrasinusoidal cells in the liver, and in the adult cat there was extensive necrosis of the parenchymal liver cells.

Trichothecenes in sublethal doses caused vomiting in cats. The vomiting dose for T-2 toxin was 0.1–0.2 mg/kg body weight and for fusarenon-X it was 0.3–0.5 mg/kg (Ueno *et al.*, 1974b).

## THE EFFECT OF T-2 TOXIN (ISOLATED FROM *F. SPOROTRICHIOIDES* NO. 921) ON MONKEYS

The best models for reproducing the same clinical signs and syndromes as those occurring in ATA in humans were cats and monkeys. For this reason we cooperated with Dr. Kriek, of the division of Nutrition Pathology, National Research Institute for Nutritional Diseases of the South African Medical Research Council, in carrying out comprehensive experiments on 19 vervet monkeys of various ages with pure T-2 toxin (Joffe and Yagen, 1977; Yagen *et al.*, 1977) from *F. sporotrichioides* No. 921 isolated from overwintered rye grains (Joffe, 1960a).

The experiments covered a wide range of toxic doses, from 10 mg/kg to 100 mg/kg, induced syndromes, and caused death within 18 hr. Long-term experiments were also carried out, with death occurring after 1 year. The effect of T-2 toxin on these primates is shown in Figures 8.20 through 8.28.

Rukmini *et al.* (1980) studied the effect of feeding T-2 toxin to rats and monkeys in India. They found that T-2 toxin showed a marked leukopenia in monkeys, and that prolonged feeding caused hemorrhage, infection of respiratory tract, and mortality. The hematological and histopathological findings also showed that T-2 toxin suppressed the immunological action in monkeys.

Jagadeesan *et al.* (1982), also in India, studied the effects of feeding T-2 toxin by gastric intubation to seven rhesus monkeys at a level of 100 µg/kg body weight/day for a period of 4–5 weeks. He also studied the hematological and immunological aspects.

At the early stage of treatment three of the animals showed vomiting, hemorrhage, and respiratory infection, which caused death. In the other monkeys the leukocyte count was reduced by more than 40%. The immunological studies showed suppression of the bactericidal activity of leukocytes and T-lymphocytes. Also B-cell number and serum immunoglobulin levels (IgG and IgM) were depressed.

After withdrawal of the toxin (during 5 mon) the affected immunity parameters were almost returned to the initial levels and resulted in the improvement of hematological and immune functions.

**Figure 8.20** Hemmorhage in skin of thorax in vervet monkey after being fed T-2 toxin. (Courtesy of N. P. J. Kriek.)

**Figure 8.21** Hemorrhage in skin of arm in vervet monkey. (Courtesy of N. P. J. Kriek.)

**Figure 8.22** Enlarged, congested, and pigmented abdominal lymph nodes in vervet monkey. (Courtesy of N. P. J. Kriek.)

## THE EFFECT OF FUSARIOTOXINS ON FISH

Very little work has been done on the effects of fusariotoxins on fish. The possibility that these toxins might contaminate fish food pellets manufactured from maize led Marasas *et al.* (1967, 1969) to investigate the effect of a toxic metabolite of *F. tricinctum* related to T-2 toxin. Rainbow trout were fed the contaminated pellets for 12 days. The LD$_{50}$ of the toxin was estimated to be 6.1 mg/kg. Fish offered pellets containing 200 mg/kg

**Figure 8.23** Liver of vervet monkey after feeding of T-2 toxin. Degeneration and marked edema of gall bladder, subserosal hemorrhages on gall bladder, and main extrahepatic bile ducts. (Courtesy of N. P. J. Kriek.)

**Figure 8.24**  Muscle (thigh). Intramuscular hemorrhages in vervet monkey. (Courtesy of N. P. J. Kriek.)

toxin or more refused the feed after the first feeding; fish given pellets containing 4 mg/kg ate the food. All fish consuming the toxin exhibited identical pathological symptoms which included shedding of intestinal mucosa and severe edema, with accumulation of fluid in body cavities and behind the eyes, causing ventral swelling and bulging eyes (Bamburg *et al.*, 1968a). When T-2 toxin isolated from *F. tricinctum* was fed to rainbow trout with various concentrations, it caused disorders in the intestinal tracts and in other organs and tissues of the fish.

Poston *et al.* (1982) studied the action of chronic toxicity in young and adult rainbow trout fingerlings which were fed a diet containing various

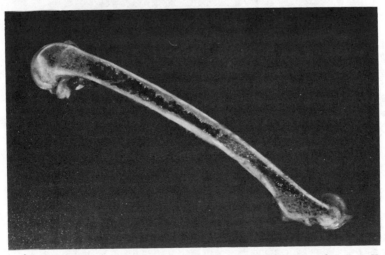

**Figure 8.25**  Bone marrow metaplasia in vervet monkeys. (Courtesy of N. P. J. Kriek.)

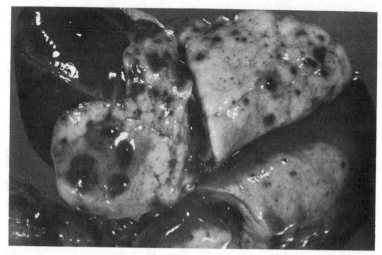

**Figure 8.26**  Lungs and heart of vervet monkey after being fed T-2 toxin. Extensive pulmonary hemorrhage, emphysema, and edema. (Courtesy of N. P. J. Kriek.)

levels of pure T-2 toxin. Feeding 15 mg/kg T-2 toxin to adult trout caused hemorrhages in the intestine and depressed the hematocrit and hemoglobin concentration. Young, growing rainbow trout fed 5 mg/kg or more of dietary T-2 toxin exhibited a significant depression in their growth and efficiency of diet.

Nowak (1973) noted that five *Fusarium* species were toxic to carp causing histopathological changes in liver, pancreas, and kidneys. Zearalenone has also been shown to affect carp. Fish fed 100 ppm F-2 in

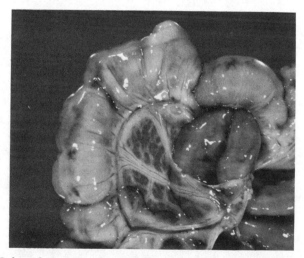

**Figure 8.27**  Colon of vervet monkey after feeding T-2 toxin. Scattered subserosal hemorrhages. (Courtesy of N. P. J. Kriek.)

**Figure 8.28** Extensive subcutaneous edema of vervet monkey head. (Courtesy of N. P. J. Kriek.)

their food suffered from severe degeneration of the epithelium of the caniculi of the testicles. However, the effects of F-2 toxin were reversible (Vanyi *et al.*, 1974a).

## THE EFFECT OF *F. POAE* ON FROGS

Frogs were tested to establish the toxicity of *F. poae* No. 60/9, isolated from overwintered prosomillet, in cooperation with Petrov and Simonov (1953). Following repeated feeding with powdered dry mycelium the animals died within 4–14 days, depending on the size of the dose. The effect of the toxin was cumulative. Postmortem dissection showed extreme hyperemia, hemorrhages, and edema of the digestive tract and of other organs and tissues.

## CONCLUSION

*Fusarium sporotrichioides* strains are the most prominent fungi of overwintered cereal grains which caused the death of thousands of people in the Orenburg district of the USSR. These strains and their compounds naturally occurring in the field are closely associated with environmental conditions and were correlated with cereal toxicoses causing the death of various experimental animals (mice, rats, guinea pigs, rabbits, dogs, cats, monkeys, and frogs) and domestic animals (such as chicks, ducklings, pigs, horses, and cattle).

The major metabolite isolate from these *Fusarium* strains was T-2 toxin and possibly other toxic compounds in smaller amounts such as diacetoxyscirpenol, neosolaniol, and HT-2 toxin.

Trichothecene mycotoxins and mainly T-2 toxin produced by fusaria of the Sporotrichiella section on a wide range of substrates have diverse toxic effects on animals. The results obtained in our experiments on cats and monkeys indicated that T-2 toxin isolated from *F. sporotrichioides* and *F. poae* was responsible for the fatal ATA disease in man.

Toxic effects of T-2 toxin were established on various laboratory and domestic animals by the author and his cooperators in acute and chronic experiments. Some researchers (Kuczuk *et al.*, 1978; Lindenfelser *et al.*, 1974; Marasas *et al.*, 1969; Nagao *et al.*, 1976; Stähelin *et al.*, 1968; Ueno, 1977b; Ueno *et al.* (1978); Wehner *et al.*, 1978) noted that chronic feeding in rodents (mice and rats) with T-2 toxin and diacetoxyscirpenol was not carcinogenic and was not mutagenic to *S. typhimurium* in the Ames test (Ames *et al.*, 1975), in contrast with the results obtained by Schoental and Joffe (1974), Schoental *et al.* (1979), and Schoental and others in many other publications (Schoental, 1975, 1977a,c, 1978, 1979a,b,c,d, 1980a,b,c, 1981a,b,c,d,e, 1983a,b,c; Schoental *et al.*, 1976, 1978a,b).

Rats surviving many months after several lethal doses of T-2 toxin can develop cardiovascular lesions; high blood pressure; skin, gastrointestinal, and neurological disorders; and tumors.

We suggest that more detailed and comprehensive studies carried out on mice, rats, and other sensitive animals (as models) are necessary in order to establish definitively whether or not T-2 toxin really induced carcinogenic effects.

The results obtained by Schoental were arrived at with the consultation and help of the famous histopathologists of Royal Veterinary College, London. The clinical and histopathological findings and the illustrating photographs confirmed that T-2 toxin caused tumors in various organs in rats. Tumors in various organs have also been reported in rats fed fusarenon-X (Ohtsubo and Saito, 1977; Saito (1983); Saito and Ohtsubo, 1974; Saito *et al.*, 1980).

# CHAPTER 9

## FUSARIOTOXICOSES IN DOMESTIC ANIMALS

Numerous outbreaks of toxicoses among domestic animals have been associated with species of the genus *Fusarium* (Bamburg and Strong, 1971; Bamburg *et al.*, 1969; Cieglar, 1978; Ueno, 1971; Ueno *et al.*, 1971a; and others). Spoiled grain was often associated with these toxicoses (McNutt *et al.*, 1928), although the exact cause was not identified in the earlier reports. Roohe and Bohsted (1930) and Roche *et al.* (1930), investigating the effect of scabby barley infected with *Gibberella saubinetii* (*G. zeae*), found that sheep and cattle with complex stomachs were not affected after eating the infected grain, but animals with simple stomachs such as horses and pigs refused the grain. When pigs did eat the contaminated grain they vomited a few minutes later. Kurmanov (1978b) also stated that animals with complex stomachs are usually less susceptible to the fusariotoxins than those with simple stomachs. Greenway and Puls (1976) reported that moldy overwintered barley infected with *Fusarium* was responsible for outbreaks of toxicoses in ducks, geese, horses, and swine in British Columbia. In the initial cases reported fatalities occurred only among the geese which, unlike ducks, lack a crop. The feed ingested by geese remains in contact with the esophagus for a much longer period of time than in other birds. This explains the advanced necrosis in the esophagus in geese, a sign that did not appear in chickens, turkeys, and pigeons. The authors noted that prolonged ingestion of palatable grains contaminated with low amounts of mycotoxin may cause lethal toxicosis in animals. It would appear that very badly infected grain was rejected as unpalatable by the animals. In fact the food refusal syndrome, often noted as a response to *Fusarium* toxins, has been explained as an instinctive defensive reaction

**345**

in animals to compounds that irritate and inflame the digestive tract (Kurmanov, 1978a).

Kriek *et al.* (1981b) stated that feeding *F. verticillioides* (*F. moniliforme* (MRC 826)) to pigs caused severe pulmonary edema, and it resulted in severe nephrosis and hepatosis when administered by rumen fistula to sheep. Feeding another less toxic isolate (*F. verticillioides* (MRC 602)) to baboons caused acute congestive heart lesions and hepatic cirrhosis.

*F. verticillioides* (MRC 826 and 602) were toxic to rats, also causing hepatic cirrhosis and intraventricular cardiac thrombosis. Kriek *et al.* (1981a,b) concluded that rats are the best model to screen the toxicity of *F. verticillioides* isolates.

Mirocha (1983) described the effects of trichothecene mycotoxins and crude *Fusarium* extracts on farm animals and summarized the results obtained by different researchers associated with the distribution and metabolism of [$^3$H]-T-2 toxin in cattle (dairy cow, chicks), and the effect of T-2 toxin in swine.

Gedek and Bauer (1983) described the occurrence and incidence of toxigenic strains of *Fusarium* species, tested for trichothecenes and zearalenone production *in vitro* in the Federal Republic of Germany, and wrote clinical reports on animal diseases (swine, poultry, horses, and cattle) connected with metabolites of *Fusarium* species. The major toxic effects of the *Fusarium* species and their toxins on the different classes of domestic animals are described in the following section.

## FUSARIOTOXICOSES IN POULTRY

Most of the natural toxicoses in agricultural animals have been associated with the larger domestic animals: pigs, horses, cattle, and so on. However, several outbreaks of disease among poultry have been traced to *Fusarium* infected feed.

Fusariotoxicoses in poultry associated with T-2 toxin were noted in the United States by Wyatt *et al.* (1972b, 1973a,b) and Chi *et al.* (1977a,b,c); in Canada by Greenway and Puls (1976); in Israel by Joffe and Yagen (1978); in Russia by Kurmanov (1978a); in Hungary by Palyusik (1977b) and Palyusik and Koplik-Kovacs (1975).

Despite the relatively few cases of natural fusariotoxicoses in poultry, considerable interest has been shown regarding the effects of various *Fusarium* species when domestic birds are fed moldy cereal grains.

As early as 1930 Mundkur and Cochran investigated the effects of feeding scabby barley infected with *G. saubinetii* to poultry. They found that mature chickens could be fed the scabby barley with little adverse effect. However, when fed to 14-day-old chicks the barley caused weight loss and "rough" plumage. Later Mundkur (1934) concluded that scabby barley is not poisonous to chicks and that chicks fed on standard mash alone gained weight, whereas those fed on a diet containing scabby barley made smaller increases in weight; artificially infected barley was more injurious than naturally contaminated barley. Roche *et al.* (1930) stated that ruminants and poultry gain weight when fed heavily scabbed grain with no apparent ill effect.

Borchers and Peltier (1947) found that moldy corn enhanced the growth of chicks when added to a poor diet. The moldy corn included unidentified species of *Fusarium* and *Aspergillus*.

Toxigenic studies were also carried out on poultry in the USSR. Alisova and Mironov (1944); Onegov and Naumov (1943); Sarkisov (1954); Sarkisov *et al.* (1948c); and Yefremov (1948) reported that chicks and ducklings, after being fed toxic overwintered prosomillet invaded by *F. sporotrichioides*, remained completely healthy.

Sergiev (1945, 1946, 1948) reported cases of mortality in chickens occurring in areas where septic angina or ATA disease were prevalent.

Spesivtseva and Kurmanov (1960) reported that feeding 2–3-month-old chicks (Leghorn and Livensky) and 6-month-old and adult chickens 25 g and 50 g, respectively, with oats contaminated with *F. poae*, caused fusariotoxicosis with a 90% mortality. The Livensky chickens refused eating the feed.

Spesivtseva (1960, 1964a,b) reported an outbreak of chicken fusariotoxicosis which was traced to *F. sporotrichiella* var. *sporotrichioides* (*F. sporotrichioides*).

According to Yarovoj (1961) grain contaminated with *F. sporotrichiella* var. *sporotrichioides* (*F. sporotrichioides*) was responsible for toxicoses among 2-month-old chicks.

Birbin (1966) reported on a massive outbreak of fusariotoxicosis in chickens on one of the farms in the Saratov district after being fed grains contaminated with *Fusarium* species.

Several cases of fusariotoxicoses among poultry in Russia are also reviewed by Kurmanov (1978a). He fed *F. poae*-infected oats to chickens, causing chronic toxicosis and death.

Prentice and Dickson (1964) suggested that in countries with acute food shortages the endosperm of lightly contaminated grain could be used as feed. They found that the factor that caused emesis in pigeons was located in the bran and aleuron layers of moderately scabbed wheat. The identity of the contaminating *Fusarium* species is of extreme importance in estimating the potential hazard of feeding contaminated grain. The toxic effects are even dependent on the particular isolate of the *Fusarium* concerned. Thus, for example, Scott (1965) found that although both of the *F. roseum* isolates he tested were toxic to ducklings, only 2 out of 10 *F. moniliforme* isolates caused toxicoses. Thirty-nine isolates of *F. equiseti* were taken from foodstuffs in the Eastern Transvaal and Swaziland. Twenty-five of them were markedly toxic when tested on ducklings (Martin *et al.*, 1971).

Extracts of *F. sporotrichiella* var. *poae* (*F. poae*) grown on barley were administered to chick embryos. Nine-day-old chick embryos were the most sensitive, and, in comparison with the rabbit skin test, the chick embryo test seemed more sensitive (Ivanov, 1968). Some extracts from *F. moniliforme* were also toxic to chick embryos (Archer, 1974).

In Czechoslovakia Vesely *et al.* (1982) studied 19 mycotoxins on White Leghorn chicken embryos incubated 40 hr by a sensitivity chick embryotoxicity screening test. Embryonic death as well as the incidence of caudal-trunk abnormalities were determined after a further 24 hr incubation. Of the mycotoxins tested, only T-2 toxin and diacetoxyscirpenol caused 100% embryonic mortality at doses as low as 0.01 µg. Abnormal development of the caudal trunk was observed after T-2 toxin and diacetoxyscirpenol (0.001 µg each) administration.

*F. tricinctum* 2016-C (*F. sporotrichioides*) cultured on maize in the laboratory was toxic to turkey poults (Christensen *et al.*, 1972a), and the severity of the toxic effect increased with the amount of infected grain fed to the test birds. A maize ration containing 1% infected corn caused 13% fatalities within 35 days. Increasing the infected maize to 2% increased the fatalities to 60–83%. If the infected maize was increased to 5, 10, or 20% of the diet, all the test birds died within 5–15 days.

Marasas and Smalley (1972) found *F. moniliforme* and *F. tricinctum* isolated from maize to be 90–100% toxic to chickens.

Fritz *et al.* (1973) observed only slight weight reductions and no major toxic effects when *F. moniliforme*-infected corn was fed to 1-day-old White Plymouth Rock chicks. Out of 15 isolates of *F. moniliforme, F. roseum,* and *F. solani* only one *F. moniliforme* and two *F. solani* isolates had any effect. These three isolates caused up to 20% weight loss in chicks, and one *F. solani* strain was lethal (Doupnick *et al.*, 1971a).

*Fusarium* species that are potential trichothecene toxin producers are often extremely toxic (especially to young poultry). Zearalenone producers are, on the whole, less toxic. Thus T-2 toxin isolated from *F. tricinctum* caused abnormal behavior of broiler chicks (Wyatt *et al.*, 1973a,c) and altered feather growth (Wyatt *et al.*, 1975b).

Puls and Greenway (1976) noted "ruffled feathers" in a test goose fed moldy overwintered barley suspected of containing T-2 toxin. Geese force-fed this barley developed leg and head tremors and died within 19 hr of feeding. Raised feathers, depression, and a tendency to stand immobile with lowered heads was noticed by Joffe and Yagen (1978) in chicks fed diets containing T-2 producing strains *F. poae* No. 958 and *F. sporotrichioides* No. 921.

Pathre *et al.* (1976) noted that *F. roseum* "Gibbosum" (*F. equiseti*) isolated from corn after growing on rice was added to a balanced diet containing 0.55–0.712 mg monoacetoxyscirpenol. When this diet was fed to turkey poults, it caused a bilateral inflammation of the beak and also gastrointestinal hemorrhaging. At higher concentrations the turkeys died within 7 days. The same contaminated diet, after being fed to laying hens, caused loss in weight gain and cessation of egg production.

According to Yoshizawa and Morooka (1974, 1977) deoxynivalenol and its monoacetate isolated from *F. roseum* were lethal when administered s.c. to Peking ducklings.

*F. tricinctum* NRRL A-23377, isolated from peanuts, produced neosolaniol monoacetate. Death of 1-day-old Dekalb cockerels usually occurred within 24 hours at higher levels of this toxin; at low levels the birds responded with increasing lethargy. $LD_{50}$ for neosolaniol monoacetate and T-2 toxin on cockerels were 0.789 and 1.84 mg/kg, respectively. For comparison of the dermal toxicity on rabbits, guinea pigs, and rats, Landsen *et al.* (1978) used 0.02 µg of neosolaniol monoacetate and T-2 toxin. Rats were slightly more sensitive to both toxins.

Chi *et al.* (1978b) examined acute toxicity of several trichothecene derivatives in 1-day-old broiler chicks by the administration of single oral doses. They found 8-acetyl-neosolaniol (3.22 mg/kg body weight) and diacetoxyscirpenol (3.8 mg/kg) to be more toxic than the others, including T-2 toxin (4.97 mg/kg). Death occurred during 8–60 hr after treating with the tested trichothecenes. Sublethal doses caused asthenia, diarrhea, and coma and decreased feed consumption and weight gain.

Robb *et al.* (1982) isolated *F. culmorum, F. tricinctum, F. nivale,* and *F. moniliforme* from maize and wheat. The trichothecenes of diacetoxyscirpenol, deoxynivalenol, and zearalenone were isolated from the maize and wheat components of the feed associated with disease in broiler chickens in Scotland in the winter of 1980–1981. Clinical symptoms included vasoconstrictive damage and dry necrosis of the toes, lack of interest in food, and loss in weight gain, whereas the weight achievement of test chicks up to 14 days old was 56% of that of the control group. Mortality in the first and second weeks was 1 and 0.5%, respectively, above normal levels, and postmortem examination revealed slightly enlarged livers and grossly distended gall bladders. The authors noted that *F. culmorum, F. moniliforme,* and *F. tricinctum* contained compounds toxic to tissue cultures of the human epithelial cell line (HEp II); zearalenone being identified in *F. culmorum* and diacetoxyscirpenol in *F. moniliforme.*

Bacon and Marks (1976) found that feeding maize contaminated with *G. zeae* (up to 30 ppm zearalenone) had no significant effect on weight gain or feed conversion rates in broiler hens and Japanese quail.

The extent of feed contamination and the feeding period are both important in gauging the toxic effects of moldy feed. Thus feed containing 62% corn showing 35% visual damage (*G. zeae*) fed to hens for 30 days had no effect on egg weight, although slight reductions in egg production and feed consumption were noted. A longer exposure (140 days) to corn showing 27% damage caused significant reductions in feed consumption and egg production, although egg weight was not affected (Adams and Tuite, 1976).

An isolate of *F. oxysporum* suspected of being toxic to cattle was shown to be toxic to turkey poults and chicks (Meronuck *et al.,* 1970).

The addition of 10% contaminated corn to the diet caused reduced weight gains in both turkeys (71% of the controls) and chicks (79%), although no internal lesions were detected. Increasing the contaminated corn to 40% of the diet caused death in all test birds (possibly due to starvation). Marasas *et al.* (1979b) also isolated a strain of *F. oxysporum* which was toxic to 1-day-old chicks. In addition they isolated strains of *F. avenaceum*, *F. culmorum*, and *F. equiseti*, which were all toxic to chicks. This sensitivity of 1-day-old chicks to fusariotoxins has led to their being used as bioassay tools in the detection of these toxins.

## The Effects of Trichothecene-Producing *Fusarium* Species on Poultry (Major Clinical Symptoms and Histological Effects Caused by *Fusarium* and Fusariotoxins)

### Oral and Upper Esophageal Lesions

One of the most easily recognized symptoms of trichothecene fusariotoxicosis in poultry is the appearance of proliferative, caseous, plague-like lesions on the beak and in the oral cavity regions. It was this symptom that led Wyatt *et al.* (1972a) to relate outbreaks of disease in broiler fowl and fancy pigeons to T-2-producing *Fusarium*. By using different feed with varying amounts of added T-2 toxin (obtained from *F. tricinctum* NRRL 3299 (*F. poae*)) it was shown that the size of these lesions was dose-related (Wyatt *et al.*, 1972b) and it was noted that these lesions bore some resemblance to those of the third phase of septic angina or ATA disease in humans. Some of the clinical symptoms of the third stage of ATA are characterized by hyperemia and hemorrhages in the mucous membranes of the oral cavity and mainly by necrotic lesions that develop in the mouth, tongue, palate, throat, and esophagus. The oral, nasal, gastric, and intestinal hemorrhages and necrosis were observed in patients during this third phase of ATA. This similarity was also noted by Joffe and Yagen (1978), and it is interesting that one of the outbreaks of fusariotoxicoses in poultry mentioned by Kurmanov (1978a) occurred in an area where septic angina had occurred in man. Inflammation of the mouth had been observed in ducklings fed oats infected by *F. sporotrichiella* var. *poae* (*F. poae*) (Spesivtseva, 1964b). Lesions have also been noted in the oral cavity in turkeys fed corn incubated with *F. tricinctum* (2061-C) (*F. sporotrichioides*) (Christensen *et al.*, 1972a) and in geese (Palyusik *et al.*, 1968). These latter researchers found that six *Fusarium* strains, including two of *F. sporotrichiella* (*F. sporotrichioides*), caused these lesions in the test geese. The yellowish-white fibrinous lesions consisted of an easily removable membranous deposit, usually at the tongue root and on the larynx cranialis. Often the underlying mucous membrane was almost sound although sometimes minor necroses were observed.

Broiler chicks' feeds appended with T-2 toxin or diacetoxyscirpenol induced necrosis and inflammation on the tongue angle of the mouth and also on the gingiva (Chi and Mirocha, 1978; Chi *et al.*, 1977a,b,c; Wyatt *et al.*, 1975a).

Chi *et al.* (1977a) studied the acute toxicity of T-2 toxin in broiler chicks and laying hens. They found that ingestion of 1 mg/kg of T-2 toxin produced necrotic lesions of the mouth, crop, and gizzard. Death in both chicks and hens occurred within 48 hr after the administration of T-2 toxin at levels of 5.03 and 6.27 mg/kg body weight, respectively.

Chi *et al.* (1977b) studied the subacute toxicity of T-2 toxin by feeding broiler chicks a diet containing 4.0 ppm T-2 toxin. Body weight and feed consumption decreased, and, after the third week, the authors observed the development of oral lesions in the bird and necrotic lesions on the mucosal layer of the gizzard. There was no evidence of hematopoietic suppression.

Hayes and Wobeser (1983) noted that young Mallard ducks (*Anas platyrhynchos*), after feeding on diets containing 20–30 ppm T-2 toxin for a period of 2 or 3 weeks, exhibited a reduction of weight gain, necrosis in oral cavity and pharynx ventriculus, and severe lymphoid depletion.

Large numbers of bacteria (*Staphylococcus epidermidis* and *Escherichia coli*) have been isolated from the oral lesions (Hamilton *et al.*, 1971; Wyatt *et al.*, 1972a). In addition to T-2 toxin (Chi *et al.*, 1977b,c; Joffe and Yagen, 1978; Hamilton *et al.*, 1971; Wyatt *et al.*, 1972b, 1973b) monoacetoxyscirpenol from *F. roseum* "Gibbosum" (*F. equiseti*) (Pathre *et al.*, 1976; Speers *et al.*, 1977) and diacetoxyscirpenol (Chi and Mirocha, 1978) have been shown to cause these oral lesions in chicks. After feeding, diacetoxyscirpenol caused the lowest weight gain in comparison with T-2 toxin and crotocin. Crotocin, at 10 ppm, had no effect on chicks. It has been suggested that these lesions are the primary effect of T-2 toxin and that by impairing the feeding ability of the birds concerned they lead to secondary symptoms such as weight loss and reduced egg production (Wyatt *et al.*, 1973a, 1975a).

### Effects on Weight Gain and Egg Production

Several researchers have shown that trichothecene toxins cause losses in weight gain and egg production, either when fed as feed contaminated with *Fusarium* or as pure toxins added to a balanced diet. The addition of as little as 1% of corn contaminated with *F. tricinctum* (2061-C) reduced feed efficiency and weight gain in turkey poults (Christensen *et al.*, 1972a), and Marasas and Smalley (1972) found that corn infected with *F. tricinctum* greatly reduced feed efficiency and weight gain in chickens even though no T-2 toxin was detected in the grain. Doses as low as 4 µg/g pure T-2 toxin in feed caused reduced feed consumption and decreased weight gains in broiler chicks (Wyatt *et al.*, 1973b). Reduced weight gain in broiler chicks has also been caused by *F. roseum* f. sp. *cereale* (Sharby *et al.*, 1973).

Sublethal doses of T-2 toxin were also observed to cause reduced weight gains in broiler chicks and laying hens. The most marked effects were in the first 10 days of a 30-day experiment (Chi *et al.*, 1977a). Chicks fed commercial feed in which wheat molded with T-2 toxin-producing *F. poae* No. 958 or *F. sporotrichioides* No. 921 (or pure T-2 toxin) was included showed reduced weight gains (Joffe, 1978a; Joffe and Yagen, 1978). Other trichothecene toxins that have been shown to have adverse effects on weight gains are diacetoxyscirpenol, neosolaniol, 8-acetylneosolaniol, HT-2 toxin, T-2 tetraol, and deacetyl HT-2 toxin (slight reductions). These toxins were also lethal at the higher doses tested (Chi *et al.*, 1978c).

The trichothecene toxins not only affect weight gain, they often reduce egg production and quality. Palyusik and Koplik-Kovacs (1975) found that toxic maize infected with *F. sporotrichioides* (3 ppm T-2 toxin) fed to 1-year-old Rhineland laying geese caused cessation of egg laying within 10 days. In addition the fertility of the eggs was greatly reduced and, although no T-2 contaminated corn was fed after the 16th day, the total egg count for the 38-day experiment was 28 (13 fertile) in the test geese as opposed to 89 (70 fertile) in the controls. It was noticed that the T-2 toxin feed also contained an additional three unidentified trichothecene toxins. Shell thickness as well as the number of eggs produced has been shown to be reduced by T-2 toxin (Wyatt *et al.*, 1975a). Thinner shells and lower egg production were also observed by Chi *et al.* (1977c) in hens fed 0.8 ppm T-2 toxin in an otherwise balanced diet.

Chi *et al.* (1977c) studied the subacute effects of T-2 toxin on the reproductive performance and health of S.C.W.L. laying hens. A diet containing 8.0 ppm toxin decreased feed consumption, egg production, and shell thickness. Hatchability was reduced by diets of 4.6 and 8.0 ppm T-2 toxin. Oral lesions in the hens and necrotic lesions in the gizzard and crop were caused by feeding with levels of up to 8.0 ppm T-2 toxin. No significant hematopoietic or pathological changes were observed, although levels of serum alkaline phosphatase, lactic dehydrogenase, and uric acid were affected.

Robison and Mirocha (1976) noted that T-2 toxin and its derivatives appeared in the gizzard (15%) and bile (9%) and also metabolized in secreta of broiler chicks into HT-2 toxin, T-2 tetraol, and neosolaniol.

Hatchability of the fertile eggs was also significantly lower than that of the control hens. White Leghorn laying hens, fed 5% corn infected with *F. tricinctum* (containing 16 ppm T-2 toxin), reduced feed intake, weight gain, and produced fewer eggs than the controls, although a 2.5% *Fusarium*-infected corn diet (8 ppm T-2 toxin) had no effect. Diets with 2.5 or 5% corn invaded with *F. roseum* "Gibbosum" (*F. equiseti*) (25 and

50 ppm monoacetoxyscirpenol) caused immediate loss in weight gain and cessation of egg production (Pathre *et al.*, 1976 and Speers *et al.*, 1977).

Allen *et al.* (1982b) stated that diacetoxyscirpenol reduced hatchability of White Leghorn females and that the embryo mortality associated with this toxin was similar to that observed by the feeding of Alaskan isolate of *F. roseum*.

Chi *et al.* (1978a) studied the transmission of radioactivity into eggs in laying hens with a single and multiple dose of tritium labeled T-2 toxin. It has been shown that radioactive T-2 toxin fed to hens accumulates in the eggs, especially in the egg white protein (Chi *et al.*, 1978a). The authors stated that within 24 hr of a single gastric intubation of 3-[$^3$H]-T-2 toxin, egg yolk and white contained 0.04 and 0.13% of the administered radioactivity, respectively. Using multiple-dosed birds intubated with eight consecutive daily doses, the radioactivity in yolk and white protein was increased; but the radioactivity accumulated in the white was greater than that in the yolk.

The radioactivity of the white protein was greater than that of the yolk after using both single and multiple doses. It was estimated that if hens were fed a diet containing 1.6 ppm T-2 toxin the resultant residue in the egg would be 0.9 µg T-2 toxin (and/or its metabolites). If the diet contained 2 ppm T-2 toxin the residue in the egg would rise to 1.1 µg T-2 toxin (and/or its metabolites).

Chi *et al.* (1978c) studied the tissue distribution and excretion of radioactivity in 6-week-old broiler chicks after a single oral dose of 3-[$^3$H]-labeled T-2 toxin. After 48 hr, the broiler chicks showed greatest radioactivity in the bile and gall bladder. T-2 toxin and/or its metabolites were excreted into the intestine through the bile duct, with the liver being the major organ of excretion of the toxin.

Studies of the distribution and excretion of radio-labeled T-2 toxin have also been noted in swine by Robison *et al.* (1979a); in poultry by Yoshizawa *et al.* (1980c); and in the lactating cow by Yoshizawa *et al.* (1981).

### Food Refusal and Emesis

One of the common effects of the trichothecene toxins in poultry is loss in weight gain, and this is usually the result of reduced food intake (e.g., Chi *et al.*, 1977a, 1978c; Christensen *et al.*, 1972b; Joffe, 1978a; Joffe and Yagen, 1978; Spears *et al.*, 1977; Wyatt *et al.*, 1972a). In some cases of experimental feeding the birds refused the moldy feed. Thus Puls and Greenway (1976) had to force-feed geese with a moldy diet suspected of containing T-2 toxin. Experimental chicks refused to eat a diet containing *F. poae* No. 958 or *F. sporotrichioides* No. 921 (Joffe and Yagen, 1978). Again the toxin in the diet was T-2 toxin.

Hamilton *et al.* (1981) noted that laying hens, after feeding on diets containing 0.35–0.70 ppm vomitoxin, consumed significantly less feed than hens on a control ration.

Hulan and Proudfoot (1982) examined the effect of vomitoxin on the mortality, feed palatability, and performance of broiler chickens. Shaver meat-type chickens were fed on diets containing various percentages of contaminated wheat containing 3 ppm vomitoxin, where the actual levels of vomitoxin ranged from 0.02 (control) to 1.87 ppm. The authors found that diets of 1.87 ppm toxin had no significant effect on mortality, weight gain, body weight, feed conversion, feed consumption, or palatability and there was no evidence of feed refusal or emesis.

These results for broiler chickens are contrary to the results obtained by Forsyth *et al.* (1977) and Hamilton *et al.* (1981).

Huff *et al.* (1981) noted that $LD_{50}$ for broiler chicks was 140 mg/kg vomitoxin per body weight. They reported that a single dose of 140 mg/kg of vomitoxin caused mortality with appearance of ecchymotic hemorrhage in the carcass, edema and irritation of gastrointestinal tract, and nervous disorders.

Moran *et al.* (1982) fed broiler cockerels corn infected with *F. graminearum* containing 800–900 mg/kg of vomitoxin for 6–11 days. Necropsy showed oral and gastric lesions, and their severity was dependent on the dietary levels of contaminated corn. Body weight and diet consumption were not significantly reduced until dietary levels of contaminated corn exceeded 116 mg/kg vomitoxin.

Tostanovskaya and Retmanskaya (1951) and Tostanovskaya (1956) reported that pigeons responded with vomiting a few hours after being fed prosomillet infected with *F. sporotrichiella* var. *poae* (*F. poae*) and *F. sporotrichiella* var. *sporotrichioides* (*F. sporotrichioides*). The authors also recommended the use of pigeons as sensitive indicators of toxins produced by the above mentioned fungi.

Emetic factors have been isolated from several species of *Fusarium*. Prentice *et al.* (1959) and Prentice and Dickson (1968) found that extracts of maize inoculated with *F. roseum* No. 162, *F. poae* No. 2518, *F. culmorum* No. 3737, *F. nivale* No. 3125, and *F. moniliforme* No. 111 all contained substances which caused emesis in pigeons when administered i.v. into a wing. Ellison and Kotsonis (1973) found that T-2 toxin isolated from *F. poae* NRRL 3287 caused emesis when tested (orally and i.v.) in pigeons. An emetic dose range which is nonlethal in pigeons was found to be 0.15 mg/kg.

Acetyl T-2 toxin isolated from the same fungi caused a slight emesis in pigeons after administration of 18 mg/kg (Kotsonis *et al.*, 1975a).

Other toxins have also been shown to cause emesis in poultry. Ueno (1971) showed that fusarenon-X (from *F. nivale* Fn-2B) caused emesis in ducklings. An s.c. dose of 2.0 mg/kg caused vomiting within 7 min. The duckling vomited and then drank water. This vomiting-drinking sequence was repeated every 2 min for a 20-min period. The symptom occurred irrespective of the mode of administration (s.c. injection or oral administration and at doses as low as 0.5 mg/kg). Other trichothecenes were also shown to cause emesis, the dose required depending on the toxin: 0.1 mg/kg for neosolaniol, T-2 toxin, and HT-2 toxin; 0.2 mg/kg for diacetoxyscirpenol; 0.4 mg/kg for diacetylnivalenol; 0.4–0.5 mg/kg for fusarenon-X; and 0.1 mg/kg for nivalenol (Ueno et al., 1974a).

Working with ducklings in the USSR, the author found that feeding them mixtures containing 2–20% wheat meal infected with *F. poae* No. 60/9 and *F. sporotrichioides* No. 347 caused pronounced swelling and inflammation in the oral cavity and also vomiting. Extensive damage to the internal organs was found on autopsy, including hemorrhages in many tissues and organs, such as the small intestine, kidneys, heart, and lungs, necrosis in the liver cells, and dilatation of the tabular cells and hemorrhages in the cortex. In the kidneys and liver blood vessels congestion was observed.

### Pathological and Histological Effects of T-2 Toxin and Other Trichothecenes

The gross clinical symptoms described previously are often accompanied by pathological and histological changes in internal organs, although the results reported are slightly conflicting. Thus in some cases (usually with mature hens as the test birds) no effect is noted on the relative sizes of internal organs such as the liver or bursa of Fabricius (e.g., Chi et al., 1977c; Hamilton et al., 1971; Wyatt et al., 1975a). Other results (usually with 1-day-old or young chicks) report changes in size and relative weights of these and other internal organs (Boonchuvit et al., 1975; Hamilton et al., 1971; Joffe, 1978a; Joffe and Yagen, 1978; Wyatt et al., 1973b).

Necrosis and hemorrhaging of internal organs are often caused by T-2 toxin intoxication (Joffe and Yagen, 1978; Palyusik, 1978; Pearson, 1978; Terao et al., 1978; Wyatt et al., 1972b).

In Israel Joffe and Yagen (1978) carried out a comprehensive study on New Hampshire chicks using two authentic toxic strains of *F. poae* No. 958 and *F. sporotrichioides* No. 921 associated with ATA in humans. The *Fusarium* cultures were grown on sterilized millet grains at 12°C for 21 days and, after drying, were added to a commercial feed in various proportions (0.5, 1, 2, 5, and 10%) and mixed until the diet was homogeneous. From these cultures the researchers also prepared ethanol extracts, after cultivating them on good quality sterilized wheat grains, and pure T-2 toxin was prepared from extracts isolated from *F. poae* No. 958 and *F. sporotrichioides* No. 921.

The experiments were performed on a total of 300 chicks. The treated chicks, which received either crude extract or mixed diet or T-2 toxin of

**Figure 9.1**   Changes in weight of 37-day-old chicks, in experimental and control groups, after feeding with 0.45 mg/kg T-2 toxin (in crude extract) isolated from *F. sporotrichioides* No. 921 (trial 1). (Joffe, unpublished results.)

*F. poae* and *F. sporotrichioides*, showed that a diet containing T-2 toxin produced severe clinical and pathological changes in organs and tissues, in addition to severe hematopoietic damage similar to that observed in humans affected by ATA. The control groups, given the same diet without toxic material, remained healthy and showed no clinical or pathological changes. The effect of different doses of T-2 toxin, derived from the above-mentioned *Fusarium* strains, on the mortality of chicks, is indicated by Joffe and Yagen (1978). Chicks given the toxic ration grew slowly and showed a reduction in weight gain. The difference between treated and control groups is shown in Figures 9.1 through 9.4.

In laying geese given 3 ppm T-2 toxin in their daily feed, postmortem examination revealed marked emaciation, cachexy, atrophy of ovary and oviduct, and severe myocardial degeneration. Microscopic examination showed premature atresia of the greater part of the ovarian follicles and hemorrhages in heart muscle and liver (Palyusik and Koplik-Kovacs, 1975).

More detailed histological effects were observed by Terao *et al.* (1978) who investigated the effect of several trichothecene toxins on the bursa of Fabricius of 1-day-old chicks. Fusarenon-X and nivalenol had similar effects on the bursa of Fabricius although they were not as toxic

**Figure 9.2** Changes in weight of 41-day-old chicks, in experimental and control groups, after feeding with 8 mg/kg pure T-2 toxin (in mixed diet) from *F. sporotrichioides* No. 921 (trial 9). (Joffe, unpublished results.)

as T-2 toxin. According to Joffe and Yagen (1978) T-2 toxin also caused swelling of the bursa of Fabricius.

According to Witlock *et al.* (1977) young broiler chickens fed 250 or 500 µg/g of citrinin in their diet exhibited changes in the morphology of the gut, compared with control birds which were fed a diet without citrinin. The authors also put 8–16 µg/g T-2 toxin in the daily ration of 1-day-old chicks for 3 weeks and noted that scanning electron microscopic examination of gut preparations showed no change in the villous structure of the gut region of chicks.

**Figure 9.3** Changes in weight of 39-day-old chicks, in experimental and control groups, after feeding with 0.6 mg/kg T-2 toxin (in crude extract) from *F. poae* No. 958 (trial 11). (Joffe, unpublished results.)

**Figure 9.4** Changes in weight of 43-day-old chicks after feeding 60 mg/kg commercial feed mixed with millet grains infected with *F. poae* No. 958 (trial 15). (Joffe, unpublished results.)

## Changes in the Blood System

Doses of up to 8 or 16 μg/g T-2 toxin had no effect on the hemoglobin, serum proteins, cholesterol or total lipids, plasma glucose, or plasma uric acid of broiler chicks (Wyatt *et al.*, 1973b). Doses of 20 μg/g in older chickens had no effect on these parameters nor on the hematocrit values, erythrocyte count, or prothrombin times (Wyatt *et al.*, 1975a). However, total plasma protein and lipid concentration and total leukocyte counts were reduced with the larger dose. In S.C.W.L. hens fed 8 ppm T-2 toxin no changes were noted in the hematocrit, hemoglobin, erythrocytes, or leukocytes of differential leukocyte counts, although the serum levels of alkaline phosphatase, lactic dehydrogenase, and uric acid were higher in hens fed T-2 toxin than in the controls (Chi *et al.*, 1977c). Doerr *et al.* (1974), however, found that although growth-inhibiting doses of T-2 toxin had no effect on whole blood clotting or recalcification, the prothrombin time was increased (cf., Wyatt *et al.*, 1975a).

In Peking ducklings fusariotoxins (from *F. "roseum"*) caused a reduction in all plasma proteins, and after 11 days the globulin was also decreased thus increasing the birds' susceptibility to bacterial infection (Sawhney, 1976, 1977). It is interesting to note that T-2 toxin was shown to increase the mortality rate of chickens infected with *Salmonella worthington*, *S. thompson*, *S. derby*, or *S. typhimurium* var. *copenhagen* (Boonchuvit *et al.*, 1975).

Joffe and Yagen (1978) showed that a mixed diet containing moldy grain contaminated with T-2 toxin caused a reduction in the circulating blood cells in young chicks, especially the leukocyte and thrombocyte counts and, to a lesser extent, the erythrocyte counts and hemoglobin values. Pure T-2 toxin at a rate of 16 mg/g feed had a slight effect on erythrocyte and hemoglobin levels but decreased the leukocyte count.

Chi et al. (1977a,b,c) did not observe, in acute and subacute toxicity with T-2 toxin, neural disturbance symptoms in broiler chicks as reported by Wyatt et al. (1973a). Therefore Chi et al. (1981) decided to study the effect of T-2 toxin (at 2.5 mg/kg body weight of 4-week-old male broiler chickens) on the neural system of chickens by determination of the concentrations of brain catecholamines and selected blood components in growing chickens. The authors stated that dopamine increased and norepinephrine concentrations decreased significantly in the brain, but concentration of brain serotonin was not affected. The authors noted that T-2 toxin affected brain function in chickens and that the changes in catecholamine metabolism may be connected with changes in motor activity and in a variety of other conditions.

The serum concentrations of glutamic oxaloacetic transamination (GOT) and lactic dehydrogenase were not significantly affected by T-2 toxin; but the serum cholesterol concentration in T-2 toxin-treated chickens was greater than in control chickens. There are also changes in the blood components. The leukocyte counts were significantly decreased during the 12–48-hr period after T-2 toxin administration; the lymphocytes were increased and the neutrophils decreased. Only the red blood cell counts remained unaffected by T-2 toxin treatment.

Bitay et al. (1981) described a case of T-2 mycotoxicosis in commercial broiler chicks. The diet, containing 2.5 ppm T-2 toxin, caused changes in the hematopoietic and lymphoid tissues and also mortality in chicks.

According to Coffin and Coombs (1980) long term levels of T-2 toxin used in conjunction with vitamin E, did not depress the hematocrit erythropoiesis or hemoglobin concentration in growing chicks.

Hoerr et al. (1981a,b) studied the clinical and pathological effects of T-2 toxin and diacetoxyscirpenol on 7-day-old chicks and compared the effect produced by these trichothecenes with those produced by a culture of F. sporotrichiella var. sporotrichioides (F. sporotrichioides). Both T-2 toxin and diacetoxyscirpenol, given in various combinations, had additive lethal effects on the chickens in single- and multiple-dose tests.

A single oral dose of T-2 toxin (2.5 mg/kg b.w.) and diacetoxyscirpenol (2.7 mg/kg) caused necrosis of lymphoid and hematopoietic tissues, necrosis of the bile duct, inflammation of the gall bladder mucosa, and necrosis of the mucosal epithelium of the ventriculus, proventriculus, and the feather epidermis.

Clinical signs and lesions were similar to those in chickens given either T-2 toxin or diacetoxyscirpenol in the diet or by crop gavage (Hoerr et al., 1981a,b, 1982a,b).

T-2 toxin and diacetoxyscirpenol given in 14 daily doses (1.5–3.5 mg/kg b.w.) caused the death of the chickens.

In the survivors body weight and hematocrit were reduced and lymphoid organs were atrophic. Lymphoid and hematopoietic lesions were more severe in T-2 toxin-treated chickens than in diacetoxyscirpenol-treated chickens in single and multiple-dose tests.

*F. sporotrichiella* var. *sporotrichioides*-infected corn from Hungary produced T-2 toxin, neosolaniol, and diacetoxyscirpenol. T-2 toxin caused the same clinical signs and histopathological findings as both T-2 toxin and diacetoxyscirpenol when given in single and multiple doses.

According to Hoerr *et al.* (1981a, 1982a) T-2 toxin was lethal to chickens and more destructive to the lymphoid and hematopoietic tissues than diacetoxyscirpenol. Hoerr *et al.* (1981b) noted that necrosis of feather epidermis occurred in chickens that were given a single dose of T-2 toxin, and Hoerr *et al.* (1982b) described the lesions of chickens by multiple oral-gavage doses of purified T-2 toxin or diacetoxyscirpenol.

T-2 toxin or diacetoxyscirpenol caused lesions in lymphocytic and hematopoietic tissues, alimentary digestive tract, liver, feathers, kidneys, and thyroid gland, and a reduction in weight gain.

Hoerr *et al.* (1982a) noted that *F. sporotrichiella* var. *sporotrichioides*, isolated from wheat in Hungary and cultivated on sterile corn at 23 °C and then at 8, 16, and 23 °C, fed in 50% of the diet to 7-day-old male broiler chickens, caused mortality.

Chickens given 0.24 mg T-2 toxin and 0.02 mg neosolaniol/kg b.w./day as 10% of their diet died and had clinical symptoms and lesions similar to those in chickens given T-2 toxin or diacetoxyscirpenol in the diet or by crop gavage (Hoerr, 1981, Hoerr *et al.*, 1981a,b). The chickens had necrosis and depletion of lymphoid and hematopoietic tissues, and other disorders were observed in the liver, digestive tract, and feather epidermis. Survivors had reduced hematopoietic cellularity in the bone marrow and necrosis of the oral and crop mucosa.

7-day-old broiler chickens fed with T-2 toxin or diacetoxyscirpenol at 4 and 16 ppm for 21 days, showed reduction of feed consumption and weight gain, and lesions with ulcers in the palate, tongue, and buccal floor (Hoerr et al., 1982c).

Six of the 10 chickens died after consuming about 10 mg/kg body weight T-2 toxin, but no deaths occurred due to diacetoxyscirpenol or a control diet. Necropsy showed necrosis of the lymphoid and hematopoietic tissues and of the digestive mucosa. Chicks that survived had atrophied lymphoid organs and were anemic.

**Estrogenic and Other Effects of
Zearalenone-Producing *Fusarium* Species**

According to Meronuck *et al.* (1970) turkeys fed a diet containing 10% of corn invaded by *F. roseum* developed swollen vents; of the eight turkeys (group J) fed the ration containing 40% of corn contaminated with *F. roseum* two died and three developed prolapsed cloacae. Some birds also had enlarged bursae of Fabricius. But turkeys and chickens which were fed a ration containing 40% of corn invaded by *F. oxysporum* died and before death greatly decreased in body weight and feed efficiency.

Zearalenone produced by *F. roseum* f. sp. *graminearum* (*F. graminearum*) caused a quadratic effect on the growth, comb weight, and ovary development in young chicks (Speers *et al.*, 1971). High doses, 1600 ppm F-2, caused a reduction in the weight of the bursa of Fabricius. Mature hens showed no significant response to the *Fusarium*-contaminated corn or corn containing pure F-2, but shell quality declined when hens were fed the *Fusarium*-contaminated corn.

Sherwood and Peberdy (1973a) found that 40 ppm zearalenone incorporated into the diet of male chicks had hardly any effect on weight gain or food conversion efficiency, but there was a significant increase in the weight of the testis and comb. Marks and Bacon (1976) observed no significant effects on body weight, egg production, weight, fertility, or hatchability in laying hens fed either corn infected with *F. roseum* "Graminearum" (supplying levels of up to 100 ppm F-2) or corn containing comparable amounts of pure F-2. Similar results were obtained by Adams and Tuite (1976).

No noticeable effect on egg production and only a slight, insignificant decline in egg fertility was observed in laying geese fed maize infected with *F. culmorum* and containing 100 ppm F-2 (Palyusik and Koplik-Kovacs, 1975). In general the estrogenic effects on poultry of *Fusarium*-infected feed are very limited, and female poultry appear to be the least susceptible of all domestic animals to these effects (Vanyi *et al.*, 1973a). However, the toxin does interfere with spermatogenesis in adult ganders and turkey-cocks (Palyusik *et al.*, 1971, 1974; Vanyi *et al.*, 1973a), but other cocks were unaffected (Vanyi *et al.*, 1973a). The sperm counts in ganders fed *F. graminearum* dropped from 1–1.2 to 0.3 million in 12 days, and in turkeys the count was reduced from 2.8–3.2 to 2.1–2.3 million in 30 days. All 17 eggs layed by a goose mated to a gander fed the contaminated maize failed to hatch. Histological examination showed that regressive changes had occurred in the testes of both species but that these were more severe in the turkeys than in the gander (Palyusik *et al.*, 1973; Vanyi *et al.*, 1973a).

Vanyi and Szailer (1974) studied the cytotoxic effect of zearalenone in various monolayer cell cultures prepared from calf, turkey, cock, and swine testicles and chicken embryo fibroblasts; secondary monolayer cultures were prepared from swine thyroid glands. Swine, turkey, and calf testicle monolayers tested with 500 to 250 μg/tube toxin were degenerated. The swine testicle cell cultures were most sensitive, followed by the turkey and calf testicle cells. The cock testicle, chick embryo fibroblast, and swine thyroid gland monolayer cell cultures were not affected even by the largest dose.

The authors concluded that the cytotoxicity test is a good model test for the rapid determination of sensitivity of various animals and tissues to the zearalenone toxin.

Chi *et al.* (1980a,b) studied the effect of zearalenone on female White Leghorn chickens. They stated that the estrogenic effects of the toxin were greater when multiple doses were used than when a single dose was administered intramuscularly or orally.

Allen *et al.* (1981a) studied the effects of zearalenone on finishing broiler chickens and young turkey poults. The authors stated that dietary zearalenone had no effect on body weight gain, feed consumption, hematopoietic system, and different organs and tissues such as weights of liver, heart, spleen, and bursa of Fabricius in both chickens and turkeys. The author concluded that the adverse effects of dietary pure zearalenone on finishing broilers and young turkey poults seem minimal.

Allen *et al.* (1981b) studied the effect of ethanol on chickens. Five mixed-sex chicks (Rhode Island Red, New Hampshire cross) were taken for each treatment. Intubation solutions were 0, 33, 50, 67, and 100% of 95% ethanol. The chicks were administered treatments via intubation into the crop with soft plastic tubing on a syringe. Birds administered 1 ml of 95% ethanol reduced weight gain and feed consumption and increased liver weight. The authors noted that crop exudates and hemorrhages were induced by all levels of ethanol. All levels of ethanol produced mild ataxia (failure of muscular coordination) within 5–10 min of dosage. Crop walls of chicks given 1 ml of undiluted ethanol were ulcerated and inflamed and also hemorrhagic, edematous, and infiltrated by heterophils and mononuclear cells. Histopathologic lesions were more severe and extensive in chicks given undiluted ethanol than in chicks given diluted ethanol.

The authors suggested that depressed growth rates of female White Leghorn chickens given zearalenone in ethanol solution as noted by Chi *et al.* (1980a) were not due to zearalenone but rather to the ethanol vehicle solution.

Chi *et al.* (1980a) studied the acute toxic effect of zearalenone on growing female White Leghorn chickens. The authors noted that all 10 chickens of the first experimental group administered a single oral dose of 15 g/kg of zearalenone using gelatin capsules survived and did not show any significant changes in the histopathological findings. In the second experimental group zearalenone was administered orally and intramuscularly at a dose of 0, 50, 200, 400, and 800 mg/kg during 7 days. The author found increased oviduct size with increasing toxin levels in 2-week-old female chicks who were administered the toxin either by oral or intramuscular injection, but by the last administration the liver weight was increased and the comb decreased. The authors concluded that the estrogenic effects of zearalenone were much greater when the toxin was administered in multiple doses rather than in a single large dose and intramuscularly rather than orally.

Chi *et al.* (1980a,b) and Allen *et al.* (1981c) reported that zearalenone increased the oviduct in growing female chickens and retarded growth of the testes in young male chickens. Allen *et al.* (1981c) found that 800 ppm of purified zearalenone had no effect on the egg production performance of White Leghorn hens.

Allen *et al.* (1982b) noted that diacetoxyscirpenol and *F. roseum* Alaska culture caused reduced hatchability after feeding to White Leghorn female chickens. The authors also found that the embryo mortality from feeding diacetoxyscirpenol was similar to that from feeding *F. roseum* Alaska culture. According to the authors *F. roseum* "Gibbosum" and *F. roseum* Alaska reduced hatchability to a similar extent, but the distribution of embryo mortality from feeding the two cultures was quite different, possibly because the primary embryo toxic agents in the *F. roseum* "Gibbosum" and *F. roseum* Alaska cultures were different.

Allen *et al.* (1983) studied the effects of *F. roseum* "Gibbosum," *F. tricinctum, F. roseum* Alaska, purified T-2 toxin, and zearalenone (F-2) on the reproductive function and immune response of Nicholas Large White female turkeys. The female turkeys were artificially inseminated and were inoculated with a killed Newcastle disease virus vaccine after 30 days of toxin feeding. The authors stated that the egg fertility and titers to Newcastle disease virus were unaffected by the dietary treatment. *F. roseum* "Gibbosum" and *F. tricinctum* caused a decrease in feed consumption, body weight, and egg producton. Zearalenone and T-2 toxin also caused a decrease in egg production.

Hatchability of fertile eggs was reduced by feeding hens *F. roseum* "Gibbosum," *F. tricinctum*, and *F. roseum* Alaska. After withdrawing the *Fusarium* toxic feed hatchability rapidly returned to control values. The authors concluded that cultures of *Fusarium* mycotoxins, other than zearalenone and T-2 toxin, were responsible for reduced hatchability in turkey females.

Dailey *et al.* (1980) studied the metabolism of intubated [$^{14}$C] zearalenone in laying hens and stated that 94% of administered [$^{14}$C]-radioactivity was isolated from excreta during 72 hr in the form of zearalenone or zearalenol. The authors also noted that 2 ppm of [$^{14}$C] zearalenone was metabolized in the egg yolk at 72 hr.

Mirocha *et al.* (1982a) described the distribution of [$^{3}$H] zearalenone and the determination of the metabolism and residue of zearalenone in excreta and edible tissue after administering it to young broiler chickens. Zearalenone and its metabolites were determined by gas chromatography-mass spectroscopy, capillary gas chromatography and by radioimmunoassay. The greatest accumulation of radioactivity occurred in the liver 30 min after administration. Zearalenone and alpha and beta zearalenols were isolated from excreta. The maximum residue was found in the muscle and contained only zearalenone.

## THE EFFECT OF TOXIC SPECIES OF *FUSARIUM* ON CATTLE

Several outbreaks of diseases among cattle have been associated with *Fusarium*-contaminated feed and feed containing fusariotoxins.

An outbreak of moldy corn poisoning in Georgia was described by Sippel *et al.* (1953). Many cows became ill after 24–48 hr but only those which were pregnant died.

Nishikado (1957a,b,c) isolated from scabbed wheat a toxigenic fungus which was identified as *G. zeae*. Nishikado (1958) noted that this wheat infected with *G. zeae*, when fed to calves and cows in Okayama in 1953, was shown to be toxic. Nishikado described in detail the distribution of wheat and barley contaminated with *F. graminearum* (*G. zeae*) in Japan. The author stated that Japan is plagued by *Fusarium* damage to a greater degree than other countries. The damage was more marked in the lee-ward regions of the seasonal winter winds than in other areas. Nishikado depicted the distribution of *Fusarium* damages in wheat and barley in his well-known map of Japan.

*F. culmorum* isolated from maize was thought to be responsible for an outbreak of mycotoxicosis in southeastern Victoria (Australia). Loss of appetite and a greatly reduced milk yield were noted and symptoms included occasional staggering among dairy cows. Symptoms disappeared as soon as the moldy food was removed (Fisher *et al.*, 1967).

Decreased fertility in cows has been linked to the consumption of moldy hay contaminated with *F. moniliforme* which was later shown to contain F-2 toxin (zearalenone) (Mirocha *et al.*, 1968b,c). Similar fertility problems were observed in dairy cows fed moldy grain contaminated with *F. graminearum* and *F. culmorum* (Roine *et al.*, 1971). Moreau (1972) reported that cows grazing on meadow rye grass (*Lolium perenne*) infected with *F. culmorum* suffered from lowered milk production, diarrhea, and premature calving.

*F. graminearum* was reported to have caused estrogenic symptoms in cattle (Danko and Toth, 1969). These estrogenic symptoms (swelling of the vulva, drop in milk production, and loss of appetite) were also associated with the occurrence of zearalenone (5–57 ppm) in the feed, although no viable *Fusarium* species were detected (Vanyi *et al.*, 1974c).

Food refusal is a common response to contaminated fodder. Cows refused feed containing 33% maize contaminated with *F. roseum* (Thomas *et al.*, 1973), and food refusal, emesis, and bloody stools are reported to have resulted from the consumption of feed contaminated with *F. roseum* (Mirocha *et al.*, 1976b). In this last case, diacetoxyscirpenol, deoxynivalenol, and zearalenone were all detected in the feed.

Fatal toxicoses in cattle have usually been associated with the trichothecene fusariotoxins. Thus T-2 toxin was detected in moldy maize used as winter fodder for a dairy herd in Wisconsin which suffered from a high abortion rate and 20% fatalities during the 5-months winter feeding period. T-2 toxin isolated from moldy corn caused bloody diarrhea, hemorrhagic syndrome, death of lactating cows (Hsu *et al.*, 1972), and several dermal necroses and disorders in the digestive tract and other organs (Joffe, 1974c, 1978b, 1983b).

Kurmonov (1968) and Kvashnina (1968) noted that *F. sporotrichioides* caused toxicosis in cattle and ruminants under experimental and field conditions with symptoms characteristic of those of ATA.

The T-2 producing species, *F. tricinctum*, has been associated with fatal hemorrhaging in calves fed contaminated silage over a period of 45 days. Death due to uncontrollable hemorrhaging after dehorning or castration occurred in the calves (Dahlgren and Williams, 1972). Bovine hemorrhagic syndrome, epistaxis, bloody diarrhea, and a 3% mortality in a 200-animal herd was attributed to trichothecene poisoning (T-2 toxin) (Hibbs *et al.*, 1974). Brewers grain contaminated with T-2 toxin was thought to be responsible for a hemorrhaging syndrome in dairy cows (Petrie *et al.*, 1977), but in spite of these reports Matthews *et al.* (1977) suggest that, although the trichothecenes are often circumstantially connected with hemorrhaging in cattle, there is little real evidence to support this. These researchers had failed to reproduce the hemorrhagic symptoms by administering T-2 toxin p.o. at rates likely to be found in naturally contaminated foods (this even when feeding was continued over a long period (79 days) and the total toxin consumed by one calf was 1.8 g). However, Kosuri *et al.* (1970) did reproduce hemorrhagic symptoms in one cow by injecting T-2 toxin. The necrosis of tail tips of some cows supported the suggestion that a butenolide toxin of *F. tricinctum* (*F. sporotrichioides*) isolated from corn and fescue grass may contribute to the syndrome observed in cattle.

Cirilli (1983) isolated diacetoxyscirpenol, nivalenol, T-2 toxin, and zearalenone from *F. graminearum*, *F. roseum*, *F. tricinctum*, and *F. poae*-contaminated corn silage, cereal grains, and corn. He reported that some dairy cattle had extensive intestinal hemorrhages after consumption of feed containing T-2 toxin. Feed contaminated by nivalenol and diacetoxyscirpenol caused a few cases of intestinal hemorrhaging in swine. *F. tricinctum*-contaminated feed caused general digestive disorders, bloody diarrhea, and hemorrhagic lesions in the stomach, intestines, and kidneys in rabbits, swine, cattle, and poultry.

Patterson (1983) described toxicoses and natural occurrence of trichothecene derivatives, mainly diacetoxyscirpenol and T-2 toxin, in

Britain. He also provided a survey of laboratory studies and animal experiments associated with *Fusarium*-produced trichothecenes and their toxicity on domestic animals, mainly cattle and pigs.

Also Szathmary (1983) reported on trichothecene toxicoses and their natural occurrence in Hungary. The author described a brief history of the trichothecene problems in Hungary, the chemistry and distribution of trichothecene-producing *Fusarium* species in different feedstuffs, and the chemical and biological assays of trichothecenes and their toxicology in animals (livestock and poultry).

Because of the severity of trichothecene poisoning symptoms in cattle it has been suggested that caution should be adopted when using poultry excreta in cattle feed. The authors concluded with the warning that care be exercised when using poultry excreta as cattle feed (Robison and Mirocha, 1976).

*F. oxysporum* (Meronuck *et al.*, 1970), *F. poae*, *F. sporotrichioides*, and *F. tricinctum*, connected with toxicoses in cattle, have all been isolated from hay associated with "respiratory symptoms" in cows (Korpinen and Ylimäki, 1972). Another respiratory disease in cattle, atypical interstitial pneumonia, has also been linked with *Fusarium* species, in this case *F. solani* (*F. javanicum*) (Wilson *et al.*, 1970, 1971).

## Experimental Toxicoses Caused by *Fusarium* Trichothecenes

Much of the work done on the effects of fusariotoxins on cattle has been hampered by the small number of test animals (often a single subject) used in each treatment. Repeated injections of T-2 toxin (0.1 mg/kg for 65 days) have been shown to cause fatal internal hemorrhaging in a steer (Grove *et al.*, 1970). Postmortem examination of a cow injected with T-2 toxin showed local abscesses at the injection site, petechial-ecchymotic hemorrhages of the epi- and endocardium, congestion of cervical lymph nodes, and hemorrhages in the small and large intestines. Before death the cow had experienced difficulty in breathing and had bloody feces.

Izmajlov (1963) was unable to produce fusariotoxicosis in cows with a very toxic culture of *F. sporotrichioides* administered with the daily feed ration. Massive, fatal hemorrhages were produced in a single cow dosed initially with 72 mg T-2 toxin i.m. per day. Loss of appetite necessitated the gradual reduction of this dose to 18 mg/day after which, with the return of appetite, the dose was increased to 35 mg i.m. until the cow died (Kosuri *et al.*, 1970).

Pier *et al.* (1976) noted that calves treated with T-2 toxin at levels of 0.32 and 0.64 mg/kg/day for 31 and 20 days, respectively, suffered severe weight loss and bloody diarrhea.

Weaver *et al.* (1980) fed a healthy pregnant Holstein cow 182 mg of pure T-2 toxin (per intubation) (0.44 mg/kg body weight) by esophageal intubation for 15 days, and a male calf (which was born) was intubated with 26.2 mg of T-2 toxin (0.6 mg/kg body weight). The calf was severely affected by the T-2 toxin, causing anorexia, hindquarter ataxia, knuckling over of the rear feet, listlessness, and depression.

In the pregnant dairy cow only a ruminal ulcer and edema of the submucosa of the reticulum, cecum, and colon was observed. At necropsy extreme congestion was detected in the gastrointestinal tract, but no hemorrhagic symptoms were found, in contrast to the results reported in natural outbreaks in cattle with T-2 toxin by Cirilli (1983); Hibbs *et al.* (1974); Hsu *et al.* (1972); and Petri *et al.* (1977), and in experiments by Grove *et al.* (1970); Kosuri *et al.* (1970); and Pier *et al.* (1976).

In addition, hemorrhaging, inflamed lips, mouth, nose, and tongue and the ulceration of the digestive tract in cattle have all been associated with T-2 poisoning associated with *F. trincinctum* (Ribelin, 1978).

The rumen appears to play an important part in neutralizing the effects of T-2 toxin. Calves, especially those with nonfunctioning rumens, are more susceptible to *F. sporotrichioides* toxin than mature cows (Kurmanov, 1978b). A diet of 500 mg/kg of fungus per day induced chronic fusariotoxicosis, clinical symptoms appearing 11–12 days after the onset of the experiment. In addition Kurmanov (1969) showed that the acidity of the feed affected its toxicity. Thus calves given acid silage (pH 3.5–3.6) infected with *F. sporotrichiella* var. *poae* (*F. poae*) developed toxicoses after 5–6 days, whereas feed of normal pH did not cause toxicosis even after 42 days (Kurmanov, 1978b).

The role of acidified food in the development of fusariotoxicosis in cattle was obvious with calves. These data were confirmed by experiments of Marchenko (1963). He studied fusariotoxicosis in sheep by feeding them silage with a low pH contaminated with *F. sporotrichioides* var. *poae*. It has been suggested that the trichothecenes may have an antibiotic effect on the ruminal microflora and thus interfere with vitamin K synthesis (Ribelin, 1978).

Stahr (1975, 1976), Stahr and Kraft (1977), and Stahr *et al.* (1978, 1979) worked out a method of analysis for animal feeds and human foods. They mainly studied animal intoxication which exhibited various symptoms such as feed refusal, mouth and lip irritation, and hemorrhage in the digestive tract and other organs and tissues. This method of analysis has been developed for detection of T-2 toxin and other scirpene toxins in feeds. The method involves extraction by acetonitrile, defatting with petroleum ether, decolorization with ferric gel, and partitioning into chloroform. The researchers recommended TLC, gas chromatography, and mass spectroscopy for a presumptive test for determination of trichothecene toxins, chicken embryo tests of extracts, field desorption, chemical ionization, and biological test on skin. Magnetic resonance and infrared spectrometry may also be used for confirmatory purposes for indication of the presence of some scirpene toxins. The recommended method is sensitive to 1 ppm for the determination of T-2 toxin, diacetoxyscirpenol, and deoxynivalenol in food and feeds described by Stahr *et al.* (1979).

Robison *et al.* (1979b) reported the transmission of T-2 toxin into a pregnant Holstein cow and into crossbred sow milk samples, collected on different days and analyzed by combined gas chromatography-mass spectrometry and selected ion monitoring mass (SIM). After intubation with 182 mg of T-2 toxin for 15 consecutive days, all milk samples on the second, fifth, tenth, and twelfth day (but not on the fourth and eighth day) contained the following concentration: 43 ppb, 160 ppb, and 30 ppb, respectively. A milk sample (58 g) from a crossbred sow, six days after parturition, contained 76 ppb of T-2 toxin. This sow contained a T-2 toxin diet at levels of 12 ppm for a period of 220 days. The minimum T-2 toxin concentration was 5 ppb to 10 ppb for cow's milk and 25 ppb to 35 ppb for sow's milk.

Yoshizawa *et al.* (1981) studied the metabolic fate of T-2 toxin in a lactating cow and found maximum levels of radioactivity in excreta and plasma, mainly in feces, urine, and milk.

Hook *et al.* (1982) noted that T-2 toxin produced by several *Fusarium* species under alternating low temperatures is immunosuppressive in laboratory animals and poultry. The authors studied the immunological, hematologic, and pathologic effect in experiments on calves using daily oral doses of T-2 toxin. The immune function of calves was compared with controls for antibody production and lymphocyte response to mitogens *Anaplasma,* and *Brucella* antigens.

## Fescue-Foot in Cattle

Tall fescue (*Festuca arudinacea* Schreb.) is a major forage grass in the United States, Australia, and New Zealand that caused outbreaks of disease in cattle called "fescue-foot." This disease is characterized by extreme loss of weight, lameness in the hind limbs, loss of the tip of the tail, rough hair coats, arched back, dry gangrene of the feet, tail tip, and ear, diarrhea, and increased respiration with a preference for shade (Cornell and Garner, 1983; Ellis and Yates, 1971; Garner and Cornell, 1978; Robbins and Russell, 1983; Yates, 1962, 1971; Yates *et al.*, 1969). Fescue-foot in cattle occurs occasionally in winter in herds grazing pastures of tall fescue.

The following researchers (Burmeister and Hesseltine, 1970; Ellis and Yates, 1971; Keyl *et al.*, 1967; Yates *et al.*, 1967, 1968, 1969, 1970) found that extracts of toxic fescue hay contaminated with *Fusarium* species caused a necrotic reaction on rabbit skin. Two of these *Fusarium* strains were identified as *F. nivale* NRRL 13318 by Keyl *et al.* (1967) and *F. nivale* NRRL 3249 by Burkhardt *et al.* (1968); Ellis and Yates (1971); Yates *et al.* (1967, 1968); Yates (1971). Later these strains were reidentified as *F. tricinctum* NRRL 3249 by Burmeister *et al.* (1971);

Ellis and Yates (1971); Ueno _et al._ (1972b, 1973d); Tookey _et al._ (1972); Yates (1971); Yates _et al._ (1969, 1970) and again identified as _F. sporotrichioides_ by Joffe and Palti (1975).

Recently our classification was confirmed by Marasas _et al._ (1984). Butenolide, T-2 toxin, and an unidentified toxin were isolated from _F. sporotrichioides_ NRRL 3249 (Burmeister _et al._, 1971, 1981; Ellis and Yates, 1971; Joffe, 1978b; Joffe and Yagen, 1977; Yates _et al._, 1967, 1968, 1969, 1970).

Neither butenolide nor T-2 toxin (both produced by _F. sporotrichioides_ NRRL 3249) have been isolated from toxic tall fescue (Yates, 1962, 1963; Yates _et al._, 1966).

_F. tricinctum_ (_F. sporotrichioides_ and _F. poae_, according to our identification—Joffe and Palti, 1975) produce diacetoxyscirpenol and T-2 toxin at 7–8°C and HT-2 toxin at 24°C (Bamburg, 1968; Bamburg _et al._, 1968a,b; Yilgan _et al._, 1966). Some strains of _F. tricinctum_ also produce butenolide. T-2 toxin and butenolide that have been used experimentally to reproduce some of the symptoms associated with fescue-foot (Grove _et al._, 1970; Kosuri _et al._, 1970; Tookey _et al._, 1972). Butenolide caused tail necrosis (Kosuri _et al._, 1970; Tookey _et al._, 1972) and tail necrosis and an arched back but no foot damage (Grove _et al._, 1970). T-2 toxin caused slight tail necrosis only and none of the other fescue-foot symptoms, although it proved to be fatal to the experimental animal (Grove _et al._, 1970).

Ellis and Yates (1971) discussed the isolation of various mycotoxins from toxic fescue hay, and Futrell _et al._ (1974) studied ruminal fungus populations.

Kalra _et al._ (1973a,b, 1977a,b) described the etiology of Degnala disease, similar to chronic ergotism and fescue toxicity among buffalo and cattle in the State of Haryene, India.

During 1972 the disease caused a mortality rate of 64% in buffalo. The outbreaks occurred in rice-growing regions and were associated with cold weather from November to March. The characteristic symptoms were necrosis and gangrene of the extremities. The isolation of toxigenic strain of _F. equiseti_ (_G. intricans_) from the toxic straw suggested the involvement of fungal metabolite(s). _F. equiseti_ was found to give a characteristic hemorrhagic reaction after application to the skin of rabbits.

We suggested that further investigations were necessary in order to understand the role of toxic rice straw in the development of the disease and also the role of toxigenic strain of _F. equiseti_ and its mycotoxin(s) in Degnala disease.

Tall fescue is the dominant forage grass in the United States for beef and dairy cattle, sheep, and horses.

In the _Proceedings of Tall Fescue Toxicosis Workshop_ (1983) the researchers (Bacon and Russell, 1983; Porter and Russell, 1983; Robbins and Russell, 1983; Siegal, 1983; and some others) revealed that there is a clear relationship between the fungus endophyte, *Epichloe typhina*, and depressed animals that were fed tall fescue seed or hay. Robbins and Russell (1983) stated that toxic samples of fescue were 100% contaminated with *E. typhina* and that the fungus is widespread throughout the fescue growing area. The authors concluded that fescue toxicosis is associated with the endophyte *E. typhina*.

On the other hand Cornell and Garner in the same *Proceedings of Tall Fescue Toxicosis Workshop* noted that "the cause of fescue foot is not clear."

Another nervous disease of sheep, cattle, horses, and deer named ryegrass staggers (RSG) was detected in many areas in New Zealand. This disease is caused by the ingestion of a fungal toxin associated with ryegrass pasture (Everest, 1983). This researcher noted that although the endophyte fungus is present in the ryegrass at all times, RSG is only severe in dry years. According to Mortimer (1983) RSG affects the previously mentioned animals grazing pastures in which perennial ryegrass (*Lolium perenne*) is dominant.

The clinical signs in sheep developed within 7–14 days after grazing on toxic pastures and caused slight trembling of the head and fasciculation of the skin, muscles of the neck, shoulder, and flank regions. As the neuromuscular disorder progresses there is headnodding, jerky limb movements, and staggering during movement.

Purified extracts of samples of ryegrass produced characteristic tremors when injected into mice. Two tremorgens named lolitrem A and B (with different chromatographic characteristics) were isolated from ryegrass neurotoxic pure fractions.

A positive association of the *Lolium* endophyte of perennial ryegrass with the lolitrems and with natural outbreaks of RGS in sheep was found. Now there is a need for convincing evidence that the lolitrem neurotoxins are responsible for causing the disease in livestock species.

### The Effects of *Fusarium* Estrogens

Cases of infertility, reduced milk production, and hyperestrogenism in cows have been associated with zearalenone or zearalenone-producing *Fusarium* species (Mirocha *et al.*, 1968c; Roine *et al.*, 1971; Vanyi *et al.*, 1974c). In general, though, cattle are less sensitive to the fusarial estrogens than other domestic animals, such as pigs, and little experimental work has been done on the estrogenic effects of fusariotoxins on cattle. In an experiment designed to investigate the "carry over" of mycotoxins into the milk and tissue of cows, it was found that no obvious clinical effects or reduced milk yield were caused by feeding a ration containing 385–1925 µg/kg zearalenone to dairy cattle for a 7-week period. No zearalenone residues were detected in the milk, serum, or urine or in the muscle, liver, and kidney tissues examined (Shreeve *et al.*, 1979).

Most of the experimental work on cattle has been centered on the growth-promoting effects of zearalenone derivative zeranol (zearalanol or RAL). This derivative has been shown to stimulate increased weight gain and feed conversion rates (Bennett *et al.*, 1974; Borger *et al.*, 1971; Brown, 1970; Davis and Mahoney, 1971; Sharp and Dyer, 1970, 1971, 1972), but Moran (1972) suggests that zearalanol may be more efficient as a growth promoter in intensive cattle rearing than in cattle kept under extensive range conditions. Perry *et al.* (1970) noted that the strongest response occurred soon after implantation, and they suggested that the implant is probably depleted after 84–112 days. Some caution must be exercised when using RAL, however, since it has been shown that levels of RAL used to increase weight gain can have adverse effects on cow fertility. It was found that heifers implanted with RAL had a lower conception rate than control animals (Nelson *et al.*, 1972).

Ralston (1978) noted that α-zearalanol exhibited anabolic properties and caused an increase in weight gain in male calves.

# FUSARIOTOXICOSES IN PIGS

Consumption of moldy grain (especially that infected with *Fusarium* species) has often been associated with losses in swine herds.

Losses incurred by swine herds have been found to be caused by two major syndromes: (1) food refusal and/or emesis; and (2) estrogenic and other reproductive disturbances in pigs caused by fusariotoxins ("hyperestrogenism," often characterized by so-called "vulvovaginitis" in swine).

## Food Refusal and Emesis

Roche *et al.* (1930) reported that pigs, horses, and dogs are very sensitive to scabby barley. Roche and Bohstedt (1930) noted that pigs were much more sensitive to scabby barley, that is, grain infected with G. *saubinetii*, than most other domestic animals, and that most of the complaints made by farmers about scabby barley in the years 1928 and 1929 referred to the fact that pigs refused to eat the grain.

Mundkur and Cochran (1930) noted that hogs fed scabbed barley in Iowa developed nausea and refused to eat the grain, and according to Raino (1932), oats infected with G. *saubinetii* are harmful to pigs and horses.

Mundkur (1934) reported food refusal and vomiting by swine in the United States when they were given barley infected by G. *saubinetii*.

Much later a similar outbreak was noted in northern Indiana in the fall and early winter of 1965. Refusal of food containing moldy maize infected with *F. graminearum* (G. *zeae*) was widely reported, although only one case of limited vomiting was recorded (Curtin and Tuite, 1966).

Korpinen and Ylimäki (1972) noted losses of appetite and "gastric symptoms" in pigs fed commercial mixes contaminated with *F. tricinctum* and *F. poae*. Much of the maize harvested in south Michigan in the autumn of 1972 was refused by pigs, and *F. graminearum* was isolated from many samples of this corn (Miller and Smith, 1975). In the winter of the same year (1972–1973) there were several outbreaks of food refusal and emesis in swine in the southern United States, and *G. zeae* was isolated from corn consumed by the animals (Futrell *et al.*, 1976). *G. zeae* was also isolated from corn in the Midwestern states in the same year (1972). In British Columbia pigs fed overwintered barley vomited, scoured, and refused to eat. These pigs returned to normal 5 days after the change of diet; suckling pigs were not affected. Once again *G. zeae* was suspected as the causal agent (Greenway and Puls, 1976). Numerous cases of feed refusal were related to maize contaminated with fusariotoxins, and *F. roseum* and *F. culmorum* were isolated from feeds associated with food refusal, emesis, abortion, and hyperestrogenism in swine (Buckle, 1983; Mirocha, 1983; Mirocha *et al.*, 1976b).

Vesonder *et al.* (1973) reported the isolation of a trichothecene, deoxynivalenol, from corn infected mainly with *F. graminearum* (*F. nivale* and *F. moniliforme* were also found). This trichothecene proved to be emetic and was given the common name vomitoxin.

According to Vesonder and Hesseltine (1980); Vesonder *et al.* (1979); and Yoshizawa and Morooka (1974) naturally occurring vomitoxin (deoxynivalenol) in cereal grains caused vomiting and refusal in swine.

In addition vomitoxin has been detected at low levels in many diets prepared from corn or soft wheat. Vomitoxin caused reduced rate of body weight gain and feed refusal in pigs. It also caused a slight decrease in egg weight when laying hens were fed contaminated wheat (Elliott, 1982).

Pepeljnjak (1983) reported trichothecene mycotoxicosis in Yugoslavia caused by toxic *Fusarium* species isolated from various sources.

Numerous experiments have demonstrated that the food refusal and emetic syndromes can be reproduced by feeding animals grain infected with several *Fusarium* species (or their perfect, *Gibberella*, stage). Early researchers mention *G. saubinetii* (Beller and Wedemann, 1929; Christensen and Kernkamp, 1936; Mundkur and Cochran, 1930; Roche and Bohstedt, 1930) or *G. zeae* (Curtin and Tuite, 1966; Tuite *et al.*, 1974) as causes of the syndrome, but also *F. roseum* (Mirocha *et al.*, 1976; Thomas *et al.*, 1973, 1975), *F. graminearum* (Ciegler and Bennett, 1980; Featherston, 1973; Forsyth, 1974; Mitchell and Beadles, 1940; Palyusik, 1973; Vesonder and Ciegler, 1979; Vesonder and Hesseltine, 1980; Vesonder *et al.*, 1973, 1976, 1979c, 1981b), *F. tricinctum, F. poae* (Korpinen and Ylimäki, 1972; Vesonder *et al.*, 1977a, 1981), *F culmorum, F. moniliforme*, and *F. nivale* (Vesonder *et al.*, 1977a) have all been shown to cause either food refusal and/or vomiting in pigs.

Attempts to detoxify the grain to make it palatable showed that the refusal factor (at least) was water soluble (Beller and Wedemann, 1929; Bennett et al., 1980, 1981a; Christensen and Kernkamp, 1936; Curtin and Tuite, 1966; Forsyth, 1974; Forsyth et al., 1976; Roche and Bohstedt, 1930). Numerous attempts have been made to determine the factors responsible for food refusal and emesis. Curtin and Tuite (1966) found that emesis was independent of food refusal and suggested that two separate toxins were involved.

Vesonder et al. (1976) and Osweiler (1982) revealed that vomitoxin isolated from F. graminearum was also responsible for food refusal in swine. They found that food refusal was the same for naturally infected corn and for healthy corn spiked with the vomitoxin fraction. They suggested that small amounts of toxin cause vomiting and larger amounts result in food refusal.

According to Mirocha et al. (1976b) deoxynivalenol and zearalenol occurred naturally in feedstuffs and were detected in all samples connected with food refusal in pigs.

Coppock et al. (1982) noted that deoxynivalenol (vomitoxin) was commonly isolated in the northern regions of the United States and Canada from animal feedstuffs and appeared together with zearalenone. Diacetoxyscirpenol could be detected in swine plasma at levels of less than 100 ppb after less than 15 min. The most pathological findings were associated with the stomach.

Schiefer and O'Ferrall (1981) described the occurrence of mycotoxicosis in swine caused by zearalenone, vomitoxin, T-2 toxin, and others.

Vesonder et al. (1978) carried out an investigation of field preharvest corn for trichothecene vomitoxin. Fifty-two preharvest corn samples were collected in mid-October 1977 from 26 farms in a four-county region of northwest Ohio. Preceding harvests in northwest Ohio were accompanied by high rainfall through September and the first 4 days of October which favored growth of G. zeae. The G. zeae contaminated kernels ranged from 2–50% for 44 corn samples. The authors isolated, from 24 corn samples, vomitoxin ranging from 0.5–10.7 μg/g, using a GLC method for vomitoxin quantitation.

Forsyth et al. (1977) also suggested that additional factors other than deoxynivalenol may be involved in the swine food refusal response. In contrast to the findings of Vesonder et al. (1976) they found that there was less tendency to refuse food when healthy corn was spiked with deoxynivalenol than when naturally infected corn samples were fed to the pigs. Further works by Vesonder et al. (1977a) also suggested that other factors were involved in food refusal. They found that swine refused corn fermented with F. culmorum (NRRL 3288), F. poae (NRRL 3287), F. moniliforme (NRRL 3197), (F. tricinctum), and F. nivale (NRRL 3289). However, only the corn incubated with F. culmorum contained deoxynivalenol. Analysis of the other refused corn failed to detect deoxynivalenol, T-2 toxins, HT-2 toxins, acetyl T-2 toxin, or fusarenon-X.

Recently Vesonder et al. (1981a) found that F. poae No. 3287, F. tri-

*cinctum* No. 3197, and *F. nivale* No. 3289 produced trichothecenes (T-2 toxin and vomitoxin) which caused refusal response in pig bioassays.

T-2 toxin and diacetoxyscirpenol have been shown to induce vomiting (when administered by injection) and food refusal in pigs (Weaver *et al.*, 1978a,b,d). Reduced weight gain is sometimes observed even when refusal or vomiting has not occurred. Thus Sharda *et al.* (1971) observed significantly reduced weight gain as a result of feeding corn contaminated with *F. roseum* Ohio isolate C. No analysis was made for toxins, but the strain was lethal to small animals. Laboratory animals fed the same strain had petechial hemorrhages in liver and lung, suggesting that a trichothecene might be responsible.

Chernov (1970) noted that toxic *F. sporotrichioides*, when fed to swine, caused vomiting and hematopoietic changes, such as an increase of the red blood cell counts and hemoglobin, decrease of thrombocytes, and increase in the number of neutrophil leukocytes in the blood. In addition there were hemorrhages in the gastrointestinal tract and significantly marked hemorrhagic diathesis, characteristic of those produced by T-2 toxin in cats (Joffe, 1971).

Sarkisov (1954) and Sarkisov *et al.* (1948c, 1972) fed a culture of *F. sporotrichioides* No. 222 isolated from overwintered millet to suckling pigs and swine. Signs of disease in suckling pigs were observed after 24 hr, and hyperemia and lesions were seen in the mucous membranes of the cavity and lips, respectively. The swine were in a bad condition: depressed, refusing to feed, and in a staggering position. The hematopoietic system of the swine was without any significant changes; only 105 days after feeding was there a decrease of leukocyte numbers in the blood (from 19800–4300 in 1 mm$^3$ of blood). Catarrhal-hemorrhagic gastroenteritis and hemorrhagic diathesis were observed at the necropsy.

## Estrogenic and Other Reproductive Disturbances in Pigs Caused by Fusariotoxins, Mainly Zearalenone

Hyperestrogenism is the most widely reported symptom in pigs. Clinical symptoms include continued swelling of the vulva which becomes firm, tense, and elevated. The lips separate and vaginal mucosa becomes visible. Swelling of the inner mucosa continues until it protrudes through the lips of the vulva. Often prolapse occurs, partially checking circulation and causing passive congestion and distention of the prolapsed organs (Mirocha *et al.*, 1971).

Vulvovaginitis was first linked to moldy corn by McNutt *et al.* (1928). They report an outbreak that began in the early summer of 1927, increased in intensity throughout the summer, and then subsided towards the winter (only a few isolated cases occurred in the winter). In this outbreak boars and barrows were affected, showing inflammation of the prepuce. In 1952 McErlean, investigating two cases of vulvovaginitis in

Ireland, showed that these symptoms were associated with the consumption of barley infected with *F. graminearum* (*G. zeae*).

Following this early work, numerous cases of vulvovaginitis in swine due to consumption of grain infected with *Fusarium* species or *G. zeae* have been reported from widely distributed locations. Outbreaks occurred in Indiana in 1957 and 1958 (Stob *et al.*, 1962) and in Minnesota in January 1964 (Christensen *et al.*, 1965). Bugeac and Berbinschi (1967) reported cases of vulvovaginitis in Rumania. They isolated *F. graminearum* from the contaminated feed. In Yugoslavia the toxin zearalenone from *F. graminearum* and *F. moniliforme* was shown to be responsible for outbreaks of vulvovaginitis and estrogenism (Ozegovic and Vukovic, 1972).

Cases of vulvovaginitis were reported in Moldavia (Kurasova *et al.*, 1973) and also in Hungary (Vanyi and Szailer, 1974; Vanyi *et al.*, 1973a) where they were associated with F-2 in feed.

In France estrogenic symptoms in pigs were recorded by Cottereau *et al.* (1974) who mention that *F. roseum* was the most prevalent species associated with this disease, but Collet *et al.* (1977) found that *F. graminearum* was the most common species isolated from maize associated with pregnancy problems in sows in southwest France. They found zearalenone in the moldy maize.

Vulvovaginitis is not the only syndrome associated with zearalenone. In Scotland an outbreak of stillbirths, neonatal mortality, small litters, and splay-legged piglets was shown to be due to this toxin (Miller *et al.*, 1973).

Kurtz (1976) and Kurtz and Mirocha (1978) noted that zearalenone caused reduction of fecundity, resorption and mummification of fetuses, and abortion in sows.

Barnikol *et al.* (1982) described syndromes of hyperestrogenism and other clinical signs in newborn and weaning piglets, in two breeding farms in the Schwarzwald-Baar region of West Germany, that were associated with *Fusarium* species isolated from wheat-ear containing 1–2.9% ergot (a fungus of the genus *Claviceps*, infecting various cereal plants and forming black sclerotia that replace many of the seeds of the host plant). T-2 toxin was found in contaminated wheat and wheat bran used as food for piglets but not zearalenone or any other trichothecene.

The authors observed defects in the joints of the corpus and torsus, mummified fetuses in each litter, and also abortion in 30–40% of the sows.

Zearalenone (100 ppm) has been shown to be estrogenic to male pigs. Reduced testes, epididymis, and vesicular glands and enlarged prostate were noticed by Palyusik (1977a), although smaller doses (6–60 ppm) had no ill effects on the reproductive potential of boars (Ruhr *et al.*, 1978). Reduced testes weight was also noticed in barrows fed *F. roseum* (Christensen *et al.*, 1972b). Other signs of hyperestrogenism in male pigs include loss of libido (Bristol and Djurickovic, 1971) and mastitis (Koen and Smith, 1945).

Zearalenone causes hyperestrogenism and infertility in swine and cows. Swine appear to be the most sensitive of all the farm animals (Mirocha, 1980).

Vanyi and Szeky (1980a,b) noted that zearalenone caused reproductive disorders and hyperestrogenism in swine and, mainly, cessation of spermiogenesis in guinea-cock and male swine. In most cases pigs returned to normal when the contaminated food was removed.

According to Chang et al. (1979) zearalenone caused infertility and reduction of litter size but was not found to be responsible for abortion in swine.

Shreeve et al. (1978) found no ill effects from feeding pregnant sows low doses (2.2 mg/g) of zearalenone. However, estrogenic effects have been shown to be transferred to unborn piglets. Almost 20% of sows fed F. graminearum (3–8 ppm zearalenone) in their diet for 15 days prior to parturition gave birth to piglets (57%) which showed signs of edema of the vulva and mammary glands, the severity of the symptoms increasing with the amount of zearalenone in the mother sow's feed (Loncarevic et al., 1977). These piglets were also heavier than "normal" piglets. However, at weaning (28 days) no differences could be observed, in appearance or weight, between the normal and the "estrogenic" piglets. The external symptoms of hyperestrogenism are accompanied by histological changes in the genital tracts and mammary glands (Kurtz et al., 1969). Ductal hyperplasia and increased mitotic index were noted in the mammary gland. In the ovary the changes varied from hypoplasia in the smallest pigs to considerable secondary follicular development and concurrent follicular atresia in the larger gilts. In the uterus edema and cellular proliferation were noted in all layers. The epithelium of the cervix showed significant changes, including metaplasia of the mucosal epithelium from a normal double layer of columnar cells to a stratified squamous cellular layer. Similar metaplasia was seen in the vagina.

Stojanovic et al. (1978) found that, although grain infected with F. graminearum contained 32 ppm zearalenone, no estrogenic symptoms were observed when this was fed to pigs for 31 days. Decreased food consumption, weight gain, and feed efficiency were noted, and the relative weights of liver, kidneys, and spleen increased, as did the leukocytes and eosinophil levels. Blood erythrocytes and total blood P, free blood Ca, and serum and liver vitamin A were decreased.

In Scotland zearalenone was suspected of causing stillbirth and splay-legged piglets (Miller et al., 1973). These symptoms were reproduced by injecting sows with 5 mg purified zearalenone (isolated from a strain of F. culmorum) i.m. once daily in the last month of pregnancy. In addition it was found that F-2 was present in the sow's milk 12 hr after the last injection but none was detected at 20–24 hr. The splaying piglets lost their uncoordinated behavior and became "normal" within 7 days of birth. Wilson *et al.* (1967) noticed the harmful effects on subsequent litters when feeding *F. roseum* Ohio strain B and C. No estrogenic effects were observed, but the sows produced weaker pigs and more mummified piglets. The toxin responsible was not identified. Similar observations were made by Sharma *et al.* (1974) who found that feeding corn infected naturally with *F. roseum* caused a reduction in the litter size and increased fetal mummification. There were no effects on the estrus cycle or weight of the live piglets. Corn artificially inoculated with *F. roseum* Ohio B and C also caused the birth of more dead and weak piglets. These researchers suggested that, although *F. roseum* has been known to produce estrogenic effects (Christensen and Kaufman, 1969; Christensen *et al.*, 1972b), the strains used in their experiments failed to do so, either due to differences in cultural conditions or strains. No analysis was made of the moldy corn so that the toxins responsible for these effects are not known. It would appear, however, that cultural conditions can influence the syndromes produced by toxic *Fusarium* species.

## Trichothecene-Induced Toxicoses in Sows

In addition to the food refusal and vomiting syndromes trichothecene toxins have also been shown to cause abortions and other reproductive disturbances in sows. *F. roseum* "Gibbosum" (*F. equiseti*) was isolated from feed implicated in cases of abortion and death in swine and cattle (Pathre *et al.*, 1976). Rice and corn artificially contaminated with this *Fusarium* were lethal to pigs and other test animals, the toxicity being similar to that of diacetoxyscirpenol (in rats). Corn cultures were shown to contain monoacetoxyscirpenol as the main toxin, although scirpentriol and traces of zearalenone were present. Diacetoxyscirpenol has been shown to be acutely toxic to pigs when administered intravenously at doses in the area of 0.4 mg/kg (LD$_{50}$ single dose—0.376 ± 0.043 mg/kg) (Weaver *et al.*, 1978d). Clinical signs became evident 20 min after injection. The pigs began to vomit and defecate frequently but still ate eagerly. However, after 60 min the pigs became lethargic and the first two symptoms continued. Modified posterior paresis developed after 6 hr and the swine were unsteady and very lethargic. After 18 hr the animals were prostrate and several died. Those that survived returned to apparent normalcy by the eighth day. Hemorrhagic bowel lesions were only observed in two of the pigs that died, although typical radiomimetic necrosis was observed in the germinal centers of the mesenteric lymph nodes and spleen of all pigs killed by the diacetoxyscirpenol injection.

Weaver *et al.* (1981) noted that, after feeding diacetoxyscirpenol to growing pigs at levels of 2, 4, 8, and 9 ppm over a period of 9 weeks, gingival, buccal, lingual lesions and a decrease in ration consumption and weight gain occurred. There was no hemorrhage or congestion in the intestinal tract and no changes in the blood count, but in a few swine acute diffusive congestion in the spleen and hyperplasia of the glandular and mucosal epithelial cell of the small intestine were observed.

The investigations of chronic toxicity of the diacetoxyscirpenol in growing pigs showed that 10 ppm caused total ration refusal. The authors stated that diacetoxyscirpenol was more potent than T-2 toxin in both acute and chronic toxicoses in growing pigs.

According to Patterson *et al.* (1979) the toxicity of pure T-2 toxin included in diets fed to pigs and calves was similar to that of toxins obtained from cultures of *Fusarium* or their extracts. The T-2 toxin was found to cause no hemorrhagic syndrome in either pigs or calves.

T-2 toxin has been shown by Weaver *et al.* (1978b) to be highly toxic to pigs, and it has been suggested that this toxin may be responsible for outbreaks of abortion and other reproductive disorders in sows. Abortion has been induced by i.v. injection of 0.41 mg/kg in two pregnant sows and of 0.21 mg/kg T-2 toxin in the third pregnant sow.

All the sows aborted, the time taken depending upon the dose (48 hr for the higher dose and 80 for the lower). The sows appeared to be clinically normal after the abortions and no internal lesions were detected.

Weaver *et al.* (1978c) studied the effect of T-2 toxin on fertility, gestation, and litter size in sows. After being fed 12 ppm T-2 toxin for 220 days the sows exhibited both reproductive and fertility abnormalities; piglets were smaller and litter size reduced.

In one sow severe liver congestion of the gall bladder mucosa was observed, while two others suffered from hyperemia of the intestinal mucosa with no effect on blood count.

Swine given dietary T-2 toxin at 24 and 32 ppm completely refused the diet. Pigs refused to eat rations containing high concentrations of T-2 toxin, and acute T-2 toxicity was therefore induced by i.v. injection of 0.13–3.2 mg/kg. Emesis, posterior paresis, lethargy, and extreme hunger were noted (Weaver *et al.*, 1978a). There was no effect on the blood chemistry or the bone marrow. Acute necrosis of the mucosal epithelial cells and the crypt cells of the jejunum and ileum were noted, in addition to severe congestion of the jejunal and ileal mucosae. Acute necrosis occurred in the lymphoid cells of the cecum, lymphoidal follicles in the spleen, and in the germinal centers of the mesenteric lymph nodes. The $LD_{50}$ was estimated as 1.21 mg/kg. These researchers showed that in studies of chronic T-2 toxicity no gross or microscopic lesions occurred and the blood and bone marrow were unaffected when pigs were fed 1–8 ppm crystalline T-2 toxin in their ration. No statistically significant differences were observed in feed consumption or weight gain, although both were lower in the animals receiving T-2 toxin. The pigs refused 16 ppm T-2 toxin in the food but tolerated 10–12 ppm. The toxin can be transferred to the sow's milk, and the toxin was detected in the milk of a sow fed 12 ppm toxin for 220 days (Robison et al., 1979b).

Robison et al. (1979a) studied the distribution and determination of levels of tritium-labeled T-2 toxin in organs, tissues, and excreta in two weanling crossbred young swine. After intubation in one swine with 0.1 mg of [$^3$H] T-2 toxin/kg body weight, the following residue levels of administered radioactivity were isolated: in muscle, 3.1 ppb; in fat, 0.49 ppb; in liver, 13.8 ppb; and in kidneys, 15.9 ppb. In the second swine, after intubation with 0.4 mg of [$^3$H] T-2/kg body weight, 11.5 ppb was found in muscle, 37.7 ppb in liver, and 61.4 ppb in kidneys. The kidneys of swine had a higher specific activity than the liver, in comparison with results in chickens obtained by Chi et al. (1978c). Close to 50% of the radioactivity was recovered in the organs and tissues of weanling swing.

Jemmali (1983) reported the trichothecene and zearalenone toxicity in France and their effects on the sexual cycle in gilts and immunological disorders in mice.

## FUSARIOTOXICOSES IN SHEEP AND GOATS

There are very few recorded cases of naturally occurring fusariotoxicoses in sheep, and few researchers have used them as experimental animals. In the Kalmyk Republic winter-pastured sheep developed a fusariotoxicosis characterized by gastrointestinal disorders and malfunctioning of the central nervous and cardiovascular systems. A loss of wool and abortions were also noted. Fusarium species isolated from grass and hay and diminished by moist and freezing weather were thought to be the cause of this outbreak (Kalmykov et al., 1967). Noskov and Kirmanov (1967) also mention the same outbreak and noted that the sheep showed decreased pain and tactile sensitivity and had enlarged livers, flaccid kidneys, swollen mesenteric glands, and often decreased spleens. Danko and Toth (1969) also mention fusariotoxicosis in sheep and noted that F. graminearum was probably responsible for this outbreak of the disease. Sheep feeding on hay infected with F. tricinctum are reported to have suffered from gastric symptoms (Korpinen and Ylimäki, 1972).

Very little experimental work has been carried out with sheep. While investigating the causes of fescue-foot in cattle, Keyl et al. (1967) found that a mixture of Cladosporium cladosporioides and a Fusarium, later identified as F. tricinctum NRRL 3249 (F. sporotrichioides), caused ruminal atony in a mature ewe when administered via a rumen fistula. Anorexia and a failure to drink were also noted. The sheep also exhibited lipidosis and central necrosis of lobules in the liver, enlarged bile ducts, uterine stroma with cellular muscle tissue, and atelectasis of the lung.

Dzhilavian (1972) and Dzhilavian and Spesivtseva (1960) induced fusariotoxicosis in sheep by feeding a single dose of 10 g/kg *F. sporo-trichiella* var. *tricinctum* (*F. sporotrichioides* var. *tricinctum*). Kur-manov (1978b) observed hemorrhages in sheep which were fed cultures of *F. sporotrichioides*. He found that *F. sporotrichioides* administered in the feed at an equivalent of 500 mg/kg live weight had no adverse effect on sheep or lambs. However, increasing the dose to an equivalent of 1 g/kg led to food refusal by the sheep after the first day, with 30% of the lambs even refusing the feed on the first day. In subsequent experiments the fungus was introduced by stomach tube. As with cattle, the pH of the contaminated feed had a great effect on the severity of the fusariotox-icosis in sheep. Both sheep and lambs were highly sensitive to sub-cutaneous administration of extracts of *F. sporotrichioides*. Symptoms appeared 1 hr after dosing with 0.05 mg/kg of the extract. The symptoms included depression and retarded responses to external stimuli, tachy-cardia, heavy mucous outflow from the nasal cavity, and diuresis. A second injection aggravated these symptoms and the animals became severely depressed, failed to rise, and, finally, became comatose. Ataxia and paralysis of the extremities were also noted. In acute fusariotoxicosis breathing often became uneven, and gastrointestinal disorders were fre-quently noted. The subacute form of the disease involved the same symptoms but in milder form. Chronic fusariotoxicosis caused weakness, depression, anorexia, progressive exhaustion and loss of mobility, in addition to edema or fission of the mouth and nose. Kurmanov (1978b) reports that sheep suffering from acute fusariotoxicosis respond to i.v. injections of a solution containing $CaCl_2$, ascorbic acid and glucose. This injection led to an improved overall condition and recovery after 2 to 3 injections. After treatment with $CaCl_2$ the sheep improved, their appe-tite was restored and the motor function of the rumen was also restored. The number of leukocytes was decreased, and the electrocardial indi-cators returned to their initial values.

Ogurcov (1960) also recommended a treatment to cure fusariotoxicosis in cattle by using i.v. injection of $CaCl_2$ (10% solution in 100–200 ml) and also an s.c. injection of caffeine.

The only other effect of *Fusarium* toxins on sheep to have been inves-tigated was the influence of the growth-promoting derivative of zearalenone: zeranol. Stob *et al.* (1962) noted improved growth rates and feeding efficiency in sheep treated with partially purified concentrates of an anabolic agent they isolated from maize infected with *G. zeae*. Brown (1970) found that 12 mg zeranol implanted into suckling and feedlot lambs improved growth and feed conversion. No adverse effects were noted. However, similar quantities implanted into male lambs were shown to cause a reduction in testis weight, which was probably due to the inhibition of gonadotropin synthesis (Reisen *et al.*, 1977). In castrated lambs 12–24 mg implants of zeranol caused mild (and 96 mg caused severe) hyperplasia and transitional and squamous transformation. Increased zeranol increased the adrenal gland weight and the adrenal gland:thyroid ratio. At 96 mg muscle and ligament structure appeared to be looser, and fat cell formation increased in the muscle along ligament muscle joints (Rothenbacker *et al.*, 1975).

Uraguchi (1971) reported convulsions and abortion in a pregnant goat fed an extract of wheat infected with *F. graminearum*. Vanyi *et al.* (1973a) reported a case of a ewe-lamb receiving 250 mg F-2 during a 7-day period and exhibiting slight swelling and reddening of the vulva and hypertrophy of the uterus. *F. nivale* fed to a goat at a rate of 2–5 mg/kg three times a week caused cell proliferation in the central nervous system and interfered with the spermatogenesis. One or two doses of 10–40 mg/kg caused death 10–60 hours after administration. Topical application resulted in a toxic skin reaction and caused skin and mucosal membrane ulceration (Kubota, 1977).

Friend *et al.* (1983b) studied the immunological and pathological effects in sheep treated daily for 21 days with an oral dose of 0.6 or 0.3 mg/kg b.w. T-2 toxin. The authors stated that lambs treated with 0.6 mg/kg b.w. of T-2 toxin suffered from leukopenia, lymphopenia, and lymphoid depletion in ruminants.

## FUSARIOTOXICOSES IN HORSES

The fusariotoxicoses most extensively recorded in horses are the diseases leukoencephalomalacia, which has been shown to be caused by maize infected with *F. moniliforme* (Badiali *et al.*, 1968; Kellerman *et al.*, 1972; Marasas *et al.*, 1977; Wilson, 1971; Wilson and Maranpot, 1971; Wilson *et al.*, 1973), and the so-called bean-hull poisoning attributed to *F. solani* (Ishii *et al.*, 1971; Ueno *et al.*, 1972a). However, numerous other toxicoses have been attributed to *Fusarium* species.

Oats and barley grain infected with *G. saubinetii* have often proved to be unpalatable to horses, who refuse to eat them (Ito, 1912a,b; Raino, 1932; Roche and Bohstedt, 1930), yet this species has been held responsible for "taumelgetreide" toxicoses in horses (as well as in other domestic animals and in humans) in Siberia (Naumov, 1916; Woronin, 1891). Tsunoda *et al.* (1968) and Tsunoda (1970) also reported cases of poisoned horses in Japan.

Red mold toxicoses attributed to *F. graminearum* have been observed to affect horses, the symptoms including hemorrhaging and food refusal (Ueno, 1977b). Mayer (1953b) found *F. tricinctum* to be responsible for an outbreak of toxicosis in horses, and Spesivtseva (1967) reported that *F. sporotrichiella* var. *poae* (*F. poae*) caused a toxicosis fatal to horses. Oats infected with *F. poae* or *F. moniliforme* caused colic and gastric symptoms in horses (Korpinen and Ylimäki, 1972), and sugar beans infected with *F. moniliforme* were associated with a neurotic syndrome and liver damage in horses (Marasas *et al.*, 1976; Van der Walt and Steyn, 1943, 1945). Bridges (1978) summarized the various syndromes caused by the different *Fusarium* species: *F. moniliforme* is associated with the leukoencephalomalacia syndrome, *F. solani* with neurological and cardiovascular disturbances, and *F. tricinctum* with stomatitis, ulcers, and edema of the lips and nares, with the toxicosis often proving fatal.

## Leukoencephalomalacia in Horses and Other Equidae

This neurotoxic syndrome has been observed in several countries for many years (Kellerman *et al.*, 1972) but was only associated with *F. moniliforme* in 1968 (Badiali *et al.*, 1968) when this fungus was isolated from moldy maize that had caused brain lesions in 11 out of the 15 donkeys that died as a result of eating the moldy maize. Wilson and Maronpot (1971) noted that the characteristic histological feature of the disease is the focal necrosis of the white matter of the cerebral hemispheres. Lesions of the visceral organs were also noted on occasion. Clinically the animals suffered first from a loss of appetite soon followed by nervous disorders. Often the animals became drowsy, unsteady, and tended to walk in aimless circles (Wilson *et al.*, 1973). Ataxia, paresis, and hypersensitivity have also been noted (Marasas *et al.*, 1976).

Cultures of *F. moniliforme* isolated from maize suspected of causing leukoencephalomalacia in South African horses were used in an attempt to reproduce the symptoms of the disease experimentally in horses and donkeys (Kellerman *et al.*, 1972). One horse and one donkey remained completely healthy. Two horses and one donkey died, but pathological examination showed that, although there were severe cardiac hemorrhages, petechiae and ecchymoses in various organs, edema, icterus, and liver damage, no brain lesions were observed apart from some small perivascular hemorrhages. In fact the symptoms were similar to those noted by Van der Walt and Steyn (1943, 1945). Further work with the same culture medium (Marasas *et al.*, 1976) succeeded in reproducing the typical clinical and pathological leukoencephalomalacia and hepatic symptoms. One horse suffered only from the leukoencephalomalacia syndrome, a second horse had the hepatic symptoms only, and a third suffered from both syndromes. This showed that either (or both) syndrome could be caused by the same fungus, and it was suggested that leukoencephalomalacia may be specifically induced by the prolonged administration of smaller doses of the toxic material. Thus doses equivalent to 0.67–1.94 kg culture material per day for 11–12 days produced the fatal hepatic syndrome (Kellerman *et al.*, 1972), whereas the LEM symptoms were caused by the equivalent of 0.33–0.44 kg/day for 90–144 days (Marasas *et al.*, 1976). However, these latter researchers pointed out that these results must be interpreted with caution, since the number of experimental animals was small and many other factors, such as age, concentration of culture material, and so on, had not been taken into consideration. Wilson (1971) noted that age may be of importance and that, at least with donkeys, older animals appear to be more susceptible to the disease.

Pienaar et al. (1981) noted that maize infected with F. verticillioides (F. moniliforme) caused field outbreaks of leukoencephalomalacia in horses.

## Bean-Hull Poisoning

Asami (1932), Oya and Morimoto (1942), and Ueno et al. (1972a) reported cases of poisoned horses associated with bean-hulls.

This disease has been recorded for many years in the horse-breeding regions of Hokkaido in Northern Japan (Ueno et al., 1972a). Bean-hulls used as fodder have been thought to be the cause of various outbreaks, including cases of abortion and still-births in pregnant mares. The most prevalent symptoms noted, however, were hepatic jaundice, impaired liver function, and disorders of the central nervous and circulatory systems. At the onset of poisoning the heart rate increased and the animals became excited and showed signs of motor irritation and clonic muscle spasms. This was followed by involuntary circular movements. Senses became dull and reflexes in skin, eyelids, cornea, and pupil diminished.

Several fungi were isolated from the bean-hulls, including F. roseum and F. solani (Ishii et al., 1971). T-2 toxin, diacetoxyscirpenol, and solaniol were all isolated from cultures of F. solani (M-1-1), but the exact role of these toxins in the horse poisoning was not determined (F. solani M-1-1 was reidentified as F. sporotrichioides by Ichinoe, 1978).

Detailed studies were carried out on two horses by the author in cooperation with Antonov, Belkin, Lukin, and Simonov (Antonov, 1951) with F. poae No. 60/9. Two series of sucrose-glucose-peptone agar medium were used for inoculation and incubation of F. poae isolated from overwintered millet grain associated with ATA in humans. The first series was grown 10 days at room temperature and then 20 days at +1°C. The second series was cultivated at low temperatures, two days at room temperature, and 45 days at low temperature, accompanied by multiple freezing and subsequent thawing.

The horse in the first series received 240 g dry culture during 23 days and the horse in the second series only a single dose of 40 g of fungus by nasopharyngeal intubation. The latter produced an acute toxicosis, the horse dying 36 hr after administration of the Fusarium culture.

The degree of toxicity depends on the environmental conditions of incubation and growth of the Fusarium culture, on size of dose, and also on the sensibility of the organism. In the previously mentioned conditions of growth F. poae No. 60/9 had a harmful effect, causing rapid death.

Clinical symptoms showed loss of appetite, severe depression, and disturbance in movement coordination. Pathological findings revealed hemorrhagic diathesis, hemorrhages of the gastrointestinal tract, liver, and kidney tissues, hyperemia and necrosis of the mucosa of small and large intestines, and a process of severe degeneration of the parenchymatous organs. Blood examination showed a slight increase in the quantity of leukocytes and lymphocytes and a drastic increase in the amount of erythrocytes and the level of hemoglobin (almost twice).

Blood examination of the horse in the first series showed reduction in white and red cell counts.

Sarkisov *et al.* (1948c) fed a horse 16.4 kg cereals infected with *F. sporotrichioides*, causing stomatitis and gingivitis only. Horses fed on herbage or whole oat grains which were toxic showed the following symptoms, according to Maisuradge (1960): (1) two days after ingestion of toxic food, hyperemia was already evident in the mouth with swelling and splitting of lips; (2) after 7 days a foul smelling slough appeared in the oral mucosa; (3) deformation of the horses' heads was observed and the animals lost weight because they refused to eat the toxic food after a while. Although one horse ate 23 kg of toxic food and another 18 kg, they recovered slowly and after a very long period returned to normal health.

Lukin and Berlin (1947) and Lukin *et al.* (1947a,b) carried out feeding tests on horses, pigs, and sheep and found that animals fed on large quantities of overwintered toxic cereals, even for as long as 39 days, did not develop ATA symptoms. They concluded that climatic conditions affect toxin production differently in different geographic regions.

## CONCLUSION

From the chemical and biological point of view, interest in fusariotoxins has concentrated upon their ability to produce mainly trichothecene metabolites and zearalenone which have caused a variety of disorders in domestic animals and humans.

Results obtained in our laboratory and in cooperation with other researchers, institutions, and laboratories in Israel, the USSR, the United States, England, and South Africa on cats, primates, mice, rats, guinea pigs, rabbits, dogs, frogs, chicks, ducklings, pigs, cattle, and horses indicate that T-2 toxin or crude extract containing T-2 toxin produces a lethal intoxication in these animals with significant clinical and histopathological syndromes.

The metabolite of diacetoxyscirpenol is also a very strong toxin. The severity of the lesions caused by diacetoxyscirpenol, for example, compared to T-2 toxin was even greater in chicks. Diacetoxyscirpenol also caused refusal of feed and severe oral lesions when fed to swine.

The high levels of vomitoxin in U.S. Midwest corn samples infected by *F. graminearum* demonstrated the importance of this toxin (Vesonder, 1973, 1979b). Vomitoxin (deoxynivalenol) and T-2 toxin are trichothecene mycotoxins which induced hemorrhages in animals (Ellison and Kotsonis, 1973); nervous system disturbances in broilers (Wyatt, 1973c); emesis, feed refusal, and decrease of weight gains in swine (Forsyth *et al.*, 1977; Schuh *et al.*, 1982; Vesonder *et al.*, 1979a, 1981); and irritation of the gastrointestinal tract in chickens (Huff *et al.*, 1981). Vomitoxin occurred in corn and cereal grains in the field in temperate climatic countries with high moisture conditions, and this indicates a problem of worldwide importance (Scott *et al.*, 1981; Sutton, 1982; Trenholm *et al.*, 1981; Neish, 1983; Vesonder *et al.*, 1982).

Zearalenone is an estrogenic mycotoxin produced by several *Fusarium*-contaminated cereals. Under certain conditions, characterized by higher moisture and alternate low temperatures, maize contaminated mainly by *F. graminearum* (*G. zeae*) produced zearalenone, a potent estrogenic toxin. This metabolite caused estrogenic syndrome (hyperestrogenism) when fed to swine, mainly in the genital system, and also emesis, refusal of feed, and diarrhea.

Zearalenone is most commonly found in maize but was also isolated from barley, oats, and mixed feeds in the United States, Japan, Europe, and other regions of the world.

Methods for analysis of zearalenone were described by various researchers. Ueno *et al.* (1977a) studied the mode of action and the fate of zearalenone in mice and rats. They stated that [$^3$H]-zearalenone was excreted mostly via feces as free zearalenone and zearalenol.

Among the zearalenone derivatives isolated from *F. graminearum* only zearalenol has been shown to be estrogenic. The presence of zearalenol in cultures is very important because it points out the possible occurrence of this compound in animal diets.

*Fusarium* trichothecene toxins, T-2 toxin, diacetoxyscirpenol, deoxynivalenol, and zearalenone are considered to be potential health hazards for farm animals and humans.

# CHAPTER 10

## TAXONOMIC PROBLEMS OF THE GENUS *FUSARIUM*

Modern taxonomy of the *Fusarium* species and varieties is based on the experimental variability of the studied isolates obtained from various substrates under various environmental conditions, as well as on morphological, cultural, and physiological properties.

### THE PRESENT STATE OF *FUSARIUM* TAXONOMY

The basic principles of the modern classification system were laid down by Appel and Wollenweber (1910) in their contribution "Grundlagen einer Monographie der Gattung *Fusarium* (Link)." Later Wollenweber and Reinking (1925, 1935a) published their taxonomy system of the *Fusarium*.

Studies on the genus *Fusarium* were also carried out by Brown (1928); Brown and Horne (1926); Coons and Strong (1931); Leonian (1929); and Mitter (1929). Coons (1928) and Coons and Strong (1928), and later Hornok (1978, 1980) and Hornok and Oras (1976) undertook an attempt to diagnose some *Fusarium* species by serological tests, but these studies found no suitable use in practice.

There are at present two systems of nomenclature and taxonomy for the identification of *Fusarium* species: first, the detailed taxonomy system of Wollenweber and Reinking (1935a), and second, the extremely simplified system of Snyder and Hansen (1940, 1941, 1945) and Snyder and Toussoun (1965).

The detailed comprehensive taxonomy system of Wollenweber (1913, 1916–1935, 1917, 1930, 1932, 1944–1945) and Wollenweber and Rein-

**386**

king (1925, 1935a, 1935b) was based on careful studies of herbarium samples and, mainly, of pure cultures isolated from various substrates. Wollenweber and Reinking (1935a) included 16 sections with 65 species, 55 varieties, and 22 forms in the genus *Fusarium,* associated with certain conidial characteristics (size, shape, length/width ratio, number of septa, structure of top and foot cells), and certain cultural features (nature of aerial mycelium and the stroma, presence of pigmentation, color of spores in mass, presence or absence of chlamydospores, sporodochia, pionnotes, and sclerotia). These authors were the first to give an orderly system to the innumerable synonyms of species and varieties of the genus *Fusarium* (Wollenweber, 1931, 1944–1945) and Wollenweber and Reinking, 1935a). Mycologists Bugnicourt (1939), Doidge (1938), Gerlach (1961, 1970, 1977, 1981), Gerlach and Ershad (1970), Gerlach and Nirenberg (1982), Jamalainen (1943a, 1943b, 1944), and Subramanian (1952, 1954, 1955) completely accepted the Wollenweber and Reinking system (1935a) because it "has been more practical and it has been possible for the research workers to know more exactly which species or subspecies is indicated" (Jamalainen, 1970). Many other mycologists and plant pathologists (Bilai, 1955, 1977; Booth, 1966a,b, 1971a,b, 1973, 1975, 1977, 1981; Carrera, 1954; Gordon, 1952; Hornok, 1975; Joffe, 1974a, 1977; Miller, 1946; Padwick, 1940; and Raillo, 1950) have proposed limited changes, the acceptability of these changes depending very much on the personal predilection or opinion of the researcher. Thus Raillo (1935, 1936, 1950) divided her taxonomy system into 17 sections, 12 subsections, 55 species, 10 subspecies, 55 varieties, and 61 forms. She took into consideration the shape of macroconidia (the shape of the apical and foot cells), the size of conidia, septation, presence or absence of microconidia, chlamydospores, sclerotia, and pigmentation of cultures.

Bilai (1952a,b,c, 1955, 1977, 1978a,b) studied the variability of morphological, cultural, and physiological characters, and the effects of temperature, moisture, and nutrient media on the growth of different *Fusarium* species. She included only 9 sections, 31 species, and 29 varieties in her system (Bilai, 1977).

Booth's (1971b) classification was based on Wollenweber and Reinking (1935a) and Gordon (1952, 1960b). He and Goos (1981) stated that the morphology of the conidiophores, and the structure of the phyalidic or polyblastic sporogenous cells that produced micro- and macroconidia, are the most important characteristics for the identification of *Fusarium* species. On this basis, for example, Booth divided the Sporotrichiella section to include only two species, *F. tricinctum* and *F. poae.* Two other species, *F. sporotrichioides* and *F. fusarioides* (*F. sporotrichioides* var. *chlamydosporum*), with polyblastic conidigenous cells, were included in the Arthrosporiella section. Booth described 44 species, 7 varieties, and 99 forms belonging to 12 sections of *Fusarium,* using the

morphology of sporogenous cells, the size of conidia and chlamydospores, their presence or absence, pigmentation, growth rate, and the
source and substrate.

Gerlach (1961, 1970, 1977, 1981) and Gerlach and Ershad (1970)
based their classification of the genus *Fusarium* solely on Wollenweber
and Reinking (1935a). Gerlach (1970) and Gerlach and Nirenberg (1982)
included about 65–70 species and 55 varieties belonging to 16 sections
in his modern *Fusarium* system.

Gordon (1935, 1939, 1944, 1952, 1954a,b, 1956a,b, 1959, 1960a,b,c,
1965) and Gordon and Sprague (1941) studied the occurrence of
*Fusarium* in Canada on various cereals and seeds of vegetables, forage,
miscellaneous crops, and soil samples. At first Gordon agreed with the
Snyder and Hansen (1940, 1941) classification concerning the species of
*F. oxysporum* and *F. solani,* but later (Gordon, 1952, 1956a,b, 1959,
1960a,b) he accepted the Wollenweber and Reinking (1935a) taxonomy
system and also recognized the Booth (1959, 1960) studies of Pyrenomycetes. Gordon described 25 species, 5 varieties, and 69 forms, including
66 forms of *F. oxysporum* (Gordon, 1965) in 14 sections in his taxonomy
system.

A second, very much simplified, system of *Fusarium* taxonomy has
been presented by Snyder and Hansen (1940, 1941, 1945) and their
school (Snyder and Toussoun, 1965; Snyder *et al.*, 1975; and Toussoun
and Nelson, 1968, 1975, 1976). This system is based on the ascomycetoid
stage, and on the shape of the macroconidia and presence or absence of
chlamydospores or microconidia.

Snyder and Hansen (1940, 1941, 1945) revised the taxonomy system of
Wollenweber and Reinking (1935a) and created a new concept in
*Fusarium* as a result of their studies of the variability exhibited in
*Fusarium* species (Snyder and Hansen, 1954; Snyder *et al.*, 1957). They
included only nine species in their taxonomy system. The authors alleged that Wollenweber and Reinking (1935a) accepted the principle
that fusaria are variable, but that this was because they did not use the
single spore culture method.

Prisiashnyuk (1932) and Raillo (1935, 1936) used the single spore
culture method as early as 1930, well before Snyder and Hansen's publication. They were studying the *Fusarium* disease of cereal crops (Prisiashnyuk, 1932) and the variability of separate conidia of varous fusaria
within the single spore culture (Raillo, 1935, 1936). They used an original method of variational statistics for conidial dimension and their variability in cultures, which revealed differences in the morphological and
cultural characters of *Fusarium* species, varieties, and forms. The taxonomy system of Raillo (1935, 1936) and her comprehensive contribution
to *Fungi of Genus Fusarium* (Raillo, 1950), published posthumously,
were based on the shape of macroconidia, the presence or absence of
chlamydospores or microconidia, and other characteristics, using the
same principles that Snyder and Hansen (1940, 1941, 1945) used later for

their new experimental concept of variability in *Fusarium* cultures. The premature death (1939) of Raillo prevented her from completing this work and her critical revision of the Wollenweber and Reinking (1935a) classification of "Die Fusarien," but she did not basically change the Wollenweber and Reinking taxonomy system, as did Snyder and Hansen (1945) and Snyder and Toussoun (1965).

Snyder and Hansen (1945) based their work on the indisputable fact that many of the *Fusarium* species possess so wide a morphological variability that distinctions between them become obscure. However, sweeping changes in *Fusarium* taxonomy, meant to make the life of plant pathologists easier, have been put forward by Snyder and Hansen (1940, 1941, 1945) and their collaborators (Snyder and Toussoun, 1965; Toussoun and Nelson, 1968, 1975, 1976). They reduced the number of all species in the genus *Fusarium* and concluded that only nine species can be justified. These authors claim that the natural variability of *Fusarium* isolates renders distinction of more numerous species unreliable, and that within the species only *formae speciales* should be distinguished according to their pathogenicity to crop plants. This reduction is exaggerated. Many mycologists and taxonomists, such as Bilai (1955), Booth (1971b, 1975), Gerlach (1970), Gerlach and Ershad (1970), Gordon (1952), Joffe (1974a, 1977), and others, studied experimentally the extent of morphological variability of isolated *Fusarium* cultures. All these workers used the single-spore culture method to examine *Fusarium* isolated from natural substrates or from subsequent cultures under a variety of environmental conditions. None of them reached the results and conclusions of Snyder and Hansen (1945), Snyder and Toussoun (1965), and Toussoun and Nelson (1968, 1975, 1976). According to their taxonomy, the only basis for the identification, separation, and delineation of *Fusarium* species is morphological characteristics, in particular the shape of macroconidia. Snyder and Hansen (1945) then combined four sections (Roseum, Arthrosporiella, Gibbosum, and Discolor, which have different morphological, cultural, and physiological characteristics) into a single species, *F. roseum*. There was no justification for including more than 50 different species, varieties, and forms, with such divergent morphological characteristics, into one species. After the publication of the species concept in *Fusarium* (Snyder and Hansen, 1945 and Snyder and Toussoun, 1965), the reaction of many mycologists and taxonomists all over the world, with the exception of some in France (Messiaen, 1959; Messiaen and Cassini, 1968) and Japan (Matuo, 1961, 1972; Matsuo, 1983), was critical. Booth (1975) stated that "there is no justification for lumping all these into one species, as did Snyder and Hansen."

Matsuo (1983) proposed a taxonomic system based on that of Snyder and Hansen (1945); however, where the latter proposed nine species, Matsuo proposed 10, the additional species being *F. splendens* Matsuo et Kaboyashi.

The illustrations of *F. roseum* species given in the pictorial guide for the identification of *Fusarium* species (Toussoun and Nelson, 1968) themselves contradict the statement that the shape of macroconidia is the major reliable and consistent criterion (Snyder and Hansen, 1945; Toussoun and Nelson, 1975). It is clear that this guide bewildered many mycologists and taxonomists. In the second issue (Toussoun and Nelson, 1976) the authors introduced corrections, connecting *F. roseum* with cultivars "Graminearum," "Culmorum," and so on (Snyder *et al.*, 1957). The term "cultivar," which the authors used, usually designates a horticulturally or agriculturally derived variety of plant and is not a suitable term to apply to fungi. The cultivars philosophy, which, according to Snyder *et al.* (1957), indicated the pathogenicity to cereals by combining various *Fusarium* species with *F. roseum*, for example, *F. roseum* f. sp. *cereale* "Graminearum," was incorrect and has nothing in common with the International Code of Botanical Nomenclature and taxonomy of the *Fusarium* species (Gerlach and Ershad, 1970). According to Booth (1975), "this rather cumbersome four-stage nomenclature has the added misleading implication that f. sp. *cereale* is specific to cereals, which is not necessarily correct."

The generally accepted modern taxonomy system should be based on morphological, cultural, and physiological (phytotoxic and toxigenic) characteristics. It is very important to classify and define precisely the *Fusarium* species that affect foods and feeds and produce toxic secondary metabolites that cause toxic syndromes known as mycotoxicoses (or fusariotoxicoses) in animals and humans. It is essential, therefore, to work out a uniform, internationally accepted taxonomy system for the identification and determination of the *Fusarium* species involved in the production of mycotoxins (including trichothecenes, zearalenone, etc.) (Kriek and Marasas, 1983; Smalley *et al.*, 1977; Ueno, 1977a). Such a uniform taxonomy system, available for biological, toxicological, chemical, and pharmacological investigations, can only be created by mycologists or mycotoxicologists who have been engaged in the classification of *Fusarium* species and who have worked with toxins (extracts) produced by these fungi.

A toxicologist, chemist, or other scientist who wishes to identify a species of *Fusarium* for toxicological purposes may have difficulty utilizing the Snyder and Hansen (1945) system since it is oversimplified. It was for this reason that I decided to create the comprehensive system of toxic *Fusarium* species and varieties presented in this chapter. I am also interested in investigating the causal factors involved in the development of trichothecene toxicoses (fusariotoxicoses) and other mycotoxicoses as associated with ecological and environmental conditions in temperate, subtropical, and tropical regions.

I criticize the Snyder and Hansen (1945) classification of *Fusarium* chiefly with regard to the distinctions made between toxigenic *Fusarium* belonging to various sections and species, the most important being the Sporotrichiella and Discolor sections.

**TABLE 10.1**  Reidentification of Sporotrichiella Strains of *Fusarium* Used in Recent Toxicological Studies in the United States

| No. and Designation of Strain | | Supplied By | Country of Origin | Reidentified As |
|---|---|---|---|---|
| NRRL 3249 | *F. tricinctum* | C. W. Hesseltine | USA | *F. sporotrichioides* |
| NRRL 5508 | *F. tricinctum* | C. W. Hesseltine | USA | *F. sporotrichioides* |
| NRRL 5509 | *F. tricinctum* | C. W. Hesseltine | USA | *F. sporotrichioides* var. *tricinctum* |
| NRRL 3299 | *F. tricinctum* | C. W. Hesseltine | USA | *F. poae* |
| NRRL 3287 | *F. poae* | C. W. Hesseltine | USA | *F. poae* |
| 2061-C | *F. tricinctum* | C. J. Mirocha | USA | *F. sporotrichioides* |
| YN-13 | *F. tricinctum* | C. J. Mirocha | USA | *F. sporotrichioides* |
| T-2 | *F. tricinctum* | W. F. O. Marasas | France | *F. poae* |

*Source:*  Adapted from Joffe and Palti, 1975.

The taxonomy of the species of the Sporotrichiella section is a matter of some dispute. Snyder and Hansen (1945) grouped various species in this section under one species, *F. tricinctum*. We have studied critically eight strains of *F. tricinctum* generally used in the United States (kindly supplied by the colleagues listed in Table 10.1) and reidentified them according to our taxonomic system (Joffe, 1974a, 1977; Joffe and Palti, 1975). This important group of *Fusarium*, which produces many toxic metabolites, has received a great deal of attention from research scientists because of its toxigenic potential and its capacity to form toxins dangerous to humans and animals.

We have therefore presented in Table 10.2 the classification of this section published by the following researchers: Bilai (1955, 1970a,b, 1977); Booth (1971a,b); Gordon (1952); Jamalainen (1943b); Joffe (1974a, 1977); Joffe and Palti (1975); Raillo (1950); Seemüller (1968); Snyder and Hansen (1945); Wollenweber and Reinking (1935a). The morphological, cultural, and toxicological characteristics of the species and varieties in the Sporotrichiella section, as established by the author, are given elsewhere (Joffe 1978b). The most prominent of these are *F. poae* and *F. sporotrichioides*, which grew on overwintered cereals in the USSR and caused the fatal disease of ATA in humans (Joffe, 1947a,b,c,d, 1950, 1956a,b, 1960a,b, 1962a, 1963a, 1965, 1969a, 1971, 1973b, 1974c, 1978b 1983b), and *F. tricinctum* (*F. sporotrichioides*, *F. poae*, *F. sporotrichioides* var. *tricinctum*, and *F. sporotrichioides* var. *chlamydosporum*), which appears in the United States on moldy corn, wheat, fescue hay, and other substrates (Bamburg and Strong, 1971; Bamburg *et al.*, 1968a,b, 1969; Booth and Morgan-Jones, 1977; Davis and Diener, 1979; Diener, 1976; Gilgan *et al.*, 1966; Hesseltine, 1974; Kosuri *et al.*, 1971; Kotsonis and Ellison, 1975; Kotsonis *et al.*, 1975a; Lansden *et al.*, 1978; Mirocha *et al.*, 1976a; Smalley *et al.*, 1970; Stahr, 1975, 1976; Stahr

**TABLE 10.2   Classification of Sporotrichiella Fusaria**

| Wollenweber and Reinking (1935a) | Jamalainen (1943b) | Snyder and Hansen (1945) | Raillo (1950) | Gordon (1952) | Bilai (1955, 1970a, 1977) | Seemüller (1968) | Joffe (1974a, 1977) | Booth[a] (1971b, 1975) |
|---|---|---|---|---|---|---|---|---|
| F. poae | F. poae F. citriforme | — | F. poae | F. poae | F. sarcochroum F. sporotrichiella var. poae | F. poae | F. poae | F. poae |
| F. sporotrichioides | F. sporotrichioides | — | F. sporotrichioides | F. sporotrichioides | F. sporotrichiella | F. sporotrichioides | F. sporotrichioides | F. sporotrichioides |
| F. sporotrichioides var. minus | — | — | F. sporotrichioides subsp. minus | — | F. sporotrichiella var. sporotrichioides | F. sporotrichioides var. minus | — | — |
| F. chlamydosporum | — | — | — | — | — | F. chlamydosporum | F. sporotrichioides var. chlamydosporum | F. fusarioides |
| F. tricinctum | — | F. tricinctum | F. sporotrichioides var. tricinctum | — | F. sporotrichiella var. tricinctum | F. tricinctum | F. sporotrichioides var. tricinctum | F. tricinctum |
| — | — | — | — | — | F. sporotrichiella var. anthophilum | — | — | — |

*Source:*   Drawn from various sources.

[a]Booth includes *F. sporotrichioides* and *F. fusarioides* in Arthrosporiella section.

and Kraft, 1977; Stahr *et al.*, 1978; Szathmary *et al.*, 1976; Yates, 1971; Yates *et al.*, 1968, 1969, 1970; Ylimäki *et al.*, 1979) producing toxic symptoms in many animals (Bamburg, 1969, 1972, 1973, 1976; Bamburg and Strong, 1969, 1971; Bamburg *et al.*, 1966, 1968a,b, 1969; Burmeister, 1971; Burmeister and Hesseltine, 1970; Burmeister *et al.*, 1972; Grove *et al.*, 1970; Mirocha and Pathre, 1973; Rabie *et al.*, 1978; Scott, 1965), and also appears in Japan (Ueno, 1977a,b, 1980a,b; Ueno *et al.*, 1972a,b, 1975; Yoshizawa *et al.*, 1980a,b,c).

Another *Fusarium* group of considerable toxicological interest is the Discolor section, mainly comprising species such as *F. graminearum* (*Gibberella zeae*), *F. sambucinum* (*G. pulicaris*), and *F. culmorum*. Different investigators proposed the system of identification of *Fusarium* species, varieties, and forms belonging to the Discolor section shown in Table 10.3. Snyder and Hansen (1945) lumped all species of this section together. They also put many other species, belonging to three other important sections (Arthrosporiella, Gibbosum, and Roseum), into one species named *F. roseum*. Grouping together 54 species, varieties, and forms, with different morphological and cultural, physiological, and toxicological characteristics under one species is completely unacceptable.

The toxigenicity of *F. graminearum* (*G. zeae*), *F. sambucinum* (*G. pulicaris*), *F. culmorum*, *F. equiseti* (*G. intricans*), *F. avenaceum* (*G. avenacea*), and others has been studied fairly extensively, especially with regard to their estrogenic and emetic effects, as observed in swine, and other symptoms of diseases in animals and plants in the United States and other countries (Brian *et al.*, 1961; Caldwell and Tuite, 1968, 1970, 1974; Caldwell *et al.*, 1970; Curtin and Tuite, 1966; Danco and Toth, 1969; Dawkins *et al.*, 1965; Grove, 1970a,b; Ishii *et al.*, 1978; Joffe, 1960a,b; Kalra *et al.*, 1977; Marasas *et al.*, 1979a; Mirocha, 1977; Mirocha and Christensen, 1974a,b; Mirocha *et al.*, 1966, 1967, 1968a,b, 1969, 1976b, 1978a; Palyusik, 1973; Pathre *et al.*, 1976; Prentice and Dickson, 1968; Prentice *et al.*, 1959; Steele *et al.*, 1974, 1976, 1977a,b; Sutton *et al.*, 1976; Ueno *et al.*, 1971c, 1972a, 1973d, 1974a,c, 1977a; Vesonder *et al.*, 1973, 1976, 1977a,b, 1978, 1979a,b; Wolf and Mirocha, 1973; Ylimäki *et al.*, 1979; Yoshizawa and Morooka, 1973, 1974; Yoshizawa *et al.*, 1979, 1980a).

Other sections which are also of great interest to toxicologists are Arachnites (*F. nivale*, *Calonectria nivalis*)), Liseola (*F. moniliforme*, *G. fujikuroi*), Lateritium (*F. lateritium*, *G. baccata*)), Eupionnotes (*F. dimerum*, *F. merismoides*), Spicarioides (*F. decemcellulare*, *C. rigidiuscula*)), Elegans (*F. oxysporum*), and Martiella (*F. solani* (*Nectria haematococca*)). Here Snyder and Hansen (1940, 1941, 1945) do not deviate so drastically, as with Discolor, from the generally accepted schemes based on Wollenweber and Reinking (1935a).

The toxigenicity of *F. nivale* (*C. nivalis*) has been widely studied in

Table 10.3  Classification of *Fusarium* Species within the Discolor Section

| Wollenweber and Reinking (1935a) | Railo (1950) | Gerlach (1970) | Bilai (1955, 1977) | Gordon (1952) | Booth (1971b) | Joffe (1974a) | Snyder and Hansen[a] (1945) |
|---|---|---|---|---|---|---|---|
| Subsection Neesiola | Subsection Eudiscolor | F. heterosporum | F. gibbosum | F. culmorum | F. trichothecioides | F. heterosporum | F. roseum |
| F. heterosporum | F. bucharicum | F. reticulatum | F. graminearum | F. graminearum | F. bucharicum | F. graminearum | |
| F. reticulosum | F. sublunatum | F. sambucinum | F. heterosporum | F. sambucinum | F. sambucinum | F. sambucinum | |
| Subsection Saubinetii | F. sambucinum | F. flocciferum | F. lateritium | | F. culmorum | F. culmorum | |
| F. sambucinum | F. bactridioides | F. trichothecioides | F. sambucinum | | F. heterosporum | F. tumidum | |
| F. bactridioides | F. culmorum | F. culmorum | F. culmorum | | F. graminearum | | |
| F. culmorum | F. tumidum | F. graminearum | F. gigas | | F. flocciferum | | |
| F. sambucinum | F. gigas | F. bucharicum | F. macroceras | | F. sulphureum | | |
| F. flocciferum | Subsection Saubinetii | F. sublunatum | | | | | |
| F. culmorum | F. macroceras | F. tumidum | | | | | |
| F. sublunatum | F. graminearum | F. macroceras | | | | | |
| F. graminearum | F. flocciferum | F. bactridioides | | | | | |
| F. tumidum | F. heterosporum | | | | | | |
| F. macroceras | | | | | | | |
| Subsection Trichothecioides | | | | | | | |
| F. trichothecioides | | | | | | | |
| F. bactridioides | | | | | | | |
| and also 5 varieties and 7 forms | and also 8 varieties and 4 forms | and also 7 varieties and 1 form | and also 7 varieties | and also 1 variety and 2 forms | and also 1 variety 1 form | and also 2 varieties | and also 1 form |

*Source:*  Drawn from various sources.

[a] According to Snyder and Hansen (1945) *F. roseum* included 54 species, varieties, and forms belonging to the following four sections: Arthrosporiella, Discolor, Gibbosum, and Roseum.

Japan (Kubota, 1977; Matsuoka *et al.*, 1979; Morooka and Tatsuno, 1970; Saito *et al.*, 1969, 1980; Tashiro *et al.*, 1979; Tatsuno, 1968, 1969, 1976; Tatsuno *et al.*, 1968a,b, 1971, 1973; Tsunoda, 1970; Ueno, 1970, 1971, 1973a,b, 1977a,b, 1980a,b, 1983a,b; Ueno and Fukushima, 1968; Ueno and Hosoya, 1970; Ueno *et al.*, 1967, 1968, 1969a,b,c, 1970a,b, 1971a,b; Yoshizawa and Morooka, 1975a,b). *F. nivale* is highly lethal to animals, has induced pathogenical changes, degeneration, and necrosis in various organs, and produced cytotoxic and strongly irritant effects on rabbit skin (Ueno *et al.*, 1970a,b,c).

   *F. moniliforme (G. fujikuroi)* is the prevalent fungus on corn in the United States (Cole *et al.*, 1973; Futrell, 1972; Melchers, 1956; Nelson and Osborne, 1956; Springer *et al.*, 1974; Tuite and Caldwell, 1971), as well as in Southern Africa (Kriek *et al.*, 1977; Martin *et al.*, 1971, 1978; Steyn *et al.*, 1979; Thiel, 1978), Israel (Joffe, 1962b, 1963c, 1967c; Joffe and Schiffmann-Nadel, 1972; Joffe *et al.*, 1973), and other parts of the world (Badiali *et al.*, 1968; Booth, 1971b; Gedek *et al.*, 1978; Ghosal *et al.*, 1978a; Kohler *et al.*, 1978; Mathur *et al.*, 1975; Nirenberg, 1976). Toxicity in different animals was reported by several investigators (Burmeister *et al.*, 1979, 1980a; Marasas and Smalley, 1972; Marasas *et al.*, 1979a,c; Mirocha *et al.*, 1968a, 1969; Scott, 1965; Stankushev *et al.*, 1977; Ueno *et al.*, 1971a; Van Rensburg *et al.*, 1971). Some metabolites of *F. moniliforme* produced a cytotoxic compound with antitumor activity (Arai and Ito, 1970), neurotic syndrome (Wilson and Maronpot, 1971; Wilson *et al.*, 1973), and necrosis in the brain by leukoencephalomalacia disease in horses (Equidae) in South Africa (Haliburton *et al.*, 1979; Kellerman *et al.*, 1972; Marasas *et al.*, 1976). *F. moniliforme* and *F. moniliforme* var. *subglutinans* produced moniliformin (Steyn *et al.*, 1978a; Thiel, 1978, 1981).

   *F. lateritium* caused irritant reactions on skin (Joffe, 1960a) and death of insects (Cole and Rolison, 1972), and *F. rigidiusculum* (*F. decemcellulare*) was lethal to mice (Ueno, 1973a; Ueno *et al.*, 1973d).

   *F. dimerum* and *F. oxysporum* caused keratitis in man (Zaias, 1966; Zapater, 1971; Zapater and Arrechea, 1975; Zapater *et al.*, 1972). *F. oxysporum* and *F. solani* are found worldwide in soil and on plants and have been widely studied as the cause of numerous destructive plant diseases. *F. oxysporum* isolated from barley caused the death of chickens (Marasas *et al.*, 1979b). Several strains of *F. oxysporum* were reported to be highly toxic to animals. They caused typical radiomimetic cellular injury and inhibited protein synthesis by rabbit reticulocytes (Ueno, 1977a,b; Ueno *et al.*, 1973b,c). According to Ueno *et al.* (1973d; 1975) monoacetylnivalenol (fusarenon-X) and diacetylnivalenol have been detected in two strains of *F. oxysporum* f. *niveum* isolated from melon, and *F. oxysporum* f. *carthami* produced fusaric acid, lycomarasmin, T-2 toxin, and diacetoxyscirpenol (Chakrabarti *et al.*, 1976; Ghosal *et al.*, 1976, 1977). *F. solani* (reidentified by Ichinoe, 1978 as *F. sporo-*

*trichioides*) on bean-hulls caused an intoxication of horses in Japan (which was characterized by disturbances of the central nervous system and hemorrhages and necrosis formation on rabbit skin) (Ishii *et al.*, 1971), produced an acute and subacute toxicity in cats by isolated T-2 toxin (Sato et al., 1975; Ueno *et al.*, 1972a), and caused death of mice, rats, and poultry (Ueno *et al.*, 1971a,b). *F. solani* was also responsible for some cases of corneal disease (keratitis) in humans (Jones *et al.*, 1969a,b,c, 1972).

## A MODERN TAXONOMY SYSTEM FOR TOXIGENIC *FUSARIUM* SPECIES AND VARIETIES

*Fusarium* species have been known to affect agricultural products under suitable environmental conditions in the field or during harvest, postharvest handling, transportation, or, in particular, storage. They produce toxic compounds which cause serious diseases in animals and humans.

During the past 50 years I have conducted a comprehensive study of *Fusarium*. Isolates were collected from the warm, semiarid climate of Israel (from a great variety of contaminated and normal (healthy) cereal grains, grasses, vegetables, horticultural plants, fruits, and soil) and also from the cold, temperate climate of the Orenburg district in the USSR and various areas of the United States. *Fusarium* cultures were also received from research institutions and culture collections (Joffe, 1974a).

Since it has been proven by the author and other researchers that *Fusarium* species and varieties cause grave disorders in humans (Bilai, 1953, 1977, Bilai and Pidoplichko, 1970; Chilikin, 1944, 1945, 1947; Joffe, 1950, 1962a, 1963a, 1965, 1971, 1974c, 1978b; Mironov, 1945a,b; Rubinstein, 1948, 1951a,b, 1953, 1956a; Yefremov, 1944a,b, 1945, 1948) and animals (Antonov *et al.*, 1951; Burmeister, 1971; Burmeister and Hesseltine, 1970; Grove *et al.*, 1970; Hsu *et al.*, 1972; Ishii *et al.*, 1971; Joffe, 1960a,b, 1971, 1974c, 1978a; Joffe and Yagen, 1978; Khrutski *et al.*, 1953; Kriek and Marasas, 1983; Lutsky and Mor, 1981; Lutsky *et al.*, 1978; Marasas *et al.*, 1967, 1969, 1971, 1978, 1979b; Mirocha and Christensen, 1974a,b; Mirocha and Kurtz, 1976; Mirocha *et al.*, 1968a, 1971, 1973, 1974; Sarkisov, 1954; Schoental, 1968, 1975, 1977a,b, 1978c, 1979a,b, 1981a,b; Schoental and Joffe, 1974; Schoental and Gibbard, 1978; Schoental and Magee, 1962; Schoental *et al.*, 1976, 1978a,b, 1979; Ueno, 1977a,b, 1980; Ueno *et al.*, 1971a,b, 1972a,b; Yagen *et al.*, 1977, 1980), the determination of the exact identity of the *Fusarium* species that produce toxins has become a matter of acute importance.

Our taxonomy of toxic *Fusarium* sp. following the generally accepted rules for isolation and identification and use of the standard media for inoculation and cultivation, is based on the study of thousands of isolates grown on a variety of substrates and under different environmental conditions (Table 10.4).

**TABLE 10.4** *Fusarium* Cultures Isolated from Various Substrates in Subtropical and Temperate Regions under Different Environmental Conditions

| Section | Total No. of Isolates | No. of Isolates Studied Morphologically from: | | | | | Total No. Studied |
|---|---|---|---|---|---|---|---|
| | | Israel | United States | USSR | Other Countries | | |
| *Arachnites* | 29 | 0 | 0 | 23 | 6 | | 29 |
| *Eupionnotes* | 149 | 39 | 13 | 15 | 27 | | 94 |
| *Sporotrichiella* | 1053 | 0 | 431 | 590 | 32 | | 1053 |
| *Spicarioides* | 10 | 2 | 0 | 0 | 8 | | 10 |
| *Macroconia* | 39 | 17 | 0 | 6 | 9 | | 32 |
| *Roseum* | 219 | 53 | 16 | 60 | 7 | | 136 |
| *Arthrosporiella* | 245 | 65 | 45 | 31 | 20 | | 161 |
| *Lateritium* | 142 | 43 | 7 | 32 | 10 | | 92 |
| *Liseola* | 2310 | 602 | 1210 | 37 | 24 | | 1873 |
| *Gibbosum* | 2224 | 576 | 74 | 88 | 32 | | 770 |
| *Discolor* | 563 | 143 | 121 | 35 | 40 | | 339 |
| *Elegans* | 4652 | 1218 | 0 | 35 | 49 | | 1302 |
| *Martiella* | 5807 | 1424 | 4 | 37 | 12 | | 1477 |
| Total | 17,442 | 4182 | 1921 | 989 | 276 | | 7368 |

*Source:* Joffe, unpublished results.

The aim of this chapter is to present a concise description of the characteristics of *Fusarium* species and varieties that produce toxins and to give researchers who work with toxigenic *Fusarium* isolates a taxonomic guide that will assist them in reaching the correct identification of this very important group of fungi.

Keys for the identification of sections and species of the genus *Fusarium* and the detailed description of *Fusarium* species and varieties are given in this chapter.

## Key for Identification of Sections

1. Microconidia present in abundance.     2.
1. Microconidia usually in chains.     3.
1. Microconidia scattered or in false heads, not in chains.     4.
1. Microconidia sparse.     5.
1. Microconidia absent.     6.
2. Microconidia in aerial mycelium, globose, ovoid, pear-shaped (pyriform); cultures rose, red.     *Sporotrichiella*
3. Macroconidia large, thick-walled with 5–11 septa; cultures intensive rose, carmine; chlamydospores absent.     *Spicarioides*
3. Macroconidia sparse, thin-walled, elongated, straight; cultures violet, white cream; chlamydospores absent.     *Liseola*
4. Macroconidia thin-walled, falcate; cultures light rose, beige, pale.     *Elegans*
4. Macroconidia thick-walled, cylindrical, fusoid with short, rounded apical cells; cultures white, cream, orange-blue to brown.     *Martiella*
5. Macroconidia thin-walled, light, straight to falcate, with beaked apical cells; cultures yellow, pale to rose to dark blue.     *Lateritium*
5. Macroconidia lanceolate with wedge-shaped basal cell and with narrow apex; cultures white, peach, ochre, yellow to red; chlamydospores present.     *Arthrosporiella*
5. Macroconidia thin-walled, falcate, dorsiventral, parabolic, or hyperbolic; cultures white to pale pink, ochre to red.     *Gibbosum*
6. Macroconidia apedicellate, curved; chlamydospores absent; cultures white, orange, rose to green.     *Arachnites*
6. Macroconidia cylindrical, lightly curved to lunate, falcate; cultures cream, orange to rose; chlamydospores present.     *Eupionnotes*
6. Macroconidia thin-walled, slender, filiform to falcate; cultures yellow, rose, purple, red, carmine; chlamydospores absent.     *Roseum*

6. Macroconidia thick-walled, fusiform, falcate; cultures purple, rose, carmine; chlamydospores present.     *Discolor*

## Brief Description of Sections

**Section Sporotrichiella Wr. em. Joffe** *(F. poae, F. sporotrichioides, F. sporotrichioides* var. *tricinctum, F. sporotrichioides* var. *chlamydosporum).* Cultures: white-red, carmine red, purple to brown. Microconidia abundant, lemon or pear shaped, globose, ellipsoid or elongate, dispersed in aerial mycelium or formed in false heads.

Macroconidia sparse, small, oblong, narrowly fusoid to falcate, pedicellate, formed in aerial mycelium or in sporodochia. Chlamydospores intercalary, terminal, in chains or knots, occasionally with plectenchymatous sclerotia. Perithecial states absent.

**Section Spicarioides Wr.** *(F. decemcellulare).* Cultures: rose, carmine red to purple. Microconidia in aerial mycelium in long chains or sometimes in false heads, globose, oval.

Macroconidia large, in sporodochia or pionnotes, thick-walled, slightly curved or straight, pedicellate, beaked at the tip with well-marked foot-cell. Chlamydospores absent.

**Section Liseola Wr.** *(F. moniliforme, F. moniliforme* var. *subglutinans).* Cultures: white-cream-brown, pale orange, violet. Microconidia in aerial mycelium, usually in long chains or small false heads, oval, fusiform, oblong, rarely pyriform.

Macroconidia thin-walled, in sporodochia, pionnotes, and aerial mycelium, sublunate, spindle-shaped, cylindrical, straight or curved, with narrow apex and base cells, typically three-septate. Chlamydospores absent.

**Section Elegans Wr.** *(F. oxysporum, F. oxysporum* var. *redolens).* Cultures: white, light rose, orange, violet to slightly purple. Mycelium felted or floccose. Microconidia scattered in mycelium or in false heads, abundant, unicellular or bicellular, variable, oval-elliptical, straight to curved, powdery, and produced from short branches of conidiophores on lateral phialides.

Macroconidia thin-walled, in aerial mycelium and sporodochia, sometimes in pionnotes, falcate, elongated, subulate, spindle- or sickle-shaped, narrowing at both ends, three to five-septate. Chlamydospores abundant, terminal, and intercalary, smooth- to rough-walled.

**Section Martiella Wr. em. Joffe** *(F. solani, F. javanicum).* Cultures: white, cream, orange-blue to brown. Microconidia abundant, oval or oblong, hyaline. Macroconidia in aerial mycelium, sporodochia or pionnotes, fusoid, cylindrical, curved or elongate, with thick walls and short, rounded apical cells and foot cells. Chlamydospores globose, oval, smooth- or rough-walled, terminal and intercalary, single or in pairs, short chains or knots.

**Section Lateritium Wr.** *(F. lateritium, F. stilboides, F. xylarioides).* Cultures: white, rose, yellow, orange, carmine-red, violet to brown-blue. Microconidia oval, elliptical, sparse, or absent. Macroconidia thin-walled, elongated, cylindrical, straight or slightly curved, with beaked apical cell and pedicellate base formed in the aerial mycelium or sporodochia, rarely in pionnotes. Chlamydospores intercalary, sparse in mycelium or macroconidia.

**Section Eupionnotes Wr.** *(F. dimerum, F. merismoides).* Cultures: pale yellow, orange, yellow-brown, rose, green. Microconidia absent. Macroconidia formed in pionnotes sporodochia on phialides singly or in groups of conidiophores, rarely in sporodochia, sublunate, cylindrical or spindle-shaped, falcate, elongated or curved, apedicellate. Aerial mycelium usually sparse, slimy. Chlamydospores and sclerotia present or rarely absent. Perithecial status present in some of the species.

**Section Gibbosum Wr. em. Joffe** *(F. equiseti* and *F. equiseti* var. *acuminatum).* Culture: white-pale pink, pale ochre, olive, carmine-red. Microconidia absent or sparse, in aerial mycelium.

Macroconidia thin-walled, formed in aerial mycelium, or from sporodochia, sometimes appear in pionnotes, falcate, narrowing at both ends with elongated apical cell, and well-developed pedicellate foot cell, dorsiventral, parabolic or hyperbolic, typically five-, rarely three-septate. Chlamydospores intercalary, abundant, smooth- or rough-walled, single or in chains and knots, yellow-brown.

**Section Arachnites Wr.** *(F. nivale).* Macroconidia in aerial mycelium, rarely in sporodochia or pionnotes, curved, apedicellate. Chlamydospores and sclerotia absent. Perithecial stage present.

**Section Arthrosporiella Wr.** *(F. semitectum, F. semitectum* Berk. and Rav. var. *majus* Wr.*).* Cultures: white, peach, ochre-yellow to carmine-red. Microconidia absent or sparse in aerial mycelium, elliptical oval, or pyriform.

Macroconidia formed from polyblastic cells of conidiophores in aerial mycelium, or mainly in sporodochia, falcate, lanceolate with wedge-shaped, pedicellate basal cell and with narrow apex, typically three to five septate. Chlamydospores intercalary, single, sometimes in chains or knots, terminal chlamydospores rare or absent.

**Section Roseum Wr.** *(F. avenaceum, F. arthrosporioides).* Cultures: yellow, ochre, carmine, purple, or red. Microconidia absent or sparse, chlamydospores absent. Macroconidia in sporodochia, pionnotes, or aerial mycelium, on phialides of branched conidiophores, sublunate, slender, almost filiform, falcate with thin walls, narrowing at both ends, pedicellate. Sclerotia white, yellow, purple to brown. Stroma yellow, rose, light red.

**Section Discolor Wr.** *(F. culmorum, F. graminearum, F. sambucinum).* Cultures: white-rose, peach, grayish, rose, red to brown. Microconidia absent. Macroconidia thick-walled, in aerial mycelium, sporodochia, and pionnotes, either broad, falcate with short apical cell and well-developed foot cell, or spindle- or sickle-shaped, with elongated, gradually narrowing apical cell, well-marked basal cell, typically five-septate. Chlamydospores intercalary, sometimes terminal, often in knots and chains. Sclerotia purple blue, brown to dark. Stroma yellow.

## Key for Identification of Species and Varieties

1. Microconidia very heterogeneous, single, in false heads, one or two-celled, generally abundant.    2.

1. Microconidia abundant, hyaline, broadly oval, elongated, singly or in false heads.    7.

1. Microconidia abundant, variable, oval-ellipsoid to kidney-shaped.    8.

1. Microconidia elongated, in chains, usually with a broad and rounded apex, or not in chains, oval to fusiform.    9.

1. Microconidia absent or rare.    10.

2. Microconidia globose, abundant, spherical with basal papillae, rarely pear-shaped.    3.

2. Microconidia pyriform to ellipsoid, globose or elongate.    4.

2. Microconidia pyriform to clavate, lemon-shaped, oval, cylindrical.    5.

2. Microconidia narrowly clavate with rounded apex, pyriform to fusiform, not spherical or globose.    6.

3. Micronidia abundant, round, or spherical with basal papillae, lemon-shaped. Macroconidia very sparse, small, curved, falcate, usually without foot cell. Typically three-septate; 19–38 × 3.5–6.8. Sporodochia and pionnotes absent.    *F. poae*

4. Microconidia pyriform, clavate. Macroconidia formed in sporodochia and in aerial mycelium, curved, falcate, elongated, and narrowly fusoid. Typically three- to five-septate; 22–45 × 3.6–5.2.    *F. sporotrichioides*

5. Microconidia more abundant than macroconidia, navicular to cylindrical. The latter formed in sporodochia, rarely in aerial mycelium, curved, spindly ellipsoid, falcate, three- to five-septate; 23–50 × 3.3–4.4. Typically three-septate.    *F. sporotrichioides* var. *tricinctum*

6.  Microconidia pyriform, slightly fusiform, or slightly clavate, formed sparsely in aerial mycelium, curved, fusoid with narrowly pointed apex and marked foot cell. Sporodochia absent or very rare, three- to five-septate; 28–42 × 3.2–4.5. Typically three-septate.     *F. sporotrichioides* var. *chlamydosporum*

7.  Microconidia broadly oval, elongated. Macroconidia elliptical, sausage-shaped, slightly curved with round apical and round basal cells on marked foot cell. Typically three-septate; 28–43 × 4.6–6.4. Cultures white, white-rose, or cream-yellow.     *F. solani*

7.  Cultures white, cream, light yellow to brown. Microconidia oval. Macroconidia slightly curved with slightly narrowing apical cell in comparison with *F. solani*. Typically three- to five-septate; 27–60 × 4.0–5.0.     *F. javanicum*

8.  Microconidia variable in shape, powdery. Macroconidia usually sparse, sickle-shaped, falcate, elongated at both ends with marked foot-cell, generally three- to five-septate. Typically three-septate. 26–42 × 2.8–4.5. Cultures white, peach to purple.     *F. oxysporum*

8.  Cultures white, cream, rose, beige, red, brown. Microconidia oval, cylindrical. Macroconidia with round apex and narrowing basal cell. Generally three- to five-septate; 25–45 × 4.0–5.5.     *F. oxysporum* var. *redolens*

9.  Microconidia in chains. Macroconidia spindly, elongated, cylindrical, straight, or slightly falcate with narrow curved apical cell and marked foot cell. Three- to five-septate; 29–67 × 2.6–4.0. Typically three-septate. Cultures peach to violet.     *F. moniliforme.*

9.  Cultures cream, rose red to dark, microconidia oval in chains. Macroconidia large, slightly cylindrical, generally seven- to eleven-septate; 50–111 × 5.6–9.0.     *F. decemcellulare*

9.  Cultures white to greyish white, purple. Microconidia oval, cylindrical to fusiform, not in chains. Macroconidia thin-walled, falcate with marked apical and basal cells. Typically three- to five-septate; 26–53 × 2.8–4.2.     *F. moniliforme* var. *subglutinans*

10.  Cultures light cream, dark yellow to brown. Microconidia rare, short, oval, spindle-shaped or clavate, zero- to two-septate. Macroconidia falcate to parabolic with more or less gradually or sharply narrowing elongated curved or noncurved apical cell and well-marked foot cell, three- to seven-septate; 27–62 × 3.5–5.5. Typically five-septate.     *F. equiseti*

10.  Cultures white-rose, peach-beige, brown, carmine red to purple. Microconidia absent or very rare. Macroconidia falcate, hyper-

bolic, dorsiventral with marked foot cell, typically five-septate; 36–54 × 3.8–4.6.    *F. equiseti* var. *acuminatum*

10.    Cultures white, yellow, violet to dark blue, brown-blue. Microconidia absent or very sparse, oval, cllliptical. Macroconidia frequently elongated or short, cylindrical, slightly curved to straight, with short, beaked apex, usually with typical foot cell, typically three- to five-septate; 20–58 × 3.0–5.2.    *F. lateritium*

10.    Cultures white, yellow to dark. Microconidia absent or rare in aerial mycelium. Macroconidia lanceolate, slightly curved with wedge-shaped foot cell and curved apex. Typically three- to five-septate; 19–39 × 2.9–4.5.    *F. semitectum*

10.    Cultures white-rose to brown. Microconidia absent or sparsely developed in aerial mycelium. Macroconidia lightly curved or almost straight, narrowing at both cnds. Typically five- to seven-septate; 38–60 × 3.6–6.0.    *F. semitectum* var. *majus*

10.    Culture sometimes whitish, mostly strongly colored, rose, red-brown, dark red. Microconidia absent or rare. Macroconidia elongated up to 90 μm, narrow, filiform or slightly hyperbolic, curved, with elongated narrowing apical cell and well-marked foot cell, three- to seven-septate; 31–92 × 2.5–4.1. Typically five-septate.    *F. avenaceum*

10.    Cultures rose, carmine to brown. Microconidia absent or sparse. Macroconidia fusiform, lanceolate, strongly dorsiventral, typically three- to five-septate. Three-septate 30–46 × 3.0–4.8; five-septate 43–62 × 3.8–5.2.    *F. arthrosporioides*

10.    Microconidia absent. Macroconidia usually in aerial mycelium, small, often sparse, one- to three-septate, curved, sometimes aseptate.    11.

11.    Cultures light-colored, white, light rose, or yellow to pale, slow-growing. Macroconidia short, elliptical, curved, sickle-shaped with pointed apex and wedge-shaped and round basal cell. Zero- to three-septate; 8–32 × 2.2–4.8.    *F. nivale.*

11.    Cultures beige, yellow to rose, slimy. Macroconidia curved, spindle- to sickle-shaped, chlamydospores present, typically one- to three-septate; 12–18 × 2.6–3.2 and 18–21 × 3.0–4.0, respectively.    *F. dimerum*

11.    Cultures cream to orange, slimy. Macroconidia cylindrical, straight, or slightly curved. Typically three-septate; 28–41 × 3.5–4.5.    *F. merismoides*

11.    Macroconidia typically five-septate, mostly abundant, slightly curved, strongly dorsiventral, with a short and wide pointed apex and well-marked foot cell.    12.

12.    Macroconidia broad; 25–68 × 4.8–8.2.    *F. culmorum*

12.  Macroconidia less broad, cylindrical, curved, strongly dorsiventral, falcate, elliptical or sickle-shaped with short, beaked apical cell or slightly curved and gradually narrowing, with elongated apical cell and well-marked foot cell.    13.

13.  Macroconidia cylindrical, curved, strongly dorsiventral, fusoid, with short beaked apical cell; 25–53 × 3.8–5.9.    *F. sambucinum*

13.  Macroconidia elongated up to 78 μm, spindle- or sickle-shaped, slightly dorsiventral, curved, with elongated apical cell; 28–78 × 3.2–5.0.    *F. graminearum*


# DETAILED DESCRIPTION OF SPECIES AND VARIETIES, THEIR MORPHOLOGY AND CULTURAL CHARACTERISTICS

## Sporotrichiella Section

### *F. poae*

This fungus is widespread in nature on various plants and soils in temperate regions. In the literature this fungus appears under the names of *F. tricinctum* and *F. tricinctum* f. *poae*, according to Snyder and Hansen (1945), and also as *F. sporotrichiella* var. *poae*, according to Bilai (1952a, 1955, 1977).

*F. poae* (Peck) Wr. No. 958 (Figs. 10.1 through 10.3).
Synonyms:

*F. citriforme* Jamal.
(Jamalainen, 1943a)
*F. tricinctum* (Cda.) Sacc. emend. Snyd. and Hans.
(Snyder and Hansen, 1945)
*F. tricinctum* (Cda.) Sacc. emend. Snyd. and Hans. *f. poae* (Pk.) Snyd. and Hans.
(Snyder and Hansen, 1945)
*F. poae* (Pk.) Wr. *f. pallens* Wr.
(Wollenweber, 1916–1935)
*F. sporotrichiella* Bilai var. *poae* (Pk.)
(Bilai, 1952a, 1955, 1977)

Cultures white, yellow, rose, carmine, or red-brown. Aerial mycelium hairy, cobwebby to felted, or somewhat powdery.

Microconidia abundant in relation to macroconidia, scattered over the mycelium, formed from small globose phialides, singly or in false heads,

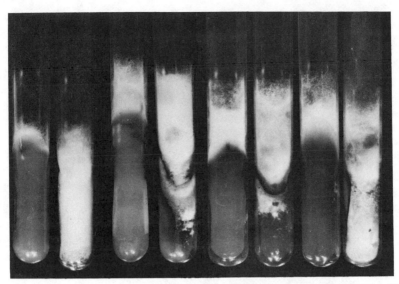

**Figure 10.1** Cultural appearance of *F. poae* No. 958, *F. sporotrichioides* No. 921, *F. sporotrichioides* var. *tricinctum* No. 1227 and *F. sporotrichioides* var. *chlamydosporum* No. 4337 on potato dextrose agar (PDA) slants (from left to right). (Adapted from Joffe, 1978b.)

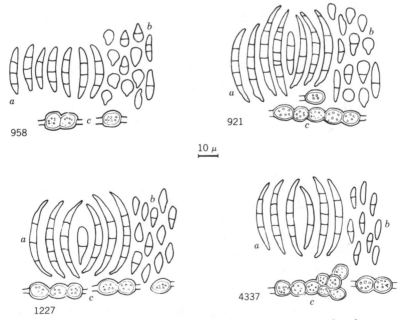

**Figure 10.2** *F. poae* No. 958; *F. sporotrichioides* No. 921; *F. sporotrichioides* var. *tricinctum* No. 1227; *F. sporotrichioides* var. *chlamydosporum* No. 4337. (*a*) macroconidia; (*b*) microconidia; (*c*) chlamydospores. (Adapted from Joffe, 1977.)

**405**

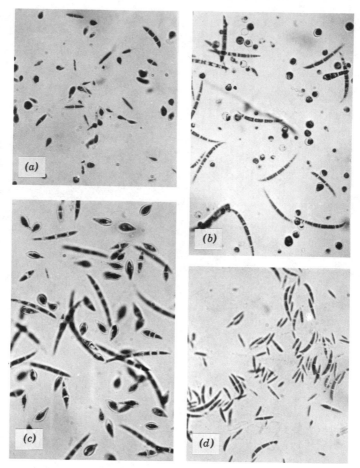

**Figure 10.3**  (*a*) *F. poae* No. 958; (*b*) *F. sporotrichioides* No. 921; (*c*) *F. sporotrichioides* var. *tricinctum* No. 1227; (*d*) *F. sporotrichioides* var. *chlamydosporum* No. 4337. (Joffe, unpublished results.)

broadly oval, round or spherical, with basal papilla, lemon-shaped, ellipsoid, rarely pear-shaped. Conidiophores well-developed and microconidia formed in lateral, small, broad phialides.

Macroconidia sparse, curved, falcate, small, only in aerial mycelium, formed on narrow, elongated phialides.

Sporodochia absent, pionnotes very rare or absent. Measurements (μm), shape, and frequency of conidia:

| | | | |
|---|---|---|---|
| zero-sept. | 91% | oval, spherical | 0.4–9 × 4.2–8.0 |
| zero-sept. | 1% | elongated, spindly | 8–14 × 2.5–4.2 |
| one-sept. | 6% | oval, pear-shaped | 10.0–15 × 4.2–7.2 |
| three-sept. | 2% | curved falcate | 17–36 × 4.0–6.6 |

Growth rate 8.0 cm

Chlamydospores rare, intercalary, single or in pairs, knots, or chains, or absent. Sclerotia absent. Stroma carmine, ochre-yellow, and sometimes violet.

## F. sporotrichioides

This fungus is widespread on various plants and soils in temperate zones. It has appeared under the name of *F. tricinctum* in papers by American and Japanese researchers, in accordance with Snyder and Hansen's (1945) classification, and has been described by Bilai as *F. sporotrichiella* and *F. sporotrichiella* var. *sporotrichioides* (1952a, 1955, 1977).

F. *sporotrichioides* Sherbakoff No. 921 (Figs. 10.1 through 10.3). Synonyms:

F. *tricinctum* (Cda.) Sacc. emend. Snyd. and Hans.
(Snyder and Hansen, 1945)

F. *tricinctum* (Cda.) Sacc. emend. Snyd. and Hans. *f. poae* (Pk.) Snyd. and Hans.
(Snyder and Hansen, 1945)

F. *sporotrichiella* Bilai
(Bilai, 1955)

F. *sporotrichiella* Bilai var. *sporotrichioides* (Sherb.) Bilai
(Bilai, 1952a, 1955, 1977)

F. *sporotrichioides* Sherb. var. *minus* Wr.
(Seemüller, 1968; Wollenweber and Reinking, 1935a)

Cultures white, white-rose, red, purple, sometimes light brown. Aerial mycelium downy, floccose.

Microconidia often as numerous as macroconidia, formed singly or in short chains, globose, pyriform, ellipsoid, lemon-shaped, elongated, slightly falcate, dispersed in aerial mycelium, formed on apical cylindrical phialides of branched conidiophores.

Macroconidia are falcate to curved, oblong, and narrowly fusoid, scattered in aerial mycelium or in sporodochia and formed on elongated phialides. Measurements (μm), shape, and frequency of conidia:

| | | | |
|---|---|---|---|
| zero-sept. | 19% | pear-shaped | 5.4–9.5 × 5.2–7.0 |
| zero-sept. | 5% | ellipsoidal | 7.5–12 × 4.4–7.5 |
| zero-sept. | 9% | elongated or sickle-shaped | 9–12 × 2.6–4.0 |
| zero-sept. | 2% | spindle-shaped | 8–15 × 2.7–3.8 |
| one-sept. | 17% | pear-shaped | 9–16.5 × 4.2–7.5 |
| one-sept. | 7% | ellipsoidal | 10–19 × 3.6–7.8 |
| one-sept. | 8% | elongated or sickle-shaped | 11–18 × 3.2–4.0 |

| one-sept. | 6% spindle-shaped | 13–24 × 2.4–4.0 |
| three-sept. | 13% sickle- or spindle-shaped | 23–34 × 3.8–5.0 |
| four-sept. | 7% sickle- or spindle-shaped | 35–42 × 3.4–4.6 |
| five-sept. | 7% sickle- or spindle-shaped | 38–47 × 4.2–5.4 |

Growth rate: 4.0 cm

Chlamydospores intercalary, rarely terminal, single or in pairs, chains or knots, globose, hyaline, or light brown, smooth-walled. Sclerotia red-brown, occasionally present. Stroma red, yellow, purple-red, or dark carmine, sometimes brown.

### *F. sporotrichioides* var. *tricinctum*

This has been classified as *F. tricinctum* by Booth (1971b); Seemüller (1968); Snyder and Hansen (1945); and Wollenweber and Reinking (1935a); and as *F. sporotrichiella* var. *tricinctum* by Bilai (1952a, 1955, 1977).

*F. sporotrichioides* var. *tricinctum* (Corda) Raillo No. 1227 (Figs. 10.1 through 10.3).

Synonyms:

*F. tricinctum* (Cda.) Sacc. emend. Snyd. and Hans.
(Snyder and Hansen, 1945)

*F. sporotrichiella* Bilai var. *tricinctum* (Cda.) Raillo
(Bilai, 1952a, 1955)

*F. tricinctum* (Cda.) Sacc.
(Seemüller, 1968; Wollenweber and Reinking, 1935a)

Cultures white, rose, red, purple, or carmine. Aerial mycelium abundant, cottony, floccose.

Microconidia more abundant than macroconidia, formed on cylindrical phialides of branched conidiophores; dispersed, usually single, in aerial mycelium or in false heads, pyriform to clavate, lemon-shaped, oval, navicular, cylindrical, ellipsoidal, slightly falcate.

Macroconidia usually formed in sporodochia on small, slightly curved phialides; falcate or elliptical, strongly curved with well-marked foot cell. Measurements (μm), shape, and frequency of conidia:

| zero-sept. | 39% pear-shaped | 8–10 × 4.8–7.4 |
| zero-sept. | 16% elongated | 7.5–15 × 3.0–3.7 |
| one-sept. | 1% pear-shaped | 9–17 × 4.5–5.2 |
| one-sept. | 8% elongated | 13–21 × 3.0–4.0 |
| three-sept. | 27% elongated, curved falcate | 22–39 × 3.4–4.2 |

| four-sept. | 3% elongated, curved falcate | $31-45 \times 3.6-4.4$ |
| five-sept. | 6% elongated, curved falcate | $37-52 \times 3.5-4.6$ |

Growth rate 4.5 cm

Chlamydospores rare, globose, smooth-walled, intercalary, single, or in chains. Sclerotia white, purple to brown, sometimes absent. Stroma carmine, purple, ochre-brown or violet, rarely colorless.

## F. sporotrichioides var. chlamydosporum

This has occurred under the names *F. fusarioides* (Booth, 1971b), *F. chlamydosporum* (Seemüller, 1968; Wollenweber and Reinking, 1935a), and *F. tricinctum* (Snyder and Hansen, 1945).

*F. sporotrichioides* var. *chlamydosporum* (Wr. and Rg.) Joffe No. 4337 (Figs. 10.1 through 10.3).

Synonyms:

*F. tricinctum* (Cda.) Sacc. emend. Snyd. and Hans.
(Snyder and Hansen, 1945)

*F. chlamydosporum* Wr. and Rg.
(Seemüller, 1968; Wollenweber and Reinking, 1935a)

*F. fusarioides* (Frag. and Cif.)
(Booth, 1971b)

Cultures: rose red, carmine, rarely white, yellow, or brown. Aerial mycelium floccose, cottony.

Microconidia usually less numerous than macroconidia, formed only singly in aerial mycelium on irregular conidiophores; they are small, narrow to clavate, lemon-shaped, usually one-celled, with rounded apex, spindle-ellipsoid or oblong, rarely pyriform or oval-ellipsoid.

Macroconidia develop on phialides of short conidiophores in the aerial mycelium. Macroconidia falcate, curved with narrowly rounded to pointed apical cell and marked foot cell. Sporodochia absent or very rare. Measurements (μm), shape, and frequency of conidia:

| zero-sept. | 31% spindle-elongated | $7.5-11 \times 2.5-3.3$ |
| one-sept. | 8% spindle-elongated | $11.5-15 \times 2.8-3.8$ |
| three-sept. | 54% curved | $28-35 \times 3.2-4.0$ |
| four-sept. | 5% curved | $34-38 \times 3.4-4.2$ |
| five-sept. | 2% curved | $38-42 \times 3.5-4.5$ |

Growth rate: 3.5 cm

Chlamydospores terminal or intercalary, abundant, single or in pairs, knots or long chains, smooth or slightly rough-walled, light brown. Sclerotia purple-brown to brown, or absent.

## Martiella Section

### *F. solani*

This occurs throughout the world and is found in abundance in soil, on various plants, and other substrates in subtropical, tropical, and temperate regions.

*F. solani* (Martius) App. and Wr. No. 990 (Figs. 10.4 through 10.6). Synonyms:

*F. argillaceum* (Fr.) Sacc.
(Saccardo, 1886; Snyder and Hansen, 1945; Wollenweber and Reinking, 1935a)

*F. solani* (Mart.) App. and Wr. var. *eumartii* (Carp.) Wr.
(Wollenweber and Reinking, 1935a; Wollenweber, 1944–45).

*F. solani* (Mart.) App. and Wr. var. *martii* (App. and Wr.) Wr.
(Wollenweber and Reinking, 1935a; Wollenweber, 1944–45)

*F. solani* (Mart.) var. *minus* (App. and Wr.) Wr.
(Wollenweber and Reinking, 1935a; Wollenweber, 1944–45)

*F. solani* (Mart.) App. and Wr. var. *striatum* (Sherb.) Wr.
(Wollenweber and Reinking, 1935a; Wollenweber, 1944–45)

*F. solani* (Mart.) App. and Wr. var. *aduncisporum* (Weim. and Hart.) Wr.
(Wollenweber and Reinking, 1935a)

*F. solani* (Mart.) App. and Wr. var. *redolens* (Wr.) Bilai
(Bilai, 1955)

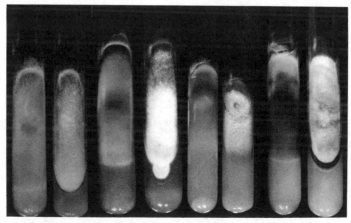

**Figure 10.4** Cultural appearance of *F. solani* No. 990, *F. javanicum* No. 300, *F. oxysporum* No. 233, and *F. oxysporum* var. *redolens* No. 916 on PDA slants (from left to right). (Joffe, unpublished results.)

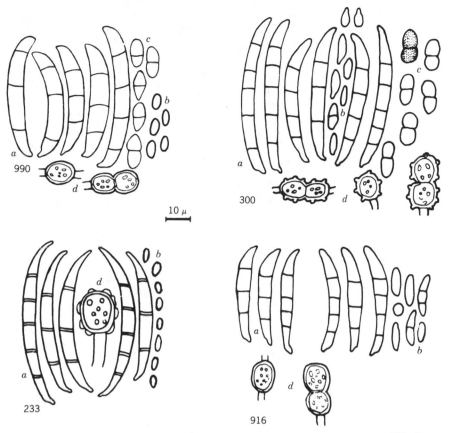

**Figure 10.5** *F. solani* No. 990; *F. javanicum* No. 300; *F. oxysporum* No. 233; *F. oxysporum* var. *redolens* No. 916. (*a*) macroconidia; (*b*) microconidia; (*c*) ascospores; (*d*) chlamydospores. (Joffe, unpublished results.)

Cultures white-grayish to brown. Aerial mycelium usually floccose, sometimes leathery.

Microconidia hyaline, cylindrical, broadly oval, oblong with one or no septum, abundantly scattered singly or often in false heads, formed on long lateral phialides of branched conidiophores.

Macroconidia in aerial mycelium and sporodochia, thick-walled, with well-marked septa, develop from short, branched conidiophores; cylindrical, sausage-shaped, elongated, slightly curved, elliptical with short, round, or sometimes beaked apical cell and round, basal cell or more or less marked foot cells. Usually three-septate. Measurements (μm) and frequency of conidia:

| | | |
|---|---|---|
| zero-sept. | 7% | 10–14 × 3.4–4.4 |
| one-sept. | 10% | 17–24 × 4.0–4.6 |

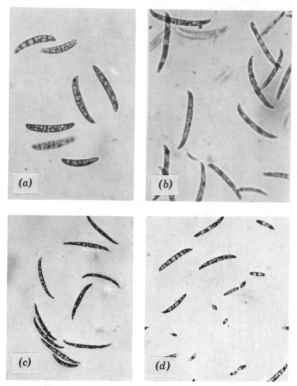

**Figure 10.6** (*a*) *F. solani* No. 990 (microculture); (*b*) *F. javanicum* No. 300; (*c*) *F. oxysporum* No. 233; *F. oxysporum* var. *redolens* No. 916. (Joffe, unpublished results.)

| | |
|---|---|
| three-sept. 71% | 28–39 × 4.8–6.6 |
| four-sept.   8% | 34–45 × 5.0–6.7 |
| five-sept.   4% | 36–52 × 5.1–6.8 |
| Growth rate: 3.5 cm | |

Chlamydospores usually abundant, terminal, or intercalary, single or in pairs, or sometimes in short chains, smooth- or rough-walled. Sclerotia rarely present. Stroma light brown to blue.

Perithecial state: *N. haematococca* Berk. and Br. Perithecia free, sparse to dense, occasionally gregarious, superficial, globose or onion-shaped; form on poorly developed pseudoparenchymatous stroma. Perithecial wall pigmented and thickened; perithecia orange, red, or light brown, 150–355 μm in diameter. Asci clavate ellipsoid to cylindrical, 65–95 × 7–11.5 with four- to eight-septate hyaline, oval, ellipsoid, sometimes pyriform ascospores, slightly constricted at septa. Perithecia are formed in 7–9-week-old cultures.

Ascospore measures (μm): one-sept. 10–17 × 4.5–7.5.

## F. javanicum

This has been reported as occurring mainly in warm regions (Joffe and Palti, 1970).

F. javanicum Koord. No. 300 (Figs. 10.4 through 10.6).

Synonyms:

F. javanicum Koord. var. theobromae (Appl. and Strk.) Wr.
(Bugnicourt, 1939; Raillo, 1950; Wollenweber and Reinking, 1935a)
F. theobromae (Appl. and Strk.) Wr. and Rg.
(Wollenweber and Reinking, 1935a)
F. javanicum Koord. var. ensiforme Wr. and Rg.
(Wollenweber and Reinking, 1935a; Bilai, 1955)

Culture white, cream, light yellow to brown. Microconidia oval, oblong, scattered in the aerial mycelium, or in false heads. Macroconidia in aerial mycelium, pionnotes, or, rarely, in sporodochia, formed on branched conidiophores. They are elliptic, slightly curved, elongate, with slightly narrowing apical cell, with or without foot cell, typically three- to five-septate. Measurements (μm) and frequency of conidia:

| | | |
|---|---|---|
| zero-sept. | 16% | 5.5–10 × 2.6–3.4 |
| one-sept. | 13% | 13–25 × 3.0–4.0 |
| two-sept. | 7% | 18–27 × 3.3–4.4 |
| three-sept. | 41% | 30–45 × 3.8–4.6 |
| five-sept. | 23% | 35–58 × 4.2–5.0 |
| Growth rate: 2 cm | | |

Chlamydospores terminal or intercalary, single or in pairs, smooth- or rough-walled. Perithecial state: *Hypomyces ipomoeae* (Hals.) Wr. Perithecia scattered single or gregarious, developed on plectenchymatous stroma, onion-shaped, 280–365 × 170–320 μm. The outer wall scaly, red with a light ostiolar papilla. The asci are cylindrical and have a rounded apex. The ascospores are ellipsoidal and slightly constricted at the centrum, septum two-celled with striate walls, 7–15 × 4–8.

## Elegans Section

### F. oxysporum

This is found worldwide in soil and on many plant species in temperate, subtropical, and tropical regions.

F. oxysporum Schlecht. No. 233 (Figs. 10.4 through 10.6).

Synonyms:

*F. angustum* Sherb.
(Sherbakoff, 1915; Wollenweber, 1931; Wollenweber and Reinking, 1935a)

*F. bostrycoides* Wr. and Rg.
(Wollenweber and Reinking, 1925; Wollenweber, 1931; Wollenweber and Reinking, 1935a)

*F. bulbigenum* Cke. and Mass.
(Wollenweber, 1931; Wollenweber and Reinking, 1935a)

*F. conglutinans* var. citrinum Wr.
(Wollenweber, 1931; Wollenweber and Reinking, 1935a)

*F. orthoceras* Appl. and Wr.
(Appel and Wollenweber, 1910; Wollenweber, 1931; Wollenweber and Reinking, 1935a)

*F. orthoceras* Appl. and Wr. var. *longius* (Sherb.) Wr.
(Wollenweber, 1917, 1931; Wollenweber and Reinking, 1935a)

*F. oxysporum* Schlecht. var. *meniscoideum* Bugn.
(Bugnicourt, 1939; Wollenweber, 1944–45)

*F. vasinfectum* (Atk.) Wr. and Rg.
(Wollenweber and Reinking, 1935a)

*F. vasinfectum* Atk. var. *lutulatum* (Sherb.) Wr.
(Wollenweber, 1931; Wollenweber and Reinking, 1935a)

*F. vasinfectum* Atk. var. *zonatum* (Sherb.) Wr.
(Wollenweber, 1931; Wollenweber and Reinking, 1935a)

Cultures white, peach, violet to purple. Microconidia in aerial mycelium or false heads, abundant, oval, elliptical, cylindrical, straight, or curved. Macroconidia thin-walled, in aerial mycelium, rarely in sporodochia or pionnotes, spindle- or sickle-shaped, falcate, elliptical, curved, subulate, or straight, with narrowed upper cell and pedicellate foot cell. Measurements (μm) and frequency of conidia:

| | | |
|---|---|---|
| zero-sept. | 59% | 6–10 × 2.4–3.4 |
| one-sept. | 17% | 11–17 × 2.6–3.6 |
| three-sept. | 19% | 28–36 × 2.8–4.9 |
| four-sept. | 14% | 30–39 × 3.5–4.4 |
| five-sept. | 1% | 33–45 × 3.7–4.8 |
| Growth rate: 4.9 cm | | |

Chlamydospores smooth or rough-walled, abundant, terminal and intercalary, single or in pairs. Sclerotia mostly present, sometimes absent.

## F. oxysporum var. redolens

This is found on a variety of plants in temperate and subtropical zones.

F. oxysporum var. redolens (Wollenw.) Gordon No. 916 (Figs. 10.4 through 10.6).

Synonyms:

F. redolens Wr.
(Wollenweber, 1913, 1931; Wollenweber and Reinking, 1935a; Raillo, 1950)
F. solani (Mart.) var. redolens (Wr.) Bilai
(Bilai, 1955)
F. redolens Wr. var. solani Sherb.
(Sherbakoff, 1915)

Cultures white, cream, rose, beige, red-brown. Microconidia oval, cylindrical, slightly curved, formed from phialides on short conidiophore branches. Macroconidia in sporodochia and pionnotes, falcate or sickle-shaped with rounded apex (as in F. solani) and narrowing basal cell, pedicellate with foot cell, typically three- to five-septate. Measurements (μm) and frequency of conidia:

| | | |
|---|---|---|
| zero-sept. | 46% | 8–14 × 3.4–4.9 |
| one-sept. | 20% | 15–23 × 3.6–4.5 |
| three-sept. | 28% | 28–36 × 3.8–5.0 |
| five-sept. | 6% | 35–48 × 4.2–5.4 |

Growth rate: 4.5 cm

Chlamydospores abundant, terminal and intercalary, globose to oval, smooth- to rough-walled. Sclerotia rare, cream or light brown or absent.

## Liseola Section

## F. moniliforme

This is distributed worldwide on a very large range of plants in subtropical, tropical, and humid temperate zones and in various cultivated and uncultivated soils in warmer rather than cooler regions of the world.

F. moniliforme Sheldon No. 1857 (Figs. 10.7 through 10.9).

Synonyms:

F. moniliforme Sheld. var. minus Wr.
(Wollenweber, 1931; Wollenweber and Reinking, 1935a)
F. lactis (Pir. and Rib.) Wr. and Rg.
(Wollenweber and Reinking, 1935a)

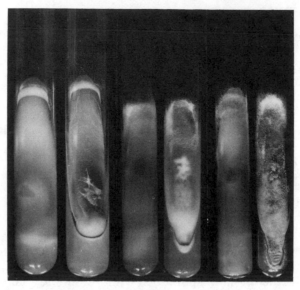

**Figure 10.7** Cultural appearance of *F. moniliforme* No. 1857, *F. moniliforme* var. *subglutinans* No. 658, and *F. decemcellulare* No. 11 on PDA slants (from left to right). (Joffe, unpublished results.)

*F. moniliforme* Sheld. var. *lactis* (Pir. and Rib.) Bilai
(Bilai, 1955)

*F. neoceras* Wr. and Rg.
(Wollenweber and Reinking, 1925; Wollenweber, 1931; Wollenweber and Reinking, 1935a)

*F. moniliforme* Sheld. var. *majus* (Wr. and Rg.) Raillo
(Raillo, 1950)

*F. verticillioides* (Sacc.) Nirenberg
(Nirenberg, 1976)

Cultures peach, pale cream, purple-violet to lilac. Aerial mycelium dense, floccose to felted, somewhat powdery with production of microconidia.

Microconidia formed on the aerial mycelium on lateral, elongated phialides in chains, ovoid, spindly oblong with a broad and rounded apex, occasionally becoming one-septate.

Macroconidia thin-walled, formed also on phialides of branched conidiophores in sporodochia or sometimes in pionnotes or scattered in aerial mycelium, spindly, falcate with elongated, narrow, slightly curved

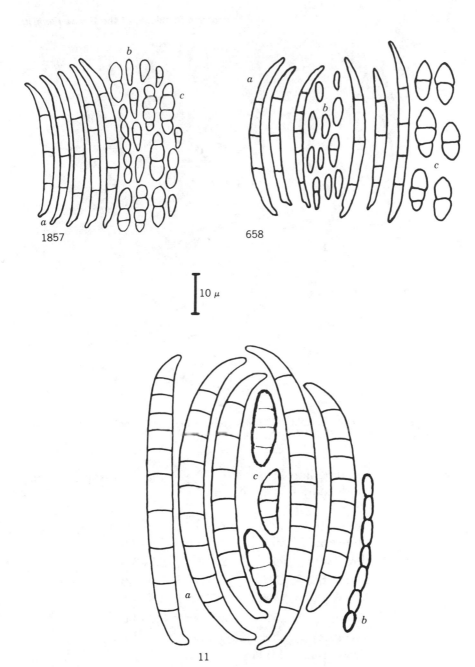

**Figure 10.8** *F. moniliforme* No. 1857; *F. moniliforme* var. *subglutinans* No. 658; *F. decemcellulare* No. 11. (*a*) macroconidia; (*b*) microconidia; (*c*) ascospores. (Joffe, unpublished results.)

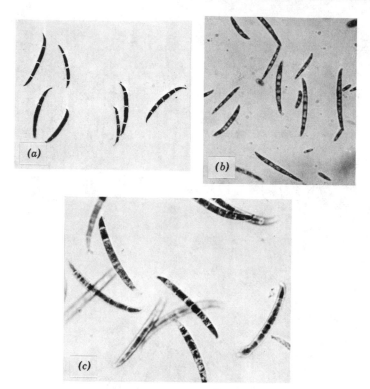

**Figure 10.9**   (a) *F. moniliforme* No. 1857; (b) *F. moniliforme* var. *subglutinans* No. 658; (c) *F. decemcellulare* No. 11. (Joffe, unpublished results.)

apical cells and pedicellate basal cells. Typically three-septate. Measurements (μm) and frequency of conidia:

| | | |
|---|---|---|
| zero-sept. | 5% | 7–13 × 2.4–3.5 |
| one-sept. | 9% | 11–12 × 2.6–3.6 |
| three-sept. | 44% | 28–45 × 2.7–3.8 |
| four-sept. | 25% | 30–50 × 3.0–4.0 |
| five-sept. | 14% | 46–54 × 3.3–4.7 |
| six-sept. | 2.5% | 50–60 × 3.0–4.0 |
| seven-sept. | 0.5% | 54–68 × 3.1–4.2 |

Growth rate: 4.9 cm

Chlamydospores absent; sclerotia dark blue; stroma wrinkled, light brown, rose, rose-red, rose-lilac.

Perithecial state: *G. fujikuroi* (Saw.) Wr. Perithecia superficial with rough outer wall, dark blue, globose or oviform, 195–380 μm in diameter, with multicellular, cylindrical paraphyses. Asci elliptical to club-shaped with four to six, rarely eight, elongated ellipsoid ascospores. Ascospore measurements (μm):

| | |
|---|---|
| one-sept. | 12–17 × 4.5–7 |
| two-sept. | 15–25 × 4.2–8.0 |
| three-sept. | 18–31 × 4.5–8.5 |

### F. moniliforme var. subglutinans

This has a wide distribution on plants and in soil in subtropical and tropical (rarely in colder) countries.

*F. moniliforme* Sheld. var. *subglutinans* Wr. and Rg. No. 658 (Figs. 10.7 through 10.9).

Synonyms:

*F. neoceras* (Wr. and Rg.) Bilai
(Bilai, 1955)

*F. neoceras* Wr. and Rg. var. *subglutinans* (Wr. and Rg.) Raillo
(Raillo, 1950; Bilai, 1955)

*F. moniliforme* Sheld. var. *lacticolor* Raillo
(Raillo, 1950; Bilai, 1955)

*F. sacchari* (Butl.) Gams var. *subglutinans* Nirenberg
(Nirenberg, 1976)

Cultures white to greyish-white, purple to dark purple. Microconidia hyaline, oval to fusiform, only in heads and later scattered in mycelium, one or two cells formed from phialides of branched conidiophores. Macroconidia thin-walled, falcate with a marked apical and pedicellate basal cell, developed in sporodochia, formed from phialides of short conidiophores. Typically three- to five-septate. Measurements (μm) and frequency of conidia:

| | | |
|---|---|---|
| zero-sept. | 8% | 9–14 × 2.0–3.2 |
| one-sept. | 16% | 13–23 × 2.5–3.6 |
| two-sept. | 3% | 20–27 × 2.7–3.7 |
| three-sept. | 55% | 28–40 × 3.0–3.8 |
| four-sept. | 12% | 34–45 × 3.2–4.0 |
| five-sept. | 6% | 41–51 × 3.6–4.2 |
| Growth rate: 4.8 cm | | |

Chlamydospores absent. Perithecial state: *G. fujikuroi* (Saw.) Wr. var. *subglutinans*. Perithecia oval with a light rough outer wall, dark blue, 200–360 × 170–385 μ, single or in groups. Asci are ellipsoid with paraphyses with four to eight hyaline ascospores, one-septate, rarely with two to three septa.

one-sept.     12–30 × 3.5–7.5
two-sept.     16–22 × 4–6
three-sept.   18–24 × 4.5–6.5

## Spicarioides Section

### *F. decemcellulare*

This occurs on many plants and in soil in tropical and subtropical areas of various countries.

*F. decemcellulare* Brick. No. 11 (Figs. 10.7 through 10.9).
Synonyms:

*F. rigidiusculum* (Brick) Snyd. and Hans.
(Snyder and Hansen, 1945)

*F. spicariae-colorantis* Sacc. and Trott.
(Wollender and Reinking, 1935a)

Culture rose, rose-red, carmine to purple. Microconidia in aerial mycelium, powdery, in chains, hyaline, oval. Macroconidia large, slightly curved, cylindrical with thick wall, in sporodochia and pionnotes, typically five- to nine-septate. Measurements (μm) and frequency of conidia:

zero-sept.     2%    6.0–10 × 3.2–4.2
one-sept.      3%    10–19 × 3.8–4.7
three-sept.    5%    28–42 × 4.6–5.4
five-sept.    16%    48–64 × 4.9–6.6
seven-sept.   51%    54–80 × 5.0–7.2
nine-sept.    17%    70–92 × 5.2–8.4
eleven-sept.   6%    85–114 × 5.4–8.6
Growth rate: 3 cm

Chlamydospores absent; stroma rose-red, purple. Perithecial state: *C. rigidiuscula* (Berk. and Br.) Sacc. Perithecia scattered or gregarious oval-round, formed on stroma, wood, cream to yellow with scaled outer wall. They measure 210–340 × 170–280 μ. Outer wall 18–32 μ wide and inner wall 12–18 μ wide. The asci (65–105 × 11–15) are curved clavate with mainly four spores, three-septate, hyaline, ellipsoid, rarely with one, two, and eight ascospores; three-sept. 20–27 × 6–11.

## Gibbosum Section

### F. equiseti

This is widespread on a variety of host plants in many climatic zones, including temperate, subtropical, and tropical regions.

F. equiseti (Corda) Sacc. No. 2385 (Figs. 10.10 through 10.12). Synonyms:

F. equiseti (Cda.) Sacc. var. *bullatum* (Sherb.) Wr.
(Gordon, 1952; Wollenweber, 1931; Wollenweber and Reinking, 1935a)

F. gibbosum App. and Wr. emend. Bilai
(Bilai, 1955)

F. gibbosum App. and Wr. emend. Bilai var. *bullatum* (Sherb.) Bilai
(Bilai, 1955)

F. scirpi Lamb. and Fautr. var. *caudatum* Wr.
(Wollenweber, 1931; Wollenweber and Reinking, 1935a)

F. scirpi Lamb. and Fautr. var. *compactum* f. 1. Wr.
(Wollenweber, 1931)

F. scirpi Lamb. and Fautr. var. *filiferum* (Preuss) Wr.
(Wollenweber, 1931; Wollenweber and Reinking, 1935a)

F. roseum Lk. emend. Snyd. and Hans.
(Snyder and Hansen, 1945)

F. roseum Lk. emend. Snyd. and Hans. f. *cereale* (Cke.)
(Snyder and Hansen, 1945)

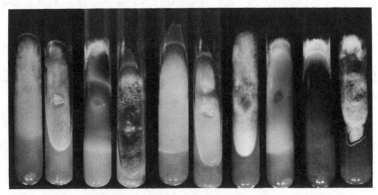

**Figure 10.10** Cultural appearance of F. equiseti No. 2385, F. equiseti var. *acuminatum* No. 194, F. *lateritium* No. 295, F. *semitectum* No. 185, and F. *semitectum* var. *majus* No. 81 on PDA slants (from left to right). (Joffe, unpublished results.)

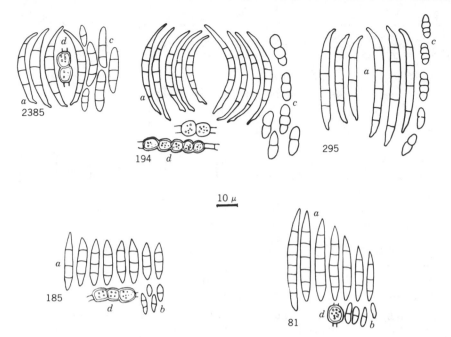

**Figure 10.11** _F. equiseti_ No. 2385; _F. equiseti_ var. _acuminatum_ No. 194; _F. lateritium_ No. 295; _F. semitectum_ No. 185; _F. semitectum_ var. _majus_ No. 81. (_a_) macroconidia; (_b_) microconidia; (_c_) ascospores; (_d_) chlamydospores. (Joffe, unpublished results.)

Cultures white, cream, pale olive to yellow-brown, with floccose aerial mycelium. Microconidia rare, spindle-shaped or oval. Zero- to two-septate.

Macroconidia narrowly falcate to parabolic and with thicker central part, elongated, thin-walled, with pointed apical cell and marked foot cell formed on phialides of branched conidiophores in aerial mycelium, sporodochia, and pionnotes. Typically five-septate. Measurements (μm) and frequency of conidia:

|              |       |                         |
| ------------ | ----- | ----------------------- |
| zero-sept.   | 0.5%  | 6.5–16 × 2.4–4.0        |
| one-sept.    | 1.5%  | 11–22 × 2.6–4.2         |
| two-sept.    | 2%    | 12–24 × 2.3–3.8         |
| three-sept.  | 22%   | 27–43 × 3.6–4.8         |
| five-sept.   | 64%   | 36–54 × 3.8–5.0         |
| seven-sept.  | 10%   | 43–62 ×4.0–5.4          |
| Growth rate: 6.0 cm |  |                    |

Chlamydospores intercalary, single or in pairs, knots or chains, occasionally terminal, globose, smooth- to rough-walled. Sclerotia rare, brown or dark blue, or absent. Stroma brown or yellow-brown to red.

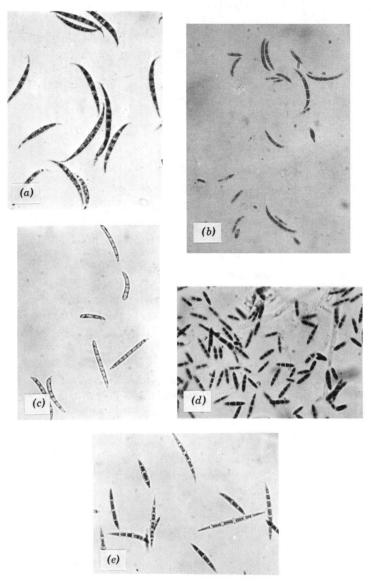

**Figure 10.12** (a) *F. equiseti* No. 2385; (b) *F. equiseti* var. *acuminatum* No. 194; (c) *F. lateritium* No. 295; (d) *F. semitectum* No. 185; (e) *F. semitectum* var. *majus* No. 81. (Joffe, unpublished results.)

Perithecial state: *G. intricans* Wr. Perithecia superficial, single or grouped, formed on poorly developed, wrinkled stroma, with weakly developed paraphyses, ovoid, 170–240 μm diameter. Asci elliptical to clavate, 64–100 × 9–14, with four to eight ascospores (rarely two, occasionally three), one to two or four to seven septa, hyaline, spindle-shaped, slightly curved or straight, slightly rounded at both ends, sometimes ellipsoidal with wedge-shaped foot cell. Ascospore measurements (μm):

| | |
|---|---|
| one-sept. | 14–22 × 3.7–5.5 |
| two-sept. | 18–30 × 3.8–6.5 |
| three-sept. | 20–34 × 4.0–5.4 |
| four-sept. | 23–36 × 3.7–6.8 |
| five-sept. | 25–38 × 4.2–7.2 |
| seven-sept. | 27–40 × 4.5–7.5 |

### *F. equiseti* var. *acuminatum*

This has appeared in the literature under the following names: *F. acuminatum* (Booth, 1971b; Butler, 1961; Gerlach, 1970, 1977; Gerlach and Ershad, 1970; Gordon, 1952, 1954a; Ylimäki, 1967), *F. gibbosum* var. *acuminatum* (Bilai, 1955, 1977), and *F. scirpi* var. *acuminatum* (Bugnicourt, 1939; Gordon, 1944; Kommedahl and Windels, 1981; Raillo, 1950; Windels *et al.*, 1976; Wollenweber and Reinking, 1935a).

*F. equiseti* (Cda.) Sacc. var. *acuminatum* (Ell. and Ev.) Joffe comb. nova. No. 194 (Figs. 10.10 through 10.12).

Synonyms:

*F. scirpi* Lamb. and Fautr. var. *acuminatum* (Ell. and Ev.) Wr.
(Wollenweber, 1931; Wollenweber and Reinking, 1935a)

*F. roseum* Lk. emend. Snyd. and Hans.
(Snyder and Hansen, 1945)

*F. roseum* Lk. f. *cereale* (Cke.) emend. Snyd. and Hans.
(Snyder and Hansen, 1945)

*F. gibbosum* App. and Wr. emend. Bilai var. *acuminatum* (Ell. and Ev.)
(Bilai, 1955)

*F. acuminatum* Ell. and Ev.
(Booth, 1971b; Wollenweber, 1917)

Cultures white-rose, peach-beige, brown, carmine-red to purple. Macroconidia broadly falcate, hyperbolic, dorsiventral with elongated apical cell and well-marked foot cell, formed on phialides of branched conidiophores, in sporodochia, sometimes pionnotes and aerial mycelium, typically five-septate. Measurements (μm) and frequency of conidia:

| | | |
|---|---|---|
| three-sept. | 19% | 20–42 × 3.3–4.4 |
| four-sept. | 15% | 26–48 × 3.6–4.8 |
| five-sept. | 56% | 34–52 × 3.8–4.8 |
| six-sept. | 4% | 42–56 × 4.0–5.0 |
| seven-sept. | 6% | 44–67 × 3.8–5.2 |
| Growth rate: 4.5 cm | | |

Chlamydospores intercalary, smooth-walled, oval, in chains and knots. Stroma plectenchymatous, red, yellow-cream to brown.

Perithecial state: *G. acuminata* Wr. Perithecia single or in groups, olive green to blue-black, grown on sterile wheat straw and found on cercal grains (barley, rye, and wheat) in the United States. They are round, oval, measure 180–200 × 250–320 μm with a smooth outer wall 20–32 μ wide, and inner walls 6–10 μ. The asci elliptical to clavate with four to eight ascospores, one to three septa, ellipsoidal to oval, hyaline, smooth.

| | |
|---|---|
| one-sept. | 12–20 × 4.8–5 |
| three-sept. | 15–26 × 5–10 |

## Lateritium Section

### *F. lateritium*

This is widespread on a variety of plants and insects in temperate, subtropical, and tropical regions.

*F. lateritium* Nees No. 295 (Figs. 10.10 through 10.12).
Synonyms:

*F. lateritium* Nees var. *longum* Wr.
(Wollenweber, 1931; Wollenweber and Reinking, 1935a)
*F. lateritium* Nees var. *majus* Wr.
(Wollenweber, 1931; Wollenweber and Reinking, 1935a)
*F. lateritium* Nees var. *minus* Wr.
(Wollenweber, 1931; Wollenweber and Reinking 1935a)
*F. lateritium* Nees var. *mori* Desm.
(Wollenweber, 1931; Wollenweber and Reinking, 1935a)
*F. lateritium* Nees var. *uncinatum* Wr.
(Wollenweber, 1931; Wollenweber and Reinking, 1935a)

*F. sarcochroum* (Desm.) Sacc.
(Wollenweber, 1931; Wollenweber and Reinking, 1935a; Bilai, 1955)

*F. lateritium* Nees var. *stilboides* (Wr.) Bilai
(Bilai, 1955)

*F. udum* Butl.
(Butler, 1961)

Cultures white, yellow, white-rose, brown, blue-black, green. Microconidia in aerial floccose or felted mycelium very sparse or absent. Macroconidia thin-walled, oblong or short, falcate to straight with beaked apex and with pedicellate foot cell formed in short, cylindrical phialides of branched conidiophores. Measurements (μm) and frequency of conidia:

| | | |
|---|---|---|
| zero-sept. | 2% | 7–15 × 2.8–4.4 |
| one-sept. | 3% | 11–24 × 2.6–4.0 |
| three-sept. | 48% | 23–42 × 2.9–4.4 |
| four-sept. | 15% | 30–46 × 3.2–4.8 |
| five-sept. | 23% | 34–60 × 3.4–5.4 |
| seven-sept. | 9% | 36–64 × 3.6–5.6 |
| Growth rate: 3.6 cm | | |

Chlamydospores rare, intercalary, in mycelium or macroconidia. Perithecial state: *G. baccata* (Wallr.) Sacc. Perithecia scattered, gregarious, developed in culture on sterile wheat straw and on stroma of wood trees and twigs of shrubs. The perithecia are black, round globose to pyriform with a narrow apical ostiole. Young perithecia soft when moist, with a smooth outer wall; when dry, wall wrinkled. They measure 160–240 × 130–220 μ, with an outer pseudoparenchymatous wall 16–27 μ wide and inner wall 9 μ.

Asci oblong or cylindrical with short stalk and round apex 60–85 × 7–10 μ. The asci with eight ascospores one-septate, and rarely with four-ascospores three-septate.

| | |
|---|---|
| one-sept. | 10–19 × 4–8 |
| three-sept. | 12–22 × 5–10 |

## Arthrosporiella Section

### *F. semitectum*

This occurs on various hosts and soil in subtropical, tropical, and temperate regions in all parts of the world.

*F. semitectum* Berk. and Rav. No. 185 (Figs. 10.10 through 10.12).
Synonyms:

*F. semitectum* Berk. and Rav. var. *majus* Wr.
(Wollenweber, 1931; Wollenweber and Reinking, 1935a)
*F. diversisporum* Sherb.
(Sherbakoff, 1915; Wollenweber, 1931; Wollenweber and Reinking, 1935a)
*F. roseum* Lk. emend. Snyd. and Hans.
(Snyder and Hansen, 1945)

Cultures white, yellow to brown. Macroconidia in aerial mycelium. Sporodochia absent. Conidia curved, lanceolate, spindle- to sickle-shaped with wedge-shaped foot cell and curved apex. Measurements (μm) and frequency of conidia:

|              |      |                       |
| ------------ | ---- | --------------------- |
| zero-sept.   | 6%   | $6–10 \times 2.5–3.6$ |
| one-sept.    | 11%  | $9–17 \times 2.6–4.0$ |
| three-sept.  | 76%  | $20–31 \times 3.1–4.4$ |
| five-sept.   | 7%   | $26–40 \times 3.3–4.7$ |
| Growth rate: 6 cm | | |

Chlamydospores sparse, globose, intercalary, single or in chains. Perfect stage unknown.

### F. semitectum var. majus

This has been found in subtropical, tropical, and temperate areas everywhere, on many plants and in soil.
    *F. semitectum* Berk. and Rav. var. *majus* Wr. No. 81 (Figs. 10.10 through 10.12).
    Synonyms:

*F. diversisporum* Sherb.
(Sherbakoff, 1915; Wollenweber, 1931; Wollenweber and Reinking, 1935a)
*F. roseum* Lk. emend. Snyd. and Hans.
(Snyder and Hansen, 1945)
*F. juglandinum* Peck
(Wollenweber and Reinking, 1935a)
*F. incarnatum* (Rob.) Sacc.
(Wollenweber and Reinking, 1935a)

Cultures white-rose to brown. Microconidia often sparse. Macroconidia formed in aerial mycelium, rarely in sporodochia, curved or almost straight, narrowing at both ends. Measurements (μm) and frequency of conidia:

|           |      |                      |
|-----------|------|----------------------|
| zero-sept. | 2%   | 11–14 × 2.3–3.2      |
| one-sept.  | 1%   | 13–22 × 2.6–3.5      |
| three-sept. | 2%  | 24–40 × 3.2–4.4      |
| four-sept.  | 8%  | 28–4.6 × 3.5–4.8     |
| five-sept.  | 61% | 35–52 × 3.6–5.0      |
| six-sept.   | 5%  | 41–56 × 3.8–5.3      |
| seven-sept. | 21% | 44–66 × 4.0–5.6      |

Growth rate: 5.2 cm

Chlamydospores sparse, intercalary, globose and smooth, single or in pairs or chains. Perfect stage unknown.

## Roseum Section

### F. avenaceum

This is widespread in temperate, subtropical, and tropical zones throughout the world on a variety of cereal crops, grasses, fruits, vegetables, and other hosts.

*F. avenaceum* (Fries) Sacc. No. 1018 (Figs. 10.13 through 10.15). Synonyms:

*F. De Tonianum* Sacc.
(Bilai, 1955)

*F. avenaceum* (Fr.) Sacc. var. *De Tonianum* (Sacc.)
(Raillo, 1950)

*F. avenaceum* (Fr.) Sacc. var. *herbarum* (Cda.) Sacc.
(Raillo, 1950)

*F. avenaceum* (Fr.) var. *pallens* Wr.
(Wollenweber, 1944–45; Wollenweber and Reinking, 1935a)

*F. avenaceum* (Fr.) Sacc. var. *volutum* Wr. and Rg.
(Wollenweber and Reinking, 1935a)

*F. graminum* Cda.
(Wollenweber, 1944–45; Wollenweber and Reinking, 1935a)

*F. avenaceum* (Fr.) Sacc. var. *graminum* Cda.
(Raillo, 1950)

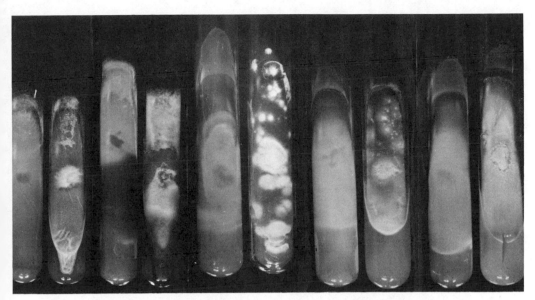

**Figure 10.13** Cultural appearance of *F. avenaceum* No. 1018, *F. arthrosporioides* No. 622, *F. nivale* No. 944, *F. dimerum* No. 53-8, and *F. merismoides* No. 8479 on PDA slants (from left to right). (Joffe, unpublished results.)

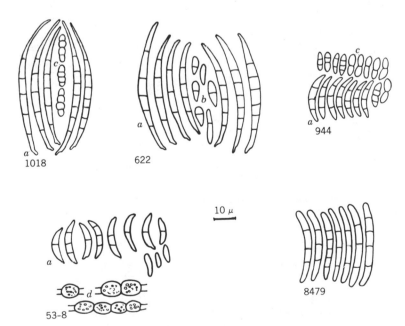

**Figure 10.14** *F. avenaceum* No. 1018; *F. arthrosporioides* No. 622; *F. nivale* No. 944; *F. dimerum* No. 53-8; *F. merismoides* No. 8479. (*a*) macroconidia; (*b*) microconidia; (*c*) ascospores; (*d*) chlamydospores. (Joffe, unpublished results.)

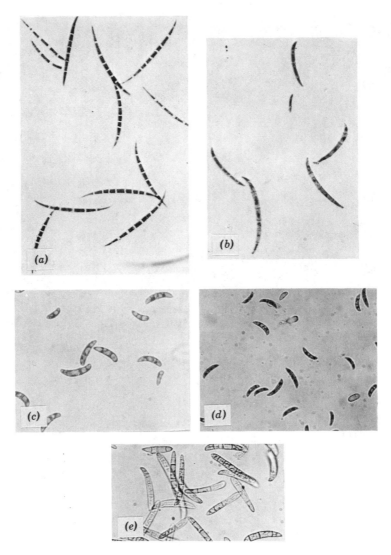

**Figure 10.15**   (*a*) *F. avenaceum* No. 1018; (*b*) *F. arthrosporioides* No. 622; (*c*) *F. nivale* No. 944; (*d*) *F. dimerum* No. 53-8; (*e*) *F. merismoides* No. 8479. (Joffe, unpublished results.)

Cultures white, yellow, or red. Aerial mycelium cobwebby to floccose.

Microconidia absent or very rarely occurring in young aerial mycelium; zero- to two-septate, elliptical, or spindle-shaped.

Macroconidia formed in sporodochia and aerial mycelium on short phialides, often also in pionnotes; they are narrowly fusoid, filiform, elongated, elliptical, rarely slightly curved or almost straight, with elongated apical cell and well-marked foot cell. Typically five-septate. Measurements (μm) and frequency of conidia:

three-sept.  10%    31–56 × 2.5–3.8
four-sept.   18%    39–72 × 3.0–4.3
five-sept.   71%    45–86 × 3.0–4.2
seven-sept.  1%     58–92 × 3.2–4.0
Growth rate: 5.2 cm

Chlamydospores generally absent. Sclerotia rare, yellow-purple or deep blue, sometimes absent. Stroma yellow, carmine, or red-brown.

Perithecial state: *G. avenaceae* R. J. Cook. Perithecia dark yellow or dark red, occurring singly or in groups, irregularly globose to pyriform, 135–240 μm diameter, with thick-walled cells. Paraphyses absent, periphyses present. Asci cylindrical to clavate, 50–95 × 8–13 μm, with eight hyaline, elliptical ascospores, rounded at both ends, constricted at the central septum, often one-, rarely three-septate. Ascospore measurements (μm):

one-sept.     12–20 × 3.8–5.0
three-sept.   16–24 × 4.5–6.2

### F. arthrosporioides

This is widespread on a variety of plants, mainly in temperate regions. *F. arthrosporioides* Sherb. No. 622 (Figs. 10.13 through 10.15). Synonyms:

*F. roseum* Lk. emend. Snyd. and Hans.
(Snyder and Hansen, 1945)

Cultures rose, carmine to brown. Microconidia absent or sparse, in aerial mycelium, pyriform to oval, formed on phialides in the aerial mycelium. Macroconidia in sporodochia on branched conidiophores which are rather like the structure of *Penicillium*. They are fusiform, lanceolate, narrowly falcate, strongly dorsiventral with marked foot cell and pointed elongated tip. Measurements (μm) and frequency of conidia:

zero-sept.    1%     8–18 × 2.4–3.8
one-sept.     3%     20–36 × 2.8–4.0
two-sept.     1%     23–40 × 3.0–4.4
three-sept.   21%    28–48 × 3.2–4.6
five-sept.    69%    45–64 × 3.6–5.0

seven-sept.  5%      62–75 × 4.2–5.2
Growth rate: 4.5 cm

Chlamydospores absent. Perithecial state unknown.

## Arachnites Section

### F. nivale

This has been isolated from cereal crops and grasses, mainly in temperate zones.
F. *nivale* (Fr.) Cesati No. 944 (Figs. 10.13 through 10.15).
Synonyms:

*F. nivale* (Fr.) Ces. var. *majus* Wr.
(Wollenweber, 1930, 1931; Wollenweber and Reinking, 1935a)
*F. nivale* (Fr.) Ces. emend. Snyd. and Hans.
(Snyder and Hansen, 1945)
*F. nivale* (Fr.) Ces. emend. Snyd. and Hans. *f. graminicola* (Berk. and
Brme.)
(Snyder and Hansen, 1945)
*F. nivale* (Fr.) Ces. var. *larvarum* (Fuckel) Bilai
(Bilai, 1955)
*F. nivale* (Fr.) Ces. var. *majus* Wr.
(Wollenweber and Reinking, 1935a; Bilai 1955)

Cultures white, rose-pale yellow to light brown, with cobwebby aerial mycelium. Microconidia absent. Macroconidia short, curbed, falcate, one- to three-septate, rarely zero-septate, with pointed apex and wedge-shaped or round base, formed in aerial mycelium and, rarely, in pionnotes or small sporodochia on broadly elliptical phialides of the conidiophore branches. Typically one- to three-septate. Measurements (μm) and frequency of conidia:

zero-sept.  1%      8–16 × 2.2–3.6
one-sept.  33%     10–22 × 2.4–4.3
two-sept.  11%     14–25 × 2.8–4.5
three-sept. 55%     15–32 × 2.6–5.0
Growth rate: 1.2 cm

Chlamydospores and sclerotia absent. Stroma pale rose or brown.
Perithecial state: *C. nivalis* Schaffnit. Perithecia initially immersed, then appear superficially on wrinkled, warty stroma, scattered singly or

in groups, globose, oviform or onion shaped, 90–275 μm in diameter with a plectenchymatous wall, composed of three cell layers. Asci elongated, rarely cylindrical, thin-walled, with six to eight ascospores, one- to three-septate, hyaline, ellipsoid, straight or slightly curved. Ascospore measurements (μm):

| | |
|---|---|
| one-sept. | 9–11 × 2.3–3.8 |
| two-sept. | 11–19 × 2.4–4.0 |
| three-sept. | 13–22 × 2.6–4.3 |

## Eupionnotes Section

### F. dimerum

This is found in various climatic zones on host plants.
   F. *dimerum* Penzig No. 53-8 (Figs. 10.13 through 10.15).
   Synonyms:

F. *dimerum* Penz. var. *pusillum* Wr.
(Wollenweber, 1931; Wollenweber and Reinking, 1935a)
F. *aquaeductuum* (Radlk. and Rabh.) Lagh. var. *dimerum* Penz.
(Bilai, 1955)
F. *episphaeria* (Tode) Snyd. and Hans.
(Bugnicourt, 1939; Snyder and Hansen, 1945)
F. *nivale* (Fr.) Ces. emend. Snyd. and Hans. *f. graminicola* (Berk. and Brme.)
(Snyder and Hansen, 1945)

Cultures orange to rose. Macroconidia in pionnotes, sporodochia, and aerial mycelium, curved, broadly lunate, spindle- to sickle-shaped, mostly one- to three-septate. Measurements (μm) and frequency of conidia:

| | | |
|---|---|---|
| zero-sept. | 3% | 8–14 × 2.2–2.8 |
| one-sept. | 10% | 12–19 × 2.6–3.0 |
| two-sept. | 23% | 14–21 × 2.8–3.4 |
| three-sept. | 64% | 18–25 × 3.1–3.6 |
| Growth rate: 2.5 cm | | |

Chlamydospores globose or oval, smooth-walled, intercalary, single or in pairs or chains. Perfect stage unknown.

## F. merismoides

This has occurred on a variety of plants and soils in different climatic regions.

*F. merismoides* Corda No. 8479 (Figs. 10.13 through 10.15).
Synonyms:

*F. merismoides* Corda var. *chlamydosporale* Wr.
(Wollenweber, 1931; Wollenweber and Reinking, 1935a)
*F. merismoides* Cda. var. *crassum* Wr.
(Wollenweber, 1931; Wollenweber and Reinking, 1935a; Raillo, 1950)
*F. episphaeria* (Tode) Snyd. and Hans.
(Snyder and Hansen, 1945)

Cultures slimy, cream, orange, slow-growing with weakly developed mycelium. Macroconidia cylindrical, straight, or slightly curved with light rounded apex with or without foot cell, formed in pionnotes on lateral phialides. Typically three-septate. Measurements (μm) and frequency of conidia.

three-sept. 79%     25–52 × 2.8–4.6
five-sept.   21%     31–58 × 3.0–4.8
Growth rate: 1 cm

Chlamydospores intercalary, single or in pairs or chains. Perfect stage unknown.

## Discolor Section

## F. culmorum

This has been isolated from a variety of host plants in all parts of the world.

*F. culmorum* (W. G. Smith) Sacc. No. 1213 (Figs. 10.16 through 10.18).
Synonyms:

*F. culmorum* (W. G. Smith) Sacc. var. *cereale* (Cke.) Wr.
(Wollenweber, 1931; Wollenweber and Reinking, 1935a)
*R. roseum* Lk. emend. Snyd. and Hans.
(Snyder and Hansen, 1945)
*F. roseum* Lk. *f. cereale* (Cke.) emend. Snyd. and Hans.
(Snyder and Hansen, 1945)

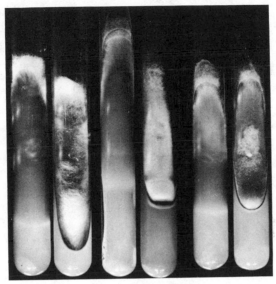

**Figure 10.16**   Cultural appearance of *F. culmorum* No. 1213, *F. sambucinum* No. 1904, and *F. graminearum* No. 765 on PDA slants (from left to right). (Joffe, unpublished results.)

Cultures red-brown, carmine, or purple. Aerial mycelium downy, dense, friable, or cobwebby. Microconidia absent. Macroconidia develop abundantly, formed from phialides of branched conidiophores in sporodochia and pionnotes, rarely in aerial mycelium, thick-walled, elliptical, slightly curved, and strongly dorsiventral, parabolic with a short, gradually, or sharply narrowing and curved apical cell and well-marked

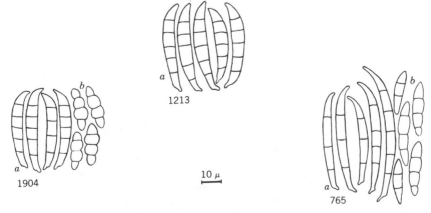

**Figure 10.17**   *F. culmorum* No. 1213; *F. sambucinum* No. 1904; *F. graminearum* No. 765. (*a*) macroconidia; (*b*) ascospores. (Joffe, unpublished results.)

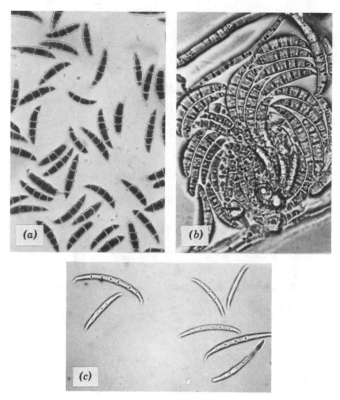

**Figure 10.18** (*a*) *F. culmorum* No. 1213; (*b*) *F. sambucinum* No. 1904 (microculture); (*c*) *F. graminearum* No. 765. (Joffe, unpublished results.)

foot cell. Typically five-septate. Measurements (μm) and frequency of conidia:

| | | |
|---|---|---|
| three-sept. | 4% | 28–40 × 5.0–6.8 |
| four-sept. | 11% | 32–44 × 5.2–7.0 |
| five-sept. | 78% | 35–50 × 5.4–7.4 |
| six-sept. | 6% | 38–5.2 × 5.5–7.6 |
| seven-sept. | 1% | 42–60 × 5.8–7.4 |
| Growth rate: 8.5 cm | | |

Chlamydospores intercalary, sometimes terminal, oval to globose, smooth- or rough-walled, single or in chains or knots. Sclerotia absent. Stroma rose, carmine-red, reddish, ochre-brown, later dark carmine or yellow-brown.

## F. sambucinum

This is widespread in many plants and soils, in temperate, subtropical, and tropical regions of the world.

F. sambucinum Fuckel No. 1904 (Figs. 10.16 through 10.18).
Synonyms:

F. sambucinum Fuckel var. *cereale* (Cke) Raillo
(Raillo, 1950)

F. sambucinum Fuckel var. *ossicola* (Berk. and Curt.) Bilai
(Bilai, 1955)

F. sambucinum Fuckel var. *minus* Wr.
(Wollenweber and Reinking, 1935a; Bilai, 1955)

F. sambucinum Fuckel var. *sublunatum* (Rg.)
(Bilai, 1955)

F. roseum Lk. emend. Snyd. and Hans.
(Snyder and Hansen, 1945)

F. roseum Lk. emend. Snyd. and Hans. *f. cereale* (Cke.)
(Snyder and Hansen, 1945)

Culture orange-rose to carmine. Aerial mycelium white, gold, yellow, rose, floccose to downy. Microconidia absent. Macroconidia formed initially in aerial mycelium and later in pionnotes or sporodochia on phialides of short conidiophores; they are falcate, sickle-shaped, slightly curved, lanceolate, strongly dorsiventral, with short, suddenly narrowing beaked apex and well-marked foot cell. Typically five-septate. Measurements (μm) and frequency of conidia:

| | | |
|---|---|---|
| three-sept. | 4% | 25–38 × 3.8–5.0 |
| four-sept. | 10% | 32–41 × 4.0–5.2 |
| five-sept. | 84% | 36–48 × 4.4–5.4 |
| six-sept. | 1% | 40–51 × 4.6–5.5 |
| seven-sept. | 1% | 44–56 × 4.8–5.6 |

Growth rate: 5.6 cm

Chlamydospores intercalary, often in chains and knots, sometimes terminal. Stroma white, yellow, brown to purple. Sclerotia dark red or brown, sometimes absent.

Perithecial state: G. pulicaris (Fr.) Sacc. Perithecia 175–250 μm in diameter, globose, scattered singly or in groups, develop on protuberant and elongated stroma; composed of several (four to five) layers of oval to globose cells. Asci 60–105 × 10–16 μm clavate, with four to eight elongated, ellipsoid, straight, or slightly curved ascospores with round ends, and slightly constricted at the septa, mostly three-septate, rarely one- to two or four- to seven-septate. Ascospore measurements ($\mu$m):

| | |
|---|---|
| one-sept. | 14–25 × 4.5–7.0 |
| two-sept. | 16–28 × 4.5–7.1 |
| three-sept. | 18–34 × 4.5–8.0 |
| four-sept. | 24–38 × 5.1–8.5 |
| seven-sept. | 29–42 × 5.3–8.8 |

### F. graminearum

This occurs worldwide on cereal crops and other hosts in temperate but rarely in subtropical or tropical regions.

*F. graminearum* Schwabe No. 765 (Figs. 10.16 through 10.18). Synonyms:

*F. roseum* Lk. emend. Snyd. and Hans. (Snyder and Hansen, 1945)

*F. roseum* Lk. emend. Snyd. and Hans. *f. cereale* (Snyder and Hansen, 1945)

Cultures variously colored: rose, white-rose, grayish rose, red to brown, carmine, and sometimes yellow. Aerial mycelium floccose, cottony or cobwebby. Microconidia absent. Macroconidia in aerial mycelium or pionnotes, rarely in sporodochia, or absent, formed on short grouped phialides of multibranched conidiophores; they are thick-walled, spindle- to sickle-shaped, falcate or dorsiventral, with slightly rounded or narrowed elongated, straight, or curved apical cell and well-marked foot cell. Typically five-septate. Measurements ($\mu$m) and frequency of conidia:

| | | |
|---|---|---|
| three-sept. | 18% | 31–52 × 3.4–4.4 |
| four-sept. | 22% | 36–58 × 3.5–4.7 |
| five-sept. | 57% | 42–72 × 3.8–4.8 |
| six-sept. | 2% | 46–74 × 4.0–5.0 |
| seven-sept. | 1% | 50–78 × 4.0–5.2 |
| Growth rate: 8.8 cm | | |

Chlamydospores present or absent, intercalary, single or in chains, rarely in knots, globose, thick-walled, hyaline to pale brown with smooth or slightly rough walls. Stroma carmine or brown, sclerotia light rose, purple, deep red, or absent.

Perithecial state: *G. zeae* (Schw.) Petch. Perithecia superficial in groups on poorly developed stroma, oval with rough tuberculare wall,

140–250 μm diameter. Asci clavate, short-stalked, with thin walls, 55–90 × 7–12 μm, usually eight, rarely four to six, hyaline, light brown or yellow, curved, fusoid ascospores with rounded ends, usually three-septate, rarely one- or four-septate. Ascospore measurements (μm):

| | |
|---|---|
| one-sept. | 13–22 × 2.4–4.8 |
| three-sept. | 15–25 × 2.7–5.2 |
| four-sept. | 19–28 × 2.6–5.5 |

## CONCLUSION

The main purpose of this chapter is to serve mycologists, mycotoxicologists, chemists, and other specialists who are interested in toxic *Fusarium* species and varieties by presenting a clear description of cultural and morphological features and a correct modern system of *Fusarium* taxonomy.

The identification of toxic *Fusarium* strains is based on my long-time experience with the natural occurrence of *Fusarium* species in the field, and also my experience with cultivating them and maintaining them under various special laboratory conditions.

The description and illustrations of all *Fusarium* species are based on thousands of single-spore strains which were isolated in the USSR and Israel.

Marasas *et al.* (1984) noted that their book includes descriptions of 203 strains of 20 *Fusarium* species belonging to 10 sections. I will deal with the progress of these authors concerning their present taxonomy.

I am very much satisfied with the authors' decision to put an end to the unacceptable taxonomy of Snyder and Hansen (1940, 1941, 1945) and their supporters and cooperators (Matuo, 1961, 1972; Messiaen, 1959; Messiaen and Cassini, 1968; Snyder and Toussoun, 1965; Toussoun and Nelson, 1968, 1975, 1976).

After 30–35 years of indefatigable criticism of Snyder and Hansen's taxonomy by numerous mycologists (Bilai, 1955, 1970, 1977; Booth, 1966, 1971, 1975; Domsch and Gams, 1972; Domsch *et al.*, 1981; Gerlach, 1970, 1978, 1981, 1982; Gerlach and Ershad, 1970; Gordon, 1952, 1954a,b, 1956a,b, 1960; Jamalainen, 1943a,b, 1944, 1970; Joffe, 1974, 1977; Joffe and Palti, 1967, 1970, 1971, 1972; Joffe *et al.*, 1964, 1965, 1967, 1972, 1973, 1974a; Nirenberg, 1976, 1981; Raillo, 1950; Seemüler, 1968; Subramanian, 1951, 1952, 1954, 1955), the cooperators of Snyder and Hansen (Drs. Marasas, Nelson and Toussoun) have understood that, at present, because of the great importance and significance of the *Fusarium* toxin problems it is not feasible to continue to use and support the simplified and unacceptable taxonomy of Snyder and Hansen and their school (Booth, 1975; Gerlach, 1970, 1981; Gerlach and Ershad, 1970; Joffe, 1974a, 1977, 1983b).

In the presentation of the taxonomy of the 203 strains by Marasas *et al.* (1984) I found many inaccuracies in their identification. For example, *F. nivale* Fn-2B was originally isolated from wheat in warm regions of Japan, and the following toxins have been isolated from this fungus: fusarenon-X; nivalenol, and diacetylnivalenol (Morooka and Tatsuno, 1970; Morooka *et al.*, 1971; Tatsuno *et al.*, 1968, 1969, 1970, 1973; Ueno, 1971; Ueno *et al.*, 1972, 1974; Yoshizawa and Morooka, 1975; Yoshizawa *et al.*, 1980b). This strain was reidentified as *F. sporotrichioides* by Nelson *et al.* (1983). I will point out that for 15–17 years this strain was referred to by these authors as a correctly determined strain of *F. nivale*.

It was rather surprising that this strain was reidentified as *F. sporotrichioides* and more so that this strain, as *F. sporotrichioides*, produced fusarenon-X, nivalenol, and diacetylnivalenol, as this strain was isolated only from warm regions of Japan. Nobody else had, up to this date, isolated *F. sporotrichioides* from tropical, subtropical, or other warm zones. We have also not been able to isolate, during 26 years in Israel, any strains belonging to the Sporotrichiella section. It is hard to believe that Japanese mycologists and researchers have classified the strain Fn-2B incorrectly. It remains only to assume that, during the years after cultivating it on various media and after multiple transfers, this strain had changed its morphological and cultural features, which is doubtful.

Because of the isolated mycotoxins (fusarenon-X, nivalenol, and diacetylnivalenol) I suggest that this strain is not identified correctly. Kubota (1977) noted that *F. nivale* Fn-2B produced fusarenon-X, nivalenol, diacetylnivalenol, and T-2 toxin.

We have identified a strain NRRL 3509 = *F. nivale* Fn-2B (according to Marasas *et al.*, 1984) as *F. sporotrichioides* var. *tricinctum* (Joffe and Palti, 1975) which produced only 0.5 mg T-2 toxin per 10 g of grain (Joffe and Yagen, 1977).

The same is true of strains such as *F. episphaeria* (*F. aquaeductuum*), *F. oxysporum* "niveum" Melon-1, *F. moniliforme* NRRL 3197, *F. poae* NRRL 3287, *F. poae* T-2, *F. poae* NRRL 3299, and many others (*F. rigidiusculum* = *F. decemcellulare*, *F. tricinctum*, *F. roseum* "Gibbosum," etc.)

Marasas *et al.* (1984) noted that our *Fusarium* cultures and extracts of cultures are "unspecified isolates of Sporotrichiella section isolated from overwintered cereals associated with ATA," or "the mycotoxicosis caused in mice by an 'unspecified' authentic isolate from overwintered grain of *F. sporotrichioides* has been described by Schoental and Joffe (1974)." This expression is groundless. Instead of groundless assertions, I recommend a cardinal revision of the taxonomy of all strains obtained from various institutions, laboratories, and researchers.

# CHAPTER 11

# ON THE QUESTION
# OF YELLOW RAIN

Since around 1975 the United States and other countries have obtained reports concerning evidence of the use of lethal chemical and biological warfare weapons against defenseless civilian populations in Southeast Asia and Afghanistan.

Reports of using chemical and biological weapons, so-called Yellow Rain, in Laos were obtained during 1975–1976. Similar reports and evidence were received from Kampuchea in 1978 and, since 1979, from Afghanistan.

Many articles and reports from UN Committees, health experts, and government representatives from the United States, Canada, and other countries, and interviews with refugees concerning chemical and biological warfare agents, have already been published. In most of these publications it was noted that the Yellow Rain weapon was associated with trichothecene mycotoxins (T-2 toxin, HT-2 toxin, deoxynivalenol, nivalenol, and diacetoxyscirpenol).

I will not debate whether there is a great deal of truth in what has been published on the subject of Yellow Rain in certain newspapers or magazines (Anderson, 1984; Anon., 1981a,b,c, 1982a,b,c; Antonov, 1982; Ashton *et al.*, 1983; Brummer, 1981; Campbell, 1981; Cullen, 1981; Dickson, 1981, 1982; Ember, 1981; Floweree, 1984; Garmon, 1981; Harmats, 1984; Harruff, 1983; Holden, 1982; Joyce, 1981a,b; Kalven, 1982; Kucewicz, 1982; Lami, 1983; Marshall, 1982a,b,c, 1983; Mirocha, 1982; Mirocha *et al.*, 1982b, 1983; Pringle, 1981, 1985; Rosen and Rosen, 1982; Schiefer, 1983a,b; Schiefer and Sutherland, 1984; Seagrave, 1981; Torrey, 1983; Wade, 1981; Watson *et al.*, 1982). Objective judgment and truth concerning the use of Yellow Rain as a chemical warfare weapon is a decision I leave to the appropriate national and international institutions.

I will however make some remarks concerning the scientific results obtained and touch upon the measures which ought to be taken when examining samples from various regions where Yellow Rain was actually used as a weapon against the population.

Due to the lack of a unified approach to the question of Yellow Rain, I think that the first measure, which is of great importance, is for an authoritative team of investigators to collect authentic samples in large enough quantities from the suspected areas of attack. The collected samples ought to be delivered safely and confidentially to various laboratories where comprehensive chemical and toxicological examinations should be undertaken. Such laboratories are available in the United States and Canada, as well as in other countries. Studies should be done by more than one investigator in more than one laboratory.

The results published by Mirocha (1982) and Mirocha *et al.* (1982b, 1983) were based on a very small number of samples obtained from contaminated vegetation, water, and from blood and urine samples taken from victims sometimes weeks after the Yellow Rain attacks. In addition organs from a soldier who had died more than one month after the first stage of Yellow Rain action were also examined.

According to my experience and investigations in this field, I am deeply sceptical as to the possibility of detection of any trichothecene mycotoxins from blood or organs of victims after a period of 48–72 hr, not to mention after a period of several weeks. I also cannot support the conclusion that the maximum detectable quantity obtained from a 140 mg sample of "yellow-green powder" containing 143 ppm T-2 toxin and 27 ppm diacetoxyscirpenol (Mirocha *et al.*, 1982b, 1983) could have caused such terrible and harmful effects and consequently such rapid death. In addition, in the majority of Yellow Rain cases since around 1975 the information received has been questionable because the samples collected were too small to verify that only trichothecene toxic metabolites were the actual tactical weapon of Yellow Rain.

I was one of the first researchers who established the etiology of ATA disease in humans in the Orenburg district of the USSR during the 1940s. I isolated, from overwintered cereal grains, hundreds of very strongly toxic strains of *Fusarium poae* and *F. sporotrichioides* and other fungi (Joffe, 1983b). Great quantities of T-2 toxin were isolated from the authentic strains responsible for outbreaks of ATA disease in humans. On the basis of our comprehensive studies we ascertained and determined the exact action of the very strong trichothecene T-2 toxin. Our investigations were carried out on animals which served as the best models for the reproduction of the clinical and histopathological syndromes resembling the ATA disease in humans. Therefore I do not believe that such low concentrations of T-2 toxin, even in the presence of other trichothecenes (DAS or HT-2 toxin), can cause the harmful effects mentioned previously.

The victims of the Yellow Rain are reported to have suffered from hemorrhaging, nausea, fever, diarrhea (bloody), vomiting of blood, breathing difficulty, itching, skin irritation, blurred vision, headache, fatigue, dizziness, and vertigo, and died within minutes or hours after exposure to Yellow Rain agents (Mirocha (1982) and Mirocha *et al.* (1982b, 1983)). In comparison ATA patients died 6–8 weeks after consumption of great quantities of toxic overwintered cereals which had been naturally contaminated with toxic fusaria mainly belonging to the Sporotrichiella section. According to my calculations, I estimated that the level which would kill a patient is at least between 40 and 75 mg of T-2 toxin and sometimes even more, depending on the toxicity, sensitivity to toxin, and age of the patient.

I may again, with full responsibility, state that T-2 toxin or other trichothecenes in levels so far isolated from Yellow Rain samples cannot cause such rapid and bizarre forms of death. I therefore do not agree with Dr. Mirocha's statement that "It is conceivable that T-2 toxin may have an etiological role in the 'sudden death toxic syndrome' of Hmong people exposed to chemical attack in Laos and Kampuchea," and also, in contrast to other scientists, that "small amounts of the toxin (T-2 toxin) and its metabolites remain bound to body tissues for much longer periods of time" (Mirocha *et al.* 1983). I think that at the present stage of investigation it is premature, and there is categorically no reason, to affirm and draw so hasty a conclusion that only trichothecenes are the etiological factors of the Yellow Rain weapon.

Even though the Yellow Rain issue has been discussed since 1975, many unanswered questions of great importance still remain.

First, is Yellow Rain associated only with trichothecene toxins? The second unsettled issue is connected with the syndromes of victims exposed to Yellow Rain, which, in the majority of cases, do not entirely resemble those symptoms caused by any of the known trichothecenes which have affected animals and humans.

Third, it is also not clear how the chemical-biological agents act on the body of the victims. To date there is a lack of detailed information as to how the lethal agents penetrate the body (e.g., by inhalation or by topical application) and why the chemical toxic compounds act so rapidly on the victims.

One way of providing an answer to these questions and furthering the understanding of the Yellow Rain issue, in my opinion, is to conduct the following experiments.

First, if Yellow Rain is indeed responsible for symptoms connected with trichothecene toxicoses (mainly T-2 toxin, HT-2 toxin, diacetoxyscirpenol, and nivalenol), then it is necessary to carry out careful investigation of the most toxic strains of *Fusarium* which produce these trichothecenes in sufficient quantities and to examine their effects on animals by various methods of administration.

Second I believe and reemphasize that there exists no other measure except that which is associated with a comprehensive examination of authentic samples gathered in great quantities from the areas of Yellow Rain attacks. From these samples it is necessary to isolate all toxic compounds and, in my opinion, all other substances (not only those belonging to trichothecenes). All these materials ought then to be studied on subhuman primates or other corresponding models (such as cats) in order to compare their reaction with that observed on the victims exposed to the Yellow Rain weapon agents.

In cases where there is difficulty in detecting compounds other than trichothecenes from authentic samples it would be advisable to add or mix an array of toxic substances, and even some gases, in order to try to reproduce the symptoms on selective test animals which resemble those observed on victims after Yellow Rain exposure.

Concerning the presence of *Fusarium* species of the Sporotrichiella section in Southeast Asia, I firmly believe that it is not possible to isolate these species from tropical regions (such as Laos or Cambodia), or subtropical areas with warm climates (such as in some regions of India or Israel), but only from cold temperate climatic regions of the world. I have isolated from Israel, from various substrates, more than 14,000 strains of *Fusarium* species and varieties belonging to different sections but have never found, during 26 years, any isolate of the Sporotrichiella section (*F. poae*, *F. sporotrichioides*, *F. sporotrichioides* var. *tricinctum*, *F. sporotrichioides* var. *chlamydosporum*). Therefore I cannot agree with Ashton *et al.* (1983) that the available evidence strongly suggests that trichothecenes isolated from Yellow Rain samples are associated with natural occurrences in Southeast Asia.

Even if some other fusaria or other fungi in the Asian tropics accidentally produced approximately 50 ppm each of three trichothecenes together with zearalenone (Rosen and Rosen, 1982), this cannot be the single factor of Yellow Rain etiology.

I believe that a more comprehensive investigation is needed to prove scientifically that the trichothecenes may indeed be the principal etiological factor of Yellow Rain.

# APPENDIX **I**

## THE EFFECT OF *FUSARIUM* CRUDE EXTRACTS ISOLATED FROM PLANTS AND SOIL IN SUBTROPICAL AREAS IN ISRAEL AND TESTED ON RABBIT SKIN

| No. of Isolate | Substrate | 6°C | 12°C | 18°C | 24°C | 30°C | 35°C |
|---|---|---|---|---|---|---|---|
| | | *F. dimerum* (Sect. Eupionnotes) | | | | | |
| A6 | Cucumber | 0 | 0 | 0 | ± | + | 0 |
| 1030 | Cucumber | 0 | 0 | 0 | + | + | 0 |
| 172a | Tomato | 0 | 0 | 0 | + | + | 0 |
| S26 | Soil | 0 | 0 | 0 | + | + | 0 |
| S39 | Soil | + | + | + | + | + | + |
| S48 | Soil | 0 | 0 | 0 | 0 | + | 0 |
| S104 | Soil | 0 | 0 | 0 | 0 | + | 0 |
| S113 | Soil | 0 | 0 | 0 | ± | + | 0 |
| S119 | Soil | 0 | 0 | 0 | + | + | 0 |
| | | *F. merismoides* | | | | | |
| 14a | Potato | 0 | 0 | 0 | ± | + | 0 |
| 853 | Potato | 0 | 0 | 0 | + | + | 0 |
| S11 | Maize | 0 | 0 | 0 | + | + | 0 |
| S17 | Maize | 0 | 0 | 0 | ± | + | 0 |
| S42 | Maize | 0 | 0 | 0 | + | ± | 0 |
| S105 | Maize | 0 | 0 | 0 | + | + | 0 |

| No. of Isolate | Substrate | 6°C | 12°C | 18°C | 24°C | 30°C | 35°C |
|---|---|---|---|---|---|---|---|
| | *F. coccidicola* (Sect. Macroconia) | | | | | | |
| 490 | | 0 | 0 | 0 | ± | ± | 0 |
| 516/2 | | 0 | 0 | 0 | + | ± | 0 |
| 517 | | 0 | 0 | 0 | + | + | ± |
| | *F. avenaceum* (Sect. Roseum) | | | | | | |
| A2b | Beet | 0 | 0 | 0 | + | 0 | 0 |
| 35/3 | Medicago | 0 | 0 | + | + | 0 | 0 |
| 47/5 | Leek | 0 | 0 | 0 | + | 0 | 0 |
| 109 | Medicago | 0 | 0 | ± | ± | 0 | 0 |
| 141 | Onion | 0 | 0 | ± | + | 0 | 0 |
| 222 | Tomato | 0 | 0 | 0 | + | 0 | 0 |
| 223 | Tomato | 0 | 0 | ± | + | 0 | 0 |
| 226 | Watermelon | 0 | 0 | 0 | ± | 0 | 0 |
| 193 | Watermelon | 0 | 0 | 0 | + | + | 0 |
| 307 | Watermelon | 0 | 0 | ± | ± | + | 0 |
| 322 | Avocado | 0 | 0 | 0 | ± | 0 | 0 |
| 339/3 | Carnation | 0 | 0 | 0 | 0 | + | 0 |
| 340 | Carnation | 0 | 0 | 0 | + | 0 | 0 |
| 327 | Tomato | 0 | 0 | 0 | ± | 0 | 0 |
| 939 | Cotton | 0 | 0 | 0 | + | 0 | 0 |
| S52 | Soil | 0 | 0 | 0 | 0 | ± | 0 |
| S59 | Soil | 0 | 0 | 0 | + | 0 | 0 |
| S67 | Soil | 0 | 0 | 0 | ± | + | 0 |
| S81 | Soil | 0 | 0 | 0 | + | + | 0 |
| S89 | Soil | 0 | 0 | 0 | + | + | 0 |
| S99 | Soil | 0 | 0 | 0 | ± | + | 0 |
| S108 | Soil | 0 | 0 | 0 | + | 0 | 0 |
| S109 | Soil | 0 | 0 | 0 | 0 | ± | 0 |
| | *F. semitectum* (Sect. Arthrosporiella) | | | | | | |
| 35 | Trifolium | 0 | 0 | 0 | ± | 0 | 0 |
| 72 | Cotton | 0 | 0 | 0 | ± | + | 0 |
| 72x | Cotton | 0 | 0 | 0 | ± | + | 0 |
| 883 | Cotton | 0 | 0 | 0 | + | + | 0 |
| 610 | Cotton | 0 | 0 | ± | + | + | 0 |
| 144/3 | Cotton | 0 | 0 | 0 | ± | + | 0 |
| 183 | Cotton | 0 | 0 | 0 | ± | + | 0 |
| 888 | Cotton | 0 | 0 | 0 | + | + | 0 |
| 889 | Cotton | 0 | 0 | 0 | + | ± | 0 |
| 904 | Cotton | 0 | 0 | + | + + | 0 | 0 |
| 906 | Cotton | 0 | 0 | + | + | 0 | 0 |
| 932 | Cotton | 0 | 0 | 0 | + | + | 0 |
| 933 | Cotton | 0 | 0 | 0 | + | ± | 0 |

| No. of Isolate | Substrate | 6°C | 12°C | 18°C | 24°C | 30°C | 35°C |
|---|---|---|---|---|---|---|---|
| 934 | Cotton | 0 | 0 | 0 | ± | ± | 0 |
| 935 | Cotton | 0 | 0 | 0 | ± | ± | 0 |
| 936 | Cotton | 0 | 0 | 0 | ± | 0 | 0 |
| 944 | Cotton | 0 | 0 | 0 | 0 | ± | 0 |
| 96 | Cucumber | 0 | 0 | 0 | ± | + | 0 |
| 292 | Cucumber | 0 | 0 | 0 | + | + | 0 |
| 293 | Cucumber | 0 | 0 | 0 | + | + | 0 |
| 171 | Sorghum | 0 | 0 | 0 | 0 | 0 | ± |
| 177 | Sorghum | 0 | 0 | 0 | 0 | 0 | ± |
| 177x | Sorghum | 0 | 0 | 0 | 0 | + | ± |
| 177/2 | Sorghum | 0 | 0 | 0 | 0 | ± | ± |
| 335 | Pepper | 0 | 0 | 0 | ± | 0 | 0 |
| 193/2 | Avocado | 0 | 0 | + | ± | ± | 0 |
| 575/1 | Avocado | 0 | 0 | 0 | + + | + | 0 |
| 578 | Tomato | 0 | 0 | + | + | + | 0 |
| 580 | Tomato | 0 | 0 | + | + + | + | 0 |
| 571/1 | Groundnut | 0 | 0 | 0 | + | 0 | 0 |
| 574 | Groundnut | 0 | 0 | + | + | 0 | 0 |
| 1758 | Avocado | 0 | 0 | 0 | ± | 0 | 0 |
| 2224 | Avocado | 0 | 0 | 0 | 0 | + | 0 |
| 2578 | Avocado | 0 | 0 | 0 | + | 0 | 0 |
| 208/1 | Melon | 0 | 0 | 0 | + | + + | 0 |
| 530/29 | Grapefruit | 0 | 0 | 0 | + | 0 | 0 |
| S7 | Soil | 0 | 0 | 0 | + + | + | 0 |
| S16 | Soil | 0 | 0 | 0 | + | 0 | 0 |
| S34 | Soil | 0 | 0 | 0 | 0 | + | 0 |
| S42 | Soil | 0 | 0 | 0 | 0 | + | 0 |
| S62 | Soil | 0 | 0 | 0 | + + | 0 | 0 |
| S74 | Soil | 0 | 0 | 0 | + | + | 0 |
| S97 | Soil | 0 | 0 | 0 | ± | + | 0 |
| S132 | Soil | 0 | 0 | 0 | ± | + | 0 |

### F. semitectum var. majus

| No. of Isolate | Substrate | 6°C | 12°C | 18°C | 24°C | 30°C | 35°C |
|---|---|---|---|---|---|---|---|
| 180a | Sorghum | 0 | 0 | 0 | ± | + | 0 |
| 876 | Cotton | 0 | 0 | 0 | ± | + | ± |
| 877 | Cotton | 0 | 0 | 0 | 0 | + | + |
| 87b | Avocado | 0 | 0 | 0 | 0 | + | 0 |
| 194b | Avocado | 0 | 0 | 0 | ± | + | 0 |
| 575 | Avocado | 0 | 0 | 0 | + | + + | 0 |
| 576a | Avocado | 0 | 0 | 0 | + | + | 0 |

### F. lateritium (Sect. Lateritium)

| No. of Isolate | Substrate | 6°C | 12°C | 18°C | 24°C | 30°C | 35°C |
|---|---|---|---|---|---|---|---|
| 81 | Tomato | 0 | 0 | ± | ± | + | 0 |
| 225 | Tomato | 0 | 0 | 0 | ± | 0 | 0 |
| 280 | Tomato | 0 | 0 | 0 | 0 | + | 0 |

| No. of Isolate | Substrate | 6°C | 12°C | 18°C | 24°C | 30°C | 35°C |
|---|---|---|---|---|---|---|---|
| 491 | Grapefruit | 0 | 0 | ± | ± | + | + + |
| 491/4 | Grapefruit | 0 | 0 | ± | ± | ± | + + |
| 492 | Grapefruit | 0 | 0 | 0 | ± | ± | + |
| 493 | Grapefruit | 0 | 0 | 0 | ± | + | + |
| 555 | Lemon | 0 | 0 | 0 | ± | + + | 0 |
| S82 | Soil | 0 | 0 | 0 | 0 | ± | 0 |
| S88 | Soil | 0 | 0 | 0 | + | + | 0 |
| S118 | Soil | 0 | 0 | 0 | ± | ± | 0 |
| S122 | Soil | 0 | 0 | 0 | + | 0 | 0 |
| S124 | Soil | 0 | 0 | 0 | + | 0 | 0 |
| S126 | Soil | 0 | 0 | 0 | + | ± | 0 |
| S129 | Soil | 0 | 0 | 0 | + | + | 0 |
| S130 | Soil | 0 | 0 | 0 | 0 | ± | + |
| S143 | Soil | 0 | 0 | 0 | + | 0 | 0 |
| S147 | Soil | 0 | 0 | 0 | + | 0 | 0 |
| S149 | Soil | 0 | + | 0 | + | + | 0 |
| S152 | Soil | 0 | + | 0 | ± | + | 0 |
| S154 | Soil | 0 | 0 | 0 | 0 | + | 0 |
| S155 | Soil | 0 | 0 | 0 | + | + | 0 |
| S158 | Soil | 0 | 0 | 0 | ± | + | 0 |

### F. moniliforme (Sect. Liseola)

| No. of Isolate | Substrate | 6°C | 12°C | 18°C | 24°C | 30°C | 35°C |
|---|---|---|---|---|---|---|---|
| A4 | Maize | 0 | 0 | ± | + | 0 | 0 |
| 11 | Maize | 0 | 0 | 0 | ± | 0 | 0 |
| 31 | Maize | 0 | 0 | 0 | ± | 0 | 0 |
| 32 | Maize | 0 | 0 | 0 | 0 | ± | 0 |
| 33 | Maize | 0 | 0 | 0 | ± | 0 | 0 |
| 34 | Maize | 0 | 0 | + + | + | ± | 0 |
| 42 | Maize | 0 | 0 | 0 | ± | ± | 0 |
| 45 | Maize | 0 | 0 | 0 | ± | + | + |
| 46-b | Maize | 0 | 0 | 0 | + | 0 | 0 |
| 48 | Maize | 0 | 0 | 0 | + | ± | ± |
| 48/2a | Maize | 0 | 0 | ± | ± | 0 | 0 |
| 49 | Maize | 0 | 0 | + | + + | ± | ± |
| 56 | Maize | 0 | 0 | 0 | 0 | ± | 0 |
| 57 | Maize | 0 | 0 | 0 | ± | 0 | 0 |
| 57/2 | Maize | 0 | 0 | 0 | + | + | 0 |
| 57/5 | Maize | 0 | 0 | 0 | ± | + | 0 |
| 58 | Maize | 0 | 0 | 0 | + | 0 | 0 |
| 63 | Maize | 0 | 0 | 0 | + + + | + | 0 |
| 64 | Maize | 0 | 0 | 0 | 0 | ± | 0 |
| 65 | Maize | 0 | 0 | 0 | + | + | 0 |
| 65a | Maize | 0 | + | + | + + | 0 | 0 |
| 65b | Maize | 0 | 0 | + | + | 0 | 0 |
| 66 | Maize | 0 | 0 | 0 | 0 | + | + |

| No. of Isolate | Substrate | 6°C | 12°C | 18°C | 24°C | 30°C | 35°C |
|---|---|---|---|---|---|---|---|
| 67 | Maize | 0 | 0 | 0 | + | + | + |
| 68 | Maize | 0 | 0 | 0 | + + | + | + |
| 70 | Maize | 0 | 0 | 0 | + | 0 | 0 |
| 79 | Maize | 0 | 0 | 0 | + | ± | ± |
| 79/1a | Maize | 0 | 0 | ± | + | + | 0 |
| 1 | Wheat | 0 | 0 | 0 | ± | ± | 0 |
| 2 | Wheat | 0 | 0 | 0 | + | 0 | 0 |
| 3 | Wheat | 0 | 0 | 0 | + | ± | 0 |
| 5 | Wheat | 0 | 0 | ± | + | + | 0 |
| 99/1/3 | Onion | 0 | 0 | 0 | + | ± | ± |
| 121/2/5 | Onion | 0 | 0 | 0 | ± | ± | 0 |
| 121/2/9 | Onion | 0 | 0 | ± | + | + | 0 |
| 127/3a | Onion | 0 | 0 | + | + | ± | 0 |
| 127/3R | Onion | 0 | 0 | 0 | + | ± | ± |
| 130/1S | Onion | 0 | 0 | 0 | ± | ± | + |
| 136/4/R | Onion | 0 | 0 | + | + | + + | + |
| 152/12/R | Onion | 0 | 0 | 0 | 0 | + | + |
| 154/3 | Onion | 0 | 0 | 0 | + | + + | 0 |
| 160/2S | Onion | 0 | 0 | 0 | 0 | + | 0 |
| 197 | Onion | 0 | 0 | ± | ± | ± | 0 |
| 330 | Onion | 0 | 0 | 0 | + | ± | 0 |
| 71 | Cotton | 0 | 0 | 0 | + | 0 | 0 |
| 288 | Cotton | 0 | 0 | 0 | ± | ± | + |
| 289 | Cotton | 0 | 0 | 0 | + + | + | 0 |
| 18 | Banana | 0 | 0 | ± | ± | 0 | 0 |
| 110 | Pepper | 0 | 0 | 0 | ± | ± | 0 |
| 344-1 | Pepper | 0 | 0 | 0 | ± | 0 | 0 |
| 178 | Sorghum | 0 | 0 | + | + + | ± | 0 |
| 179 | Sorghum | 0 | 0 | + | + + | + | + |
| 180 | Sorghum | 0 | 0 | 0 | + | ± | 0 |
| 181x | Sorghum | 0 | 0 | 0 | + | + + | 0 |
| 345 | Sorghum | 0 | 0 | 0 | + | + + | |
| 3a | Barley | 0 | 0 | 0 | + | + | 0 |
| 5a | Barley | 0 | 0 | 0 | + | ± | 0 |
| 342 | Grapefruit | 0 | 0 | 0 | + | + + | 0 |
| 342-b | Grapefruit | 0 | 0 | 0 | + | ± | ± |
| 507 | Grapefruit | 0 | 0 | 0 | 0 | ± | 0 |
| 507/1 | Grapefruit | 0 | 0 | 0 | ± | + | + |
| 510/2 | Grapefruit | 0 | 0 | + + | + | 0 | 0 |
| 511 | Grapefruit | 0 | 0 | + | + | + + | 0 |
| 512/4 | Grapefruit | 0 | 0 | 0 | ± | + | 0 |
| 512a | Grapefruit | 0 | 0 | 0 | + | 0 | 0 |
| 512/1a | Grapefruit | 0 | 0 | 0 | ± | 0 | 0 |
| 513 | Grapefruit | 0 | 0 | 0 | 0 | ± | 0 |
| 515 | Grapefruit | 0 | 0 | 0 | + | 0 | 0 |
| 520 | Grapefruit | 0 | 0 | 0 | + | + + | 0 |

| No. of Isolate | Substrate | 6°C | 12°C | 18°C | 24°C | 30°C | 35°C |
|---|---|---|---|---|---|---|---|
| 523/2 | Grapefruit | 0 | 0 | 0 | 0 | ± | 0 |
| 524 | Grapefruit | 0 | 0 | 0 | + | + + | 0 |
| 524/1x | Grapefruit | 0 | 0 | 0 | + | + | 0 |
| 651 | Mango | 0 | 0 | 0 | + | + | 0 |
| 652 | Mango | 0 | 0 | 0 | ± | + | 0 |
| 653 | Mango | 0 | 0 | 0 | + | + + | 0 |
| 654 | Mango | 0 | 0 | 0 | + | + | 0 |
| 655 | Mango | 0 | 0 | ± | + | + | 0 |
| 656 | Mango | 0 | 0 | 0 | ± | + | 0 |
| 657 | Mango | 0 | 0 | 0 | ± | + | 0 |
| 830 | Anona | 0 | 0 | 0 | + | 0 | 0 |
| 831 | Anona | 0 | 0 | 0 | ± | + | 0 |
| 801 | Castor-oil plant | 0 | 0 | 0 | + | + | 0 |
| 360/4 | Orange | | 0 | + | + | + | ± |
| 501/1 | Orange (Valencia) | 0 | 0 | 0 | ± | 0 | 0 |
| 501/2 | Orange (Valencia) | − | 0 | 0 | 0 | ± | 0 |
| 576/2 | Avocado | 0 | 0 | 0 | 0 | + | 0 |
| 2257 | Avocado | 0 | 0 | 0 | + | + + | ± |
| 110 | Avocado | 0 | 0 | 0 | + | 0 | 0 |
| 145 | Avocado | 0 | 0 | 0 | + | ± | 0 |
| 1738 | Avocado | 0 | 0 | 0 | + | 0 | 0 |
| S10 | Soil | 0 | 0 | 0 | + | ± | 0 |
| S12 | Soil | 0 | 0 | 0 | + | + | 0 |
| S14 | Soil | 0 | 0 | 0 | + | 0 | 0 |
| S43 | Soil | 0 | 0 | 0 | + | + + | 0 |
| S73 | Soil | 0 | 0 | 0 | + | + | 0 |
| S94 | Soil | 0 | 0 | 0 | + | + + | 0 |
| S95 | Soil | 0 | 0 | 0 | + | 0 | 0 |
| S107 | Soil | 0 | 0 | 0 | ± | ± | 0 |
| S119 | Soil | 0 | 0 | ± | + | + + | 0 |
| S123 | Soil | 0 | 0 | 0 | 0 | + | + |
| S125 | Soil | 0 | 0 | 0 | + | 0 | 0 |
| S131 | Soil | 0 | 0 | 0 | ± | ± | 0 |
| S137 | Soil | 0 | 0 | 0 | + | + | 0 |
| S140 | Soil | 0 | 0 | 0 | ± | + | 0 |
| S142 | Soil | 0 | 0 | 0 | + | + + | 0 |
| S151 | Soil | 0 | 0 | 0 | + | + | 0 |
| S157 | Soil | 0 | 0 | 0 | + | ± | 0 |

*F. moniliforme* var. *subglutinans*

| No. of Isolate | Substrate | 6°C | 12°C | 18°C | 24°C | 30°C | 35°C |
|---|---|---|---|---|---|---|---|
| 818 | Mango | 0 | 0 | 0 | ± | 0 | 0 |
| 819 | Mango | 0 | 0 | ± | ± | 0 | 0 |
| 820 | Mango | 0 | 0 | 0 | ± | 0 | 0 |
| 821 | Mango | 0 | 0 | 0 | 0 | + | 0 |
| 822 | Mango | 0 | 0 | 0 | + | ± | 0 |

| No. of Isolate | Substrate | 6°C | 12°C | 18°C | 24°C | 30°C | 35°C |
|---|---|---|---|---|---|---|---|
| 824 | Mango | 0 | 0 | 0 | + | 0 | 0 |
| 825 | Mango | 0 | 0 | 0 | ± | ± | 0 |
| 826 | Mango | 0 | 0 | ± | + | + | 0 |
| 874 | Mango | 0 | 0 | 0 | + | + | 0 |
| 875 | Mango | 0 | 0 | 0 | + | + | 0 |

*F. moniliforme* var. *anthophilum*

| No. of Isolate | Substrate | 6°C | 12°C | 18°C | 24°C | 30°C | 35°C |
|---|---|---|---|---|---|---|---|
| 69 | Maize | 0 | 0 | 0 | + | + + | + + |

*F. equiseti* (Sect. Gibbosum)

| No. of Isolate | Substrate | 6°C | 12°C | 18°C | 24°C | 30°C | 35°C |
|---|---|---|---|---|---|---|---|
| 62 | Onion | 0 | 0 | + | + | 0 | 0 |
| 63/1/6 | Onion | 0 | 0 | + | + + | + + | + |
| 173/3 | Onion | 0 | + | + | + + | 0 | 0 |
| 137 | Onion | 0 | 0 | 0 | ± | 0 | 0 |
| 147/5a | Onion | 0 | 0 | 0 | ± | 0 | 0 |
| 147/6b | Onion | 0 | 0 | 0 | ± | 0 | 0 |
| 153/4 | Onion | 0 | 0 | + | + | 0 | 0 |
| 153/14/R | Onion | 0 | 0 | ± | + | 0 | 0 |
| 154/3 | Onion | 0 | 0 | 0 | + | 0 | 0 |
| 154/4 | Onion | 0 | 0 | 0 | + | 0 | 0 |
| 156/2/S | Onion | 0 | 0 | + | + | 0 | 0 |
| 330/3 | Onion | 0 | 0 | + | + + | + | 0 |
| 396/3 | Onion | 0 | 0 | + | ± | 0 | 0 |
| 396/3/S | Onion | 0 | 0 | 0 | + | ± | 0 |
| 361/3 | Carrot | 0 | 0 | 0 | + + | 0 | 0 |
| 550/2 | Eggplant | 0 | 0 | ± | + | 0 | 0 |
| 61 | Tomato | 0 | 0 | + | 0 | 0 | 0 |
| 162 | Tomato | 0 | 0 | 0 | + | 0 | 0 |
| 546 | Tomato | 0 | 0 | + | + + | ± | 0 |
| 331 | Pepper | 0 | 0 | 0 | + | 0 | 0 |
| 331/2a | Pepper | 0 | 0 | ± | + + | 0 | 0 |
| 331/4 | Pepper | 0 | 0 | 0 | + | 0 | 0 |
| 334/5a | Pepper | 0 | 0 | 0 | + | ± | 0 |
| 3a | Maize | 0 | 0 | 0 | + | + | 0 |
| 38 | Maize | 0 | 0 | ± | ± | + | 0 |
| 43 | Maize | 0 | 0 | + | + | 0 | 0 |
| 44x | Maize | 0 | 0 | 0 | + | + | 0 |
| 46a | Maize | 0 | 0 | + | ± | + + | 0 |
| 38-b | Maize | 0 | 0 | 0 | + + | + | 0 |
| 48/2 | Maize | 0 | 0 | 0 | ± | 0 | 0 |
| 63a | Maize | 0 | 0 | + | + + | + | 0 |
| 79/2 | Maize | 0 | 0 | ± | + | + | 0 |
| 555a | Lemon | 0 | 0 | 0 | 0 | + | 0 |
| 555/1 | Lemon | 0 | 0 | ± | + | + | 0 |
| 555/2 | Lemon | 0 | 0 | 0 | ± | + | 0 |

| No. of Isolate | Substrate | 6°C | 12°C | 18°C | 24°C | 30°C | 35°C |
|---|---|---|---|---|---|---|---|
| 533/1 | Clemantine | 0 | 0 | 0 | ± | + | + |
| 533/2 | Clemantine | 0 | 0 | 0 | + | + + | 0 |
| 18/3 | Banana | 0 | 0 | 0 | + | + | 0 |
| 12/1 | Watermelon | 0 | 0 | 0 | + | + | 0 |
| 12/x | Watermelon | 0 | 0 | ± | + + | 0 | 0 |
| 147/2 | Watermelon | 0 | 0 | 0 | + + | 0 | 0 |
| 147/3 | Watermelon | 0 | 0 | + | + + | 0 | 0 |
| 147/3a | Watermelon | 0 | 0 | + + | + | 0 | 0 |
| 147/4 | Watermelon | 0 | 0 | ± | + + | 0 | 0 |
| 147/5 | Watermelon | 0 | 0 | + | ± | 0 | 0 |
| 147/5a | Watermelon | 0 | 0 | + | + | 0 | 0 |
| 147/5b | Watermelon | 0 | 0 | ± | + | 0 | 0 |
| 310 | Watermelon | 0 | 0 | + | ± | 0 | 0 |
| 127/1 | Melon | 0 | 0 | + | + + | 0 | 0 |
| 147/2a | Melon | 0 | 0 | 0 | ± | 0 | 0 |
| 127/5 | Melon | 0 | 0 | 0 | + + | + | + |
| 216/3 | Melon | 0 | 0 | 0 | + | 0 | 0 |
| 298a | Melon | 0 | 0 | ± | + + | + + | 0 |
| 77/1 | Cucumber | 0 | 0 | 0 | ± | + | 0 |
| 91/1 | Cucumber | 0 | 0 | 0 | + + | + + | 0 |
| 92 | Cucumber | 0 | 0 | 0 | + + | + | 0 |
| 94/1 | Cucumber | 0 | 0 | 0 | + | 0 | 0 |
| 94a | Cucumber | 0 | 0 | 0 | + + | + | 0 |
| 97/1 | Cucumber | 0 | 0 | 0 | ± | 0 | 0 |
| 97/2 | Cucumber | 0 | 0 | 0 | + | + | 0 |
| 97/5 | Cucumber | 0 | 0 | 0 | + + | 0 | 0 |
| 103 | Cucumber | 0 | 0 | + | ± | 0 | 0 |
| 124 | Cucumber | 0 | 0 | + | + + | 0 | 0 |
| 126 | Cucumber | 0 | 0 | 0 | ± | 0 | 0 |
| 128 | Cucumber | 0 | 0 | 0 | ± | ± | 0 |
| 132/1 | Cucumber | 0 | 0 | 0 | ± | + | 0 |
| 148 | Cucumber | 0 | 0 | ± | + + | + | 0 |
| 160/9 | Cucumber | 0 | 0 | ± | + + | + | 0 |
| 161 | Cucumber | 0 | 0 | + | ± | 0 | 0 |
| 212/3 | Cucumber | 0 | 0 | 0 | + | 0 | 0 |
| 212/3a | Cucumber | 0 | 0 | + + | 0 | 0 | 0 |
| 228x | Cucumber | 0 | 0 | 0 | + + | 0 | 0 |
| 230 | Cucumber | 0 | 0 | 0 | + | 0 | 0 |
| 231 | Cucumber | 0 | 0 | 0 | + | 0 | 0 |
| 311/2 | Cucumber | 0 | 0 | 0 | ± | + | 0 |
| 311/2a | Cucumber | 0 | 0 | 0 | ± | ± | 0 |
| 332 | Cucumber | 0 | 0 | 0 | ± | + + | 0 |
| 332a | Cucumber | 0 | 0 | 0 | ± | + + | 0 |
| 332/1a | Cucumber | 0 | 0 | 0 | + | + | 0 |
| 332/1b | Cucumber | 0 | 0 | 0 | ± | + | 0 |

| No. of Isolate | Substrate | 6°C | 12°C | 18°C | 24°C | 30°C | 35°C |
|---|---|---|---|---|---|---|---|
| 332/2 | Cucumber | 0 | 0 | 0 | + | ± | 0 |
| 332/1c | Cucumber | 0 | 0 | 0 | + | 0 | 0 |
| 332/3 | Cucumber | 0 | 0 | 0 | + | + | 0 |
| 357 | Cucumber | 0 | 0 | ± | ± | + | 0 |
| 625 | Strawberry | 0 | 0 | 0 | ± | + | 0 |
| 571 | Peanuts | 0 | 0 | + | + + | 0 | 0 |
| 574 | Peanuts | 0 | 0 | 0 | + + | + | 0 |
| 2923 | Peanuts | 0 | 0 | 0 | ± | 0 | 0 |
| 3037 | Peanuts | 0 | 0 | ± | + | 0 | 0 |
| 3216 | Peanuts | 0 | 0 | ± | + | 0 | 0 |
| 6922 | Peanuts | 0 | + | + | + + | ± | ± |
| 7537 | Peanuts | 0 | 0 | 0 | ± | 0 | 0 |
| 611 | Cotton | 0 | 0 | 0 | ± | ± | 0 |
| 612 | Cotton | 0 | 0 | 0 | ± | + | 0 |
| 613 | Cotton | 0 | 0 | 0 | ± | 0 | 0 |
| 220/5 | Cotton | 0 | 0 | ± | ⊥ | 0 | 0 |
| 220/7 | Cotton | 0 | 0 | 0 | ± | 0 | 0 |
| 53 | Trifolium | 0 | 0 | ± | + | + + | 0 |
| 358/1 | Carnation | 0 | 0 | ± | + | + + | 0 |
| 358/x | Carnation | 0 | 0 | 0 | + | 0 | 0 |
| 385/2 | Carnation | 0 | 0 | 0 | ± | + | 0 |
| 364x | Carnation | 0 | 0 | 0 | ± | ± | 0 |
| 364/1 | Carnation | 0 | 0 | + | + | + | 0 |
| 364a | Carnation | 0 | 0 | 0 | + | ± | 0 |
| 364b | Carnation | 0 | 0 | + | + + | + + + | 0 |
| 177 | Sorghum | 0 | 0 | 0 | + | + | 0 |
| 1173 | Avocado | 0 | 0 | 0 | ± | 0 | 0 |
| 1641 | Avocado | 0 | 0 | 0 | 0 | + | 0 |
| 1724 | Avocado | 0 | 0 | 0 | 0 | + | 0 |
| 2136 | Avocado | 0 | 0 | 0 | ± | + | 0 |
| 2269 | Avocado | 0 | 0 | 0 | + | + | 0 |
| 849 | Potato | 0 | 0 | 0 | + | 0 | 0 |
| 850 | Potato | 0 | 0 | 0 | 0 | + | 0 |
| 853 | Potato | 0 | 0 | 0 | ± | 0 | 0 |
| 855 | Potato | 0 | 0 | 0 | + | 0 | 0 |
| 954 | Wheat | 0 | 0 | 0 | ± | 0 | 0 |
| 955 | Wheat | 0 | 0 | 0 | + | 0 | 0 |
| 958 | Wheat | 0 | 0 | 0 | + | 0 | 0 |
| 960 | Wheat | 0 | 0 | 0 | ± | + | 0 |
| 997 | Millet | 0 | 0 | 0 | + | 0 | 0 |
| 998 | Millet | 0 | 0 | + | + | 0 | 0 |
| 999 | Barley | 0 | 0 | 0 | + | 0 | 0 |
| 1000 | Barley | 0 | 0 | 0 | + | 0 | 0 |
| 1001 | Barley | 0 | 0 | 0 | + | ± | 0 |
| 179/4 | Sesame | 0 | 0 | 0 | ± | + | 0 |
| 111/5 | Soil | 0 | 0 | 0 | ± | 0 | 0 |

| No. of Isolate | Substrate | 6°C | 12°C | 18°C | 24°C | 30°C | 35°C |
|---|---|---|---|---|---|---|---|
| 113/5 | Soil | 0 | 0 | 0 | + + | 0 | 0 |
| 113/20 | Soil | 0 | 0 | 0 | ± | 0 | 0 |
| 115/5 | Soil | 0 | 0 | 0 | ± | 0 | 0 |
| 115/20 | Soil | 0 | 0 | 0 | + | 0 | 0 |
| 328 | Soil | 0 | 0 | ± | 0 | 0 | 0 |
| 479/5 | Soil | 0 | 0 | 0 | + | 0 | 0 |
| 591/5 | Soil | 0 | 0 | ± | 0 | 0 | 0 |
| 695/5 | Soil | 0 | 0 | + | + | 0 | 0 |
| 841/5 | Soil | 0 | 0 | ± | ± | 0 | 0 |
| 990/5 | Soil | 0 | 0 | 0 | ± | 0 | 0 |
| 1184/20 | Soil | 0 | 0 | 0 | + | + | 0 |
| 1208 | Soil | 0 | 0 | ± | 0 | 0 | 0 |
| 1982 | Soil | 0 | 0 | 0 | ± | 0 | 0 |
| 2361 | Soil | 0 | 0 | 0 | + | 0 | 0 |
| 2375 | Soil | 0 | 0 | ± | + | + | ± |
| 2412 | Soil | 0 | 0 | ± | ± | + | 0 |
| 2472 | Soil | 0 | 0 | 0 | ± | ± | 0 |
| 3119 | Soil | 0 | 0 | ± | + | 0 | 0 |
| 2923 | Soil | 0 | 0 | + | + | 0 | 0 |
| 3037 | Soil | 0 | 0 | 0 | + | 0 | 0 |

### F. equiseti var. caudatum

| No. of Isolate | Substrate | 6°C | 12°C | 18°C | 24°C | 30°C | 35°C |
|---|---|---|---|---|---|---|---|
| 182 | Cotton | 0 | 0 | 0 | 0 | 0 | ± |
| 185 | Cotton | 0 | 0 | 0 | 0 | ± | ± |

### F. equiseti var. compactum

| No. of Isolate | Substrate | 6°C | 12°C | 18°C | 24°C | 30°C | 35°C |
|---|---|---|---|---|---|---|---|
| 891 | Cotton | 0 | 0 | 0 | + | 0 | 0 |

### F. equiseti var. acuminatum

| No. of Isolate | Substrate | 6°C | 12°C | 18°C | 24°C | 30°C | 35°C |
|---|---|---|---|---|---|---|---|
| 887 | Cotton | 0 | 0 | + | + + | 0 | 0 |
| 888/2 | Cotton | 0 | 0 | + | + | 0 | 0 |
| 890 | Cotton | 0 | 0 | + | + | 0 | 0 |
| 948 | Cotton | 0 | 0 | + | + | 0 | 0 |

### F. graminearum (Sect. Discolor)

| No. of Isolate | Substrate | 6°C | 12°C | 18°C | 24°C | 30°C | 35°C |
|---|---|---|---|---|---|---|---|
| 79 | Maize | 0 | 0 | + | + | + | 0 |
| 98 | Maize | 0 | 0 | ± | + + | + | + |
| 608a | Pepper | 0 | 0 | + | + | 0 | 0 |

### F. sambucinum

| No. of Isolate | Substrate | 6°C | 12°C | 18°C | 24°C | 30°C | 35°C |
|---|---|---|---|---|---|---|---|
| A2 | Beet | 0 | 0 | 0 | + | + | 0 |
| A2b | Beet | 0 | 0 | 0 | + | ± | 0 |
| 104 | Tomato | 0 | 0 | 0 | ± | 0 | 0 |

| No. of Isolate | Substrate | 6°C | 12°C | 18°C | 24°C | 30°C | 35°C |
|---|---|---|---|---|---|---|---|
| 864 | Dianthus | 0 | 0 | 0 | + | + | 0 |
| 866 | Dianthus | 0 | 0 | 0 | + | + | 0 |
| 868 | Dianthus | 0 | 0 | 0 | ± | + | 0 |
| 868a | Dianthus | 0 | 0 | 0 | + | 0 | 0 |
| 111 | Banana | 0 | 0 | 0 | ± | + | 0 |
| 118 | Banana | 0 | 0 | 0 | + | + | 0 |
| 122 | Carnation | 0 | 0 | 0 | + | + | 0 |
| 263 | Castor-oil plant | 0 | + + | + + + | + + + + | + + + | + |
| 405 | Onion | 0 | 0 | + | + | 0 | 0 |
| 438 | Onion | 0 | 0 | 0 | + | 0 | 0 |
| 440 | Onion | 0 | 0 | ± | + | 0 | 0 |
| 445 | Gladiolus | 0 | 0 | + | + + | 0 | 0 |
| 545/3 | Gladiolus | 0 | 0 | ± | + + | + + | 0 |
| 545/4 | Gladiolus | 0 | 0 | + | + + | + + | 0 |
| 527/1 | Orange | 0 | 0 | 0 | ± | + | 0 |
| 563 | Peanuts | 0 | 0 | ± | + | + | 0 |
| 610 | Cotton | 0 | 0 | 0 | 0 | + | 0 |
| 894 | Cotton | 0 | 0 | 0 | ± | + | 0 |
| 896 | Cotton | 0 | 0 | 0 | + | + | 0 |
| 897 | Cotton | 0 | 0 | 0 | + | ± | 0 |
| 714 | Potato | 0 | 0 | 0 | + | 0 | 0 |
| 722 | Potato | 0 | 0 | 0 | + | 0 | 0 |
| 724 | Potato | 0 | 0 | 0 | + | 0 | 0 |
| 727 | Potato | 0 | 0 | 0 | ± | 0 | 0 |
| 728 | Potato | 0 | 0 | 0 | 0 | ± | 0 |
| 845 | Potato | 0 | 0 | 0 | + | ± | 0 |
| 847 | Potato | 0 | 0 | 0 | + | ± | 0 |
| 848 | Potato | 0 | 0 | ± | + | + | 0 |
| 851 | Potato | 0 | 0 | ± | + | 0 | 0 |
| 507 | Grapefruit | 0 | 0 | 0 | 0 | ± | 0 |
| S179 | Soil | 0 | 0 | + | + + | + | 0 |
| S181 | Soil | 0 | 0 | 0 | ± | ± | 0 |
| S184 | Soil | 0 | 0 | 0 | ± | 0 | 0 |
| S191 | Soil | 0 | 0 | 0 | 0 | + | 0 |
| S193 | Soil | 0 | 0 | 0 | ± | + | 0 |
| S195 | Soil | 0 | 0 | 0 | + | + + | 0 |
| S198 | Soil | 0 | 0 | 0 | ± | + | 0 |
| S209 | Soil | 0 | 0 | 0 | + | 0 | 0 |
| S215 | Soil | 0 | 0 | 0 | 0 | + | 0 |
| S219a | Soil | 0 | 0 | 0 | + + | + | 0 |
| S226 | Soil | 0 | 0 | 0 | + | 0 | 0 |
| S228a | Soil | 0 | 0 | 0 | + | 0 | 0 |
| S244 | Soil | 0 | 0 | 0 | ± | ± | 0 |
| S254 | Soil | 0 | 0 | 0 | + + | + | 0 |
| S265 | Soil | 0 | 0 | 0 | + | 0 | 0 |

| No. of Isolate | Substrate | 6°C | 12°C | 18°C | 24°C | 30°C | 35°C |
|---|---|---|---|---|---|---|---|
| | | | | *F. culmorum* | | | |
| A10 | Cotton | 0 | 0 | 0 | + | ± | 0 |
| A10a | Cotton | 0 | 0 | 0 | + | ± | 0 |
| 118/4 | Banana | 0 | 0 | 0 | ± | + | 0 |
| 228/1 | Cucumber | 0 | 0 | 0 | + | + | 0 |
| 5a | Cucumber | 0 | 0 | 0 | + + | + | 0 |
| 208 | Cucumber | 0 | 0 | 0 | + | 0 | 0 |
| 539/1 | Grapefruit | 0 | + | + | + + | + | 0 |
| 627 | Melon | 0 | 0 | 0 | ± | 0 | 0 |
| 545 | Gladiolus | 0 | 0 | 0 | + | 0 | 0 |
| 858 | Potato | 0 | 0 | 0 | + | 0 | 0 |
| 864/1 | Carnation | 0 | 0 | 0 | + | 0 | 0 |
| 866/1 | Carnation | 0 | 0 | 0 | + | 0 | 0 |
| 868/1 | Carnation | 0 | 0 | 0 | + | 0 | 0 |
| 868/2 | Carnation | 0 | 0 | 0 | + | 0 | 0 |
| 868/3 | Carnation | 0 | 0 | 0 | + | 0 | 0 |
| 50 | Maize | 0 | 0 | ± | + | 0 | 0 |
| S67 | Soil | 0 | 0 | ± | + | 0 | 0 |
| S94 | Soil | 0 | 0 | + | + | 0 | 0 |
| S135 | Soil | 0 | 0 | + | + + | 0 | 0 |
| S256 | Soil | 0 | 0 | + | ± | 0 | 0 |
| S281 | Soil | 0 | 0 | + | + | 0 | 0 |
| S286 | Soil | 0 | 0 | + | + + | 0 | 0 |
| S289 | Soil | 0 | 0 | ± | ± | 0 | 0 |
| S301 | Soil | 0 | 0 | ± | ± | 0 | 0 |
| S306 | Soil | 0 | 0 | ± | ± | 0 | 0 |
| S312 | Soil | 0 | 0 | + | + | 0 | 0 |
| 315 | Soil | 0 | 0 | + | ± | 0 | 0 |
| | | | *F. oxysporum* (Sect. Elegans) | | | | |
| A8 | Onion | 0 | 0 | 0 | + | 0 | 0 |
| 4a/s | Onion | 0 | + | + | + + | 0 | 0 |
| 7 | Onion | 0 | 0 | 0 | + | 0 | 0 |
| 7a | Onion | 0 | 0 | ± | + | ± | 0 |
| 33/3/5 | Onion | 0 | 0 | + | + + | 0 | 0 |
| 35/8 | Onion | 0 | 0 | + | + | 0 | 0 |
| 62/3/1 | Onion | 0 | 0 | + | + | + + | 0 |
| 126/2/7 | Onion | 0 | + | + + | + + | + + + | 0 |
| 126/2/8 | Onion | 0 | 0 | + | + + | + | 0 |
| 126/10S | Onion | 0 | 0 | 0 | + | + + | 0 |
| 132/1/C | Onion | 0 | 0 | + | + | 0 | 0 |
| 132/2/S | Onion | 0 | 0 | 0 | + + | + | 0 |
| 132/4/C | Onion | 0 | 0 | ± | + | + + | 0 |
| 132/5 | Onion | 0 | 0 | 0 | ± | + | 0 |
| 132/5/C | Onion | 0 | 0 | ± | 0 | 0 | 0 |

| No. of Isolate | Substrate | 6°C | 12°C | 18°C | 24°C | 30°C | 35°C |
|---|---|---|---|---|---|---|---|
| 134/2/S | Onion | 0 | 0 | ± | + | 0 | 0 |
| 134/20/C | Onion | 0 | 0 | 0 | ± | 0 | 0 |
| 134/3/S | Onion | 0 | 0 | 0 | 0 | + | 0 |
| 134/5/S | Onion | 0 | 0 | + | + | 0 | 0 |
| 134/3a/S | Onion | 0 | 0 | + | + | 0 | 0 |
| 134/11/S | Onion | 0 | 0 | 0 | + + | + | 0 |
| 134/12/S | Onion | 0 | 0 | ± | + | + + | 0 |
| 134/15/S | Onion | 0 | 0 | 0 | + | ± | 0 |
| 139 | Onion | 0 | 0 | + | + | 0 | 0 |
| 139/1 | Onion | 0 | 0 | 0 | + | 0 | 0 |
| 139/3/C | Onion | 0 | ± | ± | + | 0 | 0 |
| 139/15/C | Onion | 0 | + | + + | + + + | + + + | 0 |
| 140/1 | Onion | 0 | 0 | ± | + | + | 0 |
| 140/2 | Onion | 0 | 0 | ± | + | 0 | 0 |
| 148/8/S | Onion | 0 | 0 | + | + + + | + + | 0 |
| 148/9/S | Onion | 0 | 0 | + | + + | ± | 0 |
| 149/1/C | Onion | 0 | 0 | + + + | + + + | + + | 0 |
| 149/4/S | Onion | 0 | 0 | + | + + + | + + | 0 |
| 150/2/S | Onion | 0 | 0 | 0 | + + | + | + |
| 151/4/S | Onion | 0 | 0 | 0 | + + | + | 0 |
| 152 | Onion | 0 | ± | + | 0 | 0 | 0 |
| 154/4/S | Onion | 0 | 0 | + | + + | + | 0 |
| 593 | Onion | 0 | 0 | + | + + | ± | 0 |
| 594 | Onion | 0 | 0 | 0 | + + | + | 0 |
| 603 | Onion | 0 | 0 | ± | + | + | 0 |
| 617 | Onion | 0 | 0 | 0 | ± | 0 | 0 |
| 618 | Onion | 0 | 0 | 0 | + | 0 | 0 |
| 361/3/S | Carrot | 0 | 0 | 0 | 0 | ± | 0 |
| 362/1 | Carrot | 0 | 0 | 0 | ± | 0 | 0 |
| 543 | Eggplant | 0 | 0 | ± | ± | 0 | 0 |
| 550 | Eggplant | 0 | ± | + | 0 | 0 | 0 |
| 556 | Eggplant | 0 | 0 | + + | + + | + | 0 |
| 557 | Eggplant | 0 | 0 | + | + + | 0 | 0 |
| 21 | Tomato | 0 | 0 | ± | + | + | 0 |
| 106 | Tomato | 0 | 0 | + | + + | + + + | 0 |
| 168 | Tomato | 0 | 0 | 0 | + + | + | + |
| 174 | Tomato | 0 | 0 | 0 | + | + | 0 |
| 222 | Tomato | 0 | 0 | + | + | + + | 0 |
| 223 | Tomato | 0 | 0 | + | + | + + | 0 |
| 224 | Tomato | 0 | 0 | + | + + | 0 | 0 |
| 225 | Tomato | 0 | 0 | 0 | + + | + | 0 |
| 275 | Tomato | 0 | 0 | + | ± | 0 | 0 |
| 276 | Tomato | 0 | 0 | 0 | + + | + | 0 |
| 277/1 | Tomato | 0 | 0 | + | + + | + | 0 |
| 278 | Tomato | 0 | 0 | 0 | + | + + | 0 |
| 280 | Tomato | 0 | 0 | 0 | + | ± | ± |

| No. of Isolate | Substrate | 6°C | 12°C | 18°C | 24°C | 30°C | 35°C |
|---|---|---|---|---|---|---|---|
| 281 | Tomato | 0 | 0 | + | + + | 0 | 0 |
| 282/1 | Tomato | 0 | ± | ± | + | + | 0 |
| 312 | Tomato | 0 | 0 | + + | + | ± | 0 |
| 316 | Tomato | 0 | 0 | 0 | + + | + | ± |
| 328 | Tomato | 0 | 0 | ± | + | 0 | 0 |
| 328/1 | Tomato | 0 | 0 | 0 | + | + + | 0 |
| 328/2 | Tomato | 0 | 0 | 0 | ± | + | 0 |
| 328/2a | Tomato | 0 | 0 | ± | + | + | 0 |
| 328/2x | Tomato | 0 | 0 | ± | + | + | 0 |
| 328/3 | Tomato | 0 | 0 | + | + | + | ± |
| 329 | Tomato | 0 | 0 | ± | + | + | 0 |
| 329/1 | Tomato | 0 | 0 | + | + + | 0 | 0 |
| 558 | Tomato | 0 | 0 | 0 | ± | + | 0 |
| 581 | Tomato | 0 | 0 | 0 | + | 0 | 0 |
| 582 | Tomato | 0 | 0 | ± | + | + | 0 |
| 584 | Tomato | 0 | 0 | + | + + | + | 0 |
| 585 | Tomato | 0 | 0 | 0 | + | 0 | 0 |
| 586 | Tomato | 0 | 0 | 0 | + | ± | 0 |
| 587 | Tomato | 0 | 0 | ± | + | 0 | 0 |
| 588 | Tomato | 0 | 0 | 0 | + + | + | + |
| 589 | Tomato | 0 | 0 | 0 | + | 0 | 0 |
| 590 | Tomato | 0 | 0 | + | + | ± | 0 |
| 591 | Tomato | 0 | 0 | 0 | + | ± | 0 |
| 592 | Tomato | 0 | 0 | 0 | + + | + | ± |
| 755 | Potato | 0 | 0 | 0 | + | + | 0 |
| 756 | Potato | 0 | 0 | + | + | 0 | 0 |
| 762 | Potato | 0 | 0 | ± | + | 0 | 0 |
| 763 | Potato | 0 | 0 | 0 | + | 0 | 0 |
| 764 | Potato | 0 | 0 | 0 | + | 0 | 0 |
| 30/2 | Pepper | 0 | 0 | 0 | 0 | + | 0 |
| 334 | Pepper | 0 | 0 | 0 | + | ± | 0 |
| 334x | Pepper | 0 | 0 | 0 | ± | + | 0 |
| 346 | Pepper | 0 | 0 | ± | + | 0 | 0 |
| 348 | Garlic | 0 | 0 | 0 | + | ± | 0 |
| 348/1 | Garlic | 0 | 0 | 0 | + + | + | 0 |
| B-6/1 | Broad bean | 0 | 0 | 0 | + | 0 | 0 |
| B-6/2 | Broad bean | 0 | 0 | 0 | + | + | 0 |
| B-6/3 | Broad bean | 0 | 0 | 0 | + | 0 | 0 |
| 47 | Leek | 0 | 0 | 0 | + | + | 0 |
| 48/2 | Maize | 0 | 0 | ± | + | 0 | 0 |
| B/1 | Barley | 0 | 0 | 0 | + | 0 | 0 |
| B/2 | Barley | 0 | 0 | 0 | + | 0 | 0 |
| B/3 | Barley | 0 | 0 | ± | + | 0 | 0 |
| 506/4 | Grapefruit | 0 | 0 | 0 | + | 0 | 0 |
| 506/4 | Grapefruit | 0 | 0 | 0 | + | 0 | 0 |

| No. of Isolate | Substrate | 6°C | 12°C | 18°C | 24°C | 30°C | 35°C |
|---|---|---|---|---|---|---|---|
| 518 | Grapefruit | 0 | 0 | ± | + | 0 | 0 |
| 518/1 | Grapefruit | 0 | 0 | ± | + | 0 | 0 |
| 518/2 | Grapefruit | 0 | 0 | ± | + | 0 | 0 |
| 518/4 | Grapefruit | 0 | 0 | 0 | ± | + | 0 |
| 519 | Grapefruit | 0 | 0 | 0 | + | + + | 0 |
| 520 | Grapefruit | 0 | 0 | 0 | ± | 0 | 0 |
| 529 | Grapefruit | 0 | 0 | 0 | + + | + | 0 |
| 529/1 | Grapefruit | 0 | 0 | 0 | + | + | ± |
| 532 | Grapefruiti | 0 | 0 | 0 | + + | + + | + |
| 499/2 | Orange (Valencia) | 0 | 0 | ± | + | + | 0 |
| 500 | Orange (Valencia) | 0 | 0 | 0 | + | + + | 0 |
| 500/2 | Orange (Valencia) | 0 | 0 | 0 | + | + | 0 |
| 504 | Orange (Valencia) | 0 | ± | + + | + + | + + + | 0 |
| 521 | Orange (Valencia) | 0 | 0 | 0 | 0 | ± | 0 |
| 526 | Orange (Valencia) | 0 | 0 | 0 | ± | + | 0 |
| 533/1 | Clemantine | 0 | 0 | ± | + | + | 0 |
| 575/2 | Avocado | 0 | 0 | 0 | ± | 0 | 0 |
| 157 | Avocado | 0 | 0 | 0 | 0 | + | 0 |
| 1278 | Avocado | 0 | 0 | 0 | 0 | + | + |
| 1758 | Avocado | 0 | 0 | 0 | ± | + | 0 |
| 2224 | Avocado | 0 | 0 | 0 | ± | ± | + |
| 135 | Watermelon | 0 | 0 | 0 | + | 0 | 0 |
| 163 | Watermelon | 0 | 0 | 0 | + + | + | 0 |
| 163/1 | Watermelon | 0 | 0 | 0 | + + | + | 0 |
| 163/6 | Watermelon | 0 | 0 | 0 | + | + | 0 |
| 226 | Watermelon | 0 | 0 | + | + + | + | 0 |
| 287x | Watermelon | 0 | 0 | + | + | + + | 0 |
| 287/1 | Watermelon | 0 | 0 | 0 | + | + | 0 |
| A5a | Melon | 0 | 0 | + | 0 | 0 | 0 |
| A7 | Melon | 0 | 0 | 0 | + | ± | 0 |
| 127/2 | Melon | 0 | 0 | 0 | ± | 0 | 0 |
| 213 | Melon | 0 | 0 | 0 | + | + | 0 |
| 229 | Melon | 0 | 0 | 0 | ± | 0 | 0 |
| 255 | Melon | 0 | 0 | + | + | 0 | 0 |
| 262 | Melon | 0 | 0 | + | + + | + + | + |
| 264 | Melon | 0 | ± | + | + + | + + + | 0 |
| 265 | Melon | 0 | 0 | + + | + + | + | 0 |
| 350/1 | Melon | 0 | 0 | ± | + | + | 0 |
| 604 | Melon | 0 | 0 | 0 | 0 | + | 0 |
| A3 | Cucumber | 0 | 0 | 0 | ± | 0 | 0 |
| A6 | Cucumber | 0 | 0 | 0 | 0 | + | 0 |
| 2 | Cucumber | 0 | + | + | ± | 0 | 0 |
| 5 | Cucumber | 0 | 0 | 0 | + | ± | 0 |
| 6 | Cucumber | 0 | 0 | ± | + | 0 | 0 |
| 10/1 | Cucumber | 0 | 0 | 0 | + + | + | 0 |

| No. of Isolate | Substrate | 6°C | 12°C | 18°C | 24°C | 30°C | 35°C |
|---|---|---|---|---|---|---|---|
| 17 | Cucumber | 0 | 0 | 0 | ± | 0 | 0 |
| 83/3 | Cucumber | 0 | 0 | 0 | + | ± | 0 |
| 89 | Cucumber | 0 | 0 | 0 | + | 0 | 0 |
| 151 | Cucumber | 0 | 0 | + | + + + | + + | 0 |
| 159 | Cucumber | 0 | 0 | + + | + + | + | 0 |
| 159/2 | Cucumber | 0 | 0 | + | + + | 0 | 0 |
| 159/3 | Cucumber | 0 | 0 | ± | + | 0 | 0 |
| 188 | Cucumber | 0 | 0 | 0 | + + | + | 0 |
| 188/2 | Cucumber | 0 | 0 | + | + | + + | 0 |
| 188/3 | Cucumber | 0 | 0 | 0 | + + | + | 0 |
| 188/4 | Cucumber | 0 | 0 | 0 | + + | + | 0 |
| 210 | Cucumber | 0 | 0 | + + | + + + | + | 0 |
| 227 | Cucumber | 0 | 0 | ± | + | 0 | 0 |
| 233 | Cucumber | 0 | 0 | ± | + + | + + | 0 |
| 258 | Cucumber | 0 | 0 | 0 | + + | + | 0 |
| 266 | Cucumber | 0 | 0 | + | + + | + + + | 0 |
| 661 | Cucumber | 0 | 0 | 0 | + + | + | + |
| 633 | Peanuts | 0 | 0 | 0 | + | + + | 0 |
| 2131 | Peanuts | 0 | 0 | ± | + | 0 | 0 |
| 3165 | Peanuts | 0 | 0 | + | + + | 0 | 0 |
| 3187 | Peanuts | 0 | 0 | 0 | ± | ± | 0 |
| R-4308 | Peanuts | 0 | 0 | 0 | ± | 0 | 0 |
| S-1277 | Peanuts | 0 | 0 | ± | 0 | 0 | 0 |
| S-1352 | Peanuts | 0 | 0 | ± | + | 0 | 0 |
| A10-a | Cotton | 0 | 0 | 0 | + | ± | 0 |
| 220/1 | Cotton | 0 | 0 | ± | + | + | 0 |
| 220/1a | Cotton | 0 | 0 | 0 | + | ± | 0 |
| 220/2 | Cotton | 0 | 0 | 0 | + | ± | 0 |
| 220/2x | Cotton | 0 | 0 | ± | + + | 0 | 0 |
| 220/3 | Cotton | 0 | 0 | ± | + | + | 0 |
| 542 | Cotton | 0 | 0 | ± | + | + | 0 |
| 901 | Cotton | 0 | 0 | 0 | + | 0 | 0 |
| 976 | Safflower | 0 | 0 | 0 | + | + | 0 |
| 977 | Safflower | 0 | 0 | 0 | ± | + | 0 |
| 978 | Safflower | 0 | 0 | 0 | ± | + | 0 |
| 979 | Safflower | 0 | 0 | 0 | + | + | 0 |
| 980 | Safflower | 0 | 0 | 0 | + | + | 0 |
| 981 | Safflower | 0 | 0 | 0 | + | 0 | 0 |
| 35/8 | Medicago | 0 | 0 | ± | + | 0 | 0 |
| 209 | Medicago | 0 | 0 | ± | + | 0 | 0 |
| 209 | Medicago | 0 | 0 | 0 | + | + | 0 |
| 175/1 | Dianthus | 0 | 0 | + | + + | 0 | 0 |
| 333 | Dianthus | 0 | 0 | + | + + | 0 | 0 |
| 333x | Dianthus | 0 | 0 | + | + + + | + + | 0 |
| 333/1 | Dianthus | 0 | 0 | + | + + | + + | + |
| 333a | Dianthus | 0 | ± | + | + + | 0 | 0 |

**460**

| No. of Isolate | Substrate | 6°C | 12°C | 18°C | 24°C | 30°C | 35°C |
|---|---|---|---|---|---|---|---|
| 358/2 | Dianthus | 0 | 0 | ± | + | + | 0 |
| 364 | Dianthus | 0 | 0 | 0 | ± | 0 | 0 |
| 364a | Dianthus | 0 | 0 | 0 | + | 0 | 0 |
| 364/2 | Dianthus | 0 | 0 | 0 | + | 0 | 0 |
| 244 | Amaranthus | 0 | 0 | ± | + | + | 0 |
| 252 | Malva | 0 | 0 | 0 | + + | + | 0 |
| S-19 | Soil | 0 | 0 | 0 | ± | 0 | 0 |
| S-33 | Soil | 0 | 0 | 0 | ± | ± | 0 |
| S-58 | Soil | 0 | 0 | 0 | ± | 0 | 0 |
| S-93 | Soil | 0 | 0 | 0 | ± | 0 | 0 |
| S-107 | Soil | 0 | 0 | 0 | ± | 0 | 0 |
| S-121 | Soil | 0 | 0 | 0 | ± | + | 0 |
| S-139 | Soil | 0 | 0 | ± | 0 | 0 | 0 |
| S-161 | Soil | 0 | 0 | 0 | ± | 0 | 0 |
| S-210 | Soil | 0 | 0 | 0 | 0 | + | 0 |
| S-211 | Soil | 0 | 0 | 0 | ± | 0 | 0 |
| S-214 | Soil | 0 | 0 | 0 | + | ± | 0 |
| S-223 | Soil | 0 | 0 | 0 | ± | 0 | 0 |
| S-231 | Soil | 0 | 0 | 0 | + | 0 | 0 |
| S-236 | Soil | 0 | 0 | ± | ± | 0 | 0 |
| S-238 | Soil | 0 | 0 | 0 | ± | + | 0 |
| S-241 | Soil | 0 | 0 | 0 | ± | 0 | 0 |
| S-246 | Soil | 0 | 0 | + | + | + + | 0 |
| S-249 | Soil | 0 | 0 | 0 | + | ± | 0 |
| S-285 | Soil | 0 | 0 | 0 | ± | 0 | 0 |
| S-366 | Soil | 0 | 0 | 0 | ± | 0 | 0 |

### F. oxysporum var. redolens

| No. of Isolate | Substrate | 6°C | 12°C | 18°C | 24°C | 30°C | 35°C |
|---|---|---|---|---|---|---|---|
| 909 | Carnation | 0 | 0 | 0 | + | ± | 0 |
| 910 | Carnation | 0 | 0 | 0 | + | + | 0 |
| 913 | Carnation | 0 | 0 | ± | + | + | 0 |
| 916 | Carnation | 0 | 0 | + | + + | 0 | 0 |
| 919 | Carnation | 0 | 0 | 0 | + | 0 | 0 |
| 921/3 | Carnation | 0 | 0 | + | + + | 0 | 0 |

### F. solani (Sect. Martiella)

| No. of Isolate | Substrate | 6°C | 12°C | 18°C | 24°C | 30°C | 35°C |
|---|---|---|---|---|---|---|---|
| 43 | Onion | 0 | 0 | + | + | 0 | 0 |
| 127/4c | Onion | 0 | 0 | 0 | + | + | 0 |
| 127/9c | Onion | 0 | 0 | 0 | + | ± | 0 |
| 127/11c | Onion | 0 | 0 | ± | ± | 0 | 0 |
| 127/12c | Onion | 0 | 0 | ± | + | 0 | 0 |
| 134/12c | Onion | 0 | 0 | ± | + | 0 | 0 |
| 140 | Onion | 0 | 0 | 0 | + + | 0 | 0 |
| 62a | Onion | 0 | 0 | + | + + | ± | 0 |
| 156 | Onion | 0 | 0 | 0 | 0 | ± | 0 |

| No. of Isolate | Substrate | 6°C | 12°C | 18°C | 24°C | 30°C | 35°C |
|---|---|---|---|---|---|---|---|
| 156/1S | Onion | 0 | 0 | 0 | + | 0 | 0 |
| 156/4c | Onion | 0 | 0 | 0 | + | 0 | 0 |
| 159/6s | Onion | 0 | 0 | 0 | 0 | + | 0 |
| 159/8S | Onion | 0 | 0 | 0 | + | 0 | 0 |
| 595 | Onion | 0 | 0 | ± | 0 | 0 | 0 |
| 547 | Carrot | 0 | 0 | 0 | + + | + | 0 |
| 37 | Tomato | 0 | 0 | 0 | + | + | 0 |
| 61/2 | Tomato | 0 | 0 | ± | 0 | 0 | 0 |
| 93 | Tomato | 0 | 0 | 0 | + + | 0 | 0 |
| 169 | Tomato | 0 | 0 | 0 | 0 | + | 0 |
| 253 | Tomato | 0 | 0 | + | + | 0 | 0 |
| 366 | Tomato | 0 | 0 | 0 | + | + | 0 |
| 757 | Potato | 0 | 0 | 0 | + | 0 | 0 |
| 758 | Potato | 0 | 0 | ± | + | 0 | 0 |
| 759 | Potato | 0 | 0 | 0 | + | 0 | 0 |
| 760 | Potato | 0 | 0 | 0 | + | ± | 0 |
| 761 | Potato | 0 | 0 | ± | + | 0 | 0 |
| 30/2 | Pepper | 0 | 0 | + | + | 0 | 0 |
| 41 | Pepper | 0 | 0 | 0 | + | + | 0 |
| 54 | Pepper | 0 | 0 | + | + + | 0 | 0 |
| 334-b | Pepper | 0 | 0 | 0 | 0 | + | 0 |
| 334/1 | Pepper | 0 | 0 | 0 | + | + | 0 |
| 341/2 | Pepper | 0 | 0 | 0 | ± | + | 0 |
| 251 | Bean | 0 | 0 | + | + + | 0 | 0 |
| 365/2 | Bean | 0 | 0 | ± | + | 0 | 0 |
| 993 | Broad bean | 0 | 0 | 0 | ± | + | 0 |
| 994 | Broad bean | 0 | 0 | 0 | 0 | + | 0 |
| 38a | Maize | 0 | 0 | + | + | 0 | 0 |
| 46 | Maize | 0 | 0 | 0 | + + | 0 | 0 |
| 79 | Maize | 0 | 0 | 0 | + | 0 | 0 |
| 79a | Maize | 0 | 0 | 0 | + | ± | 0 |
| 968 | Wheat | 0 | 0 | 0 | + | 0 | 0 |
| 969 | Wheat | 0 | 0 | 0 | + | 0 | 0 |
| 971 | Wheat | 0 | 0 | 0 | + | 0 | 0 |
| 972 | Wheat | 0 | 0 | | + | 0 | 0 |
| 530/2 | Grapefruit | 0 | 0 | 0 | ± | 0 | 0 |
| 536 | Grapefruit | 0 | 0 | 0 | + + | 0 | 0 |
| 537 | Grapefruit | 0 | 0 | 0 | ± | 0 | 0 |
| 538 | Grapefruit | 0 | 0 | + | + + | 0 | 0 |
| 480/2 | Orange (Valencia) | 0 | 0 | 0 | + + | 0 | 0 |
| 494/1 | Orange (Valencia) | 0 | 0 | 0 | + | ± | 0 |
| 495/2 | Orange (Valencia) | 0 | 0 | 0 | + | + | 0 |
| 497/3 | Orange (Valencia) | 0 | 0 | 0 | ± | ± | 0 |
| 509/2 | Orange (Valencia) | 0 | 0 | 0 | + | ± | 0 |
| 486 | Lemon | 0 | 0 | 0 | + | + | 0 |
| 486/1 | Lemon | 0 | 0 | 0 | 0 | + | 0 |

| No. of Isolate | Substrate | 6°C | 12°C | 18°C | 24°C | 30°C | 35°C |
|---|---|---|---|---|---|---|---|
| 549 | Anona | 0 | 0 | 0 | + | ± | 0 |
| 112/2 | Banana | 0 | 0 | 0 | + + | + + | + |
| 112/4 | Banana | 0 | 0 | 0 | + | + + | 0 |
| 85 | Watermelon | 0 | 0 | 0 | + | + | 0 |
| 147 | Watermelon | 0 | 0 | 0 | + | 0 | 0 |
| 147/1 | Watermelon | 0 | 0 | 0 | + + | + + | 0 |
| 164 | Watermelon | 0 | 0 | 0 | + | + | 0 |
| 164/1 | Watermelon | 0 | 0 | 0 | + | + | 0 |
| 287/3 | Watermelon | 0 | 0 | 0 | ± | + | 0 |
| 307 | Watermelon | 0 | 0 | 0 | + + | + + | 0 |
| 8 | Melon | 0 | 0 | ± | + | 0 | 0 |
| 127/3 | Melon | 0 | 0 | 0 | ± | + | 0 |
| 215 | Melon | 0 | 0 | 0 | + | ± | 0 |
| 216/2 | Melon | 0 | 0 | + | + + | ± | 0 |
| 217/4 | Melon | 0 | 0 | 0 | + | 0 | 0 |
| 219/1 | Melon | 0 | 0 | 0 | + | + | 0 |
| 219/2 | Melon | 0 | 0 | 0 | + | + | 0 |
| 219/1a | Melon | 0 | 0 | 0 | ± | + | 0 |
| 236 | Melon | 0 | 0 | 0 | + | + + | 0 |
| 219/2a | Melon | 0 | 0 | 0 | + | + | 0 |
| 309 | Melon | 0 | 0 | 0 | + | + + | 0 |
| 9 | Cucumber | 0 | 0 | 0 | + + | + + | 0 |
| 10/3 | Cucumber | 0 | 0 | 0 | + | 0 | 0 |
| 76 | Cucumber | 0 | 0 | 0 | ± | + | 0 |
| 84/1 | Cucumber | 0 | 0 | 0 | + | + | 0 |
| 94 | Cucumber | 0 | 0 | 0 | + + | 0 | 0 |
| 160 | Cucumber | 0 | 0 | + | + + | 0 | 0 |
| 160/1 | Cucumber | 0 | 0 | ± | + | + | 0 |
| 166/1 | Cucumber | 0 | 0 | 0 | + + | + + | 0 |
| 166/2 | Cucumber | 0 | 0 | ± | + + | + | 0 |
| 166/10 | Cucumber | 0 | 0 | 0 | + + | + + | 0 |
| 167 | Cucumber | 0 | 0 | 0 | + | ± | 0 |
| 167/6 | Cucumber | 0 | 0 | 0 | + | + | + |
| 198 | Cucumber | 0 | 0 | 0 | 0 | + | 0 |
| 198/14 | Cucumber | 0 | 0 | 0 | ± | 0 | 0 |
| 205/2 | Cucumber | 0 | 0 | 0 | + | 0 | 0 |
| 205/5 | Cucumber | 0 | 0 | 0 | + | + | 0 |
| 207/1 | Cucumber | 0 | 0 | 0 | + | 0 | 0 |
| 207/2 | Cucumber | 0 | 0 | 0 | + | ± | 0 |
| 242 | Cucumber | 0 | 0 | 0 | + + | 0 | 0 |
| 243 | Cucumber | 0 | 0 | 0 | ± | 0 | 0 |
| 311 | Cucumber | 0 | 0 | 0 | + | ± | 0 |
| 356 | Peanuts | 0 | 0 | 0 | ± | + | 0 |
| 356/2 | Peanuts | 0 | 0 | + | ± | 0 | 0 |
| 560/1 | Peanuts | 0 | 0 | + | 0 | 0 | 0 |
| 561 | Peanuts | 0 | 0 | 0 | + + | + | 0 |

| No. of Isolate | Substrate | 6°C | 12°C | 18°C | 24°C | 30°C | 35°C |
|---|---|---|---|---|---|---|---|
| 562/1 | Peanuts | 0 | 0 | 0 | + | 0 | 0 |
| 562 | Peanuts | 0 | 0 | + | ± | 0 | 0 |
| 564 | Peanuts | 0 | 0 | 0 | ± | 0 | 0 |
| 565 | Peanuts | 0 | 0 | + | + | 0 | 0 |
| 567/1 | Peanuts | 0 | 0 | 0 | ± | 0 | 0 |
| 568 | Peanuts | 0 | 0 | + | + + | + | 0 |
| 568/1 | Peanuts | 0 | 0 | ± | + + | 0 | 0 |
| 568a | Peanuts | 0 | 0 | + | ± | 0 | 0 |
| 569/1 | Peanuts | 0 | 0 | ± | + | 0 | 0 |
| 570/1 | Peanuts | 0 | 0 | 0 | + | ± | 0 |
| 39 | Sesame | 0 | 0 | 0 | + | + | 0 |
| 26 | Sesame | 0 | 0 | 0 | + | ± | 0 |
| 544 | Gladiolus | 0 | 0 | ± | + | + | 0 |
| 339 | Carnation | 0 | 0 | ± | + | 0 | 0 |
| 252t | Malva | 0 | 0 | 0 | + | ± | 0 |
| S1 | Soil | 0 | 0 | 0 | + | 0 | 0 |
| S2 | Soil | 0 | 0 | 0 | 0 | + | 0 |
| S11 | Soil | 0 | 0 | 0 | + | 0 | 0 |
| S13 | Soil | 0 | 0 | 0 | 0 | + | 0 |
| S17 | Soil | 0 | 0 | 0 | + | 0 | 0 |
| S23 | Soil | 0 | 0 | 0 | + | 0 | 0 |
| S37 | Soil | 0 | 0 | ± | + | 0 | 0 |
| S49 | Soil | 0 | 0 | + | + | 0 | 0 |
| S61 | Soil | 0 | 0 | 0 | + | 0 | 0 |
| S62 | Soil | 0 | 0 | ± | ± | 0 | 0 |
| S74 | Soil | 0 | 0 | 0 | + | 0 | 0 |
| S78 | Soil | 0 | 0 | 0 | + + | + | 0 |
| S97 | Soil | 0 | 0 | 0 | + | + | 0 |
| S102 | Soil | 0 | 0 | 0 | + + | 0 | 0 |
| S107 | Soil | 0 | 0 | 0 | 0 | + | 0 |
| S113 | Soil | 0 | 0 | 0 | + | 0 | 0 |
| S119 | Soil | 0 | 0 | 0 | + | 0 | 0 |
| S127 | Soil | 0 | 0 | 0 | + | 0 | 0 |
| S133 | Soil | 0 | 0 | 0 | + | + | 0 |
| S141 | Soil | 0 | 0 | ± | + + | 0 | 0 |
| S153 | Soil | 0 | 0 | 0 | + | 0 | 0 |
| S155 | Soil | 0 | 0 | 0 | + | + | 0 |
| S169 | Soil | 0 | 0 | 0 | + | + | 0 |
| S317 | Soil | 0 | 0 | 0 | + | 0 | 0 |
| S319 | Soil | 0 | 0 | 0 | + | + | 0 |
| S324 | Soil | 0 | 0 | 0 | + + | + | 0 |
| S330 | Soil | 0 | 0 | 0 | + | + | 0 |
| S334 | Soil | 0 | 0 | 0 | + | ± | 0 |
| S338 | Soil | 0 | 0 | + | + + | 0 | 0 |
| S339 | Soil | 0 | 0 | 0 | 0 | + | 0 |
| S344 | Soil | 0 | 0 | 0 | + | 0 | 0 |

| No. of Isolate | Substrate | 6°C | 12°C | 18°C | 24°C | 30°C | 35°C |
|---|---|---|---|---|---|---|---|
| S354 | Soil | 0 | 0 | 0 | + | + | 0 |
| S357 | Soil | 0 | 0 | 0 | ± | + | 0 |
| S364 | Soil | 0 | 0 | 0 | + | 0 | 0 |

*F. javanicum*

| No. of Isolate | Substrate | 6°C | 12°C | 18°C | 24°C | 30°C | 35°C |
|---|---|---|---|---|---|---|---|
| 214a | Melon | 0 | 0 | 0 | + | ± | 0 |
| 350 | Melon | 0 | 0 | 0 | + | + | 0 |
| 285 | Melon | 0 | 0 | + | + + | 0 | 0 |
| 301 | Melon | 0 | 0 | 0 | + | + | 0 |
| 308 | Melon | 0 | 0 | 0 | ± | + | 0 |
| 189 | Vegetable marrow | 0 | 0 | 0 | + | + | 0 |
| 199/1 | Vegetable marrow | 0 | 0 | 0 | ± | 0 | 0 |
| 199/10 | Vegetable marrow | 0 | 0 | 0 | + + | + | 0 |
| 206 | Vegetable marrow | 0 | 0 | + | + | + | 0 |
| 300 | Vegetable marrow | 0 | 0 | + | + + | + + | + |
| S6 | Soil | 0 | 0 | 0 | + | 0 | 0 |
| S38 | Soil | 0 | 0 | 0 | + | 0 | 0 |
| S54 | Soil | 0 | 0 | 0 | + | 0 | 0 |
| S69 | Soil | 0 | 0 | 0 | + + | + | 0 |
| S302 | Soil | 0 | 0 | 0 | ± | ± | 0 |
| S308 | Soil | 0 | 0 | 0 | + | 0 | 0 |
| S312 | Soil | 0 | 0 | 0 | ± | ± | 0 |
| S323 | Soil | 0 | 0 | 0 | + | + | 0 |
| S326 | Soil | 0 | 0 | 0 | ± | + | 0 |
| S346 | Soil | 0 | 0 | 0 | + + | + | ± |
| S359 | Soil | 0 | 0 | 0 | + | + | 0 |

# APPENDIX II

THE EFFECT OF *FUSARIUM* CRUDE EXTRACTS ISOLATED FROM PLANTS AND SOIL IN COUNTRIES OTHER THAN ISRAEL AND TESTED ON RABBIT SKIN

| No. of Strain | Substrate | Locality | 6°C | 12°C | 18°C | 24°C | 30°C | 35°C |
|---|---|---|---|---|---|---|---|---|
| | | *F. nivale* (Sect. Arachnites) | | | | | | |
| 9317 | *Secale cereale* | Germany | 0 | 0 | ++ | ± | 0 | 0 |
| 62051 | *Agrostis* | Germany | 0 | 0 | + | 0 | 0 | 0 |
| 62276 | *Secale cereale* | Germany | 0 | 0 | ± | 0 | 0 | 0 |
| 62277 | Grass | Germany | 0 | 0 | ++ | + | 0 | 0 |
| 62278 | *Secale cereale* | Germany | 0 | 0 | ++ | ± | 0 | 0 |
| 62279 | *Poa annua* | Germany | 0 | 0 | ++ | + | 0 | 0 |
| 62280 | *Triticum aestivum* | Germany | 0 | 0 | ± | 0 | 0 | 0 |
| 62281 | *Triticum aestivum* | Germany | 0 | 0 | ++ | + | 0 | 0 |
| 62282 | *Lolium perenne* | Germany | 0 | 0 | + | ± | 0 | 0 |
| 69138-b | Unknown | Finland | 0 | ± | ± | 0 | + | 0 |
| FnAm | Unknown | Finland | 0 | 0 | 0 | ++ | + | 0 |
| 131785 | Turf, LMI, Kew | England | 0 | 0 | ± | ± | 0 | 0 |
| | | *F. dimerum* (Sect. Eupionnotes) | | | | | | |
| 9375 | *Matthiola incana* | Germany | 0 | 0 | 0 | + | ± | 0 |
| | | *F. tabacinum* | | | | | | |
| 75768 | CBS, Baarn | Holland | 0 | 0 | 0 | + | + | 0 |
| | | *F. aquaeductuum* | | | | | | |
| 4251 | Rubber tube, Baarn | Holland | 0 | 0 | 0 | 0 | + | + |
| 8446 | *Fagus silvatica* | Germany | 0 | 0 | 0 | ++ | + | + |
| 15612 | Potato tuber | Australia | 0 | 0 | 0 | 0 | + | + |

| No. of Strain | Substrate | Locality | 6°C | 12°C | 18°C | 24°C | 30°C | 35°C |
|---|---|---|---|---|---|---|---|---|
| *F. merismoides* | | | | | | | | |
| 19 | CBS, Baarn | Holland | 0 | 0 | 0 | + | 0 | 0 |
| 4069 | River water | England | 0 | 0 | 0 | 0 | ± | 0 |
| 4125 | Pear decline | USA | 0 | 0 | 0 | 0 | ± | 0 |
| 5021 | Water from goldfish bowl | Holland | 0 | 0 | 0 | 0 | + | + |
| 9776 | *Triticum aestivum* | Germany | 0 | 0 | 0 | 0 | + | ± |
| 16561 | Sugar beet | England | 0 | 0 | 0 | + | ++ | 0 |
| *F. melanochlorum* | | | | | | | | |
| 9831 | *Fagus silvatica* | Austria | 0 | 0 | 0 | 0 | + | + |
| 16069 | Beech tree | Austria | 0 | 0 | 0 | 0 | ± | 0 |
| *F. sporotrichioides* (Sect. Sporotrichiella) | | | | | | | | |
| 23 | Millet | USSR | ++++ | ++++ | +++ | ++ | + | 0 |
| 60/10 | Millet | USSR | +++ | +++ | ++ | + | 0 | 0 |
| 347 | Millet | USSR | ++++ | ++++ | ++++ | ++ | + | 0 |
| 351 | Millet | USSR | ++++ | ++++ | +++ | ++ | + | 0 |
| 738 | Millet | USSR | ++++ | ++++ | +++ | ++ | + | 0 |
| 921 | Rye | USSR | +++ | ++++ | + | + | + | 0 |
| 1182 | Wheat | USSR | +++ | +++ | ++ | ± | 0 | 0 |
| 1823 | Barley | USSR | ++ | ++ | + | ± | ± | 0 |
| NRRL3249 | Tall fescue | USA | + | + | + | ± | 0 | 0 |
| NRRL5908 | Tall fescue | USA | ++ | + | ± | 0 | 0 | 0 |
| 2061-C | Corn cobs | USA | + | + | ± | ± | 0 | 0 |
| YN-13 | Corn | USA | ++ | + | ± | 0 | 0 | 0 |
| 22-26 | Unknown | Hungary | ++ | + | + | ± | 0 | 0 |

| Isolate | Substrate | Origin | | | | | | |
|---|---|---|---|---|---|---|---|---|
| 2416 | Oats | Canada | ++ | 0 | 0 | ++ | + | 0 |
| 10329 | *Malus sylvestris* | Germany | +++ | +++ | ++ | + | + | 0 |
| 10339 | *Avena sativa* | Germany | +++ | +++ | ++ | + | ± | 0 |
| 10362 | *Pinus nigra* | Germany | 0 | ++ | + | ± | + | 0 |
| 62424 | *Solanum tuberosum* | Germany | 0 | 0 | ± | 0 | 0 | 0 |
| 15661 | Scirpus | Scotland | 0 | 0 | 0 | 0 | 0 | 0 |
| *F. poae* | | | | | | | | |
| 24 | Millet | USSR | +++ | +++ | ++ | + | 0 | 0 |
| 60/9 | Millet | USSR | ++++ | ++++ | +++ | + | 0 | 0 |
| 396 | Millet | USSR | ++++ | ++++ | +++ | ++ | ± | 0 |
| 792 | Barley | USSR | +++ | +++ | ++ | ++ | 0 | 0 |
| 958 | Wheat | USSR | ++++ | ++++ | ++ | + | 0 | 0 |
| T-2 | Corn | France | ++ | + | ± | 0 | 0 | 0 |
| 3918 | Sambucus | Canada | +++ | +++ | ++ | + | + | ± |
| NRRL3287 | Unknown | USA | +++ | ++ | ± | 0 | 0 | 0 |
| NRRL3299 | Corn | USA | +++ | ++ | + | ± | 0 | 0 |
| M309 | AmTCC | USA | 0 | + | + | + | + | 0 |
| M484 | FDA | USA | 0 | 0 | ± | 0 | 0 | 0 |
| M618 | FDA | USA | + | + | 0 | 0 | 0 | 0 |
| 10317 | *Avena sativa* | Germany | +++ | ++ | + | ± | 0 | 0 |
| 15654 | Barley | USA | 0 | + | 0 | 0 | 0 | 0 |
| 16557 | Barley | Scotland | 0 | 0 | ± | 0 | 0 | 0 |
| 22-205 | Unknown | Hungary | + | ± | 0 | 0 | 0 | 0 |
| *F. sporotrichioides* var. *tricinctum* | | | | | | | | |
| M304 | FDA | USA | +++ | +++ | ++ | + | + | ± |
| 1227 | Barley | USSR | ++ | ++ | + | ± | 0 | 0 |
| 2457 | Millet | USSR | + | + | ± | ± | 0 | 0 |
| NRRL3509 | Unknown | USA | + | | | 0 | 0 | 0 |

| No. of Strain | Substrate | Locality | 6°C | 12°C | 18°C | 24°C | 30°C | 35°C |
|---|---|---|---|---|---|---|---|---|
| 22-144 | Unknown | Hungary | 0 | 0 | ± | ± | 0 | 0 |
| 7030II | Unknown | Finland | 0 | 0 | + | ± | ± | 0 |
| 7147SII5 | Unknown | Finland | 0 | 0 | ± | + | 0 | 0 |
| 71108I-22 | Unknown | Finland | 0 | 0 | 0 | + | ± | 0 |
| 62448 | Triticum aestivum | Germany | + | + | ± | 0 | 0 | 0 |
| *F. sporotrichioides* var. *chlamydosporum* | | | | | | | | |
| 1174 | Millet | USSR | +++ | +++ | ++ | ± | 0 | 0 |
| 2388 | Barley | USSR | + | + | ± | 0 | 0 | 0 |
| 4337 | Wheat | Canada | +++ | +++ | ++ | + | + | ± |
| 15615 | Soil | India | 0 | 0 | ± | + | 0 | 0 |
| 62171 | Soil | Pakistan | 0 | 0 | 0 | ± | 0 | 0 |
| 157755 | Bovine feed | Australia | 0 | 0 | 0 | ± | 0 | 0 |
| *F. decemcellulare* (Sect. Spicarioides) | | | | | | | | |
| 11 | CBS, Baarn | Holland | 0 | 0 | 0 | + | ± | 0 |
| M312 | AmTCC (FDA) | USA | 0 | 0 | 0 | 0 | ± | 0 |
| 16562 | *Kala acuminata* | Jamaica | 0 | 0 | 0 | ± | 0 | 0 |
| *F. coccidicola* (Sect. Macroconia) | | | | | | | | |
| T50 | Citrus (Orange) | Canada | 0 | 0 | 0 | + | + | ± |
| T53 | Citrus (Orange) | Canada | 0 | 0 | 0 | + | + | 0 |
| 11137 | *Prunus vulgaris* | Iran | 0 | 0 | 0 | ± | 0 | 0 |
| *F. coccophilum* | | | | | | | | |
| 11140 | *Gleditschia caspica* | Iran | 0 | 0 | 0 | + | ++ | + |
| 143094 | *Coffea arabica* | New Guinea | 0 | 0 | 0 | 0 | ± | 0 |

| Strain | Origin | Substrate | | | | | |
|---|---|---|---|---|---|---|---|
| *F. avenaceum* (Sect. Roseum) | | | | | | | |
| 7006-21 | Finland | Source unknown | 0 | ± | + | 0 | 0 |
| 10815 | Germany | *Secale cereale* | 0 | 0 | + | 0 | 0 |
| 11113 | Iran | *Paspalum dilatatum* | 0 | 0 | + | ++ | + |
| 11273 | Germany | *Betula verrucosa* | 0 | ± | + | 0 | 0 |
| *F. arthrosporioides* | | | | | | | |
| 1944 | USA | Wheat | 0 | ± | + | 0 | 0 |
| 11326 | Finland | Graminea | 0 | ± | 0 | 0 | 0 |
| 125834-b | New Zealand | Azalea | 0 | ± | 0 | 0 | 0 |
| 7005aI | Finland | Unknown | 0 | + | ++ | + | ± |
| *F. semitectum* (Sect. Arthrosporiella) | | | | | | | |
| 3365 | Tanganyika | Soybean seed | 0 | 0 | + | ± | 0 |
| 16555 | Guyana | Citrus | 0 | ± | + | 0 | 0 |
| *F. semitectum* var. *majus* | | | | | | | |
| 11129 | Iran | Coccide on prunus | 0 | + | 0 | 0 | 0 |
| *F. camptoceras* | | | | | | | |
| 9810 | Costa Rica | *Theobroma cacao* | 0 | 0 | ± | + | + |
| 16065 | Costa Rica | Cacao | ± | + | ++ | 0 | 0 |
| *F. concolor* | | | | | | | |
| 24a | Holland | CBS, Baarn | + | ± | 0 | 0 | 0 |
| 10330 | Holland | CBS, Baarn | + | + | 0 | 0 | 0 |

| No. of Strain | Substrate | Locality | 6°C | 12°C | 18°C | 24°C | 30°C | 35°C |
|---|---|---|---|---|---|---|---|---|
| *F. lateritium* (Sect. Lateritium) | | | | | | | | |
| 2526 | *Cajanus indicus* | Tanganyika | 0 | 0 | 0 | + | ± | 0 |
| 2527 | *Cajanus indicus* | Tanganyika | 0 | 0 | 0 | ± | + | 0 |
| 4005 | *Coffea excelsa* | French Equatorial Africa | 0 | 0 | 0 | ± | 0 | 0 |
| 4990 | *Coffea excelsa* | French Equatorial Africa | 0 | 0 | 0 | ± | 0 | 0 |
| 8 | CBS, Baarn | Holland | 0 | 0 | ± | ± | 0 | 0 |
| 2637 | *Hibiscus syriacus* | Canada | 0 | 0 | ± | + | + | 0 |
| 5342-SA3 | *Passiflora edulis* | New Zealand | 0 | ± | ± | + | ± | ± |
| 15632 | AmTCC | USA | 0 | 0 | 0 | + | ± | 0 |
| 62240 | *Ficus elastica* | Germany | 0 | 0 | 0 | ± | ± | 0 |
| 62242 | Dead shoot of tree | Italy | 0 | 0 | 0 | ++ | + | 0 |
| 62243 | Citrus | Iran | 0 | 0 | 0 | ± | + | 0 |
| 62245 | *Tillandsia latifolia* | Germany | 0 | 0 | 0 | + | 0 | 0 |
| 62246 | Eriosoma leaf fall on *Ulmus glabra* | Austria | 0 | 0 | 0 | ± | + | 0 |
| *F. stilboides* | | | | | | | | |
| M308 | AmTCC | USA | 0 | ± | + | + | ± | 0 |
| 4840 | *Coffea* | Malawi | 0 | 0 | 0 | ± | ± | 0 |
| 4972 | *Coffea arabica* | Zimbabwe | 0 | 0 | ± | + | 0 | 0 |
| 7780 | *Coffea* sp. | Brasilia | 0 | 0 | 0 | ± | 0 | 0 |
| *F. xylarioides* | | | | | | | | |
| 9960 | *Coffea robusta* | Guinea | 0 | 0 | 0 | ± | ± | + |
| 16070 | *Coffea robusta* | Guinea | 0 | 0 | 0 | 0 | + | ± |
| 15664 | AmTCC | USA | 0 | 0 | 0 | + | ± | 0 |

## F. moniliforme (Sect. Liseola)

| Strain | Substrate | Country | | | | | | |
|---|---|---|---|---|---|---|---|---|
| M205 | Corn, FDA | USA | 0 | + | + | ++ | 0 | 0 |
| M314 | Corn, FDA | USA | 0 | + | ± | + | 0 | 0 |
| 3104-Sa49 | Corn (Perithecia) | Canada | + | 0 | 0 | 0 | 0 | 0 |
| 3104-Sa13 | Corn | USA | 0 | ± | + | ± | 0 | 0 |
| 9272 | Musa | Libya | 0 | 0 | ± | 0 | + | 0 |
| 16564 | Fresia seed | England | + | + | + | + | 0 | 0 |
| 54 | Peanuts | Mozambique | 0 | 0 | + | 0 | + | 0 |
| 56 | Peanuts | Mozambique | 0 | ± | + | ± | 0 | 0 |

## F. moniliforme var. subglutinans

| Strain | Substrate | Country | | | | | | |
|---|---|---|---|---|---|---|---|---|
| 4410 | Corn | Canada | 0 | 0 | 0 | 0 | ± | 0 |
| 8549 | Haemanthus | Germany | 0 | 0 | + | + | + | 0 |

## F. moniliforme var. anthophilum

| Strain | Substrate | Country | | | | | | |
|---|---|---|---|---|---|---|---|---|
| 4862 | Saccharum officinarum | Nigeria | 0 | + | ± | 0 | 0 | 0 |
| 8998 | Hippeastrum | Germany | 0 | 0 | 0 | + | + | 0 |

## F. equiseti (Sect. Gibbosum)

| Strain | Substrate | Country | | | | | | |
|---|---|---|---|---|---|---|---|---|
| 21 | CBS, Baarn | Holland | 0 | 0 | ± | 0 | ± | 0 |
| 3873 | Cassava sp. | Nigeria | 0 | + | + | ++ | 0 | 0 |
| 5149 | Cereals | USA | 0 | 0 | ± | 0 | ± | 0 |
| 15622 | Cereals | USA | + | + | + | + | + | + |
| 10333 | Hordeum vulgare | Germany | + | + | + | + | 0 | 0 |

## F. equiseti var. compactum

| Strain | Substrate | Country | | | | | | |
|---|---|---|---|---|---|---|---|---|
| 4257 | Pinus elliottii seedling | Australia | 0 | + | ++ | ++ | 0 | 0 |
| 15617 | AmTCC | USA | 0 | + | + | 0 | 0 | 0 |
| 15618 | Peanut root | Tanganyika | 0 | + | + | 0 | 0 | 0 |
| 11422 | IMB | Germany | 0 | ± | ± | 0 | 0 | 0 |
| 2762-M | Peanut root | Tanganyika | 0 | ± | ± | 0 | 0 | 0 |

| No. of Strain | Substrate | Locality | 6°C | 12°C | 18°C | 24°C | 30°C | 35°C |
|---|---|---|---|---|---|---|---|---|
| | | *F. equiseti* var. *acuminatum* | | | | | | |
| 3275 | *Pinus ponderosa* | Canada | 0 | 0 | ± | + | ++ | 0 |
| 4896 | Clover rhizosphere | Australia | 0 | 0 | 0 | + | ± | 0 |
| 7885 | *Lupinus angustifolius* | Germany | 0 | 0 | + | ++ | 0 | 0 |
| 7005-SaII | Unknown | Finland | 0 | 0 | + | 0 | 0 | 0 |
| 11107 | Citrus | Iraq | 0 | 0 | + | +++ | ++ | 0 |
| 16560 | Potato tuber | Australia | 0 | 0 | 0 | ± | 0 | 0 |
| 18 | CBS, Baarn | Holland | 0 | 0 | 0 | 0 | + | ++ |
| | | *F. heterosporum* (Sect. Discolor) | | | | | | |
| 2798 | Rose twig | Canada | 0 | 0 | ± | + | 0 | 0 |
| 3396 | Raspberry | Scotland | 0 | 0 | 0 | 0 | + | 0 |
| 9949 | *Claviceps purpurea* | Germany | 0 | 0 | 0 | + | ++ | + |
| 11027 | *Lolium perenne* | Germany | 0 | 0 | 0 | 0 | ± | 0 |
| 11141 | *Zea mays* | Iran | 0 | 0 | 0 | 0 | ± | + |
| 11159 | *Pterocarya fraxinifolia* | Iran | 0 | + | 0 | 0 | 0 | 0 |
| | | *F. graminearum* | | | | | | |
| M316 | FDA | USA | 0 | 0 | ± | + | 0 | 0 |
| 9 | CBS, Baarn | Holland | 0 | 0 | 0 | 0 | ± | 0 |
| 5068 | Wheat | Australia | 0 | 0 | ± | + | ++ | 0 |
| 15624 | Corn | Canada | 0 | 0 | 0 | + | ++ | 0 |
| 22-1 | Unknown | Hungary | 0 | 0 | ± | + | 0 | 0 |
| | | *G. zeae* | | | | | | |
| M280 | Corn, FDA | USA | 0 | 0 | ± | + | 0 | 0 |
| M281 | Corn, FDA | USA | 0 | 0 | + | ++ | 0 | 0 |

| Strain | Substrate | Country | 1 | 2 | 3 | 4 | 5 |
|---|---|---|---|---|---|---|---|
| M284 | Corn, FDA | USA | 0 | + | + | 0 | 0 |
| 62050 | Avena sativa | Germany | 0 | + | ± | 0 | 0 |
| 62052 | Dianthus caryophyllus | USA | 0 | 0 | ± | 0 | 0 |
| 62056 | Dianthus caryophyllus | USA | 0 | 0 | ± | 0 | 0 |

F. sambucinum

| Strain | Substrate | Country | 1 | 2 | 3 | 4 | 5 |
|---|---|---|---|---|---|---|---|
| SA111 | Polygonum siebaldi | England | 0 | 0 | ± | + | 0 |
| 3362 | Hops | Tasmania | 0 | + | + | + | 0 |
| 4081 | Potato dry rot | Canada | 0 | +++ | +++ | + | 0 |
| 5519 | Storage rot of carrot | Nova Scotia | 0 | 0 | 0 | ± | 0 |
| 5522 | Storage rot of carrot | Nova Scotia | 0 | ± | + | + | 0 |
| 16552 | Poultry feed | Australia | 0 | 0 | 0 | ± | 0 |
| 17 | CBS, Baarn | Holland | 0 | 0 | 0 | ± | 0 |

F. sambucinum var. coeruleum

| Strain | Substrate | Country | 1 | 2 | 3 | 4 | 5 |
|---|---|---|---|---|---|---|---|
| 204-7 | Unknown | Hungary | 0 | 0 | 0 | 0 | ± |
| 4690 | Pasture soil | New Zealand | 0 | ± | 0 | 0 | 0 |
| 4917 | Strawberry | Canada | 0 | 0 | ± | 0 | 0 |
| 4932 | Sitka spruce | England | 0 | 0 | 0 | ± | 0 |

F. sambucinum var. trichothecioides

| Strain | Substrate | Country | 1 | 2 | 3 | 4 | 5 |
|---|---|---|---|---|---|---|---|
| 11125 | Solanum tuberosum | Iran | 0 | ++ | +++ | + | + |

F. culmorum

| Strain | Substrate | Country | 1 | 2 | 3 | 4 | 5 |
|---|---|---|---|---|---|---|---|
| 5322 | Soil | USA | 0 | ± | ± | 0 | 0 |
| 5205 | Barley | England | 0 | + | + | 0 | 0 |
| 16430 | AmTCC | USA | 0 | 0 | + | + | ± |
| 22 | CBS, Baarn | Holland | 0 | 0 | ± | 0 | 0 |

| No. of Strain | Substrate | Locality | 6°C | 12°C | 18°C | 24°C | 30°C | 35°C |
|---|---|---|---|---|---|---|---|---|
| | | *F. tumidum* | | | | | | |
| 16563 | *Hevea brasiliensis* | Sabah | 0 | 0 | 0 | 0 | ± | 0 |
| | | *F. oxysporum* (Sect. Elegans) | | | | | | |
| 1-7 | Unknown | France | 0 | 0 | 0 | + | 0 | 0 |
| 2-3 | Unknown | France | 0 | 0 | 0 | + | 0 | 0 |
| 3-9 | Unknown | France | 0 | 0 | + | + | 0 | 0 |
| 4-1 | Unknown | France | 0 | 0 | 0 | + | 0 | 0 |
| 6-2 | Unknown | France | 0 | 0 | + | + | 0 | 0 |
| 8-2 | Unknown | France | 0 | 0 | + | ++ | 0 | 0 |
| 9-2 | Unknown | France | 0 | 0 | 0 | + | ± | 0 |
| 9-3 | Unknown | France | 0 | 0 | ± | + | + | 0 |
| 10-2 | Unknown | France | 0 | 0 | 0 | + | 0 | 0 |
| M-310 | Corn, FDA | USA | 0 | 0 | 0 | ++ | 0 | 0 |
| 2990 | Carnation | England | 0 | 0 | 0 | + | 0 | 0 |
| 3319 | Gladiolus race 1 | Australia | 0 | 0 | 0 | ++ | + | 0 |
| 3220 | Gladiolus race 2 | Australia | 0 | 0 | 0 | ± | + | 0 |
| | | *F. oxysporum* var. *redolens* | | | | | | |
| 10326 | *Convallaria majalis* | Germany | 0 | 0 | 0 | ++ | + | 0 |

*F. solani* (Sect. Martiella)

| Strain | Substrate | Origin | 1 | 2 | 3 | 4 | 5 |
|---|---|---|---|---|---|---|---|
| M.306 | Corn, FDA | USA | 0 | + | + | 0 | 0 |
| T.25 | *Gossypium purpurescens* | Canada | 0 | + | + | 0 | 0 |
| 10 | CBS, Baarn | Holland | 0 | 0 | 0 | + | 0 |
| 14 | CBS, Baarn | Holland | 0 | 0 | ± | 0 | 0 |
| 16 | CBS, Baarn | Holland | 0 | 0 | + | 0 | 0 |
| 20 | CBS, Baarn | Holland | 0 | + | + | + | 0 |
| 7144 | *Pisum sativum* | Germany | 0 | + | ++ | 0 | 0 |
| 9800 | *Theobroma cacao* | Costa Rica | 0 | 0 | ± | 0 | 0 |

*F. solani* var. *coeruleum*

| Strain | Substrate | Origin | 1 | 2 | 3 | 4 | 5 |
|---|---|---|---|---|---|---|---|
| 4 | CBS, Baarn | Holland | 0 | 0 | 0 | ± | 0 |
| 5310 | Potato | Canada | 0 | ± | ± | 0 | 0 |

*F. solani* var. *ventricosum*

| Strain | Substrate | Origin | 1 | 2 | 3 | 4 | 5 |
|---|---|---|---|---|---|---|---|
| 8743 | Soil | Germany | 0 | 0 | ± | 0 | 0 |
| 5 | CBS, Baarn | Holland | 0 | + | + | 0 | 0 |
| 55 | Peanuts | Mozambique | 0 | ++ | + | 0 | 0 |

*F. javanicum*

| Strain | Substrate | Origin | 1 | 2 | 3 | 4 | 5 |
|---|---|---|---|---|---|---|---|
| 3 | CBS, Baarn | Holland | 0 | 0 | + | 0 | 0 |
| 6 | CBS, Baarn | Holland | 0 | 0 | ++ | ± | 0 |

# BIBLIOGRAPHY

Abbas, H. K., Mirocha, C. J., and Shier, W. T. (1984). Mycotoxins produced from fungi isolated from foodstuffs and soil: Comparison of toxicity in fibroblast and rat feeding test. *Appl. Environ. Microbiol.* **48**, 654–661.

Abramov, I. N. (1939). *Fusariosis—"Drunken Bread": The Diseases of Agricultural Plants of the Far East.* Dalgiz, pp. 78–105, Moscow.

Abramowsky, C. R., Quinn, D., Bradford, W. D., and Conant, N. F. (1974). Systematic infection by *Fusarium* in a burned child. The emergence of a saprophytic strain. *J. Pediatr.* **84**, 561–564.

Achilladelis, B., and Hanson, J. R. (1968). Studies in terpenoid biosyntheses. 7. The biosyntheses of metabolites of *Trichothecium roseum. Phytochemistry* **7**, 569–594.

Achilladelis, B., Adams, P. M., and Hanson, J. R. (1972). Studies in terpenoid biosynthesis. 8. The formation of trichothecene nucleus. *J. Chem. Soc. Perkin* 1, 1425–1428.

Adams, R. L., and Tuite, J. (1976). Feeding *Gibberella zeae* damaged corn to laying hens. *Poult. Sci.* **55**, 1991–1993.

Agrelo, C. E., and Schoental, R. (1980). Synthesis of DNA in human fibroblasts treated with T-2 toxin and HT-2 toxin (the trichothecene metabolites of *Fusarium* species) and the effects of hydroxyurea. *Toxicol. Lett.* **5**, 155–160.

Agronomov, I. A., Dounin, M. C., Bundel, A. A., Goryatchich, A. N., and Korenev, N. A. (1934). *Biochemistry and microbiology of contaminated wheat grains with Fusarium species by storage.* Snabtezhizdat, 96 pp., Moscow.

Aiiso, M. (1936). Kashin-Beck's disease in To-Hen-Do of Munchoukuo. *J. Orient. Med.* **24**, 34.

Aiiso, M., and Nayashi, N. (1937). Pathologic-anatomic studies of Kashin-Beck's disease. *J. Orient. Med.* **25**, 49.

Akhmeteli, M. A. (1977). Epidemiological features of the mycotoxicoses. *Ann. Nutr. Alim.* **31**, 957–976.

Akhmeteli, M. A., Linnik, A. B., and Cernov, K. S. (1972a). Hepatocarcinogenesis and the appearance of serum alphafetoprotein in mice treated with extract of barley grain infected with *Fusarium sporotrichioides. Bull. World Health Organ.* **47**, 663–664.

Akhmeteli, M. A., Linnik, A. B., Cernov, K. S., Voronin, V. M., and Sabad, L. M. (1972b). Study of extracts of barley grain infected with *Fusarium sporotrichioides. Bull. World Health Organ.* **47**, 123–124.

**479**

Akhmeteli, M. A., Linnik, A. B., Cernov, K. S., Voronin, V. M., Hesina, J. A., Guseva, N. A., and Sabad, L. M. (1973). Study of toxins isolated from grain infected with *Fusarium sporotrichioides*. *Pure Appl. Chem.* **35**, 209–215.

Alisova, Z. I. (1947a). General toxic action of cereal crops overwintered in the field. In Alimentary toxic aleukia. *Acta Chkalov Inst. Epidemiol. Microbiol. Second Commun.*, pp. 104–118.

Alisova, Z. I. (1947b). General action of overwintered cereal extracts and toxic fungi on laboratory animals. In Alimentary toxic aleukia. *Acta Chkalov Inst. Epidemiol. Microbiol.*, p. 192.

Alisova, Z. I. and Mironov, S. G. (1944). On toxicity of prosomillet overwintered in the field. In Data on septic angina. *Proc. First Kharkov Med. Inst. Chkalov Inst. Epidemiol. Microbiol. First Commun.*, pp. 17–36.

Alisova, Z. I. and Mironov, S. G. (1947). General toxic action of cereal crops overwintered in the field. In Alimentary toxic aleukia. *Acta Chkalov Inst. Epidemiol. Microbiol., Third Commun.*, pp. 97–103.

Allen, N. K., Mirocha, C. J., Weaver, G., Aakhus-Allen, S., and Bates, F. (1981a). Effects of dietary zearalenone on finishing broiler chickens and young turkey poults. *Poult. Sci.* **60**, 124–131.

Allen, N. K., Aakhus-Allen, S. R., and Walser, M. M. (1981b). Toxic effects of repeated ethanol intubations to chicks. *Poult. Sci.* **60**, 941–943.

Allen, N. K., Mirocha, C. J., Aakhus-Allen, S., Bitgood, J. J., Weaver, G., and Bates, F. (1981c). Effect of dietary zearalenone on reproduction of chickens. *Poult. Sci.* **60**, 1165–1174.

Allen, N. K., Peguri, A., and Mirocha, C. J. (1982a). Effects of *Fusarium* cultures, T-2 toxin and zearalenone on reproduction of turkey females. *Poult. Sci.* **61**, 1405 (Abstr.).

Allen, N. K., Jevne, R. L., Mirocha, C. J., and Lee, Y. W. (1982b). The effect of a *Fusarium roseum* culture and diacetoxyscirpenol on reproduction of White Leghorn females. *Poult. Sci.* **61**, 2172–2175.

Allen, N. K., Peguri, A., Mirocha, C. J., and Newman, J. A. (1983). Effects of *Fusarium* cultures, T-2 toxin and zearalenone on reproduction of turkey females. *Poultry Sci.* **62**, 282–289.

Aloshin, B., and Eingorn, E. (1944). Changes in blood by septic angina and experimental agranulocytosis. In Data on septic angina. *Proc. First Kharkov Med. Inst. Chkalov Inst. Epidemiol. Microbiol.*, pp. 135–172.

Aloshin, B., Burshtein, S. A., and Chernyak, B. I. (1947). Hematopoietic organs and reticuloendothelial system in aleukia. In Alimentary toxic aleukia. *Acta Chkalov Inst. Epidemiol. Microbiol. First Commun.*, pp. 125–144.

Ames, B. N., McCann, J., and Yamasaki, E. (1975). Methods for detecting carcinogens and mutagens with the salmonella/mammalian microsome mutagenicity test. *Mutat. Res.* **31**, 347–364.

Anderson, B., and Chick, E. W. (1963). Mycokeratitis: Treatment of fungal corneal ulcers with amphotericin B and mechanical debridement. *South. Med. J.* **56**, 270–274.

Anderson, B., Roberts, S. S., Gonzales, C., and Chick, E. W. (1959). Mycotic ulcerative keratitis. *A.M.A. Arch. Ophthalmol.* **62**, 169–179.

Anderson, J. (1984). Upgrading chemical warfare intelligence. *Washington Post*, November 30, TE-7.

Anon. (1981a). Fungal warfare agents used in Asia, *Chem. Eng. News*, September 21, 4.

Anon. (1981b). Yellow Rain. Soviet chemical warfare. *Time* (U.S. ed.), September 14, 22.

Anon. (1981c). Too quick on Yellow Rain. *The New York Times*, November 18, 4.

Anon. (1982a). Incriminating data. *Nature* **296**, March 25, 281.

Anon. (1982b). Mycotoxins in South-East Asia? *Nature* **296**, April 1, 379–380.

Anon. (1982c). Chemical war in Afghanistan. *The New York Times,* December 5, 6.

Antonov, N. (1982). The State Department is bluffing. *New Times* **27**, 18–20.

Antonov, N. A., Belkin, G. S., Joffe, A. Z., Lukin, A. Y., and Simonov, E. N. (1951). Feeding experiments on horses with cultures of the toxic fungi *Fusarium poae* (Peck.) Wr. and *Cladosporium epiphyllum* (Pers.) Mart. *Acta Chkalov Inst.* **4**, 47–56.

Appel, O., and Wollenweber, H. W. (1910). Grundlagen einer Monographie der Gattung *Fusarium* (Link). *Arb. Kais. Biol. Inst. Land. Forstwirtsch.* **8**, 1–207, Berlin, Verlag Paul Parey.

Arai, T., and Ito, T. (1970). Cytotoxicity and antitumor activity of fusariocins, mycotoxins from *Fusarium moniliforme.* In H. Umezawa, ed., *Progress in Antimicrobial and Anticancer Chemotherapy,* Vol. 1. Baltimore: University Park Press, pp. 87–92.

Archer, M. (1974). Detection of mycotoxins in foodstuffs by use of chick embryos. *Mycopathol. Mycol. Appl.* **54**, 453–467.

Arrechea, A., Zapater, R. C., Storero, C., and Guevara, V. (1971). Queratomicosis por *Fusarium solani. Arch. Oftalmol. Buenos Aires* **46**, 123–127.

Asami, N. (1932). Toxic encephalitis by bean-hulls in Kitami district of Hokkaido. *Bull. Appl. Vet.* **5**, 376–383.

Ashton, P. S., Meselson, M., Robinson, J. P. P., and Seeley, T. D. (1983). Origin of Yellow Rain. *Science* **222** (4622), 366–367.

Austwick, P. K. C. (1975). Mycotoxins. *Br. Med. Bull.* **31**, 222–229.

Babenkova, S. V., Zhirmunskaya, Y. A., Syroyechkovskaya, M. Y., Tsuker, M. B., and Yusevitch, Y. S. (1955). The problem of the condition of the nervous system in Urov Disease. *Clin. Med.* **33**, 48–54.

Bacon, C. W., and Marks, H. L. (1976). Growth of broilers and quail fed *Fusarium* (*Gibberella zeae*)-infected corn and zearalenone (F-2). *Poult. Sci.* **55**, 1531–1535.

Bacon, C. W., and Russell, R. B. (1983). The fungal endophyte and Tall fescue. In *Proc. Tall Fescue Toxicosis Workshop.* March 17–18, 1983, 34–42, Atlanta, Ga.

Bacon, C. W., Robbins, J. D., and Porter, J. K. (1977). Media for identification of *Gibberella zeae* and production of F-2 (zearalenone). *Appl. Environ. Microbiol.* **33**, 445–449.

Badiali, L., Abou-Youssef, M. H., Radwan, A. I., Hamdy, F. M., and Hildebrandt, P. K. (1968). Moldy corn poisoning as the major cause of an encephalomalacia syndrome in Egyptian *Equidae. Am. J. Vet. Res.* **29**, 2029–2035.

Balakrishnan, E. (1962). Mycotic keratitis. *Proc. 19th Int. Congr. Ophthalmol.* **11**, 1242, New Delhi.

Balzer, I., Ozegovic, L., Tuite, J., and Scott, P. (1977a). Panel on zearalenone. In J. V. Rodricks, C. W. Hesseltine, and M. A. Mehlman, eds., *Mycotoxins in Human and Animal Health.* Park Forest, Ill.: Pathotox, pp. 415–416.

Balzer, I., Bogdanic, C., and Muzic, S. (1977b). Natural contamination of corn (*Zea mays*) with mycotoxins in Yugoslavia. *Ann. Nutr. Alim.* **31**, 425–430.

Balzer, I., Bogdanic, C., and Pepelijnjak, S. (1978). Rapid thin-layer chromatographic method for determining aflatoxin $B_1$, ochratoxin A, and zearalenone in corn. *J. Assoc. Off. Anal. Chem.* **61**, 584–585.

Bamburg, J. R. (1969). Mycotoxins of the trichothecene family produced by cereal molds. Ph.D. dissertation, University of Wisconsin, Madison, 161 pp.

Bamburg, J. R. (1972). The biological activities and detection of naturally occurring 12,13-epoxy-$\Delta^9$-trichothecenes. *Clin. Toxicol.* **5**, 495–515.

Bamburg, J. R. (1973). Biological and physical methods for detection and identification of trichothecene mycotoxins. Abstracts of papers, No. 88. MTG, August 1973, Division of Agricultural and Food Chemistry.

Bamburg, J. R. (1976). Chemical and biochemical studies of the trichothecene mycotoxins. In J. V. Rodricks, ed., *Mycotoxins and Other Fungal Related Food Problems* (Advances in Chemistry Series No. 149). Washington, D.C.: American Chemical Society, pp. 144–162.

Bamburg, J. R., and Strong, F. M. (1969). Mycotoxins of the trichothecene family produced by *Fusarium tricinctum* and *Trichoderma liquorum*. *Phytochemistry* **8**, 2405–2410.

Bamburg, J. R., and Strong, F. M. (1971). 12,13-Epoxytrichothecenes. In S. Kadis, A. Ciegler, and S. J. Ajl, eds. *Microbial Toxins*, Vol. 7. New York: Academic, pp. 207–292.

Bamburg, J. R., Smalley, E. B., Riggs, N. V., and Strong, F. M. (1966). A toxic compound from moldy corn. *Amer. Chem. Soc. Abstr.* September (A-93), 11–16.

Bamburg, J. R., Marasas, W. F., Riggs, N. V., Smalley, E. B., and Strong, F. M. (1968a). Toxic spiroepoxy compounds from Fusaria and other Hyphomycetes. *Biotechnol. Bioeng.* **10**, 445–455.

Bamburg, J. R., and Riggs, N. V., and Strong, F. M. (1968b). The structures of toxins from two strains of *Fusarium tricinctum*. *Tetrahedron* **24**, 3329–3336.

Bamburg, J. R., Strong, F. M., and Smalley, E. B. (1969). Toxins from moldy cereals. *J. Agric. Food Chem.* **17**, 443–450.

Barer, G. L. (1947). The problem of the chemical nature of the toxic cereals. Lecture on Republic Conference about Alimentary Toxic Aleukia, January, Moscow.

Barnikol, H., Gruber, S., Thalmann, A., and Schmidt, H. L. (1982). Fusariotoxikosen beim Schwein durch Trichothezene mit Beteiligung von Mutterkorn. *Tieraerztl. Umsch.* **37**, 114–126.

Barsky, D. (1959). Keratomycosis. A report of six cases. *Arch. Ophthalmol.* **61**, 547–552.

Bart, V. V. (1960). Material on the study of the toxicity of late-gathered cereals in the Latvian S.S.R. In V. I. Bilai, ed., *Mycotoxicoses of Man and Agricultural Animals* (English Trans.). Washington, D.C.: Joint Publishers Research Service, pp. 125–131.

Bartsch, H., Camus, A. M., and Malaveille, C. (1976). Comparative mutagenicity of N-nitrosamines in a semi-solid and in a liquid incubation system in presence of rat or human tissue fractions. *Mutat. Res.* **31**, 149–162.

Bartsch, H., Malaveille, C., Camus, A. M., Martel-Planche, G., Brun, G., Hautefeuille, A., Sabadie, N., Baarbin, A., Kuroi, T., Drevon, C., Piccoli, C., and Montesano, R. (1980). Validation and comparative studies on 180 chemicals using *S. typhimurium* strains and V79 Chinese hamster cells in the presence of various metabolizing systems. *Mutat. Res.* **76**, 1–50.

Beck, E. V. (1906a). The problem of osteoarthritis deformans endemica in the Transbaikal district. Ph.D. dissertation, St. Petersburg, Medical Military Academy.

Beck, E. V. (1906b). To the problem of deforming endemic osteoarthritis in the Transbaikal district. *Russ. Physician* **3**, 74–75.

Beletskij, G. N. (1945). Measures to prevent and fight against "Septic Angina" (Alimentary Toxic Aleukia). *Hyg. a. Sanit. Moscow* **3**, 22–26.

Beller, K., and Wedemann, W. (1929). Untersuchung über Schadwirkung amerikanischer Futtergerste, *Ztschr. Infektkrankh., Hyg. Haustech.* **36**, 103–129.

Bellus, D., Fisher, H., Greuter, H., and Martin, P. (1978). Synthesen von Moniliformin, einem Mycotoxin mit Cyclobutendion-Structur. *Helv. Chim. Act.* **61**, 1784–1813.

Belt, R. J., Haas, D. C., Joseph, U., Goodwin, W., More, D., and Hoogstraten, B. (1979). Phase I study of anguidine administered weekly. *Cancer Treat. Rep.* **63**, 1993–1995.

Benjamin, R. P., Callaway, J. L., and Conant, N. F. (1970). Facial granuloma associated with *Fusarium* infection. *Arch. Derm.* **101**, 598–600.

Bennett, G. A., and Anderson, R. A. (1978). Distribution of aflatoxin and/or zearalenone in wet-milled corn products: A review. *J. Agric. Food Chem.* **26**, 1055–1060.

Bennett, G. A., and Shotwell, O. L. (1979). Zearalenone in cereal grains. *J. Am. Oil Chem. Soc.* **56**, 812–819.

Bennett, G. A., Beaumont, W. H., and Brown, P. R. (1974). Use of the anabolic agent zearanol (resorcyclic acid lactone) as a growth promoter for cattle. *Vet. Res.* **94**, 235–239.

Bennett, G. A., Peplinski, A. J., Brekke, O. L., Jackson, L. K., and Wichster, W. R. (1976). Zearalenone: Distribution in dry-milled fractions of contaminated corn. *Cereal Chem.* **53**, 299–307.

Bennett, G. A., Vandegruft, E. E., Shotwell, O. L., Watson, S. A., and Bocan, B. J. (1978). Zearalenone: Distribution in wet-milling fractions from contaminated corn. *Cereal Chem.* **55**, 455–461.

Bennett, G. A., Shotwell, O. L., and Hesseltine, C. W. (1979). Detoxification of zearalenone contaminated corn. *J. Am. Oil Chem. Soc.* **56**, 205A (Abstr.).

Bennett, G. A., Shotwell, O. L., and Hesseltine, C. W. (1980). Destruction of zearalenone in contaminated corn. *J. Am. Oil Chem. Soc.* **57**, 245–247.

Bennett, G. A., Lagoda, A. A., Shotwell, O. L., and Hesseltine, C. W. (1981a). Utilization of zearalenone-contaminated corn for ethanol production. *Am. Oil Chem. Soc.* **58**, 974–976.

Bennett, G. A., Peterson, R. E., Plattner, R. D., and Shotwell, O. L. (1981b). Isolation and purification of deoxynivalenol and a new trichothecene by high pressure liquid chromatography. *J. Am. Oil Chem. Soc.* **58**, 1002A–1005A.

Berg, B. M. (1967). Longevity studies in rats. 2. Pathology of ageing rats. In E. Cotchin and F. J. C. Roe, eds., *Pathology of Laboratory Rats and Mice*. England: Blackwell, pp. 749–786.

Berisford, Y. C., and Ayres, J. C. (1976a). Use of the insecticide naled to control zearalenone production. *J. Agric. Food Chem.* **24**, 973–975.

Berisford, Y. C., and Ayres, J. C. (1976b). Effect of insecticides on growth and zearalenone (F-2) production by the fungus *Fusarium graminearum*. *Environ. Entomol.* **5**, 644–648.

Bhat, R. V. and Tulpule, P. G. (1983). Trichothecene problems in India. In Y. Ueno, ed., *Trichothecenes, Chemical, Biological and Toxicological Aspects*. Amsterdam: Elsevier, pp. 285–289.

Bilai, V. I. (1947). *Fusarium* species on cereal crops and their toxic properties. *Microbiology* **16**, 11–17.

Bilai, V. I. (1948). The effect of extracts from toxic fungi on animal and plant tissues. *Microbiology* **17**, 142–147.

Bilai, V. I. (1952a). Taxonomy of the Sporotrichiella section of the genus *Fusarium*. *Dopov. Akad. Nauk. Ukr. SSR.* **5**, 415–419.

Bilai, V. I. (1952b). The toxicity of various *Fusarium* cultures of the Sporotrichiella section. *Mikrobiol. Zh.* **14**, 18–29.

Bilai, V. I. (1952c). Influence of various species of *Fusarium* on cereal crops. *Mikrobiol. Zh.* **14**, 58–69.

Bilai, V. I. (1953). *Toxic Fungi on the Grain of Cereal Crops*. Kiev: *Publ. Acad. Sci.*, pp. 1–93.

Bilai, V. I. (1955). *Fusaria*. Kiev: Publ. Acad. Sci.

Bilai, V. I. (1960). The specificity of toxins and antibiotics in microscopic fungi. In V. I. Bilai, ed., *Mycotoxicoses of Man and Agricultural Animals* (English Trans.) Washington, D.C.: Office of Technical Service, U.S. Dept. of Commerce, pp. 50–58.

Bilai, V. I. (1965). (Bilai, V. I., ed.) Toxins. In *Biologically Active Substances of Microscopic Fungi*. Kiev: Naukova Dumka, pp. 160–219.

Bilai, V. I. (1970a). Experimental morphogenesis in the fungi of the genus *Fusarium* and their taxonomy. *Ann. Acad. Sci. Fenn. A. IV Biol.* **168**, 7–18.

Bilai, V. I. (1970b). Phytopathological and hygienic significance of representatives of the section Sporotrichiella in the genus *Fusarium* Link. *Ann. Acad. Sci. Fenn. A., IV Biol.* **168**, 19–24.

Bilai, V. I. (1977). Fusaria. Kiev: Naukova Dumka.

Bilai, V. I. (1978a). Principles of the taxonomy and phytopathogenic structure of species of the genus *Fusarium* Lk. ex Fr. *Mikrobiol. Zh.* **40**, 148–156.

Bilai, V. I. (1978b). Mycological aspects of alimentary toxicoses. *Mikrobiol. Zh.* **40**, 205–213.

Bilai, V. I., and Pidoplichko, N. M. (1960). Primary isolation of substances through the subsurface method of cultivating toxic fungi. In V. I. Bilai, ed., *Mycotoxicoses of Man and Agricultural Animals* (English Trans.) Washington, D.C.: Joint Publishers Research Service, pp. 81–89.

Bilai, V. I., and Pidoplichko, N. M. (1970). Toxigenic microscopical fungi caused diseases of man and animals. Kiev: Naukova Dumka (in Russian).

Birbin, S. S. (1966). Fusariotoxicosis in ducks. *Veterinariya* **43**, 54–55 (Moscow).

Birone, G. M., Gyuru, F., and Feher, G. (1972). Study on the toxicity of *Fusarium* species damaging maize for white rats. *Magy. Allatorv. Lapja* **27**, 597–604.

Bitay, Z., Glavits, R., and Sellyev, G. (1979). Feeding experiments on broiler chicks with ochratoxin-A, patulin, T-2 toxin and butenolide. *Magy. Allatorv. Lapja* **34**, 417–422.

Bitay, Z., Glavits, R., Sandor, G., and Balazs, K. (1981). A case of T-2 mycotoxicosis in broiler chicks. *Magy. Allatorv. Lapja* **36**, 491–495.

Bjeldanes, I. F., and Thomson, S. V. (1979). Mutagenic activity of *Fusarium moniliforme* isolates in the *Salmonella typhimurium* assay. *Appl. Environ. Microbiol.* **37**, 1118–1121.

Bjeldanes, L. F., and Weib, L. A. (1980). Mutagenic mycotoxins from *Fusarium moniliforme*. *Environ. Mut.* **2**, 240–241.

Bjeldanes, L. F., Chang, G. W., and Thomson, S. V. (1978). Detection of mutagens produced by fungi with the *Salmonella typhimurium* assay. *Appl. Environ. Microbiol.* **35**, 1150–1154.

Bjelke, E. (1974). Epidemiological studies of the cancer of the stomach, colon and rectum, with special emphasis on the role of diet. *Scand. J. Gastroenterol.* **9**, 1–235.

Blight, M. M., and Grove, J. F. (1974). New metabolic products of *Fusarium culmorum:* Toxic trichothec-9-en-8-ones and 2-acethylquinazolin-4-(3H)-one. *J. Chem. Soc. Perkin.* **1**, 1691–1693.

Bodon, L., and Zoldac, L. (1974). Cytotoxicity studies on T-2 Fusariotoxin. *Acta Vet. Acad. Sci. Hung.* **24**, 451–455.

Boldyrev, T. E., and Shtenberg, A. I. (1950). Alimentary toxic aleukia. In *Food Hygiene, Methodological and Reference Guide for Physicians*. Moscow: Megdiz, pp. 234–236.

Bolliger, G., and Tamm, C. (1972). Vier neue Metabolite von *Gibberella zeae:* 5-formyl-zearalenon, 7′-dehydrozearalenon, 8′-hydroxy and p′-epi-Hydroxy-zearalenon. *Helv. Chim. Acta* **55**, 3030–3048.

Bonnenfant, J. L., Miller, F., and Roy, P. E. (1967). Quebec beer-drinkers cardiomyopathy. Pathological studies. *Can. Med. Assoc. J.* **97**, 910–916.

Boonchuvit, B., Hamilton, P. B., and Burmeister, H. R. (1975). Interaction of T-2 toxin with *Salmonella* infections of chickens. *Poult. Sci.* **54**, 1693–1696.

Booth, C. (1959). Studies of Pyrenomycetes. 4. *Nectria* (Part 1). Mycological Papers, No. 73, Commonwealth Mycological Institute, Kew, England, pp. 1–42.

Booth, C. (1960). Studies of Pyrenomycetes. 5. Nomenclature of some Fusaria in relation to their nectrioid perithecial stages. Mycological Papers, No. 74, Commonwealth Mycological Institute, Kew, England, pp. 1–16.

Booth, C. (1966a). Provisional key to Fusaria. Commonwealth Mycological Institute, Kew, England, pp. 1–11.

Booth, C. (1966b). Physical and biochemical techniques in the identification of Fusaria. *J. Gen. Microbiol.* **42,** 7–8 (Commonwealth Mycological Institute, Kew, England.)

Booth, C. (1971a). *Fusarium poae.* Description of pathogenic fungi and bacteria. Mycological Papers, No. 308, Commonwealth Mycological Institute, Kew, England.

Booth, C. (1971b). *The Genus Fusarium.* Commonwealth Agricultural Bureau, Farnham Royal, Bucks. Mycological Institute, Kew, England, pp. 237.

Booth, C. (1973). *Fusarium sambucinum (Gibberella pulicaris).* Descriptions of pathogenic fungi and bacteria. Mycological Papers, No. 385, Commonwealth Mycological Institute, Kew, England.

Booth, C. (1975). The present status of *Fusarium* taxonomy. *Ann. Rev. Phytopathol.* **13,** 83–93.

Booth, C. (1977). *Fusarium.* Laboratory guide to the identification of the major species. Commonwealth Mycological Institute, Kew, England.

Booth, C. (1981). Perfect states (teleomorphs) of *Fusarium* species. In P. E. Nelson, T. A. Toussoun, and R. J. Cook, eds., *Fusarium Diseases, Biology and Taxonomy.* University Park: Pennsylvania State University Press, pp. 446–452.

Booth, C., and Morgan-Jones, G. (1977). *Toxic fungi.* New York: Academic.

Borchers, R., and Peltier, G. L. (1947). Molded foodstuffs. *Poult. Sci.* **26,** 194–197.

Borger, M. L., Wilson, L. L., Sink, J. D., Ziegler, J. H., Davis, S. S., Orleg, C. F., and Rugh, M. C. (1971). Zearanol and protein effects on finishing steers. *J. Anim. Sci.* **33,** 275–276.

Borker, E., Insalata, N. F., Levi, C. P., and Witzeman, J. S. (1966). Mycotoxins in feeds and foods. *Adv. Appl. Microbiol.* **8,** 315–351.

Borut, S., and Joffe, A. Z. (1966). *Aspergillus flavus* Link and other fungi associated with stored groundnuts in Israel. *Isr. J. Bot.* **15,** 112–120.

Bottalico, A. (1975). Zearalenone production by isolates of *Fusarium* inducing foot rot disease of wheat in Italy. *Phytopathol. Mediterr.* **14,** 134–135.

Bottalico, A. (1976). La presenza di *Fusarium moniliforme* Sheld. nelle cariossidi di Granturco (*Zea mays* L.) quale problema fitopathologice e micotossicologico in Italia. 2. Aspetti micotossicologici. *Phytopathol. Mediterr.* **15,** 54–58.

Bottalico, A. (1977). Production of zearalenone by *Fusarium* species from cereals in Italy. *Phytopathol. Mediterr.* **16,** 75–78.

Bottalico, A., Lerario, P., and Frisullo, S. (1980). Presenza di aflatossine, di zearalenone e di ceppi di Aspergilli produttori di aflatossine in campioni di farina di manioca. *Zootech. Nutr. Anim.* **6,** 209–214.

Bourguignon, R. L., Walsh, A. F., Flynn, J. C., Baro, C., and Spinos, E. (1976). *Fusarium* species osteomyelitis: A case report. *J. Bone Jt. Surg.* **58A,** 722–723.

Boutibonnes, P. (1979a). Demonstration of the antibacterial activity of some mycotoxins using *Bacillus thuringiensis* (Berliner). *Mycopathology* **67,** 45–50.

Boutibonnes, P. (1979b). Antibacterial activity of zearalenone. *Can. J. Microbiol.* **25,** 421–423.

Boutibonnes, P., and Loquet, C. (1979). Antibacterial activity, DNA-attacking ability, and mutagenic ability of the mycotoxin zearalenone. *IRCS Med. Sci. Libr. Compend.* **7,** 204.

Boyd, M. R., and Wilson, B. J. (1972). Isolation and characterization of 4-ipomeanol, a lung-toxic furanoterpenoid produced by sweet potatoes (*Ipomoea batatas*). *J. Agric. Food Chem.* **20**, 428–430.

Boyd, P. A., and Wittliff, J. L. (1978). Mechanism of *Fusarium* mycotoxin action in mammary gland. *J. Toxicol. Environ. Health* **4**, 1–8.

Boyd, M. R., Wilson, B. J., and Harris, T. M. (1972). Confirmation by chemical synthesis of 4-ipomeanol, a lung-toxic metabolite of the sweet potato, *Ipomoea batatas*. *Nature (New Biology)* **236**, 158–159.

Boyd, M. R., Burka, L. T., Harris, T. M., and Wilson, B. J. (1974). Lung-toxic furanoterpenoids produced by sweet potatoes (*Ipomoae batatas*) following microbial infection. *Biochem. Biophys. Acta* **337**, 184–195.

Boyd, M. R., Burka, L. T., and Wilson, B. J. (1975). Distribution, excretion and binding of radioactivity in the rat after intraperitoneal administration of the lung-toxic furan, [$^{14}$C]-4-ipomeanol. *Toxicol. Appl. Pharmacol.* **32**, 147–157.

Boyd, M. R., Burka, L. T., Wilson, B. J., and Sasame, H. A. (1979). *In vitro* studies on the metabolic activation of the pulmonary toxin 4-ipomeanol by rat lung and liver microsomes. *J. Pharmacol. Exp. Ther.* **207**, 677–686.

Breitenstein, W., and Tamm, C. (1975). C-NMR-spectroscopy of the trichothecene derivatives verrucarol, verrucarins A and B, and roridins A, D and H. *Helv. Chim. Acta.* **58**, 1172–1180.

Breitenstein, W., and Tamm, C. (1977). Verrucarin K. The first national trichothecene derivative lacking the 12,13-Epoxy group. *Helv. Chim. Acta* **60**, 1522–1527.

Breitenstein, W., and Tamm, C. (1978). Partial Synthese von Tetrahydroverrucarin J. *Helv. Chim. Acta* **61**, 1975–1983.

Brian, P. W., Dawkins, A. W., Grove, J. F., Hemming, H. G., Love, D., and Norris, L. F. (1961). Phytotoxic compounds produced by *Fusarium equiseti*. *J. Exp. Bot.* **12**, 1–12.

Bridges, C. H. (1978). Mycotoxicoses in horses. In T. D. Wyllie and G. Morehouse, eds., *Mycotoxic Fungi, Mycotoxins, Mycotoxicoses*. New York: Marcel Dekker, pp. 173–181.

Bristol, F. M., and Djurickovic, S. (1971). Hyperestrogenism in female swine as the result of feeding moldy corn. *Can. Vet. J.* **12**, 132–135.

Brodnik, T. (1975). Influence of toxic products of *Fusarium graminearum* and *Fusarium moniliforme* on maize seed germination and embryo growth. *Seed Sci. Technol.* **3**, 691–696.

Brook, P. J., and White, E. P. (1966). Fungus toxins affecting mammals. *Ann. Rev. Phytopathol.* **4**, 171–194.

Brooks, J. R., Steelman, S. L., and Patanelli, D. J. (1971). Uterotropic and anti-implantation activities of certain resorcylic acid lactone derivatives (35521). *Proc. Soc. Exp. Biol. Med.* **137**, 101–104.

Brown, R. F. (1969). The effect of some mycotoxins on the brine shrimp (*Artemia salina*). *J. Am. Oil Chem. Soc.* **46**, 119.

Brown, R. G. (1970). An anabolic agent for ruminants. *J. Am. Vet. Med. Assoc.* **157**, 1537–1539.

Brown, W. (1928). Studies in the genus *Fusarium*. 6. General description of strains, together with a discussion of the principles at present adopted in the classification of *Fusarium*. *Ann. Bot.* **42**, 285–304.

Brown, W., and Horne, A. S. (1926). Studies in the genus *Fusarium*. 3. An analysis of factors which determine certain microscopic features of *Fusarium* strains. *Ann. Bot.* **40**, 203–221.

Brummer, A. (1981). Russia is using chemical weapons. *The Guardian*, September 15, 74.

Buckle, A. E. (1983). The occurrence of mycotoxins in cereals and animal feedstuffs. *Vet. Res. Commun.* **7**, 171–186.

Bugeac, I., and Berbinschi, C. (1967). Vulvovaginitis in sows. *Revta Zootech. Med. Vet.* **17**, 56–61, 89–90, 92–93, 95. Paul Lechevalier.

Bugnicourt, F. (1939). Les *Fusarium* et *Cylindrocarpon* de L'Indochine. Encyclopedia of Mycology. Paris VI. p. 206. Paul Lachevalier.

Bukharbayeva, A. S., and Piotrovski, S. V. (1972). Investigation of lipid mycelia from *Fusarium sporotrichiella*. In *Proc. Symp. Mycotoxins* October 3–6, 14, Kiev.

Bullerman, L. B. (1974). A screening medium and method to detect several mycotoxins in mold cultures. *J. Milk Food Technol.* **37**, 1–3.

Burda, C. D. and Fisher, E., Jr. (1959). The use of cortisone in establishing experimental fungal keratitis in rats. A preliminary report. *Am. J. Ophthalmol.* **48**, 330–335.

Burka, L. T., and Wilson, B. J. (1976). Toxic furanosesquiterpenoids from mold-damaged sweet potatoes. In J. V. Rodricks, ed., *Mycotoxins and Other Fungal Related Food Problems* (Advances in Chemistry Series No. 149). Washington, D.C.: American Chemical Society, pp. 387–399.

Burka, L. T., Kuhnert, L., Wilson, B. J., and Harris, T. M. (1974). 4-hydroxymyporone, a key intermediate in the biosyntheses of pulmonary toxins produced by *Fusarium solani* infected sweet potatoes. *Tetrahedron Lett.* **46**, 4017–4020.

Burkhardt, H. J., Lundin, R. E., and McFadden, W. H. (1968). Mycotoxins produced by *Fusarium nivale* (Fries) Cesati isolated from tall fescue (*Festuca arundinacea* Scherb.): Synthesis of 4-acetamido-4-hydroxy-2-butenoic acid-γ-lactone. *Tetrahedron* **24**, 1225–1229.

Burmeister, H. R. (1971). T-2 toxin production by *Fusarium tricinctum* on solid substrate. *Appl. Microbiol.* **21**, 739–742.

Burmeister, H. R., and Hesseltine, C. W. (1970). Biological assays for two mycotoxins produced by *Fusarium tricinctum*. *Appl. Microbiol.* **20**, 437–440.

Burmeister, H. R., Ellis, J. J., and Yates, S. G. (1971). Correlation of biological to chromatographic data for two mycotoxins elaborated by *Fusarium*. *Appl. Microbiol.* **21**, 673–675.

Burmeister, H. R., Ellis, J. J., and Hesseltine, C. W. (1972). Survey for Fusaria that elaborate T-2 toxin. *Appl. Microbiol.* **23**, 1165–1166.

Burmeister, H. R., Vesonder, R. F., and Hesseltine, C. W. (1977). Swelling of *Penicillium digitatum* conidia by *Fusarium acuminatum* NRRL 6227. *Mycopathologia* **62**, 53–56.

Burmeister, H. R., Ciegler, A., and Vesonder, R. F. (1979). Moniliformin, a metabolite of *Fusarium moniliforme* NRRL 6322, purification and toxicity. *Appl. Environ. Microbiol.* **37**, 11–13.

Burmeister, H. R., Grove, M. D., and Kwolek, W. F. (1980a). Moniliformin and butenolide: effect on mice of high-level long-term oral intake. *Appl. Environ. Microbiol.* **40**, 1142–1144.

Burmeister, H. R., Vesonder, R. F., and Kwolek, W. F. (1980b). Mouse bioassay for *Fusarium* metabolites: Rejection or acceptance when dissolved in drinking water. *Appl. Environ. Microbiol.* **39**, 957–961.

Burmeister, H. R., Ellis, J. J., and Vesonder, R. F. (1981). Survey for Fusaria that produce an antibiotic that causes conidia of *Penicillium digitatum* to swell. *Mycopathologia* **74**, 29–33.

Butko, V. S., Kalabukhov, E. P., and Adreeva, T. A. (1977). Characteristics of a number of the links in the ecosystem of an endemic Urov disease locality in Transbaikalia. *Hyg. Sanit.* **2**, 15–18.

Butler, F. C. (1961). Root and root rot diseases of wheat. *Sci. Bull. NSW Dept. Agric.* No. 77.

Caldwell, R. W., and Tuite, J. (1968). Zearalenone production among *Fusarium* species. *Phytopathology* **58**, 1046.

Caldwell, R. W., and Tuite, J. (1970). Zearalenone production in field corn in Indiana. *Phytopathology* **60**, 1696–1697.

Caldwell, R. W., and Tuite, J. (1974). Zearalenone in freshly harvested corn. *Phytopathology* **64**, 752–753.

Caldwell, R. W., Tuite, J., Stob, M., and Baldwin, R. (1970). Zearalenone production by *Fusarium* species. *Appl. Microbiol.* **20**, 31–34.

Campbell, P. (1981). Yellow Rain. Waiting for data. *Nature* **293**, October 22, 598.

Cannon, M., Jimenez, A., and Vazquez, D. (1976). Competition between Trichodermin and several other Sesquiterpene antibiotics for binding to their receptor site(s) on eukaryotic ribosomes. *Biochem. J.* **160**, 137–145.

Cannon, M., Smith, K. E., and Carter, C. J. (1976). Prevention by ribosome-bound nascent polyphenylalanine chains of the functional interaction of T-2 toxin with its receptor site. *Biochem. J.* **156**, 289–294.

Cantini, G., Scurti, J. C., and Fiussello, N. (1973). Oestrogenlike metabolites of a strain of *Gibberella zeae* (*Fusarium graminearum*) cultivated on different media. *Folia Veterinaria Latina* **3**, 203–214.

Carlson, J. R., Dyer, I. A., and Johnson, R. J. (1968). Tryptophan-induced interstitial pulmonary emphysema in cattle. *Am. J. Vet. Res.* **29**, 1983–1989.

Carlson, J. R., Dickinson, E. D., Yokoyama, M. T., and Bradley, B. (1975). Pulmonary edema and emphysema in cattle after intraruminal and intravenous administration of 3-methylindole. *Am. J. Vet. Res.* **36**, 1341–1347.

Carrera, C. J. M. (1954). The genus *Fusarium*. *Rev. Invest. Agric.* **8**, 311–456, Buenos Aires.

Carson, M. S., and Smith, T. K. (1983). Effect of feeding alfalfa and refined plant fibers on the toxicity and metabolism of T-2 toxin in rats. *J. Nutr.* **113**, 304–313.

Carter, C. J., and Cannon, M. (1978). Inhibition of eukaryotic ribosomal function by the sesquiterpenoid antibiotic fusarenon-X. *Eur. J. Biochem.* **84**, 103–111.

Carter, C. J., Cannon, M., and Smith, K. E. (1976). Inhibition of protein synthesis in reticulocyte lysates by trichodermin. *Biochem. J.* **154**, 171–178.

Casas-Campillo, C., and Bautista, M. (1965). Microbiological aspects in the hydroxylation of estrogens by *Fusarium moniliforme*. *Appl. Microbiol.* **13**, 977–984.

Chakrabarti, D. K., Basu Chaudbury, K. C., and Ghosal, S. (1976). Toxic substances produced by *Fusarium*. 3. Production and screening of phytotoxic substances of *F. oxysporum* f. sp. *carthami* responsible for the wilt disease of safflower *Carthamus tinctorius* Linn. *Experientia* **32**, 608–609.

Chandler, F. W. (1980). Color atlas and text of the histopathology of mycotoxic diseases. Chicago: Year Book Medical.

Chang, K., Kurtz, H., and Mirocha, C. J. (1979). Effects of the mycotoxin zearalenone on swine reproduction. *Am. J. Vet. Res.* **40**, 1260–1267.

Chepurov, K. V., and Cherkasova, A. V. (1954). Urov disease in farm animals and fowls. *Veterinariya* **31**, 38–42, Moscow.

Chepurov, K. V., Cherkasova, A. V., Akulov, N. M., Ostrovsky, I. I., and Martynyuk, D. F. (1955). Urov Disease. Amur Book Press, Khabarosk.

Chernikov, B. A. (1944). Problems of etiopathogenesis and therapy of "Septic Angina." In Data of septic angina. *Proc. First Kharkov Med. Inst. Chkalov Inst. Epidemiol. Microbiol.* 173–183, Orenburg, USSR.

Chernov, K. S. (1970). Mycotoxicoses of swine and their differential diagnosis. *Vet. Microbiol.* No. 16803. *All Union Inst. Exp. Vet. Med. Moscow.*

Chetvertakova, Ye. P. (1967). The clinic of Urov Disease in children. Ph.D. dissertation, Moscow. Med. Inst. Acad. Sci.

Chi, M. S., and Mirocha, C. J. (1978). Necrotic oral lesions in chickens fed diacetoxyscirpenol, T-2 toxin, and crotocin. *Poult. Sci.* **57**, 807–808.

Chi, M. S., Mirocha, C. J., Kurtz, M. J., Weaver, G., Bates, F., Shimoda, W., and Burmeister, H. R. (1977a). Acute toxicity of T-2 toxin in broiler chicks and laying hens. *Poult. Sci.* **56**, 103–116.

Chi, M. S., Mirocha, C. J., Kurtz, M. J., Weaver, G., Bates, F., and Shimoda, W. (1977b). Subacute toxicity of T-2 toxin in broiler chicks. *Poult. Sci.* **56**, 306–313.

Chi, M. S., Mirocha, C. J., Kurtz, M. J., Weaver, G., Bates, F., and Shimoda, W. (1977c). Effect of T-2 toxin on reproductive performance and health of laying hens. *Poult. Sci.* **56**, 628–637.

Chi, M. S., Robison, T. S., Mirocha, C. J., Behrens, J. C., and Shimoda, W. (1978a). Transmission of radioactivity into eggs from laying hens (*Gallus domesticus*) administered tritium labeled T-2 toxin. *Poult. Sci.* **57**, 1234–1238.

Chi, M. S., Robison, T. S., Mirocha, C. J., and Reddy, K. R. (1978b). Acute toxicity of 12,13-Epoxytrichothecenes in one-day old broiler chicks. *Appl. Environ. Microbiol.* **35**, 636–640.

Chi, M. S., Robison, T. S., Mirocha, C. J., Swanson, S. P., and Shimoda, W. (1978c). Excretion and tissue distribution of radioactivity from tritium labeled T-2 toxin in chicks. *Toxicol. Appl. Pharmacol.* **45**, 391–402.

Chi, M. S., Mirocha, C. J., Weaver, G. A., and Kurtz, H. J. (1980a). Effect of zearalenone on female white Leghorn chickens. *Appl. Environ. Microbiol.* **39**, 1026–1030.

Chi, M. S., Mirocha, C. J., Kurtz, H. J., Weaver, G. A., Bates, F., Robison, T., and Shimoda, W. (1980b). Effect of dietary zearalenone on growing broiler chickens. *Poult. Sci.* **59**, 531–536.

Chi, M. S., El-Halawani, M. E., Weibel, P. E., and Mirocha, C. J. (1981). Effect of T-2 toxin on brain catecholamines and selected blood components in growing chickens. *Poult. Sci.* **60**, 137–141.

Chi, W. O. (1982). Trichothecenes—A review. *Heterocycles* **19**, 1685–1717.

Chiba, J., Nakano, N., Morooka, N., Nakazava, S., and Watanabe, Y. (1972). Inhibitory effects of fusarenon-X, a sesquiterpene mycotoxin, on lipid synthesis and phosphate uptake in *Tetrahymena pyriformis. Japan J. Med. Sci. Biol.* **25**, 291–296.

Chilikin, V. I. (1944). *Principal Problems of Clinical Syndromes, Pathogenesis, and Therapy of Alimentary Toxic Aleukia (Septic Angina).* Kuybyshev: Regional Publishing House, p. 27.

Chilikin, V. I. (1945). Septic angina. In A. I. Nesterov, A. H. Sysin, and L. N. Karlic, eds., *Clinical Aspects and Therapy of Alimentary Toxic Aleukia (Septic Angina).* Medgiz: Publ. State Med. Lit., pp. 39–54, Moscow.

Chilikin, V. I. (1947). Peculiarities and problems concerning clinical aspects, pathogenesis and therapy of alimentary toxic aleukia. In Alimentary Toxic Aleukia. *Acta Chkalov Inst. Epidemiol. Microbiol.* 145–151, Orenburg, USSR.

Cho, B. R. (1964). Toxicity of water extracts of scabby barley to suckling mice. *Am. J. Vet. Res.* **25**, 1267–1270.

Cho, C. T., Vats, T. S., Lowman, J. T., Brandsberg, J. W., and Tosh, F. E. (1973). *Fusarium solani* infection during treatment for acute leukemia. *J. Pediatr.* **83**, 1028–1031.

Christensen, C. M. (1971). Mycotoxins. *Crit. Rev. Environ. Control* **2**, 57–80.

Christensen, C. M. (1979). Zearalenone. In W. Shimoda, ed., *Conference on Mycotoxins in*

*Animal Feeds and Grains Related to Animal Health.* Maryland: Bureau of Veterinary Food, pp. 1–79.

Christensen, C. M., and Kaufmann, H. H. (1969). Mycotoxins and grain quality. In *Grain Storage. The Role of Fungi in Quality Loss.* Minneapolis, Minn.: University of Minnesota Press, pp. 76–93.

Christensen, C. M., Nelson, G. H., and Mirocha, C. J. (1965). Effect on the white rat uterus of a toxic substance isolated from *Fusarium. Appl. Microbiol.* **13**, 653–659.

Christensen, C. M., Nelson, G. H., Mirocha, C. J., and Bates, F. (1968). Toxicity to experimental animals of 943 isolates of fungi. *Cancer Res.* **28**, 2293–2295.

Christensen, C. M., Meronuck, R. A., Nelson, G. H., and Behrens, J. C. (1972a). Effects on turkey poults of rations containing corn invaded by *Fusarium tricinctum* (Cda.) Snyd. and Hans. *Appl. Microbiol.* **23**, 177–179.

Christensen, C. M., Mirocha, C. J., Nelson, G. H., and Quast, J. F. (1972b). Effect on young swine of consumption of rations containing corn invaded by *Fusarium roseum. Appl. Microbiol.* **23**, 202.

Christensen, J. J., and Kernkamp, H. C. H. (1936). Studies on the toxicity of blighted barley to swine. *Minn. Agric. Exp. Stat. Tech. Bull.* **113**, 1–28.

Christopher, J., Carter, J., and Cannon, M. (1977). Structural requirements for the inhibitory action of 12,13-Epoxytrichothecenes on protein synthesis in eukaryotes. *Biochem. J.* **166**, 399–409.

Chu, F. S. (1977). Mode of action of mycotoxins and related compounds. In D. Perlman, ed., *Advances in Applied Microbiology,* Vol. 22. New York: Academic, pp. 83–143.

Chu, F. S., Grossman, S., Wei, W. D., and Mirocha, C. J. (1979). Production of antibody against T-2 toxin. *Appl. Environ. Microbiol.* **37**, 104–108.

Chu, F. S., Chen Liang, M. Y., and Zhang, G. S. (1984). Production and characterization of antibody against diacetoxyscirpenol. *Appl. Environ. Microbiol.* **48**, 777–780.

Chung, C. W., Trucksess, M. W., Gilles, A. L., Jr., and Reidman, K. (1974). Rabbit skin test for estimation of T-2 toxin and other skin-irritating toxins in contaminated corn. *J. Assoc. Off. Agric. Chem.* **57**, 1121–1127.

Ciegler, A. (1975). Mycotoxins: occurrence, chemistry, biological activity. *Lloydia* **38**, 21–355.

Ciegler, A. (1977). Mycotoxins as insecticides. In *Biological Regulation of Vectors. The Saprophytic and Aerobic Bacteria and Fungi* (Conference Report, DHEW Publ. No. NIH 77-1180, October 6–8, 1975). Easton, Md.: Dept. of Health, Education and Welfare, pp. 135–144.

Ciegler, A. (1978). Trichothecenes: Occurrence and toxicoses. *J. Food Prot.* **41**, 399–403.

Ciegler, A. (1979). Mycotoxins—Their biosynthesis in fungi: Biosynthesis of the trichothecenes. *J. Food Prot.* **42**, 825–828.

Ciegler, A., and Bennett, J. W. (1980). Mycotoxins and mycotoxicoses. *Bioscience* **30**, 512–515.

Ciegler, A., and Lillehoj, E. B. (1968). Mycotoxins. In W. W. Umbreit and D. Perlman, eds., *Advances in Applied Microbiology,* Vol. 10. New York: Academic, pp. 155–219.

Ciegler, A., and Vesonder, R. F. (1983). Microbial food and feed toxicants: Fungal toxins. In M. Rechcigl, Jr., ed., *Handbook of Foodborne Diseases of Biological Origin.* Boca Raton, Fla.: CRC Press, pp. 57–166.

Ciegler, A., Detroy, R. W., and Lillehoj, E. B. (1970). Patulin, penicillic acid, and other carcinogenic lactones. In S. Kadis, A. Ciegler, and J. Ajl, eds., *Microbial Toxins,* Vol. 6. New York: Academic, pp. 409–434.

Cirilli, G. (1983). Trichothecene problems in Italy. In Y. Ueno, ed., *Trichothecenes, Chemical, Biological and Toxicological Aspects.* Amsterdam: Elsevier, pp. 254–258.

Claridge, C. A., and Schmitz, H. (1978). Microbial and chemical transformations of some 12,13-epoxytrichothec-9,10-enes. *Appl. Environ. Microbiol.* **36**, 63–67.

Claridge, C. A., and Schmitz, H. (1979). Production of 3-acetoxyscirpene-4,15-diol from anguidine (4,15,diacetoxyscirpene-3-ol) by *Fusarium oxysporum* f.sp. *vasinfectum. Appl. Environ. Microbiol.* **37**, 693–696.

Claridge, C. A., Bradner, W. J., and Schmitz, H. (1978). Antitumor activity of 15-acetoxyscirpen-3,4-diol. *J. Antibiot.* **31**, 485–486.

Coffin, J. L., and Combs, G. F., Jr. (1980). Impaired vitamin E status of chicks fed T-2 toxin. *Poult. Sci.* **60**, 385–392.

Cohen, H., and Lapointe, M. R. (1980). Sephadex LH-20 clean-up, high pressure liquid chromatographic assay, and fluorescence detection of zearalenone in animal feeds. *J. Assoc. Off. Anal. Chem.* **63**, 642–646.

Cohen, H., and Lapointe, M. (1982). Capillary gas-liquid chromatographic determination of vomitoxin in cereal grains. *J. Assoc. Off. Anal. Chem.* **65**, 1429–1434.

Cole, M., and Rolinson, G. N. (1972). Microbial metabolites with insecticidal properties. *Appl. Microbiol.* **24**, 660–662.

Cole, R. J., Kirksey, J. W., Cutler, H. G., Doupnik, B. L., and Peckham, J. C. (1973). Toxin from *Fusarium moniliforme:* Effects on plants and animals. *Science* **179**, 1324–1326.

Collet, J. C., Regnier, J. M., and Marechal, J. (1977). Contamination par mycotoxines de mais conservés en cribs et visiblement alterés. *Ann. Nutr. Aliment.* **31**, 447–457.

Collins, G. J. and Rosen, J. D. (1979). Gas-liquid chromatographic-mass spectromeric screening method for T-2 toxin in milk. *J. Assoc. Off. Anal. Chem.* **62**, 1274–1280.

Collins, G. J., and Rosen, J. D. (1981). Distribution of T-2 toxin in wet-milled corn products. *J. Food Sci.* **46**, 877–879.

Collins, M. S., and Rinaldi, M. G. (1977). Cutaneous infection in man caused by *Fusarium moniliforme. Sabouraudia* **15**, 151–160.

Colvin, E. W., Malchenki, S., Raphael, R. A., and Roberts, J. S. (1973). Total synthesis of (±)-Trichodermin. *J. Chem. Soc. Perkin Trans.* **1**, 1989–1997.

Cook, P. (1971). Cancer of the oesophagus in Kenya. A summary and evaluation of the evidence for the frequency of occurrence and a preliminary indication of the possible association with the consumption of alcoholic drinks made from maize. *Br. J. Cancer* **25**, 853–880.

Coons, G. H. (1928). Some aspects of the *Fusarium* problem. Plant Pathology and Physiology in Relation to Man. *Mayo Found. Lect.* 43–92. University of Minnesota, Minneapolis.

Coons, G. H., and Strong, M. C. (1928). New methods for the diagnosis of species of the genus *Fusarium. Mich. Acad. Sci., Arts Lett.* **9**, 65–88.

Coons, G. H., and Strong, M. C. (1931). The diagnosis of species of *Fusarium* by use of growth-inhibiting substances in the culture medium. *Mich. Agric. Exp. Stat., East Lansing, Mich., Tech. Bull.* **115**, 1–78.

Coppock, R. W., Swanson, S. P., Vesonder, R., and Buck, W. (1981). Clinical signs, toxicokinetics, and pathology of DAS and DON trichothecene mycotoxins in swine. *J. Am. Vet. Med.* **181**, 276.

Cordero-Moreno, R., and Pifano, I. (1970). Queratomicosis en ojos enucleodos pó ùlcera de corneo. *Rev. Oftalmol., Venerdana* **26**, 33–44.

Cornell, C. N., and Garner, G. B. (1983). Fescue Foot. In *Proc. Tall Fescue Toxicosis Workshop*, March 17–18, Atlanta, Ga., pp. 10–14.

Cotteereau, Ph., Laval, A., Bastien, F., and Magnan, G. (1974). Une mycotoxicose oestrogenique chez le porc. *Rev. Med. Vet.* **125**, 1095–1101.

**492** Bibliography

Coulter, D. B., Wyatt, R. D., and Stewart, R. G. (1977). Electro-retinograms from broilers fed aflatoxin and T-2 toxin. *Poult. Sci.* **56**, 1435–1439.

Cox, R. H., and Cole, R. J. (1983). C NMR spectra of trichothecenes. In Y. Ueno, ed., *Trichothecenes, Chemical, Biological and Toxicological Aspects.* Amsterdam: Elsevier, pp. 39–46.

Cullen, D. (1981). Premature verdict on biological war. *The New York Times,* December 13, 6.

Cullen, D., and Smalley, E. B. (1981). New process for T-2 toxin production. *Phytopathology* **71**, 212 (Abstr.).

Cundliffe, E., and Davies, J. E. (1977). Inhibition of initiation, elongation and termination of eukaryotic protein synthesis by trichothecene fungal toxins. *Antimicrob. Agents Chemother.* **11**, 491–499.

Cundliffe, E., Cannon, M., and Davis, J. (1974). Mechanism of inhibition of eukaryotic protein synthesis by trichothecene fungal toxins. *Proc. Natl. Acad. Sci. USA* **71**, 30–34.

Cunningham, I. J. (1949). A note on the cause of tall fescue lameness in cattle. *Aust. Vet. J.* **25**, 27–28.

Curtin, T. M., and Tuite, J. (1966). Emesis and refusal of feed in swine associated with *Gibberella zeae* infected corn. *Life Sci.* **5**, 1937–1944.

Curtis, R. F., Coxon, D. T., and Levett, G. (1974). Toxicity of fatty acids in assays for mycotoxins using brine shrimp. *Food Cosmet. Toxicol.* **12**, 233–235.

Dahlgren, R. R., and Williams, D. E. (1972). Clinical report: Hemorrhagic syndrome in feed lot cattle. *Bovin. Pract.* **7**, 52–53.

Dailey, R. E., Reese, R. E., and Brouwer, E. A. (1980). Metabolism of [$^{14}$C]-zearalenone in laying hens. *Agric. Food Chem.* **28**, 286–291.

Damperov, N. I. (1939). *The Urov Kashin-Beck Disease.* Medgis: Moscow.

Danko, G., and Aldasy, P. (1968). Mycotoxin induced vulva swelling in pigs. *Magy. Allatorv. Lapja* **24**, 517–519.

Danko, G., and Toth, O. (1969). Toxicity by *Fusarium graminearum* (Schwabe) in hog, cattle and sheep. *Tudomanyas Kozlenyck* **2**, 1–18.

Davis, G. J., McLachlan, J. A., and Lucier, G. W. (1977). Fetotoxicity and teratogenicity of zearanol in mice. *Toxicol. Appl. Pharmacol.* **41**, 138–139.

Davis, G. R. F. and Schiefer, H. B. (1982). Effects of dietary T-2 toxin concentrations fed to larvae of the yellow mealworm at three dietary protein levels. *Comp. Biochem. Physiol.* **73**, 13–16.

Davis, G. R. F., and Smith, J. D. (1977). Effect of temperature on production of fungal metabolites toxic to larvae of *Tenebrio molitor. J. Invert. Pathol.* **30**, 325–329.

Davis, G. R. F., Smith, J. D., Schiefer, H. B., and Loew, F. M. (1975). Screening for mycotoxins with larvae of *Tenebrio molitor. J. Invertebr. Pathol.* **26**, 299–303.

Davis, G. R. F., Westcott, N. D., Smith, J. D., Neish, G. A., and Schiefer, H. B. (1982). Toxigenic isolates of *Fusarium sporotrichioides* obtained from hay in Saskatchewan. *Can. J. Microbiol.* **28**, 259–261.

Davis, L. W., and Mahoney, L. W. (1971). Blood urea-N and growth rate of DES or RAL treated steers. *J. Anim. Sci.* **33**, 280 (Abstr.).

Davis, N. D., and Diener, U. L. (1979). Mycotoxins. In L. R. Beuchat, ed., *Food and Beverage Mycology.* Westport, Conn.: Avi, pp. 397–444.

Davis, N. D., Wagener, R. E., Dalby, D. K., Morgan-Jones, G., and Diener, U. L. (1975). Toxic fungi in food. *Appl. Microbiol.* **30**, 159–161.

Davydova, V. L. (1947). Sensibility of human and animal skin to toxin of cereals overwintered in the field. In Alimentary toxic aleukia. *Acta Chkalov Inst. Epidemiol. Microbiol.* **2**, 94–96, Orenburg, USSR.

Davydovski, E. B., and Kestner, A. G. (1935). On so-called septic angina (morphology and pathogenesis). *Arch. Pathol. Anat.* **1**, 11–30.

Dawkins, A. W. (1966). Phytotoxic compounds produced by *Fusarium equiseti*. 2. The chemistry of diacetoxyscirpenol. *J. Chem. Soc.* C, 116–123.

Dawkins, A. W., and Grove, J. F. (1970). Phytotoxic compounds by *Fusarium equiseti*. 4. Fission of Trichothecan-3x, 4β-diols by manganese dioxide. *J. Chem. Soc.* C, 369–375.

Dawkins, A. W., Grove, J. F., and Tidd, B. K. (1965). Diacetoxyscirpenol and some related compounds. *Chem. Comm.*, 227–28.

Demakov, G. P. (1964). Fusariotoxicosis of cattle. *Vet. J.* **11**, 59–60.

DeNicola, D. B., Rebar, A. H., Carlton, W. W., and Yagen, B. (1978). T-2 toxin mycotoxicosis in the guinea pig. *Food Cosmet. Toxicol.* **16**, 601–609.

DeSimone, P. A., Greco, F. A., and Lessner, H. F. (1979). Phase I evaluation of a weekly schedule of anguidine. *Cancer Treat. Rep.* **63**, 2015–2017.

DeUriarte, L. A., Forsyth, D. M., and Tuite, J. (1976). Improved acceptance by rats of *Gibberella zeae*-damaged corn after washing. *J. Anim. Sci.* **42**, 1196–1201.

Dickson, A. D., Link, K. P., Roche, B. H., and Dickson, J. G. (1930). Report on the emetic substances in *Gibberella*-infected barley. *Phytopathology* **20**, 132 (Abstr.).

Dickson, D. (1981). U.S. chemical weapons production plan. *Nature* **293**, October 1, 327–328.

Dickson, D. (1982). Biological warfare. Soviet use. *Nature* **296**, March 25, 281–282.

Diebold, G. J., Karny, N., and Zare, R. N. (1979). Determination of zearalenone in corn by Laser fluorometry. *Anal. Chem.* **51**, 67–69.

Diener, U. L. (1976). Environmental factors influencing mycotoxin formation in the contamination of foods. *Proc. Am. Phytopathol. Soc.* **3**, 126–139.

Diggs, C. H., Scoltock, M. J., and Wiernik, P. H. (1978). Phase II evaluation of anguidine (NSC-141537) for adenocarcinoma of the colon or rectum. *Cancer Clin. Trials* **Winter 1978**, 297–299.

Dixon, S. N. (1980). Radioimmunoassay of the anabolic agent zeranol. 1. Preparation and properties of a specific antibody to zeranol. *J. Vet. Pharmacol. Ther.* **3**, 177–181.

Dixon, S. N. (1983). The efficacy, mode of action and safety of non-steroidal, non-antimicrobial growth promoters. *Vet. Res. Commun.* **7**, 51–57.

Dobrovolski, L. O. (1925). The Urov disease (endemic osteochondritis). *Vestn. Endokrinol.* **3**, 171–189, Moscow.

Doerr, J. A., Huff, W. E., Tung, H. T., Wyatt, R. D., and Hamilton, P. B. (1974). A survey of T-2 toxin, ochratoxin and aflatoxin for their effects on the coagulation of blood in young broiler chickens. *Poult. Sci.* **53**, 1728–1734.

Doerr, J. A., Hamilton, P. B., and Burmeister, H. R. (1981). T-2 toxicosis and blood coagulation in young chickens. *Toxicol. Appl. Pharmacol.* **60**, 157–162.

Doidge, E. M. (1938). Some South African Fusaria. *Bothalia* **3**, 331–483.

Doll, R. (1969). The geographical distribution of cancer. *Br. J. Cancer* **23**, 1–8.

Domaev, U. A. (1976a). Clinico-genetic study of Kashin-Beck disease. *Sov. Med.* **11**, 39–43.

Domaev, U. A. (1976b). The bioelectrical activity of the brain in Urov (Kashin-Beck) disease. *Zh. Nevropatol. Psikhiatr.* **76**, 224–228, Moscow.

Domaev, U. A. (1976c). The functional state of the neuromuscular apparatus in patients with Urov (Kashin-Beck) disease. *Zh. Nevropatol. Psikhiatr.* **76**, 659–662.

Domsch, K. H., and Gams, W. (1972). *Fungi in Agricultural Soils*. Longman.

Domsch, K. H., Gams, W., and Anderson, T. H. (1980). *Compendium of Soil Fungi*, Vol. 1, New York: Academic.

Dorell, D. G. (1971). Fatty acid composition of buckwheat seed. *J. Am. Oil Chem. Soc.* **48**, 693.

Doster, A. R., Mitchell, F. E., Farrell, R. L., and Wilson, B. J. (1978). Effects of 4-Ipomeanol, a product from mold-damaged sweet potatoes, on the bovine lung. *Vet. Pathol.* **15**, 367–375.

Dounin, M. S. (1926a). *Drunken Bread.* Moscow, pp. 3–41. Inst. Agric.

Dounin, M. S. (1926b). The fusariosis of cereal crops in European Russia in 1923. *Phytopathology* **16**, 305–308.

Doupnik, B., Jones, O., and Peckham, J. (1971a). Lack of toxicity of *Helminthosporium maydis*-invaded corn and culture filtrates to chicks and mice. *Appl. Microbiol.* **22**, 732–733.

Doupnik, B., Jones, O. H., and Peckham, J. C. (1971b). Toxic Fusaria isolated from moldy sweet potatoes involved in an epizootic of atypical interstitial pneumonia in cattle. *Phytopathology* **61**, 890.

Drabkin, B. S. (1950). The effect of extracts from some mold fungi on alcohol fermentation. *Acta Chkalov Med. Inst.* **2**, 99–102.

Drabkin, B. S., and Joffe, A. Z. (1950). The effect of extracts from toxic over-wintered cereals on *Paramaecium caudatum. Acta Chkalov Med. Inst.* **2**, 92–98.

Drabkin, B. S., and Joffe, A. Z. (1952). The protistocide effect of certain molds. *Microbiol. Acad. Sci. USSR* **21**, 700–704, Moscow.

Druckerey, H., Preussmann, R., Ivankovic, S., and Schmähl, D. (1967). Organotrope carcinogene Wirkungen bei 65 verschiedenen N-Nitroso-Verbindungen und BD-Ratten. *Z. Krebsforsch. Klin. Onkol.* **69**, 103–201.

Dudley, M. A., and Chick, E. W. (1964). Corneal lesions produced in rabbits by an extract of *Fusarium moniliforme. Arch. Ophthalmol.* **72**, 346–350.

Durachova, Z., Betina, W., Hornikova, B., and Nemec, P. (1977). Toxicity of mycotoxins and other fungal metabolites to *Artemia salina* larvae. *Zbl. Bakt. Abt. II* **132**, 294–299.

Dzhilavian, K. A. (1972). Toxicity of fungus *F. sporotrichiella* var. *tricinctum* for sheep and pigs. *Symp. Mycotoxins,* Kiev (Abstr.).

Dzhilavian, K. A., and Spesivtseva, N. A. (1960). Symptoms of fusariotoxicosis in cattle, sheep and goats. *Tr. Vses. Inst. Vet. Sanit.* **16**, 386–388.

Ehrlich, K. C., and Lillehoj, E. B. (1984). Simple method of isolation of 4-deoxynivalenol from rice inoculated with *Fusarium graminearum. Appl. Environ. Microbiol.* **48**, 1053–1054.

Ehrlich, K. C., Lee, L. S., and Ciegler, A. (1983). A simple sensitive method for detection of vomitoxin (deoxynivalenol) using reversed phase, high performance liquid chromatography. *J. Liq. Chromatogr.* **6**, 833–843.

El-Banna, A. A., Scott, P. M., Lau, P-Y., Sakuma, T., Platt, H. W., and Champbell, V. (1984). Formation of trichothecenes by *Fusarium solani* var. *coeruleum* and *Fusarium sambucinum* in potatoes. *Appl. Environ. Microbiol.* **47**, 1169–1169.

El-Kady, I. A., and El-Maraghy, S. S. (1982). Screening of zearalenone-producing *Fusarium* species in Egypt and chemically defined medium for production of the toxin. *Mycopathologia,* **78**, 25–29.

El-Kady, I. A., and Moubasher, M. H. (1982). Toxigenicity and toxins of *Stachybotrys* isolated from wheat straw samples in Egypt. *Exp. Mycol.* **6**, 25–30.

Ellestad, G. A., Lovell, F. M., Parkinson, N. A., Hargreaves, R. T., and McGahren, W. J. (1978). New zearalenone related macrolides and isocoumarins from an unidentified fungus. *J. Org. Chem.* **43**, 2339–2343.

Ellis, J. J., and Yates, S. G. (1971). Mycotoxins of fungi from fescue. *Econ. Bot.* **25**, 1–5.

Ellison, R. A. and Kotsonis, F. N. (1973). T-2 toxin as an emetic factor in moldy corn. *Appl. Microbiol.* **26**, 540–543.

Ellison, R. A., and Kotsonis, F. N. (1974). In vitro metabolism of T-2 toxin. *Appl. Microbiol.* **27**, 423–424.

Ellison, R. A., and Kotsonis, F. N. (1976). Carbon-13 nuclear magnetic resonance assignments in the trichothecene mycotoxins. *J. Org. Chem.* **41**, 576–578.

Elliott, J. R. (1982). The new problem of vomitoxin. *Assoc. Food Drug Off.*, **46**, 226–231.

Elpidina, O. K. (1945a). The etiology of septic angina and determination of grain toxicity. In Z. I. Malkin, ed., *Data on Alimentary Toxic Aleukia.* Kazan. pp. 78–82.

Elpidina, O. K. (1945b). Phytotoxin in grain causing septic angina. *Lect. Acad. Sci. USSR (DAN)* **46**, 2.

Elpidina, O. K. (1945c). *Poin Lect. Acad. Sci. (DAN)* **46**, 88.

Elpidina, O. K. (1946a). *Poin Lect. Acad. Sci. (DAN)* **51**, 163.

Elpidina, O. K. (1946b). Biological methods of determining grain toxicity causing septic angina. *Lect. Acad. Sci. USSR (DAN)* **51**, 2.

Elpidina, O. K. (1956). Biological and antibiotic properties of poin. In *Abstracts of Mycotoxicoses of Men and Agricultural Animals.* Kiev: Publ. Acad. Sci., Ukr. SSR, pp. 23–24.

Elpidina, O. K. (1958). The antiblastic properties of poin, according to experimental data. *Kazan Med. J.* **39**, 96–104.

Elpidina, O. K. (1959). Antibiotic and antiblastic properties of poin. *Antibiotics* **4**, 431–434.

Elpidina, O. K. (1960). Toxic and antibiotic properties of poin (the toxin of *Fusarium sporotrichiella* var. *poae*). In V. I. Bilai, ed., *Mycotoxicoses of Man and Agricultural Animals* (English Trans.) Washington, D.C.: U.S. Dept. of Commerce, Office of Techn. Services, pp. 73–81.

Elpidina, O. K. (1961). The action of poin, a toxin from *Fusarium sporotrichioides* v. *poae* on malignant tumor growth in experiments. *Rep. 2nd Conf. Mycotoxicoses, Kiev* (Abstr.).

Elpidina, O. K., Seminov, V. V., and Shatunova, D. G. (1972a). The mechanism of the antineoplastic activity of poin isolated from the fungus *Fusarium sporotrichiella* var. *poae* Bilai. *Symp. Mycotoxins, Kiev,* 29 (Abstr. of Lect.).

Elpidina, O. K., Shatunova, D. G., and Semionov, V. V. (1972b). The activity of poin secreted by *Fusarium sporotrichiella* var. *poae* Bilai on a tumorous (swollen) cell. *Symp. Mycotoxins, Kiev,* 30.

Ember, L. R. (1981). Scientific gaps cloud new chemical arms issue. *Chem. Eng.* December 14, 21–24.

English, M. P. (1968). Invasion of the skin by filamentous nondermatophyte fungi. *Br. J. Derm.* **80**, 282–286.

English, M. P. (1972). Observations on strains of *Fusarium solani* and *F. oxysporum* and *Candida parapsilosis* from ulcered legs. *Sabouraudia* **10**, 35–42.

English, M. P., Smith, R. J., and Harman, R. R. M. (1971). The fungal flora of ulcerated legs. *Br. J. Derm.* **83**, 567–581.

Engstrom, G. W., Richard, J. L., and Cysewski, S. J. (1977). High-pressure liquid chromatographic method for detection and resolution of rubratoxin, aflatoxin and other mycotoxins. *J. Agric. Food Chem.* **25**, 833–836.

Eppley, R. M. (1968). Screening method for zearalenone, aflatoxin and ochratoxin. *J. Assoc. Off. Anal. Chem.* **51**, 74–78.

Eppley, R. M. (1974). Sensitivity of Brine Shrimp *(Artemia salina)* to trichothecenes. *J. Assoc. Off. Anal. Chem.* **58**, 618–620.

Eppley, R. M. (1975). Methods for the detection of trichothecenes. *J. Assoc. Off. Anal. Chem.* **58**, 906–908.

Eppley, R. M. (1979). Trichothecenes and their analysis. *J. Am. Oil Chem. Soc.* **56**, 824–829.

Eppley, R. M. (1982). Recommendations of a panel on problems and solutions in trichothecene methodology. *J. Assoc. Off. Anal. Chem.* **65**, 892–893.

Eppley, R. M., and Bailey, W. J. (1973). 12,13-Epoxy$\Delta^9$-trichothecenes as the probable mycotoxins responsible for stachybotryotoxicosis. *Science* **181**, 758–760.

Eppley, R. M., and Mazzola, E. P. (1977). Structure of satrotoxin H, a metabolite of *Stachybotrys atra*. Application of proton and carbon-13 nuclear magnetic resonance. *J. Org. Chem.* **42**, 240–243.

Eppley, R. M., Joffe, A. Z., Mislivec, P., Trucksess, M., and Pohland, A. E. (1985, in press).

Eppley, R. M., Stoloff, L., and Chung, C. W. (1973). Survey of corn for *Fusarium* toxins. *87th Annu. Meet. Assoc. Off. Anal. Chem.,* 40 (Abstr.).

Eppley, R. M., Stoloff, L., Trucksess, M. W., and Chung, C. W. (1974). Survey of corn for *Fusarium* toxins. *J. Assoc. Off. Agric. Chem.* **57**, 632–635.

Eppley, R. M., Mazzola, E. P., Stack, M. E., and Dreifuss, P. A. (1980). Structure of Satratoxin F and Satratoxin G, metabolites of *Stachybotrys atra:* Application of proton and carbon-13 nuclear magnetic resonance spectroscopy. *J. Org. Chem.* **45**, 2522–2523.

Eriksen, E. (1968). Oestrogene factorer i muggent korn. Vulvovaginitis hos swin. *Nord. Vet. Med.* **20**, 369–401.

Ermakov, V. V., Kostyunina, N. A., and Kurmanov, J. A. (1978). Isolation and identification of the T-2 toxin produced by *Fusarium sporotrichiella*. *Dokl. Vses. Acad. S-Kh, Nauk.* **1**, 36–38.

Etchevers, G. C., Banasik, O. J., and Watson, C. A. (1977). Microflora of barley and its effects on malt and beer properties: A review. *Brew. Dig.* **52**, 46–53.

Eugenio, C. P. (1968a). Factors influencing the biosynthesis of the fungal estrogen (F-2) and the effect of F-2 on perithecia formation by *Fusarium* species. Ph.D. dissertation, Dept. of Plant Pathology, University of Minnesota, St. Paul.

Eugenio, C. P. (1968b). The effect of the fungal estrogen (F-2) on the production of perithecia by *Fusarium roseum* "Graminearum." *Phytopathology* **58**, 1049–1050 (Abstr.).

Eugenio, C. P., Christensen, C. M., and Mirocha, C. J. (1970a). Factors affecting production of the mycotoxin F-2 by *Fusarium roseum*. *Phytopathology* **60**, 1055–1057.

Eugenio, C. P., DeLas Casas, E., Harein, P. K., and Mirocha, C. J. (1970b). Detection of the mycotoxin F-2 in the confused flour beetle and the lesser mealworm. *J. Econ. Entomol.* **63**, 412–415.

Everest, P. G. (1983). Ryegrass staggers: An overview of the north Canterbury situation and possible costs to the farmer. *Proc. N. Z. Assoc.* **44**, 228–229.

Fazakas, A. (1958). Zusammenfassender Bericht über meine ophthalmologischen Pilzuntersuchungen. *Ophthalmologica* **126**, 91–109.

Featherston, W. R. (1973). Utilization of *Gibberella* infected corn by chicks and rats. *Poult. Sci.* **52**, 2334–2335.

Fieser, L. F., and Fieser, M. (1959). *Steroids*. New York: Reinhold, p. 727.

Fisher, E. E., Kellock, A. W., and Wellington, N. A. M. (1967). Toxic strain of *Fusarium culmorum* (W.G.Sm.) Sacc. from *Zea mays* L., associated with sickness in dairy cattle. *Nature* **215**, 322.

Floweree, C. G. (1983). Chemical weapons: A study in verification. *Arms Cent. Today* 13, April, 1.

Flury, E., Mauli, R., and Sigg, H. P. (1965). The constitution of diacetoxyscirpenol. *Chem. Comm.* 26, 27.

Foley, F. D. (1969). The burn autopsy: Fatal complications of burn. *Am. J. Clin. Pathol.* 52, 1–13.

Fontelo, P. A., Beheler, J., Bunner, D. L., and Chu, F. S. (1983). Detection of T-2 toxin by an improved radioimmunoassay. *Appl. Environ. Microbiol.* 45, 640–643.

Forgacs, J., and Carll, W. T. (1962). Mycotoxicoses. *Adv. Vet. Sci.* 7, 273–382.

Forster, R. K., and Rebell, G. (1975a). Animal model of *Fusarium solani* keratitis. *Am. J. Ophthalmol.* 79, 510–515.

Forster, R. K., and Rebell, G. (1975b). The diagnosis and management of keratomycosis. 1. Case and diagnosis. *Arch. Ophthalmol.* 93, 975–978.

Forster, R. K., and Rebell, G. (1975c). The diagnosis and management of keratomycosis. 2. Medical and surgical management. *Arch. Ophthalmol.* 93, 1134–1136.

Forsyth, D. M. (1974). Studies on *Gibberella zeae*-infected corn in diets of rats and swine. *J. Anim. Sci.* 39, 1092–1098.

Forsyth, D. M., DeUriarte, L. A., and Tuite, J. (1976). Improvement for swine of *Gibberella zeae*-damaged corn by washing. *J. Anim. Sci.* 42, 1202–1206.

Forsyth, D. M., Yoshizawa, T., Morooko, N., and Tuite, J. (1977). Emetic and refusal activity of deoxynivalenol to swine. *Appl. Environ. Microbiol.* 34, 547–552.

Foster, P. M. D., Slater, T. F. and Patterson, D. S. P. (1975). A possible enzymic assay for trichothecene mycotoxins in animal feedstuffs. *Biochem. Soc. Trans.* 3, 875–878.

Francois, J. (1968). *Les Mycoses Oculaires*, Masson, Paris.

Friend, S. C. E., Babiuk, L. A., and Schiefer, H. B. (1983a). The effects of dietary T-2 toxin on the immunological function and herpes simplex reactivation in Swiss mice. *Toxicol. Appl. Pharmacol.* 69, 234–244.

Friend, S. C. E., Hancock, D. S., Schiefer, H. B., and Babiuk, L. A. (1983b). Experimental T-2 toxicosis in sheep. *Can. J. Comp. Med.* 47, 291–297.

Friend, S. C. E., Schiefer, H. B., and Babiuk, L. A. (1983c). The effects of dietary T-2 toxin on acute herpes simplex virus type 1 infection in mice. *Vet. Pathol.* 20, 737–760.

Friedman, M. N. (1945a). Residual changes in the internal organs and blood by alimentary toxicosis with overwintered wheat. *Acta Septic Angina, Ufa, USSR* 158–161.

Friedman, Yu. M. (1945b). Prophylaxis of alimentary toxic aleukia (septic angina). In A. E. Nesterov, A. M. Sysin, and L. N. Karlic, eds., *Alimentary Toxic Aleukia (Septic Angina)*. Medgiz, Moscow: Med. Publ. House, pp. 54–57.

Frischkorn, C. G. B., Frischkorn, H. E., and Ohst, I. M. (1978). Der simultane ppm— Nachweis des anabol wirkenden Zeranols und seines Metaboliten Zearalenon in Fleisch mittels hochanflösender Flüssigkeitschromatographie (HPLC). *Z. Lebensm. Unters. Forsch.* 167, 7–10.

Fritz, J. C., Mislivec, P. B., Pla, G. W., Harrison, B. N., Weeks, C. E., and Dantzman, J. G. (1973). Toxigenicity of moldy feed for young chicks. *Poult. Sci.* 52, 1523–1530.

Fromentin, H., Salazar-Mejicanos, S., and Mariat, F. (1980). Pouvoir pathogène de *Candida albicans* pour la souris normale ou déprimée par une mycotoxine: le diacétoxyscirpénol. *Ann. Microbiol. (Inst. Pasteur)* 131B, 39–46.

Fromentin, H., Salazar-Mejicanos, S., and Mariat, F. (1981). Experimental cryptococcosis in mice treated with diacetoxyscirpenol, a mycotoxin of *Fusarium*. *Sabouraudia* 19, 311–313.

Fujimoto, Y., Morita, Y., and Tasuno, T. (1972). Recherches toxicogiques sur les substances toxiques de *Fusarium nivale:* Etude chimique des toxins principales, nivalenol, fusarenon-X et nivalenol-4, 15-di-0-acetate. *Chem. Pharm. Bull.* **20,** 1194–1203.

Fujimoto, Y., Yokura, S., Nakamura, T., Morikawa, T., and Tatsuno, T. (1974). Total synthesis of ( ± )-12,13-epoxytrichothec-9-ene. *Tetrahedron Lett.* **29,** 2523–2526.

Funnell, H. S. (1979). Mycotoxins in animal feedstuffs in Ontario, 1972–1977. *Can. J. Comp. Med.* **43,** 243–246.

Futrell, M. C. (1972). New concepts in chemical seed treatment of agronomic crops. *J. Environ. Qual.* **1,** 240–243.

Futrell, M. C., Farnell, D. R., Poe, W. E., Watson, V. H., and Coats, R. E. (1974). Fungal population in the rumen associated with fescue toxicosis. *J. Environ. Qual.* **3,** 140–143.

Futrell, M. C., Scott, G. E., and Vaughn, G. F. (1976). Problems created by fungus-infected corn shipped to swine feeders in the southern United States. *Econom. Bot.* **30,** 291–294.

Gabel, Y. O. (1947). On the chemistry of the toxic substances of overwintered millet. In *Alimentary toxic aleukia. Acta Chkalov Inst. Epidemiol. Microbiol.,* 42–44, Orenburg, USSR.

Gabrilovich, O. Y. (1906). The active principle of "drunken bread." Master's thesis, St. Petersburg. University St. Petersburg.

Gajdusek, D. C. (1953a). Alimentary toxic aleukia. In *Acute Infectious Hemorrhagic Fevers and Mycotoxicoses in the Union of Soviet Socialist Republics.* Washington, D.C.: Walter Reed Army Med. Center, pp. 82–105. Med. Sci. Publ.

Gajdusek, D. C. (1953b). Poisoning with "drunken bread." In *Acute Infectious Hemorrhagic Fevers and Mycotoxicoses in the Union of Soviet Socialist Republics.* Washington, D.C.: Walter Reed Army Med. Center, pp. 107–111. Med. Sci. Publ.

Gallo, M. A., Bailey, D. E., Babish, J. G., Cox, C. E., and Taylor, J. M. (1977). Observation on the toxicity of zearalenone in the rat. *Toxicol. Appl. Pharmacol.* **41,** 133 (Abstr.).

Gamuzov, Y. I. (1931). Results of examination of water of the Beck hydrochemical section in Eastern Transbaikal in connnection with the study of Beck's disease. *Sov. Med. Vost. Sib.* 4–5.

Garcia, N. P., Ascani, E., and Zapater, R. C. (1972). Queratomicosis por *Fusarium dimerum. Arch. Oftalmol. Buenos Aires* **47,** 332–334.

Gardner, D., Glen, A. T., and Turner, W. B. (1972). Calonectrin and 15-deacetylcalonectrin, new trichothecenes from *Calonectria nivalis. J. Chem. Soc. Perkins Trans.* **1,** 2576–2578.

Gareis, M., Bauner, J., von Montgelas, A., and Gedek, B. (1984). Stimulation of aflatoxin B and T-2 toxin production by sorbic acid. *Appl. Environ. Microbiol.* **47,** 416–418.

Garmon, L. (1981). Yellow rain riddle. *Sci. News* **120,** 250–251.

Garner, F. M., Innes, J. R. M., and Nelson, D. H. (1967). Marine neuropathology. In E. Cotchin and F. J. C. Roe, eds., *Pathology of Laboratory Rats and Mice.* Oxford: Blackwell Scientific, pp. 295–347.

Garner, G. B., and Cornell, C. N. (1978). Fescue foot in cattle. In T. D. Wyllie and L. G. Morehouse, eds., *Mycotoxic Fungi, Mycotoxins, Mycotoxicoses: An Encyclopedic Handbook,* Vol. 2. New York: Marcel Dekker, pp. 45–62.

Gedek, B., and Bauer, J. (1983). Trichothecene problems in the Federal Republic of Germany. In Y. Ueno, ed., *Trichothecenes, Chemical, Biological and Toxicological Aspects.* Amsterdam: Elsevier pp. 301–308.

Gedek, B., Hüttner, B., Kahland, D. I., Köhler, H., and Vielitz, F. (1978). Rachitis bei Masgeflügel durch Kontamination des Futters mit *Fusarium moniliforme* Sheldon. *Zbl. Vet. Med.* **B25,** 29–44.

Geimberg, V. G., and Babusenko, A. M. (1949). Dynamics of development of microflora in overwintered grain in experimental field conditions. *Hyg. Sanit.* **5**, 31–34.

Geller, G. M., Fomina, L. S., and Shlygin, G. K. (1954). A discussion of the problem of the etiology of Urov Disease. *Probl. Nutr.* **13**, 47–52.

Geminov, N. B. (1945a). Etiology of alimentary toxic aleukia. In V. Chilikin and N. Geminov, eds., *Septic Angina and Its Treatment* Kuybyshev: pp. 3–7, Inst. Epidemiology and Microbiology.

Geminov, N. B. (1945b). Epidemiology of septic angina in the Kuybyshev district in 1945. Lecture in Republic Conference of Alimentary Toxic Aleukia, Moscow, December.

Geminov, N. B. (1948). On the problem of outbreaks of alimentary toxic aleukia disease from use of acorns in food of the spring harvest. Lecture in Republic Conference of Alimentary Toxic Aleukia, Moscow, March.

Genkin, A. (1944). Condition of the upper respiratory tract in septic angina. In Data on septic angina. *Proc. 1st Kharkov Med. Inst. Chkalov Inst. Epidemiol. Microbiol.*, pp. 117–128.

Georgiyeovski, A. P. (1952). The chemical composition of the water in a focus of Urov Disease. *Hyg. Sanit.* **17**, 27–30.

Gerlach, W. (1961). *Fusarium redolens* Wr., seine Morphologie und systematische Stellung. Ein Beitrag zur Kenntnis der *Elegans*-Fusarien. *Phytopathol. Z.* **42**, 150–160.

Gerlach, W. (1970). Suggestion to an acceptable modern *Fusarium* system. *Ann. Acad. Sci. Fenn. Ser. A, IV Biol., Helsinki*, **168**, 37–49.

Gerlach, W. (1978). Critical remarks on the present situation in *Fusarium* taxonomy. In C. V. Subramanian, ed., *Proc. Int. Symp. Taxon. Fungi*, Part 1, 115–124, University Madras, India.

Gerlach, W. (1977). *Fusarium* species inciting plant diseases in the tropics. In J. Kranz, H. Schmutterer, and W. Koch, eds., *Diseases, Pests and Weeds in Tropical Crops.* Berlin: Paul Parey, pp. 210–217.

Gerlach, W. (1981). The present concept of *Fusarium* classification. In P. E. Nelson, T. A. Toussoun, and R. J. Cook, eds., *Fusarium Diseases, Biology and Taxonomy.* University Park: Pennsylvania State University Press, pp. 413–426.

Gerlach, W., und Ershad, D. (1970). Beitrag zur Kenntnis der *Fusarium* und *Cylindrocarpon*-Arten in Iran *Nova Hedwiga. Verl. J. Cramer* **20**, 725–783.

Gerlach, W., and Nirenberg, H. (1982). The genus *Fusarium*—a pictorial atlas. Berlin-Dahlem; Paul Parey.

Germanov, A. I. (1945). Some laboratory data relating to alimentary toxic aleukia. *Sov. Med.* **3**, 11–12.

Getsova, G. (1960). Material on experimental study of *Fusarium* toxicosis. In V. I. Bilai, ed., *Mycotoxicoses in Man and Agricultural Animals* pp. 110–117. (Engl. Trans.) Washington, D.C.: U.S. Dept. of Commerce, Office of Techn. Services, pp. 110–117.

Ghosal, S., Chakrabarti, D. K., and Basu Chaudhary, K. C. (1976). Toxic substances produced by *Fusarium*. 1: Trichothecene derivatives from two strains of *Fusarium oxysporum* f. sp. *carthami. J. Pharmacol. Sci.* **65**, 160–161.

Ghosal, S., Chakrabarti, D. K., and Basu Chadhary, K. C. (1977). The occurrence of 12,13-epoxytrichothecenes in seeds of safflower infected with *Fusarium oxysporum* f. sp. *carthami. Experientia* **33**, 574–575.

Ghosal, S., Biswas, K., Srivastava, R. S., Chakrabarti, D. K., and Basu Chaudhary, K. C. (1978a). Toxic substance produced by *Fusarium*. 5. Occurrence of zearalenone, diacetoxyscirpenol and T-2 toxin in moldy corn affected with *Fusarium moniliforme* Sheld. *J. Pharmacol. Sci.* **67**, 1768–1769.

Ghosal, S., Biswas, K., and Chattopadhyay, B. K. (1978b). Difference in the chemical constituents of *Mangifera india* infected with *Aspergillus niger* and *Fusarium moniliforme*. *Phytochemistry* **17**, 689–694.

Gilgan, M. W., Smalley, E. B., and Strong, F. M. (1966). Isolation and partial characterization of a toxin from *Fusarium tricinctum* on moldy corn. *Arch. Biochem. Biophys.* **114**, 1–3.

Gillespie, A. M., and Schenk, G. M. (1977). Fluorescence and phosphorescence of ochratoxins, sterigmatocystin, patulin and zearalenone: Quantitation of ochratoxins. *Anal. Lett.* **10**(2), 161–172.

Gillespie, F. D. (1963). Fungus corneal ulcer. *Am. J. Ophthalmol.* **56**, 823–825.

Gingrich, W. D. (1962). Keratomycosis. *J. Am. Med. Assoc.* **179**, 602–608.

Girotra, N. M., and Wendler, N. L. (1967). A new total synthesis of the macrolide zearalenone. *Chem. Ind., London* **2**, 1493.

Gjertsen, P., Trolle, B., and Andersen, K. (1964). Studies on gushing of weathered barley as a contributory cause of gushing in beer. *Eur. Brew. Conv. Proc. Cong., Brussels* 320–341. Amsterdam: Elsevier.

Gjertsen, P., Trolle, B., and Andersen, K. (1965). Studies on gushing. 2. Gushing caused by microorganisms specially *Fusarium* species. *Eur. Brew. Conv. Proc. Congr., Stockholm*, 428–438. Amsterdam: Elsevier.

Godtfredsen, W. O., and Vangedal, S. (1965). Trichodermin, a new sesquiterpene antibiotic. *Acta Chem. Scand.* **19**, 1088–1102.

Godtfredsen, W. O., Grove, J. F., and Tamm, C. (1967). On the nomenclature of a new class of sesquiterpenoids. *Helv. Chim. Acta* **50**, 1666–1668.

Goff, E. U., and Fine, D. H. (1979). Analysis of volatile N-Nitrosamine of alcoholic beverages. *Food Cosmet. Toxicol.* **17**, 569–573.

Goldstein, D., und Nikiforov, P. (1931). Über sogenannte Kashin-Beckische Krankheit. *Fortschr. Röentgenstr.* **43**, 321–336.

Goodwin, W., Haas, C. D., Fabian, C., Heller-Bettinger, I., and Hoogstraten, B. (1978). Phase I evaluation of anguidine (diacetoxyscirpenol, NSC-141537). *Cancer* **42**, 23–26.

Goos, R. D. (1981). Conidiogenous cells in the Fusaria. In P. E. Nelson, T. A. Toussoun, and R. J. Cook, eds. *Fusarium Diseases, Biology and Taxonomy.* University Park: Pennsylvania State University Press, pp. 453–457.

Gordon, W. L. (1935). Species of *Fusarium* isolates from field crops in Manitoba. *Proc. World's Grain Exhib. Conf., 1933* **2**, 298–299.

Gordon, W. L. (1939). *Fusarium* species associated with disease of cereals in Manitoba. *Phytopathologia* **29**, 7 (Abstr.).

Gordon, W. L. (1944). The occurrence of *Fusarium* species in Canada. 1. Species of *Fusarium* isolated from farm samples of cereal seed in Manitoba. *Can. J. Res.* **22**, 282–286.

Gordon, W. L. (1952). The occurrence of *Fusarium* species in Canada. 2. Prevalence and taxonomy of *Fusarium* species in cereal seed. *Can. J. Bot.* **30**, 209–251.

Gordon, W. L. (1954a). The occurrence of *Fusarium* species in Canada. 3. Taxonomy of *Fusarium* species in the seed of vegetable, forage, and miscellaneous crops. *Can. J. Bot.* **32**, 576–590.

Gordon, W. L. (1954b). The occurrence of *Fusarium* species in Canada. 4. Taxonomy and prevalence of *Fusarium* species in the soil of cereal plot. *Can. J. Bot.* **32**, 622–629.

Gordon, W. L. (1956a). The taxonomy and habitats of the *Fusarium* species in Trinidad, B.W.I. *Can. J. Bot.* **34**, 847–864.

Gordon, W. L. (1956b). The occurrence of *Fusarium* species in Canada. 5. Taxonomy and geographic distribution of *Fusarium* species in soil. *Can. J. Bot.* **34**, 833–846.

Gordon, W. L. (1959). The occurrence of *Fusarium* species in Canada. 6. Taxonomy and distribution of *Fusarium* species on plants, insects and fungi. *Can. J. Bot.* 37, 257–290.

Gordon, W. L. (1960a). Distribution and prevalence of *Fusarium moniliforme* Sheld. (*Gibberella fujikuroi* (Saw.) Wr.) producing substances with gibberellin-like biological properties. *Nature (London)* 186, 698–700.

Gordon, W. L. (1960b). The taxonomy and habitats of *Fusarium* species from tropical and temperate regions. *Can. J. Bot.* 38, 643–658.

Gordon, W. L. (1960c). Is *Nectria haematococca* Berk. and Br. the perfect stage of *Fusarium oxysporum* Schl. forma *pisi* (Lindf.) Snyd. and Hans. *Nature* 186, 903.

Gordon, W. L. (1965). Pathogenic strains of *Fusarium oxysporum. Can. J. Bot.* 43, 1309–1318.

Gordon, O. L. and Levitski, L. U. (1945). Peculiarities of the clinical course of alimentary toxic aleukia and the results of its therapy. Lecture at Republic Conference of Alimentary Toxic Aleukia, Moscow, December.

Gordon, W. L., and Sprague, R. (1941). Species of *Fusarium* associated with root-rots of the *Graminaceae* in the Northern Great Plains. *Plant Dis. Rep.* 25, 168–180.

Gorodijskaja, R. B. (1945). Blood changes induced by septic angina. In *Acta Septic Angina.* Ufa, USSR: Bashkir Publ. House, p. 112.

Greenhalgh, R., Neish, G. A., and Miller, D. (1983). Deoxynivalenol, acetyl deoxynivalenol, and zearalenone formation by Canadian isolates of *Fusarium graminearum* on solid substrates. *Appl. Environ. Microbiol.* 46, 625–629.

Greenway, J. A., and Puls, R. (1976). Fusariotoxicosis from barley in British Columbia. 1. Natural occurrence and diagnosis. *Can. J. Comp. Med.* 40, 12–15.

Greer, D. L., Brahim, C., and Gonzalez, L. A. (1973). Queratitis micotica en Colombia. *Trib. Med.* 74, 15–20.

Gregory, K. F., Allen, O. N., Riker, A. J., and Peterson, W. H. (1952). Antibiotics as agents for the control of certain damping-off fungi. *Am. J. Bot.* 39, 405–407.

Grinberg, G. I. (1943). Clinical aspects and pathogenesis of the so-called septic angina. *Sov. Med.* 10, 24–34.

Grollman, A. P., and Huang, M. T. (1976). Inhibitors of protein synthesis in eukaryotes. *Protein Synth.* 2, 125–127.

Gromashevski, L. V. (1945). Pathogenesis of alimentary toxic aleukia. *J. Microbiol. Epidemiol. Immunol.* 3, 65–74.

Gross, V. J., and Robb, J. (1975). Zearalenone production in barley. *Ann. Appl. Biol.* 80, 211–216.

Grove, J. F. (1969). Toxic 12,13-Epoxytrichothec-9-enes from *Fusarium* sp. *Chem. Commun.,* 1266–1267.

Grove, J. F. (1970a). Phytotoxic compounds produced by *Fusarium equiseti.* 5. Transformation products of the 4β, 15-Diacetoxy-3α, 7α-dihydroxy-12,13-epoxytrichothec-9-en-8-one and the structures of nivalenol and fusarenon. *J. Chem. Soc.* C, 375–378.

Grove, J. F. (1970b). Phytotoxic compounds produced by *Fusarium equiseti.* 6. 4β, 8α,15-Triacetoxy-12,13-epoxytrichothec-9-ene-3α, 7α-diol. *J. Chem. Soc.* C, 378–379.

Grove, J. F., and Mortimer, P. H. (1969). The cytotoxicity of some transformation products of diacetoxyscirpenol. *Biochem. Pharmacol.* 18, 1473–1478.

Grove, M. D., and Hosken, M. (1975). The larvicidal activity of some 12, 13-epoxytrichothec-9-enes. *Biochem. Pharmacol.* 24, 959–962.

Grove, M. D., Yates, S. G., Tallent, W. H., Ellis, J. J., Wolf, I. A., Kosuri, N. R., and Nichols, R. E. (1970). Mycotoxins produced by *Fusarium tricinctum* as possible causes of cattle disease. *J. Agric. Food Chem.* 18, 734–736.

Grove, M. D., Burmeister, H. R., Taylor, S. L., Weisleder, D., and Platter, R. D. (1984).

Effect of chemical modification on the epoxytrichothecene-induced feed refusal. *J. Agric. Food Chem.* **32**, 541–544.

Gubarev, E. M., and Gubareva, N. A. (1945a). Chemical nature and certain chemical properties of the toxic substances responsible for septic angina. *Biochemistry* **10**, 199–204.

Gubarev, E. M., and Gubareva, N. A. (1945b). Investigation of the structure and action on blood of inflammatory and neurotic toxins causing septic angina. Republic Conference on Alimentary Toxic Aleukia. Gen. State San. Inspection, Moscow, December 24–26, p. 6 (Abstr.).

Gugnani, H. C., Talwar, R. S., Njoku-Obi, A. N. U., and Kodilinye, H. C. (1976). Mycotic keratitis in Nigeria: A study of 21 cases. *Br. J. Ophthalmol.* **60**, 607–613.

Gurewitch, Z. A. (1944). Neuropathologic peculiarities of septic angina. In Data of septic angina. *Proc. 1st Kharkov Med. Inst. Chkalov Inst. Epidemiol. Microbiol.* 103–116, Orenburg, USSR.

Guttman, L., Chou, S. M., and Pore, R. S. (1975). Fusariosis, myasthenic syndrome and aplastic anemia. *Neurology* **25**, 922–926.

Gyongyossy-Issa, M. I. C., Christie, E. J., and Khachatourians, G. G. (1984). Charge-shift electrophoretic behavior of T-2 toxin in agarose gels. *Appl. Environ. Microbiol.* **47**, 1182–1184.

Haas, C., Goodwin, W., Leite, C., Stephens, R., and Hoogstraten, B. (1977). Phase I study of anguidine (diacetoxyscirpenol NSC # 1415377). *Proc. Am. Assoc. Cancer Res.* **18**, 296 (Abstr. C-120).

Hacking, A. (1983). Oestrogenic metabolites of Fusaria. *Vet. Res. Commun.* **7**, 161–164.

Hacking, A., Rosser, W. R., and Dervish, M. T. D. (1976). Zearalenone-producing species of *Fusarium* on barley seed. *Ann. Appl. Biol.* **84**, 7–11.

Hacking, A., Rosser, W. R., and Dervish, M. T. D. (1977). Incidence of zearalenone producing strains of *Fusarium* in barley seed. *Ann. Nutr. Aliment.* **31**, 557–562.

Hagan, S. N., and Tietjen, W. H. (1975). A convenient thin-layer chromatographic cleanup procedure for screening mycotoxins in oils. *J. Assoc. Off. Anal. Chem.* **58**, 620–621.

Hagler, W. M., and Mirocha, C. J. (1980). Biosynthesis of [$^{14}$C] zearalenone from [-$^{14}$C] acetate by *Fusarium roseum* "Gibbosum." *Appl. Environ. Microbiol.* **39**, 668–670.

Hagler, W. M., Mirocha, C. J., Pathre, S. V., and Behrens, J. C. (1979). Identification of the naturally occurring isomer of zearalenol produced by *Fusarium roseum* "Gibbosum" on rice culture. *Appl. Environ. Microbiol.* **37**, 849–853.

Hagler, W. M., Mirocha, C. J., and Pathre, S. V. (1981). Biosynthesis of radiolabeled T-2 toxin by *Fusarium tricinctum. Appl. Environ. Microbiol.* **41**, 1049–1051.

Hagler, W., Tyczkowska, K., and Hamilton, P. B. (1984). Simultaneous occurrence of deoxynivalenol, zearalenone, and aflatoxin in 1982 scabby wheat from the midwestern United States. *Appl. Environ. Microbiol.* **47**, 151–154.

Hald, B., and Krogh, P. (1983). Toxicoses and natural occurrence in Denmark. In Y. Ueno, ed., *Trichothecenes, Chemical, Biological and Toxicological Aspects.* Amsterdam: Elsevier, pp. 251–253.

Halde, C., and Okumoto, M. (1966). Ocular mycoses: A study of 82 cases. *Proc. 20th Int. Congr. Ophthalmol., Munich. Neth. Excerpta Med. Int. Congr.* Ser. 146, 705–712.

Haliburton, J. C., Vesonder, R. F., Lock, T. F., and Buck, B. (1979). Equine leucoencephalomalacia (ELEM): A study of *Fusarium moniliforme* as an etiologic agent. *Vet. Hum. Toxicol.* **21**, 348–351.

Hamilton, M. A., Knorr, M. S., and Cajori, F. A. (1953). Experimental studies of an antibiotic derived from *Fusarium bostrycoides. Antibiot. Chemother.* **3**, 853–860.

Hamilton, P. E., Wyatt, R. D., and Burmeister, H. (1971). Effect of fusariotoxin T-2 in chickens. *Poult. Sci.* **50**, 1583–1584 (Abstr.).

Hamilton, R. M. G., Thomson, B. K., and Trenholm, H. L. (1981). The effect of vomitoxin contaminated wheat on the palatability of laying diets by white Leghorn hens. *Poult. Sci.* **60**, 1665–1666. (Abstr.).

Hanson, J. R., and Achilladelis, B. (1967). The role of farnesyl pyrophosphate in the biosynthesis of trichothecin. *Chem. Ind.*, 1643–1644.

Hanson, J. R., Marten, T. and Siverns, M. (1974). Part XII. Carbon-nuclear magnetic resonance spectra of the trichothecenes and the biosynthesis of trichocolon from 2-[13]C-mevalonic acid (Part 12). *J. Chem. Soc. Perkin I*, 1033–1036.

Hanson, J. R., Hitchcock, P. B., and Myfolor, R. (1975). Cyclonerotriol [6-(3-Hydroxy-2,3-dimethylcyclopentyl)-2-methylhept-2-one-1,6-diol], a new sesquiterpenoid metabolite of *Fusarium culmorum. J. Chem. Soc. Perkin I*, 1586–1590.

Hara, Y. (1910). "Akakabibyo" (scab) of wheat and "Fushikurobyo" of rice plant. *Nogyokoku* **4**, 34–35.

Harein, P. K., DeLasCasas, E., Eugenio, C. P., and Mirocha, C. J. (1971). Reproduction and survival of confused flour beetles exposed to metabolites produced by *Fusarium roseum* var. *graminearum. J. Econom. Entomol.* **64**, 975–976.

Harruff, R. C. (1983). Chemical-biological warfare in Asia. *J. Am. Med. Assoc.* **250**, 497–498.

Hartman, G. R., Richter, H., Weiner, E. M., and Zimmermann, W. (1978). On the mechanism of action of the cytostatic drug anguidine and of the immunosuppressive agent ovalicin, two sesquiterpenes from fungi. Planta Medica: J. Med. *Plant Res.* **34**, 231–252.

Harmats, S. (1984). A chemical warfare expert who doubts the Soviets used Yellow Rain. *Washington Post*, February 26, D-1.

Harwig, J., and Scott, P. M. (1971). Brine shrimp (*Artemia salina* L.) larvae as a screening system for fungal toxins. *Appl. Microbiol.* **21**, 1011–1016.

Harwig, J., Scott, P. M., Stolz, D. R., and Blanchfield, B. J. (1979). Toxins of molds from decaying tomato fruit. *Appl. Environ. Microbiol.* **38**, 267–274.

Hayes, A. W., and Hood, R. D. (1977). Prenatal effect of mycotoxins in mammals. *Zesz. Probl. Postępów Nauk Roln.* **189**, 183–187, Warsaw.

Hayes, M. A. (1979). Morphological and toxicological studies on experimental T-2 mycotoxicosis. Ph.D. dissertation, University of Saskatchewan, Saskatoon.

Hayes, M. A., and Schiefer, H. B. (1979a). Synergistic effect of T-2 toxin and a low protein diet on erythropoiesis of mice. *Proc. 4th Int. IUPAC Symp. Mycotoxins Phycotoxins.* Park Forest, Illi.: Pathotox.

Hayes, M. A., and Schiefer, H. B. (1979b). Quantitative and morphological aspects of cutaneous irritation by trichothecene mycotoxins. *Fd. Cosmet. Toxicol.* **17**, 611–621.

Hayes, M. A., and Schiefer, H. B. (1980). Subacute toxicity of dietary T-2 toxin in mice: Influence of protein nutrition. *Can. J. Comp. Med.* **44**, 219–228.

Hayes, M. A., and Schiefer, H. B. (1982). Comparative toxicity of dietary T-2 toxin in rats and mice. *J. Appl. Toxicol.* **2**, 207–212.

Hayes, M. A., and Wobeser, G. A. (1983). Subacute toxic effects of dietary T-2 toxin in young Mallard ducks. *Can. J. Comp. Med.* **47**, 180–187.

Hayes, M. A., Bellamy, J. E. C., and Schiefer, H. B. (1980). Subacute toxicity of dietary T-2 toxin in mice: Morphological and hematological effects. *Can. J. Comp. Med.* **44**, 203–218.

Helgeson, J. P., Haberlach, G. T., and Vanderhoef, L. M. (1973). T-2 toxin decreases logarithmic growth rates of tobacco callus tissues. *Plant Physiol.* **52**, 660–662.

Herzburg, M., ed. (1970). *Proc. 1st U.S.-Japan Conf. Toxic Microorganisms.* Washington, D.C.: U.S.-Japan Coop. Prog. in Nat. Res. Panel and U.S. Dept. of Interior, pp. 490.

Hesseltine, C. W. (1974). Natural occurrence of mycotoxins in cereals. *Mycopathol. Mycol. Appl.* **53**, 141–153.

Hesseltine, C. W. (1976a). Conditions leading to mycotoxin contamination of foods and feeds. In J. V. Rodricks, ed., *Mycotoxins and Other Fungal Related Food Problems.* Washington, D.C.: American Chemical Society, pp. 1–22.

Hesseltine, C. W. (1976b). Mycotoxins other than aflatoxins. In J. M. Sharley and A. M. Kaplan, eds., *Biodeterioration of Materials,* Essex, England: Applied Science, pp. 607–623.

Hesseltine, C. W., Rogers, R. F., and Shotwell, O. (1978). Fungi, especially *Gibberella zeae* and zaeralenone occurrence in wheat. *Mycologia* **70**, 14–18.

Hibbs, C. M., Ocweiller, G. D., Buck, W. B., and MacFee, G. P. (1974). Bovine hemorrhagic syndrome related to T-2 mycotoxin. *Proc. Ann. Meet. Am. Assoc. Vet. Lab. Diag.,* **17**, 305–310.

Hidy, P. H., Baldwin, R. S., Greasham, R. L., Keith, C. L., and McMullen, J. R. (1977). Zearalenone and some derivatives: Production and biological activities. *Adv. Appl. Microbiol.* **22**, 59–82.

Hintikka, E. L. (1983). Toxicosis and natural occurrence of trichothecenes in Finland. In Y. Ueno, eds., *Trichothecenes, Chemical, Biological and Toxicological Aspects.* Amsterdam: Elsevier, pp. 221–228.

Hirayama, S., and Yomamoto, M. (1948). Biological studies on the poisonous wheat flour (1). *Bull. Natl. Inst. Hyg. Sci.* **66**, 85–98.

Hirayama, S., and Yomamoto, M. (1950). Biological studies on the poisonous wheat flour (2). *Bull. Natl. Inst. Hyg. Sci.* **67**, 117–121.

Hitokoto, H., Morozumi, S., Wauke, T., Sakai, S., and Kurata, H. (1977). Mycotoxin production of fungi on commercial foods. In J. V. Rodricks, C. W. Hesseltine, and M. A. Mehlman, eds., *Mycotoxins.* Park Forest, Ill.: Pathotox, pp. 479–487.

Hiyeda, K., Nayashi, J., and Aiiso, M. (1938). Iron as the causative agent of Kashin-Beck's disease. *J. Orient. Med.* **26**, 107.

Hobson, W., Bailey, J., and Fuller, G. B. (1977a). Hormone effects of zearalenone in non-human primates. *J. Toxic Environ. Health* **3**, 43–57.

Hobson, W., Bailey, J., Kowalk, A., and Fuller, G. (1977b). The hormonal potency of zearalenone in non-human primates. In J. V. Rodricks, C. W. Hesseltine, and M. A. Mehlman, eds., *Mycotoxins in Human and Animal Health.* Park Forest, Ill.: Pathotox, pp. 364–377.

Hoerr, F. J. (1981). Trichothecene mycotoxicosis in chickens. Ph.D. dissertation. Purdue University, West Lafayette, Ind., pp. 136–167.

Hoerr, F. J., Carlton, W. W., and Yagen, B. (1981a). The toxicity of T-2 toxin and diacetoxyscirpenol in combination for broiler chicken. *Fd. Cosmet. Toxicol.* **19**, 185–188.

Hoerr, F. J., Carlton, W. W., and Yagen, B. (1981b). Mycotoxicosis caused by a single dose of T-2 toxin or diacetoxyscirpenol in broiler chickens. *Vet. Pathol.* **18**, 652–664.

Hoerr, F. J., Carlton, W. W., Tuite, J., Vesonder, R. F., Rohwedder, W. K., and Szigeti, G. (1982a). Experimental trichothecene mycotoxicosis produced in broiler chickens by *Fusarium sporotrichiella* var. *sporotrichioides. Avian Pathol.* **11**, 385–405.

Hoerr, F. J., Carlton, W. W., Yagen, B., and Joffe, A. Z. (1982b). Mycotoxicosis produced in broiler chickens by multiple doses of either T-2 toxin or diacetoxyscirpenol. *Avian Pathol.* **11**, 369–383.

Hoerr, F. J., Carlton, W. W., Yagen, B., and Joffe, A. Z. (1982c). Mycotoxicosis caused by either T-2 toxin or diacetoxyscirpenol in the diet of broiler chickens. *Fundam. Appl. Toxicol.* **2**, 121–124.

Holden, C. (1982). "Unequivocal" evidence of Soviet toxin use. *Science* **216**, 154–155.

Holder, C. L., Nocy, C. R., and Bowman, M. C. (1977). Trace analysis of zearalenone and/or zearalanol in animal chow by high pressure liquid chromatography and gas-liquid chromatography. *J. Assoc. Off. Anal. Chem.* **60**, 272–278.

Holoday, C. E. (1980). Rapid screening method for zearalenone in corn, wheat and sorghum. *J. Am. Oil Chem. Soc.* **57**, 491A–492A.

Holzegel, K., and Kempf, H. J. (1964). *Fusarium*-mykose auf der Haut eine Verbrannten. *Dermatol. Monatschr.* **150**, 651–658.

Hood, R. D., Kuczuk, M. H., and Szczech, G. M. (1978). Effects in mice of simultaneous prenatal exposure to achratoxin A and T-2 toxin. *Teratology* **17**, 24–30.

Hook, B., Osweiler, G. D., and Buening, G. M. (1982). Effects of T-2 mycotoxin in cattle. *J. Am. Vet. Med.* **181**, 276.

Hornok, L. (1975). Occurrence of *Fusarium* species in Hungary. *Acta Phytopathol. Acad. Sci. Hung.* **10**, 347–357.

Hornok, L. (1978). On inadequacy of double diffusion test on *Fusarium* serotaxonomy. *Acta Phytopathol. Acad. Sci. Hung.* **13**, 357–363.

Hornok, L. (1980). Serotaxonomy of *Fusarium* species of the sections Gibbosum and Discolor. *Trans. Br. Mycol. Soc.* **74**, 73–78.

Hornok, L., and Oras, G. (1976). Investigations on the antigenic structure of Fusaria. 1. An electrophoretic survey of proteins, glycoprotein and lipido-proteinopolysaccharides in the mycelial extracts of *Fusarium culmorum* and *Fusarium acuminatum*. *Acta Phytopathol. Acad. Sci. Hung.* **11**, 37–43.

Howell, M. V., and Taylor, P. W. (1981). Determination of aflatoxin, ochmatoxin A, and zearalenone in mixed feeds, with detection by thin layer chromatography or high performance liquid chromatography. *J. Assoc. Off. Anal. Chem.* **64**, 1356–1363.

Hoyman, W. G. (1941). Concentration and characterization of the emetic principle present in barley infected with *Gibberella saubinetii*. *Phythopathology* **31**, 871–885.

Hsu, I. C., Smalley, C. B., Strong, F. M., and Ribelin, W. E. (1972). Identification of T-2 toxin in moldy corn associated with a lethal toxicosis in dairy cattle. *Appl. Microbiol.* **24**, 684–690.

Huff, W. E., Doerr, J. A., Hamilton, P. B., and Vesonder, R. F. (1981). Acute toxicity of vomitoxin (deoxynivalenol) in broiler chickens. *Poult. Sci.* **60**, 1412–1414.

Hulan, H. W., and Proudfoot, F. C. (1982). Effects of feeding vomitoxin contaminated wheat on the performance of broiler chickens. *Poult. Sci.* **61**, 1653–1659.

Hunt, D. C., Bourdon, A. T., and Crosby, N. T. (1978). Use of high performance liquid chromatography for the identification and estimation of zearalenone, patulin and penicillic acid in food. *J. Sci. Food Agric.* **29**, 239–244.

Hurd, R. N. (1977). Structure activity relationship in zearalenone. In J. V. Rodricks, C. W. Hesseltine, and M. A. Mehlman, eds., *Mycotoxins in Human and Animal Health*, Park Forest, Ill.: Pathotox pp. 379–391.

Hurd, R. N., and Shah, H. (1973). Total synthesis of the macrolide (R.S.)-zearalenone. *J. Med. Chem.* **16**, 543–545.

Ichinoe, M. (1978). Classification of mycotoxin-producing *Fusarium* species. *J. Jpn. Assoc. Mycotoxicology* **8**, 1–5 (in Japanese).

Ichinoe, M., and Kurata, H. (1983). Trichothecene-producing fungi. In Y. Ueno, ed., *Trichothecenes, Chemical, Biological and Toxicological Aspects*. Amsterdam: Elsevier, pp. 73–82.

Ichinoe, M., Kurata, H., Sugiura, Y., and Ueno, Y. (1983). Chemotaxonomy of *Gibberella zeae* with special reference to production of trichothecenes and zearalenone. *Appl. Environ. Microbiol.* **46**, 1364–1369.

Ide, Y., Shimbayashi, K., Obara, J., and Yonemura, T. (1967). Toxicities of wheat and rice infected with *Fusarium graminearum* and *F. nivale* for mice and chicks. *Bull. Natl. Inst. Anim. Health* **54**, 34–37.

Ikeda, Y., Omori, Y., Yoshitomo, H., Furuya, T., and Ichinoe, M. (1964). Experimental studies on some causal *Fusaria* for wheat and barley scab. 4. Feeding test in mice. *Bull. Natl. Inst. Hyg. Sci.* **82**, 130–132.

Ikediobi, C. O., Hsu, I. C., Bamburg, J. R., and Strong, F. M. (1971). Gas-Liquid chromatography of mycotoxins of the trichothecene group. *Anal. Biochem.* **43**, 327–340.

Ilus, T., Ward, P. J., Nummi, M., Adlercreutz, H., and Gripenberg, J. (1977). A new mycotoxin from *Fusarium. Phytochemistry* **16**, 1839–1840.

Ilus, T., Niku-Paavola, and Enari, T. M. (1981). Chromatographic analysis of *Fusarium* toxins in grain samples. *Eur. J. Appl. Microbiol. Biotechnol.* **11**, 244–247.

Inaba, T., and Mirocha, G. J. (1979). Preferential binding of radiolabeled zearalenone to a protein fraction of *Fusarium roseum* "Graminearum." *Appl. Environ. Microbiol.* **37**, 80–84.

Inokawa, H. (1972). Keratomycosis caused by *Fusaria. Jpn. Rev. Clin. Ophthalmol.* **66**, 258–261.

Ishibeishashi, Y. (1978). Experimental fungi keratitis due to *Fusarium.* Improvement of inoculation technique and re-examination of animal model. *Acta Soc. Ophthalmol. Japan* **82**, 643–651.

Ishii, K. (1975). Two new trichothecenes produced by *Fusarium* species. *Phytochemistry* **14**, 2469–2471.

Ishii, K. (1983). Chemistry and bioproduction of non-macrocyclic trichothecenes. In Y. Ueno, ed., *Trichothecenes, Chemical, Biological and Toxicological Aspects.* Amsterdam: Elsevier, pp. 7–19.

Ishii, K., and Mirocha, C. J. (1975). Two new trichothecenes from two strains of *Fusarium roseum. AGFD 56, ACS meet.* August (Abstr).

Ishii, K., and Ueno, Y. (1981). Isolation and characterization of two new trichothecenes from *Fusarium sporotrichioides* strain M-I-I. *Appl. Environ. Microbiol.* **42**, 541–543.

Ishii, K., Sakai, K., Ueno, Y., Tsunoda, H., and Enomoto, M. (1971). Solaniol, a toxic metabolite of *Fusarium solani. Appl. Microbiol.* **22**, 718–720.

Ishii, K., Sawano, M., Ueno, Y., and Tsunoda, T. (1974). Distribution of zearalenone-producing *Fusarium* species in Japan. *Appl. Microbiol.* **27**, 625–628.

Ishii, K., Ando, Y., and Ueno, Y. (1975). Toxicological approaches to the metabolites of *Fusaria.* 9. Isolation of vomiting factors from moldy corn infected with *Fusarium* species. *Chem. Pharmacol. Bull.* **23**, 2162–2164.

Ishii, Y., Pathre, S. V., and Mirocha, C. J. (1978). Two new trichothecenes produced by *Fusarium roseum. J. Agric. Food Chem.* **26**, 649–653.

Ito, H., Watanabe, K., and Koyama, J. (1982). The immunosuppressive effects of trichothecenes and cyclochlorotine on the antibody response in guinea pigs. *J. Pharmacol. Dyn.* **5**, 403–409.

Ito, S. (1912a). *Gibberella* disease of oats, Part 1. *Hokkaido Nokaiho* **12**, 1–7.

Ito, S. (1912b). *Gibberella* disease of oats, Part 2. *Hokkaido Nokaiho* **12**, 51–56.

Itoh, T., Tamura, T., and Matsumoto, T. (1973). Sterol composition of 19 vegetable oils. *J. Am. Oil Chem. Soc.* **50**, 122–125.

Ivanov, A. T. (1968). Study of the possibility of using chick embryos for determining the toxicity of the fungus *F. sporotrichiella* var. *poae. Trudy Vses. Inst. Vet. Sanit.* **28**, 41–47.

Iwatsu, H. (1959). Experimental studies on the etiological aspect of Kashin-Beck's disease in Japan. *Trans. Soc. Pathol. Japan* **48**, 165–167.

Izmajlov, I. A. (1963). Stachybotrytoxicosis of cattle in the western parts of USSR. Ph.D. dissertation, Lvov, Lvov. Zoovet. Inst.

Jackson, R. A. (1973). The chemistry of some derivatives of the macrolide zearalenone. Ph.D. dissertation, University of Minnesota, Minneapolis, Minn.

Jackson, R. A., Fenton, S. W., Mirocha, C. J., and Davis, G. (1974). Characterization of two isomers of 8′-hydroxyzearalenone and other derivatives of zearalenone. *J. Agric. Food Chem.* **22**, 1015–1019.

Jacobson, D. R., and Miller, W. M. (1961). Fescue toxicity. *J. Anim. Sci.* **20**, 960–961.

Jacobson, D. R., Miller, W. M., Seath, D. M., Yates, S. G., Tookey, H. L., and Wolff, I. A. (1963). Nature of fescue toxicity and progress toward identification of the toxic entity. *J. Dairy Sci.* **46**, 416–422.

Jagadeesan, V., Rukmini, C., Vijayaraghavan, M., and Tulpule, P. G. (1982). Immune studies with T-2 toxin: Effect of feeding and withdrawal in monkeys. *Fd. Chem. Toxicol.* **20**, 83–87.

Jamalainen, E. A. (1943a). Über die Fusarien Finnlands. *Publ. Finn. St. Agric. Res. Board* **122**, 1–26.

Jamalainen, E. A. (1943b). Über die Fusarien Finnlands. *Publ. Finn. St. Agric. Res. Board* **123**, 1–25.

Jamalainen, E. A. (1944). Über die Fusarien Finnlands. *Publ. Finn. St. Agric. Res. Board* **124**, 1–24.

Jamalainen, E. A. (1970). Studies on *Fusarium* fungi in Finland. *Ann. Acad. Sci. Finn. A. IV Biolog.* **113**, 54–56.

James, L. J., and Smith, T. K. (1982). Effect of alfalfa on zearalenone toxicity and metabolism in rats and swine. *J. Anim. Sci.* **55**, 110–118.

James, L. J., McGirr, L. G., and Smith, T. K. (1982). High pressure liquid chromatography of zearalenone and zearalenols in rats urine and liver. *J. Assoc. Off. Anal. Chem.* **65**, 8–13.

Jarvis, B. B., Stahly, G. P., and Pavanasasivam, G. (1980). Antileukemic compounds derived from the chemical modification of macrocyclic trichothecenes. 1. Derivatives of Verrucarin A. *Am. Chem. Soc.* **23**, 1054–1058.

Jarvis, B. B., Pavanasasivam, H., Holmlund, C. E., DeSilva, T., and Stahly, G. P. (1981). Biosynthetic intermediates to the macrocyclic trichothecenes. *J. Am. Chem. Soc.* **103**, 472–474.

Jarvis, B. B., Stahly, G. P., Pavanasasivam, G., Midiwo, J. O., DeSilva, T., Holmlund, G. E., Mazzola, E. P., and Geoghegan, R. F. (1982). Isolation and characterization of the trichoverroids and new roridins and verrucarins, *J. Org. Chem.* **47**, 1117–1624.

Jarvis, B. B., Eppley, R. M., and Mazzola, E. P. (1983). Chemistry and bioproduction of macrocyclic trichothecenes. In Y. Ueno, ed., *Trichothecenes, Chemical, Biological and Toxicological Aspects.* Amsterdam: Elsevier, pp. 20–38.

Jayaraman, S., and Parihar, D. B. (1975). Isolation of a growth promoting pigment from food grains infected with *Fusarium moniliforme. Ind. J. Exp. Biol.* **13**, 313.

Jefferys, E. G. (1970). The Gibberellin fermentation. *Adv. Appl. Microbiol.* **3**, 283–316.

Jemmali, M. (1973). The presence of an oestrogenic factor of fungal origin, zearalenone or F-2, as a natural contaminant in maize. *Ann. Microbiol.* **124B**, 109–114 (in French).

Jemmali, M. (1974). Evaluation of different mycotoxins on thin layer plates by reflectance fluorodensitometry. *Ann. Biol. Anim. Biochem. Biophys.* **14**, 845–853.

Jemmali, M. (1979). Decontamination and detoxification of mycotoxins. *Pure Appl. Chem.* **52,** 175–181.

Jemmali, M. (1983). Trichothecenes and zearalenone problems in France. In Y. Ueno, ed., *Trichothecenes, Chemical, Biological and Toxicological Aspects.* Amsterdam: Elsevier, pp. 297–300.

Jemmali, M., Ueno, Y., Ishii, K., Frayssinet, C., and Etienne, M. (1978). Natural occurrence of trichothecenes (nivalenol, deoxynivalenol,T-2) and zearalenone in corn. *Experientia* **34,** 1333–1334.

Jensen, N. P., Brown, R. D., Schmitt, S. M., Windholz, T. B., and Patchett, A. A. (1972). Chemical modification of zearalenone, Part 1. *J. Org. Chem.* **37,** 1039–1647.

Joffe, A. Z. (1947a). The mycoflora of normal and overwintered cereals in 1943–1944. In Alimentary toxic aleukia. *Acta Chkalov Inst. Epidemiol. Microbiol. First Commun.,* 28–34, Orenburg, USSR.

Joffe, A. Z. (1947b). The mycoflora of normal and overwintered cereals in 1944–1945. In Alimentary toxic aleukia. *Acta Chkalov Inst. Epidemiol. Microbiol. Second Commun.,* 35–42, Orenburg, USSR.

Joffe, A. Z. (1947c). The dynamics of toxin accumulation in overwintered cereals and their mycoflora in 1945–1946. In Alimentary toxic aleukia. *Acta Chkalov Inst. Epidemiol. Microbiol.,* p. 192, Orenburg. USSR (Abstr.).

Joffe, A. Z. (1947d). Biological properties of fungi isolated from overwintered cereals. In Alimentary toxic aleukia. *Acta Chkalov Inst. Epidemiol. Microbiol.,* p. 192 Orenburg. USSR (Abstr.).

Joffe, A. Z. (1950). Toxicity of fungi on cereals overwintered in the field (on the etiology of alimentary toxic aleukia). Ph.D. dissertation, Institute of Botany, Academy of Science, Leningrad, USSR, p. 205.

Joffe, A. Z. (1955). The antibiotic effect of molds of the genera *Fusarium* and *Penicillium* on the *Tuberculosis bacillus. Proc. Inst. Exp. Med. Sci. Lith. SSR* 3, 61–67.

Joffe, A. Z. (1956a). The etiology of alimentary toxic aleukia. In *Conference on Mycotoxicoses in Human and Agricultural Animals.* Kiev: Publ. Acad. Sci., pp. 36–38.

Joffe, A. Z. (1956b). The influence of overwintering on the antibiotic activity of several molds of the genus *Cladosporium, Alternaria, Fusarium, Mucor, Thamnidium* and *Aspergillus. Acta Acad. Sci. Lith. SSR* **Ser. B** 3, 85–95.

Joffe, A. Z. (1956c). The effect of environmental conditions on the antibiotic activity of some fungi of the genus *Penicillium. Acta Acad. Sci. Lith. SSR* **Ser.B,** 4, 101–113.

Joffe, A. Z. (1960a). Toxicity and antibiotic properties of some *Fusaria. Bull. Res. Counc. Isr.* **8D,** 81–95.

Joffe, A. Z. (1960b). The mycoflora of overwintered cereals and its toxicity. *Bull. Res. Counc. Isr.* **9D,** 101–126.

Joffe, A. Z. (1962a). Biological properties of some toxic fungi isolated from overwintered cereals. *Mycopathol. Mycol. Appl.* **16,** 201–221.

Joffe, A. Z. (1962b). *Fusarium* root rot of maize in Israel. *Plant. Dis. Rep.* **46,** 203.

Joffe, A. Z. (1963a). Toxicity of overwintered cereals. *Plant Soil* **18,** 31–44.

Joffe, A. Z. (1963b). The occurrence of *Fusarium* species in Israel. 2. Species of *Fusarium* of the section *Elegans* and a review of their taxonomy. *Annls Inst. Natl. Agron. Contrib. Serv. Mycol.* **57,** 51–61.

Joffe, A. Z. (1963c). The mycoflora of continuously cropped soil in Israel with special reference to effects of manurizing and fertilizing. *Mycologia* **55,** 271–282.

Joffe, A. Z. (1963d). Effect of manuring and fertilizing on the mycoflora of heavy soil in a crop rotation trial in Israel. *Soil Sci.* **95,** 353–355.

Joffe, A. Z. (1965). Toxin production by cereal fungi causing alimentary toxic aleukia in man. In G. N. Wogan, ed., *Mycotoxins in Foodstuffs*. Cambridge, Mass.: MIT Press, pp. 77–85.

Joffe, A. Z. (1966). Quantitative relations between some species of *Fusarium* and *Trichoderma* in citrus grove in Israel. *Soil Sci.* **102**, 240–243.

Joffe, A. Z. (1967a). Les espèces de *Fusarium* dans la mycoflore des sols en Israel. *Proc. 1st Isr. Congr. Plant Pathol.*, 32–33.

Joffe, A. Z. (1967b). La mycoflore des grains d'Arachnides en Israel. *Proc. 1st Isr. Congr. Plant Pathol.*, 54–55.

Joffe, A. Z. (1967c). The mycoflora of light soil in a citrus fertilizer trial in Israel. *Mycopathol. Mycol. Appl.* **32**, 209–230.

Joffe, A. Z. (1968a). Mycoflora of surface-sterilized groundnut kernels. *Plant Dis. Rep.* **52**, 608–611.

Joffe, A. Z. (1968b). The effect of soil inoculation with *Aspergillus flavus* on the mycoflora of groundnut soil, rhizosphere and geocarposphere. *Mycologia* **60**, 908–914.

Joffe, A. Z. (1969a). Toxic properties and effects of *Fusarium poae* (Peck.) Wr., *F. sporotrichioides* Sherb., and of *Aspergillus flavus* Link. *J. Stored Prod. Res.* **5**, 211–218.

Joffe, A. Z. (1969b). The mycoflora of groundnut rhizophere, soil and geocarposphere on light, medium and heavy soils and its relations to *Aspergillus flavus*. *Mycopathol. Mycol. Appl.* **37**, 150–160.

Joffe, A. Z. (1969c). The mycoflora of fresh and stored groundnut kernels in Israel. *Mycopathol. Mycol. Appl.* **39**, 255–264.

Joffe, A. Z. (1969d). Relationships between *Aspergillus flavus*, *A. niger* and some other fungi in the mycoflora of groundnut kernels. *Plant and Soil* **31**, 57–64.

Joffe, A. Z. (1971). Alimentary Toxic Aleukia. In S. Kadis, A. Ciegler, and S. J. Ajl, eds., *Microbial Toxins*, Vol. 7. pp. 139–189. New York: Academic, pp. 139–189.

Joffe, A. Z. (1971, 1979). Two cases of keratitis in Israel caused by *F. solani*. (Personal communication.)

Joffe, A. Z. (1972). *Fusaria* isolated from avocado, banana and citrus fruit in Israel and their pathogenicity. *Plant Dis. Rep.* **56**, 963–966.

Joffe, A. Z. (1973a). *Fusarium* species on groundnut kernels and in groundnut soil in Israel. *Plant and Soil* **38**, 339–446.

Joffe, A. Z. (1973b). *Fusarium* species of the *Sporotrichiella* section and relations between their toxicity to plants and animals. *Z. Pflanzenkr. Pflanzensch.* **88**, 92–99.

Joffe, A. Z. (1974a). A modern system of *Fusarium* taxonomy. *Mycopathol. Mycol. Appl.* **53**, 201–228.

Joffe, A. Z. (1974b). Growth and toxigenicity of *Fusarium* of *Sporotrichiella* section as related to environmental factors and culture substrates. *Mycopathol. Mycol. Appl.* **54**, 35–46.

Joffe, A. Z. (1974c). Toxicity of *Fusarium poae* and *F. sporotrichioides* and its relation to Alimentary Toxic Aleukia. In I. F. H. Purchase, ed., *Mycotoxins*. Amsterdam: Elsevier, pp. 229–262.

Joffe, A. Z. (1975). Effects of host age on infection of vegetables by *Fusarium*. *Phytoparasitica* **2**, 71 (Abstr.).

Joffe, A. Z. (1977). The taxonomy of toxigenic species of *Fusarium*. In T. D. Wyllie and L. G. Morehouse, eds., *Handbook of Mycotoxins and Mycotoxicoses*, Vol. 1. New York: Marcel Dekker, pp. 59–82.

Joffe, A. Z. (1978a). *Fusarium* toxicoses of poultry. In T. D. Wyllie and L. G. Morehouse, eds., *Handbook of Mycotoxins and Mycotoxicoses*, Vol. 2. New York: Marcel Dekker, pp. 309–321.

Joffe, A. Z. (1978b). *Fusarium poae* and *Fusarium sporotrichioides* as principal causes of alimentary toxic aleukia. In T. D. Wyllie and L. G. Morehouse, eds., *Handbook of Mycotoxins and Mycotoxicoses*, Vol. 3. New York: Marcel Dekker, pp. 21–86.

Joffe, A. Z. (1983a). *Fusarium* as field, stored and soil fungi under semiarid conditions in Israel. In Y. Ueno, ed., *Trichothecenes, Chemical, Biological and Toxicological Aspects*. Amsterdam: Elsevier, pp. 95–111.

Joffe, A. Z. (1983b). Foodborne diseases: Alimentary toxic aleukia. In M. Rechcigl, Jr., ed., *Handbook of Foodborne Diseases of Biological Origin*. CRC Press, Boca Raton, Fla: CRC Press, pp. 353–495.

Joffe, A. Z. (1983c). Environmental conditions conducive to *Fusarium* toxin formation causing serious outbreaks in animals and man. *Vet. Res. Commun.* **7**, 187–193.

Joffe, A. Z. (1984). Contamination of Arizona corn. *Science* **224**, 240.

Joffe, A. Z., and Borut, S. (1966). Soil and kernel mycoflora of groundnut fields in Israel. *Mycologia* **58**, 629–640.

Joffe, A. Z., and Lisker, N. (1968). Effect of soil fungicides on development of fungi in soil and on kernels of groundnut. *Plant Dis. Rep.* **52**, 718–721.

Joffe, A. Z., and Lisker, N. (1969). The mycoflora of fresh and subsequently stored groundnut kernels on various soil types. *Isr. J. Bot.* **18**, 77–87.

Joffe, A. Z., and Lisker, N. (1970). Effect of crop sequence and soil types on mycoflora of groundnut kernels. *Plant and Soil* **32**, 531–533.

Joffe, A. Z., and Mironov, S. G. (1944). Mycoflora of cereal crops overwintered in the field. *Lect. Publ. Lab. Septic Angina, Chkalov Inst. Epidemiol. Microbiol., Kharkov Med. Inst. Clinic. Hosp. Orenburg Railw.* 6, Orenburg, USSR (Abstr.).

Joffe, A. Z., and Mironov, S. G. (1947a). The mycoflora of normal and overwintered cereals in the field. In Alimentary toxic aleukia. *Acta Chkalov Inst. Epidemiol. Microbiol.* **2**, 28–34, Orenburg, USSR.

Joffe, A. Z., and Mironov, S. G. (1947b). The mycoflora of normal and overwintered cereals in the field. In Alimentary toxic aleukia. 2, 35–41. *Acta Chkalov Inst. Epidemiol. Microbiol.*

Joffe, A. Z. and Nadel-Schiffmann, M. (1967). Les *Fusarium* isós a partir d'avocats isolés à partir d'avocats et d'avocatiers d'Israël. *Fruits* **22**, 97–99.

Joffe, A. Z., and Palti, J. (1962). The *Fusarium* disease of maize. *Hassadeh* **42**, 703.

Joffe, A. Z., and Palti, J. (1963). *Fusaria* isolated from field crops in Israel. *Hassadeh* **43**, 505–508.

Joffe, A. Z., and Palti, J. (1964a). The occurrence of *Fusarium* species in Israel. 1. A first list of *Fusaria* isolated from field crops. *Phytopathol. Med.* **3**, 57–58.

Joffe, A. Z., and Palti, J. (1964b). Seeds as a means of spreading *Fusarium* fungi. *Hassadeh* **45**, 342–343.

Joffe, A. Z., and Palti, J. (1965). Species of *Fusarium* found associated with wilting of tomato varieties resistant to *F. oxysporum* in Israel. *Plant Dis. Rep.* **49**, 741.

Joffe, A. Z., and Palti, J. (1967). *Fusarium equiseti* (Cda.) Sacc. in Israel. *Isr. J. Bot.* **16**, 1–18.

Joffee, A. Z., and Palti, J. (1970). *Fusarium javanicum* Koord. in Israel. *Mycopathol. Mycol. Appl.* **42**, 305–314.

Joffe, A. Z., and Palti, J. (1971). The distribution of fungi of the *Fusarium solani* group in Israel. *Hassadeh* **10**, 1232–1235.

Joffe, A. Z., and Palti, J. (1972). *Fusarium* species of the *Martiella* section in Israel. *Phytopathol. Z.* **73**, 123–148.

Joffe, A. Z., and Palti, J. (1974a). *Fusaria* isolated from field crops in Israel and their pathogenicity to seedlings in glasshouse tests. *Z. Pflanzenkr. Pflanzenschutz* **81**, 196–205.

Joffe, A. Z., and Palti, J. (1974b). Relations between harmful effects on plants and on animals of toxins produced by species of *Fusarium. Mycopathol. Mycol. Appl.* **52**, 209–218.

Joffe, A. Z., and Palti, J. (1975). Taxonomic study of *Fusaria* of the *Sporotrichiella* section used in recent toxicological work. *Appl. Microbiol.* **29**, 575–581.

Joffe, A. Z., and Palti, J. (1977). Species of *Fusarium* found in uncultivated desert-type soils in Israel. *Phytoparasitica* **5**, 119–121.

Joffe, A. Z., and Schiffmann-Nadel, M. (1972). Les espèces de *Fusarium* isolées des fruits d'agrumes d'Israël. *Fruits* **27**, 117–119.

Joffe, A. Z., and Yagen, B. (1977). Comparative study of the yield of T-2 toxin produced by *Fusarium poae, F. sporotrichioides* and *F. sporotrichioides* var. *tricinctum* strains from different sources. *Mycopathologia* **60**, 93–97.

Joffe, A. Z., and Yagen, B. (1978). Intoxication produced by toxic fungi *Fusarium poae* and *F. sporotrichioides* on chicks. *Toxicon* **16**, 263–273.

Joffe, A. Z., and Yeshmantaite, N. (1955). The antibiotic effect of molds of the genera *Fusarium* and *Penicillium* on certain microbes of the enteric group *in vitro. Proc. Inst. Exp. Med. Acad. Sci. Lith. SSR* **3**, 67–72.

Joffe, A. Z., and Yeshmantaite, N. (1958). Antibiotic action of certain fungi on the agents causing intestinal infections. *Acta Inst. Exp. Med.* **3, 4**, 177–181.

Joffe, A. Z., Jackevicius, A. S., and Yeshmantaite, N. (1958). Antibiotic effect of certain *Penicillium* fungi on *Staphylococcus in vitro* and *in vivo. Acta Inst. Exp. Med.* **3, 4**, 183–192.

Joffe, A. Z., Meiri, A., and Palti, J. (1964). Les espèces de *Fusarium* en Israël. 3. *Fusarium* sur les semences de quelques végétaux. *J. Agric. Trop. Bot. Appl.* **12**, 604–606.

Joffe, A. Z., Brosh, S., and Palti, J. (1965). *Fusarium* diseases of onions and allied crops. *Hassadeh* **45**, 918–919.

Joffe, A. Z., Yaffe, Y., and Palti, J. (1967). Yield levels and mycoflora of the soil in Shamuti orange plots given various nutrient treatments. *Soil Sci.* **104**, 263–267.

Joffe, A. Z., Ausher, R., and Palti, J. (1972). Distribution and pathogenicity of *Fusarium* species associated with onions in Israel. *Phytopathol. Med.* **11**, 159–162.

Joffe, A. Z., Palti, J., and Arbel-Sherman, R. (1973). *Fusarium moniliforme* Sheld. (*Gibberella fujikuroi* (Saw.) Wr.) in Israel. *Mycopathol. Mycol. Appl.* **50**, 85–107.

Joffe, A. Z., Palti, J., and Arbel-Sherman, R. (1974). *Fusarium oxysporum* Schlecht. in Israel. *Phytoparasitica* **2**, 91–107.

Joffe, A. Z., Levin, R., and Palti, J. (1975). Preliminary studies of the effect of water stress on *Fusarium* infection of young vegetable plants. *Phytoparasitica* **3**, 71 (Abstr.).

Johnston, D. B., Sawicki, C. A., Windholz, T. B., and Patchett, H. A. (1970). Synthesis of dideoxyzearalenone and hydroxyl derivatives. *J. Med. Chem.* **13**, 941–944.

Jones, B. R. (1975). Principles in the management of oculomycosis. *Am. J. Ophthalmol.* **79**, 719–751.

Jones, B. R., Jones, D. B., Lim, A., Bron, A. J., Morgan, G., Clayton, I. M. (1969a). Corneal and intraocular infection due to *Fusarium solani. Trans. Ophthalmol. Soc. U.K.* **89**, 757–779.

Jones, D. B., Sexton, R., and Rebell, G. (1969b). Mycotic keratitis in South Florida: A review of thirty-nine cases. *Trans. Ophthalmol. Soc. U.K.* **89**, 781–797.

Jones, D. B., Wilson, L., Sexton, R., and Rebell, G. (1969c). Early diagnosis of mycotic keratitis. *Trans. Ophthalmol. Soc. U.K.* **89**, 805–813.

Jones, D. B., Forster, R. K., and Rebell, G. (1972). *Fusarium solani* keratitis treated with natamycin (pimaricin): eighteen consecutive cases. *Arch. Ophthalmol.* **88**, 147–154.

Jones, E. R. H., and Lowe, G. (1960). The biogenesis of trichothecin. *J. Chem. Soc.* 3959.

Josefsson, B. C. E., and Möller, T. E. (1977). Screening method for the detection of aflatoxin, ochratoxin, patulin, sterigmatocystin and zearalenone in cereals. *J. Assoc. Off. Anal. Chem.* **60**, 1369–1371.

Joyce, C. (1981a). New evidence of biological war in S. E. Asia. *New Sci.*, November, 480–481.

Joyce, C. (1981b). Doubts about U.S. claims on chemical warfare. *New Sci.*, September 17, 704.

Juszkiewicz, T., and Piskorska-Pliszczynska, J. (1976). Occurrence of aflatoxins $B_1$, $B_2$, $G_1$, and $G_2$, ochratoxins A and B, sterigmatocystin and zearalenone in cereals. *Med. Weter.* **32**, 617–619.

Juszkiewicz, T., and Piskorska-Pliszczynska, J. (1977). Mycotoxins in grain for animal feeds. *Ann. Nutr. Aliment.* **31**, 489–493.

Kadis, S., Ciegler, A., and Ajl, S. J., eds. (1971). *Microbial Toxins. Vol. 7. Algal and Fungal Toxins.* New York: Academic.

Kallela, K. (1978). Combined action of plant estrogens, F-2 toxin and natural estrogens. *Nord. Vet. Med.* **30**, 132–136.

Kallela, K., and Saastamoinen, I. (1979). A simple method of determining zearalenone in cereals by liquid chromatography. *Eur. J. Appl. Microbiol. Biotech.* **8**, 135–138.

Kallela, K., and Saastamoinen, I. (1981). The effect of grain preservation on the growth of the fungus *Fusarium graminearum* and on the quantity of zearalenone. *Acta Vet. Scand.* **22**, 417–427.

Kallela, K., and Vasenius, L. (1982). The effects of rumen fluid on the content of zearalenone in animals fodder. *Nord. Veterinaermed.* **34**, 336–339.

Kallela, K., Hintikka, E-L., and Ylimäki, A. (1978). Variation of F-2 toxin production on different substrates. *Nord. Vet. Med.* **30**, 424–429.

Kalmykov, S. T., Kochetov, S. T., Penkov, V. I., and Ponomareva, T. M. (1967). Fusariotoxicosis in sheep. *Veterinariya* **43**, 65–67 (Abstr.).

Kalra, D. S., Bhatia, K. C., Gautam, O. P., and Chauhan, H. V. S. (1973a). An obscure disease (possibly Degnala disease) in buffaloes and cattle. Studies on its epizootiology, pathology and etiology. *Haryana Agric. Univ. J. Res.* **2**, 256–264.

Kalra, D. S., Bhatia, K. C., Gautam, O. P., and Chauhan, H. V. S. (1973b). Chronic ergot poisoning-like syndrome (Degnala disease) in buffaloes and cattle in Haryang and Punjab States. *Indian Vet. J.* **50**, 484–486.

Kalra, D. S., Bhatia, K. C., Gautam, O. P., and Chauhan, H. V. S. (1977a). *Fusarium equiseti* associated mycotoxins as a possible cause of Degnala disease III. *Ann. Nutr. Aliment.* **31**, 745–752.

Kalra, D. S., Bhatia, K. C., Gautam, O. P., and Chauhan, H. V. S. (1977b). Pathology of Degnala disease in cattle and buffaloes. *Ann. Nutr. Alim.* **31**, 753–760.

Kalven, J. (1982). Yellow Rain: The public evidence. *Bull. At. Sci.*, May, 15–20.

Kambayashi, T., and Otake, S. (1936). Über die Tierpathogenität Pflanzenpathogener Pilze, insbesondere von *Fusarium solani* (Mart.P.P.) App. et Wr. *Z. Parasitenkd.* **8**, 611–616.

Kamimura, H., Nishijima, M., Saito, K., Takahashi, S., Ibe, A., Ochiai, S., and Naoi, Y.

(1978). Mycotoxins in foods. 8. Analytical procedure of trichothecene mycotoxins in cereals. *Jpn. J. Food Hyg. Soc.* **19**, 443–448.

Kamimura, H., Nishijima, M.,Yasuda, K., Saito, K., Ibe, A., Nayayama, T., Ushiyama, H., and Naoi, Y. (1981). Simultaneous detection of several *Fusarium* mycotoxins in cereals, grains, and foodstuffs. *J. Assoc. Off. Anal. Chem.* **64**, 1067–1073.

Kaneko, I., Schmitz, H., Essery, J. M., Rose, W., Howell, H. G., O'Herron, F. A., Nachfolger, S., Huftalen, J., Bradner, W. T., Partyka, R. A., Doyle, T. W., Davies, D., and Cundliffe, E. (1982). Structural modification of anguidine and antitumor activities of its analogues. *J. Med. Chem.* **25**, 579–589.

Kanshina, N. F. (1957). Materials on pathomorphology of experimental mycotoxicosis in dogs and white rats (to problems of etiology of Kashin-Beck disease). *Probl. Nutr.* **16**, 69–74.

Karatygin, V. M., and Rozhnova, Z. I. (1947). Vitamin insufficiency in alimentary toxic aleukia (septic angina). *Sov. Med.* **5**, 17–19.

Karlik, L. N. (1945). The history of alimentary toxic aleukia (septic angina). In A. I. Nesterov, A. H. Sysin, and L. N. Karlik, eds., *Alimentary Toxic Aleukia (Septic Angina).* Moscow: Publ. State Med. Lit., pp. 3–7.

Kashin, N. I. (1860). Urov Disease in Transbaikal. *Moskow Med. Gaz.* **7**, 53–56.

Kasirski, I. A., and Alekseyev, G. A. (1948). Alimentary toxic aleukia. In *Disease of Blood and Hematopoietic System.* Moscow: Med. Publ. House, pp. 204–217.

Kato, T., and Takitani, S. (1978). Detection and determination of trichothecene mycotoxins with 4-p-nitrobenzyl pyridine on thin-layer chromatography. *Proc. Jpn. Assoc. Mycotoxicol.* **7**, 22–23.

Kato, T., Asabe, Y., Suzuki, M., and Takitani, S. (1976). Fluorometric determination of fusarenon-X and its related mycotoxins. *Buseki, Kaga Ku,* **25**, 659–662.

Kato, T., Asabe, Y., Suzuki, M. and Takitani, S. (1977). Spectrophotometric determination of trichothecene mycotoxins with chromotropic acid. *Buseki Kaga Ku,* **26**, 422–424.

Kato, T., Asabe, Y., Suzuki, M., and Takitani, S. (1979). Spectrophotometric and fluorometric determinations of trichothecene mycotoxins with reagents for formaldehyde. *Anal. Chim. Acta* **106**, 59–65.

Katzenellenbogen, B. S., Katzenellenbogen, J. A., and Mordecai, D. (1979). Zearalenone characterization of the estrogenic potencies and receptor interactions of a series of fungal B-resorcylic acid lactones. *Endocrinology* **105**, 33–40.

Kaufman, H. E. and Wood, R. M. (1965). Mycotic keratitis. *Am. Ophthalmol.* **59**, 992–1000.

Kavetski, N. E., and Grinberg, B. M. (1945). On the problem of septic angina (alimentary toxic aleukia). *Sov. Med.* **3**, 12–17.

Kellerman, T. S., Marasas, W. F. O., Pienaar, J. G., and Naude, T. W. (1972). A mycotoxicosis of *Equidae* caused by *Fusarium moniliforme* Sheldon. A preliminary communication. *Onderstepoort J. Vet. Res.* **39**, 205–208.

Keyl, A. C., Lewis, J. C., Ellis, J. J., Yates, S. G., and Tookey, H. L. (1967). Toxic fungi isolated from tall fescue. *Mycopathol. Mycol. Appl.* **31**, 327–331.

Khabibullina, G. F. (1945). On the method of combined therapy of septic angina Bogomolt's antireticular cytotoxic serum (ACS) and sulfonamide compounds according to the data of the Ear Clinic of Bashkir Medical Institute. In *Acta Septic Angina.* Ufa, USSR: Bashkir Publ. House.

Khaikina, B. G., Uvarov, A. A., and Joffe, A. Z. (1955). Change in antibiotic properties of some *Penicillium* fungi under the influence of products of bacterial decomposition. *Proc. Chkalov Med. Inst.* **3**, 25–34.

Khera, K. S., Whalen, C., Angers, G., Vesonder, R. F., and Kuiper-Goodman, T. (1982).

Embryotoxicity of 4-Deoxynivalenol (vomitoxin) in mice. *Bull. Environ. Contam. Toxicol.* **29**, 487–491.

Kholodenko, M. I. (1947). Changes in the nervous system of toxic alimentary aleukia (so-called "septic angina"). *Neuropathol. Psychiatr.* **16**, 67–70, Moscow.

Khrutski, J. T., Joffe, A. Z., Epifanova, N. N., and Rubcova, E. B. (1953). The action of *Fusarium poae* (Peck.) Wr. on the motor activity of the stomach in dogs. *Acta Chkalov Agric. Inst.* **6**, 59–62.

Kiang, D. T., Kennedy, B. J., Pathre, S. V., and Mirocha, C. J. (1978). Binding characteristics of zearalenone analogs to estrogen receptors. *Cancer Res.* **38**, 3611–3615.

Kidd, G. H., and Wold, F. T. (1973). Dimorphism in a pathogenic *Fusarium. Mycologia* **65**, 1371–1375.

Kiessling, K. H., and Pettersson, H. (1978). Metabolism of zearalenone in rat liver. *Acta Pharmacol. Toxicol.* **43**, 285–290.

Kiessling, K. H., Pettersson, H., Sandholm, K., and Olsen, M. (1984). Metabolism of aflatoxin, ochratoxin, zearalenone and three trichothecenes by intact rumen fluid, rumen protozoa, and rumen bacteria. *Appl. Environ. Microbiol.* **47**, 1070–1073.

Kishaba, A. N., Shankland, D. L., Curtis, R. W., and Wilson, M. C. (1962). Substances inhibitory to insect feeding with insecticidal properties from fungi. *J. Econ. Entomol.* **55**, 211–214.

Knaus, E. (1978). Untersuchungen zum Nachweis der fruchtbarkeitsvermindernden Wirkung von Mykotoxinin. In *Aktuelle Probleme der landwirtschaftlichen Forschung*, Vol. 11. Linz: Veröff. Landwirtsch. Chem. Bundesversuchsanst., pp. 105–126.

Koen, J. S., and Smith, H. C. (1945). An unusual case of genital involvement in swine associated with eating moldy corn. *Vet. Med.* **40**, 131–133.

Kogan, A. M., and Jershova, O. A. (1956). The problem of biochemical changes in the bones of animals during experimental food mycotoxicosis. *Vopr. Pitan., Moscow* **15**, 56–57.

Kogan, A. M., and Vasilyeva, I. N. (1956). The mineral compositions of grain and vegetables in the regions of the Urov disease. *Probl. Nutr.* **5**, 91–92.

Köhler, H., Hüttner, B., Vielitz, E., Kahlau, D. I., and Gedek, B. (1978). Rachitis bei Mastgeflügel durch Kontamination des Futters mit *Fusarium moniliforme* Sheldon. *Zbl. Vet. Med.* **B25**, 89–109.

Kolosova, N. I. (1949). Data on chemical and toxic properties of some fatty acids isolated from grains causing alimentary toxic aleukia. Ph.D. dissertation, Moscow, Nutr. Inst. Acad. Sci. USSR.

Komissaruk, D. Z. (1957). Disputable problems of the etiology and pathogenesis of Kashin-Beck Disease. *Klin. Med.* **35**, 86–92, Moscow.

Kommedahl, T., and Windels, C. E. (1981). Root-, stalk-, and ear-infecting *Fusarium* species on corn in the U.S.A. In P. E. Nelson, T. A. Toussoun, and R. J. Cook, eds., *Fusarium, Diseases, Biology and Taxonomy*. University Park: Pennsylvania State University Press, pp. 94–103.

Konishi, T., and Ichijo, S. (1970a). Clinical studies on bean-hulls poisoning of horses. 1. Clinical and biological observations in spontaneous cases. *Res. Bull. Obihiro Zootech. Univ.* **6**, 242–257.

Konishi, T., and Ichijo, S. (1970b). Clinical studies on bean-hulls poisoning of horses. 2. Clinical and biochemical observations in experimental cases. *Res. Bull. Obihiro Zootech. Univ.* **6**, 258–273.

Kopytkova, O. I. (1948a). Differential diagnosis of alimentary toxic aleukia. *Health Kazakhstan* **7**, 39–46.

Kopytkova, O. I. (1948b). Treatment of alimentary toxic aleukia (septic angina). *Health Kazakhstan* **7**, 48–51.

Korpinen, E. L. (1972). Natural occurrence of F-2 and F-2-producing *Fusarium* strains associated with field cases of bovine and swine infertility. *IUPAC Symp. Control Mycotoxins.* August 21–22, p. 21, Kungalv, Sweden.

Korpinen, E. L., and Uoti, U. (1974). The variations in toxic effect of five *Fusarium* species in rats. *Ann. Agric. Finn.* **13**, 34–42.

Korpinen, E. L., and Ylimäki, A. (1972). Toxigenicity of some *Fusarium* strains. *Ann. Agric. Finn.* **11**, 308–314.

Kosuri, N. R., Grove, M. D., Yates, S. G., Tallent, W. H., Ellis, J. J., Wolf, I. A., and Nichols, R. E. (1970). Response of cattle to mycotoxins of *Fusarium tricinctum* isolated from corn and fescue. *J. Am. Vet. Med. Assoc.* **157**, 938–940.

Kosuri, N. R., Smalley, E. B., and Nichols, R. E. (1971). Toxicologic studies of *Fusarium tricinctum* (Corda) Snyder and Hansen from moldy corn. *Am. J. Vet. Res.* **32**, 1843–1850.

Kotik, A. N., Chemobay, V. T., Komissarenko, N. F., and Trufanova, V. A. (1979). Isolation of mycotoxin in *Fusarium sporotrichiella* and studies of its physicochemical and toxic properties. *Mikrobiol. Zh.* **41**, 636–639 (in Russian).

Kotsonis, F. N., and Ellison, R. A. (1975). Assay and relationship of HT-2 toxin and T-2 toxin formation in liquid culture. *Appl. Microbiol.* **30**, 33–37.

Kotsonis, F. N., Ellison, R. A., and Smalley, E. B. (1975a). Isolation of acetyl T-2 toxin from *Fusarium poae. Appl. Microbiol.* **30**, 493–495.

Kotsonis, F. N., Smalley, E. B., Ellison, R. A., and Gale, C. M. (1975b). Feed refusal factors in pure cultures of *Fusarium roseum* "Graminearum." *Appl. Microbiol.* **30**, 362–368.

Kovács, F., Szathmary, C., and Palyusik, M. (1975a). Data on determination of toxin F-2 (zearalenone) by high-pressure liquid, gas and thin-layer chromatography. *Acta Vet. Acad. Sci. Hung.* **25**, 223–230.

Kovács, F., Szathmary, C., and Palyusik, M. (1975b). Further data on the determination of F-2 toxin with high pressure liquid, gas and thin-layer chromatography. *Magy. Allatory Lapja* **30**, 625–628.

Kovalev, E. N. (1944). The nervous system in so-called septic angina. *Neuropathol. Psychiatr.* **13**, 75–79, Moscow.

Koza, M. A. (1945). Pathological anatomy of alimentary toxic aleukia (Septic angina). In A. E. Nesterov, A. N. Sysin, and L. N. Karlic, eds., *Alimentary Toxic Aleukia (Septic Angina).* Moscow: Med. Publ. House, pp. 21–39.

Koza, M. A., Leontiev, I. A., and Yasnitski, P. Y. (1944). *Alimentary Toxic Aleukia (Septic Angina).* Moscow: Med. Publ. House, pp. 3–43.

Kozin, N. E., and Yershova, O. A. (1945). Cultivation method for determining toxicity of cereal grain overwintered under snow (prosomillet). *Proc. Nutr. Inst. Acad. Med. Sci.* 39–48, Moscow.

Kravchenko, L. F. (1959). The main principles and tasks of conservative treatment of the Urov disease patients (Kashin-Beck). *2nd Sci. Conf. Urov Sci. Res. Stn.*, Sretensk.

Kravchenko, L. F. (1965). The Disease of Kashin-Beck (chondroosteoarthritis deformans endemica). Ph.D. dissertation, Rostov-on-Don, Rostov Univ.

Kravchenko, L. F. (1968). Urov Disease, Its Prevention and Treatment. Chita, Med. Inst.

Kraybill, H. F., and Shapiro, R. E. (1969). Implications of fungal toxicity to human health. In L. A. Goldblatt, ed., *Aflatoxin.* New York: Academic pp. 401–441.

Kretovich, V. L. (1945). Biochemistry of grain causing septic angina. Lecture in Republic

Conference of Alimentary Toxic Aleukia, General State Sanit. Insp., pp. 9–10, Moscow (Abstr.).

Kretovich, V. L., and Bundel, A. A. (1945). Investigations on the oil of toxic overwintered prosomillet. *Biochemistry* **10**, 216–224.

Kretovich, V. L., and Sosedov, N. I. (1946). Biochemical properties of toxic prosomillet. *Biochemistry* **10**, 279–284.

Kriek, N. P. J., and Marasas, W. F. O. (1983). Trichothecene research in South Africa. In Y. Ueno, ed., *Trichothecenes, Chemical, Biological and Toxicological Aspects*. Amsterdam: Elsevier pp. 273–284.

Kriek, N. P. J., Marasas, W. F. O., Steyn, P. S., Van Rensberg, S. J., and Steyn, M. (1977). Toxicity of moniliformin-producing strain of *Fusarium moniliforme* var. *subglutinans* isolated from maize. *Food Cosmet. Toxicol.* **15**, 579–587.

Kriek, N. P. J., Marasas, W. F. O., and Thiel, P. G. (1981a). Hepato- and cardiotoxicity of *Fusarium verticillioides* (*F. moniliforme*) isolates from Southern African maize. *Food Cosmet. Toxicol.* **19**, 447–456.

Kriek, N. P. J., Kellerman, T. S., and Marasas, W. F. O. (1981b). A comparative study of the toxicity of *Fusarium verticillioides* (*F. moniliforme*) to horses, primates, pigs, sheep and rats. *Onderstepoort J. Vet. Res.* **48**, 129–131.

Kubota, T. (1977). Experimental studies on *Fusarium* poisoning in goats. *Bull. Nippon Vet. Zootech. Coll.* **26**, 9–24 (in Japanese).

Kubota, T., and Ichikawa, N. (1954). On the chemical constitution of ipomeanine, a new ketone from the black-rotted sweet potato. *Chem. Ind.* **29**, 902–903.

Kucewicz, W. (1982). U.S. Army studies Yellow-Rain defences. *Wall Street J.*, April 29, 2.

Kuczuk, M. H., Benson, F. M., Heath, H., and Hayes, A. W. (1978). Evaluation of the mutagenic potential of mycotoxins using *Salmonella typhimurium* and *Saccharomyces cerevisiae*. *Mut. Res.* **53**, 11–20.

Kudryakov, V. T. (1946). Pathogenesis and therapy of alimentary toxic aleukia. *Klin. Med.* **24**, 68–69, Moscow.

Kuo, C., Taub, D., Hoffsommer, R. D., and Wendler, N. L. (1967). The resolution of (±)-zearalenone. Determination of the absolute configuration of the natural enantiomorph. *Chem. Commun.* 761–762.

Kurasova, V. V., Leonov, A. N., and Golban, D. M. (1973). Fusariotoxicosis causing vulvovaginitis in pigs. *Vetarinariya* **11**, 98–99, Moscow.

Kurata, H. (1978). Current scope of mycotoxin research from the viewpoint of food mycology. In K. Uraguchi and M. Yamazaki, eds., *Toxicology, Biochemistry and Pathology of Mycotoxins*. New York: Wiley, pp. 13–64.

Kurata, H., Udagawa, S., and Sakabe, F. (1964). Experimental studies on some causal *Fusarium* of wheat and barley scab, I-III. *Bull. Natl. Inst. Hyg. Sci.* **82**, 123–130.

Kurata, M. I., Sugiura, Y., and Ueno, Y. (1983). Chemotaxonomy of *Gibberella zeae* with special reference to production of trichothecenes and zearalenone. *Appl. Environ. Microbiol.* **46**, 1364–1369.

Kuratra, M. S., and Singh, A. (1973). Experimental reproduction of gangrenous syndrome in buffaloes (*Bos bubalus*). *Zbl. Vet. Med. B.* **20**, 481–489.

Kurbatova, T. G. (1948). Alimentary toxic aleukia (review of the literature). *Acta Kirovsk Inst. Epidemiol. Microbiol.* **2**, 71–75.

Kurmanov, I. A. (1961). Fusariotoxicosis of sheep in the Stavropol District. *Vet. J.* **11**, 30–31.

Kurmanov, I. A., (1968). Some questions of fusariotoxicosis in animals. *Veterinariya*. **45**, Moscow. 53–56.

Kurmanov, I. A. (1969). Investigations on *Fusarium* toxicosis in animals. *Vet. Bull.* **39**, 330 (Abstr. 1957).

Kurmanov, I. A. (1971). Pathogenesis of *Fusarium* toxicosis in animals. *Veterinariya* **7**, 92–93, Moscow.

Kurmanov, I. A. (1978a). Fusariotoxicosis in chickens in the USA. In T. D. Wyllie and L. G. Morehouse, eds., *Mycotoxic Fungi, Mycotoxins, Mycotoxicoses—An Encyclopedic Handbook*, Vol. 2. New York: Marcel Dekker, pp. 322–326.

Kurmanov, I. A. (1978b). Fusariotoxicosis in cattle and sheep. In T. D. Wyllie and L. G. Morehouse, eds., *Mycotoxic Fungi, Mycotoxins, Mycotoxicoses—An Encyclopedic Handbook*, Vol. 2. New York: Marcel Dekker, pp. 85–110.

Kuroda, H., Mori, T., Nishioka, C., Okasaki, H., and Takagi, A. (1979). Studies on gas chromatographic determination of trichothecene mycotoxins in food. *J. Food Hyg. Soc. Japan* **20**, 137–142.

Kurtz, H. J. (1976). Estrogenic syndrome caused by zearalenone (F-2) mycotoxicosis. *Proc. Int. Pig Vet. Soc.*, p. 10, Ames, Iowa.

Kurtz, H. J., and Mirocha, C. J. (1978). Zearalenone (F-2) induced estrogenic syndrome in swine. In T. D. Wyllie and L. G. Morehouse, eds., *Mycotoxic Fungi, Mycotoxins, Mycotoxicoses—An Encyclopedic Handbook*, Vol. 2. New York: Marcel Dekker, pp. 216–268.

Kurtz, H. J., Nairn, M. E., Nelson, G. H., Christensen, C. M., and Mirocha, C. J. (1969). Histologic changes in the genital tracts of swine fed estrogenic mycotoxin. *Am. J. Vet. Res.* **30**, 551–556.

Kvashnina, E. S. (1948). Mycoflora of cereal crops overwintered in the field. In *Data on Cereal Crops Wintered under Snow*. Moscow, USSR: Publ. Minis. Agric., A. Kh. Sarkisov, pp. 41–49.

Kvashnina, E. S. (1968). Toxic and biological properties of *Fusarium* species associated with mycotoxicoses in animals. *All-Union Inst. Exp. Vet.* **35**, 341–349.

Kvashnina, E. S. (1976). Physiological and ecological characteristics of *Fusarium* species of the Sporotrichiella Section. *Mikrobiol. Fitopatol.* **10**. 275–282 (in Russian).

Kvashnina, E. S. (1978). Physiological and ecological characteristics of *Fusarium* species of the Sporotrichiella Section. *Bull. Vses. Ordena Lenina Inst. Eksp.* **32**, 42–45.

Kvashnina, E. S. (1979). Morphological and cultural characteristics of species of the genus *Fusarium* Section Sporotrichiella and its distribution in the U.S.S.R. *Mikrobiol. Fitopatol.* **13**, 3–10 (in Russian).

Kvashnina, E. S., and Gabrilova, D. A. (1956). Fusariotoxicosis of sheep and cattle. *Bull. Sci. Tech. Inf. VNIVS* **1**, 65.

Lafarge-Frayssinet, C., Lespinats, G., Lafont, P., Loisillier, R., Mousset, S., Rosenstein, V., and Frayssinet, C. (1979). Immunosuppressive effects of *Fusarium* extracts and trichothecenes: Blastogenic response of murine splenic and thymic cells to mitogens (40439). *Proc. Soc. Exp. Biol. Med.* **160**, 302–311.

Lafarge-Frayssinet, C., Decloitre, F., Mousset, S., Martin, M., and Frayssinet, C. (1981). Induction of DNA single-stranded breaks by T-2 toxin, a trichothecene metabolite of *Fusarium*. Effect on lymphoid organs and liver. *Mutat. Res.* **88**, 115–123.

Lafont, P., and Lafont, J. (1980). Contaminations du mais per des mycotoxines. *Bull. Acad. Vet. Fr.* **53**, 533–538.

Lafont, P., Lafarge-Frayssinet, C., Lafont, J., Bertin, G., and Frayssinet, C. (1977). Métabolites toxiques de *Fusarium* agents de l'aleucemie toxique alimentaire. *Ann. Microbiol. (Inst. Pasteur)* **128B**, 215–220.

Lami, L. (1983). Yellow Rain: The conspiracy of closed mouths. *Commentary* **76** (4), 60–61.

Lando, Y. K. (1935). On the pathologic anatomy of septic angina. *Med. J. Kazakhstan* **4, 5,** 88–91.

Lando, Y. K. (1939). Material relating to the pathologic anatomy of septic angina. *Sov. Publ. Health Serv., Kirghiz SSR* **6,** 72–86.

Lansden, J. A., Clarkson, R. J., Neely, W. C., Cole, R. J., and Kirksey, J. W. (1974). Spectroanalytical parameters of fungal metabolites. 4. Moniliformin. *J. Assoc. Off. Anal. Chem.* **57,** 1392–1396.

Lansden, J. A., Cole, R. J., Dorner, J. W., Cox, R. H., Cutler, H. G., and Clark, J. D. (1978). A new trichothecene mycotoxin isolated from *Fusarium tricinctum. J. Agric. Food Chem.* **26,** 246–249.

Lappe, U., and Barz, W. (1978). Degradation of pisatin by fungi of the genus Fusarium. *Z. Naturforsch.* **33,** 301–302.

Lásztity, R., and Wöller, L. (1975a). Effect of zearalenone and some derivatives on animals fed on contaminated fodder. *Acta Aliment.* **4,** 189–197.

Lásztity, R., and Wöller, L. (1975b). Toxinbildung bei *Fusarium*-Arten und vorkommen der Toxin in Landwirtschaftlichen Producten. *Nahrung* **19,** 537–546.

Lásztity, R., and Wöller, L. (1977). Investigation of the effect of zearalenone toxin produced by *Fusarium* fungi and its derivatives on animal organisms fed with infected fodder. In *Mycotoxins in Food.* Zeszyty Problemowe Postępów Nauk Rolniczych. Warsaw: pp. 193–200.

Lásztity, R., Tamás, K., and Wöller, L. (1977). Occurrence of *Fusarium* mycotoxins in some Hungarian corn crops and the possibilities of detoxication. *Ann. Nutr. Aliment.* **31,** 495–498.

Laverde, C., Moncada, L. H., Restrepo, A., Diaz, F., and Vera, C. (1972). Estudio micrologico y bacteriologico de las ulceras de cornea en pacientes del Hospital Universitario San Vincente de Paul. *Rev. Soc. Colomb. Oftalmol.* **3,** 175–190.

Laverde, C. L., Moncada, L. H., Restrepo, A., and Vera, C. L. (1973). Mycotic keratitis: 5 cases caused by unusual fungi. *Sabouraudia,* **11,** 119–123.

Lazarus, J. A., and Schwarz, L. H. (1948). Infection of urinary bladder with unusual fungus strain (*Fusarium*). *Urol. Cutaneous Rev.* **52,** 185–189.

Lee, S., and Chu, F. S. (1981a). Radioimmunoassay of T-2 toxin in corn and wheat. *J. Assoc. Off. Anal. Chem.* **64,** 156–161.

Lee, S., and Chu, F. S. (1981b). Radioimmunoassay of T-2 toxin in biological fluids. *J. Assoc. Off. Anal. Chem.* **64,** 684–688.

Lee, Y-W., and Mirocha, C. J. (1984). Production of nivalenol and fusarenon-X by *Fusarium tricinctum* Fn-2B on a rice substrate. *Appl. Environ. Microbiol.* **48,** 857–858.

Lelievre, J., Bremond, J., and Rebour, J. (1962). Enzootic de vulvovaginite chez la truite. *Bull. Soc. Vet. Prat. Fr.* **46,** 18–19.

Leonian, L. H. (1929). Studies on the variability and dissociations in the genus *Fusarium. Phytopathology* **19,** 753–867.

Levin, I. I. (1945). Clinic and therapy of alimentary toxic aleukia. Abstract in Republic Conference of Alimentary Toxic Aleukia, General State Sanit. Insp., Moscow, December 24–26, pp. 10–12.

Levin, I. I. (1946). Treatment of alimentary toxic aleukia with sulfamid compound preparations. *Clin. Med.* **7,** 54–59.

Levitski, L. M. (1948). The use of penicillin in alimentary toxic aleukia. Lecture at Republic Conference of Alimentary Toxic Aleukia, Moscow, March.

Lew, H. (1978). Zearalenon und Trichothecene. In *Aktuelle Probleme der Landwirtschaftlicher Forschung,* Vol. 11. Linz: Veröf. Landwirtsch-Chem. Bundesversuchsanst., pp. 127–143.

Ley, A. P., and Sanders, T. E. (1956). Fungus Keratitis: A report of 3 cases. *A.M.A. Arch. Ophthalmol.* **56**, 257–264.

Li, K. H., Kao, J. C., and Wu, Y. K. (1962). A survey of the prevalence of carcinoma of the esophagus in North China. In *Selected Papers on Cancer Research*. Shanghai: Scientific and Technical Publ., pp. 215–221.

Li, M. H., Lu, S. H., Ji, C., Wang, M. Y., Cheng, S. J., and Jin, C. L. (1979). Formation of carcinogenic N-nitroso compound in corn-break inoculated with fungi. *Sci. Sin.* **22**, 471–477.

Li, M. H., Lu, S. H., Ji, C., Wang, Y. L., Wang, M. Y., Cheng, S. J., and Tiang, G. Z. (1980). Experimental studies on the carcinogenicity of fungus-contaminated food from Lin-xian county. In H. V. Gelboim, ed., *Genetic and Environmental Factors in Experimental and Human Cancer*. Tokyo: Japan Science Press, pp. 139–148.

Liao, L. -L., Grollman, A. P., and Horwitz, S. B. (1976). Mechanism of action of the 12,13-epoxytrichothecene, anguidine, an inhibitor of protein synthesis. *Biochim. Biophys. Acta* **454**, 273–284.

Lieberman, J. R., and Mirocha, C. J. (1970). Biosynthesis of F-2 (zearalenone) by *Fusarium roseum* "Graminearum." *Phytopathology* **60**, 1300 (Abstr.).

Lieberman, J. R., Wolf, J. C., Rao, R. G., and Harein, P. K. (1971). Inhibition of F2 (zearalenone) biosynthesis and perithecial production in *Fusarium roseum* "Graminearum." *Phytopathology* **61**, 900 (Abstr.).

Lieberman, T. W., Ferry, A. P., and Botlone, E. J. (1979). *Fusarium solani* endophthalmitis without primary corneal involvement. *Am. J. Ophthalmol.* **88**, 764–767.

Lillehoj, E. B. (1973). Feed sources and conditions conducive to production of aflatoxin, ochratoxin, *Fusarium* toxins and zearalenone. *J. Am. Vet. Med. Assoc.*, **163**, 1281–1284.

Lin, C. C. (1959). Studies on the Kashin-Beck disease in Taiwan. *Trans. Soc. Pathol. Japan* **48**, 827.

Lin, P., and Tang, W. (1980). Zur Epidemiologie und Ätiologie des Oesophaguscarcinoms in China. *J. Can. Res. Clin. Oncol.* **96**, 121–130.

Lindenfelser, L. A., Lillehoj, E. B., and Burmeister, H. R. (1974). Aflatoxin and trichothecene toxins: Skin tumor induction synergistic acute toxicity in white mice. *J. Natl. Cancer Inst.* **52**, 113–116.

Lindenfelser, L. A., Ciegler, A., and Hesseltine, C. W. (1978). Wild rice as fermentation substrate for mycotoxin production. *Appl. Environ. Microbiol.* **35**, 105–108.

Loncarević, A., Jovanović, M., Lješević, Z., Stanov, Z., Bogetić, V., and Tosevski, J. (1977). Appearance of vulvovaginitis in newborn piglets originated from sows fed diet contaminated with *Fusarium graminearum*. *Acta Vet.* **27**, 151–157.

Lopatin, G. M. (1946). Alimentary toxic aleukia (septic angina) in children. *Pediatrica* **1**, 27–30.

Louria, D. B., Smith, J. K., and Finkel, G. C. (1968). Mycotoxins other than aflatoxin: tumor producing potential and possible relation to human disease. *Ann. N.Y. Acad. Sci.* **174**, 583–591.

Lovelace, C. E. A., and Nyathi, C. B. (1977). Estimation of the fungal toxins, zearalenone and aflatoxin, contaminating opaque maize beer in Zambia. *J. Sci. Food Agric.* **28**, 288–292.

Lovla, D. S. (1944). Septic angina in the Chkalov district. In *Alimentary toxic aleukia. Acta Chkalov Inst. Epidemiol. Microbiol.*, p. 1, Orenburg, USSR.

Lozanov, N. N., and Tsareva, V. Y. (1944). *Septic Angina*. Kazan: Tatar ASSR Publ. House, pp. 3–15.

Lu, S., Li, M., Ji, C., Wang, M., Wang, Y., and Huang, L. (1979). A new N-nitro compound,

N-3-methylbutyl-N-1-methylacetonyl-nitrosamine, in corn-bread inoculated with fungi. *Sci. Sin.* **22**, 601–607.

Lu, S. H., Camus, A. M., Ji, C., Wang, Y. L., Wang, M. Y., and Bartsch, H. (1980a). Mutagenicity in *Salmonella typhimurium* of N-3-methylbutyl-N-1-methyl-acetonyl-nitrosamine and N-methyl-N-benzylnitrosamine, N-nitrosation products isolated from cornbread contaminated with commonly occurring moulds in Linshien county, a high incidence area for oesophageal cancer in Northern China. *Carcinogenesis* **1**, 867–870.

Lu, S. H., Wang, Y. L., and Li, M. H. (1980b). Effect of fungi on the formation of carcinogenic nitrosamines and their precursors in food. *Acta Acad. Med. Sin.* **2**, 24–28.

Lukin, A. Y., and Berlin, M. G. (1947). Toxic influences of overwintered prosomillet on the organs of horses and pigs. *Proc. Chkalov Agric. Inst.* **3**, 65–71.

Lukin, A. Y., Antonov, N. A., and Simonov, I. N. (1947a). Feeding tests with toxic cereals on pigs and sheep. *Proc. Chkalov Agric. Inst.* **3**, 78–86.

Lukin, A. Y., Antonov, N. A., and Simonov, I. N. (1947b). Feeding tests with toxic cereals on horses. *Proc. Chkalov Agric. Inst.* **3**, 93–106.

Lutsky, I., and Mor, N. (1981a). Experimental alimentary toxic aleukia in cats. *Lab. Anim. Sci.* **31**, 43–47.

Lutsky, I., and Mor, N. (1981b). Alimentary toxic aleukia (Septic angina, endemic panmyelotoxicosis, alimentary toxic aleukia). T-2 toxin-induced intoxication of cats. *Am. J. Pathol.* **104**, 189–191.

Lutsky, J., Mor, N., Yagen, B., and Joffe, A. Z. (1978). The role of T-2 toxin in experimental alimentary toxic aleukia: A toxicity study in cats. *Toxicol. Appl. Pharmacol.* **43**, 111–124.

Lyass, M. A. (1940). *Agranulocytosis.* Vitebsk, USSR: Medical Institute, pp. 3–95.

Lynn, J. R. (1964). *Fusarium* keratitis treated with cycloheximide. *Am. J. Ophthalmol.* **58**, 637–641.

Machida, Y., and Nozoe, S. (1972a). Biosynthesis of trichothecin and related compounds. *Tetrahedron* **28**, 5113–5117.

Machida, Y., and Nozoe, S. (1972b). Biosynthesis of trichothecin and related compounds. *Tetrahedron Lett.* 1969–1971.

Magee, P. N., and Barnes, J. M. (1967). Carcinogenic nitroso compounds. *Adv. Cancer Res.* **10**, 163–246.

Main, C. E., and Hamilton, P. B. (1972). Animal toxicity of phytopathogenic fungi. *Appl. Microbiol.* **23**, 193–195.

Mains, E. B., Vestal, C. M., and Curtis, F. B. (1930). Scab of small grains and feeding trouble in Indiana in 1928. *Proc. Indiana Acad. Sci.* **39**, 101–110.

Maisuradge, G. I. (1953). The role of fungi in developing toxicosis in horses by feeding them germinating oats, Abstr. Ph.D. Dissertation, Moscow, p. 1–15, Nutr. Inst. Acad. Sci.

Maisuradge, G. I. (1960). Clinical and pathomorphological data of poisoning horses with germinating oats. *Vses. Nauchn. Isl. Inst. Vet. Sanit.* **16**, 407–414.

Malaiyandi, M., Barrette, J. P., and Wavrock, P. L. (1976). Bis-Diazotized benzidine as a spray reagent for detecting zearalenone on thin layer chromoplates. *J. Assoc. Off. Anal. Chem.* **59**, 959–962.

Malkin, Z. I., and Odelevskaja, N. N. (1945). Clinical aspects and treatment of alimentary toxic aleukia (septic angina). In Alimentary Toxic Aleukia, p. 19, Kazan.

Manburg, E. M. (1944). Clinical aspects of septic angina. *Proc. 1st Kharkov Med. Inst. Chkalov Inst. Epidemiol. Microbiol.* **1**, 85–102.

Manburg, E. M., and Rachalski, E. A. (1944). Clinic and treatment of septic angina.

*Chkalov Inst. Epidemiol. Microbiol., Kharkov Med. Inst. Clin. Hospital Orenburg Railw.,* pp. 18–21.

Manburg, E. M., and Rachalski, E. A. (1947). Clinical aspects and therapy of alimentary toxic aleukia. *Acta Chkalov Inst. Epidemiol. Microbiol.* pp. 152–163, Orenburg, USSR.

Mann, D. D., Buening, G. M., Hook, B. S., and Osweiler, G. D. (1982). Effect of T-2 toxin and the bovine immune system. *Humoral Factors Infect. Immunol.* 36, 1249–1252.

Manoilova, O. S. (1947). Chemical diagnosis of toxic overwintered cereals. Lecture in Republic Conference of Alimentary Toxic Aleukia, Moscow, January.

Marasas, W. F. O., and Smalley, E. B. (1972). Mycoflora, toxicity and nutritive value of moldy maize. *Onderstepoort J. Vet. Res.* 39, 1–10.

Marasas, W. F. O., Smalley, E. B., Degurse, P. E., Bamburg, J. R., and Nichols, R. E. (1967). Acute toxicity to rainbow trout (*Salmo gairdnerii*) of a metabolite produced by the fungus *Fusarium tricinctum. Nature* 214, 817–818.

Marasas, W. F. O., Bamburg, J. R., Smalley, E. B., Strong, F. M., Regland, W. L., and Degurse, P. E. (1969). Toxic effects on trout, rats and mice of T-2 toxin produced by the fungus *Fusarium tricinctum* (Cd.) Snyd. and Hans. *Toxicol. Appl. Pharmacol.* 15, 471–482.

Marasas, W. F. O., Smalley, E. B., Bamburg, J. R., and Strong, F. M. (1971). Phytotoxicity of T-2 toxin produced by *Fusarium tricinctum. Phytopathology* 61, 1488–1491.

Marasas, W. F. O., Kellerman, T. S., Pienaar, J. G., and Maudé, T. W. (1976). Leukoencephalomalacia: A mycotoxicosis of Equidae caused by *Fusarium moniliforme* Sheldon. *Onderstepoort J. Vet. Res.* 43, 113–122.

Marasas, W. F. O., Kriek, N. P. J., van Rensburg, S. J., Steyn, M., and van Schalkwyk, G. C. (1977). Occurrence of zearalenone and deoxynivalenol, mycotoxins produced by *Fusarium graminearum* Schwabe, in maize in South Africa. *S. Afr. J. Sci.* 73, 346–349.

Marasas, W. F. O., Kriek, N. P. J., Steyn, M., van Rensburg, S. J., van Schalkwyk, D. J. (1978). Mycotoxicological investigations on Zambian maize. *Food Cosmet. Toxicol.* 16, 39–45.

Marasas, W. F. O., Kriek, N. P. J., Wiggins, V. M., Steyn, P. S., Towers, D. K., and Hastie, T. J. (1979a). Incidence, geographical distribution and toxigenicity of *Fusarium* species in South African corn. *Am. Phytopathol. Soc.* 69, 1181–1185.

Marasas, W. F. O., Leistner, L., Hofman, G., and Eckardt, C. (1979b). Occurrence of toxigenic strains of *Fusarium* in maize and barley in Germany. *Eur. J. Appl. Microbiol. Biotechnol.* 7, 289–305.

Marasas, W. F. O., van Rensburg, S. J., and Mirocha, C. J. (1979c). Incidence of *Fusarium* species and the mycotoxins deoxynivalenol and zearalenone in corn produced in the oesophageal cancer areas in Transkei. *J. Agric. Food Chem.* 27, 1108–1112.

Marasas, W. F. O., Wehmer, F. C., van Rensburg, S. J., and van Schalkwyk, D. J. (1981). Mycoflora of corn produced in human esophageal cancer in Transkei, Southern Africa. *Phytopathology,* 71, 792–794.

Marks, H. L., and Bacon, C. W. (1976). Influence of *Fusarium*-infected corn and F-2 on laying hens. *Poult. Sci.* 55, 1864–1870.

Marasas, W. F. O., Nelson, P. E., and Toussoun, T. A. (1984). Toxigenic *Fusarium* species identity and mycotoxicology. University Park: Pennsylvania State University Press.

Marchenko, G. F. (1963). Experimental fusariotoxicosis of sheep. *Vet. J.* 3, 100.

Marianashvili, M. H. (1964). Fusariotoxicosis of cattle. *Vet. J.* 4, 46.

Marshall, E. (1979). Carcinogens in Scotch. *Science* 205, 768–769.

Marshall, E. (1982a). More on Yellow Rain. *Science* 216, May 28, 1982, 967.

Marshall, E. (1982b). Yellow Rain: Filling in the gaps. *Science* **217**, July 2, 31–34.

Marshall, E. (1982c). The Soviet Elephant grass theory. *Science* **217**, July 2, 32.

Marshall, E. (1983). Yellow Rain experts battle over corn mold. *Science* **221**, August 5, 526–529.

Marth, E. H., and Calanog, B. G. (1976). Toxigenic fungi. In M. P. Defiqueiredo and D. F. Foesser, eds., *Food Microbiology: Public Health and Spoilage Aspects.* Westport, Conn.: Avi, pp. 210–256.

Martin, P. M. D., and Keen, P. (1978). The occurrence of zearalenone in raw and fermented products from Swaziland and Lesotho. *Sabouaaudia* **16**, 15–32.

Martin, R. A., and Johnston, H. W. (1982). Effects and control of *Fusarium* diseases of cereal grains in the Atlantic provinces. *Can. J. Plant Pathol.* **4**, 210–216.

Martin, P. M. D., Gilman, G. A., and Keen, P. (1971). The incidence of fungi in foodstuffs and their significance based on a survey in the eastern Transvaal and Swaziland. In I. F. Purchase, ed., *Mycotoxins in Human Health.* London: MacMillan, pp. 281–290.

Martin, P. M. D., Horwitz, K. B., Ryan, D. S., and McGuire, W. L. (1978). Phytoestrogen interaction with estrogen receptors in human breast cancer cells. *Endocrinology* **103**, 1860–1867.

Masuda, E., Takemoto, T., Tatsumo, T., and Obra, T. (1982). Immunosuppressive effect of a trichothecene Fusarenon-X in mice. *Immunology* **45**, 743–749.

Masuko, H., Ueno, Y., Otokawa, M., and Yaginuma, K. (1977). The enhancing effect of T-2 toxin on delayed hypersensitivity in mice. *Japan J. Med. Sci. Biol.* **30**, 159–163.

Mathur, S. K., Mathur, S. B., and Neergaard, P. (1975). Detection of seed-borne fungi in sorghum and location of *Fusarium moniliforme* in the seed. *Seed Sci. Technol.* **3**, 683–690.

Matsumoto, T. (1972). Keratomycosis. *Jpn. J. Med. Mycol.* **13**, 122–124.

Matsumoto, T., and Soejima, N. (1976). Keratomycosis. *Mykosen* **19**, 217–222.

Matsumoto, H., Ito, T., and Ueno, Y. (1978). Toxicological approaches to the metabolites of Fusaria. 12. Fate and distribution of T-2 toxin in mice. *Jpn. J. Exp. Med.* **48**, 393–399.

Matsumoto, M., Minato, H., Tori, K., Ueyama, M. (1977). Structures of isororidin E, epoxy-isororidin E, and epoxy- and diepoxyroridin H, new metabolites isolated from *Cylindrocarpon* species determined by Carbon-13 and hydrogen-1 NMR spectroscopy. *Tetrahedron Lett.* **47**, 4093–4096.

Matsuo, T. (1983). *Fusarium* as plant pathogens. In Y. Ueno, ed., *Trichothecenes, Chemical, Biological and Toxicological Aspects.* Amsterdam: Elsevier, pp. 83–94.

Matsuoka, Y., and Kubota, K. (1981). Studies on the mechanisms of diarrhea induced by fusarenon-X, a trichothecene mycotoxin from *Fusarium* species. *Toxicol. Appl. Pharmacol.* **57**, 293–301.

Matsuoka, Y., and Kubota, K. (1982). Studies on mechanisms of diarrhea induced by fusarenon-X, a trichothecene mycotoxin from *Fusarium* species; characteristics of increased intestinal absorption rate induced by fusarenon-X. *J. Pharmacol. Dyn.* **5**, 193–199.

Matsuoka, Y., Kubota, K., and Ueno, Y. (1979). General pharmacological studies of fusarenon-X, a trichothecene mycotoxin from *Fusarium* species. *Toxicol. Appl. Pharmacol.* **50**, 87–94.

Matthews, J. G., Patterson, D. S. P., Roberts, B. A., and Shreeve, B. J. (1977). T-2 toxin and haemorrhagic syndrome of cattle. *Vet. Rec.* **101**, 391.

Matuo, T. (1961). On the classification of Japanese *Fusaria. Ann. Phytopathol. Soc. Japan* **26**, 43–47.

Matuo, T. (1972). Taxonomic studies of phytopathogenic *Fusaria* in Japan. *Rev. Plant Prot. Res. Tokyo* **5**, 34–45.

Mayer, C. F. (1953a). Endemic panmyelotoxicosis in the Russian grain belt. 1. The clinical aspects of alimentary toxic aleukia (ATA): A comprehensive review. *Mil. Surg.* **113**, 173–189.

Mayer, C. F. (1953b). Endemic panmyelotoxicosis in the Russian grain belt. 2. The botany, phytopathology and toxicology of Russian cereal food. *Mil. Surg.* **113**, 295–315.

McCann, J., Choi, E., Yamasaki, E., and Ames, B. N. (1975). *Salmonella* microsome test: Assay of 300 chemicals. *Proc. Natl. Acad. Sci. USA* **72**, 5135–5139.

McErlean, B. A. (1952). Vulvovaginitis of swine. *Vet. Rec.* **64**, 539–540.

McLaughlin, C. S., Vaughan, M. H., Campbell, I. M., Wei, C. M., Stafford, M. E., and Hansen, B. S. (1977). Inhibition of protein synthesis by trichothecenes. In J. V. Rodricks, C. W. Hesseltine, and M. A. Mehlman, eds., *Mycotoxins in Human and Animal Health*. Park Forest, Ill., Pathotox, pp. 263–273.

McNutt, S. H., Purwin, P., and Murray, C. (1928). Vulvovaginitis in swine. *J. Am. Vet. Med. Assoc.* **73**, 484–492.

Melchers, L. R. (1956). Fungi associated with Kansas hybrid seed corn. *Plant Dis. Rep.* **40**, 500–506.

Meronuck, R. A., Garren, K. H., Christensen, C. M., Nelson, G. H., and Bates, F. (1970). Effects on turkey poults and chicks of rations containing corn invaded by *Penicillium* and *Fusarium* species. *Am. J. Vet. Res.* **31**, 551–555.

Messiaen, C. M. (1959). La systématique du genre *Fusarium selen* Snyder et Hansen. *Rev. Pathol. Vég. Entomol. Agric. Fr.* **38**, 253–266.

Messiaen, C. M., et Cassini, R. (1968). Recherches sur les fusarioses. 4. La systématique des *Fusarium*. *Ann. Epiphyt.* **19**, 387–454.

Messiaen, C. M., and Cassini, R. (1981). Taxonomy of *Fusarium*. In P. E. Nelson, T. A. Toussoun, and R. J. Cook, eds., *Fusarium Diseases, Biology and Taxonomy*. University Park: Pennsylvania State University Press, pp. 427–445.

Miller, J. D., Yong, J. C., and Trenholm, H. L. (1983). *Fusarium* toxins in field corn. 1. Time course of fungal growth and production of deoxynivalenol and other mycotoxins. *Can. J. Bot.* **61**, 3080–3087.

Miessner, H., and Schoop, G. (1929). Über den Pilzbefall amerikanischer "Giftgerste," *Dtsch. Tierärztl. Wschr.* **37**, 167–170.

Mikami, R., and Stemmerman, G. N. (1958). Keratomycosis caused by *Fusarium oxysporum. Am. J. Clin. Path.* **29**, 257–262.

Mikhailovski, S. V. (1945). The first experience in therapeutic application of Bogomolt's antireticular cytotoxic serum (ACS) in so-called septic angina. *Acta Septic Angina Ufa*, p. 94.

Milama, N. and Lelievre, H. (1979). Contribution à l'identification et au dosage de deux 12-13 epoxytrichothecenes (T-2 toxine et diacetoxyscirpenol) par chromatographie en phase gazeuse. *Analysis* **7**, 232–235.

Miller, E. E., and Smith, N. A. (1975). Moldy wheat refusal by swine. *Michigan Agr. Exp. Stn.* **289**, 90–92.

Miller, J. J. (1946). Cultural and taxonomical studies on certain *Fusaria* in culture. 1. Mutation in culture. *Can. J. Res.* **C24**, 188–211.

Miller, J. K., Hacking, A., Harrison, J., and Gross, V. J. (1973). Stillbirths, neonatal mortality and small litters in pigs associated with the ingestion of *Fusarium* toxin by pregnant cows. *Vet. Rec.* **93**, 555–559.

Mills, J. T., and Frydman, C. C. (1980). Mycoflora and condition of grains from overwintered fields in Manitoba. 1877–78. *Can. Plant Dis. Surv.* **60**, 1–7.

Minato, H., Katayama, T., and Tori, K. (1975). Vertisporin, a new antibiotic from *Verticimonosporium diffractum, Tetrahedron Letts.* **30**, 2579–2582.

Ming, Y. N., and Yu, T. F. (1966). Identification of a *Fusarium* species, isolated from a corneal ulcer. *Acta Microbiol. Sinica* **12**, 1180–1186.

Miranda, H., Fernandez, W., and Sanchez, V. (1969). Drei Fälle von Keratomykose verursacht durch *Fusarium* sp. *Mycopathol. Mycol. Appl.* **37**, 179–185.

Mirocha, C. J. (1977). Biological activity of various derivatives of zearalenone. *Acta Microbiol. Acad. Sci. Hung.* **24**, 106.

Mirocha, C. J. (1979). Trichothecene mycotoxins. In W. Shimoda, ed., *Conference on Mycotoxins in Animal Feeds and Grains Related to Animal Health* (Report No. FDA/BVM-79-139). Washington D.C.: Food and Drug Administration, pp. 289–373.

Mirocha, C. J. (1980). Pharmacological and toxicological studies on zearalenone in food-producing animals. Final report of contract 223-77-7211 submitted to the FDA, U.S. Bureau of Veterinary Medicine, Washington, D.C.

Mirocha, C. J. (1982). Hazards of scientific investigation: Analysis of samples implicated in biological warfare. *Toxicol.-Toxin Rev.* **1**, 199–203.

Mirocha, C. J. (1983). Effect of trichothecene mycotoxins on farm animals. In Y. Ueno, ed., *Trichothecenes, Chemical, Biological and Toxicological Aspects.* Amsterdam: Elsevier, pp. 177–194.

Mirocha, C. J., and Christensen, C. M. (1974a). Fungus metabolites toxic to animals. *Ann. Rev. Phytopathol.* **12**, 303–330.

Mirocha, C. J., and Christensen, C. M. (1974b). Estrogenic mycotoxins synthesized by *Fusarium.* In I. F. M. Purchase, ed., *Mycotoxins.* Amsterdam: Elsevier, pp. 129–148.

Mirocha, C. J., and Christensen, C. M. (1982). Mycotoxins. In C. M. Christensen, ed., *Storage of Cereal Grains and Their Products.* St. Paul, Minn.: American Association of Cereal Chemists, pp. 241–280.

Mirocha, C. J., and Pathre, S. V. (1973). Identification of the toxic principle in a sample of poaefusarin. *Appl. Microbiol.* **26**, 719–724.

Mirocha, C. J., and Pathre, S. V. (1979). Mycotoxins—their biosynthesis in fungi: Zearalenone biosynthesis. *J. Food Prot.* **42**, 821–824.

Mirocha, C. J., and Swanson, S. P. (1983). Regulation of perithecia production of *Fusarium roseum* by zearalenone. *J. Food Saf.* **5**, 41–53.

Mirocha, C. J., Christensen, C. M., and Nelson, G. H. (1966). An estrogenic metabolite of *Fusarium graminearum.* In R. I. Mateles and G. N. Wogan, eds., *Biochemistry of Some Foodborne Microbiol. Toxins.* Cambridge, Mass.: MIT Press, pp. 119–130.

Mirocha, C. J., Christensen, C. M., and Nelson, G. H. (1967). Estrogenic metabolite produced by *Fusarium graminearum* in stored corn. *Appl. Microbiol.* **15**, 497–503.

Mirocha, C. J., Christensen, C. M., and Nelson, G. H. (1968a). Physiologic activity of some fungal estrogens produced by *Fusarium. Cancer Res.* **28**, 2319–2332.

Mirocha, C. J., Christensen, C. M., and Nelson, G. H. (1968b). Toxic metabolites produced by fungi implicated in mycotoxicoses. *Biotech. Bioeng.* **10**, 469–482.

Mirocha, C. J., Harrison, J., Nichols, A. A., and McClintock, M. (1968c). Detection of fungal estrogen (F-2) in hay associated with infertility in dairy cattle. *Appl. Microbiol.* **16**, 797–798.

Mirocha, C. J., Christensen, C. M., and Nelson, G. H. (1969). Biosynthesis of the fungal estrogen F-2 and a naturally occurring derivative (F-3) by *Fusarium moniliforme. Appl. Microbiol.* **17**, 482–483.

Mirocha, C. J., Christensen, C. M., and Nelson, G. H. (1971). F-2 (zearalenone), estrogenic mycotoxin from *Fusarium.* In S. Kadis, A. Ciegler, and S. J. Ajl, eds., *Microbiological Toxins,* Vol. 7. New York: Academic, pp. 107–138.

Mirocha, C. J., Christensen, C. M., Davis, G., and Nelson, G. H. (1973). Detection of diethylstilbestrol in swine feedstuff. *J. Agric. Food Chem.* **31**, 135–138.

Mirocha, C. J., Schauerhamer, B., and Pathre, S. V. (1974). Isolation, detection and quantitation of zearalenone in maize and barley. *J. Assoc. Off. Anal. Chem.* **57**, 1104–1110.

Mirocha, C. J., Pathre, S. V., and Behrens, J. (1976a). Substances interfering with the gas liquid chromatographic determination of T-2 mycotoxin. *J. Assoc. Off. Anal. Chem.* **59**, 221–223.

Mirocha, C. J., Pathre, S. V., Schauerhamer, B., and Christensen, C. M. (1976b). Natural occurrence of *Fusarium* toxins in feedstuff. *Appl. Environ. Microbiol.* **32**, 553–556.

Mirocha, C. J., Pathre, S. V., and Christensen, C. M. (1977a). Zearalenone. In J. V. Rodricks, H. W. Hesseltine, and M. A. Mehlman, eds., *Mycotoxins in Human and Animal Health.* Park Forest, Ill.: Pathotox, pp. 345–364.

Mirocha, C. J., Pathre, S. V., and Christensen, C. M. (1977b). Chemistry of *Fusarium* and *Stachybotrys* mycotoxins. In T. D. Wyllie and L. G. Morehouse, eds., *Mycotoxic Fungi, Mycotoxins, Mycotoxicoses, An Encyclopedic Handbook,* Vol. 1. New York: Marcel Dekker, pp. 305–420.

Mirocha, C. J., Pathre, S. V., Behrens, J. C., and Schauerhamer, B. (1978). Uterotropic activity of *cis* and *trans* isomers of zearalenone and zearalenol. *Appl. Environ. Microbiol.* **35**, 986–987.

Mirocha, C. J., Schauerhamer, B., Christensen, C. M., and Kommedahl, T. (1979a). Zearalenone, deoxynivalenol and T-2 toxin associated with stalk rot in corn. *Appl. Environ. Microbiol.* **38**, 557–558.

Mirocha, C. J., Schauerhamer, B., Christensen, C. M., Niku-Paavola, M.-L., and Nummi, M. (1979b). Incidence of zearalenol (*Fusarium* mycotoxin) in animal feed. *Appl. Environ. Microbiol.* **38**, 749–750.

Mirocha, C. J., Pathre, S. V., and Christensen, C. M. (1980). Mycotoxins. In Y. Pomeranz, ed., *Advances in Cereal Science and Technology,* Vol. 3. St. Paul, Minn.: American Association of Cereal Chemists, pp. 159–225.

Mirocha, C. J., Pathre, S. V., and Robison, T. S. (1981). Comparative metabolism of zearalenone and transmission into bovine milk. *Food Cosmet. Toxicol.* **19**, 25–30.

Mirocha, C. J., Robison, T. S., Pawlovsky, R. J., and Allen, N. K. (1982a). Distribution and residue determination of [³H] zearalenone in broilers. *Toxicol. Appl. Pharmacol.* **66**, 77–87.

Mirocha, C. J., Watson, S., and Hayes, W. (1982b). Occurrence of trichothecenes in samples from Southeast Asia associated with Yellow Rain. *Proc. 5 Int. IUPAC Symp. Mycotoxins and Phycotoxins.* pp. 130–133, Vienna, Austria, September.

Mirocha, C. J., Pawlovsky, R. A., Chatterjee, K., Watson, S., and Hayes, W. (1983). Analysis for *Fusarium* toxins in various samples implicated in biological warfare in Southeast Asia. *J. Assoc. Off. Anal. Chem.* **66**, 1485–1499.

Mironov, S. G. (1944). To the problems of septic angina. Lecture Publ. Lab. Septic Angina, *Chkalov Inst. Epidemiol. Microbiol.,* Orenburg, USSR, pp. 4–5 (Abstr.).

Mironov, S. (1945a). Etiology of septic angina (alimentary toxic aleukia) and measures for its prevention. *M. Microbiol. Epidemiol. Immunol.* **6**, 70–77.

Mironov, S. G. (1945b). New data on the etiology of alimentary toxic aleukia (septic angina). Lecture in Republic Conference of Alimentary Toxic Aleukia, Moscow, December 24–26, pp. 3–6 (Abstr.).

Mironov, S. G., and Alisova, Z. I. (1947). Detection of toxic compounds of fungal derivation in overwintered cereals by means of immunity tests. *Acta Chkalov Inst. Epidemiol. Microbiol.* **2**, 192.

Mironov, S. G., and Davydova, V. L. (1947). Sensibility of man and animal skin to toxins of cereals overwintered in the field. *Acta Chkalov Inst. Epidemiol. Microbiol.* **2**, 89–93.

Mironov, S. G., and Fok, R. (1944). Toxicity of overwintered prosomillet in the field. *Proc. 1st Kharkov Med. Inst. Chkalov Inst. Epidemiol. Microbiol.* 1, 37–46.

Mironov, S. G., and Fok, R. A. (1947a). Skin test on rabbits for determination of toxicity of overwintered cereals. *Acta Chkalov Inst. Epidemiol. Microbiol.* 2, 80–82.

Mironov, S. G., and Fok, R. A. (1947b). Skin test on rabbits for determination of toxicity of overwintered cereals. *Acta Chkalov Inst. Epidemiol. Microbiol.* 2, 83–88.

Mironov, S. G., and Joffe, A. Z. (1947a). The dynamics of toxin accumulation in overwintered cereals in 1943–1944. *Acta Chkalov Inst. Epidemiol. Microbiol.* 2, 19–22.

Mironov, S. G. and Joffe, A. Z. (1947b). The dynamics of toxin accumulation in overwintered cereals in 1944–1945. *Proc. Chkalov Inst. Epidemiol. Microbiol.* 2, 23–27.

Mironov, S. G., and Myasnikov, V. A. (1947). Characteristics of toxins from overwintered cereals in the field. *Acta Chkalov Inst. Epidemiol. Microbiol.* 2, 61–65.

Mironov, S. G., Soboleva, R., Fok, R., and Yudenich, V. (1944). Phytopathological analysis of overwintered prosomillet samples in the field. *Proc. 1st Kharkov Med. Inst. Chkalov Inst. Epidemiol. Microbiol.* 1, 47–52.

Mironov, S. G., Joffe, A. Z., Bakbardina, M. K., Fok, R., and Davydova, V. L. (1947a). Phytopathological analysis of toxic overwintered prosomillet samples. *Acta Chkalov Inst. Epidemiol. Microbiol.* 2, 11–18.

Mironov, S. G., Strukov, A. I., and Fok, R. A. (1947b). Skin test on rabbits for determination of toxicity of overwintered cereals. *Acta Chkalov Inst. Epidemiol. Microbiol.* 2, 73–79.

Mishustin, E. N., Kretovich, V. L., and Bundel, A. A. (1946). The fermentation test as a diagnostic method for toxic overwintered prosomillet. *Hyg. Sanit.* 11, 32–35.

Misiurenko, I. P. (1972). Formation of toxin by deep cultivation of *Fusarium sporotrichiella* Bilai. *Proc. Symp. Mycotoxins, Acad. Sci. Ukr. SSR, Kiev,* October 3–9, p. 17.

Mitchell, H. H., and Beadles, J. R. (1940). The impairment in nutritive value of corn grain damaged by specific fungi. *Agric. Res.* 61, 135–142.

Mitchell, J. S., and Attleberger, M. H. (1973). *Fusarium* keratomycosis in the horse. *Vet. Med. Small Anim. Clin.* 68, 1257–1260.

Mitter, J. H. (1929). Studies in the genus *Fusarium.* 7. Saltation in the section *Discolor. Ann. Bot.* 43, 379–410.

Miyake, I. (1909). Study on the fungi growing on rice-plant in Japan. *Bot. Mag.* 8, 89.

Mizuno, S. (1975). Mechanism of inhibition of protein synthesis initiated by diacetoxyscirpenol and fusarenon-X in the reticulocyte lysate system. *Biochem. Biophys. Acta* 383, 207–214.

Moerck, K. E., McElfresh, P., Wohlman, A., and Hilton, B. W. (1980). Aflatoxin destruction in corn using sodium bisulfite, sodium hydroxide and aqueous ammonia. *J. Food Prot.* 43, 571–574.

Möller, J. M., Thalmann, A., and Hausmann, M. (1978). Über das Vorkommen von Fusarien und zearalenon in Futtermitteln. *Landwirtsch. Forschung* 31, 38–44.

Möller, T. E., and Josefsson, E. (1978). High pressure liquid chromatography of zearalenone in cereals. *J. Assoc. Off. Anal. Chem.* 61, 789–792.

Montesano, R., and Bartsch, H. (1976). Mutagenic and carcinogenic N-nitroso compounds: Possible environmental hazards. *Mutat. Res.* 32, 179–228.

Moran, E. T., Jr., Hunter, B., Ferket, P., Young, L. G., and McGirr, L. G. (1982). High tolerance of broilers to vomitoxin from corn infected with *Fusarium graminearum. Poult. Sci.* 61, 1828–1831.

Moran, J. B. (1972). The effect of zearalenol on seasonal growth rates of cattle in a dry monsoonal environment. *Austrian J. Exp. Agric. Anim. Husb.* 12, 345–347.

Moreau, C. (1972). Mycotoxicose chez des vaches laitières liée au developpement du

*Fusarium roseum* var. *culmorum* sur l'herbe d'une prairie. *Comp. Rend. Acad. Agric. Fr.* **58**, 383–387.

Moreau, C. (1974a). *Moisissures Toxigènes dans l'Alimentation.* Paris: Ed. Masson.

Moreau, C. (1974b). Trois cas de paralysie chez des porcs et des volailles vraisemblablement liés a une action toxique de *Fusarium moniliforme* Sheld. *Bull. Soc. Mycol. Fr.* **90**, 201–208.

Moreau, C. (1979). Troubles nerveus et digestifs liés à la consommation, par les animaux, d'aliments contaminés par des *Aspergillus, Pencillium* et *Fusarium. Rev. Mycol.* **43**, 227–238.

Morel-Chany, E., Burtin, P., Trineal, G., and Frayssinet, C. (1980). Cytostatic effects of T-2 toxin on cultured human neoplastic cells of intestinal origin. *Bull. Cancer* **67**, 149–154.

Morimoto, S. (1936). The strange disease called bean-hulls poisoning. *Bull. Appl. Vet.* **9**, 139–141 (in Japanese).

Morin, Y. L., and Daniel, P. (1967). Quebec beer-drinkers cardiomyopathy. Etiological considerations. *Can. Med. Assoc. J.* **97**, 926–928.

Morooka, N., and Tatsuno, T. (1970). Toxic substances (fusarenon and nivalenol) produced by *Fusarium nivale. Proc. 1st U.S.-Japan Conf. Toxic Microorganisms, Washington, D.C.*, pp. 114–119.

Morooka, N., Nakano, N., Nakazawa, S., and Tsunoda, H. (1971). On the chemical properties of Fusarenon and related compounds obtained from toxic metabolites of *Fusarium nivale. Nippon Nogeikagaku Kaishi* **45**, 151–155.

Morooka, N., Uratsuchi, N., Yoshizawa, T., and Yamamoto, H. (1972). Studies on the substances in barley infected with *Fusarium. J. Food Hyg. Soc. Japan* **13**, 368–375.

Mortimer, P. H. (1983). Ryegrass staggers: Clinical, pathological and aetiological aspects. *Proc. N.Z. Assoc.* **44**, 230–233.

Muchkin, N. I. (1967). Urov disease or Kashin-Beck disease in animals. *Veterinariya* **3**, 74–75.

Muchkin, N. I. (1968). Etiology of Urov disease. *Vest. Sel'khoz* **3**, 138–141.

Muller, B., Achini, R., and Tamm, C. (1975). Biosynthesis der verrucarine und roridine. 3. Der Einbau von (3R)-[5-$^{14}$C]-[2-$^{14}$C] und an C (2) stereospezifisch tritiertem mevalonat in verrucarol. *Helv. Chim. Acta* **58**, 471–482.

Muller, B., and Tamm, C. (1975). Biosynthesis of the verrucarins and roridins. 4. The mode of incorporation of (3R)-[(5R)-5-$^3$H]-mevalonate into verrucarol. Evidence for the identity of the C (II)- hydrogen atom of the trichothecene skeleton with the (5R)- hydrogen atom of (3R)- mevalonic acid. *Helv. Chim. Acta* **58**, 483–488.

Müller, H. M. (1978). Zearalenon-ein östrogenwirksamer Mykotoxin. *Übers. Tierernahr.* **6**, 265–300.

Mundkur, B. B. (1934). Some preliminary feeding experiments with scabbed barley. *Phytopathology* **24**, 1237–1243.

Mundkur, B. B., and Cochran, B. L. (1930). Some feeding tests with scabbey barley. *Phytopathology* **20**, 132.

Murashinskij, K. E. (1934). On the study of fusariosis of cereal crops. Species of genus *Fusarium* on cereal crops in Siberia. *Proc. Siber. Agric. Acad. Omsk.* **3**, 87–114.

Murphy, W. K., Livingston, R. B., Gottlieb, J. A., Burgess, M. A. and Rawson, R. W. (1976). Phase I evaluation of anguidine. *Proc. Am. Assoc. Cancer Res.* **17**, 90 (Abstr. 358).

Mutton, K. J., Lucas, T. J., and Harknes, J. L. (1980). Disseminated *Fusarium* infection. *Med. J. Aust.* **2**, 624–625.

Myasnikov, A. L. (1935). Clinical aspects of alimentary hemorrhagic aleukia. In *Alimentary Hemorrhagic Aleukia (Septic Angina).* Novosibirsk: *West. Sib. Territ. Publ. Health Serv.*

Myasnikov, V. A. (1947). Experiments in destruction and neutralization of toxic material from overwintered prosomillet. In Alimentary Toxic Aleukia. *Acta Chkalov Inst. Epidemiol. Microbiol.* pp. 55–57, Chkalov.

Myasnikov, V. A. (1948). Characteristics of extracts of various fungi treated by the barium method, compared to extracts of prosomillet exposed in winter. *Chkalov Inst. Epidemiol. Microbiol.* (MS).

Nagao, M., Honda, M., Hamasaki, T., Natori, S., Ueno, Y., Yamasaki, M., Seino, Y., Yahagi, T., and Sugimuri, T. (1976). Mutagenicities of mycotoxins on *Salmonella. Proc. Japan Assoc. Mycotoxicol.* **3, 4,** 41–43.

Naik, D. M., Busch, L. V., and Barron, G. L. (1979). Influence of temperature on the strain of *Fusarium graminearum* Schwabe in zearalenone production. *Can. J. Plant Sci.* **58,** 1095–1098.

Nakamura, Y., Takeda, S., and Ogasawara, K. (1951b). Studies on the poisoning of the damaged wheat with the scab. 2. On toxicity test. *Hokkaido Eisei Kenkyujoho* **2,** 47–50.

Nakamura, Y., Takeda, S., Ogasawara, K., Karashimada, T., and Ando, K. (1951a). Studies on the food poisoning of the damaged wheat with the scab. 1. Growth conditions of the fungus. *Hokkaido Eisei Kenkyujoho* **2,** 35–46 (in Japanese).

Nakamura, Y., Ohta, M., and Neno, Y. (1977). Reactivity of 12,13-epoxytrichothecenes with epoxy hydrolase, glutathione-S-transferase and glutathione. *Chem. Pharmacol. Bull.* **25,** 3410–3414.

Nakano, N. (1968). Inhibitory effects of fusarenon on multiplication of *Tetrahymena pyriformis. Jpn. J. Med. Sci. Biol.* **21,** 351–354.

Nakano, N., Kunimoto, T., and Aibara, K. (1973). Chemical and biological assays of fusarenon-X and diacetoxyscirpenol in cereal grains. *J. Food Hyg. Soc.* **14,** 56–64.

Nakhapetov, M. I. (1944). *Septic Angina.* Leningrad: Medgiz, p. 14.

Naoi, Y. (1983). Clean-up procedures and GLC analysis. In Y. Ueno, ed., *Trichothecenes, Chemical, Biological and Toxicological Aspects.* Amsterdam: Elsevier pp. 121–124.

Naoi, Y., Kazama, E., Saito, K., Ogawa, H., Shimura, K., and Kimura, Y. (1974). Studies on mycotoxins in food. Analytical procedure for T-2 toxin and diacetoxyscirpenol in powdered wheat. *Tokyo Toritsu Eisei Kenkyusho Kenkyu Nempo* **25,** 203–206.

Naumann, G., Green, W. R., and Zimmerman, L. E. (1967). Mycotic keratitis, a histopathologic study of 73 cases. *Am. J. Ophthalmol.* **64,** 668–682.

Naumov, N. A. (1916). Intoxicating bread. *Trudy Biuro Mikol. Fitopat.* **2,** 1–216.

Neish, G. A., Farnworth, E. R., and Cohen, H. (1982). Zearalenone and trichothecene production by some *Fusarium* species associated with Canadian grain. *Can. J. Plant Pathol.* **4,** 191–194.

Neish, G. A., Farnworth, E. R., Greenhalgh, R., and Young, J. C. (1983). Observations on the occurrence of the *Fusarium* species and their toxins in corn in Eastern Ontario. *Can. J. Plant Pathol.* **5,** 11–16.

Nelson, R. R. (1971). Hormonal involvement in sexual reproduction in the fungi with special reference to F-2, a fungal estrogen. In S. Akai and S. Ouchi, eds., *Morphological and Biochemical Events in Plant-Parasite Interaction.* Tokyo, Japan: Phytopathological Society, pp. 181–205.

Nelson, G. H., Christensen, C. M., and Mirocha, C. J. (1966a). Feeds, fungi and animal health: Three Minnesota researchers report the effects of toxins in feedstuffs. *Minnesota Sci.* **23,** 12–13.

Nelson, G. H., Christensen, C. M., and Mirocha, C. J. (1966b). Work on mycotoxicoses at the University of Minnesota. *Proc. 64th Annu. Meet. U.S. Livest. Sanit. Assoc.,* pp. 614–618.

Nelson, G. H., Christensen, C. M., and Mirocha, C. J. (1973). *Fusarium* and estrogenism in swine. *J. Am. Vet. Med. Assoc.* **11**, 1276–1277.

Nelson, L. A., Perry, T. W., Stob, M., and Huber, D. A. (1972). Effect of DES and RAL on reproduction of heifers. *J. Anim. Sci.* **35**, 250 (Abstr.).

Nelson, R. R., and Osborne, J. C. (1956). The relative prevalence and geographic distribution of fungi associated with moldy corn in eastern North Carolina in 1955. *Plant Dis. Rep.* **40**, 225–227.

Nelson, P. E., Toussoun, T. A., and Marasas, W. F. O. (1983). *Fusarium Species: An Illustrated Manual for Identification.* University Park: Pennsylvania State University Press.

Nesterov, A. I. (1945). Alimentary Toxic Aleukia. Megiz, Moscow, pp. 3–54.

Nesterov, A. I. (1964). The clinical course of Kashin-Beck disease. *Arthritis Rheum.* **7**, 29–40.

Nesterov, V. S. (1948). The clinical aspects of septic angina. *Clin. Med.* **7**, 34.

Newberne, P. M. (1974b). The new world of mycotoxins—Animal and human health. *Clin. Toxicol.* **7**, 161–177.

Newberne, P. M. (1974a). Mycotoxins: Toxicity, carcinogenicity and the influence of various nutritional conditions. *Environ. Health Perspect.* **9**, 1–32.

Newmark, E., Ellison, A. C., and Kaufman, H. E. (1970). Primaricin therapy of *Cephalosporium* and *Fusarium* keratitis. *Am. J. Ophthalmol.* **69**, 458–466.

Niku-Paavola, M.-L., and Nummi, M. (1977). Research into toxic metabolites of fungi. *Kemia-Kemi* **4**, 151–153 (in Finnish).

Niku-Paavola, M.-L., Ilus, T., Ward, P. J., and Enati, T. M. (1976). Quantitative analysis of *Fusarium* toxins. *Proc. 3rd Int. IUPAC Symp. Mycotoxins Foodst.*, September 16–18, Paris (Abstr.).

Niku-Paavola, M. L., Illus, T., Ward, P. J., and Nummi, M. (1977). Thin layer analysis of *Fusarium* toxins in grain samples. *Arch. Inst. Pasteur Tunis* **3**, 4, 264–278.

Nirenberg, H. (1976). Untersuchungen über die morphologische und biologische Differenzierung in der *Fusarium*—Section *Liseola*. *Mitt. Biol. Bundesanst.* **169**.

Nirenberg, H. (1981). A simplified method for identifying *Fusarium* spp. occurring in wheat. *Can. J. Bot.* **59**, 1599–1609.

Nishikado, Y. (1957a). Studies on the wheat scab, caused by *Gibberella zeae* (Schw.) Petch and its control. *Nogaku Kenkyu* **45**, 59–86.

Nishikado, Y. (1957b). Studies on the control of *Fusarium* scab disease of wheat. 2. *Agric. Inst. Okayama Univ.*, **45**, 141–158.

Nishikado, Y. (1957c). Studies on the control of *Fusarium* scab disease of wheat. 3. *Agric. Inst. Okayama Univ.*, **45**, 159–220.

Nishikado, Y. (1958). Studies on the control of *Fusarium* scab disease of wheat. *Data on Agricultural New Technics. Agr. Inst. Okayama.* **45**, 159–220.

Nishikado, Y. (1958). Studies on the control of *Fusarium* scab disease of wheat. *Data on Agricultural New Technics.* **97**, 59–220 (in Japanese).

Noskov, A. I., and Kurmanov, I. A. (1967). Fusariotoxicosis in sheep. *Veterinariya* **43**, 65.

Nowak, T. (1973). Anatomical and pathological changes in carp fed on maize infected with fungi of the genus *Fusarium*. *Med. Vet.* **29**, 420–422 (in Polish).

Nummi, M. (1977). Recent studies on *Fusarium* toxins, *Zesz. Probl. Postępów Nauk Roln.* **189**, 189–192.

Nummi, M., Niku-Paavola, M.-L., and Enari, T.-M. (1975). Der Einfluss eines *Fusarium*-Toxins auf die Gersten-Vermälzung. *Bruwissenschaft* **28**, 130–134.

Ogasawara, K. (1965). On the food poisoning of wheat and barley caused by *Fusarium* scab disease. *J. Food Hyg. Soc. Jap.* **6**, 81–82 (in Japanese).

Ogurcov, A. F. (1960). A case of fusariotoxicosis of cattle. *Vet. J.* **9**, 70.

Ohira, I. (1938). Clinical investigation on bean-hulls poisoning of horses. *Bull. Appl. Vet.* **11**, 433–437 (in Japanese).

Ohta, M., Ishii, T., and Ueno, Y. (1977). Metabolism of trichothecene mycotoxins. 1. Microsomal deacetylation of T-2 toxin in animal tissues. *J. Biochem.* **82**, 1591–1598.

Ohta, M., Matsumoto, H., Ishii, K., and Ueno, Y. (1978). Metabolism of trichothecene mycotoxins. 2. Substrate specificity of microsomal deacetylation of trichothecenes. *J. Biochem.* **84**, 697–706.

Ohtsubo, K. (1973). Mycotoxins in foodstuffs and their biological action on mammals. *Beitr. Path. Bd.* **148**, 218–219.

Ohtsubo, K. (1983). Chronic toxicity of trichothecenes. In Y. Ueno, ed., *Trichothecenes, Chemical, Biological and Toxicological Aspects.* Amsterdam: Elsevier, pp. 171–176.

Ohtsubo, K., and Saito, M. (1970). Cytotoxic effect of scirpene compounds fusarenon-X, produced by *Fusarium nivale*, dihydronivalenol and dihydrofusarenon-X on HeLa cells. *Jpn. J. Med. Sci. Biol.* **23**, 217–225.

Ohtsubo, K., and Saito, M. (1977). Chronic effects of trichothecene toxins. In J. V. Rodricks, H. W. Hesseltine, and M. A. Mehlman, eds., *Mycotoxins in Human and Animal Health.* Park Forest, Ill.: Pathotox, pp. 255–262.

Ohtsubo, K., Yamada, M.-A., and Saito, M. (1968). Inhibitory effect of nivalenol, a toxic metabolite of *Fusarium nivale*, on the growth cycle and biopolymer synthesis of HeLa cells. *Jpn. J. Med. Sci. Biol.* **21**, 185–194.

Ohtsubo, K., Kaden, P., and Mittermayer, C. (1972). Polyribosomal breakdown in mouse fibroblasts (L-cells) fusarenon-X, a toxic principle isolated from *Fusarium nivale*. *Biochim. Biophys. Acta* **287**, 520–525.

Okubo, K., Isoda, M., Senboku, T., Satta, K., and Yukisada, S. (1966a). Pathological studies on the poisoning by *Fusarium nivale*. I. Pathological findings of acute poisoning of guinea pigs by peroral administration of crude sample. *Japan J. Vet. Sci.* **28**, 381.

Okubo, K., Isoda, M., Senboku, T., Kawamura, H., Kuniya, H., and Kobayashi, T. (1966b). Studies on the pathology of poisoning of *Fusarium nivale*. 2. Experimental acute poisoning of mice. *Japan J. Vet. Sci.* **28**, 489.

Okubo, K., Isoda, M., Nakamura, T., Morooka, S., and Nakano, N. (1969). Studies on the essential nature of the red scab, *Fusarium nivale*. 2. Histological changes in mice administered intraperitoneally with toxic chemical substances isolated during the fractionation process of "Fusarenon," the metabolite of *Fusarium nivale*. *Bull. Nippon Vet. Zootech. Coll.* **18**, 1–14.

Okuniev, N. V. (1948). New data on the chemical nature of the initial toxicity of cereal grains causing alimentary toxic aleukia. Lecture in Republic Conference of Alimentary Toxic Aleukia, Moscow p. 16 (Abstr.).

Oldham, J. W., Allred, L. E., Milo, G. E., Kindig, O., and Capen, C. C. (1980). The toxicological evaluation of the mycotoxins T-2 and T-2 tetraol using normal human fibroblasts in vitro. *Toxicol. Appl. Pharmacol.* **52**, 159.

Olifson, L. E. (1955a). Chemical nature of water soluble toxic substances from overwintered millet. *Proc. Orenburg State Med. Inst.* **4**, 79–82.

Olifson, L. E. (1955b). New chemical methods of determining toxicity of cereal crops. *Publ. Acad. Sci. USSR*, Moscow, 58–59.

Olifson, L. E. (1956a). The influence of physico-chemical factors on process accumulation of toxic substances in prosomillet. *D.J. Mendeleyev Chem. Soc.* **6**, 67–73.

Olifson, L. E. (1956b). Methods of harmless grains infected by fungi of *Fusarium sporo-trichiella*. *Proc. 3rd Conf. Chem. Agric. Orenburg*, 76–82.

Olifson, L. E. (1956c). The chemical activity of *Fusarium sporotrichiella*. *Abstr. Mycotoxicoses Human Agric. Anim.* Kiev: Publ. Acad. Sci. 21–22.

Olifson, L. E. (1957a). Toxic substances isolated from overwintered cereals and their chemical nature. *D. J. Mendeleyev Chem. Soc.* **7**, 37–46.

Olifson, L. E. (1957b). Chemical action of some fungi on overwintered cereals. *D. J. Mendeleyev Chem. Soc.* **7**, 21–35.

Olifson, L. E. (1960). The chemical activity of some species of fungi which affect the grain of cereals. In V. I. Bilai, ed., *Mycotoxicoses of Man and Agricultural Animals* (English Trans.). Washington, D.C.: Joint Publishers Research Service, pp. 58–66.

Olifson, L. E. (1962). Spectrographic investigations of biologically active compounds from *Fusarium sporotrichioides*. *Ann. Rep. D. J. Mendeleyev Chem. Soc.* **7**, 109.

Olifson, L. E. (1965a). Chemical methods for determination of the toxicity of grains and feeds infected by fungi of *Fusarium sporotrichiella*. Lecture 9, Mendeleyev Conference of General Applied Chemistry, Section on Analytical Chemistry, Moscow, pp. 51–53.

Olifson, L. E. (1965b). Chemical and biological properties of toxic materials derived from grain infected with the fungus *Fusarium sporotrichiella*. Ph.D. dissertation, *Moscow Tec. Inst. Ind. Nutr.* (Abstr.).

Olifson, L. E. (1972). The problems of toxic steroids in microscopic fungi of *Fusarium sporotrichiella*. *Proc. Symp. Mycotoxins Acad. Sci. Ukr. SSR*, Kiev, October, 12.

Olifson, L. E., and Joffe, A. Z. (1954). On the changes in some chemical constants of millet oil under the influence of fungi developing on millet. *D. J. Mendeleyev Chem. Soc.* **5**, 61–65.

Olifson, L. E., Joffe, A. Z., and Drabkin, B. S. (1949). Chemical method for determining of toxicity of millet overwintered under snow. *D. J. Mendeleyev Chem. Soc.* **3**, 133–137.

Olifson, L. E., Drabkin, B. S., and Joffe, A. Z. (1950). The influence of *Fusarium* fungi on millet oil. *Acta Chkalov Med. Inst.* **2**, 103–107.

Olifson, L. E., Kenina, S. M., Zhilin, A. N., and Perepelitsina, P. N. (1969). Biological methods for determination of toxicity of grain infected with toxigenic fungi. *Vopr. Pitan.* **28**, 58–61.

Olifson, L. E., Kenina, S. M., and Kartashova, V. L. (1972). Chromatographic method to identify toxicity of grain of cereals (wheat, rye, millet and others) affected by a toxigenic strain of *Fusarium sporotrichiella*. Instruction on how to identify the toxicity of the grain. *D. J. Mendeleyev Chem. Soc.* p. 15, 3–8.

Olifson, L. E., Kenina, S. M., Kartashova, V. L., and Galkovich, K. G. (1975). Chromatographic determination of the toxicity of grain infected with *Fusarium sporotrichiella*. *Vopr. Pitan.* **2**, 83–86.

Olifson, L. E., Kenina, S. M., and Kartashova, V. L. (1978). Study of the fractional composition of the lipid complex of groats infected by the microscopic fungus *Fusarium sporotrichiella* Bilai. *Prikl. Biokhim. Mikrobiol.* **14**, 630–634.

Olsen, M., Pettersson, H., and Kiessling, K-H (1981). Reduction of zearalenone to zearalenol in female rat liver by 3α-hydroxysteroid dehydrogenase. *Acta Pharmacol. Toxicol.* **48**, 157–161.

Onegov, A. P., and Naumov, V. A. (1943). Toxicity of overwintered cereals. *Proc. Kirov, Zootekh. Vet. Inst.* **5**, 110–119.

Oparin, S. V. (1939). The phosphorus-calcium metabolism in patients with Urov diseases. *Med. Bull.* **7**, 2.

Osweiler, G. D. (1982). Wheat scab toxins. A threat to Missouri livestock. *Vet. Med. Rev.*, July/August, 2–3. Published by College of Veterinary Medicine, Univ. of Missouri, Columbia.

Otokawa, M. (1983). Immunological disorders. In Y. Ueno, ed., *Trichothecenes, Chemical, Biological and Toxicological Aspects.* Amsterdam: Elsevier pp. 163–170.

Otokawa, M., Shibahara, Y., and Egashira, Y. (1979). The inhibitory effect of T-2 toxin on tolerance induction of delayed-type hypersensitivity in mice. *Japan J. Med. Sci. Biol.* **32**, 37–45.

Oya, T., and Morimoto, S. (1942). On the bean-hulls poisoning in Tokachi-Schimizu district of Hokkaido. *Bull. Appl. Vet.* **15**, 21–32.

Ožegović, L. (1970). Moldy maize poisoning in swine (F-2 zearalenone-*Fusarium* toxicosis). *Veterinariya* **19**, 525–531, Sarajevo.

Ožegović, L., and Vuković, V. (1972). Zearalenone-(F-2)-*Fusarium* Toxicose der Schweine in Jugoslawien. *Mykosen* **15**, 171–174.

Padwick, G. W. (1940). The genus *Fusarium.* 5. *Fusarium udum* Butler, *F. vasinfectum* Atk., and *F. lateritium* Nees var. *uncinatum* Wr. *Indian J. Agric. Sci.* **10**, 863–878.

Paita, C. (1962). Alcuni casi di vulvovaginite comparsi in scrofette hel friceli. *Vet. Ital.* **4**, 195–198.

Palchevski, N. A. (1891). *Diseases of Cultivated Cereals in South Ussuri District.* Leningrad SPB. pp. 1–79.

Palti, J. (1978). Toxigenic fusaria, their distribution and significance as causes of disease in animals and man. *Acta Phytomedica.* Berlin: Paul Parey.

Palti, J., and Joffe, A. Z. (1971). Causes of *Fusarium* wilts of cucurbits in Israel and conditions favouring their development. *Phytopathol. Zeitsch.* **70**, 31–42.

Palyusik, M. (1973). Experimental vulvooedema of the porcine vagina. Fusariotoxicosis caused by *Fusarium graminearum. Magy. Allatorv. Lapja* **28**, 297–303.

Palyusik, M. (1977a). Effect of zearalenone *Fusarium* toxin on the prostate gland of swine. *Acta Microbiol. Acad. Sci. Hung.* **24**, 104 (Abstr.).

Palyusik, M. (1977b). Experimental fusariotoxicosis of swine and geese. In *Mycotoxins in Food,* Warsaw: pp. 209–212, Proc. 2nd Intern. Symposium IUPAC a. PAN.

Palyusik, M. (1978). Die Bedeutung der Mykotoxine für den Gesundheitzustand der Haustiere. *Veröff. Landwirtsch. Chem.* **11**, 11–32. Linz: Bundesversuchsanstalt.

Palyusik, M., and Koplik-Kovács, E. (1975). Effect on laying geese of feeds containing the fusariotoxins T-2 and F-2, *Acta Vet. Acad. Sci. Hung.* **25**, 363–368.

Palyusik, M., Szép, I., and Szoke, F. (1968). Data on susceptibility to mycotoxins of day-old goslings. *Acta Vet. Acad. Sci. Hung.* **18**, 363–372.

Palyusik, M., Koplik-Kovács, E., and Guszal, E. (1971). Effect of *Fusarium graminearum* on the semen production in ganders and turkey cocks. *Magy. Allatorv. Lapja* **26**, 300–303.

Palyusik, M., Koplik-Kovacs, E., and Guzal, E. (1973). Die Wirkung von *Fusarium graminearum* auf die Fruchtbarkeit der Gänseriche. *Dermatol. Monatsschr.* **159**, 384–385.

Palyusik, M., Nagy, G., and Zoldag, L. (1974). Wirkung von verschiedenen *Fusarium*—Spiezies auf die Spermiogenesis. *Magy. Allatorv. Lapja* **29**, 551–553.

Pareles, S. R., Collins, G. J., and Rosen, J. D. (1976). Analysis of T-2 toxin (and HT-2 toxin) by mass fragmentography. *J. Agric. Food Chem.* **24**, 872–875.

Parker, J. C., and Klintworth, G. K. (1971). Miscellaneous uncommon diseases attributed to fungi and actinomycetes. In R. D. Baker, ed., *The Pathologic Anatomy of Mycoses.* Springer: pp. 953–1012, New York.

Pathre, S. V., and Mirocha, C. J. (1976). Zearalenone and related compounds. In J. V. Rodricks, ed., *Mycotoxins and Other Fungal Related Food Problems* (Advances in Chemistry Series No. 149). Washington, D.C.: American Chemical Society, pp. 178–227.

Pathre, S. V., and Mirocha, C. J. (1977). Assay methods for trichothecenes and review of their natural occurrence. In J. V. Rodricks, H. W. Hesseltine, and M. A. Mehlman, eds., *Mycotoxins in Human and Animal Health*. Park Forest, Ill.: Pathotox, pp. 229–253.

Pathre, S. V., and Mirocha, C. J. (1978). Analysis of deoxynivalenol from cultures of *Fusarium* species. *Appl. Environ. Microbiol.* **35**, 992–994.

Pathre, S. V., and Mirocha, C. J. (1979). Trichothecenes: Natural occurrence and potential hazard. *Amer. Oil Chem. Soc.*, **56**, 820–823.

Pathre, S. V., Behrens, J., and Mirocha, C. J. (1974). A new mycotoxin produced by *Fusarium roseum* Gibbosum. *168th Natl. Meet. Am. Chem. Soc.* Paper No. 56. Atlantic City, N.J. (Abstr.).

Pathre, S. V., Mirocha, C. J., Christensen, C. M., and Behrens, J. (1976). Monoacetoxyscirpenol: A new mycotoxin produced by *Fusarium roseum* "Gibbosum." *J. Agric. Food Chem.* **24**, 97–103.

Pathre, S. V., Mirocha, C. J., Fenton, S. W., and Riddle, R. M. (1978). Biosynthesis of zearalenone: Incorporation of I-[$^{13}$C] and I,2-[$^{13}$C] acetate. *Proc. Am. Chem. Soc.*, September 11–15, Miami, Fla.

Pathre, S. V., Mirocha, C. J., and Fenton, S. W. (1979). Criteria of purity for zearalenone standards. *J. Assoc. Off. Anal. Chem.* **62**, 1268–1273.

Pathre, S. V., Fenton, S. W., and Mirocha, C. J. (1980). 3'-Hydroxyzearalenones, two metabolites produced by *Fusarium roseum*. *Agric. Food Chem.* **28**, 421–424.

Patterson, D. S. P. (1973). Mycotoxins: Metabolism and liver injury. *Biochem. Soc. Trans.* **11**, 917–922.

Patterson, D. S. P. (1978). Mycotoxins in animal feeds. *J. Sci. Food Agric.* **29**, 78.

Patterson, D. S. P., Shreeve, B. J., and Roberts, B. A. (1980). Mycotoxin residues in body fluids and tissues of food-producing animals. *Zbl. Bact. Parasit. Infek. Hyg.* **1980**, 321–328.

Patterson, D. S. P. (1983). Trichothecenes: Toxicoses and natural occurrence in Britain. In Y. Ueno, ed., *Trichothecenes, Chemical, Biological and Toxicological Aspects*. Amsterdam: Elsevier pp. 259–264.

Patterson, D. S. P., and Roberts, B. A. (1979). Mycotoxins in animal feedstuffs: Sensitive thin-layer chromatographic detection of aflatoxin, ochratoxin A, sterigmatocystin, zearalenone and T-2 toxin. *J. Assoc. Off. Anal. Chem.* **62**, 1265–1267.

Patterson, D. S. P., Roberts, B. A., Shreeve, B. J., Wrathall, A. E., and Gitter, M. (1977). Aflatoxin, ochratoxin and zearalenone in animal feedstuffs: Some clinical and experimental observations. *Ann. Nutr. Aliment.* **31**, 643–650.

Patterson, D. S. P., Mathews, J. G., Shreve, B. J., Roberts, B. A., McDonald, S. M., and Hayes, A. W. (1979). The failure of trichothecene mycotoxins and whole cultures of *Fusarium tricinctum* to cause experimental hemorrhagic syndromes in calves and pigs. *Vet. Rec.* **105**, 252–255.

Pearson, A. W. (1978). Biochemical changes produced by *Fusarium* T-2 toxin in the chicken. *Res. Vet. Sci.* **24**, 92–97.

Peckham, J. C., Mitchell, F. E., Jones, O. H., and Doupnik, B. (1972). Atypical interstitial pneumonia in cattle fed moldy sweet potatoes. *J. Am. Vet. Med. Assoc.* **160**, 169–172.

Peguri, A., Allen, N. K., and Mirocha, C. J. (1982). Effect of *Fusarium* cultures, T-2 toxin and zearalenone on reproduction of turkey males. *Poult. Sci.* **61**, 1524–1525. (Abstr.).

Pentman, I. S. (1935). Pathologic anatomy of alimentary hemorrhagic aleukia (septic angina). *West-Siberian Reg. Health,* 17–29, Novosibirsk.

Pepeljnjak, S. (1983). Trichothecene problems in Yugoslavia. In Y. Ueno, ed., *Trichothecenes, Chemical, Biological and Toxicological Aspects.* Amsterdam: Elsevier pp. 265–272.

Peregud, G. M. (1947). Clinical aspects and treatment of oral cavity and upper respiratory tract in cases of alimentary toxic aleukia. *Acta Chkalov Inst. Epidemiol. Microbiol.,* pp. 170–175.

Perez, M. (1966). *Fusarium nivale* as a case of corneal mycosis. *Klin. Oczna* **36,** 609–612.

Perkel, N. V. (1957). Toxicity of some forms of the fungus *Fusarium sporotrichiella* isolated from the grain of Eastern Siberia. *Vopr. Pitar.* **16,** 64–69.

Perkel, N. V. (1960). Study of the toxicity of Transbaikal strains of *Fusarium sporotrichiella* Bilai in connection with the etiology of Urov disease (Kashin-Beck disease). In V. I. Bilai, ed., *Mycotoxicoses of Man and Agricultural Animals.* (English Trans.), Washington, D.C.: Joint Publishers Research Service, pp. 117–125.

Perlman, D. P., and Peruzzotti, G. P. (1970). Microbial metabolites as potentially useful pharmacologically active agents. In D. P. Perlman, ed., *Advances in Applied Microbiology.* New York: Academic. pp. 277–294.

Perry, T. W., Stob, M., Huber, D. A., and Peterson, R. C. (1970). Effect of subcutaneous implantation of resorcylic acid lactone on performance of growing and finishing beef cattle. *J. Anim. Sci.* **31,** 789–793.

Perstneva, A. P. (1949). On some peculiarities of the clinical aspects of alimentary hemorrhagic aleukia (septic angina). *Klin. Med.* **27,** 85–89.

Pestka, J. J., Lee, S. C., Lau, H. P., and Chu, F. S. (1981). Enzyme-linked immunosorbent assay for T-2 toxin. *J. Am. Oil. Chem. Soc.* **58,** 940A–944A.

Peters, C. A. (1972). Photochemistry of zearalenone and its derivatives. *J. Med. Chem.* **15,** 867–868.

Peterson, J. E., and Baker, T. J. (1959). An isolate of *Fusarium roseum* from human burns. *Mycologia* **51,** 453–456.

Petrie, L., Robb, L. J., and Stewart, A. F. (1977). The identification of T-2 toxin and its association with hemorrhagic syndrome in cattle. *Vet. Rec.* **101,** 326.

Petrov, V. E., and Simonov, E. N. (1953). Toxicity of *Fusarium poae* and its effect on the blood of frogs. *Acta Andreyev Inst. Agr. Sci., Orenburg* **5,** 48–52.

Pidoplichko, N. M. (1953). *The Mycoflora of Rough Fodder.* Kiev: Publ. Acad. Sci. Ukr. SSR.

Pidoplichko, N. M., and Bilai, V. I. (1946). Toxic fungi on cereal grains. Kiev: Publ. Acad. Sci. Ukr. SSR.

Pidoplichko, N. M., and Bilai, V. I. (1960). Toxic fungi which develop in food products and fodder. In V. I. Bilai, ed., *Mycotoxicoses of Man and Agricultural Animals.* (English Trans.) Washington, D.C.: Joint Publishers Research Service, pp. 3–36.

Pienaar, J. G., Kellerman, T. S., and Marasas, W. F. O. (1981). Field outbreaks of leukoencephalomalacia in horses consuming maize infected by *Fusarium verticillioides* in South Africa. *J. S. Afr. Vet. Assoc.* **52,** 21–24.

Pier, A. C., Cysewski, S. J., Richard, J. L., Baetz, A. L., and Mitchell, L. (1976). Experimental mycotoxicosis in calves with aflatoxin, ochratoxin, rubratoxin and T-2 toxin. *Proc. U.S. Anim. Health Assoc.* **80,** 130–148.

Polack, F. G., Kaufman, E., and Newmark, E. (1971). Keratomycosis: Medical and surgical management. *Arch. Ophthalmol.* **85,** 410–416.

Poliantseva, A. I. (1945). Residual changes in the internal organs and blood in convalescents from alimentary toxic aleukia caused by ingestion of overwintered millet and rye. *Acta Septic Angina, Ufa,* 154–158.

Pomasski, A. (1915). Regarding the changes in chemical composition of rye resulting from the activity of certain *Fusarium* forms. *Mat. Mikol. Fitopatol. Ross.* **1,** 77–106, Leningrad.

Pomeranz, Y. (1964). Formation of toxic compounds in storage-damaged foods and feedstuffs. *Cereal Sci. Today* **9,** 93–96.

Popova, A. A. (1948). Alimentary toxic aleukia in some counties of Kirovsk district in 1942–1946. *Acta Kirovsk Inst. Epidemiol. Microbiol.* **2,** 77–80.

Popp, M. (1930). Untersuchungen über die amerikanische Giftgerste. *Chem. Ztschr.* **54,** 715.

Popp, M., and Contzen, J. (1930). Investigation of poisoned barley from the United States. *Tierernähr.* **2,** 315–355.

Porter, J. K., and Russell, J. R. (1983). Chemical constituents of tall fescue associated with toxicity. In *Proc. Tall Fescue Toxicosis Workshop,* March 17–18, 1983, 27, Atlanta, Ga.

Poston, H. A., Coffin, J. L., and Combs, G. F., Jr. (1982). Biological effects of dietary T-2 toxin on rainbow trout, *Salmo gairdneri. Aquatic Toxicol.,* **2,** 79–88.

Poznanski, A. S. (1947). Neurophysical disturbances in alimentary toxic aleukia. *Acta Chkalov Inst. Epidemiol. Microbiol.,* Second Commun., Orenburg, USSR. 176–178.

Prasad, N. (1949). Variability of the cucurbit root-rot fungus, *Fusarium (Hypomyces) solani* f. *cucurbitae. Phytopathology* **39,** 133–141.

Prentice, N., and Dickson, A. D. (1964). Emetic material in scabbed wheat. *Cereal Chem.* **41,** 548–550.

Prentice, N., and Dickson, A. D. (1968). Emetic material associated with *Fusarium* species in cereal grains and artificial media. *Biotech. Bioeng.* **10,** 413–427.

Prentice, N., Dickson, A. D., and Dickson, J. G. (1959). Production of emetic material by species of *Fusarium. Nature* **184,** 1319.

Prior, M. G. (1981). Mycotoxins in animal feedstuffs and tissues in Western Canada, 1975–1979. *Can. J. Comp. Med.* **45,** 116–119.

Pringle, P. (1981). Haig "Yellow Rain" charge challenged. *The Observer,* September 20th.

Pringle, P. (1985). Yellow Rain: The cost of chemical arms control. *SAIS Rev.* **5,** November 1, pp. 151–162.

Prisiashnyuk, A. A. (1932). Contribution to the study of cereal crops fusariosis. *Bull. Plant Prot.* **5,** 173–200.

*Proc. Tall Fescue Toxicosis Workshop,* March 17–18, 1983, Atlanta, Ga.

Pullar, E. M., and Lerew, W. M. (1937). Vulvovaginitis in swine. *Aust. Vet. J.* **13,** 28.

Puls, R., and Greenway, J. A. (1976). Fusariotoxicosis from barley in British Columbia. 2. Analysis and toxicity of suspected barley. *Can. J. Comp. Med.* **40,** 16–19.

Pulsford, M. F. (1950). A note on lameness in cattle grazing on Tall Meadow Fescue (*Festuca arundinacea*) in South Australia. *Aust. Vet. J.* **26,** 87–88.

Purchase, I. F. H., ed. (1974). *Mycotoxins.* Amsterdam: Elsevier, p. 443.

Purchase, I. F. H., Tustin, R. C., and van Rensburg, S. J. (1975). Biological testing of food grown in the Transkei. *Food Cosmet. Toxicol.* **13,** 639–647.

Putanna, S. T. (1967). Primary keratomycosis. *J. All-India Ophthalmol. Soc.* **17,** 171–200.

Qin, A., Zhaug, B.-S., Wang, Z-S., Wang, G., Fy, Z., and Wang, R. (1981). Studies on the "Sore foot disease" of cattle. *Acta Vet. Zootech. Sin.* **12,** 137–143 (in Chinese).

Rabie, C. J., Lübben, A., Louw, A. I., Rathbone, E. B., Steyn, P. S., and Vleggaar, R. (1978).

Moniliformin, a mycotoxin from *Fusarium fusarioides*. *J. Agric. Food Chem.* **26**, 375–379.

Rabie, C. J., Marasas, W. F. O., Thiel, P. G., Lubben, A., and Vleggaar, R. (1982). Moniliformin production and toxicity of different *Fusarium* species from Southern Africa. *Appl. Environ. Microbiol.* **43**, 517–521.

Radkevich, P. E. (1952). Poisoning with overwintered grain. In *Veterinary Toxicology*. Moscow: State Publ. Agric. Lit., p. 119.

Raillo, A. I. (1935). The diagnostic estimation of morphological and cultural signs of the species of the genus *Fusarium*. *Publ. All-Union Acad. Agric. Sci.* Ser. **2**, **7**, 1–98, Leningrad.

Raillo, A. I. (1936). Systematics and methods for the determination of the species of the genus *Fusarium*. *Acta Bot. Inst. Acad. Sci. USSR* Ser. **2**, **3**, 803–857.

Raillo, A. I. (1950). *Fungi of the Genus Fusarium*. Moscow: Pub. State Agr. Lit.

Raino, A. J. (1932). Punahome *Fusarium roseum* Link-*Gibberella saubinetii* (Mont.) Sacc. ja sen Aiheuttamat mykset kaurassa. *Valt. Maalolouskoetoiminnan Julk.* **50**, 45 (Abstr.).

Rakieten, N., Gordon, B. S., Beatty, A., Cooney, D. A., Davis, R. D., and Schein, P. S. (1971). Pancreatic islet cell tumors produced by the combined action of streptozotocin and nicotinamide. *Proc. Soc. Exp. Biol. Med.* **137**, 280–283.

Ralston, A. T. (1978). Effect of zearalanol on weaning weight of male calves. *J. Anim. Sci.* **47**, 1203–1206.

Rao, H. R. G., Eugenio, C., Christensen, C. M., De Las Casas, E., and Harein, P. K. (1971). Survival and reproduction of confused flour beetles exposed to fungus metabolites. *J. Econ. Entomol.* **64**, 1563–1565.

Razumov, M. I., and Rubinstein, Y. I. (1951). Experimental food mycotoxic enchondral osteodystrophy (on the etiology of Kashin-Beck Disease). *Probl. Nutr.* **1**, 227–246.

Rebell, G. (1981). *Fusarium* infections in human and veterinary medicine. In P. E. Nelson, T. A. Toussoun, and R. J. Cook, eds., *Fusarium Diseases, Biology and Taxonomy*. University Park: Pennsylvania State University Press, pp. 210–220.

Reisen, J. W., Beeler, B. J., Abenes, F. B., and Woody, C. O. (1977). Effects of zearanol on the reproductive system of lambs. *J. Am. Anim. Sci.* **45**, 293–298.

Reisler, A. V. (1943). *Septic Angina*. Moscow: Publ. Med. House, p. 19.

Reisler, A. V. (1952). Alimentary toxic aleukia. In *Hygiene of Nutrition*, Moscow: Med. State Publ. House, pp. 402–405.

Reiss, J. (1972). Vergleichende Untersuchungen über die Toxizität einiger Mykotoxine gegenüber den Larven des Salinenkrebses. *Zbl. Bakt. Hyg. I.* **155**, 531–534 (Abstr.).

Reiss, J. (1975). Insecticidal and larvicidal activities of the mycotoxins aflatoxin B, rubratoxin B, patulin and diacetoxyscirpenol towards *Drosophila melanogaster*. *Chem. Biol. Interact.* **10**, 339–342.

Renault, L., Goujet, M., Mouin, A., Boutin, G., Palisse, M., and Alamagny, A. (1979). Suspicion de mycotoxicose provoquee par les trichothecenes les poulets de char. *Bull. Acad. Vet. Fr.* **52**, 181–188.

Ribelin, W. E. (1978). Trichothecene toxicosis in cattle. In T. D. Wyllie and L. G. Morehouse, eds., *Mycotoxic Fungi, Mycotoxins, Mycotoxicoses—An Encyclopedic Handbook*, Vol. 2. New York: Marcel Dekker, pp. 36–45.

Richard, J. L., Cysewski, S. D., Pier, A. C., and Booth, G. D. (1978). Comparison of effects of dietary T-2 toxin on growth, immunogenic organs, antibody formation and pathogenic changes in turkeys and chickens. *Am. J. Vet. Res.* **39**, 1674–1678.

Ripperger, H., Seifert, K., Römer, A., Rullkötter, J. (1975). Isolierung von diacetoxyscirpenol aus *Fusarium solani* var. *coeruleum*. *Phytochemistry* **14**, 2298–2299.

Richardson, K. E., Hagler, W. M., Jr., and Hamilton, P. B. (1984a). Bioconversion of $\alpha$-[$^{14}$C] Zearalenol and $\beta$-[$^{14}$C] Zearalenol into [$^{14}$C] Zearalenone by *Fusarium roseum* "Gibbosum." *Appl. Environ. Microbiol.* **47**, 1206–1209.

Richardson, K. E., Hagler, W. M., Jr., and Hamilton, P. B. (1984b). Method for detecting production of zearalenone, zearalenol, T-2 toxin, and deoxynivalenol by *Fusarium* isolates. *Appl. Environ. Microbiol.* **47**, 643–646.

Robb, J., Kirkpatrick, K. S., Norval, M. (1982). Association of toxin-producing fungi with disease in broilers. *Vet. Rec.*, October 23, *111*, 389–390.

Robbins, J. D., and Russell, R. B. (1983). The tall fescue toxicosis problem. In *Proc. Tall Fescue Toxicosis Workshop*, March 17–18, 1983, p. 1–4, Atlanta, Ga.

Roberts, B. A., and Patterson, D. S. P. (1975). Detection of twelve mycotoxins in mixed animal feedstuffs, using a novel membrane cleanup procedure. *J. Assoc. Off. Anal. Chem.* **58**, 1178–1181.

Robison, T. S., and Mirocha, C. J. (1976). Metabolism of the T-2 toxin in poultry. *172nd Am. Chem. Soc. Natl. Meet., Div. Agric. Food Chem.* (Abstr. 61.).

Robison, T. S., Mirocha, C. J., Kurt, H. J., Behrens, J. C., Weaver, G. A., and Chi, M. S. (1979a). Distribution of tritium-labeled T-2 toxin in swine. *J. Agric. Food Chem.* **27**, 1411–1413.

Robison, T. S., Mirocha, C. J., Kurtz, H. J., Behrens, J. C., Chi, M. S., Weaver, G. A., and Nystrom, S. D. (1979b). Transmission of T-2 toxin into bovine and porcine milk. *J. Dairy Sci.* **62**, 637–641.

Robison, T. S., Fenton, S. W., and Mirocha, C. J. (1979c). Isolation of 4′,5′-dihydroxyzearalenone-A zearalenone-related metabolite of *Fusarium roseum* "Gibbosum." *178th Am. Chem. Soc. Natl. Meet. Div. Agric. Food Chem.* (Abstr. 80).

Roche, B. H., and Bohstedt, G. (1930). Scabbed barley and oats and their effect on various classes of livestock. *Am. Soc. Anim. Prod.* **23**, 219–222.

Roche, B. H., Bohstedt, G., and Dickson, J. G. (1930). Feeding scab-infected barley. *Phytopathology* **20**, 132 (Abstr.).

Rodricks, J. W., and Eppley, R. M. (1974). *Stachybotrys* and stachybotryotoxicosis. In I. F. H. Purchase, ed., *Mycotoxins*. New York: Elsevier, pp. 181–197.

Rodricks, J. V., Hesseltine, C. W., and Mehlman, M. A., eds. (1977). *Mycotoxins in Human and Animal Health*. Park Forest, Ill., Pathotox.

Roine, K., Korpinen, E. L., and Kallela, L. (1971). Mycotoxicoses as probable cause of infertility in dairy cows. *Nord. Vet. Med.* **23**, 628–633.

Rokhlin, D. G. (1938). The problem of geographical prevalence of the Beck disease and rudimentary Beck's changes. *News Röentgenol. Radiol.* **21**, 62–71, Moscow.

Roll, R. (1970). Toxische und Kazerogene Wirkung der Mycotoxine von Mikroorganismen. *Bundesgesundheitsblatt* **13**, 157–165.

Romanova, E. D. (1947). Clinical observation and therapy of alimentary toxic aleukia. *Acta Chkalov Inst. Epidemiol. Microbiol.* Second Commun. Orenberg, USSR. 164–169.

Romer, T. (1976). Methods of detecting mycotoxins in mixed feed and feed ingredients. *Feedstuffs* **48**, 18–46.

Romer, T. R. (1977). Analytical approaches to trichothecene mycotoxins. *Cereal Foods World* **22**, 521–523.

Romer, A. and Rullkotter, J. (1975). Isolierung von Diacetoxyscirpenol aus *Fusarium solani* var. *coeruleum*. *Phytochemistry* **14**, 2298–2299.

Romer, T. R., Boling, T. M., and MacDonald, J. L. (1978). Gas-liquid chromatographic determination of T-2 toxin and diacetoxyscirpenol in corn and mixed feeds. *J. Anal. Off. Anal. Chem.* **61**, 801–808.

Rosen, R. T., and Rosen, J. D. (1982). Presence of four *Fusarium* mycotoxins and synthetic material in "Yellow Rain." Evidence for the use of chemical weapons in Laos. *Biomed. Mass Spectrosc.* **9**, 443–450.

Rosenstein, Y., Lafont, P., and Frayssinet-Lafarge, G. (1978). Effets immunosuppresseurs des toxines de *Fusarium*. *Collect. Med. Leg. Toxicol. Med.* **107**, 51–57.

Rosenstein, Y., Lafarge-Fayssinet, C., Lespinats, G., Loisillier, F., Lafont, P. and Frayssinet, C. (1979). Immunosuppressive activity of *Fusarium* toxins. Effect on antibody synthesis and skin grafts of crude extracts, T-2 toxin and diacetoxyscirpenol. *Immunology* **36**, 111–118.

Rosenstein, Y., Kretschmer, R. R., and Lafarge-Frayssinet, C. (1981). Effect of *Fusarium* toxins, T-2 toxin and diacetoxyscirpenol on murine T-independent immune responses. *Immunology* **44**, 555–560.

Rossi, C. (1959). An episode of intoxication by *Festuca arundinacea* in cows of Val d'Aosta. *Vet. Ital.* **10**, 853–864.

Rothenbacker, H., Wiggins, J. P., and Wilson, L. L. (1975). Pathologic changes in endocrine glands and certain other tissues of lambs implanted with the synthetic growth promotant zeranol. *Am. J. Vet. Res.* **36**, 1313–1317.

Rowlatt, U. F. (1967). Pancreatic neoplasms in rats and mice. In E. Cotchin and F. J. C. Roe, eds., *Pathology of Laboratory Rats and Mice.* Oxford: Blackwell, pp. 85–101.

Rowsey, J. J., Teaceers, D. L., Smith, J. A., Newson, D. L., and Rodriquez, J. (1979). *Fusarium oxysporum* endophthalmitis. *Arch. Ophthalmol.* **97**, 103–105.

Rozov (1889). *Trip Around the World, from Moscow to Amur and Siberia.* Chapter 4, pp. 125–168, Moscow Publ. Selchozgiz.

Rubinstein, Y. I. (1948). Microflora of cereals overwintered under snow. *Proc. Nutr. Inst. Acad. Med. Sci. USSR, Moscow* pp. 29–38.

Rubinstein, Y. I. (1949). Toxic cultures of *Fusarium* from Transbaikal cereals (on the etiology of Kashin-Beck disease). *Hyg. Sanit.* **14**, 35–39.

Rubinstein, Y. I. (1950a). Biochemical properties of *Fusarium* cultures (Section Sporotrichiella). *Microbiology* **19**, 438–443.

Rubinstein, Y. I. (1950b). Experimental food mycotoxical enchondral dystrophy (on etiology and pathogenesis of Kashin-Beck disease). *Collect.—Jubilee Sess. Inst. Nutr. Acad. Med. Serv. USSR*, p. 80. Moscow.

Rubinstein, Y. I. (1951a). On the problem of digestive mycotoxicoses. *Nov. Med. Moscow* **2**, 30–36.

Rubinstein, Y. I. (1951b). Some properties of *Fusarium sporotrichioides* toxin. *Acta Acad. Med. Sci. USSR Nutr. Probl.* **13**, 247–253.

Rubinstein, Y. I. (1953). The etiology of Urov (Kashin-Beck) disease. *Nutr. Probl.* **12**, 73–81.

Rubinstein, Y. I. (1956a). Mycotoxicoses of men and the problems of their further investigation. *Publ. Acad. Sci. Ukr. SSR, Kiev*, 10–13.

Rubinstein, Y. I. (1956b). Actual problems in study of fusariotoxicoses. *Nutr. Probl.* **15**, 8.

Rubinstein, Y. I. (1956c). The fluorescence method of determining the toxicity of *Fusarium sporotrichiella*. *Microbiologiya* **25**, 171–174.

Rubinstein, Y. I. (1960a). The effect of the cultivation conditions on toxin formation in *Fusarium sporotrichiella*. In V. I. Bilai, ed., *Mycotoxicoses of Man and Agricultural Animals* (English Trans.). Washington, D.C.: Joint Publishers Research Service, pp. 66–73.

Rubinstein, Y. I. (1960b). Human mycotoxicoses, food fusariotoxicoses. In V. I. Bilai, ed., *Mycotoxicoses of Man and Agricultural Animals* (English Trans.) Washington, D.C.: Joint Publishers Research Service, pp. 89–110.

Rubinstein, Y. I., and Lyass, L. S. (1948). On the etiology of alimentary toxic aleukia (septic angina). *Hyg. Sanit.* **7**, 33–38.

Rubinstein, Y. I., Kukel, Y., and Kudinova, G. (1961). New experimental chronic toxicosis caused by *Fusarium sporotrichiella*. Lecture No. 2 in Conference of Mycotoxicology, Kiev, p. 16 (Abstr.).

Rubinstein, Y. I., Kukel, J. P., and Kudinova, G. P. (1967). Papillomatosis with hyper-keratosis of the proventriculus in rats following their feeding on grain infected with the fungus *Fusarium sporotrichiella* var. *poae* (Pk.) Bilai. *Vopr. Pitan.* **26**, 57–61.

Ruddick, J. A., Scott, P. M., and Harwig, J. (1976). Teratological evaluation of zearalenone administered orally to the rat. *Bull. Environ. Contamin. Toxicol.* **15**, 678–681.

Ruhr, L. P., Osweiler, G. D., and Foley, C. W. (1978). The effect of zearalenone on fertility in the boar. *Proc. Annu. Meet. Am. Assoc. Vet. Lab. Diagn.* **21**, 127–133.

Rukmini, C., and Bhat, R. V. (1978). Occurrence of T-2 toxin in *Fusarium* infected sorghum from India. *J. Agric. Food Chem.* **26**, 647–649.

Rukmini, C., Prasad, J. S., and Rao, K. (1980). Effect of feeding T-2 toxin to rats and monkeys. *Food Cosmet. Toxicol.* **18**, 267.

Rush-Munro, F. M., Black, H., and Dingley, J. M. (1971). Onychomycosis caused by *Fusarium oxysporum*. *Aust. J. Derm.* **12**, 18–29.

Ruzsas, C., Mess, B., Biro-Gosztonyi, M., and Woller, L. (1978). Effect of pre- and perinatal administration of the fungous $F_2$-toxin on the reproduction of the albino rat. *Dev. Endocrinol.* **3**, 57–60.

Ryazanov, V. A. (1947). Lecture in Republic Conference of Alimentary Toxic Aleukia, Moscow, December.

Ryazanov, V. A. (1948). Lecture in Republic Conference on Alimentary Toxic Aleukia, Moscow, March.

Ryegrass staggers. *Annu. Proc. N.Z. Grassl. Assoc.* **44**, 227–266.

Saccardo, P. A. (1886). *Sylloge Fungorum* **4**, 718.

Saito, M. (1983). Acute and chronic biological effect of trichothecene toxins. In M. Rech-cigl, Jr., ed., *Handbook of Foodborne Diseases of Biological Origin*. Boca Raton, Fla.: CRC Press, pp. 239–245.

Saito, M., and Ohtsubo, K. (1974). Trichothecene toxins of *Fusarium* species. In I. F. H. Purchase, ed., *Mycotoxins*. Amsterdam: Elsevier, pp. 263–281.

Saito, M., and Okubo, K. (1970). Studies on the target injuries in experimental animals with mycotoxins of *Fusarium nivale*. In M. Herzberg, ed., *Toxic Microorganisms*. Washington, D.C.: pp. 82–95. UJNR Joint Panel and Toxic Micro-organisms and the U.S. Department of the Interior.

Saito, M., and Tatsuno, T. (1971). Toxins of *Fusarium nivale*. In S. Kadis, A. Ciegler, and S. J. Ajl, eds., *Microbial Toxins*, Vol. 7. New York: Academic, pp. 293–316.

Saito, M., Enomoto, M., and Tatsuno, T. (1969). Radiomimetic biological properties of the new scirpene metabolite of *Fusarium nivale*. *GANN* **60**, 599–603.

Saito, M., Ohtsubo, K., Umedo, M., Enomot, M., Kurata, H., Udagawa, S., Sakabem, F., and Ichinoe, M. (1971). Screening tests using HeLa cells and mice for detection of myco-toxin-producing fungi isolated from foodstuffs. *Jpn. J. Exp. Med.* **41**, 1–20.

Saito, M., Ishiko, T., Enomoto, M., Ohtsubo, K., Umeda, M., Kurata, H., Udagawa, S., Taniguchi, S., and Sekita, S. (1974). Screening test using HeLa cells and mice for detection of mycotoxin-producing fungi isolated from foodstuffs. *Jpn. J. Exp. Med.* **44**, 63–82.

Saito, M., Horiuchi, T., Ohtsubo, K., Hatanaka, Y., and Ueno, Y. (1980). Low tumor inci-dence in rats with long-term feeding of fusarenon-X, a cytotoxic trichothecene pro-duced by *Fusarium nivale*. *Jpn. J. Exp. Med.* **50**, 293–302.

Salazar, S., Fromentin, H., and Mariat, F. (1980). Action du diacetoxyscirpénol sur la candidose expérimentale de la souris. *C. R. Acad. Sci. Paris,* **Ser. D, 290,** 877–878.

Salceda, S. R. (1976). Fungi and the human eye. *Kaliklasau* **5,** 143–174.

Salceda, S. R., Valenton, M. J., Nievera, L-Fc., and Abendanio, R. M. (1974). Corneal ulcer: A study of 110 cases. *J. Philipp. Med. Assoc.* **49,** 493–508.

Sarkisov, A. Kh. (1944). Method of determining toxicity of cereal crops overwintered in the field. *Hyg. Sanit.* **9,** 19–22.

Sarkisov, A. Kh. (1945). On the etiology of so-called septic angina. Abstract in Republic Conference on Alimentary Toxic Aleukia, Moscow, December 24–26, pp. 7–8.

Sarkisov, A. Kh. (1948). Data on toxicity of cereals over-wintered under snow. *Govt. Ed. Min. Agric. USSR,* pp. 22–40, Moscow.

Sarkisov, A. Kh. (1950). Etiology of septic angina in man. *J. Microbiol. Epidemiol. Immunol.* **1,** 43–47.

Sarkisov, A. Kh. (1954). Mycotoxicoses. *Govt. Ed. Min. Agr. USSR,* pp. 216, Moscow.

Sarkisov, A. Kh. (1960). General characterization of the alimentary mycotoxicoses of agricultural animals. In V. I. Bitai, ed., *Mycotoxicoses of Man and Agricultural Animals* (English Trans.) Washington, D.C.: Joint Publishers Research Service, pp. 155–163.

Sarkisov, A. Kh. (1961). Fusariotoxicosis of cattle and sheep. *Proc. 2nd Conf. Mycotoxicoses Man Agric. Anim., Kiev.* pp. 13–15.

Sarkisov, A. Kh., and Kvashnina, E. S. (1948). Toxicobiological properties of *Fusarium sporotrichioides. Govt. Publ. Min. Agric. USSR,* pp. 89–92, Moscow.

Sarkisov, A. Kh., Kvashnina, E. S., and Koroleva, V. P. (1948a). The exposition of toxic fungi. *Govt. Ed. Min. Agr. USSR,* pp. 49–54, Moscow.

Sarkisov, A. Kh., Kornejev, N. E., Kazakov, M. V., Kvashnina, E. S., Gerasimova, P. A., and Akulova, N. S. (1948b). Comparative effect of fungi isolated from overwintered cereal crops on laboratory animals and cattle. *Govt. Ed. Min. Agric. USSR,* pp. 54–80, Moscow.

Sarkisov, A. Kh., Korneyev, N. E., Kvashnina, E. S., Koroleva, V. P., Gerasimova, P. A., and Akulova, N. S. (1948c). The harm caused by overwintered cereal crops to livestock and poultry. *Govt. Ed. Min. Agric. USSR,* pp. 10–21, Moscow.

Sarkisov, A. Kh., Djilavyan, Kh. A., Rukhlyada, V. V., and Chernov, K. S. (1972). Differential diagnosis of mycotoxicoses in farm animals. *Bull. All Union Order of Lenin Inst. Exp. Vet. Sci.* **Ser. 12,** 57–61.

Sarudi, I. Jr., Kasa, I., and Bajonoczy, H. (1976). Über die spektrofluorometrische Bestimmung des Zearalenon (F₂)-Toxins in Maiskorn. *Dtsch. Lebensm. Rdsch.* **72,** 96–99.

Sarudi, I., Kupai, J., Muleczki, J., Galambos, A., Czukas, B., Ormos, Z., Pataki, K., and Horvath, L. (1979). Treatment of fodder contaminated with zearalenone ($F_2$ toxin). *Hung. Teljes.* **16,** 20–38.

Sass, B., Rabstein, L. S., Madison, R., Nimms, R. M., Peters, R. L., and Kelloff, G. J. (1975). Incidence of spontaneous neoplasms in F344 rats throughout the natural life-span. *J. Natl. Cancer Inst.* **54,** 1449–1465.

Sato, N., and Amano, R. (1976). Herstellung von T-2 toxin unter Einsatz von *Fusarium solani* in ruhender Kultur. *Fleischwirtschaft* **56,** 1354–1355.

Sato, N., and Ueno, Y. (1977). Comparative toxicities of trichothecenes. In J. V. Rodricks, H. W. Hesseltine, and M. A. Mehlman, eds., *Mycotoxins in Human and Animal Health.* Park Forest, Ill.: Pathotox, pp. 295–307.

Sato, N., Ueno, Y., and Enomoto, M. (1975). Toxicological approaches to the toxic metabolites of fusaria. 8. Acute and subacute toxicities of T-2 toxin in cats. *Japan J. Pharmacol.* **25,** 263–270.

Sato, N., Ito, Y., Kumada, H., Ueno, Y., Asano, K., Saito, M., Ohtsubo, K., Ueno, I., and Hatanaka, Y. (1978). Toxicological approaches to the metabolites of fusaria. 13. Hematological changes in mice by a single and repeated administrations of trichothecenes. *J. Toxicol. Sci.* **3**, 335–356.

Sawhney, D. S. (1976). Biochemical responses of duckling to dietary Fusariotoxins. *Toxicol. Appl. Pharmacol.* **37**, 141 (Abstr.).

Sawhney, D. C. (1977). Changes in plasma protein of ducklings from dietary fusariotoxins. *Toxicol. Appl. Pharmacol.* **41**, 168 (Abstr.).

Sayer, J. M., and Emery, T. F. (1968). Structures of the naturally occurring hydroxamic acids Fusarinines A and B. *Biochemistry* **7**, 184–190.

Schappert, K. T., and Khachatourians, G. G. (1984). Influence of membrane on T-2 toxicity in *Saccharomyces* spp. *Appl. Environ. Microbiol.* **47**, 681–684.

Schiefer, H. B. (1983a). The possible use of chemical warfare agents in Southeast Asia. *Conflict Q.* **3**, 32–41. Fredericton, New Brunswick, Canada.

Schiefer, H. B. (1983b). Facts, not rhetoric on Yellow Rain. Nature **304**, 10 (letter to the editor).

Schiefer, H. B., and O'Ferrall, B. K. (1981). Alleged mycotoxicosis in swine: Review of a court case. *Can. Vet. J.* **22**, 134–139.

Schiefer, H. B., and Sutherland, R. G. (1984). Problems associated with verification of alleged CBW use in Southeast Asia. *Proc. 1st World Cong. New Compds. Biol. Chem. Warf.: Toxicol. Eval.*, Ghent, Belgium, May 21–23.

Schmidt, R., Ziegenhagen, E., and Dose, K. (1981a). High performance liquid chromatography of trichothecenes. 1. Detection of T-2 toxin and HT-2 toxin. *J. Chromatogr.* **212**, 370–373.

Schmidt, R., Bieger, A., Ziegenhagen, E., and Dose, K. (1981b). Bestimmung von T-2 Toxin in pflanzlichen Nahrungsmitteln. 1. T-2 Toxin in verschimmeltem Reis und Mais. *Fresenius Z. Anal. Chem.* **308**, 133–136.

Schoental, R. (1968). Experimental induction of gastrointestinal tumours in rodents by N-alkyl-N-nitrosourethanes and certain related compounds. *Gann Monogr.* **3**, 61–71.

Schoental, R. (1975). Mycotoxicoses "by proxy." *Int. J. Environ. Stud.* **8**, 171–175.

Schoental, R. (1977a). Health hazards due to oestrogenic mycotoxins in certain foodstuffs. *Int. J. Environ. Stud.* **11**, 149–150.

Schoental, R. (1977b). The role of nicotinamide and the certain other modifying factors in diethylnitrosamine carcinogenesis. Fusaria mycotoxins and "spontaneous" tumours in animals and man. *Cancer* **40**, 1833–1840.

Schoental, R. (1977c). Health hazards due to T-2 toxin. *Vet. Rec.* **101**, 473–474.

Schoental, R. (1978). Mycotoxins in food and the variations in tumor incidence among laboratory rodents. *Nutr. Cancer* **1**, 12–13.

Schoental, R. (1979a). The role of *Fusarium* mycotoxins in the etiology of tumours of the digestive tract and of certain other organs in man and animals. *Front. Gastrointest. Res.* **4**, 17–24.

Schoental, R. (1979b). *Fusarium* mycotoxins in the aetiology of neonatal abnormalities. Cardiovascular and sexual disorders and tumours in man and animals. *Toxicon* **17** (Suppl. 1), 164.

Schoental, R. (1979c). Possible public health significance of *Fusarium* toxins. In Proc. 3rd Meeting on Mycotoxins in Animal Disease, pp. 67–70. Ministry of Agriculture, Fisheries and Food Agricultural Development and Advisory Service (U.K. Weybridge).

Schoental, R. (1979d). Pellagra and *Fusarium* mycotoxins. *Int. J. Environ. Stud.* **13**, 327–330.

Schoental, R. (1980a). Relationships of *Fusarium* mycotoxins to disorders and tumors associated with alcoholic drinks. *Nutr. Cancer* **2**, 88–92.

Schoental, R. (1980b). Mouldy grain and the aetiology of pellagra. The role of toxic metabolites of *Fusarium. Biochem. Soc. Trans.* **8**, 147–150.

Schoental, R. (1980c). Cancer in digestive tract and mycotoxins of *Fusarium* species. *The Lancet,* September 13, p. 593.

Schoental, R. (1981a). Relationships of *Fusarium* toxins to tumours and other disorders in livestock. *J. Vet. Pharmacol. Therap.* **4**, 1–6.

Schoental, R. (1981b). Mycotoxin and fetal abnormalities. *Int. J. Environ. Stud.* **17**, 25–29.

Schoental, R. (1981c). *Fusarium* mycotoxins and the effects of high-fat diets. *Nutr. Cancer* **3**, 57–62.

Schoental, R. (1981d). Carcinogenic mycotoxins. In R. Schoental and T. A. Connors, eds., *Dietary Influences on Cancer: Traditional and Modern.* Boca Raton, Fla.: CRC Press, pp. 109–148.

Schoental, R. (1981e). Carcinogenic contaminants of food. In R. Schoental and T. A. Connors, eds., *Dietary Influences on Cancer; Traditional and Modern.* Boca Raton, Fla.: CRC Press, pp. 149–183.

Schoental, R. (1983a). Environmental factors and cancer of the oesophagus. The role of seleniferous soil and trichothecene mycotoxins—a hypothesis. *Int. J. Environ. Stud.* **21**, 159–164.

Schoental, R. (1983b). Health hazards of secondary metabolites of *Fusarium. Microbiol. Aliments Nutr.* **1**, 101–107.

Schoental, R. (1983c). Chronic, including teratogenic and carcinogenic effects of trichothecenes: A short review. *Vet. Res. Commun.* **7**, 165–170.

Schoental, R., and Gibbard, S. (1978). Increased excretion of urinary porphyrins by white rats given intragastrically the chemical carcinogens diethylnitrosamine, monocrotaline T-2 toxin, and ethylmethane sulphonate. *Trans. Biochem. Soc.* **7**, 127–129.

Schoental, R., and Joffe, A. Z. (1974). Lesions induced in rodents by extracts from cultures of *Fusarium poae* and *F. sporotrichioides. J. Pathol.* **112**, 37–42.

Schoental, R., and Magee, P. N. (1962). Induction of squamous carcinoma of the lung and of the stomach and oesophagus by diazomethane and N-methyl-N-nitroso-urethane, respectively. *Br. J. Cancer* **16**, 92–100.

Schoental, R., Joffe, A. Z., and Yagen, B. (1976). Chronic lesions in rats treated with crude extracts of *Fusarium poae* and *F. sporotrichioides.* The role of moldy food in the medicine of oesophagus, mammary and certain other abnormalities and tumour in livestock and man. *Br. J. Cancer* **34**, 310 (Abstr.).

Schoental, R., Joffe, A. Z., and Yagen, B. (1978a). Irreversible depigmentation of dark mouse hair by T-2 toxin (a metabolite of *Fusarium sporotrichioides*) and by calcium pantothenate. *Experientia* **34**, 763–764.

Schoental, R., Joffe, A. Z., and Yagen, B. (1978b). The induction of tumours of the digestive tract and of certain other organs in rats given T-2 toxin, a secondary metabolite of *Fusarium sporotrichioides. Br. J. Cancer* **38**, 171 (Abstr.).

Schoental, R., Joffe, A. Z., and Yagen, B. (1979). Cardiovascular lesions and various tumours found in rats given T-2 toxin, a trichothecene metabolite of *Fusarium* species. *Cancer Res.* **39**, 2279–2289.

Schroeder, H. W., and Hein, H. (1975). A note on zearalenone in grain sorghum. *Cereal Chem.* **52**, 751–752.

Schuh, M., Leibetseder, J., and Glawischung, E. (1982). Chronic effects of different levels of deoxynivalenol on weight gain, feed consumption, blood parameters, pathological as well as histopathological changes in fattening pigs. *Proc. 5th Int. IUPAC Symp. Mycotoxins Phycotoxins. Tech. Univ., Vienna*, pp. 273–276. Oxford: Pergamon Press.

Scott, D. B. (1965). Toxigenic fungi isolated from cereal and legume products. *Mycopathol. Mycol. Appl.* **25**, 213–222.

Scott, P. M. (1978). Mycotoxins in feeds and ingredients and their origin. *J. Food Prot.* **41**, 385–398.

Scott, P. M. (1982). Assessment of quantitative methods for detection of trichothecenes in grains and grain products. *J. Assoc. Off. Anal. Chem.* **65**, 876–883.

Scott, P. M. (1983). Trichothecene problems in Canada. In Y. Ueno, ed., *Trichothecenes, Chemical, Biological and Toxicological Aspects.* Amsterdam: Elsevier pp. 218–220.

Scott, P. M., Lawrence, J. W., and van Walbeek, W. (1970). Detection of mycotoxins by thin-layer chromatography: Application to screening of fungal extracts. *Appl. Microbiol.* **20**, 839–842.

Scott, P. M., van Walbeek, W., Kennedy, B., and Anyeti, D. (1972). Mycotoxins and toxigenic fungi in grains and other agricultural products. *J. Agric. Food Chem.* **20**, 1103–1109.

Scott, P. M., Panalaks, T., Kanhere, S., and Miles, W. F. (1978). Determination of zearalenone in cornflakes and other corn-based foods by thin-layer chromatography, high pressure liquid chromatography, and gas-liquid chromatography/high resolution mass spectrometry. *J. Assoc. Off. Anal. Chem.* **61**, 593–600.

Scott, P. M., Harwig, J., and Blanchfield, B. J. (1980). Screening *Fusarium* isolates from overwintered Canadian grains for trichothecenes. *Mycopathologia* **72**, 175–180.

Scott, P. M., Lau, P-Y., and Kanhere, S. R., (1981). Gas chromatography with electron capture and mass spectrometric detection of deoxynivalenol in wheat and other grains. *J. Assoc. Off. Anal. Chem.* **64**, 1364–1371.

Scott, P. M., Kanhere, S. R., Lau, P-Y., Dexter, J. E., and Greenhalgh, R. (1983). Effects of experimental flour milling and breadbaking on retention of deoxynivalenol (vomitoxin) in hard red spring wheat. *Cereal Chem.* **60**, 421–424.

Scott, P. M., Lawrence, G. A., Telli, A., and Iyengar, J. R. (1984). Preparation of deoxynivalenol (vomitoxin) from field inoculated corn. *J. Assoc. Off. Anal. Chem.* **67**, 32–34.

Scott, P. M., Nelson, K., Kanhere, S. R., Karpinski, K. F., Hayward, S., Neish, G. A., and Teich, A. H., (1984). Decline in deoxynivalenol (vomitoxin) concentrations in 1983 Ontario winter wheat before harvest. *Appl. Environ. Microbiol.* **48**, 884–886.

Schwartzman, M. I. (1937). So-called "Urov disease" or endemic deforming osteoarthritis in Zabaikalye. *Sov. Med. J.* **3**, 181–184.

Seagrave, S. (1981). *Yellow Rain: A Journey through the Terror of Chemical Warfare.* New York: M. Evans.

Seaman, W. L. (1982). Epidemiology and control of mycotoxigenic fusaria on cereal grains. *Can. J. Plant Pathol.* **4**, 187–190.

Seemüller, E. (1968). Untersuchungen über die morphologische und biologische Differenzierung in der *Fusarium* Section *Sporotrichiella. Mitt. Biol. Bundesanst.* **127**, 1–97, Berlin-Dahlem.

Segal, R., Milo-Goldzweig, I., Joffe, A. Z., and Yagen, B. (1983). Trichothecene-induced hemolysis. 1. The hemolytic activity of T-2 toxin. *Toxic. Appl. Pharmacol.* **70**, 343–349.

Seitz, L. M., and Mohr, H. E. (1976). Simple method for simultaneous detection of aflatoxin and zearalenone in corn. *J. Assoc. Off. Anal. Chem.* **59**, 106–109.

Serafimov, B. N. (1945). Mental disorder caused by alimentary toxic aleukia. *Neuropathol. Psychiatry, Moscow* **5**, 25–30.

Serafimov, B. N. (1946). Mental symptomatology of alimentary toxic aleukia. *Neuropathol. Psychiatry, Moscow* **6**, 50–53.

Sergiev, P. G. (1945). Epidemiology of alimentary toxic aleukia. In A. I. Nesterov, A. N. Sysin, and L. N. Karlik, eds., *Alimentary Toxic Aleukia (Septic Angina)*. Moscow: Medgiz, pp. 7–11.

Sergiev, P. G. (1946). Harvesting losses and overwintered grain in the field—Source of disease. *Kolkhozn. Prod.* **7**, 1–16.

Sergiev, P. G. (1948). *Septic angina*. In *Artic. Lect. Conversat. on Sanit. Instructive Subjects*. Barnaul, pp. 62–65.

Sergiyevski, F. P. (1948). *Urov Kashin-Beck Disease*. Chita: Chitgiz, p. 62.

Sergiyevski, F. P. (1952). *Urov (Kashin-Beck) Disease*. Chita: Chitgiz.

Shannon, G. M., Shotwell, O. L., Lyons, A. J., White, D. G., and Garcia-Aguirre, G. (1980). Laboratory screening for zearalenone formation in corn hybrids and inbreds. *J. Assoc. Off. Anal. Chem.* **63**, 1275–1277.

Sharby, T. F., Templeton, G. E., Beasley, J. N., and Stephenson, E. L. (1973). Toxicity resulting from feeding experimentally molded corn to broiler chicks. *Poult. Sci.* **52**, 1007–1014.

Sharda, D. P., Wilson, R. F., Williams, L. E., Swiger, L. A., and Gross, R. F. (1971). Mold toxicity in swine and laboratory animals. Effect of feeding corn inoculated with pure cultures of *Fusarium roseum* Ohio isolate C. *J. Anim. Sci.* **32**, 1169–1173.

Sharma, V. D., Wilson, R. F., and Williams, L. E. (1974). Reproductive performance of female swine fed corn naturally molded or inoculated with *Fusarium roseum*, Ohio isolates B and C. *J. Anim. Sci.* **38**, 598–602.

Sharp, G. D., and Dyer, I. A., (1968). The effects of zearalenol on performance of heifers. *Proc. West. Sect., Am. Soc. Anim. Sci.* **19**, 127.

Sharp, G. D., and Dyer, I. A. (1970). Metabolic responses to zearalenol implants. *Proc. West. Sect., Am. Soc. Anim. Sci.* **21**, 147–152.

Sharp, G. D., and Dyer, I. A. (1971). Effect of zearalenol on the performance and carcass composition of growing-finishing ruminants. *J. Anim. Sci.* **33**, 865–871.

Sharp, G. D., and Dyer, I. A. (1972). Zearalenol metabolism in steers. *J. Anim. Sci.* **34**, 176–179.

Shchipatchev, V. G. (1928). The etiology of Kashin-Beck Disease (arthritis deformans endemica). In *Transbaikal Collect. Works Irkutsk State Univ.* **14**, 109–131.

Shemshelevitch, S. B., and Dubniakova, A. M. (1945). Clinicohematological characteristics of alimentary toxic aleukia. Abstract in *Republic Conference on Alimentary Toxic Aleukia*, Moscow, December 24–26, pp. 13–14.

Sherbakoff, C. D. (1915). Fusaria of potatoes. *Mem. Cornell Univ. Agric. Exp. Stn.* **6**, 85–270.

Sherwood, R. F., and Peberdy, J. F. (1972a). Fusaria in the ration. Effects of mycotoxins in feeds. *Feed Farm Suppl.* **19**, 19–21.

Sherwood, R. F., and Peberdy, J. F. (1972b). Zearalenone production by *Fusarium* spp. in grain. *J. Gen. Microbiol.* **73**, 23.

Sherwood, R. F., and Peberdy, J. F. (1972c). Factors affecting the production of zearalenone by *Fusarium graminearum* in grain. *J. Stored Prod. Res.* **8**, 71–75.

Sherwood, R. F., and Peberdy, J. F. (1972d). Effects of mycotoxins in feeds. *Feed Farm Suppl.* **10**, 9–11.

Sherwood, R. F., and Peberdy, J. F. (1973a). Effects of zearalenone on the developing male chicken. *Br. Poult. Sci.* **14**, 127–129.

Sherwood, R. F., and Peberdy, J. F. (1973b). Aflatoxin and zearalenone occurrence in stored corn. *Cereal Chem.* 687–697.

Sherwood, R. F., and Peberdy, J. F. (1974a). Production of the mycotoxin, zearalenone, by *Fusarium graminearum* growing on stored grain. 1. Grain storage at reduced temperatures. *J. Sci. Food Agric.* **25**, 1081–1087.

Sherwood, R. F., and Peberdy, J. F. (1974b). Production of the mycotoxin, zearalenone, by *Fusarium graminearum* growing on stored grain. 2. Treatment of wheat grain with organic acids. *J. Sci. Food Agric.* **25**, 1089–1093.

Shimuzu, T., Nakano, N., Matsui, T., and Aibara, K. (1979). Hypoglycemia in mice administered with fusarenon-X. *Japan J. Med. Sci. Biol.* **32**, 189–198.

Shipchandler, M. T. (1975). Chemistry of zearalenone and some of its derivatives. *Heterocycles* **3**, 471–520.

Shirlaw, J. F. (1939). Degnala disease of buffaloes. An account of the lesions and pathology. *Indian J. Vet. Sci.* **9**, 173–177.

Shklovskaja, R. S., and Brodskaja, F. P. (1944). *The so-called "septic angina."* In G. S. Lurie, ed., *Data Scient.—Pract. Works of Town Sterlitamak Physicians.* Bashkiria: Sterlitamak, pp. 5–31.

Shotwell, O. L. (1977a). Mycotoxins—corn-related problems. *Cereal Food World* **22**, 524–527.

Shotwell, O. L. (1977b). Assay methods for zearalenone and its natural occurrence. In J. V. Rodricks, H. W. Hesseltine, and M. A. Mehlman, eds., *Mycotoxins in Human and Animal Health.* Park Forest, Ill.: Pathotox, pp. 403–413.

Shotwell, O. L., Hesseltine, C. W., Goulden, M. L., and Vandegraft, E. E. (1970). Survey of corn for aflatoxin, zearalenone, and ochratoxin. *Cereal Chem.* **47**, 700–707.

Shotwell, O. L., Hesseltine, C. W., Vandegraft, E. E., and Goulden, M. L. (1971). Survey of corn from different regions for aflatoxin, ochratoxin and zearalenone. *Cereal Sci. Today* **16**, 266–273.

Shotwell, O. L., Goulden, M. L., Bothast, R. J., and Hesseltine, C. W. (1975). Mycotoxins in hot spots in grains. 1. Aflatoxin and zearalenone occurrence in stored corn. *Cereal Chem.* **52**, 687–697.

Shotwell, O. L., Goulden, M. L., and Bennett, G. A. (1976). Determination of zearalenone in corn: Collaborative study. *J. Assoc. Off. Anal. Chem.* **59**, 666–670.

Shotwell, O. L., Goulden, M. L., Bennett, G. A., Plattner, R. D. and Hesseltine, C. W. (1977). Survey of 1975 wheat and soybeans for aflatoxin, zearalenone and ochratoxin. *J. Assoc. Off. Anal. Chem.* **60**, 778–783.

Shotwell, O. L., Bennett, G. A., Goulden, M. L., Plattner, R. D., and Hesseltine, C. W. (1980). Survey for zearalenone, aflatoxin, and ochratoxin in U.S. grain sorghum from 1975 and 1972 crops. *J. Assoc. Off. Anal. Chem.* **63**, 922–926.

Shreeve, B. J., and Patterson, D. S. P. (1975). Mycotoxicosis. *Vet. Rec.* **97**, 279–280.

Shreeve, B. J., Patterson, D. S. P., and Roberts, B. A. (1975). Investigation of suspected cases of mycotoxicosis in farm animals in Britain. *Vet. Rec.* **97**, 275–278.

Shreeve, B. J., Patterson, D. S. P., Roberts, B. A., and Wrathall, A. E. (1978). Effect of moldy feed containing zearalenone on pregnant sows. *Br. Vet. J.* **134**, 421–427.

Shreeve, B. J., Patterson, D. S. P., and Roberts, B. A. (1979). The carry-over of aflatoxin, ochratoxin and zearalenone from naturally contaminated feed to tissues, urine and milk of dairy cows. *Food Cosmet. Toxicol.* **16**, 151–152.

Sidorik, O. T., Misyurenko, I. P., Shevchenko, I. N., and Kindzelsky, L. P. (1974). Effects of fungus *Fusarium sporotrichiella* toxin on hemopoiesis in healthy animals and those with leukemia. *Microbiol. J.* **36**, 641–643.

Siegal, M. R. (1983). Mode of transmission of the fescue endophyte. In *Proc. Tall Tescue Toxicosis Workshop.* March 17–18, 1983, p. 48, Atlanta, Ga.

Siegfried, R. (1977). Fusarium-Toxine. *Naturwiss.* **64**, 274.

Siegfried, R. (1978). Qualitative und quantitative Bestimmung eines *Fusarium*-toxins (Diacetoxyscirpenol) in Futtermais. *Landwirtsch. Forsch.* **31**, 141–144.

Siegfried, R. and Frank, K. (1978). Beitrag zur Analytik der *Fusarium*—Toxine (Trichothecen—Toxine). *Z. Lebensm. Unters. Forsch.* **166**, 363–367.

Sigg, H. P., Mauli, R., Flury, E., and Hauser, D. (1965). Die Konstitution von diacetoxyscirpenol. *Helv. Chim. Acta* **48**, 962–988.

Sigtenhorst, M. L., and Gingrich, W. D. (1957). Bacteriologic studies of keratitis. *South. Med. J.* **50**, 346–350.

Singh, G., and Malik, S. R. K. (1972). Therapeutic keratoplasty in fungal corneal ulcers. *Br. J. Ophthalmol.* **56**, 41–45.

Sinha, R. N. (1966). Development and mortality of *Tribolium castaneum* and *T. confusum* (*Coleoptera: Tenebrionidae*) on seedborne fungi. *Ann. Entomol. Soc. Am.* **59**, 192–201.

Sippel, W. L., Burnside, J. E., and Atwood, M. B. (1953). A disease of swine and cattle caused by eating moldy corn. *Proc. 90th Ann. Meet. Am. Vet. Med. Assoc.*, pp. 174–181.

Siraj, M. Y., Phillips, T. D., and Hayes, A. W. (1978). Interaction of mycotoxins with copper-folin reagent. *J. Food Prot.* **41**, 370–372.

Sirotinina, O. N. (1945). Toxicity of cereals overwintered in the field. Ph.D. dissertation, Saratov, p. 232. Saratov University.

Sloey, W., and Prentice, N. (1962). Effects of *Fusarium* isolates applied during malting on properties of malt. *Am. Soc. Brew. Chem. Proc.*, pp. 24–29.

Slonevski, S. I. (1945). Alimentary toxic aleukia (septic angina) in Udmurt, ASSR. *Hyg. Sanit.* **6**, 23.

Smalley, E. B. (1973). T-2 toxin. *J. Am. Vet. Med. Assoc.* **163**, 1278–1281.

Smalley, E. B., and Strong, F. M. (1974). Toxic trichothecenes. In I. F. H. Purchase, ed., *Mycotoxins*. Amsterdam: Elsevier, pp. 199–228.

Smalley, E. B., Marasas, W. F. O., Strong, F. M., Bamburg, J. R., Nichols, R. E., and Kosuri, N. R. (1970). Mycotoxicoses associated with moldy corn. In M. Herzberg, ed., *Toxic Microorganisms*. Washington, D.C.: U.S. Dept. of the Interior, pp. 163–173.

Smalley, E. B., Joffe, A. Z., Palyusik, M., Kurata, H., and Marasas, W. F. O. (1977). Panel on trichothecene toxins. In J. V. Rodricks, C. W. Hesseltine, and M. A. Mehlman, eds., *Mycotoxins in Human and Animal Health*. Park Forest, Ill.: Pathotox pp. 337–340.

Smirnova, V. A. (1945). Long-term consequences of septic angina in laryngeal organs. *Data Septic Angina, Ufa*, p. 145.

Smith, K. E., Cannon, M., and Cundliffe, E. (1975). Inhibition at the initiation level of eukaryotic protein synthesis by T-2 toxin. *Febs Lett.* **50**, 8–12.

Smith, T. K. (1980a). Influence of dietary fiber, protein and zeolite on zearalenone toxicosis in rats and swine. *J. Anim. Sci.* **50**, 278–285.

Smith, T. K. (1980b). Effect of dietary protein, alfalfa and zeolite on excretory patterns of $5',5',7',7'$-[$^3$H] zearalenone in rats. *Can. J. Physiol. Pharmacol.* **58**, 1251–1255.

Snyder, W. C., and Hansen, H. N. (1940). The species concept in *Fusarium. Am. J. Bot.* **27**, 64–67.

Snyder, W. C., and Hansen, H. N. (1941). The species concept in *Fusarium* with reference to section Martiella. *Am. J. Bot.* **28**, 738–742.

Snyder, W. C., and Hansen, H. N. (1945). The species concept in *Fusarium* with reference to Discolor and other sections. *Am. J. Bot.* **32**, 657–666.

Snyder, W. C., and Hansen, H. N. (1954). Variation and speciation in the genus *Fusarium. Ann. N.Y. Acad. Sci.* **60**, 16–23.

Snyder, W. C., and Toussoun, T. A. (1965). Current status of taxonomy in *Fusarium* species and in their perfect stages. *Phytopathology* **55**, 833–837.

Snyder, W. C., Hansen, H. N., and Oswald, W. (1957). Cultivars of the fungus *Fusarium. J. Madras Univ.* **B27**, 182–187.

Snyder, W. C., Georgopoulos, S. G., Webster, R. K., and Smith, S. N. (1975). Sexuality and genetic behavior in the fungus *Hypomyces (Fusarium) solani* f. sp. *cucurbitae. Hilgardia* **43**, 161–185.

Sorokin, N. (1890). *To Some Diseases of Cultured Plants of South Ussuri Land, Seaside District.* Kazan: Imperial Univ. Press.

Speers, G. M., Meronuck, R. A., Barnes, D. M., and Mirocha, C. J. (1971). Effect of feeding *Fusarium roseum* sp. *graminearum* contaminated corn and the mycotoxin F-2 on the growing chick and laying hen. *Poult. Sci.* **50**, 627–633.

Speers, G. M., Mirocha, C. J., and Christensen, C. M. (1973). Effect of feeding *F. tricinctum* and *F. roseum* isolate "oxyrose" invaded corn and the purified T-2 mycotoxin on S.C.W.L. hens. *Poult. Sci.* **52**, 2088 (Abstr.).

Speers, G. M., Mirocha, C. J., Christensen, C. M., and Behrens, J. C. (1977). Effects on laying hens of feeding corn invaded by two species of *Fusarium* and pure T-2 mycotoxin. *Poult. Sci.* **56**, 98–102.

Speigelhalder, B., Eisenbrand, G., and Preussmann, R. (1979). Contamination of beer with trace quantities of N-Nitrosodimethylamine. *Food Cosmet. Toxicol.* **17**, 29–31.

Spesivtseva, N. A. (1960). *Mycoses and Mycotoxicoses*, 1st ed. Moscow: Kolos.

Spesivtseva, N. A. (1964a). *Mycoses and Mycotoxicoses*, 2nd ed. Moscow: Kolos.

Spesivtseva, N. H. (1964b). Fusariotoxicosis in poultry. In *Mycoses and Mycotoxicoses*, 2nd ed. Moscow: Kolos, p. 435–447.

Spesivtseva, N. H. (1967). Fusariotoxicosis in horses. *Proc. All-Union Inst. Vet. Sanit.* **28**, 11–13.

Spesivtseva, M. A., and Kurmanov, I. A. (1960). Mikotoksikozy. *Zhivotnykh Vet. J.* **10**, 351–361 Moscow.

Sphon, J. A., Driefuss, P. A., and Schulten, H.-R. (1977). Field desorption mass spectrometry of mycotoxins and mycotoxin mixtures and its application as a screening technique for foodstuffs. *J. Assoc. Off. Anal. Chem.* **60**, 73–82.

Springer, J. P., Clardy, J., Cole, R. J., Kirksey, J. W., Hill, R. K., Carlson, R. M., and Isidor, J. L. (1974). Structure and synthesis of moniliformin, a novel cyclobutane microbial toxin. *J. Am. Chem. Soc.* **96**, 2267–2268.

Stähelin, H., Kalberer-Rusch, M. E., Singer, E., and Lazary, S. (1968). Über einige biologische Wirkungen des Cytostaticum Diacetoxyscirpenol. *Arzneim. Forsch.* **18**, 989–994.

Stahl, C., Vanderhoef, L. N., Siegel, N., and Helgeson, J. P. (1973). *Fusarium tricinctum* T-2 toxin inhibits auxin-promoted elongation in soybean hypocotyl. *Plant Physiol.* **52**, 663–666.

Stahr, H. M. (1975). *Mycotoxins in Methods Manual.* Ames: Iowa State University, pp. 170–178.

Stahr, H. M. (1976). Investigation of analytical methods for scirpene toxins in food and feed. Ph.D. dissertation, Iowa State University, Ames.

Stahr, H. M., and Kraft, A. A. (1977). Analysis of some scirpene toxins in foods and feeds. *IFT 37th Ann. Meet. Philadelphia*, p. 113 (Abstr. 154.)

Stahr, H. M., Ross, P. F., Hyde, W., and Obioha, W. (1978). Scirpene toxin analyses of feed associated with animal intoxication. *Appl. Spectrosc.* **32**, 174–183.

Stahr, H. M., Kraft, A. A., and Schuh, M. (1979). The determination of T-2 toxin, diacetoxyscirpenol and deoxynivalenol in foods and feeds. *Appl. Spectrosc.* **33**, 294–297.

Stamatovic, W., Ljesevic, Z., and Djurickovic, S. (1963). On a fungal alimentary intoxication in swine (vulvovaginitis suum). *Veterinarski Glas.* **6**, 507–510.

Stanford, G. K., Hood, R. D., and Hyes, A. W. (1975). Effect of prenatal administration of T-2 toxin to mice. *Res. Commun. Chem. Pathol. Pharmacol.* **10**, 743–746.

Stankushev, K., Duparinova, M., Dzhurov, A., Vranska, T., and Tikhova, D. (1977). Parasitic *Fusarium* species in corn and attempted isolation of phytoestrogens from them. *Vet. Med. Nauk.* **14**, 33–39.

Steele, J. A., Lieberman, J. R., and Mirocha, C. J. (1974). Biogenesis of zearalenone (F-2) by *Fusarium roseum* "Graminearum." *Can. J. Microbiol.* **20**, 531–534.

Steele, J. A., Mirocha, C. J., and Pathre, S. V. (1976). Metabolism of zearalenone by *Fusarium roseum* "Graminearum." *J. Agric. Food Chem.* **24**, 89–97.

Steele, J. A., Mirocha, C. J., and Pathre, S. V. (1977a). Metabolism of zearalenone by *Fusarium roseum. J. Toxicol. Environ. Health* **3**, 35–42.

Steele, J. A., Mirocha, C. J., and Pathre, S. V. (1977b). Metabolism of zearalenone by *Fusarium roseum.* In J. V. Rodricks, C. W. Hesseltine, and M. A. Mehlman, eds., *Mycotoxins in Human and Animal Health.* Park Forest, Ill.: Pathotox, pp. 393–401.

Steinberger, G. K., Britt, R. H., Enzmann, D. R., Finlay, J. L., and Arvin, A. M. (1983). *Fusarium* brain abscess. Case report. *J. Neurosurg.* **58**, 598–601.

Steyn, M., Thiel, P. G., and van Schalkwyk, G. C. (1978a). Isolation and purification of moniliformin. *J. Assoc. Off. Anal. Chem.* **61**, 578–580.

Steyn, P. S. (1981). Multimycotoxin Analysis. *Pure Appl. Chem.* **53**, 891–902.

Steyn, P. S., Vleggaar, R., Rabie, C. J., Krick, N. P. J., and Harington, J. S. (1978b). Trichothecene mycotoxins from *Fusarium sulphureum. Phytochemistry* **17**, 949–951.

Steyn, P. S., Wessels, P. L., and Marasas, W. F. O. (1979). Pigments from *Fusarium moniliforme* Sheldon. *Tetrahedron* **35**, 1551–1555.

Stipanovic, R. D., and Schroeder, H. W. (1975). Zearalenone and 8'-hydroxyzearalenone from *Fusarium roseum. Mycopathologia* **57**, 77–78.

Stob, M. (1973). Estrogen in foods. In *Toxicants Occurring Naturally in Foods.* Washington, D.C.: National Academy of Science, pp. 550–557.

Stob, M., Baldwin, R. S., Tuite, J., Andrews, F. N., and Gillette, K. G. (1962). Isolation of an anabolic, uterotrophic compound from corn infected with *Gibberella zeae. Nature* **196**, 1318.

Stojanovic, Z., Popovic, M., Kovcin, S., and Gagrein, M. (1978). Effect of moldy corn contaminated with *Fusarium* toxins on the performance and health of pigs. *Savrem Poljopr.* **26**, 5–18.

Stoloff, L. (1971, 1972, 1975–1978). Reports on Mycotoxins. *J. Assoc. Off. Anal. Chem.* **54**, 305–309; **55**, 260–265; **58**, 213–217; **59**, 317–323; **60**, 348–353; **61**, 340–346.

Stoloff, L. (1972). Analytical methods for mycotoxins. *Clin. Toxicol.* **5**, 465–494.

Stoloff, L. (1973). Fluorodensitometry of Mycotoxins. In J. C. Touchstone, ed., *Quantitative Thin-Layer Chromatography.* New York: Wiley pp. 95–122.

Stoloff, L. (1976). Occurrence of mycotoxins in food and feeds. In J. V. Rodricks, ed., *Mycotoxins and their Fungal Related Food Problems* (Advances in Chemistry Series No. 149). Washington, D.C.: American Chemical Society, pp. 24–50.

Stoloff, L., and Dalrymple, B. (1977). Aflatoxin and zearalenone in dry-milled corn products. *J. Assoc. Off. Anal. Chem.* **60**, 579–582.

Stoloff, L., and Francis, O. J., Jr. (1980). Survey for aflatoxins and zearalenone in canned and frozen sweet corn. *J. Assoc. Off. Anal. Chem.* **63**, 180–181.

Stoloff, L., Beckwith, A. C., and Cushmac, M. E. (1968). TLC spotting solvent for aflatoxin. *J. Assoc. Off. Anal. Chem.* **51**, 65–67.

Stoloff, L., Nesheim, S., Yin, L., Rodricks, J. V., Stack, M., and Campbell, A. D. (1971). A multimycotoxin detection method for aflatoxins, ochratoxins, zearalenone, sterigmatocystin and patulin. *J. Assoc. Off. Anal. Chem.* **54**, 91–97.

Stoloff, L., Henry, S., and Francis, O. J., Jr. (1976). Survey of aflatoxin and zearalenone in 1973 corn crop stored on farms and in country elevators. *J. Assoc. Off. Anal. Chem.* **59**, 118–121.

Strukov, A. I. (1947). Pathological changes in animal tissues caused by toxin from overwintered cereals. *Acta Chkalov Inst. Epidemiol. Microbiol. 2nd Commun.*, pp. 117–119.

Strukov, A. I., and Mironov, S. G. (1944). Toxic products from overwintered cereals tested on rabbit skin reaction and mucous membranes. *Chkalov Inst. Epidemiol. Microbiol.*, p. 10.

Strukov, A. I., and Mironov, S. G. (1947). Pathological changes in animal tissues caused by toxin from overwintered cereals. *Acta Chkalov Inst. Epidemiol. Microbiol. 1st Commun.*, pp. 109–116.

Strukov, A. I., and Tischchenko, T. (1944). Pathomorphology of septic angina. *Proc. 1st Kharkov Med. Inst. Chkalov Inst. Epidemiol. Microbiol.*, pp. 53–84.

Strukov, A. I., and Tischchenko, M. A. (1947). To some suppositions on pathomorphology and pathogenesis of alimentary toxic aleukia. *Acta Chkalov Inst. Epidemiol. Microbiol.*, pp. 120–124.

Subramanian, C. V. (1951). Is there a "wild type" in genus *Fusarium*? *Proc. Natl. Inst. Sci. India* **17**, 403–411.

Subramanian, C. V. (1952). Fungi isolated and recorded from Indian Soils. *J. Madras Univ.* **22B**, 206–222.

Subramanian, C. V. (1954). Studies on South Indian fusaria. 3. Fusaria isolated from some crop plants. *J. Madras Univ.* **B, 24**, 21–46.

Subramanian, C. V. (1955). The ecological and taxonomic problems in fusaria. *Proc. Indian Acad. Sci.* **41B**, 102–109.

Suga, K. (1972). Diagnosis and therapy of mycotic keratitis. *Jpn. J. Clin. Ophthalmol.* **25**, 1859–1866.

Sugimoto, T., Miniamizawa, M., Takano, K., and Furukawa, Y. (1975). Zearalenon: Gas liquid chromatographic essay of zearalenone in grain. *J. Food Hyg. Soc. Japan* **17**, 12–18.

Sugimura, T., Sato, S., Nagao, M., Yahagi, T., Matsushima, T., Seino, Y., Takeuchi, M., and Kawachi, T. (1976). Overlapping of carcinogens and mutagens. In P. N. Magee *et al.*, eds., *Fundamentals in Cancer Prevention*. Baltimore, Md.: University Park Press, pp. 191–215.

Sukroongreung, S., Schapper, K. T., and Khachatourians, G. (1984). Survey of sensitivity of twelve yeast genera toward T-2 toxin. *Appl. Environ. Microbiol.* **48**, 416–419.

Sullivan, J. D., Jr., and Ikawa, M. (1972). Variations on inhibition of growth of five *Chlorella* strains by mycotoxins and other toxic substances. *J. Agric. Food Chem.* **20**, 291–292.

Suringa, D. W. R. (1970). Treatment of superficial onychomycosis with topically applied glutaraldehyde. *Arch. Derm.* **102**, 163.

Sutton, J. C. (1982). Epidemiology of wheat head blight and maize ear rot caused by *Fusarium graminearum*. *Can. J. Plant Pathol.* **4**, 195–209.

Sutton, J. C., Baliko, W., and Funnell, H. S. (1976). Evidence for translocation of zearalenone in corn plants colonized by *Fusarium graminearum*. *Can. J. Plant Sci.* **56**, 7–12.

Sutton, J. C., Baliko, W., and Liu, H. J. (1980a). Fungal colonization and zearalenone accumulation in maize ears injured by birds. *Can. J. Plant Sci.* **60**, 453–461.

Sutton, J. C., Baliko, W., and Funnell, H. S. (1980b). Relation of weather variables to incidence of zearalenone in corn in Southern Ontario. *Can. J. Plant Sci.* **60**, 149–155.

Suzuki, T., Hushino, Y., Kurisu, M., Nose, N., and Watanabe, A. (1978). Gas chromatographic determination of zearalenone in *Fusarium* culture and cereal. *Shokuhin Eiseigaku Zasshi* **19**, 201–207.

Suzuki, T., Kurisu, N., Nose, N., and Watanabe, A. (1981). Determination of butenolide (*Fusarium* toxin) by gas chromatography with an electron-capture detector. *J. Food Hyg. Soc. Japan* **22**, 197–202 (in Japanese).

Svojskaya, V. S. (1947). Experiments in isolating toxic substances from overwintered prosomillet. *Acta Chkalov Inst. Epidemiol. Microbiol.* pp. 45–54.

Swanson, S. P., Ramswamy, V., Beasley, V. R., Buck, W. B., and Burmeister, H. R. (1983). Gas-liquid chromatographic determination of T-2 toxin in plasma. *J. Assoc. Off. Anal. Chem.* **66**, 909–912.

Szathmary, C. I., Mirocha, C. J., Palyusik, M., and Pathre, S. V. (1976). Identification of mycotoxins produced by species of *Fusarium* and *Stachybotrys* obtained from Eastern Europe. *Appl. Environ. Microbiol.* **32**, 579–584.

Szathmary, C. I., Mirocha, C. J., Palyusik, M., and Pathre, S. V. (1977). Chemical and biological studies on the toxins of *Fusarium* and *Stachybotrys* strains. *Magy. Allatorv. Lapja* **32**, 455–461.

Szathmary, C., Galacz, J., Vida, L., and Alexander, G. (1980). Capillary gas chromatographic-mass spectrometric determination of some mycotoxins causing fusariotoxicoses in animals. *J. Chromatogr.* **191**, 327–331.

Szathmary, C. I. (1983). Trichothecene toxicosis and natural occurrence in Hungary. In Y. Ueno, ed., *Trichothecenes, Chemical, Biological and Toxicological Aspects.* Amsterdam: Elsevier, pp. 229–250.

Takeda, Y., Morishita, E., and Shibota, S. (1968). Metabolic products of fungi. 30. The structure of fuscofusarin. *Chem. Pharm. Bull.* **16**, 2213–2215.

Takeda, S., Ogasawara, K., Oh-hara, H., and Konishi, T. (1953). Study on the damaged wheat with *Fusarium* scab disease. 4. Animal experiments on toxicity of the fungus. *Proc. Hokkaido Inst. Hyg. Sci.* **4**, 22–28.

Takitani, S., Asabe, Y., Kato, T., Suzuki, M., and Lleno, Y. (1979). Spectrodensitometric determination of trichothecene mycotoxins with 4-(p-nitrobenzyl) pyridine on silica gel thin-layer chromatograms. *J. Chromatogr.* **172**, 335–342.

Takitani, S., and Asabe, Y. (1983). Thin-layer chromatographic analysis of trichothecene mycotoxins. In Y. Ueno, ed., *Trichothecenes, Chemical, Biological and Toxicological Aspects.* Amsterdam: Elsevier, pp. 113–120.

Takizawa, N., Noguchi, K., Iwatsu, H., and Ishibashi, M. (1956). Studies on Kashin-Beck's disease in Japan (Report 1). *Acta Pathol. Japon* **6**, 341.

Takizawa, N., Iwatsu, H., Ishibashi, M., and Noguchi, K. (1957). Studies on the Kashin-Beck's disease in Japan (Report 2). *Acta Pathol. Japon* **7**, 594–595.

Takizawa, N., Shimada, T., Kanisawa, M., and Noguchi, K. (1958). Studies on Kashin-Beck's disease in Japan (Report 3). *Acta Pathol. Japon* **8**, 431.

Takizawa, N., Noguchi, K., Suga, S., Matsumura, Y., Aiso, K., Ide, G., Kanisawa, M., Kayama, T., Terao, K., and Chia Chin Lin (1959). Studies on Kashin-Beck's disease in Japan (Report 4). The new endemic districts of this disease in Japan and Taiwan and the experimental studies of the injection of the extracts of plants and drinking water of the endemic district into rats. *Acta Pathol. Japon* **9**, 485.

Takizawa, N., Noguchi, K., Suga, S., Matsumura, Y., Aiso, K., Ide, G., Kanisawa, M., Terao, K., Suko, A., and Ooyabu, T. (1960). Studies on Kashin-Beck's disease in Japan (Report

5). Upon the new endemic districts in Japan and experimental research for the cause of this disease. *Acta Pathol. Japon* **10**, 250–251.

Takizawa, N., Noguchi, K., Matsumura, Y., Aiso, K., Kawase, O., Ide, G., Kanisawa, M., and Terao, K. (1961). Studies on Kashin-Beck's disease in Japan (Report 6). New endemic districts in Japan and experimental research for the cause of the disease. *Acta Pathol. Japon* **11**, 185–186.

Takizawa, N., Noguchi, K., Ichikuni, M., Kobo, K., Tachikawa, R., Terao, K., Ogata, M., Suzuki, J., Tateiwa, M., Ishikawa, Y., Kawamura, B., Yoshida, K.,Takahashi, M., Koyama, R., and Irone, K. (1963). Studies on Kashin-Beck's disease in Japan (Report 7). New prevalent districts and experimental studies on the cause of this disease. *Acta Pathol. Japon* **13**, 56.

Takizawa, N., Noguchi, K., Kobo, K., Miyaki, K., Suzuki, J., Kawamura, B., Ogata, M., Ide, G., Terao, K., and Iwazaki, I. (1964). Studies on Kashin-Beck's disease in Japan (Report 8). On the newly found prevalent districts in Japan and on the cause of Kashin-Beck's disease. *Acta Pathol. Japon* **14**, 246.

Talayev, B. T., Mogunov, B. I., and Sharbe, E. N. (1936). Alimentary hemorrhagic aleukia. *Nutr. Probl.* **5, 6**, 27–35.

Tamm, C. (1974). Antibiotic complex of the verrucarins and roridins. In W. Herz, H. Grisebach and G. W. Kirby, eds., *Progress in the Chemistry of Organic Natural Products*, Vol. 31. Springer Verlag, Vienna, pp. 64–117.

Tamm, C. (1977). Chemistry and biosynthesis of trichothecenes. In J. V. Rodricks, C. W. Hesseltine, and M. A. Mehlman, eds., *Mycotoxins in Human and Animal Health*. Park Forest, Ill.: Pathotox, pp. 209–228.

Tanaka, K., Amano, R., Kawada, K., and Tanabe, H. (1974). Gas-liquid chromatography of trichothecene mycotoxins. *J. Food Hyg. Soc. Japan* **15**, 195–200.

Tanaka, K., Minamisawa, M., Manabe, M., and Matuura, S. (1975). Biological test using brine shrimp. 1. The influence of mycotoxins on the brine shrimp. *Rep. Natl. Food Res. Inst.* **30**, 43–48, Tokyo.

Tannenbaum, S. R., Kraft, P., Baldwin, J., and Branz, S. (1977). The mutagenicity of methylbenzylnitrosamine and its a-acetoxy derivatives. *Cancer Lett.* **2**, 305–310.

Tashiro, F., Hairai, K., and Ueno, Y. (1979). Inhibitory effects of carcinogenic mycotoxins on deoxyribonucleic acid dependent ribonucleic acid polymerase and ribonuclease. *Appl. Environ. Microbiol.* **38**, 191–196.

Tashiro, F., Kawabata, Y., Naoi, M., and Ueno, Y. (1980). Zearalenone-estrogen receptor interaction and RNA synthesis in rat uterus. *Med. Mycol., Zbl. Bact.* Suppl. 8, 311–320. Gustav Fischer Verlag, Stuttgard, New York.

Tatarinov, D. I. (1945). Problems to be discussed in the clinical aspect and therapy of alimentary toxic aleukia. *Acta Septic Angina, Ufa*, p. 12.

Tatsuno, T. (1968). Toxicologic research on substances from *Fusarium nivale*. *Cancer Res.* **28**, 2393–2396.

Tatsuno, T. (1969). Biochemical aspects of *Fusarium* toxicosis. *Japan Biochem. Soc.* **41**, 153–171.

Tatsuno, T. (1976). Recherches chimiques et biologiques sur les mycotoxines des *Fusariums*. *Ann. Pharmacol. Fr.* **34**, 25–29.

Tatsuno, T. (1983). Chemical synthesis of trichothecenes. In Y. Ueno, ed., *Trichothecenes, Chemical, Biological and Toxicological Aspects*. Amsterdam: Elsevier, pp. 47–59.

Tatsuno, T., Morooko, N., Saito, M., Enomoto, M., Umeda, M., Okhubo, K., and Tsunoda, H. (1966). Toxicological studies on the toxic metabolites of *Fusarium nivale*. *Folia Pharmacol. Jpn.* **62**, 26–27.

Tatsuno, T., Saito, M., Enomato, M., and Tsunoda, H. (1968a). Nivalenol, a toxic principle of *Fusarium nivale*. *Chem. Pharmacol. Bull.* **16**, 2519–2520, Tokyo.

Tatsuno, T., Ueno, Y., Neno, I., Morito, Y., Hosoya, M., Saito, M., Enomoto, M., Okubo, K., Morooko, S., and Makano, N. (1968b). Toxicological studies on the reddish wheat by *Fusarium* sp. (*F. Nivale*). *Folia Pharmacol. Jpn.* **64**, 121–122.

Tatsuno, T., Fujimoto, Y., and Morita, Y. (1969). Toxicological research on substances from *Fusarium nivale*. 3. The structure of nivalenol and its monoacetate. *Tetrahedron Lett.* **33**, 2823–2826.

Tatsuno, T., Morita, Y., Tsunoda, H., and Umeda, M. (1970). Recherches toxicologiques des substances metaboliques du *Fusarium nivale*. 7. La troisieme substance metabolique de *F. nivale*, le diacetate de nivalenol. *Chem. Pharmacol. Bull.* **18**, 1485–1487, Tokyo.

Tatsuno, T., Saito, M., Kubota, Y., and Tsunoda, H. (1971). Recherches toxicologiques des substances metaboliques du *Fusarium nivale*. 8. La quatrieme substance metabolique de *Fusarium nivale*. *Chem. Pharmacol. Bull.* **19**, 1498–1500, Tokyo.

Tatsuno, T., Ohtsubo, K., and Saito, M. (1973). Chemical and biological detection of 12,13-epoxytrichothecenes isolated from *Fusarium* species. *Pure Appl. Chem.* **35**, 309–313.

Taub, D., Girotra, N. N., Hoffsommer, R. D., Kuo, C. H., Slates, H. L., Weber, S., and Wendler, N. L. (1968). Total synthesis of the macrolide zearalenone. *Tetrahedron* **24**, 2443–2461.

Taylor, T. F., and Watson, W. H. (1976). The structure of hydrated 8′-hydroxyzearalenone, $C_{18}H_{22}O_6 \cdot H_2O$. An estrogenic syndrome producing microtoxin. *Acta Cryst.* **B32**, 710–714.

Tenk, I., Fodor, E., and Szathmary, C. (1982). The effect of pure *Fusarium* toxins (T-2, F-2, DAS) on the microflora of the gut and plasma glucocorticoid levels in rat and swine. *Zbl. Bakt. Hyg., I. Abt. Orig. A* **252**, 384–393.

Terao, K. (1983). The target organella of trichothecenes in rodents and poultry. In Y. Ueno, ed., *Trichothecenes, Chemical, Biological and Toxicological Aspects*. Amsterdam: Elsevier, pp. 147–162.

Terao, K., and Ueno, Y. (1978). Morphological and functional damage to cells and tissues. In K. Uraguchi and M. Yamazaki, eds., *Toxicology, Biochemistry and Pathology of Mycotoxins*. New York: Wiley, pp. 189–238.

Terao, K., Kera, K., and Yazima, T. (1978). The effects of trichothecene toxins on the bursa of Fabricius in day-old chicks. *Virchows Arch. B. Cell. Path.* **27**, 359–370.

Teregulov, G. N. (1945). Clinical aspects and treatment of alimentary toxic aleukia by propaedeutic-therapy material of Bashkir Med. Inst. *Data on Septic Angina, Ufa*, p. 30.

Thiel, P. G. (1978). A molecular mechanism of the toxic action of moniliformin, a mycotoxin produced by *Fusarium moniliforme*. *Biochem. Pharmacol.* **27**, 483–486.

Thiel, P. G. (1981). The determination of moniliformin by high-performance liquid chromatography. In P. Krogh, ed., *Proc. 4th IUPAC Symp. Mycotoxins Phycotoxins, 24th–31st Aug., 1979, Lausanne, Switzerland*. Park Forest, Ill.: Pathotox.

Thomas, F., Eppley, R. M., and Trucksess, M. W. (1975). Rapid screening method for aflatoxin and zearalenone in corn. *J. Assoc. Off. Anal. Chem.* **58**, 114–116.

Thomas, J. W., Miller, E. R., and Marshall, S. (1973). Moldy corn for pigs, rats, mice, rabbits and cows. *J. Anim. Sci.* **37**, 292 (Abstr.).

Thouvenot, D. R., and Morfin, R. F. (1979). Quantitation of zearalenone by gas-liquid chromatography on capillary glass columns. *J. Chromatogr.* **170**, 165–175.

Thouvenot, D. R., and Morfin, R. F. (1980). Interferences of zearalenone, zearalenol or estradiol-17B with steroid-metabolizing enzymes of the human prostate gland. *J. Steroid Biochem.* **13**, 1337–1345.

Thouvenot, D. R., and Morfin, R. F. (1983). Radioimmunoassay for zearalenone and zearalanol in human serum: Production, properties, and use of porcine antibodies. *Appl. Environ. Microbiol.* **45**, 16–23.

Tidd, B. K. (1967). Phytotoxic compounds produced by *Fusarium equiseti.* 3. Nuclear magnetic resonance spectra. *J. Chem. Soc.* C, 218–220.

Tikhonov, V. A. (1977). Role of the hereditary factors in the etiology of endemic Kashin-Beck disease. *Klin. Med.* **55**, 142–143, Moscow.

Tochina, Y. (1933a). On scab of cereals and grains. 1. *Byochugai Zashi* **20**, 106–114.

Tochina, Y. (1933b). On scab of cereals and grains. 2. *Byochugai Zashi* **20**, 179–188.

Tomina, M. V. (1948). Distribution of *Fusarium* fungal toxin from overwintered cereal in animal organs. Abstract in Republic Conference on Alimentary Toxic Aleukia, Moscow, p. 10.

Tookey, H. L., Yates, S. G., Ellis, J. J., Grove, M. D., and Michols, R. E. (1972). Toxic effects of a butenolide mycotoxin and *Fusarium tricinctum* in cattle. *J. Am. Vet. Med. Assoc.* **160**, 1522–1526.

Torrey, L. T. (1983). Yellow Rain: Is it really a weapon? *New Sci.* **99**, 350–351.

Tostanovskaya, A. A. (1956). Toxicological characteristics of grain infected with *Fusarium sporotrichiella* var. *poae* and other fungal species from cereal crops in the USSR. *Publ. Acad. Sci. Ukr. SSR, Kiev,* p. 33.

Tostanovskaya, A. A., and Retmanskaya, U. M. (1951). A study in toxic grains infected with some fungi of the genus *Fusarium. Microbiol. J. Acad. Sci. Ukr. SSR* **1**, 13.

Toussoun, T. A., and Nelson, P. E. (1968). *Fusarium: A Pictorial Guide to the Identification of Fusarium Species According to the Taxonomic System of Snyder and Hansen.* University Park: Pennsylvania State University Press.

Toussoun, T. A., and Nelson, P. E. (1975). Variation and speciation in the fusaria. *Ann. Rev. Phytopathol.* **13**, 71–82.

Toussoun, T. A., and Nelson, P. E. (1976). *Fusarium: A Pictorial Guide to the Identification of Fusarium Species According to the Taxonomic System of Snyder and Hansen,* 2nd ed. University Park: Pennsylvania State University Press.

Trenholm, H. L., Warner, R., and Farnworth, E. R. (1980). Gas chromatographic detection of the mycotoxin zearalenone in blood serum. *J. Assoc. Off. Anal. Chem.* **63**, 604–611.

Trenholm, H. L., Cochrane, W. P., Cohen, H., Elliot, J. I., Farnworth, E. R., Friend, D. W., Hamilton, M. G., Neish, G. A., and Standish, J. F. (1981a). Survey of vomitoxin contamination of the 1980 white winter wheat crop in Ontario (Canada). *J. Am. Oil Chem. Soc.* **58**, 992A–994A.

Trenholm, H. L., Warner, R. M., and Farnworth, E. R. (1981b). High performance liquid chromatographic method using fluorescence detection for quantitative analysis of zearalenone and α-zearalenol in blood plasma. *J. Assoc. Off. Anal. Chem.* **64**, 302–310.

Trolle, B. (1969). Danish grain quality research. In *Eur. Brew. Conv. Proc. 12th Congr.* pp. 99–106. Amsterdam: Elsevier.

Trucksess, M. W., Nesheim, S., and Epply, R. M. (1984). Thin layer chromatographic determination of deoxynivalenol in wheat and corn. *J. Assoc. Off. Chem.* **67**, 40–43.

Tseng, T.-C. (1983). Natural occurrence of trichothecenes in Taiwan. In Y. Ueno, ed., *Trichothecenes, Chemical, Biological and Toxicological Aspects.* Amsterdam: Elsevier, pp. 290–296.

Tsubouchi, H., Yamamoto, K., Moriyama, S., and Sakabe, Y. (1976). Effects of mycotoxins on hatching of brine shrimp eggs. *Nagoya-shi Eisei Kenkyush Ho* **22**, 50–53.

Tsunoda, H. (1970). Microorganisms which deteriorate stored cereals and grains. *Proc. 1st U.S.-Japan Conf. Toxic Microorganisms,* Washington, D.C., pp. 143–162.

Tsunoda, H., Tsuruto, O., Matsunami, S., and Ishii, S. (1957). Researches on the microorganisms which deteriorate the stored cereals and grains. 14. Studies on rice parasitic molds, *Gibberella* and *Fusarium. Proc. Food Res. Inst.* **12,** 27–31.

Tsunoda, H., Tsuruta, O., and Matsunami, S. (1958). Researches on the microorganisms which deteriorate the stored cereals and grains. 16. Studies on rice parasitic molds, *Gibberella and Fusarium.* Proc. Food Res. Inst. **13,** 26–28.

Tsunoda, M., Toyazaki, N., Morooka, N., Nakano, N., Yoshiyawa, H., Okuba, K., and Isoda, M. (1968). Researches on the microorganisms which deteriorate the stored cereals and grain. 34. Detection of injurious strains and properties of their toxic substances of scab, *Fusarium* blight grown on the wheat. *Proc. Food Res. Inst.* **23,** 89–116.

Tsvetkov, I. I. (1946). On the incidence of alimentary toxic aleukia in urban settlements. *Jubilee Proc. Astrakhansk Lunacharsk State Med. Inst.* **8,** 142–146.

Tuite, J., and Caldwell, R. (1971). Infection of corn seed with *Helminthosporium maydis* and other fungi in 1970. *Plant Dis. Rep.* **55,** 387–389.

Tuite, J., Shaner, G., Rombo, F., Foster, J., and Caldwell, W. (1974). The *Gibberella* ear rot epidemics of corn in Indiana in 1965 and 1972. *Cereal Sci. Today* **19,** 238–241.

Turner, G. V., Phillips, T. D., Heidelbaugh, N. D., and Russell, L. H. (1983). High pressure liquid chromatographic determination of zearalenone in chicken tissues. *J. Assoc. Off. Anal. Chem.* **66,** 102–104.

Tuyns, A. J. (1979). Epidemiology of alcohol and cancer. *Cancer Res.* **39,** 2840–2843.

Udagawa, S. I., Ichinoe, M., and Kurata, H. (1970). Occurrence and distribution of mycotoxin producers in Japanese foods. In M. Herzberg, ed., *Toxic Microorganisms.* Washington, D.C.: UJNR Joint Panels on Toxic Microorganisms and U.S. Dept. of the Interior, pp. 174–184.

Ueda, A., Ono, T., Yamagiwa, S. (1967). Neuropathological studies on bean-hulls poisoning of horses. *Res. Bull. Obihiro Univ.* **5,** 149–152.

Ueno, Y. (1970). Inhibition of protein synthesis in animal cells by nivalenol and related metabolites; toxic principles of rice infected with *Fusarium nivale.* In M. Herzberg, ed., *Toxic Microorganisms.* Washington, D.C.: UJNR Joint Panels on Toxic Microorganisms and Dept. of the Interior, pp. 76–79.

Ueno, Y. (1971). Toxicological and biological properties of fusarenon-X, a cytotoxic mycotoxin of *Fusarium nivale* Fn-2B. In J. V. Rodricks, C. W. Hesseltine, and M. A. Mehlman, eds., *Mycotoxins in Human and Animal Health.* Park Forest, Ill.: Pathotox, pp. 163–178.

Ueno, Y. (1973a). Akakabi toxins (*Fusarium* toxins). 1. *J. Food Hyg. Soc. Japan* **14,** 403–414.

Ueno, Y. (1973b). Akakabi toxins (*Fusarium* toxins). 2. *J. Food Hyg. Soc. Japan* **14,** 501–510.

Ueno, Y. (1977a). Trichothecenes: Overview address. In J. V. Rodricks, C. W. Hesseltine, and M. A. Mehlman, eds., *Mycotoxins in Human and Animal Health* Park. Forest, Ill.: Pathotex, pp. 189–207.

Ueno, Y. (1977b). Mode of action of trichothecene mycotoxins. *Pure Appl. Chem.* **49,** 1737–1745.

Ueno, Y. (1980a). Toxicological evaluation of trichothecene mycotoxins. In D. Eaker and T. Wadström, eds., *Natural Toxins.* New York: Pergamon, pp. 663–671.

Ueno, Y. (1980b). Trichothecene mycotoxins mycology, chemistry and toxicology. *Adv. Nutr. Res.* **3,** 301–353.

Ueno, Y. (1980c). Trichothecene mycotoxins, mycology, chemistry and toxicology. In H. H.

Draper, ed., *Advances in Nutritional Research*, Vol. 3. New York: Plenum, pp. 301–353.

Ueno, Y. (1983a). Historical background of trichothecene problems. In Y. Ueno, ed., *Trichothecenes, Chemical, Biological and Toxicological Aspects.* Amsterdam: Elsevier, pp. 1–6.

Ueno, Y. (1983b). Biological detection of trichothecenes. In Y. Ueno, ed., *Trichothecenes, Chemical, Biological and Toxicological Aspects.* Amsterdam: Elsevier, pp. 125–133.

Ueno, Y. (1983c). General toxicity. In Y. Ueno, ed., *Trichothecenes, Chemical, Biological and Toxicological Aspects.* Amsterdam: Elsevier, pp. 135–146.

Ueno, Y., and Fukushima, K. (1968). Inhibition of protein and DNA synthesis in Ehrlich ascites tumour by nivalenol, a toxin principle of *Fusarium nivale*-growing rice. *Experientia* **24,** 10, 1032–1033.

Ueno, Y., and Hosoya, M. (1970). Bioassay of toxic principles of rice infected with *Fusarium nivale* employing rabbit reticulocytes. *Proc. 1st U.S. Japan Conf. Toxic Microorganisms,* Washington, D.C., pp. 80–81.

Ueno, Y. and Kubota, K. (1976). DNA-attacking ability of carcinogenic mycotoxins in recombination-deficient mutant cells of *Bacillus subtilis. Cancer Res.* **36,** 445–451.

Ueno, Y., and Matsumoto, H. (1975a). Mode of action of trichothecene mycotoxin. *Proc. 1st Int. Congr. IAMS* **4,** 314–323, Tokyo.

Ueno, Y., and Matsumoto, H. (1975b). Inactivation of some thiol-enzymes by trichothecene mycotoxins from *Fusarium* species. *Chem. Pharmacol. Bull.* **23,** 2439–2442.

Ueno, Y., and Ohta, M. (1977). Microsomal biotransformation of trichothecene mycotoxins. *Acta Microbiol. Acad. Sci. Hung.* **24,** 105–106.

Ueno, Y., and Shimada, M. (1974). Reconfirmation of the specific nature of reticulocytes bioassay system to trichothecene mycotoxins of *Fusarium* spp. *Chem. Pharmacol. Bull.* **22,** 2744–2746.

Ueno, Y., and Ueno, I. (1978). Toxicology and biochemistry of mycotoxins. In K. Uraguchi and M. Yamazaki, eds., *Toxicology, Biochemistry and Pathology of Mycotoxins.* Tokyo: Kadanscha LTD, pp. 107–188.

Ueno, Y., and Yagasaki, S. (1975). Toxicological approaches to the metabolites of fusaria. 10. Accelerating effect of zearalenone on RNA and protein synthesis in the uterus of ovariectomized mice. *Jpn. J. Exp. Med.* **42,** 199–205.

Ueno, Y., and Yamakawa, H. (1970). Antiprotozoal activity of scirpene mycotoxins of *Fusarium nivale* Fn 2B. *Jpn. J. Exp. Med.* **40,** 385–390.

Ueno, Y., Hosoya, M., and Tatsuno, T. (1967). Effect of toxin of *Fusarium nivale* on the polymer synthesis of reticulocytes. *Biochemistry* **39,** 708 (in Japanese).

Ueno, Y., Hosoya, M., Morita, Y., Ueno, I., and Tatsuno, T. (1968). Inhibition of the protein synthesis in rabbit reticulocyte by nivalenol, a toxic principle isolated from *Fusarium nivale*-growing rice. *J. Biochem.* **64,** 479–485.

Ueno, Y., Hosoya, M., and Ishikawa, Y. (1969a). Inhibitory effects of mycotoxins on the protein synthesis in rabbit reticulocytes. *J. Biochem.* **66,** 419–422.

Ueno, Y., Ishikawa, Y., Amakai, K., and Tsunoda, H. (1969b). An investigation on the culture condition of *Fusarium nivale* for the production of toxic metabolites. *J. Jpn. Biochem. Soc.* **41,** 505.

Ueno, Y., Ueno, T., Tatsuno, T., Ohokubo, K., and Tsunoda, H. (1969c). Fusarenon-X a toxic principle of *Fusarium nivale*-culture filtrate. *Experientia* **25,** 1062.

Ueno, Y., Ishikawa, Y., Amakai, K., Nakajima, M., Saito, M., Enomot, M., and Ohtsubo, K. (1970a). Comparative study on skin-necrotizing effect of scirpene metabolites of *Fusaria. Jpn. J. Exp. Med.* **40,** 33–38.

Ueno, Y., Saito, K., and Tsunoda, H. (1970b). Isolation of toxic principles from the culture filtrate of *Fusarium nivale*. *Proc. 1st U.S. Jap. Conf. Toxic Microorganisms*, Washington, D.C., p. 120.

Ueno, Y., Ishikawa, Y., Saito-Amakai, K., and Tsunoda, H. (1970c). Environmental factors influencing the production of fusarenon-X, a cytotoxic mycotoxin of *Fusarium nivale* Fn2B. *Chem. Pharmacol. Bull.* **18**, 304–312.

Ueno, Y., Ishikawa, Y., Nakajima, M., Sakai, K., Ishii, K., Tsunoda, H., Saito, M., Enomoto, M., Ohtsubo, K., and Umedo, M. (1971a). Toxicological approaches to the metabolites of fusaria. 1. Screening of toxic strains. *Jpn. J. Exp. Med.* **41**, 257–272.

Ueno, Y., Ueno, T., Amakai, K., Ishikawa, Y., Tsunoda, H., Okubo, K., Saito, M., and Enomoto, M. (1971b). Toxicological approaches to the metabolites of fusaria. 2. Isolation of fusarenon-X from the culture filtrate of *Fusarium nivale* Fn2B. *Jpn. J. Exp. Med.* **41**, 507–519.

Ueno, Y., Ueno, T., Iitoy, Y., Tsunoda, M., Enomoto, M., and Ohtsubo, K. (1971c). Toxicological approaches to the metabolites of fusaria. 3. Acute toxicity of fusarenon-X. *Jpn. J. Exp. Med.* **41**, 521–539.

Ueno, Y., Ishii, K., Sakai, K., Kanaeda, S., Tsunoda, H., Tanaka, T., and Enomoto, M. (1972a). Toxicological approaches to the metabolites of fusaria. 4. Microbial survey on "Bean-hulls poisoning of horses" with the isolation of toxic trichothecenes, neosolaniol and T-2 toxin of *Fusarium solani* M-1-1. *Jpn. J. Exp. Med.* **42**, 187–203.

Ueno, Y., Sato, N., Ishii, K., Sakai, K., and Enomoto, M. (1972b). Toxicological approaches to the metabolites of fusaria. 5. Neosolaniol, T-2 toxin and butenolide, toxic metabolites of *Fusarium sporotrichioides* NRRL 3510 and *Fusarium poae* 3287. *Jpn. J. Exp. Med.* **42**, 461–472.

Ueno, Y., Ishii, K., Sato, N., Shimada, N., Tsunoda, H., Sawano, M., and Enomoto, M. (1973a). Screening of trichothecene-producing fungi and the comparative toxicity of isolated mycotoxins. *Jpn. J. Pharmacol.* **23**, 133 (Abstr.).

Ueno, Y., Nakajima, M., Sakai, K., Ishii, K., Sato, N., and Shimada, N. (1973b). Comparative toxicology of trichothecene mycotoxins: Inhibition or protein synthesis in animal cells. *J. Biochem.* **74**, 285–296.

Ueno, Y., Sato, N., Ishii, K., Shimada, N., Tokito, K., Enomoto, N., Saito, M., Ohtsubo, K., and Ueno, I. (1973c). Subacute toxicity of trichothecene mycotoxins of *Fusarium* to cats and mice. *Jpn. J. Pharmacol.* **23**, 133 (Abstr.)

Ueno, Y., Sato, N., Ishii, K., Sakai, K., Tsunoda, H., and Enomoto, M. (1973d). Biological and chemical detection of trichothecene mycotoxins of *Fusarium* species. *Appl. Microbiol.* **25**, 699–704.

Ueno, Y., Ishii, K., Sato, N., and Ohtsubo, K. (1974a). Toxicological approaches to the metabolites of fusaria. 6. Vomiting factor from moldy corn infected with *Fusarium* spp. *Jpn. J. Exp. Med.* **44**, 123–127.

Ueno, Y., Sato, N., and Ishii, K. (1974b). Trichothecene compounds—vomiting factor in *Fusarium* toxicoses. *Jpn. J. Pharmacol.* **245**, 40 (Abstr.).

Ueno, Y., Shimada, N., Yagasaki, S., and Enomoto, M. (1974c). Toxicological approaches to the metabolites of fusaria. 7. Effects of zearalenone on the uteri of mice and rats. *Chem. Pharmacol. Bull.* **22**, 2830–2835, Tokyo.

Ueno, Y., Sawano, M., and Ishii, K. (1975). Production of trichothecene mycotoxins by *Fusarium* species in shake culture. *Appl. Microbiol.* **30**, 4–9.

Ueno, Y., Ayaki, S., Sato, N., and Ito, T. (1977a). Fate and mode of action of zearalenone. *Ann. Nutr. Aliment.* **31**, 935–948.

Ueno, Y., Ishii, K., Sawano, M., Ohtsubo, K., Motsuda, Y., Tanaka, T., Kurata, H., and Ichinoe, M. (1977b). Toxicological approaches to the metabolites of *Fusaria*. 11.

Trichothecenes and zearalenone from *Fusarium* species isolated from river sediments. *Jpn. J. Exp. Med.* **47**, 177–184.

Ueno, Y., Kubata, K., Ito, T., and Nakamura, Y. (1978). Mutagenicity of carcinogenic mycotoxins in *Salmonella typhimurium*. *Cancer Res.* **38**, 536–542.

Uglov, V. A. (1913). *Wheat and Rye Grains of Ussuri and Amur District in Connection with their Chemical Composition and Analysis in General*. Leningrad, pp. 17–24.

Umedaa, M., Yamamoto, T., and Saito, S. (1972). DNA-strand breakage in HeLa cells induced by several mycotoxins. *Jpn. J. Exp. Med.* **42**, 527–535.

Uraguchi, K. (1971). Mycotoxins of fusaria. In M. Raskova, ed., *Pharmacology and Toxicology of Naturally Occurring Toxins*, Section 71, Vol. 2. International Encyclopaedia of Pharmacology and Therapeutics. Elmosford, N.Y.: Pergamon, pp. 247–298.

Uraguchi, K. and Yamazaki, M., eds. (1978). *Toxicology: Biochemistry and Pathology of Mycotoxins*. New York: Wiley.

Uraguchi, K., Sakai, Y., Tatsuno, T., Tsunoda, H., and Wakamatsu, H. (1958). Toxicological approach to metabolites of *Fusarium roseum* and other "red" molds growing on rice, wheat and other cereal grains. *Folia Pharmacol. Jpn.* **54**, 127–128.

Urakura, U. (1933). On wheat damage (8). *Byochugai Zashi* **20**, 390–397.

Urry, W. H., Wehrmeister, H. L., Hodge, E. B., and Hidy, F. H. (1966). The structure of zearalenone. *Tetrahedron Lett.* **27**, 3109–3114.

Vanderheuvel, W. J. H. (1968). The gas-liquid chromatographic behavior of the zearalenone, a new family of biologically active natural products. *Sep. Sci.* **3**, 151–163.

Van der Walt, S. J., and Steyn, D. G. (1943). Recent investigations into the toxicity of plants in the Union of South Africa, 13. *Onderstepoort J. Vet. Sci. Anim. Ind.* **18**, 207–224.

Van der Walt, S. J., and Steyn, D. G. (1945). Recent investigations into the toxicity of plants in the Union of South Africa, 15. *Onderstepoort J. Vet. Sci. Anim. Ind.* **21**, 45–55.

Van Rensburg, S. J., Purchase, I. F. H., and Van der Walt, S. J. (1971). Hepatic and renal pathology induced in mice by feeding fungal cultures. In I. F. H. Purchase, ed., *Symp. Mycotoxins Hum. Health*. London: Macmillan, pp. 153–161.

Vanyi, A., and Romvaryne, S. E. (1974). Fusariotoxicoses. 3. Study on cytotoxic effect of $F_2$-toxin (zearalenone) in monolayer tissue cultures. *Magy. Allatorv. Lapja* **29**, 191–194.

Vanyi, A., and Szailer, E. (1974). Investigation on the cytotoxic effect of F-2 toxin (zearalenone) in various monolayer cell cultures. *Acta Vet. Acad. Sci. Hung.* **24**, 407–412.

Vanyi, A. and Szeky, A. (1980a). Fusariotoxicoses. 6. The effect of F-2 toxin (zearalenone) on the spermatogenesis of male swine. *Magy. Allatorv. Lapja* **35**, 242–246.

Vanyi, A. and Szeky, A. (1980b). Fusariotoxicoses. 7. Disturbed spermatogenesis caused by zearalenone (F-2 fusariotoxin) and by imperfect illumination in guinea-cocks. *Magy. Allatorv. Lapja* **35**, 247–252.

Vanyi, A., Danko, G., Aldasy, P., Erös, T., and Szigeti, G. (1973a). Fusariotoxicosis. 1. A laboratory method for demonstration of $F_2$-toxin in vitro. *Magy. Allatorv. Lapja* **28**, 303–307.

Vanyi, A., Erös, T., Aldasy, G., Danko, G., and Szigeti, G. (1973b). Fusariotoxicosis. 2. A laboratory method for demonstration of $F_2$-toxin in vitro. *Magy. Allatorv. Lapja* **28**, 308–309.

Vanyi, A., Buza, L., and Szeky, A. (1974a). Fusariotoxicosis. 4. The effect of $F_2$ toxin (zearalenone) on the spermiogenesis of the carp. *Magy. Allatorv. Lapja* **29**, 457–461.

Vanyi, A., Szeky, A., and Romvaryne, Szailer, E. (1974b). Fusariotoxicoses. 5. The effect of $F_2$ toxin on the sexual activity of female swine. *Magy. Allatorv. Lapja* **29**, 723–730.

Vanyi, A., Szeremeredi, G., Quarini, L., and Romvaryne Szailer, E. (1974c). Fusariotoxicosis on a cattle farm. *Magy. Allatorv. Lapja* **29**, 544–546.

Vea, E. V., and Wright, V. F. (1973). Effect of F-2 and T-2 mycotoxins on the ATPase enzyme system of *T. confusum*. *Tribolium Inf. Bull.* **16**, 110.

Veindrach, G. M., and Fadeyeva, S. V. (1937). The blood-picture in septic granulocytic angina. *Kazan Med. J.* **9**, 1065–1072.

Vertinski, K. I., and Adutkevitch, V. A. (1948). Pathomorphological studies of cats who died of experimental alimentary toxic aleukia. In *Cereal Crops Overwintered Under Snow*. Moscow: USSR Ministry of Agriculture, pp. 80–86.

Veselá, D. and Vesely, D. (1983). The toxinogenic vomitoxin-producing fusaria isolated from wheat. *Vet. Med. Praha* **28**, 687–692.

Veselá D., Vesely, D., and Adamkova, A. (1981). The occurrence of zearalenone and zearalenone-producing fusaria in feeds. *Vet. Med. Praha* **26**, 737–741.

Veselá, D., Vesely, D., and Fassatová, O. (1982). Findings of toxinogenic fungus strains from the genera *Fusarium* and *Alternaria* on stored wheat. *Sov. UVTIZ-Ochr. Rostlin*, **18**, 253–258.

Vesely, D., Veselá, D., and Jelinek, R. (1982). Nineteen mycotoxins tested on chicken embryos. *Toxicol. Lett.* **13**, 239–245.

Vesonder, R. F. (1983). Natural occurrence in North America. In Y. Ueno, ed., *Trichothecenes, Chemical, Biological and Toxicological Aspects*. Amsterdam: Elsevier pp. 210–217.

Vesonder, R. F., and Ciegler, A. (1979). Natural occurrence of vomitoxin in Austrian and Canadian corn. *Eur. J. Appl. Microbiol. Biotechnol.* **8**, 237–240.

Vesonder, R. F., and Hesseltine, C. W. (1980). Vomitoxin: Natural occurrence on cereal grains and significance as a refusal and emetic factor to swine. *J. Process Biochem.* **16**, 12–15.

Vesonder, R. F., and Hesseltine, C. W. (1981). Metabolites of Fusarium. In P. E. Nelson, T. A. Toussoun, and R. J. Cook, eds., *Fusarium Diseases, Biology and Taxonomy*. University Park: Pennsylvania State University Press, pp. 350–364.

Vesonder, R. F., Ciegler, A., and Jensen, A. H. (1973). Isolation of the emetic principle from *Fusarium*-infected corn. *Appl. Microbiol.* **26**, 1008–1010.

Vesonder, R. F., Ciegler, A., Jensen, A. H., Rohwedder, W. K., and Weisleder, D. (1976). Co-identity of the refusal and emetic principle from *Fusarium*-infected corn. *Appl. Environ. Microbiol.* **31**, 280–285.

Vesonder, R. F., Ciegler, A., and Jensen, A. H. (1977a). Production of refusal factors by *Fusarium* strains on grains. *Appl. Environ. Microbiol.* **34**, 105–106.

Vesonder, R. F., Tjarks, L. W., Ciegler, A., Spencer, G. F., and Wallen, L. L. (1977b). 4-acetamido-2-butenoic acid from *Fusarium graminearum*. *Phytochemistry* **16**, 1296–1297.

Vesonder, R. F., Ciegler, A., Rogers, R. F., Burbridge, K. A., Bothast, R. J., and Jensen, A. H. (1978). Survey of 1977 crop year preharvest corn for vomitoxin. *Appl. Environ. Microbiol.* **36**, 885–888.

Vesonder, R. F., Ciegler, A., Burmeister, H. R., and Jensen, A. H. (1979a). Acceptance by swine and rats of corn amended with trichothecene. *Appl. Environ. Microbiol.* **38**, 344–346.

Vesonder, R. F., Ciegler, A., Rohwedder, W. K., and Eppley, R. (1979b). Re-examination of 1972 midwest corn for vomitoxin. *Toxicon* **17**, 658–660.

Vesonder, R. F., Tjarks, L. W., Rohwedder, W. K., Burmeister, H. R., and Laugal, J. A. (1979c). Equisetin, an antibiotic from *Fusarium equiseti* NRRL 5537, identified as derivative of N.methyl-2,4,-pyrollidone. *J. Antibiot.* **32**, 759–761.

Vesonder, R. F., Ellis, J. J., and Rohwedder, W. K. (1981a). Swine refusal factors elaborated by *Fusarium* strains and identified as trichothecenes. *Appl. Environ. Microbiol.* **41,** 323–324.

Vesonder, R. F., Ellis, J. J., and Rohwedder, W. K. (1981b). Elaboration of vomitoxin and zearalenone by *Fusarium* isolates and the biological activity of *Fusarium*-producing toxins. *Appl. Environ. Microbiol.* **42,** 1132–1134.

Vesonder, R. F., Ellis, J. J., Kwolek, W. F., and DeMarini, D. J. (1982). Production of vomitoxin on corn by *Fusarium graminearum* NRRL 5883 and *Fusarium roseum* NRRL 6101. *Appl. Environ. Microbiol.* **43,** 967–970.

Vinogradov, A. P. (1949). The causes of occurrence of Urov epidemics. *Proc. Biogeochem. Lab. Acad. Sci. USSR Press,* **9,** 17–24.

Wallace, E. M., Pathre, S. V., Mirocha, C. J., Robison, T. S., and Fenton, S. W. (1977). Synthesis of radiolabeled T-2 toxin. *J. Agr. Food Chem.* **25,** 836–838.

Walshe, M. M., and English, M. P. (1966). Fungi in nails. *Br. J. Derm.* **78,** 198.

Ware, G. M., and Thrope, C. W. (1978). Determination of zearalenone in corn by high pressure liquid chromatography and fluorescence detection. *J. Assoc. Off. Anal. Chem.* **61,** 1058–1062.

Warwick, G. R., and Harington, J. S. (1973). Some aspects of the epidemiology and etiology of oesophageal cancer with particular emphasis on the Transkei, S. Africa. *Adv. Cancer Res.* **17,** 81–229.

Watson, S. A., Mirocha, C. J., and Hayes, A. W. (1982). "Yellow Rain; What is it?" *Spec. Symp. ASPET/SOT Joint Meet.,* August 15–19, Louisville, Ky.

Weaver, G. A., Kurtz, H. J., Bates, F. Y., Chi, M. S., Mirocha, C. J., Behrens, J. C., and Robinson, T. S. (1978a). Acute and chronic toxicity of T-2 mycotoxin in swine. *Vet. Rec.* **103,** 531–535.

Weaver, G. A., Kurtz, H. J., Mirocha, C. J., Bates, F. Y., Behrens, J. C., Robison, T. S., and Gipp, W. F. (1978b). Mycotoxin-induced abortions in swine. *Can. Vet. J.* **19,** 72–74.

Weaver, G. A., Kurtz, H. J., Mirocha, C. J., Bates, F. Y., and Behrens, J. C. (1978c). Effect of T-2 toxin on porcine reproduction. *Can. Vet. J.* **19,** 310–314.

Weaver, G. A., Kurtz, H. J., Mirocha, C. J., Bates, F. Y., and Behrens, J. C. (1978d). Acute toxicity of the mycotoxin diacetoxyscirpenol in swine. *Can. Vet. J.* **19,** 267–271.

Weaver, G. A., Kurtz, H. J., Mirocha, C. J., Bates, F. Y., Behrens, J. C., Robison, T. S., and Swanson, S. P. (1980). The failure of T-2 mycotoxin to produce hemorrhaging in dairy cattle. *Can. Vet. J.* **21,** 210–213.

Weaver, G. A., Kurtz, H. J., and Bates, F. Y. (1981). Deacetoxyscirpenol toxicity in pigs. *Res. Vet. Sci.* **31,** 131–135.

Wehner, F. C., Marasas, W. F. O., and Thiel, P. G. (1978). Lack of mutagenicity to *Salmonella typhimurium* of some *Fusarium* mycotoxins. *Appl. Environ. Microbiol.* **35,** 659–662.

Wehrmeister, H. L., and Robertson, D. E. (1968). Total synthesis of the macrocyclic lactone, dideoxyzearalone. *J. Org. Chem.* **33,** 4173–4176.

Wei, C.-M., and McLaughlin, C. S. (1974). Structure-function relationship in the 12,13-epoxytrichothecenes. Novel inhibitors of protein synthesis. *Biochem. Biophys. Res. Commun.* **57,** 838–844.

Wei, R.-D., Strong, F. M., Smalley, E. B., and Schnoes, H. K. (1971). Chemical interconversion of T-2 and HT-2 toxins and related compounds. *Biochem. Biophys. Res. Commun.* **45,** 396–401.

Wei, R.-D., Smalley, E. B., and Strong, F. M. (1972). Improved skin test for detection of T-2 toxin. *Appl. Microbiol.* **23,** 1029–1030.

Wei, C. M., Hansen, B. S., Vaughan, M. H., Jr., and McLaughlin, C. S. (1974). Mechanism of action of the mycotoxin trichodermin, a 12,13-epoxytrichothecene. *Proc. Natl. Acad. Sci. USA,* **71,** 713–717.

Wells, J. M., and Payne, J. A. (1976). Toxigenic species of *Penicillium, Fusarium* and *Aspergillus* from Weevil damaged pecans. *Can. J. Microbiol.* **37,** 881–885.

Wentworth, B. C., Mashaly, M., Birrenkott, G., Zimmerman, N., and Wineland, M. J. (1979). The performance of growing turkeys and ducks implanted and fed zearanol. *Poult. Sci.* **58,** 1122.

Wheeler, M. S., McGinnis, M. R., Schell, W. A., and Walker, D. H. (1981). *Fusarium* infection in burned patients. *Am. J. Clin. Pathol.* **75,** 304–311.

White, E. P. (1967). Isolation of (±)-2-acetamido-2.5-dihydro-5-oxofuran from *Fusarium equiseti. J. Chem. Soc.* **C,** 346–347.

Wilenz, S. L., and Sproul, E. E. (1938a). Spontaneous cardiovascular disease in the rat. 1. Lesions of the heart. *Am. J. Pathol.,* **14,** 177–200.

Wilenz, S. L., and Sproul, E. E. (1938b). Spontaneous cardiovascular disease in the rat. 2. Lesions of the vascular system. *Am. J. Pathol.* **14,** 201–216.

Willemart, J. P., and Bouffault, J. C. (1983). A ral compound as an anabolic in cattle. *Vet. Res. Commun.* **7,** 35–44.

Williams, M., Shaffer, S. R., Garner, G. B., Yates, S. G., Tookey, H. L., Kinter, L. D., Nelson, S. L., and McGinty, J. T. (1975). Induction of foot syndrome in cattle by fractionated extracts of toxic fescue hay. *Am. J. Vet. Res.* **36,** 1356–1357.

Wilson, B. J. (1971). Recently discovered metabolites with unusual toxic manifestations. In I. F. H. Purchase, ed., *Symp. Mycotoxins Hum. Health.* London: Macmillan, pp. 223–229.

Wilson, B. J. (1973a). Toxicity of mold-damaged sweet potatoes. *Nutr. Rev.* **31,** 73–78.

Wilson, B. J. (1973b). 12,13-epoxytrichothecenes: Potential toxic contaminants of foods. *Nutr. Rev.* **31,** 169–172.

Wilson, B. J. (1978). Hazards of mycotoxins to public health. *J. Food Prot.* **41,** 375–384.

Wilson, B. J., and Hayes, A. W. (1973). Microbial toxins. In F. M. Strong, ed., *Toxicants Occurring Naturally in Foods.* Washington, D.C.: National Academy of Science, pp. 372–423.

Wilson, B. J., and Maronpot, R. R. (1971). Causative fungus agent of leucoencephalo-malacia in equine animals. *Vet. Res.* **88,** 484–486.

Wilson, B. J., Yang, D. T. C., and Boyd, M. B. (1970). Toxicity of mold-damaged sweet potatoes (*Ipomoea batatas*). *Nature* **227,** 521–522.

Wilson, B. J., Boyd, M. R., Harris, T. M., and Yang, D. T. C. (1971). A lung oedema factor from moldy sweet potatoes (*Ipomoea batatas*). *Nature* **231,** 52–53.

Wilson, B. J., Maronpot, R. R., and Hindebrandt, P. K. (1973). Equine leucoencephalo-malacia. *J. Am. Vet. Med. Assoc.* **163,** 1293–1294.

Wilson, C. A., Everard, D. M., and Schoental, R. (1982). Blood pressure changes and cardiovascular lesions found in rats given T-2 toxin, a trichothecene secondary metabolite of certain *Fusarium* microfungi. *Toxicol. Lett.* **10,** 35–40.

Wilson, D. M., Tabor, W. H., and Trucksess, M. W. (1976). Screening method for the detection of aflatoxin, ochratoxin, zearalenone, penicillic acid and citrinin. *J. Assoc. Offic. Anal. Chem.* **59,** 125–127.

Wilson, L. A., Kuehne, J. W., Hall, S. W., and Ahearn, D. G. (1971). Microbial contamina-tion in ocular cosmetic. *Am. J. Ophthalmol.* **71,** 1298–1302.

Wilson, R. F., Sharma, V. D., Williams, L. E., Sharda, D. P., and Teague, H. S. (1967). Effect of feeding *Fusarium roseum* mold to rats and hogs. *J. Anim. Sci.* **26,** 1479–1480.

Wimmer, J. (1978). Das Auftreten von Kolbenfusariosen beim Mais. In *Aktuelle Probleme der Landwirtschaftlichen Forschung*, Vol. 11. Linz, Veröff. Landwirtsch.-Chem. Bundesversuchsanstalt, pp. 77–105.

Windels, C. E., Windels, M. B., and Kommedahl, T. (1976). Association of *Fusarium* species with picnic beetles on corn ears. *Phytopathology* **66**, 328–331.

Windholz, T. B., and Brown, R. D. (1972). Chemical modifications of zearalenone. 2. *J. Org. Chem.* **37**, 1647–1651.

Witlock, D. R., Wyatt, R. D., and Ruff, M. D. (1977). Morphological changes in the avian intestine induced by citrinin and lack of effect of aflatoxin and T-2 toxin as seen with scanning electron microscopy. *Toxicon* **15**, 41–44.

Wogan, G. W. (1975). Mycotoxins. *Ann. Rev. Pharmacol.* **15**, 437–451.

Wolf, J. C., and Mirocha, C. J. (1971). Regulation of the perfect stage in *Fusarium roseum* "Graminearum" by F-2 (zearalenone). *Phytopathologia* **61**, 918.

Wolf, J. C., and Mirocha, C. J. (1973). Regulation of sexual reproduction in *Gibberella zeae* (*Fusarium roseum* "Graminearum") by F-2 (zearalenone). *Can. J. Microbiol.* **19**, 725–734.

Wolf, J. C., and Mirocha, C. J. (1977). Control of sexual reproduction in *Gibberella zeae* (*Fusarium roseum* "Graminearum") *Appl. Environ. Microbiol.* **33**, 546–550.

Wolf, J. C., Lieberman, J. R., and Mirocha, C. J. (1972). Inhibition of F-2 (zearalenone) biosynthesis and perithecium production in *Fusarium roseum* "Graminearum." *Phytopathology* **62**, 937–939.

Wollenweber, H. W. (1913). Studies on the *Fusarium* problem. *Phytopathology* **3**, 24–48.

Wollenweber, H. W. (1916–1935). "Fusaria autographice delineata." Berlin: Selbstverlag.

Wollenweber, H. W. (1917). Fusaria autographice delineata. *Ann. Mycol.* **15**, 1–56.

Wollenweber, H. W. (1930). Fusaria autographice delineata. No. 882. Berlin.

Wollenweber, H. W. (1931). *Fusarium*-Monographie. Fungi parasitici et saprophytici. *Z. Parasitkde.* **3**, 269–516.

Wollenweber, H. W. (1932). *Hyphomycetes (Die Gattung Fusarium). Handbuch d. Pflanzenkrankheiten*, Band 2, pp. 577–819. Verlag Paul Parey, Berlin.

Wollenweber, H. W. (1944–1945). *Fusarium*—Monographie 2. Fungi parasitici et saprophytici, *Zentr. Bakt. Parasitenk, Abt. 2.* **106**: 104–135; 171–202.

Wollenweber, H. W., and Reinking, O. A. (1925). Aliquot fusaria tropicalia, nova vel revisa. *Phytopathology* **15**, 155–169.

Wollenweber, H. W., and Reinking, O. A. (1935a). *Die Fusarien, Ihre Beschreinbung, Schadwirkung und Bekampfung*. Berlin: Paul Parey.

Wollenweber, H. W. and Reinking, O. A. (1935b). *Die Verbreitung der Fusarien in der Natur*. Berlin: Friedländer.

Woolley, D. W. (1948). Studies on the structure of lycomarasmin. *J. Biol. Chem.* **176**, 1291–1298.

Woronin, M. (1891). Über das "Taumel-Getreide" in Süd-Ussurien. *Bot. Z.* **49**, 81–93.

Wright, V. F., DeLas, Casas, E., and Harein, P. K. (1976). The response of *Tribolium confusum* to the mycotoxins zearalenone (F-2) and T-2 toxin. *Environ. Entomol.* **5**, 371–374.

Wright, V. F., Vesonder, R. F., and Ciegler, A. (1982). Mycotoxins and other fungal metabolites as insecticides. In E. Kurstak, ed., *Microbiology and Viral Pesticides*. New York: Marcel Dekker, pp. 559–583.

Wyatt, R. D., Harris, J. R., Hamilton, P. B., and Burmeister, H. R. (1972a). Possible outbreak of fusariotoxicosis in avia. *Avian Dis.* **16**, 1123–1130.

Wyatt, R. D., Weeks, B. A., Hamilton, P. B., and Burmeister, H. R. (1972b). Severe oral lesions in chickens caused by ingestion of dietary fusariotoxin T-2. *Appl. Microbiol.* **24**, 251–257.

Wyatt, R. D., Colwell, W. M., Hamilton, P. B., and Burmeister, H. R. (1973a). Neural disturbances in chickens caused by dietary T-2 toxin. *Appl. Microbiol.* **26**, 757–761.

Wyatt, R. D., Hamilton, P. B., and Burmeister, H. R. (1973b). The effect of T-2 toxin in broiler chickens. *Poult. Sci.* **52**, 1853–1859.

Wyatt, R. D., Colwell, W. M., Hamilton, P. B., and Burmeister, H. R. (1973c). Neurotoxicity of T-2 toxins in broilers. *Poult. Sci.* **52**, 2105 (Abstr.).

Wyatt, R. D., Doerr, J. A., Hamilton, P. B., and Burmeister, H. R. (1975a). Egg production, shell thickness, and other physiological parameters of laying hens affected by T-2 toxin. *Appl. Microbiol.* **29**, 641–645.

Wyatt, R. D., Hamilton, P. B., and Burmeister, H. R. (1975b). Altered feathering of chicks caused by T-2 toxin. *Poult. Sci.* **54**, 1042–1045.

Wyllie, T. D., and Morehouse, L. G., eds. (1977–1978). *Mycotoxic Fungi, Mycotoxins, Mycotoxicoses. An Encyclopedic Handbook.* Vol. 1, *Mycotoxic Fungi and Chemistry of Mycotoxins;* Vol. 2, *Mycotoxicoses of Domestic and Laboratory Animals;* Vol. 3, *Mycotoxicoses of Man and Plants: Mycotoxin Control and Regulatory Practices.* New York: Marcel Dekker.

Yachevski, A. A. (1904). *Rye. "Inebriant Bread"; Yearly Information of Diseases and Damages, Cultivated and Wild Useful Plants.* St. Peterburg, pp. 16–17. Publ. Bureau for Mycology.

Yagen, B., and Joffe, A. Z. (1976). Screening of toxic isolates of *Fusarium poae* and *F. sporotrichioides* involved in causing alimentary toxic aleukia. *Appl. Microbiol.* **32**, 423–427.

Yagen, B., Joffe, A. Z., Horn, P., Mor, N., and Lutsky, I. I. (1977). Toxins from a strain involved in alimentary toxic aleukia. In J. V. Rodricks, C. W. Hesseltine, and M. A. Mehlman, eds., *Mycotoxins in Human and Animal Health.* Park Forest, Ill.: Pathotox, pp. 327–336.

Yagen, B., Horn, P., Joffe, A. Z., and Cox, R. H. (1980). Isolation and structural elucidation of a novel sterol metabolite of *Fusarium sporotrichioides* 921. *J. C. S. Perkin* **1**, 2914–2917.

Yahagi, T., Nagao, M., Seino, Y., Matsushima, T., Sugimura, T., and Okada, M. (1977). Mutagenicities of N-nitrosamines on *Salmonella. Mutat. Res.* **48**, 121–130.

Yamazaki, M. (1978). Chemistry of mycotoxins. In K. Uraguchi and M. Yamazaki, eds., *Toxicology, Biochemistry and Pathology of Mycotoxins.* Tokyo: Kadanscha TLD, pp. 65–106.

Yang, C. S. (1980). Research on esophageal cancer in China: A review. *Cancer Res.* **40**, 2633–2644.

Yap, H-Y., Murphy, W. K., DiStefano, A., Blumenschein, G. R., and Bodey, G. P. (1979). Phase II study of anguidine in advanced breast cancer. *Cancer Treat. Rep.* **63**, 789–791.

Yarovoj, P. F. (1961). Mycotoxicosis of chicks. *Proc. Kazakh. NIVI* **1**, 447–449.

Yates, S. C. (1962). Toxicity of tall fescue forage: A review. *Econ. Bot.* **16**, 295–303.

Yates, S. G. (1963). Paper chromatography of alkaloids of tall fescue hay. *J. Chromatogr.* **12**, 423–426.

Yates, S. G. (1971). Toxin-producing fungi from fescue pasture. In S. Kadis, A. Ciegler, and S. J. Ajl, eds., *Microbial Toxins.* Vol. 7. New York: Academic, pp. 191–206.

Yates, S. G., and Tookey, H. L. (1965). Festucine, an alkaloid from tall fescue (*Festuca arundinacea* Schreb.): Chemistry of the functional groups. *Aust. J. Chem.* **18**, 53–60.

Yates, S. G., Tookey, H. L., and Ellis, J. J. (1966). Toxic butenolide produced by *Fusarium nivale* isolated from tall fescue (*Festuca arundinacea* Schreb.). *Am. Chem. Soc.* **152**, 81.

Yates, S. G., Tookey, H. L., Ellis, J. J., and Burkhardt, H. J. (1967). Toxic butenolide produced by *Fusarium nivale* (Fries) Cesati isolated from fescue (*Festuca arundinaceae* Schreb). *Tetrahedron Lett.* **7**, 621–625.

Yates, S. G., Tookey, H. L., Ellis, J. J., and Burkhardt, H. J. (1968). Mycotoxins produced by *Fusarium nivale* isolated from tall fescue (*Festuca arundinaceae* Schreb.). *Phytochemistry* **7**, 139–146.

Yates, S. G., Tookey, H. L., Ellis, J. J., Tallent, W. H., and Wolff, I. A. (1969). Mycotoxins as a possible cause of fescue toxicity. *J. Agric. Food Chem.* **17**, 437–442.

Yates, S. G., Tookey, H. L., and Ellis, J. J. (1970). Survey of tall-fescue pasture: Correlation of toxicity of *Fusarium* isolates to known toxins. *Appl. Microbiol.* **19**, 103–105.

Yefremov, V. V. (1944a). On the so-called alimentary toxic aleukia (septic angina). *Sov. Med.* **1, 2**, 19–21.

Yefremov, V. V. (1944b). Alimentary toxic aleukia (septic angina). *Hyg. Sanit.* **7, 8**, 18–45.

Yefremov, V. V. (1945). Etiology and pathogenesis of alimentary toxic aleukia (septic angina). In *Alimentary Toxic Aleukia (Septic Angina)* Moscow: Medgiz, pp. 18–21.

Yefremov, V. V. (1948). *Alimentary Toxic Aleukia.* Moscow: Publ. Med. Lit.

Ylimäki, A. (1967). Root rot as a cause of red clover decline in leys in Finland. *Ann. Agric. Finn.* **6**, 1–59.

Ylimäki, A., Koponen, H., Hintikka, E.-L., Nummi, M., Niku-Paavola, M.-L., Ilus, T., and Enari, T.-M. (1979). Mycoflora and occurrence of *Fusarium* toxins. *Tech. Res. Cent. Finl.* **21**, 3–28.

Yoshizawa, T. (1978). Toxic metabolites of fusaria invaded field crops. *Proc. Jap. Assoc. Mycotoxicol.* **8**, 6–11.

Yoshizawa, T. (1983a). *Fusarium* metabolites other than trichothecenes. In Y. Ueno, ed., *Trichothecenes, Chemical, Biological and Toxicological Aspects.* Amsterdam: Elsevier, pp. 60–71.

Yoshizawa, T. (1983b). Red-mold diseases and natural occurence in Japan. In Y. Ueno, ed., *Trichothecenes, Chemical, Biological and Toxicological Aspects.* Amsterdam: Elsevier, pp. 195–209.

Yoshizawa, T., and Hosokawa, H. (1983). Natural occurrence of deoxynivalenol and nivalenol, trichothecene mycotoxins in commercial foods. *Jpn. J. Food Hyg.* **24**, 413–415.

Yoshizawa, T., and Morooka, N. (1973). Deoxynivalenol and its monoacetate: New mycotoxins from *Fusarium roseum* and moldy barley. *Agr. Biol. Chem.* **37**, 2933–2934.

Yoshizawa, T., and Morooka, N. (1974). Studies on the toxic substances in the infected cereals. 3. Acute toxicities of new trichothecene mycotoxins: Deoxynivalenol and its monoacetate. *J. Food Hyg. Soc. Japan* **15**, 261–269.

Yoshizawa, T., and Morooka, N. (1975a). Biological modification of trichothecene mycotoxins: Acetylation and deacetylation of deoxynivalenol by *Fusarium* spp. *Appl. Microbiol.* **29**, 54–58.

Yoshizawa, T., and Morooka, N. (1975b). Comparative studies on microbial and chemical modifications of trichothecene mycotoxins. *Appl. Microbiol.* **30**, 38–43.

Yoshizawa, T., and Morooka, N. (1977). Trichothecenes from mold-infected cereals in Japan. In J. V. Rodricks, C. W. Hesseltine, and M. A. Mehlman, eds., *Mycotoxins in Human and Animal Health.* Park Forest, Ill.: Pathotox, pp. 309–321.

Yoshizawa, T., and Sakamoto, T. (1982). *In vitro* metabolism of T-2 toxin and its derivatives in animal livers. *Proc. Jpn. Assoc. Mycotoxicol.* **14,** 26–28.

Yoshizawa, T., Tsuchiya, Y., Teraura, M., and Morooka, N. (1976). Studies on the toxic substances in the infected cereals. 4. Trichothecene from *Fusarium*-infected wheat in the crop field. *Proc. Jpn. Assoc. Mycotoxicol.* **2,** 30–32.

Yoshizawa, T., Shirota, T., and Morooka, N. (1978). Deoxynivalenol and its acetate as feed refusal principles in rice cultures of *Fusarium roseum* No. 117 (ATCC 28114). *J. Food Hyg. Soc. Japan* **19,** 178–184.

Yoshizawa, T., Matsuura, Y., Tsuchiya, Y., Morooka, N., Kitani, K., Ichinoe, M., and Kurata, H. (1979). On the toxigenic fusaria invading barley and wheat in Southern Japan. *J. Food Hyg. Soc. Japan* **20,** 21–26.

Yoshizawa, T., Onomot, C., and Morooka, N. (1980a). Microbial acetyl conjugation of T-2 toxin and its derivatives. *Appl. Environ. Microbiol.* **39,** 962–966.

Yoshizawa, T., Swanson, S. P., and Mirocha, C. J. (1980b). In vitro metabolism of T-2 toxin in rats. *Appl. Environ. Microbiol.* **40,** 901–906.

Yoshizawa, T., Swanson, S. P., and Mirocha, C. (1980c). T-2 metabolites in the excreta of broiler chickens administered 3H-labeled T-2 toxin. *Appl. Environ. Microbiol.* **39,** 1172–1177.

Yoshizawa, T., Mirocha, C. J., Behrens, J. C., and Swanson, S. P. (1981). Metabolic fate of T-2 toxin in a lactating cow. *Food Cosmet. Toxicol.* **19,** 31–39.

Yoshizawa, T., Sakamoto, T., Ayano, Y. and Mirocha, C. J. (1982a). 3'-Hydroxy T-2 and 3'-Hydroxy HT-2 toxins: New metabolites of T-2 toxin, a trichothecene mycotoxin, in animals. *Agric. Biol. Chem.* **46,** 2613–2615.

Yoshizawa, T., Sakamoto, T., Ayano, Y., and Mirocha, C. J. (1982b). Chemical structure of new metabolites of T-2 toxin. *Proc. Jpn. Assoc. Mycotoxicol.* **15,** 13–15.

Yoshizawa, T., Sakamoto, T., and Okamoto, K. (1984). *In vitro* formation of 3'-Hydroxy T-2 and 3'-Hydroxy HT-2 toxins from T-2 toxin by liver homogenates from mice and monkeys. *Appl. Environ. Microbiol.* **47,** 130–134.

Young, C. N., and Meyers, A. M. (1979). Opportunistic fungal infection by *Fusarium oxysporum* in a renal transplant patient. *Sabouraudia* **17,** 219–223.

Young, N. A., Kwon-Chung, K. J., Kubota, T. T., Jennings, A. E., and Fisher, R. I. (1978). Disseminated infection by *Fusarium moniliforme* during treatment for malignant lymphoma. *J. Clin. Microbiol.* **7,** 589–594.

Yudenitch, V., Mironov, C., Soboleva, R., and Fok, R. (1944). Septic angina in the Chkalov district in 1944. *Proc. 1st Kharkov Med. Inst. Chkalov Inst. Epidemiol. Microbiol.,* pp. 5–16.

Zagrafski, B., Terziev, F., Tolev, I., and Beltchev, M. (1957). Study of Urov Kashin-Beck Disease in Korean People's Democratic Republic. *Clin. Med.* **35,** 92–95.

Zaias, N. (1966). Superficial white onychomycosis. *Sabouraudia* **5,** 99–109.

Zaias, N. (1971). Onychomycosis. *Arch. Dermatol.* **105,** 263–274.

Zapater, R. C. (1971). Keratomycose por *Fusarium*. *Proc. 5th Congr. Int. Soc. Hum. Anim. Mycol., Paris,* pp. 187–188.

Zapater, R. C., and Arrechea, A. (1975). Mycotic keratitis by *Fusarium*. *Ophthalmologica, Basel* **170,** 1–12.

Zapater, R. C., Arrechea, A., and Guevara, V. H. (1972). Queratomycosis por *Fusarium dimerum*. *Sabouraudia* **10,** 274–275.

Zapater, R. C., Brunzini, M. A., Albesi, E. V., and Silicardo, C. A. (1976). El genero *Fusarium* como agente etiologico de micosis oculares: Presentacion de 7 cases. *Arch. Oftalmol. Buenos Aires* **51,** 279–286.

Zavyalova, A. P. (1946). Chemical and toxicological characteristics of lipoproteins isolated from wheat, causing alimentary toxic aleukia. Ph.D. dissertation, Med. Inst., Kuybyshev.

Zodzishki, B. Y., (1933). Data on the study of clinical aspects, pathogenesis and alimentary panhematopathy treatment in so-called septic angina (arpanulocytosis, hemorrhagic aleukia, panmyelopthisis). Ph.D. dissertation, Med. Inst., Novosibirsk.

Zhukhin, V. A. (1945). Data on pathogenic anatomy and pathogenesis in so-called septic angina (alimentary toxic aleukia). *Acta Septic Angina. Ufa*, pp. 82–86.

# INDEX